Yuri Bahturin

Identical Relations in Lie Algebras

De Gruyter Expositions in Mathematics

Volume 68

Yuri Bahturin

Identical Relations in Lie Algebras

2nd edition

DE GRUYTER

Mathematics Subject Classification 2010
Primary: 17B01, 17B65, 17B70; Secondary: 16R10, 20E10, 20F40

Author
Prof. Yuri Bahturin
Memorial University of Newfoundland
Department of Mathematics & Statistics
St. John's NL A1C 5S7
Canada
bahturin@mun.ca

ISBN 978-3-11-056557-7
e-ISBN (PDF) 978-3-11-056665-9
e-ISBN (EPUB) 978-3-11-056570-6
ISSN 0938-6572

Library of Congress Control Number: 2021938532

Bibliographic information published by the Deutsche Nationalbibliothek
The Deutsche Nationalbibliothek lists this publication in the Deutsche Nationalbibliografie; detailed bibliographic data are available on the Internet at http://dnb.dnb.de.

Typesetting: VTeX UAB, Lithuania
Printing and binding: CPI books GmbH, Leck

MIX
Papier aus verantwortungsvollen Quellen
FSC® C083411
www.fsc.org

www.degruyter.com

To my family

Preface to the second edition

The second edition of "Identical relations in Lie algebras" is appearing more than 30 years after the first edition and more than 40 years after my first book "Lectures on Lie Algebras," Akademie-Verlag, Berlin, 1978, on the same topic. Since then, the area has experienced a considerable progress. In this second edition, we try to reflect its modern state. Since most of the material in the first edition remains relevant, we decided to keep it in its original form (with few inevitable corrections) and to add several additional chapters dedicated to the later development. The first edition included almost all of the results in the area of Lie algebras with identical relations, given with full proofs, a notable exception being Engel Lie algebras. The amount of the material which has appeared since then makes it impossible to include all existing results. Moreover, we were not able to include full proofs in many cases. We decided to avoid including proofs that appear in other books, such as A. I. Kostrikin's "Around Burnside" [181], Yu. P. Razmyslov's "Identities of Algebras and their Representations" [313], A. Giambruno and M. V. Zaicev "Polynomial Identities and Asymptotic Methods" [123], Yu. A. Bahturin, A. A. Mikhalev, V. M. Petrogradsky, and M. V. Zaicev "Infinite-dimensional Lie superalgebras" [46]. We also mention a few relevant books, such as L. A. Bokut' and G. P. Kukin "Algorithmic and Combinatorial Algebra" [78], A. A. Mikhalev and A. A. Zolotykh "Combinatorial aspects of Lie superalgebras" [227], C. Reutenauer "Free Lie algebras" [320], and M. R. Vaughan-Lee, "The Restricted Burnside Problem" (2nd ed.) [357].

In the second edition, we put our stress on the topics which have been especially active in the recent couple of decades. One of such topics is the growth: the codimension growth and the growth of algebras with polynomial identities. While studying the codimension growth, we restrict ourselves to the field of characteristic zero. In this case, we can deal only with multilinear identities. One of the main tools here is the representation theory of the symmetric group. In distinction with the codimension growth in the associative algebras, which is always at most exponential, in the case of Lie algebras we have varieties of polynomial, exponential, and overexponential growth. Although many varieties of Lie algebras have overexponential growth, the majority of the most meaningful examples have growth which is at most exponential. Such are all finite-dimensional algebras, Cartan-type simple algebras, Kac–Moody affine algebras and others. S. P. Mishchenko showed that there are no algebras whose growth is intermediate between polynomial and exponential, and finite-dimensional algebras with polynomial growth are solvable.

So the varieties with exponential growth play a special role and get the most extensive treatment in this book; see Chapter 9. Among the results we give with (partial) proof is M. V. Zaicev's theorem saying that the exponent of the codimension growth of a finite-dimensional Lie algebra is an integer. In the case of associative algebras, A. Giambruno and M. V. Zaicev proved that the exponent of any associative algebra is an integer. In this book we give the proof of the first example, due to S. P. Mishchenko

https://doi.org/10.1515/9783110566659-201

and M. V. Zaicev, showing that the varieties of Lie algebras do not need to have an integral exponent. There were many more examples of nonintegral exponent for Lie algebras, finite-dimensional Lie superalgebras, and very striking examples of other algebras based on the deep facts in combinatorics and number theory. Unfortunately, we were unable to give many proofs or even a survey of all these results, some were beyond the scope of identities in Lie algebras.

Varieties of overexponential growth, explored mostly by V. M. Petrogradsky after the pioneering works of I. B. Volichenko, are presented in Section 9.2. I am deeply indebted to Professor Petrogradsky for the material which I used while writing several sections of the book, devoted to the hierarchy of varieties of overexponential codimension growth, the growth of solvable Lie algebras, and examples of nonnilpotent Engel Lie algebras.

One more topic in the theory of infinite-dimensional algebras, which became prominent in the recent decades, is the growth of Lie algebras. Growth functions, in particular, Hilbert series, serve as a replacement of the dimension or codimension in infinite-dimensional spaces, the number of generators, when this number is infinite, and so on. In Chapter 10, we give a survey of the work of V. M. Petrogradsky on the growth of free algebras in certain varieties of Lie algebras. The techniques, necessary to prove these deep results, go as far as the theory of functions of complex variables. The remainder of the chapter is devoted to the expansion of the theory of free Lie algebras with the help of the notion of the growth and Hilbert series. One of the central results is an analogue, by V. M. Petrogradsky, of Nielsen–Schreier formula for the number of generators for the subgroup of a free group to the case of free Lie algebras. Several sections are devoted to the study, by Y. Bahturin and A. Olshanskii, of the growth of subalgebras and subideals in free Lie algebras, with application to the standard theory of free Lie algebras. We also discuss the notion of distortion for the embeddings of Lie algebras.

One more topic that became popular in a couple of recent decades is the theory of graded Lie algebras, Lie superalgebras, and Lie algebras endowed with an additional structure, such as the action of a Hopf algebra. One of the reasons is that during this period of time we learned a lot about the classification of group gradings on many important algebras, for example, associative matrix algebras, simple Lie algebras, and others. This additional structure on algebras makes it easier to study them in the absence of definitive results about the identities of ungraded algebras. For example, the determination of all identities of associative or Lie matrix algebras remains a very difficult problem beyond matrices of order 2(!). At the same time, the determination of *graded identities* for "popular" gradings is often much easier. Moreover, if two algebras graded by the same group have the same graded identities then their ordinary identities are the same, too!

We study identities of graded Lie algebras and algebras with the action of Hopf algebras in Chapters 11 and 12. We start with a popular theme of the study of Lie algebras whose identity component or invariant subalgebra is a PI-algebra. A basic result here a theorem by Y. Bahturin and M. V. Zaicev confirming that if a *Lie-type algebra* has this property then it satisfies a nontrivial identity, as a whole. Then we give a far-reaching generalization of this result by V. Linchenko. One of Linchenko's results, in conjunction with some deep results by E. Zelmanov, became quite popular in the solution of some problems in the group theory: if the subalgebra of invariants of a Lie-type algebra under the action of a finite group satisfies a nontrivial identity, is the same true for the whole algebra? We then mention the results of A. S. Gordienko about the codimension growth of algebras with the action of Hopf algebras or other additional structure.

We also include some results by Y. Bahturin and F. Yasumura, expanding famous Yu. P. Razmyslov's results about the isomorphism of simple algebras with the same polynomial identities to the case of algebras with additional structure. Here we recall Razmyslov's approach via multioperator, or Ω-algebras, and use it to prove, for example, that two finite-dimensional graded simple Lie algebras, having the same graded identities, are graded isomorphic. So the polynomial identities are not only able to recover algebras but also their gradings.

In Chapter 13 we treat Engel Lie algebras. Although general Engel algebras do not need to be algebras with polynomial identities, they definitely belong to the same realm of algebras. We give a simplified proof by A. Olshanskii of E. S. Golod's example of a finitely generated nonnilpotent Engel Lie algebra, which was the first of this kind. We also treat an example by V. M.Petrogradsky and other authors, based on the idea of self-similarity first suggested by R. I. Grigorchuk for the solution of the Burnside and Milnor's problems in group theory.

Then we give a short survey of the famous results of A. I. Kostrikin and E. Zelmanov on the restricted Burnside problem. The proofs in this area remain to be technically complicated, but luckily there is a monograph of A. I. Kostrikin "Around Burnside" [181] where the reader can find the proofs of many key results in this area.

There are many other topics in the theory, some of them we mention in the last Chapter 14: Razmyslov's theory of identities in simple Lie algebras, special (or SPI) Lie algebras, and the automorphisms of varietally free Lie algebras. We give some of these results with full proofs, like the author's joint results with V. Shpilrain and S. Nabiev about the automorphisms, and for many important and technically complex results, mainly by V. Drensky, we give statements and references to Professor Drensky's surveys and original papers.

A couple of remarks about the notation. In some parts of the book, following the traditions of nonassociative algebra, we denote the product of two elements x, y simply as xy. In other parts, where the connection with associative algebras is inherent, we use the commutator notation for the product, $[x, y]$. The product, or commutator, of more than two elements $x_1, x_2, \ldots, x_n$ depends on the placement of brackets. When we write $x_1x_2\cdots x_n$, in the nonassociative notation, it usually means one of two things: right-normed product defined by induction as $x_1\cdots x_n = x_1(x_2\cdots x_n)$ or left-normed product $x_1\cdots x_n = (x_1\cdots x_{n-1})x_n$. Actually, almost half of the people call the first product left-normed and the second right-normed. When we quoted results of various authors, we did our best to indicate to what party a particular author belongs (occasionally, they change sides!). Similar is the situation with the left- and right-normed commutators.

Another remark is about the notation for the base ring/field of coefficients. In the chapters coming from the first edition, we used lowercase k. This created a lot of confusion in the situations where k should be naturally used as an index: $i, j, k, \ldots$ So in the remainder of the book, we have used $\mathbb{F}$ to denote the field/ring of coefficients.

Several people helped me with the second edition. I cordially thank Vesselin Drensky, Sergei Mishchenko, Alexander Olshanskii, Victor Petrogradsky, Gunnar Traustason, Felipe Yukihide Yasumura and Mikhail Zaicev. Special thanks to Felipe who did an invaluable job of digitizing the first edition and to Victor (Mikhailovich) who prepared portions of the text related to his work.

Finally, I would like to thank my wife Galina who helped me with some editing, but mostly for her support during this time of total uncertainty. Thank you, Galya!

Preface to the first edition

An important property of an algebraic system is to satisfy an identical relation. "Although simplest closed propositions of the language of logic, the identities form a language rich enough to express quite fine properties of algebraic systems" (Malcev [219]). Lie algebras satisfying identical relations have been at the focus of investigations since the very beginning of the theory, more than a century ago. Indeed, a large portion of usually studied classes of Lie algebras can be singled out by imposing certain identical relations. Of these one can readily name such classes as abelian, nilpotent, solvable Lie algebras and some others. Moreover, recent investigations have shown that in some important classes of Lie algebras their members are completely determined by identities. This is the case, for instance, for finite-dimensional simple algebras over any fixed field which is either finite (Bahturin and Olshanskii [50]) or algebraically closed (Razmyslov [309]). Many classical results admit reformulation in the language of identities. Of course, a reformulation of this kind is rather formal as it concerns such theorems as those of Lie and Engel. However, the course of evolution has led to problems where identities become the core of the subject. An impressive illustration for this is supplied by A. I. Kostrikin's work on Burnside's problem. Here one of the principal points is the study of Lie algebras satisfying the Engel identities.

It is from the 1950s, when this and some other works were published, that we date the beginning of a regular study of identical relations in Lie algebras. A basis for much of the research was formed by the work of A. I. Shirshov who proved the theorem on subalgebras of free Lie algebras and gave exceptionally useful constructions of linear bases in these algebras (Shirshov [335, 338, 340]). Bokut' [76] has studied polynilpotent Lie algebras, Latyshev [200, 201] has introduced standard identities and initiated the study of special Lie algebras. Shmelkin [341] has applied the methods of Lie algebras to the study of varieties of groups. An important Engel identity has been studied by Cohn [90], Higgins [141], and some others.

The next stage in the development of the theory, the study of systems of identities without their immediate connection to the particular types of algebras, as well as the study of varieties of Lie algebras as classes determined by such systems, began in the fall of the 1960s and the beginning of the 1970s. To a considerable extent, this work was facilitated by the achievements in the theory of varieties of groups (H. Neumann [247]). Thus, for some years, the former theory was "in chase" of the latter one. As a result of this we find examples of a variety of Lie algebras without finite basis for identities, the theorem on the freeness of the semigroup of varieties of Lie algebras over an infinite field, some other analogues of group-theoretic results over infinite fields (Parfenov [255]; Vaughan-Lee [354]; Bahturin [21, 22]).

It was realized very soon, however, that this new theory had many specific features. For instance, in contrast to the group theory case, the multiplication of varieties over arbitrary fields need not be associative (Bahturin [22]). Another distinction is that

https://doi.org/10.1515/9783110566659-202

the product of finitely based varieties of Lie algebras over an infinite field is necessarily finitely based [22]. The behavior of varieties of Lie algebras and groups differs in such questions due to the structure of lattices of varieties (Sheina [332]), basic and axiomatic ranks (Bahturin [29]), and many others. These results gave the theory its present appearance, with sufficiently many attractive problems of its own.

In this book we sought to treat the modern state of the theory of Lie algebras satisfying identical relations, as well as some of its applications. The book is designed in such a way that its initial chapters should be accessible to a general reader with only a university background in algebra. Although we tried to keep our text as elementary and self-contained as possible, the subsequent chapters need a higher degree of sophistication.

There are eight chapters in this book. In the first we give principal definitions and prove some basic theorems, with an eye to our future needs. A particular feature of our treatment is trying to prove theorems, wherever possible, in the setting of algebras over arbitrary commutative rings with unity. The second chapter is devoted to the study of free Lie algebras, with an emphasis on bases and subalgebras. We also sketch some other topics such as restricted Lie algebras, free products, and cohomology. Here our main hero, Lie algebras satisfying identical relations, is introduced. In Chapter 3 we continue our preparation work and consider the action of the symmetric and the general linear groups on nonassociative polynomials. We will not make full use of the techniques developed here, but recent investigations showed that the main applications are still ahead.

Chapter 4 is introductory to the varieties of Lie algebras and it gives an exposition of the most traditional parts of the theory. We treat some results on multihomogeneous varieties, operations on the varieties, as well as such important topics as the basic rank and Schreier varieties of Lie algebras. We also discuss wreath products, an important tool, due to Shmelkin [342]. In Chapter 5 we deal with the finite basis problem. We start with some general results and then sketch the important techniques of partially well-ordered sets, due to Higman [142]. These techniques are applied to the proof of the finite basis property of some solvable varieties of Lie algebras. Then we give examples of varieties and Lie algebras admitting no finite bases for their identities. Finally, we describe some more recent techniques which have led to the proof of the finite basis property for finite-dimensional solvable Lie algebras, as well as for any Lie algebras satisfying the identities of the matrix algebra of order 2 over a field of characteristic zero.

Chapter 6 is devoted to the study of special Lie algebras, an extensive class of not necessarily finite-dimensional Lie algebras satisfying identical relations, which inherit many of the usual properties of finite-dimensional Lie algebras. We give a detailed discussion of these algebras, including their structure and identities. This culminates in the description of Lie algebras (and Lie superalgebras) whose universal envelope algebra is an associative algebra satisfying a nontrivial polynomial identity

(PI-algebra). Finally, we apply this to the description of Lie algebras, all of whose irreducible representations are of finite bounded degree.

In Chapter 7 we discuss identities of finite Lie rings and give the solution of a problem, due to Malcev, on the finite basis for identities satisfied by such rings. The techniques developed here are subsequently used for various questions on identities in locally finite varieties of Lie algebras. These techniques are known to have been applied in other classes of nonassociative algebras.

Chapter 8 deals with applications to the group theory. We first present some classical results of Hall, Magnus, Witt, and Malcev on the connection between free and nilpotent groups on the one side and Lie algebras on the other. Then we discuss a correspondence between varieties of groups and Lie algebras. This is applied to the study of free groups of product varieties. We also give an application of powerful techniques due to Razmyslov which allowed him to solve difficult problems of the theory of varieties.

It was not our aim to embrace all the results in the area considered. The most important omission is the theory of Lie algebras satisfying the Engel conditions. The main reason for this is that, in our opinion, this topic merits a special monograph. We also remark that the theory of varieties of Lie algebras to a certain extent has been considered in the book of Amayo and Stewart [5], as well as in the author's book (Bahturin [26]). In the present book we borrowed a lot from the latter publication.

We have used here research results of many mathematicians, but, without any doubt, the names of A. Yu. Olshanskii, Yu. P. Razmyslov, and A. L. Shmelkin deserve special mention. I cordially thank these people. I am indebted to every member of the Department of Algebra of Moscow University with which I have been connected for nearly 20 years. I am grateful to M. V. Zaicev, A. Yu. Olshanskii, and Yu. P. Razmyslov, who have read some extracts of the manuscript, for much helpful advice.

Contents

1 Basic theory of Lie algebras

1.1 Main definitions

1.1.1 Rings, modules, and algebras

We assume that the reader is familiar with the main definitions and standard theorems concerning principal algebraic structures: groups, rings, and modules (Jacobson [151]; Herstein [140]; Kostrikin [178]; Bahturin [38]). Herstein, I. The only possible difference with our approach is that the rings are not necessarily associative. Thus *the ring* is any set R with operations of addition and multiplication such that with respect to the addition R is an abelian group and the two operations are bound by the distributivity laws

$$x(y + z) = xy + xz, \quad (x + y)z = xz + yz$$

where x, y, z are arbitrary elements in R. At the same time our concept of the module is a more limited one since we consider only unitary modules over rings with unity. Thus an abelian group M over an associative ring k with unit 1 is called a *left k-module* if for all $\lambda \in k$ and all $x \in M$ there exists an element $\lambda x \in M$, and the following laws hold:

$$(\lambda + \mu)x = \lambda x + \mu x, \quad (\lambda\mu)x = \lambda(\mu x), \quad \lambda(x + y) = \lambda x + \lambda y, \quad 1x = x$$

where λ, μ are arbitrary in k and x, y are arbitrary in M. A ring R which is at the same time a left module over an associative commutative ring k is called an *algebra over k* (or a k-algebra) if for all $\lambda \in k$ and $x, y \in R$ we have

$$\lambda(xy) = (\lambda x)y = x(\lambda y).$$

All basic notions and results in the ring theory may easily be translated into the algebra language. For instance, a *subalgebra* S of an algebra R over a ring k is a subring in R which is a k-submodule. The notion of *homomorphism* is defined for algebras over the same ring k as a homomorphism of rings which is a homomorphism of k-modules. If X is a subset in an algebra R then $\mathrm{alg}(X)$ denotes a subalgebra generated by X and $\mathrm{id}_R(X)$ denotes an ideal in R generated by X as an ideal. If $X = \{x\}$, then we write $\mathrm{alg}(x)$, $\mathrm{id}_R(x)$. We will now formulate two standard theorems in which the following notation is used. Given two k-algebras R_1 and R_2, we write $R_1 \cong R_2$ if R_1 and R_2 are isomorphic as k-algebras; if S, T are two submodules in a k-module R then $S+T$ denotes a k-submodule in R whose elements are of the form $s + t$, where $s \in S$, $t \in T$.

Theorem 1.1 (The homomorphism theorem). *Let φ be a surjective homomorphism of a k-algebra R_1 onto a k-algebra R_2. Then the kernel $\ker\varphi$ of this homomorphism is an ideal of R_1 and*

$$R_1/\ker\varphi \cong R_2.$$

https://doi.org/10.1515/9783110566659-001

Theorem 1.2 (The isomorphism theorem). *Let R be a k-algebra, S, T its ideals, and U its subalgebra.*

(i) *If $S \supset T$ then S/T is an ideal in R/T and*

$$(R/T)/(S/T) \cong R/S.$$

(ii) *$S \cap U$ is an ideal in U, $U + S$ is a subalgebra in R and*

$$U/(S \cap U) \cong (U + S)/S.$$

1.1.2 The anticommutativity law

A ring R is called *anticommutative* if the square of each element in R is zero. Given arbitrary x, y in such a ring, we have

$$0 = (x + y)^2 = x^2 + xy + yx + y^2 = xy + yx.$$

Thus, in an anticommutative ring, in addition to the defining identical relation

$$x^2 = 0, \tag{1.1}$$

we always have an identity of the form

$$xy = -yx. \tag{1.2}$$

Simple examples show that equation (1.1) is not a consequence of equation (1.2). Namely, there exists a ring S in which for all $x, y \in S$ we have equation (1.2) but, for some $x \in S$, equation (1.1) does not hold (see Exercise 1.12.1).

1.1.3 The Jacobi identity. Lie ring

Suppose that for any x, y, z in a ring R we have

$$x(yz) + y(zx) + z(xy) = 0. \tag{1.3}$$

Then we say that R satisfies the *Jacobi identity*. The polynomial on the left-hand side of equation (1.3) is called the *Jacobian* of x, y, z. It is denoted by $J(x, y, z)$.

An anticommutative ring satisfying the Jacobi identity is called a *Lie ring*. If a Lie ring is an algebra over a commutative ring k then we are given a *Lie algebra over k*.

1.1.4 Some other classes of rings

A ring R satisfying the associativity law $x(yz) = (xy)z$, x, y, z being arbitrary elements in R, is called *associative*. An *alternative* ring is a ring in which each pair of elements generates an associative subring. A *commutative* ring R is a ring in which $xy = yx$ for all $x, y \in R$. A commutative ring R with additional relation $(x^2y)x = x^2(yx)$ for all $x, y \in R$ is called a *Jordan ring*.

1.1.5 Right-normed products

Let R be a not necessarily associative ring, and consider $x_1, x_2, \dots, x_n \in R$. The *right-normed product* of the above elements is defined by induction as

$$x_1x_2\cdots x_n = x_1(x_2\cdots x_n) \quad \text{if } n > 1.$$

If $n = 1$, we simply put x_1. Moreover, we write, for short,

$$\underbrace{x\cdots x}_{m}y = x^m y.$$

1.1.6 Derivation

Let R be a k-algebra. An endomorphism δ of a k-module R is called a *derivation* if, for any $x, y \in R$, we have

$$\delta(xy) = \delta(x)y + x\delta(y).$$

1.1.7 Extension of the base ring

Let k' be a commutative ring containing k as a subring with the same unity as in k'. If R is a k-algebra then $R_{k'} = k' \otimes_k R$ is an algebra over k'. In this case, we say that $R_{k'}$ is obtained by extending the base ring. If we assume that R is a free k-module with a basis $\{e_\alpha \mid \alpha \in I\}$ then $R_{k'}$ is a free k-module with a basis $\{1 \otimes e_\alpha \mid \alpha \in I\}$. Suppose that

$$e_\alpha e_\beta = \sum_\gamma c_{\alpha\beta}^\gamma e_\gamma \quad c_{\alpha\beta}^\gamma \in k.$$

Then

$$(1 \otimes e_\alpha)(1 \otimes e_\beta) = \sum_\gamma c_{\alpha\beta}^\gamma (1 \otimes e_\gamma).$$

In particular, this is true when k is a field (Lang [199, Chapter XVI]; Kostrikin [178, Chapter 8]).

1.1.8 Finite-dimensional algebras

Let R be a k-algebra. We suppose that R is finitely generated as a k-module; that is, there exist $e_1, \ldots, e_n \in R$ such that for any $x \in R$ we have

$$x = \lambda_1 e_1 + \cdots + \lambda_n e_n,$$

with $\lambda_1, \ldots, \lambda_n$ being some elements in k. In this case we say that R is a *finite-dimensional k-algebra*. More restrictive is the notion of finite length. To be precise, we admit that a k-module R possesses an unrefinable series of k-submodules of the form

$$\{0\} = R_0 \subset R_1 \subset \cdots \subset R_s = R,$$

in which, for every $i = 1, 2, \ldots, s$, we have $R_{i-1} \neq R_i$. Then s is independent of the particular choice of the series and it is called the *length* of the k-module R. An algebra R over k which has finite length as a k-module is called a *k-algebra of finite length s*. We then write $s = l_k(R)$. Given a subalgebra S in an algebra R of finite length, we have

$$l_k(R) = l_k(S) + l_k(R/S).$$

This relation enables us to proceed by induction while proving theorems on algebras with finite length. It should be noticed that, in the case where k is a field, the length of a k-algebra is merely its dimension as a vector space over k. All the proofs of the above assertions can be found in Lang [199, Chapter IV].

1.2 Examples of Lie algebras

1.2.1 Lie algebra of an associative algebra

Let A be an associative k-algebra. For all $x, y \in A$, we set

$$[x, y] = xy - yx.$$

This element of A is called the *commutator* of x and y. Let $[A]$ denote the k-module A with operations in which multiplication is replaced by taking the commutator of two elements. We will show that $[A]$ is a Lie algebra. Firstly, + and $[\cdot,\cdot]$ are bound by the distributivity laws. In fact, given $x, y, z \in [A]$, we have

$$\begin{aligned}[x, (y + z)] &= x(y + z) - (y + z)x = xy + xz - yx - zx \\ &= (xy - yx) + (xz - zx) = [x, y] + [x, z].\end{aligned}$$

Similarly, $[(x+y), z] = [x, z] + [y, z]$. Now if $x \in [A]$, then $[x, x] = xx - xx = 0$. Hence $[A]$ is anticommutative. To prove the Jacobi identity, we compute the Jacobian with respect to the commutator $[\cdot,\cdot]$:

$$[x, [y, z]] = [x, yz - zy] = xyz - xzy - yzx + zyx.$$

Permutating the letters cyclically, we get

$$[y,[z,x]] = yzx - yxz - zxy + xzy,$$
$$[z,[x,y]] = zxy - zyx - xyz + yxz.$$

Summing up the left- and right-hand sides of the three equations, we find $J(x,y,z) = 0$. Finally, if $\lambda \in k$, $x,y \in [A]$ then

$$\lambda[x,y] = \lambda(xy - yx) = (\lambda x)y - y(\lambda x) = [(\lambda x), y].$$

Similarly, $\lambda[x,y] = [x,\lambda y]$. Now we see that $[A]$ satisfies all the axioms of a Lie algebra over k. The right-normed product in the Lie ring A will be called the right-normed commutator and it will be written in the form $[x_1, x_2, \ldots, x_n]$ where $x_1, x_2, \ldots, x_n \in A$. We will also be writing

$$[\underbrace{x,\ldots,x}_{n},y] = [nx, y]$$

(if there is no risk of confusion with multiplication by elements in k).

1.2.2 Lie algebra of endomorphisms of a *k*-module

Let M be an arbitrary k-module. By $\operatorname{End}_k M$ we mean the (associative) algebra of all homomorphisms of a k-module M into itself (i. e., k-endomorphisms). If φ, ψ are two such endomorphisms, $x \in M$, $\lambda \in k$, then, by definition,

$$(\varphi + \psi)(x) = \varphi(x) + \psi(x),$$
$$(\varphi\psi)(x) = \varphi(\psi(x)),$$
$$(\lambda\varphi)(x) = \lambda\varphi(x).$$

We will denote the Lie algebra $[\operatorname{End}_k M]$ of the associative algebra $\operatorname{End}_k M$ by $\mathfrak{gl}(M,k)$. If M is a finite-dimensional vector space over a field k with a basis $E = \{e_1, e_2, \ldots, e_n\}$ then with each endomorphism φ we may uniquely associative a matrix A_φ which is the matrix of the linear operator φ in the basis E. The mapping $\varphi \mapsto A_\varphi$ is an isomorphism of the associative algebra $\operatorname{End}_k M$ and the algebra k_n of all $n \times n$ square matrices with entries from k under the usual operations of addition and multiplication. The Lie algebra $[k_n]$ will be denoted by $\mathfrak{gl}(n,k)$. With the mapping $\varphi \mapsto A_\varphi$ obviously preserving the commutator of two elements, we have the isomorphism of Lie algebras

$$\mathfrak{gl}(M,k) \cong \mathfrak{gl}(n,k).$$

An important example of Lie algebras is the subalgebra $\mathfrak{sl}(n,k)$ of $\mathfrak{gl}(n,k)$ consisting of all matrices X with trace zero. The set of such matrices being closed under commutator follows from the equation

$$\operatorname{tr}(XY) = \operatorname{tr}(YX)$$

and the fact that tr is a linear function. It should be recalled that given a matrix $X = (x_{ij})$, $1 \leq i,j \leq n$, its *trace* is defined by

$$\operatorname{tr} X = \sum_{i=1}^{n} x_{ii}.$$

Another example is the Lie algebra $\mathfrak{so}(n,k)$ of skew-symmetric matrices of order n over k, that is, matrices X such that ${}^tX = -X$, ${}^tX$ being the transpose of X. Using well-known properties of passing to the transpose, we have

$$\begin{aligned} {}^t[X,Y] &= {}^t(XY - YX) = {}^t(XY) - {}^t(YX) = {}^tY{}^tX - {}^tX{}^tY \\ &= YX - XY = -[X,Y]. \end{aligned}$$

1.2.3 Lie algebras of derivations

Let R be an algebra over a ring k. We denote by $\operatorname{Der}_k(R)$ the subset in $\mathfrak{gl}(R,k)$ consisting of all derivations of R. We will verify that $\operatorname{Der}_k(R)$ is a subalgebra of $\mathfrak{gl}(R,k)$. Indeed, given $\delta_1, \delta_2 \in \operatorname{Der}_k(R)$, $x_1, x_2 \in R$, we have

$$\begin{aligned} [\delta_1,\delta_2](x_1x_2) &= (\delta_1\delta_2 - \delta_2\delta_1)(x_1x_2) \\ &= \delta_1(\delta_2(x_1)x_2 + x_1\delta_2(x_2)) - \delta_2(\delta_1(x_1)x_2 + x_1\delta_1(x_2)) \\ &= \delta_1\delta_2(x_1)x_2 + \delta_2(x_1)\delta_1(x_2) + \delta_1(x_1)\delta_2(x_2) + x_1\delta_1\delta_2(x_2) \\ &\quad - \delta_2\delta_1(x_1)x_2 - \delta_1(x_1)\delta_2(x_2) - \delta_2(x_1)\delta_1(x_2) - x_1\delta_2\delta_1(x_2). \end{aligned}$$

Collecting like terms, using distributivity and the definition of the commutator, we arrive at

$$[\delta_1,\delta_2](x_1x_2) = [\delta_1,\delta_2](x_1)x_2 + x_1[\delta_1,\delta_2](x_2).$$

Hence $[\delta_1,\delta_2] \in \operatorname{Der}_k(R)$. It is obvious that $\operatorname{Der}_k(R)$ is a k-submodule in $\mathfrak{gl}(R,k)$. Now we see that $\operatorname{Der}_k(R)$ is a Lie algebra. It is called the *Lie algebra of derivations* of a k-algebra R. An important particular case is the Lie algebra $W_n(k)$ of all derivations of the polynomial ring $k[x_1,x_2,\dots,x_n]$ over a field k. Each element of W_n can be written in the form

$$\delta = f_1\frac{\partial}{\partial x_1} + f_2\frac{\partial}{\partial x_2} + \cdots + f_n\frac{\partial}{\partial x_n}$$

where $f_1, f_2, \dots, f_n$ are some polynomials in $x_1, x_2, \dots, x_n$ (the images of x_i under δ) and $\partial/\partial x_1, \partial/\partial x_2, \dots, \partial/\partial x_n$ are usual partial derivatives.

1.3 Derivation algebra of a Lie algebra

1.3.1 Inner derivations

Let L be a Lie algebra over a ring k, $x \in L$. By $\operatorname{ad} x$ we will be denoting an endomorphism of k-module L whose value on an element $y \in L$ is given by

$$\operatorname{ad} x(y) = xy.$$

Let us verify that $\operatorname{ad} x$ is a derivation. Since L is anticommutative and satisfies the Jacobi identity $J(x, y, z) = 0$, we have

$$\begin{aligned}\operatorname{ad} x(yz) = x(yz) = -y(zx) - z(xy) = y(xz) + (xy)z \\ = \operatorname{ad} x(y)z + y \operatorname{ad} x(z).\end{aligned}$$

Any derivation of the form $\operatorname{ad} x$, $x \in L$, is called an *inner derivation* determined by x. The set of all inner derivations of a Lie algebra L is a subset $\operatorname{ad} L$ in Lie algebra $\operatorname{Der}_k(L)$ of all derivations of L. There is a natural mapping $\operatorname{ad} : L \to \operatorname{Der}_K(L)$ that associates with each $x \in L$ an inner derivation $\operatorname{ad} x$ determined by x.

Theorem 1.3. *The above mapping* ad *is a Lie algebra homomorphism. The image* $\operatorname{ad} L$ *under this homomorphism is an ideal of* $\operatorname{Der}_k L$.

Proof. It is obvious that ad is a homomorphism of k-modules. Now let $x, y \in L$. Then

$$\begin{aligned}\operatorname{ad}(xy)(z) &= (xy)z = -z(xy) = x(yz) + y(zx) \\ &= x(yz) - y(xz) = (\operatorname{ad} x \operatorname{ad} y)(z) - (\operatorname{ad} y \operatorname{ad} x)(z) \\ &= [\operatorname{ad} x, \operatorname{ad} y](z).\end{aligned}$$

Hence $\operatorname{ad}(xy) = [\operatorname{ad} x, \operatorname{ad} y]$, proving the first part of the theorem.

Now we will show that $\operatorname{ad} L$ is an ideal of $\operatorname{Der}_k L$. For this we compute $[\delta, \operatorname{ad} x]$, where $\delta \in \operatorname{Der}_k L$, $x \in L$. Given an arbitrary $y \in L$, we have

$$\begin{aligned}[\delta, \operatorname{ad} x](y) &= (\delta \operatorname{ad} x - \operatorname{ad} x\delta)(y) \\ &= \delta(xy) - \operatorname{ad} x(\delta(y)) \\ &= (\delta x)(y) + x(\delta(y)) - x(\delta(y)) \\ &= (\delta x)y = \operatorname{ad} \delta(x)(y).\end{aligned}$$

Thus $[\delta, \operatorname{ad} x] = \operatorname{ad} \delta(x) \in \operatorname{ad} L$. Hence $\operatorname{ad} L$ is indeed an ideal of $\operatorname{Der}_k L$. □

1.3.2 Center of a Lie algebra

Let L be a Lie algebra over a commutative ring k. The set $Z(L)$ of all elements $z \in L$ such that for any $x \in L$ we have $xz = 0$ is called the *center* of L.

Corollary of Theorem 1.1. The center of a Lie algebra is an ideal of this algebra.

Proof. It is obvious from the definition of the center and of the mapping ad that for any Lie algebra L we have $Z(L) = \ker \operatorname{ad}$. By the homomorphism theorem in Section 1.1.1, the kernel of a homomorphism of algebras is an ideal. □

We notice that, according to the homomorphism theorem,

$$L/Z(L) \cong \operatorname{ad} L.$$

An important conclusion can be made in the case where the center of a Lie algebra L is equal to zero. In this case L is isomorphic to a subalgebra of a Lie algebra of the form $[A]$, A being an associative algebra (cf. Section 1.2.1). It will be seen later (Section 2.5.3) that this assertion is true for any Lie algebras which are free k-modules.

A Lie algebra L is called *abelian* if it coincides with its center. In this case $xy = 0$ for all $x, y \in L$.

1.4 Semidirect product

1.4.1 Cartesian product of rings

Let $(R_\alpha)_{\alpha\in I}$ be a nonempty family of algebras over the same ring k. We define the Cartesian product

$$R = \coprod_{\alpha\in I} R_\alpha$$

as an algebra whose elements are all functions

$$f : I \to \bigcup_{\alpha\in I} R_\alpha$$

such that $f(\alpha) \in R_\alpha$. The k-algebra structure is given by

$$\begin{aligned}(\lambda f)(\alpha) &= \lambda f(\alpha),\\ (f+g)(\alpha) &= f(\alpha) + g(\alpha),\\ (fg)(\alpha) &= f(\alpha)g(\alpha),\end{aligned}$$

where $\lambda \in k$, $f, g \in R$, and α is an arbitrary element of the index set I. It is easy to verify that all the axioms defining k-algebras are satisfied. Moreover, the Cartesian product of k-algebras satisfying any of the laws mentioned in Sections 1.1.2–1.1.4 is an algebra with the same property. In particular, the Cartesian product of Lie algebras over k is a Lie algebra over k.

A homomorphism $\pi_\alpha : R \to R_\alpha$ such that $\pi_\alpha(f) = f(\alpha)$, $\alpha \in I$, is called a *projection* of R onto R_α. A subalgebra S of the Cartesian product R is called a *sub-Cartesian product* in R if for every $\alpha \in I$ we have $\pi_\alpha(S) = R_\alpha$.

1.4.2 Direct product

We preserve the notation from Section 1.4.1. Let $f \in R$. The *support* supp f of a function f is the set of all indices $\alpha \in I$ such that $f(\alpha) \neq 0$. The subset S of a Cartesian product R of a family $(R_\alpha)_{\alpha\in I}$ of k-algebras whose elements are functions of f with finite support is a subalgebra in R and it is called the *direct product* of the above family. The direct product is denoted by

$$S = \prod_{\alpha\in I} R_\alpha.$$

In the case where I is finite, say, $I = \{1, 2, \dots, n\}$, the notions of the Cartesian and of the direct product coincide, and we write simply

$$R = \prod_{i\in I} R_i = R_1 \times R_2 \times \cdots \times R_n.$$

It is convenient to represent the elements of R as vectors with n components of the form $(r_1, r_2, \dots, r_n)$, where $r_i \in R_i$, $i = 1, 2, \dots, n$. All the operations are then defined componentwise. The concept of the subdirect product is defined along the same lines as that of the sub-Cartesian product.

1.4.3 Direct sum

Let S be an algebra and $S_1, S_2, \dots, S_n$ a collection of its ideals. We say that S is the *direct sum* of its (nonzero) ideals $S_1, S_2, \dots, S_n$ if each element x in S has precisely one representation of the form

$$x = x_1 + x_2 + \cdots + x_n, \tag{1.4}$$

where $x_i \in S$, $i = 1, 2, \dots, n$. In this case we write $S = S_1 \oplus \cdots \oplus S_n$. If y is another element in S and $y = y_1 + y_2 + \cdots + y_n$ then

$$xy = x_1y_1 + x_2y_2 + \cdots + x_ny_n \tag{1.5}$$

is the representation for xy and

$$\lambda x = \lambda x_1 + \lambda x_2 + \cdots + \lambda x_n \tag{1.6}$$

is the representation for λx, $\lambda \in k$. Indeed, if $i \neq j$ then x_iy_j is both in S_i and in S_j. From the uniqueness of representations of the form in equation (1.4), we see that $S_i \cap S_j = \{0\}$. Hence $x_iy_j = 0$. Now

$$\begin{aligned} xy &= (x_1 + x_2 + \cdots + x_n)(y_1 + y_2 + \cdots + y_n) \\ &= x_1y_1 + x_2y_2 + \cdots + x_ny_n + \sum_{i\neq j} x_iy_j = x_1y_1 + x_2y_2 + \cdots + x_ny_n. \end{aligned}$$

From $x_iy_i \in S_i$ we see that equation (1.5) is the desired representation for xy. Using equations (1.5) and (1.6), we find that by associating an element of the form $x = s_1 + \cdots + s_n$, $s_i \in S_i$, with an element $(s_1, s_2, \ldots, s_n) \in S_1 \times S_2 \times \cdots \times S_n$ we obtain an isomorphism of the direct product and the direct sum

$$S_1 \times S_2 \times \cdots \times S_n \cong S_1 \oplus S_2 \oplus \cdots \oplus S_n.$$

1.4.4 Semidirect product of Lie algebras

Let L and M be Lie algebras over the same ring k. We assume that there is a homomorphism φ of L into the derivation algebra $\operatorname{Der}_k M$ of M. The set of all pairs (x, y), $x \in L$, $y \in M$, becomes an algebra if we put

$$\begin{aligned}(x_1, y_1) + (x_2, y_2) &= (x_1 + x_2, y_1 + y_2),\\ \lambda(x, y) &= (\lambda x, \lambda y),\\ (x_1, y_1)(x_2, y_2) &= (x_1x_2, \varphi(x_1)(y_2) - \varphi(x_2)(y_1) + y_1y_2).\end{aligned}$$

It is obvious that N (our notation for the above set of pairs) is a k-module and the multiplication in N is anticommutative. Let us verify the validity of the Jacobi identity. We remark that the Jacobian is a multilinear alternating function of its arguments; moreover, $(x, y) = (x, 0) + (0, y)$. Thus the equation $J(u, v, w) = 0$ should be verified in the four cases below:

(i) $u = (x_1, 0)$, $v = (x_2, 0)$, $w = (x_3, 0)$;
(ii) $u = (x_1, 0)$, $v = (x_2, 0)$, $w = (0, y)$;
(iii) $u = (x, 0)$, $v = (0, y_1)$, $w = (0, y_2)$;
(iv) $u = (0, y_1)$, $v = (0, y_2)$, $w = (0, y_3)$.

In Case (i), $J(u, v, w) = (J(x_1, x_2, x_3), 0) = 0$ since L is a Lie algebra. Similarly, in Case (iv), $J(u, v, w) = (0, J(y_1, y_2, y_3)) = 0$, since M is a Lie algebra.

In Case (ii), we have

$$\begin{aligned}u(vw) &= (x_1, 0)((x_2, 0)(0, y)) = (x_1, 0)(0, \varphi(x_2)(y))\\ &= (0, \varphi(x_1)\varphi(x_2)(y)),\\ v(wu) &= (x_2, 0)((0, y)(x_1, 0)) = (x_2, 0)(0, -\varphi(x_1)(y))\\ &= (0, -\varphi(x_2)\varphi(x_1)(y)),\\ w(uv) &= (0, y)((x_1, 0)(x_2, 0)) = (0, y)(x_1x_2, 0)\\ &= (0, -\varphi(x_1x_2)(y)).\end{aligned}$$

Therefore,

$$\begin{aligned}J(u, v, w) &= (0, \varphi(x_1)\varphi(x_2)(y) - \varphi(x_2)\varphi(x_1) - \varphi(x_1x_2)(y))\\ &= (0, ([\varphi(x_1), \varphi(x_2)] - \varphi(x_1x_2))(y)) = 0.\end{aligned}$$

The final equation is true since φ is a homomorphism of Lie algebras, that is,

$$\varphi(x_1x_2) = [\varphi(x_1), \varphi(x_2)].$$

In Case (iii), we have

$$\begin{aligned} u(vw) &= (x,0)((0,y_1)(0,y_2)) = (x,0)(0,y_1y_2) \\ &= (0,\varphi(x)(y_1y_2)), \\ v(wu) &= (0,y_1)((0,y_2)(x,0)) = (0,y_1)(0,-\varphi(x)(y_2)) \\ &= (0,-y_1\varphi(x)(y_2)), \\ w(uv) &= (0,y_2)((x,0)(0,y_1)) = (0,y_2)(0,\varphi(x)(y_1)) \\ &= (0,y_2(\varphi(x)(y_1))). \end{aligned}$$

Therefore,

$$\begin{aligned} J(u,v,w) &= (0,\varphi(x)(y_1y_2) - y_1(\varphi(x)(y_2)) + y_2(\varphi(x)(y_1))) \\ &= (0,\varphi(x)(y_1y_2) - y_1(\varphi(x)(y_2)) - (\varphi(x)(y_1)y_2)) \\ &= 0. \end{aligned}$$

The final equation holds since for each $x \in L$ the endomorphism $\varphi(x)$ is a derivation of M.

Thus N is a Lie algebra over k called the *semidirect product* of L by M corresponding to φ. We will denote this algebra by $L \rtimes_\varphi M$, or simply by $L \rtimes M$, if φ is fixed. In the case where φ is the zero map, we arrive at the direct product of L by M.

1.4.5 Split extension of algebras

A k-algebra R is called an *extension* of a k-algebra S with quotient algebra T if R has an ideal $S' \cong S$ such that $R/S' \cong T$. This extension is called a *split extension* if R has a subalgebra T' such that $T' \cap S' = \{0\}$ and $R = T' + S'$ (k-module sum). In this case $T' \simeq T$. Indeed, using the isomorphism theorem (Section 1.1.1), we find

$$T \cong R/S' \cong T' + S'/S/' = T'/T' \cap S' = T'/\{0\} \cong T'.$$

We notice that if a Lie algebra R is a split extension of S by T then for an appropriate homomorphism $\varphi : T \to \operatorname{Der}_k S$ we have

$$R \cong T \rtimes_\varphi S.$$

Indeed, we may identify S and S', as well as T and T', by the above isomorphisms. Then $R = T \oplus S$. For each $x \in T$, let $\varphi(x)$ denote the restriction to S of an inner derivation ad x. Since ad is a homomorphism (Theorem 1.3), the same is true of φ. Now we can compute the product of any pair of elements in R as follows:

$$\begin{aligned}(x_1 + y_1)(x_2 + y_2) &= x_1x_2 + x_1y_2 - x_2y_1 + y_1y_2\\ &= x_1x_2 + \operatorname{ad} x_1(y_2) - \operatorname{ad} x_2(y_1) + y_1y_2\\ &= x_1x_2 + \varphi(x_1)(y_2) - \varphi(x_2)(y_1) + y_1y_2.\end{aligned}$$

This shows that the mapping $(x, y) \mapsto x + y$ from $T \rtimes_\varphi S$ into R is compatible with multiplication in both algebras. The other operations being preserved and the bijectivity of this mapping are obvious. Hence we obtain the asserted isomorphism.

1.5 Series of ideals in Lie algebras

1.5.1 Products of ideals

Given a k-algebra R and its subsets S an T, we define their *product* ST as a k-module generated by all products of the form xy, $x \in S$, $y \in T$. In the case where R is a Lie algebra or an associative algebra, the product ST of two ideals S, T is itself an ideal. Indeed, each element r in ST can be written in the form

$$r = x_1y_1 + \cdots + x_my_m, \quad \text{where } x_i \in S, y_i \in T, i = 1, \ldots, m,$$

with m being a natural number. Therefore it is sufficient to show that, given $u \in R$, $x \in S$, $y \in T$, we have $u(xy), (xy)u \in ST$. In the case of associative algebras, these products are equal to $(ux)y$ and $x(yu)$, respectively, and all is true since S and T are ideals. In the case of Lie algebras,

$$-(xy)u = u(xy) = -x(yu) - y(ux) = xy' + x'y,$$

where $x' = ux \in S$, $y' = y - yu \in T$. It should be remarked that, for Lie algebras, we always have $ST = TS$.

1.5.2 Lower central series

Let L be a Lie algebra over a ring k. We define L^n, $n = 1, 2, \ldots$, using induction by setting $L^1 = L$ and $L^{n+1} = L^nL$. According to Section 1.5.1, we obtain a descending series of ideals of the form

$$L = L^1 \supseteq L^2 \supseteq \cdots \supseteq L^n \supseteq L^{n+1} \supseteq \cdots.$$

This is called the *lower central series* of a Lie algebra L. In fact, this notion is a particular case of the notion of a descending central series which is defined as follows.

A series

$$L = L_1 \supseteq L_2 \supseteq \cdots \supseteq L_n \supseteq L_{n+1} \supseteq \cdots,$$

in which all L_i are ideals of L, is called a *descending central series* if for each natural $n \geq 1$ the ideal L_n/L_{n+1} of L/L_{n+1} belongs to the center of the latter algebra. This is equivalent to the inclusion $LL_n \subset L_{n+1}$, $n = 1, 2, \ldots$ It is obvious that the lower central series is a descending central series and that given an arbitrary descending central series as above we have $L_n \supseteq L^n$. Indeed, $L_1 = L^1$, and if L_n is shown to contain L^n, we have $L_{n+1} \supseteq LL^n = L^{n+1}$ which, by induction, proves our assertion.

1.5.3 Derived series

Again let L be a Lie algebra over a ring k. We define $L^{(n)}$, $n = 0, 1, 2, \ldots$, by induction if we set $L^{(0)} = L$ and $L^{(n+1)} = L^{(n)}L^{(n)}$. According to Section 1.5.1, we get as a result a descending chain of ideals in L of the form

$$L = L^{(0)} \supseteq L^{(1)} \supseteq \cdots \supseteq L^{(n)} \supseteq L^{(n+1)} \supseteq \cdots.$$

It is called the *derived series* of L. The first term $L^{(1)}$ is called the *derived algebra* and it is often denoted by L'. Similarly, we often write $L^{(2)} = L''$. It is clear that $L^1 = L'$. If $L^2 = \{0\}$ then the product of any two elements in L is zero, that is, L is an abelian Lie algebra (see Section 1.3.2). The derived series is solvable in the following sense. Let

$$L = L_0 \supseteq L_1 \supseteq \cdots \supseteq L_n \supseteq L_{n+1} \supseteq \cdots$$

be a series of ideals in L such that, for all $n = 0, 1, \ldots$, L_n/L_{n+1} is an abelian ideal of L/L_{n+1} (this is equivalent to writing $L_nL_n \subseteq L_{n+1}$). We call any such series *solvable*. It is obvious that the derived series is solvable and that, given a solvable series as above, we have $L_n \supseteq L^{(n)}$ for all $n \geq 0$.

1.5.4 Centralizer

Let L be a Lie algebra over a ring k, and S, T be nonempty subsets in L. We define the *centralizer* $C_L(S, T)$ as a k-submodule in L generated by all $x \in L$ such that $Sx \subset T$. We remark that if S, T are ideals of L then $C_L(S, T)$ enjoys the same property. Indeed, assume that $sx = t \in T$ for all $s \in S$. Then given a $y \in L$, we have

$$s(xy) = (sx)y + x(sy) = ty + xs' \in T.$$

Here $s' = sy$ is in S since S is an ideal of L. Now $ty \in T$ since T is an ideal of L, and $xs' \in T$ since $x \in C_L(S, T)$. If, more generally, T is an ideal of L then, similarly, $C_L(S, T)$

is a subalgebra of L. If $T = \{0\}$, then $C_L(S,T)$ is denoted by $C_L(S)$ and is called the *centralizer of S in L*. If $S = L$ then $C_L(S)$ is merely the center of L (see Section 1.3.2), and we have $C_L(L) = Z(L)$.

Proposition 1.4. *Let S, T be ideals in a Lie algebra L such that $l_k(S/T) = n$. Then*

$$l_k(L/C_L(S,T)) \le n^2.$$

Proof. We consider the algebra $\mathrm{End}_k(S/T)$ of all k-endomorphisms of S/T. Mapping $\mathrm{ad} : x \mapsto \mathrm{ad}\, x$ induces a homomorphism $\varphi : L \to \mathrm{End}_k(S/T)$ such that for each $s \in S$ we have

$$\varphi(x)(s + T) = xs + T.$$

The kernel of this homomorphism is equal to $C_L(S,T)$. Hence $L/C_L(S,T)$ is isomorphic to a k-submodule in $E = \mathrm{End}_k(S/T)$. It is sufficient to show that $l(E) \le n^2$. But it is more appropriate to prove that, given k-modules M, N of lengths m, n, the length of the k-module $H = \mathrm{Hom}_k(M,N)$ of all homomorphisms from M into N is at most mn. Indeed, if $m = n = 1$ then either $H = M$ for $M \cong N$ or $H = \{0\}$ for $M \ncong N$ (Schur's lemma). Now we apply induction over the length of N and then over the length of M. If N_1 is a submodule of N then with each $f \in H$ we may associate a homomorphism $\bar{f} \in \mathrm{Hom}(M, N/N_1)$ such that $\bar{f}(m) = f(m) + N_1$. This gives rise to a homomorphism $\alpha : H \mapsto \mathrm{Hom}(M, N/N_1)$ whose kernel is $\mathrm{Hom}(M, N_1)$. Then

$$l(H) = l(\mathrm{Hom}(M, N_1)) + l(\mathrm{Hom}(M, N/N_1)).$$

Now if M_1 is a submodule of M then we may construct a homomorphism from H into $\mathrm{Hom}(M_1, N)$ by mapping an $f \in H$ into its restriction $\bar{f}$ on M_1. The kernel of this homomorphism may naturally be identified with a submodule in $\mathrm{Hom}(M/M_1, N)$ and the induction applies. From the above procedure it is clear that $l_k(H) \le mn$. □

1.5.5 Upper central series

We define an ascending chain of ideals $Z_n(L)$, $n = 0, 1, 2, \ldots$, by setting $Z_0(L) = \{0\}$, and $Z_{n+1}(L) = C_L(L, Z_n(L))$. According to a remark in Section 1.5.4, for each $n \ge 0$, $Z_n(L)$ is indeed an ideal. Obviously, $Z_1(L) = Z(L)$ is the center of L. Thus we have a series of the form

$$\{0\} = Z_0(L) \subseteq Z_1(L) \subseteq \cdots \subseteq Z_n(L) \subseteq Z_{n+1}(L) \subseteq \cdots.$$

We call this the *upper central series* of a Lie algebra L. Again, a more general notion is that of an *ascending central series*, that is, of a series

$$\{0\} = L_0 \subseteq L_1 \subseteq \cdots \subseteq L_n \subseteq L_{n+1} \subset \cdots$$

such that L_{n+1}/L_n is in the center of L/L_n, $n = 0, 1, 2, \dots$ Moreover, given an ascending central series as above, we have $L_n \subseteq Z_n(L)$. Indeed, $L_0 = Z_0(L)$ and if $L_n \subseteq Z_n(L)$ we have $L_{n+1}L \subseteq L_n \subseteq Z_n(L)$, hence $L_{n+1} \subseteq C(Z_n(L), L) = Z_{n+1}(L)$.

1.6 Modules

1.6.1 Representation of a Lie algebra

Let L be a Lie algebra over a ring k, and V a k-module. We consider the associative algebra $\mathrm{End}_k\, V$ of all endomorphisms of V and its associated Lie algebra $\mathfrak{gl}(V) = \mathfrak{gl}(V, k)$ (see Section 1.2.2). Any homomorphism $\rho : L \to \mathfrak{gl}(V)$ is called a *representation* of a Lie algebra *L by endomorphisms of V*. If the kernel of ρ is 0, we say that ρ is a faithful representation. Mapping $\mathrm{ad} : L \to \mathrm{Der}\, L \subset \mathfrak{gl}(L)$ is a homomorphism (Theorem 1.3) and its kernel is the center of L. Thus we have a representation of L called the *adjoint representation*. Adjoint representation is faithful if and only if the center of a Lie algebra is zero.

Given a representation ρ of a k-algebra L in a k-module V and a k-submodule W of V such that $\rho(L)(W) \subset W$, the restriction $\sigma = \rho\,|_W : x \mapsto \rho(x)\,|_W$ is a representation σ of L by endomorphisms of W We call it a *subrepresentation* of ρ. We may also define a representation $\tau = \rho/\sigma : L \to \mathfrak{gl}(V/W)$ by setting

$$\tau(x)(v + W) = \tau(x)(v) + W.$$

The definition is unambiguous since W is invariant, that is, $\rho(L)(W) \subset W$. Thus we obtain a representation τ of L, called the *quotient representation* of ρ over σ. The notions of subrepresentation and quotient representation are particular cases of the notion of *subquotient representation* which can be defined both as a subrepresentation of a quotient representation or a quotient representation of a subrepresentation. Finally, we say that a representation is *irreducible* if it has no subrepresentations different from the zero representation and the representation itself.

1.6.2 Module over a Lie algebra

Let L be a Lie algebra over a ring k and V a k-module. We say that V is a module over L, or an L-module, if there exists a k-bilinear map $L \times V \to V$ satisfying the following axiom:

$$(xy)v = x(yv) - y(xv). \tag{1.7}$$

Here $x, y \in L$, $v \in V$, yv is the image of (y, v) under the above bilinear map. Let V_1, V_2 be two L-modules. The mapping $\varphi : V_1 \to V_2$ is called the *homomorphism* of L-modules if φ is a k-homomorphism and for each $v \in V_1$ we have $x\varphi(v) = \varphi(xv)$. By way of generalizing to the case of L-modules of the concept of the right-normed products, we write $g_1g_2 \cdots g_nv$ in place of $g_1(g_2 \cdots (g_nv) \cdots)$, where $g_1, g_2, \dots, g_n \in L$ and $v \in V$.

The notion of the L-module is equivalent to that of representation. Indeed, if V is an L-module, we may define a mapping $\rho : L \to \mathfrak{gl}(V, k)$ by setting $\rho(x)(v) = xv$, where $x \in L$, $v \in V$. In terms of ρ, equation (1.7) can be rewritten as

$$(xy)(v) = \rho(x)(\rho(y)(v)) - \rho(y)(\rho(x)(v)).$$

In other words, for all $x, y \in L$, we have

$$\rho(xy) = [\rho(x), \rho(y)]$$

in $\mathfrak{gl}(V)$. This shows that $\rho : L \to \mathfrak{gl}(V)$ is a homomorphism, or a representation of L in the k-module V ($\rho(x)$ being a k-endomorphism easily follows from the bilinearity of the map defining L-module). The inversion of this argument proves that the k-module of a representation ρ of a Lie algebra L becomes an *L-module* if we put $xv = \rho(x)(v)$ for all $x \in L$, $v \in V$.

Similarly to Section 1.6.1, we define the concepts of *submodule, quotient module*, and *subquotient module* of an L-module as companions to the concepts of subrepresentation, quotient representation, and subquotient representation of a representation of a Lie algebra L. Thus a k-submodule V/W becomes an L-module if we put $x(v + W) = xv + W$ for all $x \in L$, $v \in V$. The result is an L-quotient module V/W. Given L-submodules $U \supseteq W$ of an L-module V, we say that U/W is an L-subquotient (or L-factor) of V. If an L-module V has no nontrivial submodules different from V then it is called a *simple (irreducible)* L-module. A direct sum of simples L-modules is called a *semisimple (completely reducible)* L-module.

In the case where an L-module V is defined via a representation $\rho : L \to \operatorname{End}_k V$, we say that V is the module of ρ. In the case of the adjoint representation ad, we see that a Lie algebra is an L-module with respect to the multiplication on the left. The module of the adjoint representation (i. e., L itself) is called the *adjoint module*. Its submodules are the ideals of L. If the adjoint L-module is simple then L is called a *simple Lie algebra*.

If H is a subalgebra of a Lie algebra L and V is an L-module then, naturally, V is an H-module. Thus, having an L-module V, we may speak about H-submodules, H-quotient modules, etc. In particular, given a subalgebra H of a Lie algebra L, we may view L as an H-module with respect to the multiplication of the elements of L by the elements of H on the left. Given ideals $U \supseteq W$ of L, we turn U/W into an H-module by putting

$$x(u + W) = xu + W,$$

where $x \in H$, $u \in U$. Finally, if $L \subseteq [A]$ and A an associative algebra then the (associative) multiplication by the elements of L on the left makes A an L-module, $x \circ a = xa$.

Indeed, $[x, y]a = xya - yxa = x(ya) - y(xa)$. In this case we speak about the (left) *regular module* (representation).

1.6.3 Chief series

Note that if V is an L-module, L being a Lie algebra, then, having constructed the corresponding representation $\rho : L \to \operatorname{End} V$ and having taken the associative subalgebra A in $\operatorname{End} V$ generated by $\rho(L)$, we will see that V becomes an A-module. Moreover, within V, the concepts of an L-module, L-quotient module, etc., are identical to those of an A-module, A-quotient module, etc. This enables us to translate into the language of modules over a Lie algebra many of the results concerning modules over usual associative rings without even minor changes. This we do, and the reader is referred to Lang [199, Chapters III, IV, XVII].

Our terminology as concerns the Lie algebra case will deviate somewhat from that used in the associative algebra case. Thus we will say *chief subquotient* (factor) in place of simple *subquotient* (factor) and *chief series* in place of *composition series*. The chief series of a Lie algebra is a chief series of the adjoint L-module. Alternatively, the chief series is a series of ideals in which all factors are chief and which, starting from the zero subalgebra, reaches the algebra itself in finitely many steps.

Theorem 1.5 (Jordan–Hölder). *If an L-module V has at least one chief series*

$$\{0\} = V_0 \subset V_1 \subset \cdots \subset V_s = V$$

(of finite length) then each other chief series has the same length. If U/V is a chief factor of V then, for a suitable choice of i, we have $U/W \cong V_i/V_{i-1}$.

1.6.4 Semisimple series

Let V be an L-module. A series

$$\{0\} = V_0 \subset V_1 \subset \cdots \subset V_s = V$$

is called *semisimple* if, for each $i = 0, 1, \ldots, s$, the factor V_i/V_{i-1} is a semisimple L-module. In this and in the next subsections, we discuss the behavior of chief and semisimple series while passing to subalgebras.

Theorem 1.6. *Let L be a Lie algebra and M its subalgebra, both possessing finite chief series. We consider a chief series in L of the form*

$$\{0\} = L_0 \subset L_1 \subset \cdots \subset L_{i-1} \subset L_i \subset \cdots \subset L_n = L. \tag{1.8}$$

Then:
(i) *Every chief factor of L is isomorphic to a factor of series* (1.8).
(ii) *Every chief factor of M (as an M-module) is isomorphic to an M-factor of an M-module L_i/L_{i-1} of suitable $i = 1, \dots, s$.*
(iii) *If M is an ideal of L and if L possesses a semisimple series of length m as the L-module then L/M possesses a semisimple series of length m at most.*

Proof. (i) is a reformulation of Section 1.6.3. To prove (ii), we must refine the series

$$\{0\} = M \cap L_0 \subseteq M \cap I_1 \subseteq \cdots \subseteq M \cap L_{i-1} \subseteq M \cap L_i \subseteq \cdots \subseteq M \cap L_n = M$$

(cf. proof of Jordan–Hölder's theorem). But the factors of this series are isomorphic (as M-modules) to submodules of the factors in the series (1.8). Indeed,

$$M \cap L_i/M \cap L_{i-1} = M \cap L_i/(M \cap L_i) \cap L_{i-1} \cong M \cap L_i + L_{i-1}/L_{i-1} \subseteq L_i/L_{i-1}.$$

(iii) Since M belongs to the annihilator of the L-module L/M, the existence of a semisimple series in L/M is equivalent to the existence of such a series in the L-module L/M. Let

$$\{0\} = S_0 \subset S_1 \subset \cdots \subset S_{j-1} \subset S_j \subset \cdots \subset S_m = L$$

be an arbitrary semisimple series of length m in the L-module L. We consider a series of L-submodules in L/M of the form

$$\begin{aligned} 0 = M/M \subseteq S_1 + M/M \subseteq \cdots \subseteq S_{i-1} + M/M \\ \subseteq S_i + M/M \subseteq \cdots \subseteq S_m + M/M = L/M. \end{aligned}$$

An arbitrary factor of this series has the form

$$\begin{aligned} (S_i + M/M)/(S_{i-1} + M/M) &\cong S_i + M/S_{i-1} + M \\ &= S_i + (S_{i-1} + M)/S_{i-1} + M \\ &\cong S_i/S_i \cap (S_{i-1} + M) = S_i/S_{i-1} + S_i \cap M \\ &\cong (S_i/S_{i-1})(S_{i-1} + S_i \cap M)/S_{i-1}. \end{aligned}$$

Thus, the factor in sight is isomorphic to a homomorphic image of a semisimple L-module S_i/S_{i-1} and, according to a theorem in Lang [199, Chapter XVII], is itself semisimple as L-module (hence as L/M-module). □

Theorem 1.7. *Let L be a Lie algebra and V a simple L-module of finite length over the ground ring k. Suppose we are given an ideal M of L. Then any two chief factors of M-module V are isomorphic.*

Proof. Let W be a minimal M-submodule. We denote by T a maximal M-submodule of M-module V all of whose chief factors are isomorphic to W. Let us show that T is an L-submodule. Suppose this is not the case. Then there is an $x \in L$ such that $xT \not\subseteq T$. Now the mapping $t \mapsto xt + T$, $t \in T$, is an M-homomorphism from T into V/T. Indeed, for an arbitrary $y \in M$, we have

$$yt \mapsto x(yt) + T = yxt + (xy)t + T = y(xt) + T.$$

The latter equation follows since xy is in M as soon as $y \in M$ and since $MT \subseteq T$. We notice that the image of M-submodule T under this map is equal to $xt + T/T$ so that $xT + T/T$ is an M-submodule in V/T. Now let S/T be a minimal nonzero M-submodule in $xT + T/T$. It does exist since $xT + T \supseteq xT$ but $xT \not\subseteq T$. Since $xT + T/T$ is a homomorphic image of T, by the homomorphism theorem and Section 1.6.3, every chief factor, and S/T among them, is isomorphic to a chief factor of M-module T. Hence $S/T \cong W$. Now in the M-submodule S there is a series of M-submodules

$$\{0\} = T_0 \subset T \subset S.$$

Having refined this series to a chief one, we obtain a series

$$\{0\} = T_0 \subset T_1 \subset \cdots \subset T_r = T \subset S$$

all of whose factors are isomorphic to the M-module W. Once more using Section 1.6.3, we find that each chief factor of M-module S is isomorphic to W, which contradicts our choice of T. Now T is shown to be a nonzero L-submodule of V, that is, $T = V$. □

1.6.5 Modules and extensions

An ideal M of a Lie algebra L over a ring k is a submodule of the adjoint L-module. Moreover, if $C = C_L(M)$ and if N is any ideal of L contained in C then M becomes an (L/N)-module if we put

$$(x + N) \circ m = xm.$$

In particular, if M is abelian then $M \subseteq C$, that is, M is a module over $G = L/M$ with respect to the action of the form $(x+M)\circ m = xm$. Obviously, the extension of an abelian ideal M by $G = L/M$ is not completely determined by the structure of G-module on M. However, this takes place for the case of split extensions (see Section 1.4.5). Conversely, given a G-module M, we may, viewing M as an abelian Lie algebra, construct a split extension L of M by G as the set of pairs (x, m), $x \in G$, $m \in M$, with multiplication

$$(x, m)(y, n) = (xy, xm - yn). \tag{1.9}$$

By comparison with Section 1.4.4, we can see that L is isomorphic to the semidirect product $G \rtimes_\varphi M$, where $\varphi(x)(m) = xm$.

1.6.6 Lie algebra cohomology. Nonsplit extensions

Let G be a Lie algebra over a ring k, M a certain G-module. Let $C^n(G,M)$ denote the set of all k-multilinear skew-symmetric mappings from $\underbrace{G\times\cdots\times G}_{n}$ into M, $n=1,2,\ldots$ The elements of $C^n(G,M)$ are called the *n-dimensional (M-valued) cochains*. The mapping $\partial=\partial_n : C^n(G,M)\to C^{n+1}(G,M)$ given by

$$\begin{aligned}(\partial f)(x_1,\ldots,x_{n+1}) &= \sum_{i<j}(-1)^{i+j}f(x_ix_j,x_1,\ldots,\hat{x}_i,\ldots,\hat{x}_j,\ldots,x_{n+1})\\ &\quad+\sum_{i=1}^{n}(-1)^{i+1}x_if(x_1,\ldots,\hat{x}_i,\ldots,x_n)\end{aligned} \tag{1.10}$$

is called the *coboundary* operator (the letters under hats in equation (1.10) should be omitted). An element $f\in C^n(G,M)$ is called an *n-dimensional cocycle* if $\partial f=0$ and it is called an *n-dimensional coboundary* if there exists a cochain $g\in C^{n-1}(G,M)$ such that $f=\partial g$. The set of all n-dimensional cocycles is denoted by $Z^n(G,M)$, while the set of n-dimensional coboundaries is denoted by $B^n(G,M)$. From $\partial^2=0$ it follows that $Z^n(G,M)\supseteq B^n(G,M)$. Now the nth cohomology module $H^n(G,M)$ is defined by

$$H^n(G,M)=Z^n(G,M)/B^n(G,M).$$

Given a 2-cocycle f, we may construct a *standard extension* L_f of abelian ideal M by G as k-module $G\oplus M$ with multiplication

$$(x+m)*(y+n)=xy+xn-ym+f(x,y). \tag{1.11}$$

The anticommutativity follows from f being skew-symmetric. The verification of the Jacobi identity $J(u,v,w)=0$ (see Section 1.4.4) reduces to the cases where $u=x$, $v=y$, $w=z$ or $u=x$, $v=y$, $w=m$ ($x,y,z\in G$, $m\in M$). We write

$$x(yz)=x(f(y,z))+f(x,yz).$$

By permuting cyclically and using equation (1.10) together with the Jacobi identity, we get

$$\begin{aligned}J(x,y,z)&=xf(y,z)+yf(z,x)+zf(x,y)+f(x,yz)+f(y,zx)+f(z,xy)\\ &=(\partial f)(x,y,z)=0.\end{aligned}$$

Further

$$\begin{aligned}J(x,y,m)&=x*y*m+y*m*x+m*x*y\\ &=xym-yxm+m*(xy+f(x,y))\\ &=xym-yxm-(xy)m=0\end{aligned}$$

since M is both a G-module and an abelian algebra. Hence L_f is a Lie algebra. It is obvious that M is an abelian ideal of L_f and that $L_f/M \cong G$. Now if 2-cocycles f_1 and f_2 are *cohomologous cocycles* (that is, $f_1 + B^2(G,M) = f_2 + B^2(G,M)$) then L_{f_1}, as an extension, is *equivalent* to L_{f_2} in the following sense. There exists an isomorphism $\varphi : L_{f_1} \to L_{f_2}$ making the following diagram commutative:

$$\begin{array}{ccccc} M & \xrightarrow{\mu_1} & L_{f_1} & \xrightarrow{\varepsilon_1} & G \\ \downarrow{\scriptstyle 1_M} & & \downarrow{\scriptstyle \varphi} & & \downarrow{\scriptstyle 1_G} \\ M & \xrightarrow{\mu_2} & L_{f_2} & \xrightarrow{\varepsilon_2} & G \end{array} \quad . \tag{1.12}$$

Here μ_1 and μ_2 are natural imbeddings and $\varepsilon_1, \varepsilon_2$ are natural epimorphisms. The commutativity means that $\varphi\mu_1 = \mu_2 1_M$, $1_G\varepsilon_1 = \varepsilon_2\varphi$. To prove this, we choose the 1-cocycle h with $f_1 = f_2 + \partial h$. Then we may write

$$f_1(x,y) = f_2(x,y) - h(xy) + xh(y) - yh(x), \tag{1.13}$$

for $x, y \in G$. Now the desired isomorphism will be defined if we set $\varphi(x + m) = x + h(x) + m$. The commutativity follows from equation (1.11), by immediate calculations. Conversely, if L_{f_1} and L_{f_2} are equivalent as extensions then the diagram being commutative yields the existence of a linear map $h : G \to M$ such that $\varphi(x+m) = x + h(x) + m$. Since φ preserves products, we immediately arrive at equation (1.13). Hence $f_1 - f_2 = \partial h$. Thus we have proved the second statement of Theorem 1.8 below.

Theorem 1.8. *Let G be a Lie algebra over a ring k and M a G-module. For any extension L of an abelian ideal M by G, which is k-split and in which the adjoint G-module structure on M is as above, there exists a 2-cocycle f such that L is equivalent to a standard extension L_f. Any two standard extensions L_{f_1} and L_{f_2} are equivalent if and only if f_1 and f_2 are cohomologous.*

Proof. To prove the first statement, we represent L in the form $L = \bar{G} \oplus M$ where $\bar{G} \cong G$ as k-modules. Write $\bar{x}$ for the image of $x \in G$ under this isomorphism. Then $f(x,y) = \bar{x}\bar{y} - \overline{xy}$ is an element of M. If $b = \bar{y} + n$ then (cf. Section 1.6.5)

$$(\bar{x} + m)(\bar{y} + n) = \bar{x}\bar{y} + xn + ym - \overline{xy} + xn - ym + f(x,y). \tag{1.14}$$

Since L satisfies the Jacobi identity and the anticommutativity law, we immediately find that $f \in Z^2(G,M)$. By comparison of equations (1.11) and (1.14), we see that L is equivalent to L_f, proving the first statement of the theorem. The second part was verified previously. □

Corollary 1.9 (Corollary of Theorem 1.8). *Let G be a Lie algebra over a field k and M a G-module. Any extension of M by G with prescribed adjoint action of G on M splits if and only if $H^2(G,M) = 0$.*

Proof. In the case of a field the condition of being k-split is vacuous. It remains to notice that L_0 is a split extension and that any extension equivalent to a split one is itself split. □

1.6.7 Basis of a Lie module

Let G be a Lie algebra and E a well-ordered set of its generators as a k-module (according to Zermelo's axiom, any set can be well-ordered). Let M be a G-module generated by a nonempty set T. It is important to have an explicit form of a generating set for M as a k-module. Since any module is the sum of cyclic ones, we restrict ourselves to this particular case (by a *cyclic module* we mean a module generated by a single element).

Proposition 1.10. *Let M be a cyclic G-module generated by an element t and G be a Lie algebra over k with a linearly ordered generating set E as k-modules. Then M is generated over k by the union of the one-element set $\{t\}$ and the set of all elements of the form*

$$e_1 \cdots e_n t, \text{ where } e_1, \dots, e_n \in E, \text{ such that } e_1 \geq \cdots \geq e_n \text{ and } n \geq 1. \tag{1.15}$$

Proof. Obviously, M is the span of t and the set of all elements of the form $f_1 \cdots f_m t$ where $f_1, \dots, f_m \in E$ and the row $(f_1, \dots, f_m)$ is not necessarily decreasing. We will show that $f_1 \cdots f_m t$ is a k-combination of the elements of the form (1.15). We will proceed by induction over m and, for m fixed, over the number of inversions in the row $(f_1, \dots, f_m)$. We admit that there is an inversion in the row $(f_1, \dots, f_m)$, that is, for some i we have $f_i < f_{i+1}$. Then we may write

$$f_1 \cdots f_i f_{i+1} \cdots f_m t = f_1 \cdots f_{i+1} f_i \cdots f_m t + f_1 \cdots (f_i f_{i+1}) \cdots f_m t.$$

The number of inversions in $(f_1, \dots, f_{i+1}, f_i, \dots, f_m)$ is less than that in the row in sight, hence we may apply the induction hypothesis to $f_1 \cdots f_{i+1} f_i \cdots f_m t$. Now $f_i f_{i+1}$ is a linear combination of elements in E. By distributivity then $f_1 \cdots (f_i f_{i+1}) \cdots f_m t$ is a k-combination of elements, to which we may apply induction (this time over m). □

An important particular case is that of a *free G-module M freely generated by a set T* (basic set, or *basis*). By this we mean a G-module such that any map $\varphi : T \to N$, N being any G-module, extends to a homomorphism of G-modules $\bar{\varphi} : M \to N$.

Theorem 1.11. *Let G be a Lie algebra over a ring k which is a free k-module with a free k-basis E which is well-ordered. Then there exists a free G-module with any prescribed free generating set T and this is a free k-module freely generated by T along with the set of all elements of the form*

$$e_1 e_2 \cdots e_n t, \quad \text{where } t \in T, e_1, e_2, \dots, e_n \in E, e_1 \geq e_2 \geq \cdots \geq e_n,$$

with n running through all natural numbers.

Proof. We introduce a free k-module F whose basis is formed by the elements of the form Pt, P being a decreasing row (possibly empty) $(e_1, e_2, \ldots, e_n)$ of elements in E (that is, $e_1 \geq e_2 \geq \cdots \geq e_n$), $t \in T$. It will be convenient to assume that E is well-ordered with respect to decreasing order. This enables us to proceed by inverse induction with respect to the given order. We write $|P| = n$ and also $e \geq P$ if $e \geq e_1$. If the latter holds, we write $P' = eP$, where $P' = (e, e_1, \ldots, e_n)$. Our next purpose is to define the action of G on F using the direct induction over $|P|$ and the inverse induction over the order of E. In doing so, it is required that the resulting element $e * Pt$ be a k-combination of the elements of the form $P't$ such that $|P'| \geq |P| + 1$.

Now if $|P| = 0$, we put $e * Pt = et$. If $|P| = n \geq 1$ and $e \geq P$, we put $e * (Pt) = (eP)t$. Otherwise we write $P = e_1P'$, $e_1 > e$, and then we put

$$e * (Pt) = e_1 * (e * P't) + (ee_1) * P't. \tag{1.16}$$

The definition is correct since the right-hand side is well-defined by induction. We notice that, according to our definition, it is obvious that

$$e * Pt = \bar{P}t + \sum_{\alpha} P_\alpha t,$$

where $\bar{P}$ results by putting the elements of $\{e\} \cup P$ in the proper order and $|P_\alpha| \leq |P|$. Now we must show that the above definition turns F into a G-module, that is,

$$(ef) * Pt = e * (f * Pt) - f * (e * Pt). \tag{1.17}$$

We proceed by induction on $|P|$. Since equation (1.17) is trivial for $e = f$, we may choose $e > f$. If $e \geq P$ then the second term of the right-hand side of equation (1.17) can be transformed by equation (1.16) (where P is replaced by eP). Then $-f * (ePt) = e * (f * Pt) - (fe) * Pt$, proving (1.17). Thus we arrive at the toughest case where $e_1 > e > f$. This is dealt with by a calculation below in which we apply the induction over $|P|$ and the Jacobi identity:

$$\begin{aligned}(ef) * Pt &= (ef) * e_1 * P't = +e_1 * (ef) * P't - e_1(ef) * P't\\ &= +e_1 * e * f * P't - e_1 * f * e * P't - e(e_1f) * P't - f(ee_1) * P't\\ &= +e_1 * e * f * P't - e_1 * f * e * P't - e * (e_1f) * P't\\ &\quad + (e_1f) * e * P't - f * (ee_1) * P't + (ee_1) * f * P't\\ &= \{+e_1 * e * f * P't - e * e_1 * f * P't + (ee_1) * f * P't\}\\ &\quad + \{-e_1 * f * e * P't + f * e_1 * e * P't + (e_1f) * e * P't\}\\ &\quad + \{e * f * Pt - f * e * Pt\}.\end{aligned}$$

We are going to show that the first two summands of the resulting expression are zero. For this we consider $f * P't$. As is noticed previously,

$$f * P't = \bar{P}t + \sum_{\alpha} P_\alpha t,$$

where $\bar{P}$ is composed of elements of the set $P \cup \{f\}$ and $|P_\alpha| \le |P'|$. In this case $e_1 \ge \bar{P}$ and, applying equation (1.16), we have

$$e * e_1 * \bar{P}t = e * e_1\bar{P}t = e_1 * e * \bar{P}t + (ee_1) * \bar{P}t.$$

By induction on $|P|$, we have $e * e_1 * P_\alpha t = e_1 * e * P_\alpha t + (ee_1) * P_\alpha t$, hence the first of the two summands in sight is zero. The same argument can be applied to the remaining summand, proving equation (1.17). Thus we see that F is a G-module.

Now we can show that F is a free G-module with free generating set T. First we note that $Pt = e_1e_2\cdots e_nt$, proving the assertion concerning k-bases. Let N be an arbitrary G-module and $\varphi : T \to N$ an arbitrary map. If $P = (e_1, e_2, \ldots, e_n)$ is a descending row then we put

$$\bar{\varphi}(Pt) = e_1e_2\cdots e_n\varphi(t).$$

According to equation (1.16), $\bar{\varphi}$ is a G-homomorphism extending φ. □

1.7 Nilpotent Lie algebras

1.7.1 Equivalent definitions of nilpotency

We employ here the notation and some results from Section 1.5.

Theorem 1.12. *The following conditions on a Lie algebra L over a ring k are equivalent (c is a natural number):*

(i) $L^c \ne \{0\}, L^{c+1} = \{0\}$.

(ii) $Z_{c-1}(L) \ne L, Z_c(L) = L$.

(iii) *L possesses a finite central series of length c and has no series of this form with length less than c.*

(iv) *Any product of c + 1 elements of L is equal to zero and some product of c elements is nonzero.*

(v) *L satisfies the identity*

$$x_1x_2\cdots x_{c+1} = 0,$$

but it does not satisfy any such identity of smaller degree.

Proof. We remark first that for any natural r and s, it is true that $L^sL^r \subset L^{r+s}$. For $L^rL^1 = L^{r+1}$ by definition. Proceeding by induction on r, we write

$$\begin{aligned} L^sL^{r+1} &= L^s(L^rL^1) \subset (L^sL^r)L^1 + L^r(L^sL^1) \\ &= L^{s+r}L^1 + L^rL^{s+1} = L^{s+r+1}. \end{aligned}$$

What we use here is the Jacobi identity and the induction hypothesis, along with the definition of the lower central series.

Now to prove the equivalence of (i)–(iii), we notice that if L has a central series of length c,

$$L = L_1 \supseteq L_2 \supseteq \cdots \supseteq L_{c+1} = \{0\}, \tag{1.18}$$

then, according to properties of lower and upper central series in Sections 1.5.2 and 1.5.5, we may write for every natural $\ell \leq c$ that

$$L^\ell \subseteq L_\ell \quad \text{and} \quad Z_\ell(L) \supseteq L_{c-\ell+1}.$$

Therefore, in this case, $L^{c+1} = \{0\}$ and $Z_c(L) = L$. Now choosing, as appropriate, the lower central series or the upper central series as a particular case of (1.18), we arrive at any of the implications (i) $\iff$ (ii) $\iff$ (iii).

Next we prove (i) $\iff$ (iv). First we show that given any $x_1, \ldots, x_{\ell+1} \in L$ we always have $a = (x_1 x_2 \cdots x_{\ell+1})_\sigma \in L^{\ell+1}$, with σ denoting an arrangement of brackets in the associative word $x_1 x_2 \cdots x_{\ell+1}$. The basis of an induction over ℓ is obvious. Suppose the assertion proven for smaller values of ℓ. We remark that, for some $s < \ell$ we may write $a = (x_1 \cdots x_s)_\tau (x_{s+1} \cdots x_{\ell+1})_\pi$, with τ, π being some arrangements of brackets. By the induction hypothesis, it is true that $(x_1 \cdots x_s)_\tau \in L^s$ and that $(x_{s+1} \cdots x_{\ell+1})_\pi \in L^{\ell-s+1}$. According to the very beginning of the proof, we have $a \in L^s L^{\ell-s+1} \subset L^{\ell+1}$. Hence, assuming that (i) holds, we have for any $x_1, \ldots, x_{c+1} \in L$ and any arrangement of brackets that

$$(x_1 \cdots x_{c+1})_\sigma = 0.$$

Now it is immediate from the definition that, for each ℓ, L^ℓ is a k-submodule generated by all right-normed products of the form

$$x_1 x_2 \cdots x_\ell, \quad x_i \in L, i = 1, 2, \ldots, \ell.$$

Therefore if at least one of the right-normed products of the form $x_1 x_2 \cdots x_c$ is nonzero, we are forced to conclude $L^c \neq \{0\}$. These remarks are sufficient to prove the remaining implications. □

A Lie algebra L satisfying the equivalent conditions listed in Theorem 1.12 is called *c-step nilpotent*.

1.7.2 Frattini subalgebra

Let L be a Lie algebra over a ring k. A subalgebra M of L is called *maximal* if $M \neq L$ and M is not contained in a subalgebra of L different from L and M itself. Let $\{M_\alpha \mid \alpha \in I\}$ be the (possibly empty) set of all maximal subalgebras in L. Then we define the *Frattini subalgebra* $\Phi(L)$ as L itself if $I = \varnothing$ and as the intersection

$$\Phi(L) = \bigcap_{\alpha \in I} M_\alpha$$

otherwise. An element $x \in L$ is called a *nongenerator* if L can be generated by any set S such that L is generated by the union $S \cup \{u\}$.

Theorem 1.13. *For any Lie algebra L, its Frattini subalgebra $\Phi(L)$ is precisely the set of all nongenerators of L. If, in addition, L is a nilpotent Lie algebra then $\Phi(L)$ is an ideal of L containing the derived algebra L^2.*

Proof. We single out an $x \in L \setminus \Phi(L)$. Then there is a maximal subalgebra M such that $x \notin M$. By maximality, L is generated by M and x. Now it is clear that L is not generated by M, since M is already a subalgebra different from L. Thus x is not a nongenerator. Conversely, suppose $x \in \Phi(L)$ and let S be a set generating L if united with x. Suppose S is not a generating set of L. Then, by Zorn's lemma, there exists a subalgebra T which is maximal with the property that it contains S but avoids x. Then T is a maximal subalgebra of L. Indeed, if a subalgebra P properly contains T then $P \supseteq \{S, x\}$, that is, $P = L$. Now by the definition of the Frattini subalgebra, $x \in \Phi(L) \subseteq T$ whence $T \supseteq \{S, x\}$, that is, $T = L$. This contradiction shows that, in fact, S generates L and, therefore, x is a nongenerator.

Now we restrict ourselves to the case where L is a nilpotent algebra. By the above, we have to prove that any $x \in L^2$ is a nongenerator. In fact, a stronger assertion holds, namely any subset S in L^2 is a nongenerator subset. It is sufficient to show that L^2 is such a set. Now let S be a set of elements in L which generates L along with L^2. Using induction on the step of nilpotence of L, we will prove that S generates L. The base of induction at $c = 1$ is obvious. For the induction step, we consider L/L^c. Since $(L/L^c)^2 = L^2/L^c$, we have L/L^c generated by its derived algebra along with the set of the cosets of the form $x + L^c$, $x \in S$. By induction hypothesis and taking inverse images, we see that L is generated by $S \cup L^c$. Now it remains to show that L^c is included in the subalgebra R generated by S. This is the case since, given an $x \in L$, we may write $x = t + c$ where $t \in T$ and $c \in L^c$. Now L^c is a k-span of the elements of the form $x_1 x_2 \cdots x_c$. If $x_i = t_i + c_i$, $t_i \in T$, $c_i \in L^c$ then

$$x_1 x_2 \cdots x_c = t_1 t_2 \cdots t_c \in T.$$

Thus $L^c \subseteq T$, that is, $T = L$. Hence S generates L without L^2. To finish the proof, it is sufficient to note that any subalgebra of L containing the whole of L^2 is always an ideal of L.

□

1.7.3 Lie algebras of nilpotent endomorphisms

If L is a Lie algebra of step c then, for any $x \in L$, the inner derivation $\operatorname{ad} x$ is nilpotent of index at most c, i. e., $(\operatorname{ad} x)^c = 0$. The question of the conditions on a Lie algebra which ensure that the nilpotence of every inner derivation implies the nilpotence of the whole of algebra is among central in the theory of Lie algebras of arbitrary dimension. In the case of Lie algebras of linear operators, this question is immediately related to that of simultaneously reducing a collection of operators to the triangular form. The main result for the finite-dimensional case is the following.

Theorem 1.14 (Engel's theorem). *Let L be a Lie algebra of nilpotent endomorphisms of a k-module V such that $l_k(L) < \infty$. Then either $V = \{0\}$ or else there is a nonzero $v_0 \in V$ such that $x(v_0) = 0$ for all $x \in L$.*

Proof. We use induction on $l_k(L)$ with obvious basis when $l_k(L) = 0$. Now suppose $l_k(L) > 0$ and $V \neq \{0\}$. Our first remark is that for any $x \in L$ the inner derivation $\operatorname{ad} x$ is nilpotent. Indeed, we know that, given such an x, there exists a natural number n such that $x^n = 0$. Then let y be an element in L. We may write

$$(\operatorname{ad} x)^{2n-1}(y) = \sum_{i=0}^{2n-1} \beta_i x^i y x^{2n-1-i}, \quad \text{where } \beta_i \in k.$$

For each i, $0 \leq i \leq 2n-1$, it is true that either $i \geq n$ or $2n-1-i \geq n$. Hence each summand of the right-hand side is zero.

Our second remark is that given a subalgebra M in L, either $M = L$ or M is an ideal in a subalgebra $M' \subset M$ such that $M' \neq M$.

To prove this, we will view the k-module L/M as an M-module with respect to the adjoint action (Section 1.6.2), that is,

$$x(y + M) = [x, y] + M, \quad x \in M, y \in L.$$

Let $\pi : M \to \operatorname{End}(L/M)$ be the corresponding representation. Then $\bar{M} = \pi(M)$ is a Lie algebra of endomorphisms of L/M. If $x^n = 0$ then

$$\pi(x)^{2n-1}(y + M) = (\operatorname{ad} x)^{2n-1}(y) + M = M.$$

Thus all $\pi(x)$, $x \in M$, are nilpotent endomorphisms. Besides, $l_k(M) < l_k(L)$. Then by the induction hypothesis there exists a $y \notin M$ such that $\pi(x)(y + M) = M$ for all $x \in M$. In other words, what we have found is an element $y \notin M$ such that $[y, M] \subset M$. Now we put $M' = ky + M$. From the above property of y we see that M' is a subalgebra and that M is an ideal of M'.

Now let M be a maximal (different from L) subalgebra in L. By the above, M is an ideal of L. Choose $y \notin M$. Then $L = ky + M$. Let W denote the k-submodule in V whose elements are all w with $xw = 0$, x being arbitrary in M. By the induction hypothesis,

W is a nonzero k-module. Note that $L(W) \subset W$. Indeed, given $w \in W$, for any $x \in M$ we have

$$x(y(w)) = (xy)(w) = (yx)(w) + [x,y](w) = y(x(w)) = 0.$$

Here $[x,y](w) = 0$ since M is an ideal and then $[x,y] \in M$.

To finish the proof, we note that the restriction $\bar{y}$ of y to W is nilpotent, that is, there exists a natural m such that $\bar{y}^m = 0$ and $\bar{y}^{m-1} \neq 0$. Thus for some w_0 in W, we have $\bar{y}^m(w_0) = 0$ but $\bar{y}^{m-1}(w_0) \neq 0$. Setting $v_0 = y^{m-1}(w_0)$, we obtain a nonzero element in V such that $y(v_0) = 0$ and $M(v_0) = \{0\}$ (since $v_0 \in W$). Thus v_0 is annihilated by the whole of L, that is, v_0 is the desired element. □

Corollary 1.15. *Let L be a Lie algebra of finite length over k and for any $x \in L$ the inner derivation* $\operatorname{ad} x$ *is nilpotent. Then L is a nilpotent Lie algebra.*

Proof. We apply Engel's theorem to the Lie algebra $\operatorname{ad} L$ of endomorphisms of the k-module L. By our hypotheses, this consists of nilpotent endomorphisms and its length is finite since $\operatorname{ad} L \simeq L/Z(L)$. Hence either $L = \{0\}$ or else there is an element v_0 in L such that for any $x \in L$ we have $xv_0 = 0$. This proves that $Z_1(L) = Z(L) \neq L$. Factoring over $Z_1(L)$ and repeating the procedure, we will find that either $L = Z_1(L)$ or else $Z_1(L) \neq Z_2(L)$. Since $l(L) < \infty$, this process terminates, that is, for some natural m we have $Z_m(L) = L$. According to Section 1.7.1, L is a nilpotent Lie algebra. □

1.7.4 Triangular form

Here we restrict ourselves to a classical case where k is a field and $\dim V < \infty$. In this case, given a basis of V, one may associate with every element of $L \subset \operatorname{End}_k V$ an $n \times n$ matrix, n being the dimension of V.

Corollary 1.16. *Let L be a Lie algebra of nilpotent linear operators of a finite-dimensional vector space V over a field k, $n =$* $\dim V$*. Then there is a basis of V in which the matrices of all operators in L have the triangular form*

$$\begin{pmatrix} 0 & a_{12} & a_{13} & \dots & a_{1,n-1} & a_{1n} \\ 0 & 0 & a_{23} & \dots & a_{2,n-1} & a_{2n} \\ \vdots & \vdots & \ddots & \ddots & \vdots & \vdots \\ 0 & 0 & 0 & \ddots & 0 & a_{n-1,n} \\ 0 & 0 & 0 & \dots & 0 & 0 \end{pmatrix},$$

with a_{ij} being some elements in k.

Proof. It is sufficient to find a basis $\{e_1, e_2, \dots, e_n\}$ in V such that $Le_i \subseteq ke_1 + \dots + ke_{i-1}$, $i = 2, \dots, n$, and $Le_1 = \{0\}$. Now the existence of e_1 is immediate from Engel's theorem.

To find the remaining vectors, one has to pass to V/ke_1 and to a Lie algebra of linear operators of the latter space which is induced by L. Then the induction on n applies. □

1.7.5 Some sufficient conditions for the nilpotency

The two theorems below give sufficient conditions for a Lie algebra to be nilpotent. Theorem 1.17 applies to Lie algebras of finite length, whereas Theorem 1.19 requires no such restrictions. In a corollary of Theorem 1.17, we give a property of the *Frattini ideal* $\varphi(L)$; by this we mean the greatest ideal of a Lie algebra L which is contained in its Frattini subalgebra $\Phi(L)$.

Theorem 1.17 (D. W. Barnes). *Let L be a Lie algebra of finite length over a ring k, with M, N being its ideals such that $N \subseteq \Phi(L)$. Suppose that $N \subseteq M$ and that M/N is a nilpotent Lie algebra. Then M is nilpotent.*

Proof. By the hypotheses, there exists a nonnegative integer c such that $M^{c+1} \subseteq N$. We consider an arbitrary $x \in M$. Our purpose is to show that $\varphi = \operatorname{ad} x$ is a nilpotent endomorphism of L. For this we build two series of k-submodules of the form

$$\operatorname{Im}\varphi \supseteq \operatorname{Im}\varphi^2 \supseteq \cdots \quad \text{and} \quad \ker\varphi \subseteq \ker\varphi^2 \subseteq \cdots,$$

with L being of finite length. Then there exists a t, $t > c$, such that $\operatorname{Im}\varphi^{t+s} = \operatorname{Im}\varphi^t$ and $\ker\varphi^{t+s} = \ker\varphi^t$ for all $s \geq 0$. Hence $L = \ker\varphi^t + \operatorname{Im}\varphi^t$, $\ker\varphi^t$ being a subalgebra. Indeed, pick any y in L. Then there exists a z in L such that $\varphi^t(y) = \varphi^{2t}(z)$. Accordingly, $u = y - \varphi^t(z) \in \ker\varphi^t$, i. e., $y = u + \varphi^t(z) \in \ker\varphi^t + \operatorname{Im}\varphi^t$. Now if $u, v \in \ker\varphi^t$ then

$$\varphi^{2t}(uv) + \sum_{i=0}^{2t} \alpha_i \varphi^i(u)\varphi^{2t-i}(v) = 0$$

since either $i \geq t$ or $2t - i \geq t$. Now $\ker\varphi^{2t} = \ker\varphi^t$, therefore $uv \in \ker\varphi^t$.

Now we recall that $t > c$, hence $\operatorname{Im}\varphi^t \subset N \subset \Phi(L)$. Thus we have

$$L = \ker\varphi^t + \Phi(L).$$

With L being a module of finite length, we may write $\Phi(L) = ku_1 + \cdots + ku_m$, where $u_1, \dots, u_m$ are some nongenerators. Casting them away one after another and having in mind that $\ker\varphi^t$ is a subalgebra, we finally arrive at $L = \ker\varphi^t$. Thus $\operatorname{ad} x$, and, moreover, $\operatorname{ad} x \mid_M$ are nilpotent for each $x \in M$. Applying the corollary of Engel's theorem shows that M is a nilpotent Lie algebra. □

Corollary 1.18. *Let L be a Lie algebra of finite length over k, $\phi(L)$ its Frattini ideal. Then $\phi(L)$ is nilpotent.*

Proof. Immediate from Theorem 1.17 if we set $M = N = \phi(L)$. □

Theorem 1.19. *Let L be a Lie algebra over k and M a nilpotent ideal such that L/M^2 is a nilpotent Lie algebra. Then L is itself nilpotent.*

Proof. We proceed by induction over the nilpotent step c of M, the assertion being trivial for $c = 1$. It should be noticed that M^c is an ideal of L (Section 1.5.1). We may apply the induction hypothesis to L/M^c. Thus, there exists a natural d such that $(L/M^c)^{d+1} = \{0\}$, i. e., $L^{d+1} \subseteq M^c$. Since $c > 1$, any $a \in M^c$ can be written in the form

$$a = \sum_i m_i n_i, \quad m_i, n_i \in L.$$

Let us compute the product $x_{2d-1}x_{2d-2}\cdots x_2x_1a$ (right-normed product, with $x_1, \ldots, x_{2d-1} \in L$). Since $\operatorname{ad} x$ for $x \in L$ is a derivation, we may write

$$x_{2d-1}\cdots x_2x_1a = \sum_i \sum (x_{i_1}\cdots x_{i_s}m)(x_{j_1}\cdots x_{j_{2d-1-s}}n_i)$$

where $i_1 > \cdots > i_s$, $j_1 > \cdots > j_{2d-1-s}$, $\{i_1, \ldots, i_s, j_1, \ldots, j_{2d-1-s}\} = \{2d-1, \ldots, 2, 1\}$. For every i, one of the two factors of each summand on the right-hand side has length exceeding d, that is, it is an element of M^c. Another factor being in M, we finally see that $x_{2d-1}\cdots x_2x_1a = 0$. Hence L is a nilpotent algebra whose nilpotent step is at most $3d - 1$. □

1.8 Solvable Lie algebras

We define a nonassociative monomial $\delta_\ell(x_1, \ldots, x_{2^\ell})$ by setting $\delta_0(x) = x$ and, using induction over ℓ,

$$\delta_{\ell+1}(x_1, \ldots, x_{2^{\ell+1}}) = \delta_\ell(x_1, \ldots, x_{2^\ell})\delta_\ell(x_{2^\ell+1}, \ldots, x_{2^{\ell+1}}).$$

Theorem 1.20. *Let L be a Lie algebra over a commutative ring k. Then the following conditions are equivalent:*

(i) *For some integer $\ell \geq 1$, we have $L^{(\ell-1)} \neq \{0\}$ but $L^{(\ell)} = \{0\}$.*

(ii) *L possesses a finite solvable series of ideals*

$$L = L_0 \supseteq L_1 \supseteq \cdots \supseteq L_\ell = \{0\}$$

(that is, in which L_i/L_{i+1} is abelian) and ℓ is the shortest length of such a series.

(iii) *L possesses a series of subalgebras of the form*

$$L = M_0 \supseteq M_1 \supseteq \cdots \supseteq M_\ell = \{0\}$$

where $M_i^2 \subseteq M_{i+1}$ and ℓ is the least length of such a series.

(iv) *The following relation holds identically in L:*

$$\delta_\ell(x_1,\ldots,x_{2^\ell}) = 0,$$

this being false for $\delta_{\ell-1}$.

Proof. It has been noted in Section 1.5.3 that the derived series is characterized by the property that for each solvable series, as in (ii), and each $n \geq 0$ we always have $L_n \supseteq L^{(n)}$. Hence (i) and (ii) are equivalent. Now if we start with (iii) then, by induction, using $M_i^2 \subseteq M_{i+1}$, we deduce $L^{(i)} \subseteq M_i$, hence $L^{(\ell)} = \{0\}$. Therefore (i) holds: if $L^{(\ell-1)} = \{0\}$ then the derived series has shorter length than a series in (iii), contradicting the hypothesis. Now (i) and (iii) are equivalent. The implication (i) $\Rightarrow$ (iv) is simple, since the inductive definition of $\delta_\ell(x_1,\ldots,x_{2^\ell})$ shows that, given any $v_1,\ldots,v_{2^\ell} \in L$, we have $\delta_\ell(v_1,\ldots,v_{2^\ell}) \in L^{(\ell)}$. To prove the converse, we need to show that $L^{(\ell)}$ is a k-span of the above elements. This is obvious for $\ell = 0$. Arguing by induction, we may assume that $L^{(\ell-1)}$ is a k-span of the elements of the form $\delta_{\ell-1}(v_1,\ldots,v_{2^{\ell-1}})$, $v_1,\ldots,v_{2^{\ell-1}} \in L$. By definition then $L^{(\ell)}$ is a linear span of the products of the form

$$\delta_{\ell-1}(v_1,\ldots,v_{2^{\ell-1}})\delta_{\ell-1}(v_{2^{l-1}+1},\ldots,v_{2^\ell}) = \delta_\ell(v_1,\ldots,v_{2^\ell}).$$

This proves the implication (iv) $\Rightarrow$ (i) and completes the proof of the theorem. □

A Lie algebra L satisfying the equivalent conditions (i)–(iv) is called *l-step solvable.*

1.8.1 Examples and properties of solvable Lie algebras

(i) The most typical example of solvable Lie algebras over a commutative ring k is that of Lie algebras of (upper) triangular matrices of the same order under the commutator of matrices, the entries of all matrices being in k. We remark that the derived algebra L^2 of such an L consists of niltriangular matrices and hence is nilpotent (see Section 1.8). This particular fact is a rule for finite-dimensional Lie algebras over algebraically closed fields of characteristic zero.

(ii) Any extension (see Section 1.4.5) of a solvable Lie algebra by a solvable one is itself solvable: this is trivial by Theorem 1.20(iii). A Lie algebra which is an extension of an abelian algebra by an abelian one is called *metabelian.* Thus metabelian Lie algebras are of solvable step at most 2.

(iii) The property of being solvable is inherited while passing from a Lie algebra to its subalgebras and quotient algebras. Given a family of Lie algebras which are solvable of step at most ℓ, their Cartesian product is solvable with the same bound ℓ on its solvable step. The most natural approach to this is via Theorem 1.20(iv).

Theorem 1.21 (Lie's theorem). *Let $\rho : L \to \mathfrak{gl}(V)$ be an irreducible representation of a finite-dimensional solvable Lie algebra L by linear operators of a finite-dimensional vector space V over a field k of characteristic zero. Then $\rho(L^2) = \{0\}$.*

Proof. The core of the proof is in the case where k is algebraically closed. Having imposed this restriction, we will be proving that $\dim V = 1$. We proceed by induction over $\dim L$. The basis of induction ($\dim L = 0$) is obvious. If $\dim L > 0$ then there is an ideal M in L such that $\dim M + 1 = \dim L$. Indeed, by solubility, $L^2 \neq L$ and every subspace $M \supseteq L^2$ is an ideal of L. By Theorem 1.7, the M-module V has a chief series

$$\{0\} = V_0 \subset V_1 \subset \cdots \subset V_{i-1} \subset V_i \subset \cdots \subset V_n = V,$$

each factor of which V_i/V_{i-1} is isomorphic to V, which is a simple M-module. By induction, $\dim V_1 = 1$ and then $n = \dim V$. Besides, for each $y \in M$ there is a $\lambda(y) \in k$, such that, given an i, $0 < i \leq n$ and any v in V_i we have

$$\rho(y)(v + V_{i-1}) = \lambda(y)v + V_{i-1}.$$

As a result

$$\operatorname{tr}\rho(y) = n\lambda(y). \tag{1.19}$$

If $y \in L^2 \subseteq M$ then $\lambda(y) = 0$. For, in this case $\rho(y)$ is a linear combination of commutators of linear operators, hence $\operatorname{tr}\rho(y) = 0$, hence, by (1.19), $\lambda(y) = 0$. Now let V' denote the M-submodule of V whose elements v are such that for every $y \in M$ we have $\rho(y)(v) = \lambda(y)v$. This submodule is nonzero because $V' \supseteq V_1$. We will show that $\rho(L)(V') \subseteq V'$. Let $x \in L$, $v \in V'$, $y \in M$. Then we may write

$$\begin{aligned}\rho(y)\rho(x)(v) &= \rho(x)\rho(y)(v) - \rho(xy)(v)\\ &= \rho(x)(\lambda(y)v) - \lambda(xy)(v) = \lambda(y)\rho(x)(v).\end{aligned}$$

The latter equation is true since $xy \in L^2$ and $v \in V'$. Since V is irreducible, we have $V' = V$. Now choose an x_0 in L outside M. Then $L = kx_0 \oplus M$, and if v_0 is a nonzero eigenvector for $\rho(x_0)$ in V', we may write $\rho(L)(kv_0) \subseteq kv_0$. Hence $kv_0 = V$, proving $\dim V = 1$.

Coming back to arbitrary zero characteristic fields, we denote by k^* a finite Galois extension of k which contains all characteristic roots of operators in L. Let $L^* = k^* \otimes_k L$. Then $V^* = k^* \otimes_k V$ is a natural L^*-module, and this module is semisimple.[1] In each

1 Let W^* be a minimal nonzero L^*-submodule in V^*, and Γ the Galois group of k^*/k. Since the subspace $P = \sum_{\gamma\in V} \gamma W^*$ is an L^*-submodule and it is invariant under all automorphisms in Γ, there exists an L-submodule W in V such that $P = k^* \otimes W$. But V is a simple L-module. Therefore, $W = V$ and then $V^* = P$. Thus we see that V^* is a sum of simple L^*-modules. As usual, this is equal to a direct sum, showing that L is semisimple.

simple direct summand of this module $(L^*)^2$ acts by zero operators. This is a trivial modification of the first part of the proof since the only thing we needed there was the existence of eigenvectors. Now $(L^*)^2(V^*) = \{0\}$. Obviously, then we have $\rho(L^2)(V) = \{0\}$ as well. □

Corollary 1.22. *The derived algebra of a finite-dimensional solvable Lie algebra over a field of characteristic zero is nilpotent.*

Proof. We consider the adjoint L-module L. Let

$$\{0\} = L_0 \subset L_1 \subset \cdots \subset L_m = L$$

be a chief series of this module. Applying Lie's theorem to all simple L-modules of the form L_i/L_{i-1}, we see that L^2 annihilates them all. Thus, given an x in L^2, we have $(\operatorname{ad} x)^m = 0$. By the corollary of Engel's theorem, L^2 is nilpotent Lie algebra, as desired. □

In Exercises 1.12.32 and 1.12.33, we give examples showing that Lie's theorem and its corollary do not remain true in the case of infinite-dimensional representations or fields of nonzero characteristic.

Corollary 1.23. *Let L be a solvable Lie algebra of linear operators of a finite-dimensional vector space V over an algebraically closed field k of characteristic 0. Then, there exists a basis of V in which all the matrices of operators in L have the triangular form.*

Proof. By the first part of the proof of Lie's theorem, the operators in L possess a common eigenvector v_0. This enables us to pass to the L-module V/kv_0 where the induction applies. To make this more rigorous, the reader is referred to the proof of the corollary in Section 1.7.4. □

1.8.2 Engel Lie algebras

This subsection and three subsequent ones are devoted to Lie algebras in which all inner derivations are nilpotent. Such algebras are called the *Engel* Lie algebras. According to the corollary of Engel's theorem, any Engel Lie algebra of finite length over the base ring is nilpotent. Here we start with an infinite-dimensional Lie algebra over an arbitrary ring which is Engel but not nilpotent. For this we should consider a Lie algebra

$$L = \prod_{i=1}^{\infty} L_i,$$

where L_i is the Lie algebra of all niltriangular matrices of order $i + 1$. Immediate verification shows that $\mathfrak{nt}_{i+1}(k)$ (our notation for the latter algebra) is i-step nilpotent.

Since each element x in L is a function with finitely many nonzero values, each inner derivation is nilpotent. Now L is not nilpotent since for each natural c it contains a subalgebra L_{c+1} such that $L_{c+1}^{c+1} \neq \{0\}$. It should be remarked, however, that for each c as above L possesses an element x_c such that $(\operatorname{ad} x_c)^c \neq 0$, for instance, e_{12} is such an element of $L_c = \mathfrak{nt}_{c+1}(k)$. Therefore, it is of interest to consider Lie algebra L in which $(\operatorname{ad} x)^n = 0$, with n being a fixed number.

1.8.3 Example (P. M. Cohn)

Let k be a commutative ring of simple additive exponent p. There exists a nonnilpotent metabelian Lie algebra L over k in which $(\operatorname{ad} x)^{p+1} = 0$ for all $x \in L$.

Let V denote a free k-module whose basis is formed by the set $\{e_f \mid f \in \mathbb{Z}_p^{\mathbb{N}}\}$, with $\mathbb{Z}_p^{\mathbb{N}}$ being the set of all functions of natural argument, valued in the residue class ring $\mathbb{Z}_p$. For each natural m, we define an endomorphism δ_m of V which maps e_f into $e_{\bar{f}}$ where $\bar{f}$ is given by

$$\bar{f}(i) = \begin{cases} f(i), & \text{if } i \neq m, \\ f(i) + 1, & \text{if } i = m. \end{cases}$$

It is obvious that $\delta_m \delta_n = \delta_n \delta_m$ for all n, m and that $\delta_m^p = \varepsilon$ (ε being the identity endomorphism of V). We set $x_i = \delta_i - \varepsilon$, $i = 1, 2, \ldots$ Then

$$\begin{aligned} x_i x_j &= -(\delta_i - \varepsilon)(\delta_j - \varepsilon) = \delta_i \delta_j - \delta_i - \delta_j + \varepsilon \\ &= \delta_j \delta_i - \delta_j - \delta_i + \varepsilon = (\delta_j - \varepsilon)(\delta_i - \varepsilon) = x_j x_i. \end{aligned}$$

Therefore the k-span of $\{x_i \mid i \in \mathbb{N}\}$ is an abelian Lie algebra L and V is an L-module. We denote by T the semidirect product $T = L \rtimes V$ (where V is viewed as an abelian Lie algebra, cf. Section 1.6.5). Let us show first that for all $t \in T$ we have $(\operatorname{ad} t)^{p+1} = 0$. It should be remarked that using the binomial decomposition and the additive exponent of k, it is easy to prove that

$$x_i^p = (\delta_i - \varepsilon)^p = \delta_i^p - \varepsilon^p = 0.$$

In other words, $(\operatorname{ad} x_i)^p(e_f) = 0$ for all $i \in \mathbb{N}$, $f \in \mathbb{Z}_p^{\mathbb{N}}$. Now, let $t = \sum_s \alpha_s x_s + \sum_f \beta_f e_f$. For all $u \in T$, obviously, $tu \in V$, hence $(\operatorname{ad} e_f)(tu) = 0$. Now

$$\begin{aligned} (\operatorname{ad} t)^{p+1}(u) &= (\operatorname{ad} t)^p(tu) = \left(\operatorname{ad} \sum_s \alpha_s x_s\right)^p (tu) \\ &= \sum_s \alpha_s^p (\operatorname{ad} x_s)^p (tu) = 0. \end{aligned}$$

It remains to verify that T is a nonnilpotent algebra. For this we define a function f_m, where m is an integer, taking value 1 at the points $1, 2, \ldots, m$, and 0 otherwise. Then

$$x_m x_{m-1} \cdots x_1 e_{f_0} = e_{f_m} \neq 0.$$

Thus T is a nonnilpotent Lie algebra satisfying the condition $(\operatorname{ad} x)^{p+1} = 0$ identically. To complete the construction, we note that T is an extension of an abelian ideal by an abelian algebra, i. e., T is metabelian.

It is interesting that similar examples over fields of characteristic zero or over fields of characteristic $p > n$ are no longer possible. This will follow from a theorem below for which we need the following.

Lemma 1.24. *Let A be a subgroup of the additive group of an associative k-algebra B and n a natural number such that for all $v \in A$ we have $v^n = 0$. Then for any $v_1, v_2, \ldots, v_n \in A$, the following relation holds in B:*

$$\sum_{\sigma \in S_n} v_{\sigma(1)} v_{\sigma(2)} \cdots v_{\sigma(n)} = 0, \qquad (1.20)$$

where S_n is the symmetric group of order n.

Proof. We define, for every collection $\varepsilon_1, \varepsilon_2, \ldots, \varepsilon_n$ of numbers 0, 1, the elements $a(\varepsilon_1, \varepsilon_2, \ldots, \varepsilon_n)$ by setting

$$a(\varepsilon_1, \varepsilon_2, \ldots, \varepsilon_n) = \varepsilon_1 v_1 + \varepsilon_2 v_2 + \cdots + \varepsilon_n v_n.$$

Since $a(\varepsilon_1, \epsilon_2, \ldots, \varepsilon_n)^n = 0$ and denoting the left-hand side of (1.20) by w, we may write

$$0 = (\varepsilon_1 v_1 + \varepsilon_2 v_2 + \cdots + \varepsilon_n v_n)^n = r_0 + \varepsilon_1 r_1 + \varepsilon_1 \varepsilon_2 r_2 + \cdots + \varepsilon_1 \varepsilon_2 \cdots \varepsilon_n w, \qquad (1.21)$$

where r_0 is the linear combination of monomials without entries of v_1, $\epsilon_1 r_1$ is the linear combination of monomials outside r_0 without entries of v_2, $\varepsilon_1 \varepsilon_2 r_2$ is the linear combination of monomials outside $r_0 + \varepsilon_1 r_1$ without entries of v_3, and so on. Since (1.21) is true for all values of coefficients, setting $\varepsilon_1 = 0$ gives $r_0 = 0$, for all $\varepsilon_1, \varepsilon_2, \ldots, \varepsilon_n$. Setting $\varepsilon_1 = 1$, we arrive at a relation of the form

$$r_1' + \varepsilon_2 r_2' + \cdots + \varepsilon_2 \cdots \varepsilon_n w = 0,$$

where r_i' is equal to r_i at $\varepsilon_1 = 1$. Proceeding similarly, we finally get $w = 0$. □

Theorem 1.25 (P. J. Higgins). *Let L be a solvable Lie ring of solvable step ℓ and n an integer such that $(\operatorname{ad} x)^n = 0$ for all $x \in L$. Suppose the additive group of L has no elements of order dividing $n!$. The L is a c-step nilpotent ring, where*

$$c \leq \frac{n^\ell - 1}{n - 1}.$$

Proof. We set $M = L^2$. Since all M^i are ideals of L, $i = 1, 2, \ldots$, each factor M^i/M^{i+1} becomes a natural L-module under the adjoint action

$$x(m + M^{i+1}) = xm + M^{i+1},$$

where $x \in L$, $m \in M^i$. By our convention and by the definition of the lower central series, we have

$$L^2(M^i/M^{i+1}) = M(M^i/M^{i+1}) = \{0\}.$$

Hence, for all $x, y \in L$, we may write

$$xy(m + M^{i+1}) = yx(m + M^{i+1}).$$

For the completion of the proof, we proceed by induction over the solvable step ℓ of L. Since $M = L^2$ is $(\ell-1)$-step solvable, by the induction hypothesis, we may write $M^{c'+1} = \{0\}$, where

$$c' = \frac{n^{\ell-1} - 1}{n-1}.$$

To apply Lemma 1.24, we set $A = \operatorname{ad} L$ and $B = \operatorname{End}_k L$. Then, for all $x_1, x_2, \ldots, x_n \in L$, we have

$$\sum_{\sigma \in S_n} \operatorname{ad} x_{\sigma(1)} \operatorname{ad} x_{\sigma(2)} \cdots \operatorname{ad} x_{\sigma(n)} = 0.$$

Applying the left-hand side of the latter equation to M^i, we will have

$$\Big(\sum_{\sigma \in S_n} \operatorname{ad} x_{\sigma(1)} \operatorname{ad} x_{\sigma(2)} \cdots \operatorname{ad} x_{\sigma(n)}\Big)(M^i) \subseteq M^{i+1}.$$

By the commutativity of action of L in M^i/M^{i+1} mentioned above, we may rewrite

$$n!(\operatorname{ad} x_n \operatorname{ad} x_{n-1} \cdots \operatorname{ad} x_1)(M^i) \subseteq M^{i+1}.$$

Since $(\operatorname{ad} x)(L) \subset L^2 = M$, we arrive at

$$(n!)^{c'} (\operatorname{ad} x_{c'n} \cdots \operatorname{ad} x_1 \operatorname{ad} x_0)(L) \subseteq M^{c'+1} = \{0\}.$$

From the hypothesis concerning the additive torsion and the above, we finally get

$$\operatorname{ad} x_{c'n} \cdots \operatorname{ad} x_1 \operatorname{ad} x_0 = 0.$$

Thus L is nilpotent and its nilpotence step is at most

$$c'n + 1 = \frac{n^{l-1} - 1}{n-1} + 1 = \frac{n^l - 1}{n-1}. \qquad \square$$

The restriction on the torsion of L can be replaced by a condition on k in the case where L is a k-algebra. For instance, we might assume that $n!1$ is an invertible element in k, or else that k is a field of characteristic 0 or $p > n$. Further information concerning Engel Lie algebras can be found in Chapter 8.

1.9 Frattini theory for Lie algebras of finite length

1.9.1 An example

We recall that the Frattini ideal $\varphi(L)$ of a Lie algebra L over a ring k is the greatest ideal of L which is contained in its Frattini subalgebra $\Phi(L)$. Here we give an example of a Lie algebra in which $\varphi(L) \neq \Phi(L)$. This should be compared with the group theory case where the Frattini subgroup is always normal. Now let k be the two-element field $\mathbb{Z}_2$ and L a 3-dimensional simple Lie algebra over k with a basis $\{e_1, e_2, e_3\}$ such that $e_i e_j = e_l$ where $\{i, j, l\} = \{1, 2, 3\}$. Since $e_i e_j e_l = 0$ for pairwise different i, j, l, we have $J(e_i, e_j, e_l) = 0$. This is enough for the Jacobi identity to be satisfied. Now all maximal subalgebras in L are 2-dimensional and they are the spans of the elements of the form $e_i, e_j + e_l$, where $\{i, j, l\} = \{1, 2, 3\}$. Each such subalgebra contains $a = e_1 + e_2 + e_3$. Hence $\Phi(L) = k(e_1 + e_2 + e_3)$. Since L is a simple Lie algebra and $\varphi(L) \subseteq \Phi(L)$, we have $\varphi(L) = \{0\}$. Thus $\varphi(L) \neq \Phi(L)$.

However, in the case of solvable Lie algebras of finite length we always have $\varphi(L) = \Phi(L)$. Before we formulate Theorem 1.26 due to D. W. Barnes concerning this, we remark that a minimal ideal of a solvable Lie algebra is abelian. Indeed, given a solvable Lie algebra M, for some $l > 0$ we have $M^{(l-1)} \neq \{0\}$ and $M^{(l)} = \{0\}$. According to Section 1.5.1, $M^{(l-1)}$ is an ideal of L, that is, $M = M^{(l-1)}$ and $M^2 = (M^{(l-1)})^2 = M^{(l)} = \{0\}$, as required.

Theorem 1.26 (D. W. Barnes). *Let L be a solvable k-algebra of finite length. Then the Frattini subalgebra $\Phi(L)$ is a nilpotent ideal of L.*

Proof. It is sufficient to show that $\Phi(L)$ is an ideal of L and then to use the corollary of Theorem 1.17. We will proceed by induction over $d = l_k(L)$ with an obvious basis $d = 1$. Now suppose $d > 1$. We single out a minimal ideal M in L. Let Φ_M be the intersection of all maximal subalgebra of L which include M. Since $\Phi(L/M) = \Phi_M/M$, we see by induction that Φ_M is an ideal of L. Thus, if $M \subseteq \Phi(L)$, we are done. Therefore we may assume that $M \nsubseteq \Phi(L)$. Now let N be an arbitrary maximal subalgebra of L such that $N \nsupseteq M$. We will show that $L = M \oplus N$. It is enough to show $M' = M \cap N = \{0\}$. Indeed, if $x \in L$, then $x = m + n$, $m \in M$, $n \in N$. Taking a y in M', we may write

$$M \ni xy = (m + n)y = 0 + ny \in N.$$

Since M is abelian, we have $C_L(M) \supseteq M$. Assume first that $C = C_L(M) \neq M$. Then, given a subalgebra $N \nsupseteq M$, we always have that $C_L(M) \cap N$ is a nonzero ideal of L contained in N. This follows from $L = M + N$ and $C \supset M$. Let C_N be a minimal ideal of L which is a subset of $C_L(M) \cap N$. Then in every maximal subalgebra N there is a minimal ideal (either M or C_N). Hence $\Phi(L)$ can be represented as the intersection of ideals of the form Φ_M, with M running through the set of all minimal ideals of L. Therefore $\Phi(L)$ is an ideal of L.

It remains to consider the case of a minimal ideal $M \not\subseteq \Phi(L)$ such that $C_L(M) = M$. We will show that, in this case, $\Phi(L) = \{0\}$. Assuming the converse, we may choose a nonzero $x \in \Phi(L)$. Let N be a maximal subalgebra with $N \not\supseteq M$. Then $x \in N \setminus M$ and for some $m \in M$ we have $xm \neq 0$. Since $(\operatorname{ad} m)^2 = 0$, the linear operator $1 + \operatorname{ad} m$ is an automorphism of L (its inverse is $1 - \operatorname{ad} m$). Set $x' = (1+\operatorname{ad} m)(x) = x+mx$. The definition of the Frattini subalgebra shows that, given an automorphism α of L, we always have $\alpha(\Phi(L)) = \Phi(L)$. Hence $x' \in \Phi(L)$, that is, $x' \in N$. Now for each $m' \in M$, we have

$$m'x' = m'x + m'(mx) = m'x, \quad \text{i. e., } m'(x' - x) = 0.$$

Therefore $x' - x \in C_L(M) \cap N = \{0\}$, giving $x = x'$. But then $mx = 0$, contradicting the choice of m. Hence $\Phi(L) = \{0\}$. □

1.9.2 Solvable and nilpotent radicals. Socle

Let L be a Lie algebra over a ring k. The greatest solvable ideal of L (if this exists) is called the *solvable radical* of L and it is denoted by $S(L)$. Similarly, we define the *nilpotent radical* $N(L)$. The sum $\operatorname{Soc}(L)$ of all minimal ideals (which is the direct sum of some of them) of a Lie algebra L is called its *socle*. The sum $S_0(L)$ of all abelian minimal ideals of L is called its *abelian socle*. In the remaining subsections of this section, we will prove some results describing the structure of a Lie algebra of finite length in terms of the notions introduced just above.

Proposition 1.27. *Let L be a Lie algebra of finite length over a ring k. Then the solvable and nilpotent radicals exist and coincide with the sums of all solvable and, respectively, nilpotent ideals of L. Moreover, $S(L/S(L)) = \{0\}$.*

Proof. Since the property of being solvable is preserved under extensions, the sum $U + V$ of two solvable ideals U, V is itself solvable. Indeed,

$$(U + V)/V \cong U/U \cap V$$

and V, $U/U \cap V$ are solvable. From the finiteness of the length, the sum of all solvable ideals is, in fact, the sum of finitely many of them, that is solvable.

Now let U, V be two nilpotent ideals of L. To show the nilpotency of $L_1 = U + V$, we apply the induction over the sum of the nilpotent steps of U and V. If both U and V are abelian, then $U \cap V$ is central in L_1 and $L_1^2 \subset U \cap V$. Thus $L_1^3 = \{0\}$, as needed. If one of U and V (say, V) is nonabelian then V^2 is a nonzero ideal of L_1, and for L_1/V^2 we can apply induction. Using Theorem 1.19, we derive that L_1 is nilpotent. □

Lemma 1.28. *Let L be a Lie algebra over a ring k, M its subalgebra, K an ideal of L contained in $\Phi(M)$. Then $K \subseteq \Phi(L)$.*

Proof. We assume $K \not\subseteq \Phi(L)$. Then there is a maximal subalgebra P of L such that $P + K = L$. Using the modular law, we find that $P \cap M + K = M$. Since $K \subseteq \Phi(M)$ we have, in fact, $M = P \cap M$, or $K \subseteq M = P \cap M \subseteq P$. This contradicts our assumption. □

Lemma 1.29. *Let L be a Lie algebra of finite length over k with Frattini ideal $\varphi(L) = \{0\}$, M an abelian ideal of L. Then M is complemented in L, that is, there exists a subalgebra K such that $L = M \oplus K$.*

Proof. Let K denote a subalgebra L which is minimal with respect to the property $L = M + K$. We consider $M \cap K$. Since $M(M \cap K) = \{0\}$ and $K(M \cap K) \subset M \cap K$, $(M \cap K)$ is an ideal of L. If $M \cap K \subseteq \Phi(K)$ then, by Lemma 1.28, we have $M \cap K \subseteq \Phi(L)$. With $M \cap K$ being an ideal, we have, in fact, $M \cap K \subseteq \varphi(L) = \{0\}$. Thus it can be assumed that $M \cap K \not\subseteq \Phi(K)$. In this case, K has a maximal subalgebra P such that $K = M \cap K + P$. Thus we write

$$L = M + K = M + (M \cap K + P) = (M + M \cap K) + P = M + P,$$

contradicting the minimality of K with respect to the property of being $L = M + K$. Thus $L = M \oplus K$ and K is the desired complement. □

Theorem 1.30 (D. Towers). *Let L be a Lie algebra of finite length over a commutative ring k, $\Phi(L) = \Phi$, $\varphi(L) = \varphi$, $N(L) = N$. Then $\varphi(L/\varphi) = \{0\}$, and*
(i) $\Phi \cap N = \varphi \supseteq N^2$;
(ii) $N(L/\varphi) = N/\varphi = S_0(L/\varphi)$;
(iii) *N/φ is complemented in L/φ, that is, for some subalgebra M/φ we have $L/\varphi = N/\varphi \oplus M/\varphi$.*

Proof. The triviality of the Frattini ideal $\varphi(L/\varphi)$ is an immediate consequence of the definition. To prove (i), we recall (Section 1.7.2) that for a nilpotent Lie algebra N its derived algebra is included in $\Phi(N)$. Since N^2 is an ideal of L, by Lemma 1.28, we have $N^2 \subseteq \Phi(L) = \Phi$. Now by the corollary of Theorem 1.17, we have $\varphi \subseteq \Phi \cap N$. We set $\Phi_1 = \Phi \cap N$ and prove $\Phi_1 = \varphi$. Assume that Φ_1 is not an ideal of L. Then $L\Phi_1 \not\subseteq \Phi$. Let P be a maximal subalgebra of L such that $L\Phi_1 \not\subseteq P$. From $L\Phi_1 \subseteq N$ it follows that $L = N + P$. This latter relation implies

$$L\Phi_1 = (N + P)\Phi_1 \subseteq N^2 + P \subseteq \Phi(L) + P \subseteq P.$$

The resulting inclusion $L\Phi_1 \subseteq P$ contradicts the choice of P. Hence Φ_1 is an ideal of L, hence $\varphi = \Phi$. Thus the proof of (i) is complete.

Now the first of the equations in (ii) is a consequence of Section 1.7.5. Indeed, if S/φ is a nilpotent ideal of L/φ then, according to Section 1.7.5, S is a nilpotent ideal of L, that is, $S \subseteq N$ and $S/\varphi \subseteq N/\varphi$. Thus $N(L/\varphi) \subseteq N/\varphi$, the converse inclusion being obvious. Further, it is clear that $S_0 = S_0(L/\varphi) \subseteq N(L/\varphi)$. Let M/φ be a complement for S_0 in L/φ (its existence follows from Lemma 1.29). Then $L/\varphi = S_0 \oplus M/\varphi$. Applying the modular law, we deduce $N/\varphi = S_0 + (M/\varphi \cap N/\varphi)$. Since $N^2 \subseteq \varphi$ (see (i)) the ideal $M/\varphi \cap N/\varphi$ of L/φ is abelian. Now this ideal, if nonzero, must intersect $S_0 = S_0(L/\varphi)$. This is true since, on the one hand, $S_0(L/\varphi)$ is the sum of all minimal abelian ideals of L/φ and, on the other hand, $M/\varphi \cap N/\varphi$, being a nonzero ideal, must include a minimal (abelian) ideal. Thus $M/\varphi \cap N/\varphi = \{0\}$, hence $N/\varphi = S_0$.

As for (iii), this is immediate from Lemma 1.29 and, in fact, it has been proved in the preceding paragraph. □

1.10 Graded algebras

1.10.1 Definition of graded algebras

Let R be a k-algebra, k being a commutative ring with unity and Ξ a commutative semigroup (written additively). A collection of k-submodules $\{R_\xi \mid \xi \in \Xi\}$ is called a Ξ-*grading* of R if

$$\text{(i)}\quad R = \bigoplus_{\xi \in \Xi} R_\xi; \qquad \text{(ii)}\quad R_\xi R_\eta \subseteq R_{\xi+\eta}.$$

Usually, the role of Ξ will be played by the semigroup of natural numbers with ordinary addition, or the group of all integers under addition, or the group of residue classes $\mathbb{Z}_m$ modulo m.

A nonzero $r \in R_\xi$ is called *homogeneous of degree* ξ. Suppose next that Ξ is a linearly ordered semigroup, that is, Ξ is endowed with a linear order $\le$ such that $\xi \le \eta$ implies $\xi + \zeta \le \eta + \zeta$ for any $\xi, \eta, \zeta \in \Xi$. Now if r is an arbitrary nonzero element in R such that $r = r_{\xi_1} + r_{\xi_2} + \cdots + r_{\xi_t}$, where $\xi_1 < \xi_2 < \cdots < \xi_t$, $r_{\xi_i} \in R_{\xi_i}$, then r_{ξ_t} is called the *leading term* of r and r_{ξ_1} the lowest one.

1.10.2 Examples

(i) The polynomial ring $R = k[x_1, x_2, \ldots, x_n]$ is $\mathbb{N}'$-graded, where $\mathbb{N}' = \mathbb{N} \cup \{0\}$, for each $n \in \mathbb{N}'$ the submodule R_n being the k-span of the monomials of total degree n.

(ii) Lie algebra $L = W_n$ possesses a $\mathbb{Z}$-grading in which $L_k = \{0\}$ for $k \le -2$ and L_{s-1} is the k-span of all derivations of the form $f(\partial/\partial x_i)$, f being a homogeneous polynomial of degree s.

(iii) A k-algebra R with $\mathbb{Z}_2$-grading $R = R_0 \oplus R_1$ is called a Lie *superalgebra* if the following relations hold in R:

$$x_\alpha x_\beta + (-1)^{\alpha\beta} x_\beta x_\alpha = 0,$$
$$x_\alpha(x_\beta y) = ((x_\alpha x_\beta) y) + (-1)^{\alpha\beta}(x_\beta(x_\alpha y)),$$

where $x_\alpha \in R_\alpha$, $x_\beta \in R_\beta$, $y \in R$. Thus R_0 is a Lie algebra, R_1 is an R_0-module, and the multiplication induces a symmetric bilinear map from $R_1 \times R_1 \to R_0$ such that

$$a(bc) + b(ac) + c(ab) = 0,$$
$$(ga)b + a(gb) = g(ab),$$

where $g \in R_0$, $a, b, c \in R_1$.

1.10.3 Homogeneous subalgebras

An additive subgroup S of a graded algebra R with index semigroup Ξ is called *homogeneous* if

$$S = \bigoplus_{\xi\in\Xi}(S \cap R_\xi).$$

Alternatively, with each $r = r_{\xi_1} + \cdots + r_{\xi_t}$ in S, all its homogeneous components r_{ξ_i}, $i = 1, \ldots, t$, are in S. So let S be a homogeneous ideal. Then putting $S_\xi = S \cap R_\xi$, $\xi \in \Xi$, we may write

$$R/S = \bigoplus_{\xi\in\Xi} R_\xi / \bigoplus_{\xi\in\Xi} S_\xi \cong \bigoplus_{\xi\in\Xi}(R_\xi/S_\xi).$$

Thus the quotient algebra R/S is endowed with the structure of Ξ-graded algebra, its homogeneous component $(R/S)_\xi$ being the image of R_ξ under natural epimorphism, and it is isomorphic to R_ξ/S_ξ as k-modules. Thus, for instance, we may speak about the quotient Lie superalgebra over a homogeneous ideal.

1.10.4 Filtration

Again let Ξ be a linearly ordered semigroup. A family of k-modules $\{P^\xi \mid \xi \in \Xi\}$ of a k-algebra P is an (ascending) filtration if:
(i) $P = \bigcup_{\xi\in\Xi} P^\xi$;
(ii) from $\xi \le \eta$ it follows that $P^\xi \subseteq P^\eta$;
(iii) $P^\xi P^\eta \subseteq P^{\xi+\eta}$;
(iv) $\bigcap_{\xi\in\Xi} P^\xi = \{0\}$.

Similarly, we define the *descending filtration*. Given an algebra P with ascending filtration as above, we define a graded algebra $R = \operatorname{gr} P$, associated with P as follows. For any $\xi \in \Xi$, we set $\operatorname{gr}_\xi P = P^\xi/\bar{P}^\xi$, where $\bar{P}^\xi = \bigcup_{\eta<\xi} P^\eta$. Then the k-module direct sum

$\mathrm{gr} = \bigoplus_{\xi\in\Xi} \mathrm{gr}_\xi P$ becomes a Ξ-graded k-algebra if, given two homogeneous elements $a = x_\xi + \bar{P}^\xi$, $b = y_n + \bar{P}^\eta$, we put

$$ab = xy + \bar{P}^{\xi+\eta}.$$

If Q is a k-submodule in P then the k-span of the elements of the form $\mathrm{gr}\, x = x + \bar{P}^\xi$, $x \in P^\xi \cap Q$, is obviously a homogeneous submodule in $\mathrm{gr}\, P$.

Moreover, if Q is a subalgebra, then so is $\mathrm{gr}\, Q$. Finally, if $R = \bigoplus_{\xi\in\Xi} R_\xi$ is a graded algebra then, setting $R^\xi = \bigoplus_{\eta\leq\xi} R_\eta$ makes R into a filtered algebra. Clearly, $\mathrm{gr}_\xi R \cong R_\xi$. This enables us to identify $\mathrm{gr}\, R$ and R. In this case, given a k-submodule Q in R, the submodule $\mathrm{gr}\, Q$ is already homogeneous, and this is the k-span of the set of leading terms of elements in Q.

Lemma 1.31. *Let k be a field and $\mathrm{gr}\, Q$ the k-span of leading terms of elements in Q. If Ξ is well-ordered and $\dim L/Q = m < \infty$ then also $\dim(L/\mathrm{gr}\, Q) = m$.*

Proof. We note that if $Q \neq R$ then there exists an index ξ such that $\mathrm{gr}_\xi Q \neq R_\xi$. Indeed, assuming the contrary, we proceed as follows. Let $r \in R$ be a homogeneous element of minimal degree with $r \notin Q$. Since $\mathrm{gr}_\eta Q = R_\eta$, there exists $q \in Q$ such that $r - q \in \bigoplus_{\xi<\eta} R_\xi$. We have $R_\xi \subseteq Q$ for all $\xi < \eta$. Hence $r - q \in Q$, that is, $r \in Q$, a contradiction. Now let the subspace E_χ in R_χ be such that $R_\chi = E_\chi \oplus \mathrm{gr}_\xi Q$. If $E = \bigoplus_{\xi\in\Xi} E_\xi$ then $R = E \oplus \mathrm{gr}\, Q$. It is clear from the above that $R = Q + E$. Assume that $Q \cap E \neq 0$. Let q be an element of the least degree η in $Q\cap E$. Then $\mathrm{gr}\, q \in \mathrm{gr}_\xi Q \cap \mathrm{gr}_\xi E = \mathrm{gr}_\xi Q \cap E_\xi = \{0\}$, a contradiction. Hence $R = Q \oplus E$, making obvious the equation $\dim R/Q = \dim(R/\mathrm{gr}\, Q) = \dim E$. □

1.11 Restricted Lie algebras

1.11.1 Example

Let k be a ring of prime additive exponent p, for instance, a field of characteristic $p > 0$. Let A be an associative k-algebra and $[A]$ its Lie algebra. The unary operation of raising an element to the pth power has the following properties:

(i) $(\lambda x)^p = \lambda^p x^p$, where $\lambda \in k$, $x \in A$;

(ii) $\mathrm{ad}\, x^p = (\mathrm{ad}\, x)^p$, where $x \in A$;

(iii) $(x + y)^p = x^p + y^p + \sum_i s_i(x, y)$,

where $is_i(x, y)$ is the coefficient of t^{i-1} in the polynomial

$$\mathrm{ad}(tx + y)^{p-1}(x) \in A[t], \quad x, y \in A.$$

Property (i) is obvious. To prove (ii), we set, for all $x, a \in A$, $l_x(a) = xa$ and $r_x(a) = ax$. Then $l_x, r_x \in \mathrm{End}_k A$ and $l_x r_y = r_y l_x$ $(y \in A)$. Applying the binomial decomposition, we write

$$(\operatorname{ad} x)^p = (l_x - r_x)^p = (l_x)^p - (r_x)^p + \sum_{i=1}^{p-1} (-1)^i \binom{p}{i} l_x^i r_x^{p-i}$$
$$= (l_x)^p - (r_x)^p = l_{x^p} - r_{x^p} = \operatorname{ad} x^p.$$

We have used here that for each $i = 1, \ldots, p-1$ the binomial coefficient

$$\binom{p}{i} = \frac{p!}{i!(p-i)!}$$

is divisible by p.

In proving (iii), we start with the following equation with indeterminate coefficients in the polynomial ring $A[t]$:

$$(tx + y)^p = t^p x^p + y^p + \sum_{i=1}^{p-1} t^i s_i(x, y).$$

Taking the derivative of both sides with respect to t, we get

$$\sum_{i=0}^{p-1} (tx + y)^i x (tx + y)^{p-1-i} = \sum_{i=1}^{p-1} i t^{i-1} s_i(x, y).$$

Now we consider $\operatorname{ad}(tx + y)^{p-1}(x)$. Arguing as in the proof of (ii), we may write

$$\operatorname{ad}(tx + p)^{p-1} = (l_{tx+y} - r_{tx+y})^{p-1}$$
$$= \sum_{i=0}^{p-1} (-1)^i \binom{p-1}{i} l_{tx+y}^i r_{tx+y}^{p-1-i}.$$

Now the proof will be complete if we remark that, for any $i = 0, 1, \ldots, p-1$, we always have $\binom{p-1}{i} = (-1)^i \pmod p$. But this can be easily proved by induction on i using the relation

$$\binom{p}{i} = \binom{p-1}{i} + \binom{p-1}{i-1}.$$

Now it should be noticed that, given $x, y \in [A]$, we have $s_i(x, y) \in \operatorname{alg}(x, y)$. This makes correct the following definition.

1.11.2 Definition of restricted Lie algebra

A Lie algebra L over a ring k of prime additive exponent $p > 0$ endowed with additional unary operation $x \mapsto x^p$ which satisfies the axioms (i), (ii), and (iii) is called a *restricted*

Lie algebra, or a p-algebra. Any k-submodule M in L which is closed under multiplication and exponentiation is called a p-subalgebra. Similarly, we define a p-ideal. The quotient algebra over a p-ideal is naturally a p-algebra. The least p-subalgebra of a p-algebra L containing a subset X in L is called a p-span of X and it is denoted by $\mathrm{alg}_p(X)$. We write inductively $x^{p^r} = (x^{p^{r-1}})^p$.

Proposition 1.32. *Let L be a p-algebra and M a Lie subalgebra of L (that is, not necessarily a p-subalgebra) which is the k-span of an X. If $P = \mathrm{alg}_p(M)$ is the p-span of M then P is the k-span of a subset Y whose elements are of the form x^{p^r}, $x \in X$, $r = 0, 1, 2, \ldots$*

Proof. It is enough to show that the set of all k-combinations of the form

$$\lambda_1 x_1^{p^{r_1}} + \lambda_2 x_2^{p^{r_2}} + \cdots + \lambda_s x_s^{p^{r_s}}, \quad \lambda_1, \lambda_2, \ldots, \lambda_s \in k,$$

where $x_1, x_2, \ldots, x_s \in X$ and $r_1, r_2, \ldots, r_s$ are some nonnegative integers, is a p-subalgebra in L. This is really the case. Firstly, given $x, y \in X$, we have

$$\begin{aligned} x^{p^u} y^{p^v} &= (\mathrm{ad}\, x)^{p^u}(y^{p^v}) = -(\mathrm{ad}\, x)^{p^{u-1}}(y^{p^v} x) \\ &= -(\mathrm{ad}\, x)^{p^{u-1}}(\mathrm{ad}\, y)^{p^v}(x). \end{aligned}$$

The element on the right-hand side is in M, that is, it is equal to a linear combination of elements in X, proving that the set under consideration is multiplication closed. Secondly, using induction over $s \geq 1$ and what has been just shown, we may write

$$\begin{aligned} (\lambda_1 x_1^{p^{r_1}} + \lambda_2 x_2^{p^{r_2}} + \cdots + \lambda_s x_s^{p^{r_s}})^p &= \lambda_1^p x_1^{p^{r_1+1}} + (\lambda_2 x_2^{p^{r_2}} + \cdots + \lambda_s x_s^{p^{r_s}})^p \\ &\quad + \sum_{i=1}^{p-1} s_i(\lambda_1 x_1^{p^{r_1}}, \lambda_2 x_2^{p^{r_2}} + \cdots + \lambda_s x_s^{p^{r_s}}). \end{aligned}$$

All s_i are the products in the sense of Lie algebras and, by the above, they are in M. □

1.11.3 Examples

(i) Given any k-algebra R, $\mathrm{char}\, k = p > 0$, its derivation algebra $\mathrm{Der}_k(R)$ is a p-algebra, since for any derivation $\delta : R \to R$, its p-power (in the common sense) is also a derivation.

(ii) Lie algebra $\mathfrak{sl}(n, k)$, $\mathrm{char}\, k = p > 0$, is a p-subalgebra in the algebra k_n of all $n \times n$ matrices with respect to usually raising a matrix to the pth power.

(iii) Any Lie k-algebra which can be embedded in a Lie algebra of the form $[A]$, A being an associative algebra over a ring of characteristic $p > 0$, can be also embedded in a p-algebra. This is an immediate corollary of Proposition 1.32.

1.12 Exercises

1.12.1. Any commutative ring, in which $x+x=0$, satisfies the identical relation (1.2). Give an example of a ring in which equation (1.2) holds but equation (1.1) is not satisfied.

1.12.2. In any anticommutative ring R, the Jacobian $J(x,y,z)$ is an alternating function of its arguments. In particular, for all $x, y \in R$, we have $J(x,y,x) = J(x,x,y) = J(y,x,x) = 0$.

1.12.3. Let k be a commutative ring without zero divisors and I its ideal. Then k-module I becomes a Lie algebra over k only if the multiplication is zero.

1.12.4. Let A be a commutative algebra, δ its derivation. Set $[a,b] = a\delta(b) - b\delta(a)$. Then A becomes a Lie algebra.

1.12.5. Prove that if δ is a nilpotent derivation of an algebra G over a field of characteristic zero then

$$e^{\delta} = 1 + \sum_{n=1}^{\infty} \frac{\delta^n}{n!}$$

is an automorphism of G.

1.12.6. Determine the center of $\mathfrak{gl}(n,k)$.

1.12.7. Determine, with respect to isomorphism, all semidirect products of a 1-dimensional algebra by a 1-dimensional one. Describe all 2-dimensional algebras over a field.

1.12.8. Prove that, given three ideals A, B, C of a Lie algebra L, we always have

$$A(BC) \subseteq B(CA) + C(AB).$$

1.12.9. Give an example of a Lie algebra in which the lower central series is strictly decreasing and the upper central series is finite.

1.12.10. An ideal P of a Lie algebra G is called *prime* if under the condition $HK \subseteq P$, $H, K \lhd G$, it follows that either $H \subseteq P$ or $K \subseteq P$. Prove that in the case of Lie algebras this condition is equivalent to the following: for any $a, b \in G$ from $a\,\mathrm{id}_L(b) \subseteq P$ is follows that either $a \in P$ or $\mathrm{id}_L(b) \subset P$.

1.12.11. Let the center of a Lie algebra G over a field of have finite codimension. Then $\dim G^2 < \infty$.

1.12.12. Let G be a Lie algebra over a field and let G be finitely generated. Suppose $\dim G^2 < \infty$. Then $\dim(G/Z(G)) < \infty$. Is it true for Lie algebras that are not finitely generated?

1.12.13. Let V, W be two G-modules, G being a Lie algebra over a ring k, $\mathrm{Hom}_k(V,W)$ the k-module of k-endomorphisms from V into W. Set $(gf)v = gf(v) - f(gv)$, where $g \in G$, $f \in \mathrm{Hom}_k(V,W)$, $v \in V$. Prove that in this way $\mathrm{Hom}_k(V,W)$ becomes a G-module. A homomorphism f is a G-invariant, i. e., $Gf = \{0\}$, if and only if this is a homomorphism of G-modules.

1.12.14. Let V be a G-module, G being a Lie algebra over k, $B(V)$ the space of bilinear functions on V with values in k. Set $(gb)(v, w) = b(gv, w) + b(v, gw)$. Prove that in this way $B(V)$ becomes a G-module. Show that if G is finite-dimensional over a field k, then the function $f(g_1, g_2) = \operatorname{tr}(\operatorname{ad} g_1 \operatorname{ad} g_2)$ on the adjoint G-module is invariant, that is, $Gf = \{0\}$.

1.12.15. Determine all chief factors of $\operatorname{gl}(2, k)$ over any field k of arbitrary characteristic.

1.12.16. Let $G = \mathfrak{sl}(2, k)$, $\operatorname{char} k = 0$, $H = kh$, where

$$h = \begin{pmatrix} 1 & 0 \\ 0 & -1 \end{pmatrix}.$$

Is G a semisimple H-module with respect to the adjoint action?

1.12.17. Let $A = k_2$ be the matrix algebra of order 2 over a field k of characteristic 0, $L \subseteq [A]$, $L = \mathfrak{sl}(2, k)$. Is it true that the left regular L-module A is semisimple?

1.12.18. We use the notation of Section 1.6.6. Let $C^n(G, M)$ denote the space of all k-multilinear maps from G into M. Given any $g \in G$ and any $f \in C^n(G, M)$, we define

$$(\theta(g)f)(x_1, \ldots, x_n) = g(f(x_1, \ldots, x_n)) - \sum_{i=1}^{n} f(x_1, \ldots, gx_i, \ldots, x_n).$$

Show that θ is a representation of G and that $Z^n(G, M)$ is stable under $\theta(G)$.

1.12.19. (see Exercise 1.12.18). For any $x \in G$, we define $i(x) : Z^n(G, M) \to Z^{n-1}(G, M)$ by setting $(i(x)f)(x_1, \ldots, x_{n-1}) = f(x, x_1, \ldots, x_{n-1})$. Prove that $i(x)^2 = 0$ and $\theta(x)i(y) - i(y)\theta(x) = i(xy)$.

1.12.20. (see Exercise 1.12.19). Using induction over the dimension of cochains, prove that, given any $x \in G$, we always have

$$\partial i(x) + i(x)\partial = \theta(y), \quad \partial\theta(y) = \theta(y)\partial.$$

1.12.21. (see Exercise 1.12.20). Prove that ∂^2 commutes with $i(x)$ for any $x \in G$. Proceeding by induction on the dimension of cochains, derive $\partial^2 = 0$.

1.12.22. Let G be a Lie algebra, V a G-module, $V^G = \{v \mid Gv = \{0\}\}$. Prove that $H^0(G, M) \cong V^G$.

1.12.23. Consider the adjoint G-module, with G being a Lie algebra. Prove that $H^1(G, G) \cong \operatorname{Der} G / \operatorname{ad} G$.

1.12.24. Let $G = ka \oplus kb$, $M = kv$, $ab = 0$, $GM = \{0\}$. Compute $H^2(G, M)$.

1.12.25. Let G be a nilpotent Lie algebra over a commutative ring k, N a nontrivial ideal of G, and $Z(G)$ the center of G. Show that $N \cap Z(G) \neq \{0\}$.

1.12.26. Let $N_G(H) = \{g \mid gH \subseteq H\}$ denote the normalizer of a subalgebra H in a Lie algebra G. Show that $N_G(H)$ is a subalgebra and that H is an ideal in $N_G(H)$. If G is a nilpotent Lie algebra over a ring k and $H \neq G$ then also $N_G(H) \neq H$.

1.12.27. Let G be a finitely generated nilpotent Lie ring. Prove that any one of its subrings is finitely generated.

1.12.28. Let G be a finitely generated nilpotent Lie ring. Suppose that $|Z(G)| < \infty$. Prove that then also $|G| < \infty$.

1.12.29. Suppose that the additive group of a finite nilpotent Lie ring is a p-group, with p being a prime number. Prove that then $\Phi(G) = G^2 + pG$.

1.12.30. Suppose that a simple Lie algebra over k is locally nilpotent. Prove that then G is a simple k-module.

1.12.31. Let G be a 2-dimensional metabelian Lie algebra with a basis $\{x, y\}$ such that $[x, y] = y$, and $V = k[t]$ the polynomial ring over the base field k. Make M into a G-module by setting

$$xf = tf(t), \quad yf = f(t-1).$$

Prove that, in the case $\operatorname{char} k = 0$, V is a simple G-module.

1.12.32. Let G be as above, and k a field of characteristic $p > 0$ and V a p-dimensional space over k with a basis $e_1, \dots, e_p$. Set $xe_i = e_{i+1}$, $1 \le i < p$, $xe_p = e_1$, $ye_j = (j-1)e_j$, $1 \le j \le p$. Prove that V becomes a faithful irreducible G-module.

1.12.33. Apply the construction of the above problem and semidirect products to exhibiting a finite-dimensional solvable Lie algebra over any field of characteristic $p > 0$ whose derived algebra is not nilpotent.

1.12.34. Prove that any metabelian Lie algebra over a field of characteristic 0 satisfying the identical relation $(\operatorname{ad} x)^n = 0$ is c-step nilpotent, $c \le n$.

1.12.35. Prove that a Lie algebra with the law $(\operatorname{ad} x)^2 = 0$ is nilpotent of the nilpotent step at most 2, provided 3 is invertible in the base ring.

1.12.36. Prove that any Lie algebra with the law $(\operatorname{ad} x)^2 = 0$ is at most 3-step nilpotent.

1.12.37. Is it true that any Lie algebra satisfying the identical relation $(\operatorname{ad} x)^2 = \operatorname{ad} x$ is nilpotent?

1.12.38. Let G be a metabelian Lie algebra of linear operators of a vector space V over a field k, $\operatorname{char} k = 0$. Suppose that there exists a natural number l such that $g^l(v) = 0$. Prove that, in this case, G is a nilpotent algebra.

1.12.39. Prove that, given a Lie algebra L and a nilpotent ideal M of L, algebra L is Engel if and only if such is L/M^2.

1.12.40. Let G be a 3-dimensional nonabelian Lie algebra over a field of characteristic $p > 0$. Endow G with the structure of a p-algebra. How many pairwise nonisomorphic structures of a p-algebra might one introduce on G? Give an example of a nilpotent Lie algebra over a field of characteristic $p > 0$ which cannot be given such a structure.

1.12.41. Describe the structure of p-algebras on 2-dimensional Lie algebras over a field of characteristic $p > 0$.

1.12.42. Verify the assertions in Section 1.11.3.

1.12.43. Is there a structure of a p-algebra on the Lie algebra of skew-symmetric matrices?

1.12.44. Prove that any Lie algebra L can be embedded in a Lie algebra $\bar{L}$ in such a way that an ideal M of L is an ideal of $\bar{L}$ if and only if M is characteristic, that is, M is left-invariant with respect to all derivations of L.

1.13 Comments

The bulk of this chapter is formed by classical results usually included in all books on Lie algebras. The main distinction is that we attempted to impose no restrictions on the base commutative ring. On the other hand, we left aside the important notion of a semisimple algebra. A short summary of the results concerning this important topic will be given in Section 6.2, for our immediate purposes there. Those in need of deeper information on this and many other branches of Lie algebras are referred to the following (incomplete) list of books: of Bourbaki [81–83], Jacobson [150], Serre [328, 329] as well as to Kaplansky [159], Postnikov [293], Goto and Grosshans [130], Humphreys [145], Gilmore [126], Hochschild [144], Seligman [326], Wam Zue-Xian [368], and Winter [370].

Our theory of extensions in Section 1.6 is somewhat sketchy. For more details, we refer to Jacobson [150] and Séminaire "Sophus Lie" [327].

The Frattini theory which we need principally for studying identities in finite Lie algebras is based on the papers of Barnes [63], Barnes and Newell [64], and Towers [349]. The sources of Example 1.8.3 and Theorem 1.25 are Cohn [90] and Higgins [141], respectively.

2 Free Lie algebras

2.1 Lie algebras satisfying identical relations

2.1.1 Nonassociative polynomials

Let $X = \{x_1, x_2, \dots, x_n, \dots\}$ be a countable set which we call the set of *variables*. A *nonassociative monomial* of degree 1 is any element in X. Given a natural number $n > 1$, a *nonassociative monomial* of degree n is an expression of the form $(u)(v)$, where u is a nonassociative monomial of degree i and v a nonassociative monomial of degree $n-i$. Thus, by induction, we have defined nonassociative monomials of arbitrary natural degree. To simplify our notation, we do not write brackets in the expression $(u)(v)$, around u (and/or v) if u (and/or v) is a monomial of degree 1. A *nonassociative polynomial* over a ring k is a formal linear combination of nonassociative monomials with coefficients in k. If f includes no variables except $x_1, x_2, \dots, x_n$, and $a_1, a_2, \dots, a_n$ is a collection of elements of a k-algebra R then $f(a_1, a_2, \dots, a_n)$ is an element of R which results by applying the sequence of operations forming f to $a_1, a_2, \dots, a_n$ in place of $x_1, x_2, \dots, x_n$.

2.1.2 Algebras satisfying an identity (PI-algebras)

Let $\mathbf{V}$ denote a class of algebras over a fixed ring k and $R \in \mathbf{V}$. We say that R is an algebra of $\mathbf{V}$ satisfying a (nontrivial) identity (or a PI-algebra) if there exists a nonassociative polynomial such that for any $a_1, a_2, \dots, a_n \in R$ we have $f(a_1, a_2, \dots, a_n) = 0$, but there is an algebra S in $\mathbf{V}$ and some elements $b_1, b_2, \dots, b_n \in S$ such that $f(b_1, b_2, \dots, b_n) \neq 0$. The expression of the form $f(x_1, x_2, \dots, x_n) \equiv 0$ is then called an *identical relation*, and we say that R satisfies this identical relation (or identity) within $\mathbf{V}$. Thus, having taken for $\mathbf{V}$ any of the classes of commutative, anticommutative, associative, alternative, Jordan or Lie algebras, we arrive at the concept of a commutative, anticommutative, associative, etc., algebra satisfying an identical relation. Any Lie ring is an algebra satisfying an identity in the class of all anticommutative rings since there exist rings which are anticommutative but do not satisfy the Jacobi identity. The synonymic terminology of PI-algebras is applied, as a rule, to denote the associative algebras satisfying a nontrivial relation (polynomial identity algebras).

2.1.3 Lie algebras with identities. Examples

(i) Any abelian Lie algebra is a Lie algebra with identity. A polynomial annihilating any abelian Lie algebra is $f(x, y) = xy$.

https://doi.org/10.1515/9783110566659-002

(ii) Any nilpotent Lie algebra is an algebra with identity (cf. Section 1.7.1). The nontriviality of the identity $x_1 x_2 \cdots x_c \equiv 0$ is implied by the existence of nilpotent Lie algebras which are not c-step nilpotent (e. g., $\mathfrak{nt}_{c+1}(k)$).

(iii) Solvable Lie algebras of fixed solvable step l are Lie algebras satisfying the nontrivial identical relation $\delta_1(x_1, x_2, \dots, x_{2^l}) \equiv 0$ (cf. Section 1.8, Theorem 1.20). This identity is nontrivial which follows from the existence of Lie algebras L with $L^2 = L$: any nonabelian simple Lie algebra is such.

(iv) We consider a right-normed nonassociative polynomial of the form

$$s_n(x_1, \dots, x_{n+1}) = \sum_{\sigma \in S_n} \epsilon_\sigma x_{\sigma(1)} x_{\sigma(2)} \cdots x_{\sigma(n)} x_{n+1}, \tag{2.1}$$

where S_n is the symmetric group of order n, and $\epsilon_\sigma = 1$ if σ is an even permutation and $\epsilon_\sigma = -1$ otherwise. Then, given a Lie algebra of dimension less than n, we easily see that $s_n(x_1, \dots, x_{n+1}) \equiv 0$ is an identity in L. This follows, since given a k-generating set $E = \{e_1, \dots, e_{n-1}\}$ for L, any value of equation (2.1) is a linear combination of its values in the elements of E. Then at least two variables of $x_1, \dots, x_n$ take the same value. Since these variables alternate, we finally get zero. To show the nontriviality of the identity, we consider $\mathfrak{gl}(n+1, k)$ and put

$$x_{n+1} = e_{11}, \quad x_n = x_{21}, \quad x_{n-1} = x_{32}, \quad \dots, \quad x_1 = e_{n+1,n},$$

where e_{ij} is a matrix unit. Using the rules for multiplying matrix units ($e_{ij} e_{rs} = \delta_{jr} e_{is}$), we easily see that the only nonzero summand in the value of the right-hand side of equation (2.1) corresponds to $\sigma = 1$ and it is equal to $e_{n+1,1}$. Hence $s_n(x_1, \dots, x_{n+1}) \equiv 0$ is not an identity of $\mathfrak{gl}(n+1, k)$.

(v) Let W_m denote the Lie algebra of derivations of $k[x_1, \dots, x_m]$ (Section 1.2.3). For each $m = 1, 2, \dots$, this is a Lie algebra satisfying an identical relation. We will show that $s_n(x_1, \dots, x_{n+1}) \equiv 0$ is an identity of W_m with $n = m^2 + 2m + 2$. We recall that W_m is a graded Lie algebra $W_m = L_{-1} \oplus L_0 \oplus L_1 \oplus \cdots$ and

$$u = \sum_{i=1}^{m} f_i \frac{\partial}{\partial x_i} \in L_s$$

iff for every $i = 1, \dots, m$, either $f_i = 0$ or f_i is homogeneous of degree $s+1$. If $s_n(x_1, \dots, x_{n+1}) \equiv 0$ is not an identity then, the polynomial being multilinear, there exist homogeneous $u_1, \dots, u_{n+1}$ such that $u = s_n(u_1, \dots, u_{n+1}) \neq 0$. Now, for some s, we have $u \in L_s$. If $s \geq 0$, then supposing $f_1 \neq 0$ we can find a j such that $\partial f_1 / \partial x_j \neq 0$. Therefore

$$\left[\frac{\partial}{\partial x_j}, u\right] = \frac{\partial f_1}{\partial x_j} \frac{\partial}{\partial x_1} + g_2 \frac{\partial}{\partial x_2} + \cdots + g_m \frac{\partial}{\partial x_m} \neq 0.$$

On the other hand, applying the rules for a derivation, we have

$$\left[\frac{\partial}{\partial x_j}, u\right] = s_n\left(\left[\frac{\partial}{\partial x_j}, u_1\right], \dots, u_{n+1}\right) + \cdots + s_n\left(u_1, \dots, \left[\frac{\partial}{\partial x_j}, u_{n+1}\right]\right). \tag{2.2}$$

Hence one of the summands in equation (2.2) is different from zero and it is of degree strictly less than s. Now we remark that $\dim L_{-1} = m$ and equation (2.1) is skew-symmetric over all arguments but x_{n+1}. Thus, if we substitute elements of a basis of L_{-1} into equation (2.1), at most $m + 1$ elements may take value in L_{-1}. Since the resulting value of $s_n(x_1, \ldots, x_{n+1})$ must be in L_{-1} at most m variables may take value in a basis of the space $\bigoplus_{t>0} L_t$. Now we see that at least $n - (2m + 1) = m^2 + 1$ variables must be valued in L_0. Since $\dim L_0 = m^2$, we finally see that any substitution gives zero, proving our assertion.

(vi) Suppose that L is a Lie subalgebra in a Lie algebra of the form $[A]$ (see Section 1.2.1), A being an associative PI-algebra. Then L is a Lie algebra with identity. Indeed, according to Jacobson [151, Chapter X], there exists a natural number n such that if $f \equiv 0$ is an identity in the associative algebra k_n of all square matrices of order n, then, for suitable $l > 0$, $f^l \equiv 0$ is an identity in A. Since k_n is an algebra of dimension n^2, an identity of the form

$$s_{n^2+1}(x_1, \ldots, x_{n^2+1}, x_{n^2+2}) \equiv 0$$

holds in this algebra. Then, in A, we have

$$(s_{n^2+1}(x_1, \ldots, x_{n^2+1}, x_{n^2+2}))^l \equiv 0.$$

Proceeding as in the proof of Engel's theorem, Section 1.7, we derive

$$(\operatorname{ad} s_{n^2+1}(x_1, \ldots, x_{n^2+2}))^{2l-1} \equiv 0.$$

Thus, in L, we have

$$(\operatorname{ad} s_{n^2+1}(x_1, \ldots, x_{n^2+2}))^{2l-1}(x_{n^2+3}) \equiv 0.$$

The verification of this identity being nontrivial is rather a delicate task. The best tool for this and many other similar problems is using the developed theory of free Lie algebras. This theory will be dealt with in the remaining sections of this chapter. An alternative approach to the above particular result is used in Section 6.3.

2.2 Free Lie algebra

2.2.1 Free groupoid

Let X be a nonempty set, $\Gamma = \Gamma(X)$ the set of all nonassociative monomials in the variables comprising X. This set is considered along with the operation $*$ of juxtaposition, that is, given $u, v \in \Gamma(X)$, their product is the monomial $(u)(v)$ (as in Section 2.1.1, a pair of brackets surrounding u (and/or v) is omitted if u (and/or v) is of degree 1). A set with one binary operation is called a *groupoid*. Then $\Gamma(X)$ is a *free groupoid* in the following sense.

Proposition 2.1. *Let G be a groupoid. For any map $\varphi : X \to G$, there exists a unique homomorphism $\bar{\varphi} : \Gamma(X) \to G$ such that $\bar{\varphi}\,|_X = \varphi$.*

Proof. We define $\bar{\varphi}$ using induction over the degree of monomials. For the monomials of degree 1, that is, the elements of X, we set $\bar{\varphi}(x) = \varphi(x)$. If now $w = (u)(v)$ then set set $\bar{\varphi}(w) = \bar{\varphi}(u)\bar{\varphi}(v)$ (multiplication on the right-hand side is performed in G). All the required properties of $\bar{\varphi}$ are obvious. □

2.2.2 Multidegree

If, in the above, we set $G = \mathbb{N}$, the groupoid of natural numbers under addition, then the mapping $x \to 1$ for all $x \in X$ extends to a homomorphism $d : \Gamma(X) \to \mathbb{N}$. It is obvious that $d(v)$ is the degree of the monomial v.

Now let $G = \mathbb{Z}^X$ be the set of functions from X into $\mathbb{Z}$. We set $(\alpha+\beta)(x) = \alpha(x)+\beta(x)$. For any $x \in X$, let α_x denote the function such that $\alpha_x(x) = 1$ and $\alpha_x(y) = 0$ if $y \neq x$. The map $x \to \alpha_x$ for all $x \in X$ extends to a homomorphism $m : \Gamma(X) \to \mathbb{Z}^X$. Given a monomial $v \in \Gamma(X)$, its image $m(v)$ under M is called the *multidegree* of v. For instance,

$$d(((xy)y)(xy)) = 5, \quad m(((xy)y)(xy)) = 2\alpha_x + 3\alpha_y.$$

Now given a natural n, we denote by $\Gamma_n(X)$ (or simply Γ_n) the set of all monomial $v \in \Gamma(X)$ such that $d(v) = n$. For any $\alpha \in \mathbb{Z}^X$, we denote by $\Gamma_\alpha(X)$ the set of all $v \in \Gamma(X)$ such that $m(v) = \alpha$. It is clear that $\Gamma_\alpha(X)$ is nonempty if and only if α is a function of *finite support* $\operatorname{supp}\alpha$, where

$$\operatorname{supp}\alpha = \{x \in X \mid \alpha(x) \neq 0\}$$

and, in addition, for all $x \in \operatorname{supp}\alpha$ we have $\alpha(x) > 0$. The set of all functions of this sort is denoted by Φ. Given a function $\alpha \in \Phi$, we define $|\alpha| = \sum_{x\in X} \alpha(x)$. Thus

$$\Gamma(X) = \bigcup_{n=1}^{\infty} \Gamma_n(X), \quad \Gamma(X) = \bigcup_{\alpha\in\Phi} \Gamma_\alpha(X), \quad \Gamma_n = \bigcup_{|\alpha|=n} \Gamma_\alpha,$$

in all cases the union is being disjoint.

2.2.3 Free algebra

Now suppose we are given a commutative ring k with unity and a nonempty set X. Let $\Gamma = \Gamma(X)$ be the free groupoid on X. We introduce a free k-module whose basis is Γ, denoted by $F(X)$. We will define multiplication in $F(X)$ to turn it into a k-algebra. So let $f, g \in F(X)$. Then

$$f = \sum_{u\in\Gamma} \lambda_u u, \quad g = \sum_{v\in\Gamma} \mu_v v.$$

We set

$$f * g = \sum_{u,v \in \Gamma} \lambda_u \mu_v (u)(v).$$

It is obvious that then $F(X)$ becomes a k-algebra and, if we naturally identify Γ with the set of all monomials in $F(X)$, then Γ becomes a subgrouppoid in $F(X)$. The following proposition is analogous to what was proved in Section 2.2.1.

Proposition 2.2. *Let R be an arbitrary k-algebra and φ an arbitrary map from X into R. Then φ uniquely extends to a homomorphism $\bar{\varphi} : F(X) \to R$ such that $\bar{\varphi}\mid_X = \varphi$.*

Now let n be a natural number and $\alpha \in \Phi$ (see Section 2.2.2). We denote by $F_n = F_n(X)$ the k-span of $\Gamma_n(X)$ and by $F_\alpha = F_\alpha(X)$ the k-span of $\Gamma_\alpha(X)$. The elements in F_n are called *homogeneous of degree n*, and those in F_α are called *multihomogeneous of multidegree α*. If $\alpha(x) = 1$ or 0 for all $x \in X$, then we say that the elements of F_α are multilinear in the variables $X' = \operatorname{supp} \alpha$. The following are the direct decompositions as k-modules:

$$F(X) = \bigoplus_{n=1}^{\infty} F_n(X), \quad F(X) = \bigoplus_{\alpha \in \Phi} F_\alpha(X). \tag{2.3}$$

It is important to remark that, for any $\alpha \in \Phi$, the k-module $F_\alpha(X)$ is of finite rank. Furthermore, for any natural n, m and any $\alpha, \beta \in \Phi$, we may write

$$F_n * F_m \subseteq F_{n+m}, \quad F_\alpha * F_\beta \subseteq F_{\alpha+\beta}.$$

2.2.4 Free Lie algebras

It is obvious that any homomorphism from $F(X)$ into a Lie algebra sends to zero the elements of either of the two forms, $f * f$ and $J(f, g, h) = f * (g * h) + g * (h * f) + h * (f * g)$, where $f, g, h \in F(X)$. Let I be the least ideal of $F(X)$ containing all these elements. Then the quotient algebra $L(X) = F(X)/I$ is called the *free Lie algebra with the set X of free generators*. The multiplication in $L(X)$ will be denoted by juxtaposition with the same convention about brackets as in Section 2.2.1. Given an arbitrary Lie algebra G, we know that any map $\varphi : X \to G$ extends to a homomorphism $\bar{\varphi} : F(X) \to G$ and, as we saw just now, $K = \ker \bar{\varphi} \supseteq I$. By the homomorphism theorem (Section 1.1), part (i), there exists a homomorphism $\tilde{\varphi} : L(X) = F(X)/I \to G$ whose kernel is K/I. In particular, if G is an abelian Lie algebra which is a free k-module with basis X, then an appropriate $\tilde{\varphi}$ maps the k-submodule of L/X generated by all $x + I$, $x \in X$, onto the free k-module generated by X in such a way that $x + I \mapsto x$. This enables us to identify X with its image in $F(X)$, and we may say that kX is a free k-submodule in $L(X)$. With this convention, we have proved the following.

Proposition 2.3. *Let G be any arbitrary Lie algebra over k and φ a mapping from X into G. Then φ uniquely extends to a homomorphism $\bar{\varphi} : L(X) \to G$ such that $\bar{\varphi}\,|_X = \varphi$.*

Proposition 2.4. *Ideal I is homogeneous with respect to any of the decompositions in equation* (2.3).

Proof. We have to show that

$$I = \bigoplus_{n=1}^{\infty} (I \cap F_n) \quad \text{and} \quad I = \bigoplus_{\alpha \in \Phi} (I \cap F_\alpha).$$

The two cases being similar, we restrict ourselves to proving the second decomposition. So, we must show that given an element in I of the form $\sum_\alpha \lambda_\alpha f_\alpha$ with $m(f_\alpha) = \alpha$, then all f_α are in I. By definition, all elements in I are some linear combinations of elements of the form

$$u_1 * \cdots * u_s * f * f * u_{s+1} * \cdots * u_t,$$
$$v_1 * \cdots * v_p * J(f, g, h) * v_{p+1} * \cdots * v_q,$$

with some brackets and where all u_i, v_j are monomials. Since every monomial is multihomogeneous and the product of multihomogeneous elements is itself multihomogeneous, it is sufficient to restrict ourselves to the case where a has the form of $J(f, g, h)$ of $f * f$. We note that if $f = \sum \lambda_\alpha f_\alpha$, $g = \sum \mu_\beta g_\beta$, $h = \sum \eta_\gamma h_\gamma$, where all $f_\alpha, g_\beta, h_\gamma$ are multihomogeneous then

$$J(f, g, h) = \sum \lambda_\alpha \mu_\beta \eta_\gamma J(f_\alpha, g_\beta, h_\gamma),$$

where $J(f_\alpha, g_\beta, h_\gamma) \in F_{\alpha+\beta+\gamma} \cap I$, and all is done. Now if

$$f = \sum_{i=1}^{t} \lambda_{\alpha_i} f_{\alpha_i},$$

we may write

$$f * f = \sum_{i,j=1}^{t} \lambda_{\alpha_i} \lambda_{\alpha_j} f_{\alpha_i} * f_{\alpha_j} = \sum_{i=1}^{t} \lambda_{\alpha_i}^2 f_{\alpha_i} f_{\alpha_i} + \sum_{i<j} \lambda_{\alpha_i} \lambda_{\alpha_j} (f_{\alpha_i} f_{\alpha_j} + f_{\alpha_j} f_{\alpha_i}).$$

Here $f_{\alpha_i} * f_{\alpha_i} \in F_{2\alpha_i} \cap I$. Besides,

$$f_{\alpha_i} f_{\alpha_j} + f_{\alpha_j} f_{\alpha_i} = (f_{\alpha_i} + f_{\alpha_j}) * (f_{\alpha_i} + f_{\alpha_j}) - f_{\alpha_i} * f_{\alpha_i} - f_{\alpha_j} * f_{\alpha_j} \in I \cap F_{\alpha_i + \alpha_j}.$$

Hence the assertion is true for $f * f$, too. Now I is homogeneous. □

Let us set $L_n(X) = F_n(X) + I/I$ and $L_\alpha(X) = F_\alpha(X) + I/I$, $n \in \mathbb{N}$, $\alpha \in \Phi$. Since I is homogeneous, using Section 1.10.3, we derive that

$$L(X) = \bigoplus_{n=1}^{\infty} L_n(X) \quad \text{and} \quad L(X) = \bigoplus_{\alpha \in \Phi} L_\alpha(X).$$

The elements in $L_n(X)$ are called *homogeneous of degree n*, and those in $L_\alpha(X)$ *multihomogeneous of multidegree α*. If for every x we have $\alpha(x) = 0$ or 1 then the elements of $L_\alpha(X)$ are called *multilinear* (polynomials) in the variables of supp α.

We will see in Section 2.3.3 that all $L_\alpha(X)$ and $L_n(X)$ are free k-modules.

2.2.5 Free associative algebra

Let k be a commutative ring with unit 1, X a nonempty set, $W(X)$ the set of all (associative) words in the alphabet X with operation of juxtaposition. The empty word is denoted by 1 and it is playing the role of unity in the semigroup $W(X)$ (by associativity no brackets are needed now). Let $A(X)$ denote the free k-module with basis $W(X)$. The multiplication of $A(X)$ is uniquely determined by the requirements of $A(X)$ becoming a k-algebra and $W(X)$ becoming a subsemigroup in $A(X)$. Thus we obtain the free associative algebra with 1 and with the free generating set X.

Proposition 2.5. *Let S be an associative algebra with* 1 *over k and let φ be an arbitrary map: $X \to S$. Then there exists a unique homomorphism $\varphi : A(X) \to S$ such that $\bar{\varphi}(1) = 1$ and $\bar{\varphi}\mid_X = \varphi$.*

Proof is left to the reader.

Similarly to what has been done in Section 2.2.3, we may write

$$W(X) = \bigcup_{n=0}^{\infty} W_n(X), \quad W(X) = \bigcup_{\alpha\in\Phi} W_\alpha(X),$$

$$A(X) = \bigoplus_{n=1}^{\infty} A_n(X), \quad A(X) = \bigoplus_{\alpha\in\Phi} A_\alpha(X).$$

Then we may introduce the notions of homogeneous, multihomogeneous, and multilinear elements in the same way as previously (the only difference is in the appearance of $W_0(X) = \{1\}$ and $A_0(X) = k \cdot 1$).

2.2.6 Identities

The above construction enables us to effectively define the concept of the *identity of Lie algebra* over a ring k (see Section 2.1). This is any equation of the form $f \equiv 0$, f being a nonzero element in some $L(X)$, a free Lie algebra over k. This identity is satisfied by an algebra G if and only if for any homomorphism $\varphi : L(X) \to G$ we have $\varphi(f) = 0$. The "nontriviality" part of the definition of the identity follows since $f \neq 0$ in $L(X)$. Similarly, using $A(X)$ we may effectively define the concept of the identity of associative algebras. We say that an identity $g \equiv 0$ is a *consequence* of $f \equiv 0$ if the former is satisfied by every algebra which satisfies the latter. The identities which are consequences of each other are called *equivalent*.

Given a subset V in $L(X)$ and a Lie algebra G, we denote by $V(G)$ the least ideal of G which is generated by all elements of the form $\pi(v)$, where π is an arbitrary homomorphism from $L(X)$ into G and $v \in V$. If $V = \varnothing$ then, clearly, $V(G) = \{0\}$. The ideal $V(G)$ is called the *verbal ideal of G determined by the set of nonassociative polynomials V*. Now the quotient algebra $G/V(G)$ satisfies all the identities of the form $v \equiv 0$, $v \in V$. The ideal $V(L(Y))$ is the set of all consequences of the system of identities $\{v \equiv 0 \mid v \in V\}$ depending on the variables in Y. Similarly, one may define the concept of the verbal ideal of associative algebras which is usually called the *T-ideal.*

2.3 Bases of free Lie algebras

2.3.1 Basic family

Let X be a nonempty set and $\Gamma = \Gamma(X)$ the free groupoid over X (see Section 2.2.1). The *basic* family is any subset $R = R(X) \subseteq \Gamma(X) = \Gamma$ with well-ordering $\leq$ and satisfying the following conditions:

(R1) $X \subseteq R$;

(R2) A monomial $w = u * v$ is in R if and only if:
(i) $u, v \in R$, (ii) $u < v$, (iii) if $v = v_1 * v_2$ then $u \geq v_1$;

(R3) $u * v > u$.

We set $R_n = R \cap \Gamma_n(X)$, $R_\alpha = R \cap \Gamma_\alpha(X)$, where $n \in \mathbb{N}$, $\alpha \in \Phi$ (see Section 2.2.2).

An example of a basic family can be obtained as follows. We let $R^1 = X$ and introduce any well-ordering in this set. The sets R^n, $n = 2, 3, \ldots$, will be defined by induction in such a way that R^{n+1} contain R^n and all the monomials of degree $n+1$ from $R^n * R^n$ which satisfy (ii) and (iii). Then we extend the ordering of R^n to R^{n+1} arbitrarily but make sure that for any $u \in R^{n+1} \setminus R^n$ and $v \in R^n$ we have $u > v$. Then we set $R = \bigcup_{n=1}^{\infty} R^n$. Any basic family of this sort is called *Hall's basic family*.

Our principal result of this section is that the image of any basic family under the natural homomorphism from $\Gamma(X)$ into $L(X)$ (see Section 2.2.1) is a basis of the k-module $L(X)$, which is thus free. Let ε denote this homomorphism. So, given $u \in \Gamma(X)$, $\varepsilon(u)$ is always an element of $L(X)$. According to our convention in Section 2.2.4, if $x \in X$ then $\varepsilon(x)$ is simply x. The monomials of the form $\varepsilon(u)$, $u \in R$, will be called *basic*. We write $\varepsilon(u) \geq \varepsilon(v)$ if $u \geq v$.

Lemma 2.6. *Let R be a basic family in $\Gamma(X)$. Then for any $\alpha \in \Phi$ we have $L_\alpha = k\varepsilon(R_\alpha)$.*

Proof. We recall that for any $\alpha \in \Phi$ it is true that $L_\alpha = k\varepsilon(\Gamma_\alpha)$. Then we proceed by induction on $|\alpha|$. If $|\alpha| = 1$ then $R_\alpha = \Gamma_\alpha$, and all is done. Now let $|\alpha| > 1$. Any element in L_α is a linear combination of the elements of the form $w = uv$, $u \in L_\beta$, $v \in L_\gamma$, where $\beta + \gamma = \alpha$. By induction, u and v are represented as linear combinations of basic monomials. By distributivity, one may restrict oneself to the case of $w = uv$, where u and v are basic. Applying anticommutativity, we obtain that any element in L can be represented as a linear combination of elements w of the form $w = uv$, where $u \in \varepsilon(R_\beta)$, $v \in \varepsilon(R_\gamma)$, $\beta + \gamma = \alpha$, and $u < v$. Now we will prove, using induction over $|\gamma|$, that w as above can be represented as a linear combination of basic monomials $w_1 = u_1v_1, \dots, w_s = u_sv_s$, where $w_i > u$, $i = 1, 2, \dots, s$. The basis of induction is obvious: if $|\gamma| = 1$ then for $w = uv$ we have (R2) (since v is not representable as a product), that is, $w \in \varepsilon(R)$. Besides, by (R3), we have $w > u$.

So, we embark on proving the main part with $|\gamma| \geq 2$. Then $v = v_1v_2$, where $v_1 < v_2$. If $u \geq v_1$ then $w \in \varepsilon(R)$ and all is done. If, otherwise, $u < v_1$, we apply the Jacobi identity

$$w = u(v_1v_2) = -v_1(v_2u) - v_2(uv_1) = v_1(uv_2) - v_2(uv_1).$$

By induction, the monomials uv_2 and uv_1 ($u < v_1 < v_2$) can be written as linear combinations of basic monomials u' such that $u' > u$. Thus we have represented $w = uv$ as a linear combination of elements of the form v_1u' and v_2u' where $v_1, v_2, u' > u$. Hence v_1u' and v_2u' are in $\varepsilon(\Gamma_\alpha) \setminus \{uv\}$. Now, if further $v_1 < u'$ and $u' = u_1'u_2'$ with $v_1 < u_1'$ then proceeding similarly we may express v_1u' through monomials of the form $u_1'v_1'$ and $u_2'v_1'$ such that $v_1', u_1', u_2' > v_1$, that is, through monomials in $\varepsilon(\Gamma_\alpha) \setminus \{uv, v_1u'\}$. Note that Γ_α being finite, the above procedure cannot be continued infinitely. Thus, after several steps, the procedure terminates yielding the desired expression for w. At each step of the procedure, as we have seen, $w = uv$, $u < v$, is representable as a linear combination of $w_i = u_iv_i$ with $u_i, v_i > u$. Now if w_i becomes basic then by (R3) we may write $w_i > u_i > u$, as required. □

2.3.2 Canonical homomorphism from $L(X)$ into $A(X)$

Let $A(X)$ be a free associative algebra with free generating set X (Section 2.2.5). We will view $A(X)$ as a Lie algebra under commutation, denoted by $A(X)$ (Section 1.2.1). By universality of free Lie algebras (see Proposition 2.3), the identity map $i : X \to X$ extends to a homomorphism $\bar{i} : L(X) \to [A(X)]$. We will see later that $\bar{i}$ is injective, that is, the subalgebra in $[A(X)]$ generated by X is a free Lie algebra. We start with an assertion which will be repeatedly used throughout the book.

Proposition 2.7. *Let B denote a Lie subalgebra in $[C]$, C being an associative k-algebra with 1. Suppose that B is the k-span of a well-ordered set $E \subseteq B$, and that C, as an associative algebra with 1, is generated by B. Then C is the k-span of 1 and of the set of all ordered products of the form*

$$e_1e_2\cdots e_n, \quad \text{where } e_1 \geq e_2 \geq \cdots \geq e_n, \quad e_1, e_2, \dots, e_n \in E, \quad n = 1, 2, \dots$$

Proof. We consider C as a regular B-module (cf. the end part of Section 1.6.2). Then C is a cyclic module generated by 1. Now Section 1.6.7 applies, giving what was desired. □

In the particular case, i. e., where $B = \bar{\imath}(L(X))$, $C = A(X)$, $E = \bar{\imath}\varepsilon(R(X))$, we write the above as follows.

Corollary 2.8. *For any $\alpha \in \Phi$, $\alpha \neq 0$, the k-submodule $A_\alpha(X)$ is the k-span of the ordered products of the form*

$$w_1 w_2 \cdots w_s, \quad \text{where } w_i = \bar{\imath}(u_i), \quad u_1 \geq \cdots \geq u_s, u_i \in \varepsilon(R_{\alpha_i}), \quad \alpha_1 + \cdots + \alpha_s = \alpha. \tag{2.4}$$

Proof. Immediate from Lemma 2.6, Proposition 2.7, and from the definition of the k-module $A_\alpha(X)$ (Section 2.2.5). □

2.3.3 Arranging brackets on words

We know, by definition, that the k-module $A_\alpha(X)$ is the k-span of all words of the form

$$x_1 x_2 \cdots x_n, \quad \text{where } m(x_1 x_2 \cdots x_n) = \alpha \tag{2.5}$$

and that these elements form a basis of A as a free k-module. We will show that for any $\alpha \in \Phi$, $\alpha \neq 0$, the number of elements in equations (2.4) and (2.5) is the same.

Lemma 2.9. *Every associative monomial admits precisely one arrangement of brackets which turns it into an element of the form of equation* (2.4).

Proof. We use induction over the length n of the word $w = x_1 x_2 \cdots x_n$ with obvious basis at $n = 1$. We will call any arrangement as in equation (2.4) a correct one.

(i) *Existence of an arrangement.* Consider $w' = x_1 x_2 \cdots x_{n-1}$. By the induction hypothesis, we may arrange the brackets on w' to obtain an element of the form $w_1 w_2 \cdots w_s$, where $w_i = \bar{\imath}(u_i)$, $u_i \in \varepsilon(R)$, $i = 1, 2, \ldots, s$, $u_1 \geq u_2 \geq \cdots \geq u_s$. If $x_n \leq u_s$ then we put $u_{s+1} = x_n$, $w_{s+1} = \bar{\imath}\varepsilon(u_{s+1})$, and then $w_1 w_2 \cdots w_s w_{s+1}$ is the element in $A(X)$ such that $u_1 \geq u_2 \geq \cdots \geq u_s \geq u_{s+1}$ and this element is the result of an arrangement of brackets on w.

Now let us assume that $u_s < x_n$ and that p is the least of the natural numbers for which $u_p(u_{p+1}(u_{p+2}(\cdots(u_s x_n)\cdots)))$ is a basic monomial, that is, $u_p < u_{p+1}(u_{p+2}(\cdots(u_s x_n)\cdots))$. Then the monomial

$$u_p' = u_p(u_{p+1}(u_{p+2}(\cdots(u_s x_n)\cdots)))$$

satisfies all the requirements in (R1) and (R2), that is, $u_p' \in \varepsilon(R)$. Set $w_p' = \bar{\imath}(u_p')$. Then the product

$$w_1 w_2 \cdots w_{p-1} w_p' \text{ (if } p \neq 1) \text{ or } w_1' \text{ (if } p = 1)$$

results by arranging brackets on w and, if $p \neq 1$, $u_{p-1} \geq u'_p$, proving the existence part in Lemma 2.9.

(ii) *Uniqueness of arrangement.* We consider arrangements of brackets on the word $w = x_1x_2\cdots x_n$. Suppose we have two such arrangements, namely, $w_1w_2\cdots w_s$ and $w'_1w'_2\cdots w'_r$. The monomials u_s and u'_r with $\bar{\imath}(u_s) = w_s$ and $\bar{\imath}(u'_r) = w'_r$, according to (R2), can be uniquely written in the form

$$u_s = a_1(a_2\cdots(a_lx_n)\cdots) \quad \text{and} \quad u'_r = b_1(b_2(\cdots(b_mx_n)\cdots)),$$

where $a_1 \geq a_2 \geq \cdots \geq a_l$ and $b_1 \geq b_2 \geq \cdots \geq b_m$ are some elements in $\bar{\imath}\varepsilon(R)$. If $\bar{\imath}(a_i) = \bar{a}_i$, $\bar{\imath}(b_j) = \bar{b}_j$, $i = 1,\ldots,l$; $j = 1,\ldots,m$, then the elements

$$w_1w_2\cdots w_{s-1}\bar{a}_1\bar{a}_2\cdots\bar{a}_l \quad \text{and} \quad w'_1w'_2\cdots w'_{r-1}\bar{b}_1\bar{b}_2\cdots\bar{b}_m$$

are correct arrangements of brackets on $w' = x_1x_2\cdots x_{n-1}$. To prove this, it is sufficient to show that $u_{s-1} \geq a_1$ and $u'_{r-1} \geq b_1$. This is the case: for instance, applying (R3), we may write $u_{s-1} \geq u_s \geq a_1$, and, similarly, $u'_{r-1} \geq u'_r \geq b_1$. By the induction hypothesis applied to w', we may derive

$$w_1 = w'_1 = v_1,\ldots,\bar{a}_l = \bar{b}_m = v_t. \tag{2.6}$$

However, it should be noted that $l = m$, which follows from a property of both a_1 and b_1: in the above sequence of equal monomials, either of a_1 or b_1 is the leftmost of the monomials $\bar{v}_p = \bar{\imath}(v_p)$ for which $v_p(v_{p+1}(\cdots(v_tx_n)\cdots))$ is a basic monomial. Hence $l = m$, $s = r$, and then $w_1 = w'_1,\ldots,w_{s-1} = w'_{s-1}$, $a_1 = b_1,\ldots,a_l = b_l$. Hence, also $w_s = w'_s$. □

Theorem 2.10. *For any nonempty set X and any commutative ring k with* 1, *the free Lie algebra L(X) is a free k-module. For any basic family R(X), the set of basic monomials is a free basis of this module.*

Proof. In the free k-module $A_\alpha(X)$, there are a free basis whose elements are all of the form (2.5) and a generating set whose elements have the form (2.4). By Lemma 2.21, the number of elements of either of the forms (2.4) and (2.5) is the same. Now this proves that the set in (2.4) is a basis. To make this clear, we denote the elements of the basis as $a_1,\ldots,a_q$, and those in (2.4) by $b_1,\ldots,b_q$. Let $S = (\sigma_{ij})$, $T = (\tau_{ij})$ be two square matrices of order q such that

$$a_i = \sum_{j=1}^{q} \sigma_{ij}b_j, \quad b_i = \sum_{j=1}^{q} \tau_{ij}a_j.$$

Then

$$a_i = \sum_{j=1}^{q} \sigma_{ij}b_j = \sum_{j=1}^{q}\sum_{l=1}^{q} \sigma_{ij}\tau_{jl}a_l.$$

Since $\{a_1, \dots, a_q\}$ is a basis, we must have $\sum \sigma_{ij}\tau_{jl} = \delta_{il}$, the Kronecker's delta. That is, ST is the unit matrix I_q of order q. In this case $\det S \cdot \det T = 1$. Let $\widehat{T}$ denote the adjoint of T. Then $T\widehat{T} = \widehat{T}T = \det T \cdot I_q$. We write $T(ST)\widehat{T} = TI_q\widehat{T} = \det T \cdot I_q$. Hence, $TS \cdot \det T \cdot I_q = \det T \cdot I_q$. Multiplying both sides by $\det S$, we arrive at $TS = I_q$. Now suppose $\{b_1, \dots, b_q\}$ is not a basis, that is, there exist $\lambda_1, \lambda_2, \dots, \lambda_q \in k$ such that at least one λ_i is nonzero and for which

$$\lambda_1 b_1 + \lambda_2 b_2 + \cdots + \lambda_q b_q = 0.$$

Writing this down in the matrix form gives $(\lambda_1, \lambda_2, \dots, \lambda_q)T = 0$. Multiplying by S on the right, we get $(\lambda_1, \lambda_2, \dots, \lambda_q)TS = 0$. Since $TS = I_q$, we obtain $\lambda_1 = \lambda_2 = \cdots = \lambda_q = 0$, a contradiction.

So the elements in (2.4) form a basis of $A_\alpha(X)$. Since $A(X)$ is the direct sum of such submodules, we see that the elements of the form $\bar{i}(u_1)\bar{i}(u_2)\cdots\bar{i}(u_s)$, where $u_1 \geq u_2 \geq \cdots \geq u_s$, $u_i \in \varepsilon(R)$, are linearly independent. A subsystem whose elements all have $s = 1$, that is, have the form $\bar{i}(u)$, $u \in \varepsilon(R)$ is then a free generating set (as k-modules) for the images of $L(X)$ in $A(X)$ under $\bar{i}$. Thus $R(X)$ is linearly independent in $L(X)$. □

Corollary 2.11. *The subalgebra of $[A(X)]$, generated by X under the commutation, is a free Lie algebra freely generated by X.*

2.3.4 Hall's basis

This is a particular case of the general scheme in which (R3) is replaced by a stronger condition

(R3′) If $d(u) > d(v)$ then $u > v$.

A part of one such basis up to degree 6 is given below (all the products are right-normed):

$$\begin{aligned} x < y < xy < x^2y < yxy < x^3y < yx^2y < y^2xy < x^4y < yx^3y < y^2x^2y < y^3xy \\ < (xy)(x^2y) < (xy)(yxy) < x^5y < yx^4y < y^2x^3y < y^3x^2y < y^4xy \\ < (xy)(x^3y) < (xy)(yx^2y) < (xy)(y^2xy) < (x^2y)(yxy)\cdots. \end{aligned}$$

2.3.5 Shirshov's basis

Let X be a well-ordered set and $W(X)$ the semigroup of words in X ordered lexicographically. Thus we write $x_1x_2\cdots x_s > x_1'x_2'\cdots x_r'$ if there exist no nonnegative p such that $x_{s-p} < x'_{r-p}$ and, at the same time, $x_{s-p+1} = x'_{r-p+1}, \dots, x_s = x'_r$. For instance, if $x_1 < x_2 < x_3$ then

$$x_2x_3 > x_1x_2^2x_3 > x_2x_1x_3.$$

An *associative word* $a \in W(X)$ is called *basic* if for any decomposition $a = a_1a_2$, $a \neq a_1, a_2$, we have $a > a_2a_1$. With each monomial $w \in \Gamma(X)$, we associate a word $\bar{w} \in W(X)$ by omitting brackets on w. Now let $\check{R}$ denote the set of nonassociative monomials defined by the following conditions:
($\check{R}$1) $X \subseteq \check{R}$;
($\check{R}$2) a monomial of the form $w = u * v$ is in $\check{R}$ if and only if (i) $u, v \in \check{R}$, (ii) $\bar{u} < \bar{v}$, and (iii) if $v = v_1 * v_2$ then $\bar{u} \geq \bar{v}_1$.

Theorem 2.12 (A. I. Shirshov). *The image of $\check{R}$ under the natural map of $\Gamma(X)$ into $L(X)$ (see Section 2.3.1) is a basis of the free k-module $L(X)$.*

Proof. It is sufficient to show that the relation $w_1 < w_2 \iff \bar{w}_1 < \bar{w}_2$ is a well-ordering of $\check{R}$ satisfying (R3) in Section 2.3.1. To prove the first part of this assertion, it is sufficient to show that any $w \in \check{R}$ is uniquely determined by $\bar{w}$. For this we proceed by induction on the degree of w. Let $\bar{w} = ax_1b$, where x_1 is the least letter of X occurring here and where b has no occurrences of x_1. It is clear that $b \neq 1$, moreover, $b = x_2b'$ for some letter $x_2 > x_1$. The monomial (x_1x_2) is a factor of w. Assuming the opposite, we would find a factor of w of the form $(w' * x_1) * (x_2 * w'')$. But then $w' * x_1 \in \check{R}$, that is, $w' * x_1$ is simply x_1. By (R2), then $x_1 > x_2$, which is impossible. Now we enlarge our alphabet by adjoining $y = x_1x_2$ with its ordering. Then replacing in w each occurrence of (x_1x_2) by y we obtain a monomial w_1 in the variables $x_1, y, x_2, \dots$ We notice that $\bar{w}_1$ is a basic associative word of shorter length; that is, uniquely determines w_1. Arguing similarly, we see that any basic word a is of the form $a = \bar{w}$, where $w \in \check{R}$. Now it remains to verify the validity of (R3). So let $u, v \in \check{R}$. We want to prove $\bar{u} < \overline{uv}$. Let the converse hold, i. e., $\bar{u} > \overline{uv}$. If u is not a basic subword in $\overline{uv}$ then $\bar{v}\bar{u} > \bar{u}\bar{v}$, which is impossible since uv is basic. Now $\bar{u}$ is not a right subword of $\bar{v}$ since $\bar{u} < \bar{v}$. Hence $\bar{u} = p\bar{v}$ for some $p \in W(X)$, $p \neq 1$. Now we have $p\bar{v} > \overline{uv}$, whence $p > \bar{u} = p\bar{v}$. As before, for suitable $q \in W(X)$, $\bar{u} = qp$. Furthermore, $\bar{u} = qp = p\bar{v} > \bar{v}p$. This gives $q > \bar{v}$. Since the lengths of q and $\bar{v}$ are the same, this inequality leads to $\bar{u} = qp > pq > p\bar{v} = \bar{u}$, a contradiction. Thus our former assumption is not true, that is, in fact, $\bar{u} < \overline{uv}$. □

We write down the elements of R in the case of $L(x, y)$ whose degrees are at most 6:

$$\begin{aligned} x &< x^5y < x^4y < x^3y < x^2y < (x^3y)(xy) < (x^2y)(xy) \\ &< xy < (x^4y)y < (x^3y)y < (xy)((x^2y)y) < (x^2y)y < ((x^2y)(xy))y \\ &< (xy)y < ((x^3y)y)y < (x^2y)y < ((xy)((xy)y))y \\ &< ((xy)y)y < (((x^2y)y)y)y < (((xy)y)y)y < ((((xy)y)y)y)y < y. \end{aligned}$$

We conclude the section by a proposition which easily follows by induction over the degree of the basic monomial.

Proposition 2.13. *We consider $L(X)$ as a Lie subalgebra in $[A(X)]$. Let $u = \varepsilon(w)$ be a basic monomial $w \in \check{R}$. We represent u as a linear combination of associative words. Then the leading term of this combination with respect to the above order of $W(X)$ is equal to $\bar{w}$.*

2.4 Subalgebras of free Lie algebras

2.4.1 Example

A well-known theorem in the group theory due to Nielsen–Schreier says that every subgroup of a free group is itself free. This result has no full counterpart in the theory of Lie algebras over arbitrary commutative rings. For instance, let $L(\{x, y\})$ be a free Lie ring, which is a Lie algebra over integers. Further, let H denote its subring generated by $2x, y, xy$. We write

$$L = L_1 \oplus L_2 \oplus \cdots \oplus L_n \oplus \cdots, \qquad H_i = H \cap L_i.$$

Since H is generated by homogeneous elements, it is homogeneous, that is,

$$H = H_1 \oplus H_2 \oplus \cdots \oplus H_n \oplus \cdots.$$

Here $H_1 = \langle 2x\rangle \oplus \langle y\rangle$. If H^2 is the derived algebra then $(H^2)_1 = \{0\}$, $(H^2)_2 = \langle 2xy\rangle, \ldots$ Suppose that H is a free Lie ring. Then H/H^2 must be a free abelian group. But H/H^2 contains an element $xy + H$ of additive order 2, a contradiction. Hence, H is not a free Lie ring.

However, if H is generated by a subset of kX which can be included into a basis of this free module, then H is a free Lie algebra. This is an easy consequence of an auxiliary lemma, which follows.

Lemma 2.14. *Let $L(X)$ be a free Lie algebra over a commutative ring k, Y being a basis of the free k-module kX. Then Y is the free generating set for $L(X)$.*

Proof. Let φ denote an automorphism of the free module $kX = kY$ sending X into Y, and ψ its inverse. The restriction $\varphi|_X$ and $\psi|_X$ extend to homomorphisms $\bar{\varphi}, \bar{\psi} : L(X) \to L(X)$. Thus we have

$$(\bar{\varphi}\bar{\psi})|_{kX} = \bar{\varphi}|_{kX} \cdot \bar{\psi}|_{kX} = \varphi|_{kX} \cdot \psi|_{kX} = \varphi\psi|_{kX} = 1_{kX}.$$

Similarly, $(\bar{\psi}\bar{\varphi})|_{kX} = 1_{kX}$. Hence $\bar{\varphi}\bar{\psi} = \bar{\psi}\bar{\varphi} = 1_{L(X)}$, proving that $\bar{\varphi}$ is an automorphism of $L(X)$ transforming X into Y. Since X is a free generating set, the same is true for Y. □

2.4.2 Free generators for some ideals

In this subsection we are going to prove a proposition which will give the freeness of some special subalgebras of free Lie algebras. This will enable us to prove the main theorem concerning the freeness of homogeneous subalgebras in the case where the base ring satisfies some rather mild restrictions (all projective modules over this ring must be free). Then, as corollaries, we will derive well-known theorems of Shirshov and Witt in which the base ring is respectively a field or the ring of integers.

We will require some more notation. Let X be a nonempty set and $\Gamma = \Gamma(X)$ the free groupoid over X. A subset S is called a *left ideal* of Γ if for all $u \in \Gamma$, $v \in S$, we have $u * v \in S$. Suppose that R is a basic family satisfying the following condition: for any $u \in R \setminus S$, $v \in R \cap S$, we have $u \leq v$. An element $w \in S$ is called *S-reducible* if it can be represented in the form $w = u * v$, where $u, v \in S$. All the other elements $w \in S$ are called S-irreducible.

In the following proposition we identify the elements of $\Gamma(X)$ with their canonical images in $L(X)$.

Proposition 2.15. *Let S be a left ideal in $\Gamma(X)$, Y the set of S-irreducible elements in $R \cap S$ (R as above). Then the k-submodule M of the free Lie algebra $L(X)$ generated over k by $R \cap S$ is a free Lie algebra with Y as a free generating set.*

Proof. Let Y' denote a set which is in a bijective correspondence β with Y, $\beta : Y' \to Y$. We consider $\Gamma' = \Gamma(Y')$ and let $\bar{\beta} : \Gamma(Y') \to \Gamma(Y)$ be the extension of β to the isomorphism of groupoids. We will be constructing a basic family R' in $\Gamma' = \Gamma(Y')$ as follows. We set $R_1' = Y'$ and, using β^{-1}, we transfer to Y' the ordering of Y (which is a subset of the well-ordered set R). Suppose that we managed to construct $R_1', R_2', \ldots, R_{m-1}'$ so that $\bar{\beta}(R_i') \subseteq R \cap S$. We also suppose that $R_1' \cup R_2' \cup \cdots \cup R_{m-1}'$ is a linearly ordered set in such a way that $\bar{\beta}|_{R_1' \cup R_2' \cup \cdots R_{m-1}'}$ is a homomorphism of ordered sets. We will construct R_m' from these data as the set of all monomials $u * v$ such that $u \in R_t'$, $v \in R_{m-t}'$, $u < v$, and if $v = v_1 * v_2$ then $u \geq v_1$. Let us show first that $\bar{\beta}(R_m') \subset R \cap S$. Indeed, $\bar{\beta}(u), \bar{\beta}(v) \in R \cap S$ and $\bar{\beta}(u) < \bar{\beta}(v)$. If $v = v_1 * v_2$, $v_1, v_2 \in \bar{\beta}^{-1}(S)$, then $\bar{\beta}(v) = \bar{\beta}(v_1) * \bar{\beta}(v_2)$ and, by induction $u \geq v_1$, that is, $\bar{\beta}(u) \geq \bar{\beta}(v_1)$. Hence, $w \in R \cap S$. If v is irreducible then $v \in Y'$. Now the image of Y' is equal to Y and then from $\bar{\beta}(v) = w_1 * w_2$ we have $w_1 \in R \setminus S$ and $w_2 \in R \cap S$. But $\bar{\beta}(u)$ is an element of $R \cap S$, hence, $\bar{\beta}(u) \geq w_1$. Therefore $\bar{\beta}(u * v)$ is in R as required. Now we transfer to $R_1' \cup R_2' \cup \cdots \cup R_m'$ the ordering of the set $\bar{\beta}(R_1' \cup R_2' \cup R_m')$. Setting $R' = \bigcup_{m=1}^{\infty} R_m'$, we obtain the set of monomials in Γ with some ordering satisfying the conditions (R1), (R2), (R3) from Section 2.3.1. Hence R' becomes a basis of the free k-module $L(Y')$. The image of $L(Y')$ under a homomorphism $\tilde{\beta}$, induced by β, is a subalgebra H of $L(X)$. This coincides with $k(R \cap S)$, which is true since S is an ideal in Γ. The elements of R', which is a basis of the k-module L', map injectively into the elements of the k-basis R of $L(X)$. Thus $\tilde{\beta} : L(Y) \to H$ is an isomorphism of the free Lie algebra $L(Y')$ and the subalgebra $k(R \cap S)$ mapping Y' into Y. Hence Y is a set of free generators for $k(R \cap S)$. □

Corollary 2.16. *Let $L(X)$ be a free Lie algebra over an arbitrary commutative ring k.*
(i) *Suppose that $X = X' \cup X''$, where $X' \cap X'' = \varnothing$, and $X' \neq \varnothing$. Let R'' denote a basic set for $L(X'')$. Then the ideal T of L generated by X' is a free Lie algebra, and we may choose a free generating set for T as the union of X' and the set of all right-normed products of the form*

$$v_1 v_2 \cdots v_{s-1} x_s, \quad \text{where } x_s \in X',\ v_1 \geq v_2 \geq \cdots \geq v_{s-1},\ v_i \in R''. \tag{2.7}$$

(ii) *The derived algebra $L(X)^2$ is a free Lie algebra provided $|X| > 1$, and its free generating set consists of all right-normed products of the form*

$$x_1 x_2 \cdots x_{s-1} x_s, \quad \text{where } x_i \in X,\ x_1 \geq x_2 \geq \cdots \geq x_{s-1} < x_s,\ s \geq 2. \tag{2.8}$$

Proof. In either case we have to choose an ideal and then to apply the proposition.

(i) In this case S is the set of all monomials u in $\Gamma(X)$ which are in $\Gamma_\alpha(X)$, where α is an arbitrary function in Φ such that $\alpha(X') \neq 0$. It is obvious that S is a two-sided ideal in $\Gamma(X)$. We will be building a basic family as in Section 2.3.1. As it was remarked there, in defining the ordering on $P = R_1 \cup R_2 \cup \cdots \cup R_n$ there is a certain freeness. We will introduce ordering in three steps. Firstly, we require that, given $u \in P \cap S$ and $v \in P \setminus S$, we always have $u > v$. Secondly, we require that for any $u, v \in P$ with $uv \in P$ we have $u < uv$. Thirdly, we extend this partial ordering to the linear one. While performing all the above we see that the second step is a complement of the first one for if one member of the pair u, uv is in S and the other is outside S then, necessarily, $u \notin S$ and $uv \in S$, so that, according to the first step, we already have $u < uv$. Finally, we set $R = \bigcup_{i=1}^{\infty} R_i$. Now $R'' = R \cap \Gamma(X'')$ is a basic family of $L(X'')$ and the irreducible elements of $R \cap S$ have the form as in (2.7).

(ii) In this case we set $S = \Gamma \setminus \Gamma_1$ and choose for R any of Hall's bases. Then the irreducible elements in $R \cap S$ will be precisely of the form as in (2.8) and $k(R \cap S) = L^2$. □

Before we state our main theorem we recall that the direct summands of free modules are called *projective* modules.

Theorem 2.17. *Let k be a commutative ring over which all projective modules are free. Suppose that $L = L(X)$ is a free Lie algebra with some $\mathbb{N}$-grading in which X is homogeneous and H is a homogeneous subalgebra which is a direct summand of the k-module L. Then H is a free Lie algebra with a homogeneous set of free generators.*

Proof. By our hypotheses, we may write $L = \bigoplus_{t=1}^{\infty} L_t$ where L_t is the homogeneous component of L of degree t, $t = 1, 2, \ldots$ Moreover, $H = \bigoplus_{t=1}^{\infty} H_t$, where $H_t = H \cap L_t$, $t = 1, 2, \ldots$, and, for some k-submodule K we have $L = K \oplus H$. By definition, H, K are projective, that is, free. Now $K = L/H = \bigoplus_{t=1}^{\infty} L_t/H_t$. Hence, all L_t/H_t are projective, thus free, too. Therefore, for each t, there exists a k-submodule L_t' such that $L_t = H \oplus L_t'$ where H_t, L_t' are all free as k-modules. We represent H as the intersection of the decreasing sequence $\{H^{(s)}, s = 0, 1, 2, \ldots\}$, $H^{(0)} = L$ and

$$H^{(s)} = H_1 \oplus \cdots \oplus H_s \oplus L_{s+1} \oplus L_{s+2} \oplus \cdots \tag{2.9}$$

if $s = 1, 2, \ldots$ Suppose we have constructed the homogeneous subsets $B^{(s)}$, $s = 1, 2, \ldots$, such that $B^{(s)}$ is a free generating set for $H^{(s)}$, and $B_t^{(s)} = B^{(s)} \cap L_t = B_t^{(s-1)}$, $t = 1, 2, \ldots, s-1$. Then $B = \bigcup_{t=1}^{\infty} B_t^{(t)}$ is a homogeneous free generating set for H. Indeed, let R be a basic family of monomials over B. Any linear dependence among the basic monomials in R is, in fact, a linear dependence among the basic monomials in $B^{(s)}$ for suitable s. But this latter is a free generating set for $H^{(s)}$, a contradiction. Thus R is a basis of a k-module generated by it. Arguing as in the proof of Proposition 2.15, we can see that the natural map of $L(B)$ into a subalgebra generated by B is an isomorphism. Obviously, $H = \text{alg}(B)$, i. e., H is a free Lie algebra for which B is a free generating set.

So to prove Theorem 2.17, we should construct a free generating set for $H^{(s)}$ which includes the free generators of $H^{(s-1)}$ of degree at most $s-1$ with respect to the grading of L. In this case we have to consider a subalgebra H with $H_t = L_t$ for all $t \neq s$, and for $t = s$ we have $L_s = H_s \oplus L_s'$. If we put $X_t = X \cap L_t$, $t = 1, 2, \ldots$, then our purpose is to construct a free homogeneous generating set Y for H such that $Y_1 = X_1, \ldots, Y_{s-1} = X_{s-1}$. Let G be a homogeneous subalgebra in L generated by $X_1 \cup X_2 \cup \cdots \cup X_{s-1}$. Since X is a free generating set, we have $L_s = kX_s \oplus G_s$. Obviously, $G_s \subseteq H_s$. Applying the modular law, we get $H_s = (kX_s \cap H_s) \oplus G_s$. Moreover,

$$kX_s/kX_s \cap H_s \cong kX_s + H_s/H_s \cong kX_s + G_s/H_s \cong L_s/H_s \cong L_s'.$$

Now L_s' being projective as a k-module, there exists a k-submodule $L_s'' \subset L_s$ such that $kX_s = (kX_s \cap H_s) \oplus L_s''$. Thus, in the free k-module kX_s one may choose a basis $Z_s = Z_s' \cup Z_s''$ with $kX_s \cap H = kZ_s'$ and $L_s'' = kZ_s''$. Setting $Z_t = X_t$ for $t \neq s$, we arrive at a new basis of kX of the form $Z = \bigcup_{t=1}^{\infty} Z_t$. Using Lemma 2.14, we may replace our free generating set by a new one, namely Z, for $L(X)$. For this set we may write $kZ_s = (kZ_s \cap H_s) \oplus kZ_s''$ and $kZ_s \cap H = kZ_s'$. Without any loss of generality, we may think, from the very beginning, that $X_s = X_s' \cup X_s''$ and $kX_s \cap H_s = kX_s'$, $L_s'' = kX_s''$. Let S denote the set of the monomials in $\Gamma(X)$ which are not elements in X_s''. It is clear that S is an ideal, and that for the k-span T of S the conditions of part (i) of the corollary to Proposition 2.15 are satisfied. Now the free generating set Y for T is the union of $X \setminus X_s''$ and the set of all products of the form (2.7), each of which is an element of $\bigoplus_{j \geq s+1}^{\infty} L_j$. Since $T = H$, we have found for H a free generating set with $Y_t = X_t$ for all $t = 1, 2, \ldots, s-1$. □

The hypotheses of Theorem 2.17 are satisfied by any principal ideal domain (e. g., by any field), by the polynomial rings over fields, etc. In some cases it is possible to abandon restrictions imposed on the subalgebra or weaken them somewhat.

Theorem 2.18 (A. I. Shirshov). *Let H be a nonzero subalgebra of a free Lie algebra $L(X)$ over a field k. Then H is a free Lie algebra.*

Proof. We consider the natural grading of $L = L(X)$ (with respect to degrees in X). Then (see Section 1.10.4) with each subalgebra H we can associate a homogeneous subalgebra $\operatorname{gr} H$ generated by leading terms of elements in H. Since k is a field, by Theorem 2.17, $\operatorname{gr} H$ is a free Lie algebra freely generated by a free generating homogeneous set of the form $\operatorname{gr} Y$ for suitable $Y \subseteq H$, the natural mapping $Y \to \operatorname{gr} Y$ associating with each element its leading term being bijective. Obviously, Y is a free generating set for $\operatorname{alg}(Y)$, the subalgebra generated by Y. Indeed, this mapping induces a k-endomorphism from $\operatorname{alg}(Y)$ into $\operatorname{gr}(H)$ which brings the basic family $R(Y)$ into the basic family $R(\operatorname{gr} Y)$. The same argument shows that $kR(Y) = H$, that is, $\operatorname{alg}(Y) = H$. □

In what follows we will need the following.

Corollary 2.19. *Let L be a free Lie algebra over a field k and H a proper nonzero ideal in L. Then H is a free Lie algebra of infinite rank.*

Proof. Since $H \neq L$ there exists $x \in L \setminus H$. We set $G = H \oplus kx$. Since $G^2 \subseteq H$, a free generating set Y for G can be chosen using Lemma 2.14 so that $x \in Y$ and $Y \setminus \{x\} \subseteq H$. Using the first part of the corollary of Proposition 2.15, we find that a free generating set for H can be chosen to include all products of the form

$$(\operatorname{ad} x)^s(y), \quad S = 0, 1, 2, \dots, \ y \in Y \setminus \{x\}.$$

By the invariance of the rank of a free Lie algebra (see Exercise 2.8.17), H cannot be generated by a finite set. □

Theorem 2.20 (E. Witt). *Let L be a free Lie ring and H its homogeneous subring. If the abelian group L/H is free then H is a free Lie ring.*

Proof. We apply natural grading of $L(X)$. By our hypotheses, H is the direct summand of the abelian group $L(X)$. Each projective module over $\mathbb{Z}$ being free follows using the theorem about subgroups of abelian groups. Thus Theorem 2.17 applies. □

2.5 Universal enveloping algebra

2.5.1 Definition

Let G be a Lie algebra over a commutative ring k with unit 1. An associative k-algebra $U = U(G)$ with 1 is called the *universal enveloping algebra* for G if

(i) there exists a homomorphism, called *canonical*, of Lie algebras $\varepsilon : G \to [U]$;

(ii) for every associative algebra B with 1 over k and any homomorphism $\varphi : G \to [B]$, there exists unique homomorphism of associative algebras with 1, $\psi : U \to B$, such that $\psi\varepsilon = \varphi$ (ψ being a homomorphism of associative algebras with 1 means $\psi(1) = 1$).

It follows from the definition that $U(G)$ is unique up to isomorphism. Its existence can be shown as follows. We represent G as $G = L(X)/R$, for an appropriate free Lie algebra $L(X)$ and its ideal R (Section 2.2.4). By the universal property of free Lie algebras and free associative algebras, it follows that $A(X)$ is the universal enveloping algebra for $L(X)$ and that the canonical map $i : L(X) \to A(X)$ is the homomorphism extending the identity map of X. Let S be the least ideal of $A(X)$ containing $i(R)$. Then $U = A(X)/S$ is the universal enveloping algebra for G. The canonical map will be obtained if we put $\varepsilon = \pi_2 i \pi_1^{-1}$, where $\pi_1 : L(X) \to L(X)/R$ and $\pi_2 : A(X) \to A(X)/S$ are canonical maps of an algebra onto its quotient algebra. The details of necessary verifications are left to the reader.

2.5.2 Basis for the universal enveloping algebra

Let G be a subalgebra in a Lie algebra of the form $[B]$, B being an associative algebra. We will say that B is the *enveloping algebra* for G if B is generated by G as an associative algebra. As has been shown in Proposition 2.7, a k-generating set for the enveloping algebra B can be chosen in the form

$$e_1 e_2 \cdots e_n, \quad e_1 \geq \cdots \geq e_n, \tag{2.10}$$

where $e_i \in E$ is an element of an ordered basis for G. A refinement of this assertion reads as follows.

Lemma 2.21. *Let k be a commutative ring with 1 over which all projective modules are free. Suppose that A is an associative k-algebra generated by its Lie subalgebra L. Let E and F be two linearly ordered bases of the k-free module L. Let S_E denote the set of all ordered monomials as in (2.10). If S_E is a basis of A as a free k-module then similar set S_F is a free basis of A.*

Proof. For brevity, the elements of the form $d_1 d_2 \cdots d_t$, $d_i \in D \subseteq L$ will be called *D-monomials of degree t*. Let A_t denote the k-submodule of A generated by all L-monomials of degree at most t. Then $A = \bigcup_{t=1}^{\infty} A_t$. Modulo A_{t-1} the k-module A_t is generated by all ordered E-monomials of degree t. The proof of Lemma 2.21 will be complete if we manage to show that the images of ordered F-monomials in A_t/A_{t-1} form a basis of this free k-module. Moreover, it suffices to restrict ourselves to F'-monomials, F' being a finite subspace in F. Let E' be a finite subset in E such that $kE' \supseteq F'$, and let F'' be a finite subset in F containing F' and such that $kF'' \supseteq kE'$. Then $kF'' = kF' \oplus T$, T being a free k-module. Using modular law, we may write

$$kE' = kE' \cap (kF' \oplus T) = kF' \oplus (kE' \cap T).$$

Since $kE' \cap T$ is a direct summand of the free k-module kE', it is free. Let E'' be the union of F' and of a basis for $kE' \cap T$. Then $kE'' = kE'$, hence $|E'| = |E''|$ (see Theorem 2.10). Now the number of ordered E' and E''-monomials is the same (the ordering of E'' extends that of F'). Applying the same argument as in the final part of Theorem 2.10, we may conclude that the images of ordered E''-monomials of degree t in A_t/A_{t-1} are k-independent. Since F'-monomials form a part of the set of E''-monomials, our proof is complete. □

Theorem 2.22 (Poincare–Birkhoff–Witt (PBW)). *Let G be a Lie algebra over a commutative ring k with unit and $U(G)$ its universal enveloping algebra. If G is a free k-module and E a well-ordered basis for G then the canonical map $\varepsilon : G \to U(G)$ is injective and $U(G)$ is a free k-module freely generated by 1 and by the images of all ordered monomials of the form*

$$e_1 e_2 \cdots e_n, \quad e_1 \geq e_2 \geq \cdots \geq e_n,\ n \geq 1. \tag{2.11}$$

Proof. We will give two different proofs of PBW, the first in the case where all projective k-modules are free. In Section 2.5.4, we will give another proof, which is free of this restriction.

Proof I. Consider the case where $G = L(X)$ and $E = \varepsilon(R)$ is a set of basic monomials. Then $A(X) = k1 \oplus \sum_{n\geq 1} A_n(X)$ and, according to the proof of Theorem 2.10, a basis of $A(X)$ is formed by 1 together with the set of all ordered E-monomials. If F is another linearly ordered basis for $L = L(X)$, the same conclusion follows using Lemma 2.21. We warn the reader that we have made use of the lemma where it had been assumed that all projective modules over k had had to be free. Now we are continuing our proof. Let G be an arbitrary Lie algebra which is free as a k-module. We represent G in the form $G = L(X)/R$ and, using the corollary of Theorem 2.10, we identify $L = L(X)$ with its canonical image in $A(X)$. Then $U(G) = A(X)/S$ where S is a two-sided ideal of $A = A(X)$ generated by R. Since G is a free k-module, we may represent the k-module L in the form $L = G' \oplus R$, where $G' \cong G$. By our hypotheses on the ring, we see that R is a free k-module. Let E' be a basis of G' whose image under the natural homomorphism $\pi_1 : L \to L/R$ coincides with E. Let E'' be a k-basis for R. We introduce a linear ordering on $E' \cup E''$ by setting $e' > e''$, $e' \in E'$, $e'' \in E''$, and so that $\pi_1|_E$ is the isomorphism of ordered sets. Then by the first part of this proof a basis of $U(A)$ can be chosen in the form of the union of $\{1\}$ and of the set of all ordered $(E' \cup E'')$-monomials

$$e_1' e_2' \cdots e_s' e_1'' e_2'' \cdots e_l'', \quad e_i' \in E',\ e_j'' \in E'', \quad e_1' \geq \cdots \geq e_s' \geq e_1'' \geq \cdots \geq e_l''. \tag{2.12}$$

We remark that the submodule in $A(X)$ generated by the monomials of the form (2.12) such that $l > 0$ is a two-sided ideal in $A(X)$. This ideal is contained in S, hence it is equal to S. Applying π_2 (see Section 2.5.1), we derive the conclusion about the basis of $U(G)$. According to this conclusion, the images of elements in E under the canonical map are k-independent, that is, ε is injective. □

In Section 6.1 we will give an example of a Lie algebra for which the canonical map is not injective (of course, then it is not a free k-module).

2.5.3 Filtration on $U(G)$

Let U^n, $n \geq 1$, denote the k-submodule of $U(G)$ spanned by all G-monomials of degrees $1, 2, \ldots, n$ and by 1. We also set $U^0 = k1$ and $U^{-1} = \{0\}$. Then $U^n U^m \subseteq U^{n+m}$, $n, m \geq -1$ so that $(U^n)_{n \geq -1}$ is a filtration on $U(G)$. We may form a graded k-module by setting

$$\operatorname{gr} U(G) = \sum_{n=-1}^{\infty} U^{n+1}/U^n.$$

As in Section 1.10.4, this can be turned into an algebra if we set

$$(g + U^{n-1})(h + U^{m-1}) = gh + U^{n+m-1}, \quad g \in U^n,\ h \in U^m.$$

Then $\operatorname{gr} U(G)$ is generated by the elements of the form $g + U^0$, $g \in G$. We may write

$$(g_1 + U^0)(g_2 + U^0) = g_1 g_2 + U^1 = g_2 g_1 + [g_1, g_2] + U^1 = g_2 g_1 + U^1$$

since $[g_1, g_2] \in G \subseteq U^1$. Thus $\operatorname{gr} U(G)$ is a commutative algebra.

Proposition 2.23. *We adopt the hypotheses of the PBW theorem. Then the identity map $E \to E$ extends to an isomorphism of the polynomial ring $k[E]$ and $\operatorname{gr} U(G)$.*

Proof. Since $\operatorname{gr} U(G)$ is commutative such a homomorphism does exist. It is surjective since $\operatorname{gr} U(G)$ is generated by the elements of the form $e + U^0$, $e \in E$. Its injectivity is immediate from the PBW theorem. □

2.5.4 G-modules and $U(G)$-modules

Suppose we are given a G-module M, G being a Lie algebra over k. Then M naturally becomes a $U(G)$-module such that the restriction of the action to G coincides with the original action of G on M. Indeed, let $\rho : G \to \mathfrak{gl}(M)$ be the corresponding representation of G in M (see Section 1.6.2). We denote by A the enveloping associative algebra for the set $\{1\} \cup \rho(G)$ of k-endomorphisms of M. Then, according to the definition of the universal enveloping algebra (Section 2.5.1), there exists a unique homomorphism $\psi : U(G) \to A$ extending ρ and mapping 1 to 1. Thus M is a $U(G)$-module with the above restriction to G. It is obvious that any $U(G)$-module is naturally a G-module. Under this correspondence, the notions of submodule, quotient module, and direct sum of modules are the same with respect to either action. Thus we may say that the universal enveloping algebra for a Lie algebra is an analogy of the group ring for a group.

Proof II of PBW theorem. Let M denote the free cyclic G-module with a free generator t (Section 1.6.7). According to preceding remarks, this is a cyclic $U(G)$-module. Thus the mapping $1 \mapsto t$ extends to an epimorphism of $U(G)$-modules from $U(G)$ onto M. Under this epimorphism the monomials in the set (2.11) are mapped to the elements of the form $e_1 e_2 \cdots e_n t$ ($e_1 \geq e_2 \geq \cdots \geq e_n$). By Section 1.6, Theorem 1.11, the set of these elements is k-free. The same conclusion then holds for the set (2.11). That this set generates $U(G)$ is immediate since the inverse map $t \mapsto 1$ also extends to an epimorphism from M onto $U(G)$. □

2.5.5 Derivations of free Lie algebras

Let $L = L(X)$ be a free Lie algebra and $\delta : X \to L$ an arbitrary map from X onto L. We will show that δ uniquely extends to a derivation $\bar{\delta} \in \operatorname{Der} L$. It is convenient to view L as a Lie subalgebra of $A(X)$ (see the corollary of Theorem 2.10). If we construct $\tilde{\delta} \in \operatorname{Der} A(X)$ such that $\tilde{\delta}|_X = \delta$ then $\bar{\delta} = \tilde{\delta}|_{L(X)}$ is the desired derivation. We recall that a basis of $A(X)$ is formed by 1 and all possible words in X. Now we set $\tilde{\delta}(1) = 0$ and

$$\tilde{\delta}(x_1 x_2 \cdots x_n) = \delta(x_1) x_2 \cdots x_n + x_1 \delta(x_2) x_3 \cdots d_n + \cdots + x_1 x_2 \cdots \delta(x_n).$$

A simple verification using induction over n shows that $\tilde{\delta}$ is indeed a derivation of $A(X)$. The uniqueness of $\bar{\delta}$ is clear since $L(X)$ is generated by X.

In a particular case where $\delta = 1_X$, we arrive at a derivation δ such that $\bar{\delta}(w) = d(w)w$ whenever w is a homogeneous polynomial of degree $d(w)$. Now $\bar{\delta}$ can be obtained from $\operatorname{Der} A(X)$ as above. However, this can be obtained as a restriction of another map $\pi : A(X) \to A(X)$ given on the homogeneous elements as follows:

$$\pi(1) = 0, \quad \pi(x_1) = x_1, \quad \ldots, \quad \pi(x_1 x_2 \cdots x_n) = [x_1, x_2, \ldots, x_n], \quad \ldots$$

Obviously, π maps $A(X)$ into $L(X)$. To verify $\pi|_{L(X)} = \bar{\delta}$, it is sufficient to show that $\pi|_{L(X)}$ is a derivation of $L(X)$. Since $\pi|_X = 1_X$, this will give $\pi|_{L(X)} = \bar{\delta}$. For the proof we consider the adjoint L-module (see Section 1.6.1). Since $A(X)$ is the universal enveloping algebra for $L(X)$, this latter becomes an $A(X)$-module. Let $*$ denote the action of $A(X)$ on $L(X)$. If $u \in L(X)$, $v \in L(X)$ then $u * v = (\operatorname{ad} u)(v) = [u, v]$. We consider $\pi(ab)$, a, b being homogeneous elements in $A(X)$, $b \neq 1$. We may write $\pi(ab) = a * \pi(b)$. Hence, given $u, v \in L(X)$, we have

$$\begin{aligned} \pi([u,v]) &= \pi(uv) - \pi(vu) = u * \pi(v) - v * \pi(u) \\ &= [u, \pi(v)] - [v, \pi(u)] = [\pi(u), v] + [u, \pi(v)]. \end{aligned}$$

Thus $\pi|_{L(X)}$ is indeed a derivation, that is, $\pi|_{L(X)} = \bar{\delta}$. We can formulate what we have proved in the following way.

Theorem 2.24 (Specht–Wever criterion). *Let $A(X)$ be a free associative algebra over a field k of characteristic zero and $L(X)$ its free Lie subalgebra freely generated by X with respect to the operation of bracketing. A homogeneous element $w \in A(X)$ of degree $n \geq 1$ is an element of $L(X)$ if and only if $\pi(w) = nw$.*

Proof. If $\pi(w) = nw$ then $w = (1/n)\pi(w) \in L(X)$. If, conversely, $w \in L(X)$ then, as we have shown above, $\pi(w) = \bar{\delta}(w) = nw$. □

2.5.6 Injective modules

Let R be a ring and Q a left R-module. We say that Q is an *injective* R-module if, for any R-module M and its submodule N, every R-homomorphism $\varphi : N \to Q$ extends to a homomorphism $\tilde{\varphi} : M \to Q$. In particular, if G is a Lie algebra over a ring k, $R = U(G)$ its universal enveloping algebra then we arrive at the concept of an *injective module over a Lie algebra G*. The following proposition is well-known (cf. Curtis and Reiner [95] or Lambek [198]).

Proposition 2.25. *Any R-module can be embedded in an injective R-module.*

The main aim of this section is the proof of the following.

Theorem 2.26. *Let a Lie algebra L over a ring k be an extension of an abelian ideal M by a quotient algebra G such that M is an injective G-module. If G is a free k-module then L is a split extension.*

By Section 1.6, the corollary of Theorem 1.8, it is sufficient to show that $H^2(G, M) = 0$. Let $U = U(G)$. We consider a U-module of the form $F = U \otimes_k E(G)$, where $E(G)$ is the exterior (Grassmann's) algebra of the free k-module G (see Lang [199, Chapter XVI]). A basis of $E(G)$ over k is formed by 1 and all tensors of the form

$$e = e_1 \wedge e_2 \wedge \cdots \wedge e_t, \quad e_1 < e_2 < \cdots < e_t, \tag{2.13}$$

where $e_1, \ldots, e_t$ are the elements of a well-ordered basis of the free k-module G. These elements (more precisely, tensors of the form $1 \otimes e$) form a basis of the free k-module F. Now let E_t denote the k-span of the tensors of the form in equation (2.13) where t is fixed. We set $F_t = U \otimes E_t$. Then the alternating functions $f : G \times \cdots \times G \to M$ can be identified with U-homomorphisms from F into M, namely, we associate with a function f a homomorphism $\bar{f}$ such that $\bar{f}(e) = f(e_1, \ldots, e_t)$, $e \in E_t$. Now we define a U-homomorphism $\delta_t : F_t \to F_{t-1}$ by setting

$$\begin{aligned}\delta_t(1 \otimes e_1 \wedge \cdots \wedge e_t) = &\sum_{i=1}^{t} (-1)^{i+1} e_i \otimes e_1 \wedge \cdots \wedge \hat{e}_t \wedge \cdots \wedge e_t \\ &+ 1 \otimes \sum_{1 \leq i < j \leq k} (-1)^{i+j} [e_i, e_j] \wedge e_1 \wedge \cdots \wedge \hat{e}_i \wedge \cdots \wedge e_t.\end{aligned}$$

Here ˆ shows that the symbol under it should be omitted. Using this notation we may rewrite the coboundary formula (2.10) in Section 1.6 in the following way:

$$(\partial_{t-1}f)(e_1,\dots,e_t) = \bar{f}(\delta_t e)$$

(here, for clarity, we explicitly show the indices at the operators ∂ and δ). An important property of the sequence

$$\cdots \to F_{t+1} \xrightarrow{\delta_{t+1}} F_t \xrightarrow{\delta_t} F_{t-1} \to \cdots \to F_0 \to 0$$

is its exactness, that is, for all t, $t = 0,1,\dots,$ we have $\ker\delta_t = \operatorname{Im}\delta_{t+1}$. This sequence is called a free resolution of a U-module F_0. Its property of being exact needs rather difficult verification (see, e. g., Séminaire "Sophus Lie" [327]). Using the above identification $f \mapsto \bar{f}$ of t-dimensional cochains with U-homomorphisms from F_t into M, it is sufficient to verify the exactness of the following sequences:

$$\cdots \to \operatorname{Hom}_U(F_{t-1},M) \xrightarrow{\partial_{t-1}} \operatorname{Hom}_U(F_t,M) \xrightarrow{\partial_t} \operatorname{Hom}_U(F_{t+1},M) \to \cdots.$$

As we have remarked in Section 1.6.6, for any $t \geq 1$, we have

$$\operatorname{Im}\partial_{t-1} = B^t(G,M) \subseteq Z^t(G,M) = \ker\delta_t.$$

We will show the converse inclusion. Let f be in $\ker\delta_t$. Then for any $x \in F_{t+1}$ we have $\bar{f}(\delta_{t+1}x) = 0$. Since $\operatorname{Im}\delta_{t+1} = \ker\delta_t$, the map $f \circ \delta_t^{-1}$ is a correctly defined U-homomorphism of $\operatorname{Im}\delta \subseteq F_{t-1}$ into M. By the definition of the injective module, it extends to a U-homomorphism $\bar{g} : F_{t-1} \to M$ such that $\bar{g}|_{\operatorname{Im}\delta_t} = \bar{f} \circ \delta_t^{-1}$. Thus for any $y \in F_t$, we have $\bar{g}(\delta_t y) = \bar{f}(y)$. By a property of the bijection $f \mapsto \bar{f}$, we find $\partial_{t-1}g = f$, that is $B^t(G,M) = Z^t(G,M)$ for all $t \geq 1$. Hence $H^t(G,M) = \{0\}$.

2.6 Free product

2.6.1 Presenting algebras in terms of generators and defining relations

Let X be a nonempty set, $L(X)$ a free Lie algebra freely generated by X, and R a subset in $L(X)$ (possibly empty). By $(X|R)$ we denote the quotient algebra $G = L/S$ of $L = L(X)$ by $S = \operatorname{id}_L(R)$, the least ideal of L containing R. If $X = \{x_1, x_2, \dots\}$, $R = \{r_1, r_2, \dots\}$ then we may replace $(X|R)$ by $(x_1, x_2, \dots | r_1 = 0, r_2 = 0, \dots)$. In either case, we say that G is a Lie algebra given by the generating set X and by the set of defining relations $\{r = 0 \mid r \in R\}$. For instance, the Lie algebra G_1 of all matrices of the form

$$\left\{\begin{pmatrix} \alpha & 0 \\ 0 & \beta \end{pmatrix} \middle| \alpha, \beta \in k\right\}$$

can be presented as $(x_1, x_2 \mid x_1x_2 = 0)$. The Lie algebra G_2 of all matrices of the form

$$\left\{\begin{pmatrix} \alpha & \beta \\ 0 & 0 \end{pmatrix} \middle| \alpha, \beta \in k\right\}$$

can be presented as $(x_1, x_2 \mid x_1x_2 = x_2)$. A Lie algebra G is called *finitely generated Lie algebra* (resp., *finitely related* or *finitely presented*) if G has a presentation $(X \mid R)$ in which $|X| < \infty$ (resp., $|R| < \infty$ or $|X|, |R| < \infty$). A free Lie algebra has a presentation of the form $L(X) = (X \mid \varnothing)$.

Lemma 2.27. *Let $G = (X \mid R)$ and H be a Lie algebra over k. We consider a map $\varphi : X \to H$ such that for each $r = r(x_1, x_2, \dots, x_n)$ in R we have $r(\varphi(x_1), \varphi(x_2), \dots, \varphi(x_n)) = 0$. Then the map of the form $\tilde{\varphi}(w(x_1, \dots, x_n) + S) = w(\varphi(x_1), \dots, \varphi(x_n))$ is a correctly defined unique homomorphism from G to H such that $\tilde{\varphi}(x + S) = \varphi(x)$ for all $x \in X$.*

Proof. An obvious consequence of the defining property of a free Lie algebra (Section 2.2.3) and the definition as above. □

2.6.2 Free product of Lie algebras

Let $(G_\alpha)_{\alpha \in I}$ be a family of Lie algebras over k with presentations $(X_\alpha \mid R_\alpha)$, where for all $\alpha \neq \beta$ we have $X_\alpha \cap X_\beta = \varnothing$. We set $X = \bigcup_\alpha X_\alpha$, $R = \bigcup_\alpha R_\alpha$. Then we say that $G = (X \mid R)$ is the *free product* of the family $(G_\alpha)_{\alpha \in I}$ and we write

$$G = \coprod_{\alpha \in I} G_\alpha.$$

If $I = \{1, 2, \dots, n\}$, we simply write $G = G_1 * G_2 * \dots * G_n$. We note that for any $\alpha \in I$ there exists a homomorphism $i_\alpha : G_\alpha \to G$, extending the identity map of X_α into X.

Proposition 2.28. *Let H be a Lie algebra and $(\varphi_\alpha)_{\alpha \in I}$ a family of homomorphisms $\varphi_\alpha : G_\alpha \to H$. Then there exists precisely one homomorphism $\varphi : G \to H$ such that $\varphi i_\alpha = \varphi_\alpha$.*

Proof. Let $\varepsilon_\alpha : L(X_\alpha) \to G_\alpha$ be a natural homomorphism from $L(X_\alpha)$ into G_α. Then $R_\alpha \subseteq \ker \varphi_\alpha \varepsilon_\alpha$. We define $\psi : L(X) \to H$ by putting $\psi(x_\alpha) = \varphi_\alpha \varepsilon_\alpha$ for all $\alpha \in I$. This map extends to a homomorphism $\bar{\psi} : L(X) \to H$ such that $\ker \bar{\psi} \supseteq R_\alpha$ for any $\alpha \in I$. Hence $\ker \bar{\psi} \subseteq R$ and $\ker \bar{\psi} \supseteq S$, where S is the least ideal of $L(X)$ containing R. Applying Lemma 2.27, we represent $\bar{\psi}$ in the form $\bar{\psi} = \varphi\varepsilon$ where ε is the natural homomorphism $\varepsilon : L(X) \to G$. Now it is obvious that φ is as desired. Its uniqueness follows since G is generated by its subalgebras $i_\alpha(G_\alpha)$. □

Corollary 2.29. *For any α the mapping i_α is injective, that is, $i_\alpha(G_\alpha)$ is an isomorphic copy of G_α in G.*

Proof. Apply Proposition 2.28 with $H = G_\alpha$, $\varphi_\beta = 0$ for all $\beta \neq \alpha$, $\varphi_\alpha = 1_{G_\alpha}$. □

An alternative definition of the free product can now be given as follows: $G = \coprod_{\alpha\in I} G_\alpha$ is the *free product* of the family $(G_\alpha)_{\alpha\in I}$ if it is unique up to isomorphism Lie algebra containing for each $\alpha \in I$ a subalgebra G_α and such that any family $(\varphi_\alpha)_{\alpha\in I}$ of homomorphisms $\varphi_\alpha : G_\alpha \to H$ uniquely extends to a homomorphism $\varphi : G \to H$ such that $\varphi|_{G_\alpha} = \varphi_\alpha$.

2.6.3 Universal enveloping algebra of the free product of Lie algebras

Let $(R_\alpha)_{\alpha\in I}$ be a family of associative k-algebras with pairwise nonintersecting bases $(B_\alpha)_{\alpha\in I}$. Let F be a free k-module with a basis B whose elements are the words of the form

$$b_1 b_2 \cdots b_s, \quad b_i \in B_{\alpha_i}, \ \alpha_i \neq \alpha_{i+1}, \ i = 1, 2, \ldots, s. \tag{2.14}$$

We define an operation in F on the elements of B as follows. If $w_1 = b_1 b_2 \cdots b_s$, $w_2 = b_1' b_2' \cdots b_r'$ and $\alpha_s \neq \beta_1$, then we put $w_1 \circ w_2 = w_1 w_2$. Otherwise, that is, when $\alpha_s = \beta_1 = \alpha$, we note that $b_s b_1' = \sum_i \lambda_i b_i''$, $\lambda_i \in k$. In this case we put

$$w_1 \circ w_2 = \sum_i \lambda_i b_1 b_2 \cdots b_{s-1} b_i'' b_2' \cdots b_r'.$$

Each member of this sum is a k-multiple of an element of the form in (2.14). It is easy to verify that this operation is associative. Thus F is an associative k-algebra called the free product $F = \coprod_{\alpha\in I} R_\alpha$ of the family of algebras $(R_\alpha)_{\alpha\in I}$. Now if $(U_\alpha)_{\alpha\in I}$ is the family of universal enveloping algebras for a family of Lie algebras $(G_\alpha)_{\alpha\in I}$ which are free k-modules then we define R_α as an associative subalgebra in U_α generated by G_α. Thus $U_\alpha = k \cdot 1 + R_\alpha$. We set $U = k \cdot 1 + \coprod_{\alpha\in I} R_\alpha$.

Proposition 2.30. *The above algebra U is the universal enveloping algebra for the free product of Lie algebras $G = \coprod_{\alpha\in I} G_\alpha$ (here, as before, each G_α is a free k-module).*

Proof. Let ε_α be the natural imbedding of G_α into U_α (in fact, into R_α). By Proposition 2.28, the family $(\varepsilon_\alpha)_{\alpha\in I}$ extends to a homomorphism $\varepsilon : G \to [U]$. Suppose now that φ is a homomorphism from G into a Lie algebra of the form $[P]$ where P is an associative k-algebra with 1. Let $i_\alpha : G_\alpha \to G$ be the imbedding as in the corollary of Proposition 2.28. Then φi_α is a homomorphism from G_α into $[P]$. By definition, there exists a homomorphism $\psi_\alpha : G_\alpha \to P$ such that $\psi_\alpha \varepsilon_\alpha = \varphi i_\alpha$ and such that $\psi_\alpha(1) = 1$. The family of maps $(\psi_\alpha)_{\alpha\in I}$ naturally extends to a homomorphism of associative algebras $\psi : U \to P$ mapping 1 to 1 and such that $\psi(w) = \psi_{\alpha_1}(b_1) \cdots \psi_{\alpha_s}(b_s)$, where w is an element of the form (2.14). Now ψ is the desired homomorphism. Indeed, a property of $\psi\varepsilon$ is that $\psi\varepsilon i_\alpha = \psi_\alpha \varepsilon_\alpha = \varphi i_\alpha$ for any α. By the uniqueness of φ (Proposition 2.28), we derive $\psi\varepsilon = \varphi$. □

2.6.4 Basis for free product

Suppose all algebras G_α, $\alpha \in I$, are free k-modules with bases $E_\alpha = \{e_{\alpha\beta} \mid \beta \in I_\alpha\}$ and that all sets I_α, $\alpha \in I$, and I itself are well-ordered. We set $E = \bigcup_{\alpha\in I} E_\alpha$ and introduce an ordering on E by putting $e_{\alpha\beta} \geq e_{\alpha'\beta'}$ if either $\alpha > \alpha'$, or $\alpha = \alpha'$ and then $\beta > \beta'$. Let $\check{R} = \check{R}(E)$ denote the Shirshov's basis in E (see Section 2.3.5). A monomial w in R is called *special* if its underlying associative word $\bar{w}$ has no subword of the form $e_{\alpha i}e_{\alpha j}$ with $i < j$. Let R_s denote the set of images of all special monomials in $G = \prod^*_{\alpha\in I} G_\alpha$.

Theorem 2.31 (A. I. Shirshov). *The set R_s is a basis of the free product $G = \coprod_{\alpha\in I} G_\alpha$.*

Proof. We note first that by Proposition 2.30 a basis of $U(G)$ is formed by 1 and all products of the form

$$\begin{gathered} e_{\alpha_1 i}\cdots e_{\alpha_1 j}e_{\alpha_2 s}\cdots e_{\alpha_2 r}\cdots e_{\alpha_t n}\cdots e_{\alpha_t m}, \quad t \geq 1,\ \alpha_i \in I, \\ i \geq \cdots \geq j,\ s \geq \cdots \geq r,\ n \geq \cdots \geq m, \quad \alpha_i \neq \alpha_{i+1},\ i = 1,\dots,t. \end{gathered} \tag{2.15}$$

Let $\varepsilon : G \to U$ denote the natural homomorphism from G into U. We choose $w \in R_s$ and write $\varepsilon(w)$ in the form of a linear combination of basic elements (2.15) of U. Let $\check{w}$ be the leading term of this combination with respect to the lexicographic ordering. We note that $\check{w}$ arises in an alternative way. Namely, we may consider w as an element of the free associative algebra $A(E)$ and write it as an associative polynomial in E. By Proposition 2.13, its lexicographically greatest word is equal to the word $\check{w}$ resulting by forgetting brackets on w. Under the natural map of $A(E)$ onto U, $\check{w}$ is mapped to w because $\bar{w}$ has no subwords of the form $e_{\alpha i}e_{\alpha j}$, $i < j$. By linear independence of $\bar{w}$, for different $w \in R$ we derive linear independence of words w, hence that of special monomials forming R_s.

Now we are going to prove that special monomials span G. It is sufficient to verify that $\varepsilon(\check{R})$ is in the linear span of the set of special monomials. Let w be an element in $\check{R}$. Suppose $\bar{w}$ has a subword of the form $e_{\alpha i}e_{\alpha j}$, $i < j$. Apply induction on the degree and, within a given degree, on lexicographical ordering. In w as above, there is a factor of the form

$$(w_t \cdots w_2 w_1 e_{\alpha i})e_{\alpha j} \tag{2.16}$$

where $w_t \geq \cdots \geq w_2 \geq w_1 < e_{\alpha i}$ (it is possible that $t = 0$). Having expanded monomial (2.16) according to derivation rules, we will see, as in Section 2.3.5, that $\varepsilon(w)$ is a linear combination of $\varepsilon(w')$, where w' results from w by replacing (2.16) with

$$w_t \cdots w_2 w_1 e_{\alpha i} e_{\alpha j} \tag{2.17}$$

and some basic monomials which are strictly less than $\varepsilon(w)$. Since in the image of equation (2.17) $e_{\alpha i}e_{\alpha j}$ can be replaced with its product in G,

$$e_{\alpha i}e_{\alpha j} = \sum_p c_{ij}^p e_{\alpha p}, \quad c_{ij}^p \in k,$$

we may apply our induction hypothesis. □

2.6.5 Subalgebras of free products

In a paper by Shirshov [339], a negative answer is given to the question whether the subalgebras of free products of Lie algebras have structure similar to their ambient products (this is the case for groups, see Kurosh [196]). A complete solution of the problem has been given by Kukin [191] who described these subalgebras in terms of generators and defining relations. In many cases such subalgebras are free. One example of this kind is given below, and we will need it in Chapter 4.

Proposition 2.32. *Let L be a free Lie algebra over a ring k and G a Lie algebra over k, which is free as a k-module. Let F be their free product, $F = L * G$. Then the ideal $J = \mathrm{id}_F L$ generated in F by L is a free Lie algebra.*

Proof. Let $\{x_i \mid i \in I\}$ be a free generating set for L and $\{e_\alpha \mid \alpha \in M\}$ a basis for G over k, M being well-ordered. We will show that a free generating set for J may be chosen in the form of the set of all monomials

$$e_{\alpha_1} e_{\alpha_2} \cdots e_{\alpha_n} x_i, \quad \alpha_1, \alpha_2, \dots, \alpha_n \in M,\ i \in I,\ \alpha_1 \geq \cdots \geq \alpha_n,\ n \geq 0. \tag{2.18}$$

For the proof we consider a free Lie algebra $\bar{L}$ freely generated by the set of all pairs of the form (m, x_i), where m is a monomial in the canonical basis for $U(G)$ (see the PBW theorem), $i \in I$. We will need a homomorphism $g \mapsto \bar{d}_g$ from G into $\operatorname{Der} \bar{L}$. To define it, we set

$$d_g((m, x_i)) = \sum_{j=1}^{t} \beta_j (m_j, x_i) \quad \text{if } gm = \sum_{j=1}^{t} \beta_j m_j \text{ in } U(G).$$

According to Section 2.5.5, d_g uniquely extends to a derivation $\bar{d}_g$ of $\bar{L}$. The action of d_g on the first component of the free generator (m, x_i) is the restriction of the regular representation of G in $U(G)$ (see the end of Section 2.5.4). Thus $g \mapsto d_g$, hence $g \mapsto \bar{d}_g \in \operatorname{Der} \bar{L}$ is a homomorphism of Lie algebras. Let $S = G \ltimes \bar{L}$ be the semidirect product with respect to $g \mapsto \bar{d}_g$ (Section 1.4.4). By Proposition 2.28, there exists a homomorphism $\chi : F \to S$, extending the identity map of G into G, and the map $x_i \mapsto (1, x_i)$ from L into $\bar{L}$. Under χ the images of elements in (2.18) become free generators for $\bar{L}$, hence the set of these elements is a free generating set for a subalgebra which it generates. However, it is a simple matter that J is generated by the G-submodules whose generators are x_i, $i \in I$. From Section 1.6, Proposition 1.10, the k-generators of these modules have the form as in (2.18). □

2.7 Free restricted Lie algebra

2.7.1 Definition and construction

If k is a field of characteristic $p > 0$ and X a nonempty set then the *free restricted Lie algebra* (p-algebra) $L_p(X)$ can be defined as a Lie p-algebra generated by X and such that any map $\varphi : X \to G$, where G is a Lie p-algebra, uniquely extends to a p-homomorphism $\bar{\varphi} : L_p(X) \to G$ (see basic definitions in Section 1.11).

Proposition 2.33. *Let $A = A(X)$ be a free associative algebra over k with free generating set X, $L = L(X)$ a free Lie algebra freely generated by X in A with respect to bracketing. If R is a basis for L then the set*

$$R_1 = \{w^{p^t} \mid w \in R, t = 0, 1, \dots\}$$

is linearly independent and it is a basis for a Lie p-algebra $L_p(X)$ which is a free p-algebra.

Proof. Since R_1 is a part of a basis for $A(X)$, given in the PBW theorem, it is linearly independent. By Proposition 1.32, Section 1.11, the linear span of R_1 is a Lie p-algebra generated by X. Any element in $L_p(X)$ has the form

$$a = \lambda_1 w_1^{p^{t_1}} + \lambda_2 w_2^{p^{t_2}} + \dots + \lambda_s w_s^{p^{t_s}}.$$

If $\varphi : X \to N$, where N is a Lie p-algebra, then φ extends to a homomorphism of Lie algebras $\varphi_1 : L(X) \to N$. We set

$$\bar{\varphi}(a) = \lambda_1 \varphi_1(w_1)^{p^{t_1}} + \lambda_2 \varphi_1(w_2)^{p^{t_2}} + \dots + \lambda_s \varphi_1(w_s)^{p^{t_s}}.$$

Using identities in Section 1.11.1, it is easy to verify that $\bar{\varphi}$ is a homomorphism of Lie p-algebras. □

The elements in R_1 are called *p-basic monomials*.

A special feature of free Lie p-algebras is that they are similar to both free Lie algebras and free groups (of course, in a very limited sense!). For instance, here, in addition to an analogue of Theorem 2.31 (Shirshov) on subalgebras, we also have an analogue of Schreier's formula for ranks of subgroups of free groups. This latter analogy is our next purpose.

Using the above proposition, we will be applying to $L_p(X)$ all the notions developed in the case of $L(X)$ and $A(X)$.

2.7.2 Elementary transformations

Let $S = \{s_\alpha \mid \alpha \in I\}$ be a subset in a free Lie p-algebra $L_p(X)$. An *elementary transformation* of S is as map $\omega : S \to L_p(X)$ such that $\omega(s_\alpha) = s_\alpha$ for all $\alpha \in I$ except one of them, say β, and

$$\omega(s_\beta) = \lambda s_\beta + w(s_{\alpha_1}, \dots, s_{\alpha_t})$$

where $\lambda \neq 0$, $\lambda \in k$, $\alpha_1, \dots, \alpha_t \neq \beta$ and $w(x_{\alpha_1}, \dots, x_{\alpha_t}) \in L_p(x_1, x_2, \dots)$.

Lemma 2.34. *Elementary transformations of a free generating set induce automorphisms of $L_p(X)$.*

Proof. If ω is an elementary transformation as above then ω' such that $\omega'(s_\alpha) = s_\alpha$ and $\omega'(s_\beta) = \lambda^{-1}s_\beta - \lambda^{-1}w(s_{\alpha_1}, \dots, s_{\alpha_t})$ in composition with ω gives the identity transformation of S. In the case where $S = X$, let $\bar{\omega}$ and $\bar{\omega}'$ denote the corresponding endomorphisms of $L_p(X)$. Then under both $\bar{\omega}\bar{\omega}'$ and $\bar{\omega}'\bar{\omega}$, X is mapped identically, i. e., $\bar{\omega}$ is an automorphism. □

2.7.3 *p*-Reduced set

A set T of elements in $L_p(X)$ is called *p-reduced* if for any $f \in T$ its homogeneous component of the highest degree (leading term) is not an element of the subalgebra generated by leading terms of remaining elements in T.

Lemma 2.35. *Any p-subalgebra B of a free Lie p-algebra $L_p(X)$ has a p-reduced set of generators.*

Proof. We will be constructing T as a union of the form $T = \bigcup_{j=1}^{\infty} T_j$ where T_j is either empty or consists of elements of degree j. We set $T_0 = \varnothing$. Proceeding by induction, we assume that $T_0, T_1, \dots, T_i$ have already been defined. Let S_i denote the set of elements of degree at most $i+1$ in the subalgebra generated by $T_0 \cup T_1 \cup \cdots \cup T_i$. Then T_{i+1} is defined as a maximal subset of elements in B of degree $i+1$ which is linearly independent modulo S_i. It is obvious that B is generated by T constructed as just above. Now let $f \in T$ be obtained on the $(i+1)$st step of this procedure. Suppose $\bar{f}$ being an element of the subalgebra C generated by leading terms of all remaining elements in T. C being homogeneous, $\bar{f}$, in fact, is an element of the subalgebra generated by the leading terms of elements in $\bigcup_{j=0}^{i+1} T_j$. Then

$$\bar{f} = \alpha_1\bar{f}_1 + \cdots + \alpha_q\bar{f}_q + w(\bar{f}_{q+1}, \dots, \bar{f}_s)$$

where $f_1, \dots, f_q \in T_{i+1}$ and $f_{q+1}, \dots, f_s \subseteq \bigcup_{j\leq i} T_j$.

From the above relation, we derive a relation of the form

$$f = \alpha_1 f_1 + \cdots + \alpha_q f_q + w(f_{q+1}, \dots, f_s) + g, \tag{2.19}$$

where g is an element of degree at most i. It follows form equation (2.19) that $g \in B$, that is, $g \in S_i$. Hence equation (2.19) contradicts the definition of T_{i+1}. Thus T is a p-reduced set. □

Theorem 2.36 (E. Witt). *Let k be a field of prime characteristic p, $L_p(X)$ a free Lie p-algebra over k, H an arbitrary p-subalgebra in $L_p(X)$. Then H is a free restricted Lie p-algebra.*

Proof. As in Theorem 2.18, we restrict ourselves to the case where H is a homogeneous subalgebra. Then, using Section 2.7.3, we will find in H a p-reduced homogeneous generating set T. Any relation among the elements of T is a relation among the elements of a finite subset of T. Thus we may restrict ourselves to the case where X is finite and T its finite p-reduced subset. Let $T_1 = \{f_1, \ldots, f_l\}$. The elements in T_1 are linearly independent and are of degree 1. Hence, obviously, $|X| \geq l$. The map $\varphi : x_1 \mapsto f_1, \ldots, x_l \mapsto f_l, x_i \mapsto x_i (i \geq l)$ is a product of elementary transformations. According to Section 2.7.2, it is sufficient to show that $\varphi(T)$, which is still a p-reduced set, is a free generating set for the p-subalgebra $\varphi(H)$ which it generates. So, we may think that $T_1 \subseteq X$. The cases $T \subseteq X$ and $T_1 = X$ being trivial, we assume that there exists an element $x_{l+1} \in X \setminus T$ which enters nontrivially the elements of T. In this case T becomes a subset of the p-ideal I generated by $(X \setminus \{x_{l+1}\}) \cup \{x_{l+1}^p\}$. We note that I is a free Lie p-algebra freely generated by

$$X = \{x_{l+1}^p, x_{l+1}^s x_\alpha \mid \alpha \neq l+1, 0 \leq s < p\}.$$

A proof of this can be given applying an argument from the corollary of Proposition 2.15; using identity (ii) in Section 1.11.1, we easily observe that I is generated by X_1. Its being a free p-generating set can be derived from the linear independence of p-basic monomials constructed on X_1. Indeed, usual basic monomials in X_1 except x_{l+1}^p form a part of the set of all basic monomials in the elements of the form (2.7) where $X'' = \{x_{l+1}\}$. From their linear independence, it follows that X_1 freely generates a free subalgebra with respect to the operations of Lie algebra and a free Lie p-algebra with respect to the operations of a Lie p-algebra.

Since I is homogeneous, T is still a p-reduced set in the free Lie p-algebra I. Since the elements of T all depend on x_{l+1}, if we rewrite them on the free generators of I then the sum of the degrees of elements in T decrease. Applying induction over this parameter, we conclude that T generates a free Lie p-subalgebra. □

In the case of Lie p-algebras (contrary to the case of Lie algebras), a p-subalgebra of finite codimension in a finitely generated p-algebra is itself finitely generated. The precise value of the rank of a free p-subalgebra is given below.

Theorem 2.37. *Let B be a subalgebra of free Lie p-algebra $L(X)$ of rank n, i. e., $|X| = n$. If $\dim L/B = j$ then B is a free Lie p-algebra of rank*

$$N(n, j) = p^j(n-1) + 1.$$

Proof. Let $\bar{B}$ be a p-subalgebra generated by the leading terms of elements in B. A p-reduced set of generators for $\bar{B}$ is merely the set $\bar{T}$ of leading terms of elements of a p-reduced set for B. By the proof of Theorem 2.36, then the rank of $\bar{B}$ is the same as that for B. To finally pass to the case of homogeneous subalgebras, we recall that, according to Section 1.10.4, we have

$$\dim L_p(X)/B = \dim L_p(X)/\bar{B}.$$

Thus we may restrict ourselves to the case where B is a homogeneous p-subalgebra and T is a homogeneous p-reduced set. As in the proof of Theorem 2.36, we may think that the elements of degree 1 in T are included in X as its proper part. If $x_{l+1} \notin T$ then we may pass to the p-ideal I generated by $(X \setminus \{x_{l+1}\}) \cup \{x_{l+1}^p\}$. This ideal is a free Lie p-algebra or rank $p(n-1)+1$. Furthermore, $\dim I/B = j-1$. Applying the induction hypothesis over j, we see that B is a free Lie p-subalgebra of rank

$$N(n(p-1)+1, j-1) = p^{j-1}(p(n-1)) + 1 = p^j(n-1) + 1 = N(n,j). \qquad \square$$

Corollary 2.38. *Let B be a p-subalgebra of a finitely generated Lie p-algebra A such that* $\dim(A/B) < \infty$. *Then B is a finitely generated Lie p-algebra.*

Proof. It is sufficient to represent A in the form of a quotient algebra of a free Lie p-algebra $L_p(X)$ of appropriate rank and then choose the full inverse image for B in $L_p(X)$. □

2.8 Exercises

2.8.1. Prove that $\mathfrak{gl}(n,k)$ does not satisfy any identity of degree less than $2n$.

2.8.2. Prove that W_1 does not satisfy any identity of degree less than or equal to 4.

2.8.3. Let $f(z_1,\dots,z_m)$ be a multilinear associative polynomial, G a Lie algebra over a ring k. We define a k-submodule G_f as the set of all elements $x \in G$ such that for all $x_1,\dots,x_m \in G$ we have $f(\operatorname{ad} x_1,\dots,\operatorname{ad} x_m)(x) = 0$. Prove that G_f is an ideal of G.

2.8.4. Suppose that any algebra with identical relation $v(x_1,\dots,x_n) \equiv 0$ is solvable and f is as in Exercise 2.8.2. If any Lie algebra with identity $f(\operatorname{ad} x_1,\dots,\operatorname{ad} x_m)(x) \equiv 0$ is solvable then any Lie algebra with identity

$$f(\operatorname{ad} x_1,\dots,\operatorname{ad} x_m)v(x_{x+1},\dots,x_{m+n}) = 0$$

is solvable.

2.8.5. Derive the recurrence formula for the number a_n of possible arrangements of brackets on a word of length n. Find a_n for $n = 1,\dots,10$.

2.8.6. The same question as in Exercise 2.8.4 in the case of free commutative groupoids.

2.8.7. Is it true that the free alternative and free Jordan algebras are graded with respect to their free generating sets?

2.8.8. Let a, b be elements of a free Lie algebra over a ring k such that $ab = 0$. Then a and b are proportional.

2.8.9. In the free Lie algebra $L = L(x_1, x_2)$ find a free generating set for a subalgebra H such that
(i) $H = L^n$;
(ii) $H = (L^n)^m$;
(iii) $H = \mathrm{id}_L(x_2 - x_1x_2)$.

2.8.10. Let

$$L_1 = (x, y \mid (\operatorname{ad} x)^2(y) = (\operatorname{ad} y)^2(x) = 0),$$
$$L_2 = (x, y \mid (\operatorname{ad} x)^2(y) = (\operatorname{ad} y)^3(x) = 0),$$
$$L_3 = (x, y \mid (\operatorname{ad} x)^3(y) = (\operatorname{ad} y)^2(x) = 0).$$

Which of L_i, $i = 1, 2, 3$, is nilpotent or solvable?

2.8.11. Let L_1, L_2 be two Lie algebras over a field k and $F = L_1 * L_2$ their free product. Let φ denote the natural homomorphism of F onto $D = L_1 \times L_2$ extending the identity maps of L_1 and L_2. Prove that the kernel C of this homomorphism (this is called *Cartesian subalgebra*) is a free Lie algebra with a free generating set of the form

$$S = \{e_1 \cdots e_n f_1 \cdots f_m e_{n+1}, \quad \text{where } e_i, f_j \text{ are arbitrary elements of some bases for } L_1, L_2 \text{ and } e_1 \geq \cdots \geq e_n, f_1 \geq \cdots \geq f_m\}.$$

Hint: Consider free Lie algebra $L(S)$. Taking into account Shirshov's bases for $L(S)$ and F, prove that the identity map of S extends to the isomorphism of $L(S)$ onto C.

2.8.12. Let L_1, L_2 be two finite-dimensional Lie algebras over a field k. Prove that their free product $F = L_1 * L_2$ is residually finite (see Section 6.6.3).

2.8.13. Let L_1 and L_2 be nilpotent Lie algebras over a field k. Prove that $F = L_1 * L_2$ is residually nilpotent (see Section 6.6.5).

2.8.14. Let u, v be two elements of a free product of Lie algebras L_1 and L_2. If $uv = 0$ then either u, v are the elements of one of the subalgebras L_1, L_2 or u and v are proportional.

2.8.15. Prove that no Lie algebra can be a free and at the same time a direct product of its nontrivial subalgebras.

2.8.16. Give another proof of Shirshov's theorem using the argument from Section 2.7.

2.8.17. Find a free generating set for the p-ideal J of a free Lie p-algebra $L_p(x_1, x_2)$ generated by x_1^p, x_2^p.

2.8.18. Let X, Y be nonempty sets such that $L(X) = L(Y)$. Prove that then $|X| = |Y|$.

2.8.19. Let L be a free Lie algebra of countable rank. Prove that for any c the derivation algebra $\operatorname{Der} L$ contains a nilpotent subalgebra of nilpotent step precisely c.

2.8.20. Embed a 3-dimensional nilpotent Lie algebra $L = kx \oplus ky \oplus kz$ with center kz and quotient algebra $G = L/kz$ in a Lie algebra of the form $\bar{L} = G \rtimes M$ so that $M \cap L = kz$, M being an abelian ideal of $\bar{L}$. Is it possible to have $\bar{L}$ nilpotent?

2.9 Comments

The origins of the theory of free Lie algebras lie in the paper of P. Hall [139], Magnus [213, 214], and Witt [371] devoted to the structure of free and nilpotent groups. An explicit form of a basis in a free Lie algebra (Hall's basis) is given in a paper by M. Hall [136]. Theorems on subalgebras of free Lie algebras have been proved in Shirshov [336] and Witt [372]. A. I. Shirshov has given miscellaneous constructions of bases of free Lie algebras [338, 340]. This topic is considered also in a separate issue of Springer Lecture Notes (Viennot [359]). The notion of the free product of a free Lie algebra has been introduced and studied by Shirshov [339]. As a matter of fact, in that paper a basis is constructed for the free product with amalgamated subalgebras. This construction has been subsequently studied in detail by Kukin [191]. We left, by necessity, beyond the limits of the present volume a part of the theory concerning Lie algebras presented in terms of generators and defining relations. Many brilliant results in this area are due to algebraists of Shirshov's school. In particular, a negative solution has been obtained of the word problem in the class of all Lie algebras (see Bokut' [77]), as well as within some classes of solvable algebras (see Kukin [193, 195]). We mention few more papers of Kukin devoted to free Lie algebras [188, 194, 192].

Additional information concerning injective modules and the question of splitting the extension with injective kernel may be found in Lambek [198] and Séminaire "Sophus Lie" [327]. The theorem on subalgebras of free restricted Lie algebras is due to Witt [372]. Its present version is due to Kukin [190]. The same source is for Theorem 2.37. We also mention a book by Reutenauer [319] totally devoted to the free Lie algebras.

3 Homogeneous structure of free Lie algebras

3.1 Introduction

3.1.1 $L(X)$ as GL- and S_n-module

Let $L = L(X)$ be a free Lie algebra with a free generating set X over a ring k. Let also $V = L(X)$ be a (free) submodule generated by X in L, $G = \mathrm{GL}(V)$ the automorphism group of this k-module. The map $x \mapsto g(x)$, $x \in X$, $g \in G$, extends to a Lie algebra automorphism of $L(X)$. In this way $L(X)$ becomes a G-module. It is important for us here that, given a k-algebra P, the set T of all elements $f \in L(X)$ such that $f \equiv 0$ in P is a G-submodule in L. A considerable simplification of the G-module structure on $L(X)$ is due to the fact that

$$L = \bigoplus_{n=1}^{\infty} L_n,$$

all homogeneous components L_n being G-invariant. In particular, if $m = |X| < \infty$ then we are dealing with k-modules of finite dimension. If, in addition, k is a field, we arrive at finite-dimensional modules over a classical group $\mathrm{GL}(m,k)$. The classical approach to studying homogeneous representations of degree n is via representations of the symmetric group S_n. A good theory of the latter exists in the case where k is a field of characteristic zero. As we will see in Chapter 4, in this case the ideal of identities of an arbitrary algebra is homogeneous. Thus the determination of the homogeneous structure of a G-module $L_n(X)$ is more or less equivalent to the study of identities in $m = |X|$ variables. If we restrict ourselves to homogeneous identities of some fixed degree n in m variables, the problem of determining G-submodules in $L_n(X)$ is equivalent to that of determining all systems of identities up to equivalence. The most important case is where $m = n$, the case of multilinear identities. Here it is sufficient to determine the S_n-structure in P_n, the space of all multilinear polynomials in $L_n(X)$ where S_n is naturally considered as a subgroup in $\mathrm{GL}(n,k)$.

The remaining part of this section will be devoted to deriving a formula, due to Witt, giving the dimension of $L_n(X)$. In Section 3.2 we meet Young diagrams, the main tool for describing irreducible representations of S_n. Here the base field is of characteristic 0. Then, in Section 3.3, we study some basic notions concerning Frobenius algebras. This will be applied to describing homogeneous representations of $\mathrm{GL}(V)$, as well as to some other purposes in Chapter 8. In the final Section 3.4, we study L_n as a functor in the category of $\mathrm{GL}(V)$-modules over a wider class of rings. In the case where k is a field, we arrive at the multiplicity formula for irreducible $\mathrm{GL}(V)$-submodules in the $\mathrm{GL}(V)$-module $L_n(V)$. This section is more challenging since it requires closer acquaintance with the representation theory.

https://doi.org/10.1515/9783110566659-003

3.1.2 Möbius function

We define a function $\mu : \mathbb{N} \to \{-1, 0, 1\}$ by setting $\mu(1) = 1$, $\mu(n) = 0$ if n is divisible by a square different from 1, and $\mu(n) = (-1)^q$ if $n = p_1 \cdots p_q$, $p_1, \ldots, p_q$ being pairwise different primes. This function is called *Möbius function.* From the definition we easily conclude that if $n = p_1^{s_1} \cdots p_q^{s_q} \geq 2$ then

$$\sum_{d|n} \mu(d) = \sum_{t=0}^{q} \binom{q}{t} (-1)^t = 0, \tag{3.1}$$

where, as above, $p_1, \ldots, p_q$ are pairwise different prime numbers.

Möbius inversion formula. Let f, g be functions of the natural argument. If

$$f(n) = \sum_{m|n} g(m) \quad \text{then } g(n) = \sum_{m|n} \mu(m) f\left(\frac{n}{m}\right).$$

Applying induction over n with basis $n = 1$, we may write

$$f(n) = g(n) + \sum_{\substack{m|n \\ m \neq n}} \sum_{p|m} \mu\left(\frac{m}{p}\right) f(p).$$

Putting $\frac{m}{p} = \delta$, we rewrite

$$g(n) = f(n) - \sum_{\substack{p|n \\ p \neq n}} \sum_{\substack{\delta|(n/p) \\ \delta \neq (n/p)}} \mu(\delta) f(p).$$

Applying equation (3.1), we finally get

$$g(n) = \mu(1) f(n) + \sum_{\substack{p|n \\ p \neq n}} \mu\left(\frac{n}{p}\right) f(p) = \sum_{p|n} \mu\left(\frac{n}{p}\right) f(p).$$

Obviously, this formula is equivalent to that we are proving.

Example 3.1. Let $p(n)$ denote the sum of all primitive roots of 1 of degree n, $s(n)$ the sum of all roots of degree n. Then, obviously,

$$s(n) = \sum_{m|n} p(m).$$

By the just proved formula, we have

$$p(n) = \sum_{m|n} \mu(m) s(n/m).$$

But $s(k) = 0$ if $k \neq 1$ and $s(1) = 1$. Hence $p(n) = \mu(n)$.

3.1.3 Witt's formula

As we have learnt in Section 2.3.3, a free Lie algebra $L(X)$ is a free module over the base ring k. Moreover, $L(X) = \bigoplus_{n=1}^{\infty} L_n(X)$ and all $L_n(X)$ are free k-modules, too. If X is a finite set then $L_n(X)$ is a free k-module of finite rank. Here we compute this rank and put $|X| = d$. Since the definition of basic monomials (forming a basis in $L(X)$) does not depend on the base ring, we may restrict ourselves to the case where k is a field. Let $l_d(n)$ denote the rank (in the particular case the dimension) of the k-module $L_n(X)$, $|X| = d$.

Theorem 3.2.

$$l_d(n) = \frac{1}{n} \sum_{m|n} \mu(m) d^{m/n}.$$

Proof. Let $\{w_1, w_2, \ldots\}$ be a homogeneous basis in $L(X)$, the degree of w_i being equal to d_i. According to Section 2.3, $L(X)$ is embedded in $A(X)$ and a basis for $A(X)$ can be chosen in either of the two following ways:

(i) The set of all words of length n in X, their number is, obviously, d^n.

(ii) The set of all products of the form

$$w_s^{e_s} w_{s-1}^{e_{s-1}} \cdots w_1^{e_1}$$

where e_i are some nonnegative numbers, w_i basic monomials as above, $n = e_1 d_1 + \cdots + e_{s-1} d_{s-1} + e_s d_s$.

Thus d^n is equal to the number of all finite sequences $(e_1, \ldots, e_s)$ such that

$$n = \sum_{i=1}^{s} e_i d_i.$$

In other words, the coefficients of the following two power series are the same:

$$A(t) = \prod_{i=1}^{\infty} \frac{1}{1 - t^{d_i}}, \quad B(t) = \frac{1}{1 - dt}.$$

Indeed,

$$\prod_{i=1}^{\infty} \frac{1}{1 - t^{d_i}} = \prod_{i=1}^{\infty} (1 + t^{d_i} + t^{2d_i} + \cdots)$$

and

$$\frac{1}{1 - dt} = 1 + dt + d^2 t^2 + \cdots.$$

Now the number of degrees d_i equal to the same number m is precisely $l_d(m)$. Hence

$$\prod_{m=1}^{\infty} \frac{1}{(1-t^m)^{l_d(m)}} = \frac{1}{1-dt}.$$

Taking logarithms of both sides (a more detailed study of this is given in Chapter 8), we obtain

$$\sum_{m,v} \left(\frac{1}{v}\right) l_d(m) t^{mv} = \sum_{n=1}^{\infty} \left(\frac{1}{n}\right) d^n t^n,$$

whence

$$\sum_{mv=n} \left(\frac{1}{v}\right) l_d(m) = \frac{1}{n} d^n.$$

Finally,

$$d^n = \sum_{m|n} m l_d(m).$$

Now all that remains is to apply Möbius inversion formula. □

3.2 Representations of the symmetric group

3.2.1 Necessary prerequisites

When treating the theory of irreducible representations of the symmetric group, we will repeatedly use some basic facts which can be found in many sources, for instance, in the following: Chapters 7–9 of Kostrikin [178], Chapters 2–4 of Skornyakov [345], or Kerber [164].

Now let S_n denote the symmetric group on $\{1, 2, \ldots, n\}$ and $A = kS_n$ its group algebra over an algebraically closed field of characteristic zero. Then

(i) Each finite-dimensional A-module is semisimple, that is, the direct sum of irreducible (left) A-modules.

(ii) Algebra A is equal to the direct sum $A = \varepsilon_1 A + \varepsilon_2 A + \cdots + \varepsilon_r A$ of two-sided ideals $\varepsilon_i A$, $i = 1, 2, \ldots, r$, where r is the number of the conjugacy classes of S_n. Every irreducible S_n-module is isomorphic to a minimal left ideal of A which is contained in precisely one two-sided ideal $\varepsilon_i A$. Central elements ε_i have the property $\varepsilon_i^2 = \varepsilon_i$; they are called *idempotents*. Each minimal ideal I is generated by a (not necessarily central) idempotent. If $I \subseteq \varepsilon_i A$ then $\dim \varepsilon_i A = (\dim I)^2$.

(iii) The conjugacy classes of S_n are in one-to-one correspondence with the partitions of n of the form

$$n = n_1 + n_2 + \cdots + n_t \quad \text{where } n_1 \geq n_2 \geq \cdots \geq n_t$$

labeling the conjugacy classes of permutations of the form

$$\sigma = \sigma_1\sigma_2\cdots\sigma_t,$$

with $\sigma_1, \sigma_2, \ldots, \sigma_t$ being independent cycles of lengths $n_1, n_2, \ldots, n_t$, respectively.

Thus, according to these assertions, the isomorphism classes of S_n-modules are in one-to-one correspondence with the decompositions of n as in (iii). Our aim is to give the explicit form and dimension of a simple module corresponding to the given partition. In the process of doing this, we will see that the arising idempotents are in $\mathbb{Q}S_n$. Thus all the theory works over an arbitrary field of characteristic zero.

3.2.2 Constructing the idempotents

With each decomposition $d = \{n_1, n_2, \ldots, n_t\}$, $n_1 \geq n_2 \geq \cdots \geq n_t$, of n we associate a diagram with t rows, the ith row containing n_i boxes. For instance, $d = \{5, 3, 1, 1\}$ is a decomposition of 10 then the associated diagram has the form

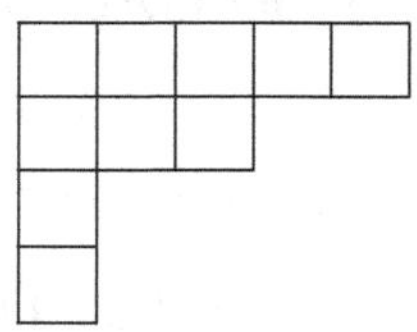

We will designate this diagram by the same letter d and this will be called the *Young diagram* associated with partition d. If $\tau \in S_n$ then τd will designate the *Young tableau*, that is, the Young diagram d stuffed with the numbers $1, 2, \ldots, n$ in the following way. We consider the sequence $\tau(1), \ldots, \tau(n)$ and, preserving the order, insert its elements into the boxes of the diagram from above downwards and from the left to the right. For instance, if d is as above and

$$\tau = \begin{pmatrix} 1 & 2 & 3 & 4 & 5 & 6 & 7 & 7 & 9 & 10 \\ 2 & 1 & 10 & 7 & 6 & 4 & 5 & 3 & 9 & 8 \end{pmatrix}$$

then

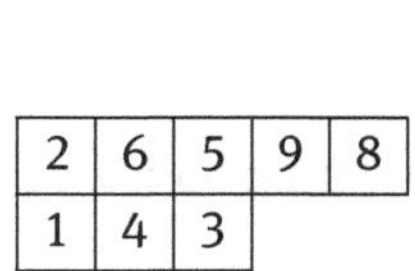

If $\tau = 1$ then we write $\tau d = d$ so that d is also a tableau corresponding to the trivial arrangement of numbers in the diagram d. As a result, we obtain the action of S_n on the set of all Young tableaux corresponding to a fixed diagram d given by $\tau(\sigma d) = (\tau\sigma)d$. For a fixed tableau τd, let $C = C_{\tau d}$ denote a subgroup of S_n whose elements are permutations leaving invariant the sets of symbols in each column of τd. Similarly, $R = R_{\tau d}$ is a subgroup of S_n whose elements leave invariant the rows of τd. Now we set

$$e_{\tau d} = \sum_{\substack{\rho \in R \\ \sigma \in C}} \varepsilon_\sigma \rho\sigma.$$

Here ε_σ stands for the sign of permutation σ, that is, 1 if σ is even and -1 is σ is odd. The above element is nonzero since $C \cap R = \{1\}$ and, in this case, $pq = p'q'$ if and only if $p = p', q = q'$.

We will show that the left ideals $Ae_{\tau d}$ are minimal and that left S_n-modules $Ae_{\tau d}$ and $A_{\tau' d'}$ are isomorphic if and only if $d = d'$. In this way, we will obtain representatives of all isomorphism classes of S_n-modules.

Lemma 3.3. *Any element $\gamma \in S_n$ can be written in the form $\gamma = \rho\sigma, \rho \in R_{\tau d}, \sigma \in C_{\tau d}$ if and only if any two symbols in the same row of τd are in different rows of $\gamma\tau d$.*

Proof. Let a, b be two symbols in the same row of τd and $\gamma = \rho\sigma$. Then a and b are in the same row of $\rho\tau d$. Now

$$\gamma\tau d = \rho\rho^{-1}\gamma\rho^{-1}\rho\tau d = \rho\sigma\rho^{-1}\rho\tau d.$$

It is obvious that if $\sigma \in C_{\tau d}$ then $\rho\sigma\rho^{-1} \in C_{\rho\tau d}$. Hence a, b, which are in different columns of $\rho\tau d$, remain in different columns of $\gamma\tau d$, as required.

Conversely, let any two symbols in the same row of τd be in the same row of $\gamma\tau d$. Then all the symbols of the first column of $\gamma\tau d$ are in different rows of τd. Hence there exists a permutation $\rho_1 \in R_{\tau d}$ such that all these symbols are in the first column of $\rho_1\tau d$. Repeating this for the remaining columns, we will find a permutation $\rho \in R_{\tau d}$ such that all the sets of elements in columns of $\rho\tau d$ and $\gamma\tau d$ coincide. Then for a permutation $\sigma' \in C_{\rho\tau d}$ we will have $\gamma\tau = \sigma'\rho\tau$. But then $C_{\rho\tau d} = \rho C_{\tau d}\rho^{-1}$, whence $\sigma' = \rho\sigma\rho^{-1}$, $\sigma \in C_{\tau d}$. In this case $\gamma\tau = \rho\sigma\tau$ and $\gamma = \rho\sigma$, where $\rho \in R_{\tau d}$, $\sigma \in C_{\tau d}$. □

3.2.3 Lexicographic ordering

We introduce the lexicographic ordering on the set of all decompositions (and diagrams). Thus,

$$d = \{n_1, n_2, \ldots, n_u\} > \{n_1', n_2', \ldots, n_v'\} = d'$$

if, given the first s with $n_s \neq n_s'$, we have $n_s > n_s'$.

Lemma 3.4. *Suppose $d > d'$. Then for all τ, τ' we have $e_{\tau' d'} e_{\tau d} = 0$.*

Proof. We will show first that there exist two symbols which are in the same row of τd and the same column of $\tau' d'$. If this is not the case, then all n_1 symbols in the first row of τd are in different columns (their number is n_1') of $\tau' d'$. Thus $n_1 \leq n_1'$, hence $n_1 = n_1'$. Replacing τ' by $\sigma' \tau'$, where $\sigma' \in C_{\tau' d'}$, we arrive at a tableau $\sigma' \tau' d'$ whose first row is the same as in τd and, as previously, each pair of symbols in the same row of τd is not in the same column of $\sigma' \tau' d'$. The same holds then for the tableaux resulting from the two given by discarding their first rows. Applying the induction hypothesis, we will find that $u = v$ and $n_1 = n_1'$, $n_2 = n_2'$, ..., $n_u = n_u'$, a contradiction.

So, let a, b be two symbols in the same row of τd and in the same column of $\tau' d'$. Let $\pi = (a, b) \in S_n$. Then $\pi \in R_{\tau d}$, $\pi \in C_{\tau' d'}$. We note that if $\rho \in R_{\tau d}$ and $\sigma \in C_{\tau' d'}$ then $\rho e_{\tau d} = e_{\tau d}$ and $e_{\tau' d'} \sigma = \varepsilon_\sigma e_{\tau' d'}$. In this case

$$e_{\tau' d'} e_{\tau d} = e_{\tau' d'} \pi^{-1} \pi e_{\tau d} = -e_{\tau' d'} e_{\tau d}.$$

Hence $e_{\tau' d'} e_{\tau d} = 0$. □

Lemma 3.5. *Let κ be an element in A such that $\rho \kappa \sigma = \varepsilon_\sigma \kappa$ for all $\sigma \in R_{\tau d}$ and $\sigma \in C_{\tau d}$. Then $\kappa = r e_{\tau d}$ where $r \in k$.*

Proof. We represent κ in the form $\kappa = \sum_{\gamma \in S_n} a_\gamma \gamma$, where $a_\gamma \in k$. Then for any $\rho \in R_{\tau d}$, $\sigma \in C_{\tau d}$, we will have

$$\begin{aligned} \kappa = \varepsilon_\sigma \rho^{-1} \kappa \sigma^{-1} &= \varepsilon_\sigma \sum_\gamma a_\gamma (\rho^{-1} \gamma \sigma^{-1}) \\ &= \varepsilon_\sigma \sum_\chi a_{\rho \chi \sigma} \chi. \end{aligned}$$

Hence, for all $\rho \in R_{\tau d}$, $\sigma \in C_{\tau d}$, we have

$$a_\gamma = \varepsilon_\sigma a_{\rho \gamma \sigma}. \tag{3.2}$$

For $\gamma = 1$, we obtain $a_{\rho \sigma} = \varepsilon_\sigma a_1$, $\sigma \in C_{\tau d}$. Suppose γ has no form $\rho \sigma$ where $\rho \in R_{\tau d}$, $\sigma \in C_{\tau d}$. By Lemma 3.3, there exist symbols a, b in the same row of τd and in the same column of $\gamma \tau d$. We consider $\pi = (a, b) \in S_n$. Then $\pi \in R_{\tau d}$, $\pi \in C_{\gamma \tau d}$. For a suitable $\delta \in C_{\tau d}$, we may then write $\pi = \gamma \delta \gamma^{-1}$. By equation (3.2),

$$a_\gamma = \varepsilon_\sigma a_{\pi \gamma \delta^{-1}} = \varepsilon_\pi a_\gamma = -a_\gamma.$$

Hence $a_\gamma = 0$. □

Corollary 3.6. *$e_{\tau d}^2 = y e_{\tau d}$, where y is a nonzero rational number.*

Proof. As we remarked already, $\rho e_{\tau d} = e_{\tau d}$ and $e_{\tau d} \sigma = \varepsilon_\sigma e_{\tau d}$, where $\rho \in R_{\tau d}$, $\sigma \in C_{\tau d}$. Then

$$\rho e_{\tau d}^2 \sigma = \rho e_{\tau d} e_{\tau d} \sigma = \varepsilon_\sigma e_{\tau d}^2.$$

By Lemma 3.5, $e_{\tau d}^2 = \gamma e_{\tau d}$ where γ is an integer coefficient of 1 in $e_{\tau d}^2$. Let us show that $\gamma \neq 0$. For this we consider a linear operator $\varphi : A \to A$ such that $\varphi(x) = xe_{\tau d}$. We compute its trace in the natural basis $\sigma_1 = 1, \sigma_2, \dots, \sigma_{n!}$ of A. If $e_{\tau d} = a_1\sigma_1 + \cdots$ then $\sigma_i e_{\tau d} = \cdots + a_1\sigma_i + \cdots$. Hence $\operatorname{tr}\varphi = a_1 n!$. Moreover, $a_1 = 1$ since $\sigma_1 = 1$ enters the above representation of $e_{\tau d}$ with coefficient 1. We remark that $Ae_{\tau d}$ is the linear span of elements of the form $\mu e_{\tau d}$, $\mu \in S_n$. Thus, by a transformation with rational coefficients, we may pass to a new basis $\{v_1, \dots, v_n\}$ such that m first elements $v_1, \dots, v_m$ form a basis of $Ae_{\tau d}$. Obviously, $m \geq 1$. We compute $\operatorname{tr}\varphi$ in this new basis. Since for any $\kappa \in Ae_{\tau d}$ we have $\kappa e_{\tau d} = \gamma\kappa$ and since $Ae_{\tau d}$ is a left ideal of A, we get $\operatorname{tr}\varphi = m\gamma$. Thus $m\gamma = n!$ and $\gamma \neq 0$. □

Theorem 3.7. *For any partition d and any tableau τd, the left ideal $Ae_{\tau d}$ is a minimal left ideal of $A = kS_n$. Two S_n-modules $Ae_{\tau d}$ and $Ae_{\tau' d'}$ are isomorphic if and only if $d = d'$. Any simple S_n-module is isomorphic to a module of the form $Ae_{\tau d}$.*

Proof. We put $e = \gamma^{-1}e_{\tau d}$ (see Lemma 3.5). Then $e^2 = e$. We will show that any ideal of the form $Ae = Ae_{\tau d}$ is minimal. Assuming to the contrary, we may write Ae as the direct sum $Ae_1 \oplus Ae_2$. Indeed, by Maschke's theorem, $Ae = A_1 \oplus A_2$. Let $e_1 \in A_1$ and $e_2 \in A_2$ be such that $e = e_1 + e_2$. Then $e_1 = ae$, hence $e_1 e = e_1$. Similarly, $e_2 e = e_2$. Applying this and the equation

$$e = e^2 = e(e_1 + e_2) = ee_1 + ee_2,$$

we obtain $e_1 = ee_1$ and $e_2 = ee_2$. Hence $e_1e_2 = e_1(e - e_1) = e_1 - e_1^2 \in Ae_1$. Clearly, $e_1e_2 \in Ae_2$. Therefore $e_1e_2 = 0$. Thus we have succeeded in finding nonzero elements $a = ee_1e$ and $b = ee_2e$ in eAe such that $ab = e_1e_2 = 0$. Now we remark that $D = eAe = e_{\tau d}Ae_{\tau d}$ is a skew field. Indeed, suppose $x = e_{\tau d}ue_{\tau d}$ is a nonzero element in D. Then for any $\rho \in R_{\tau d}$, $\sigma \in C_{\tau d}$, we may write

$$\rho x\sigma = \rho e_{\tau d}ue_{\tau d}\sigma = e_{\tau d}ue_{\tau d}\varepsilon_\sigma = \varepsilon x.$$

By Lemma 3.5, $x = \lambda e_{\tau d}$ where $\lambda \in k$. Hence $D \cong ke_{\tau d} = ke$, that is, D is isomorphic to k, a contradiction. Thus $Ae_{\tau d}$ is a minimal left ideal in A, i. e., a simple S_n-module.

Now we prove that $d \neq d'$ implies $Ae_{\tau d} \not\cong Ae_{\tau' d'}$. Assuming the opposite, there exists an isomorphism φ from $Ae_{\tau d}$ into $Ae_{\tau' d'}$, $d < d'$. Then $\varphi(e_{\tau d}) = a$ and, $e_{\tau d}$ being a rational multiple of an idempotent, we have $Ae_{\tau' d'} = Ae_{\tau d}a$, in particular, $e_{\tau' d'} = be_{\tau d}a$. Furthermore,

$$e_{\tau' d'} = \gamma^{-1}e_{\tau' d'}^2 = \gamma be_{\tau d}ae_{\tau' d'}.$$

We will arrive at a contradiction if we show that $e_{\tau d}ae_{\tau' d'} = 0$. Now this is easy. We restrict ourselves to the case where $a \in S_n$. Then $e_{\tau d}ae_{\tau' d'}a^{-1}a = e_{\tau d}e_{a\tau' d'}a$. By Lemma 3.4, we have $e_{\tau d}e_{a\tau' d'} = 0$. □

3.2.4 Decomposing kS_n

The theorem just proved gives the complete description of isomorphism classes of simple S_n-modules over arbitrary fields of characteristic zero. Obviously, the description of minimal left ideals in the group algebra kS_n is far more difficult. We will exhibit here a subset Y in the set of all Young tableaux τd such that

$$kS_n = \bigoplus_{\tau d \in Y} kS_n e_{\tau d}. \tag{3.3}$$

A Young tableau is called *standard* if each of its rows and each of its columns contains its entries in the increasing order. We say that $\tau d > \tau' d'$ if $(\tau(1), \dots, \tau(n))$ is lexicographically greater than $(\tau'(1), \dots, \tau'(n))$, that is, the first nonzero number in the sequence $\tau(i) - \tau'(i)$, $i = 1, \dots, n$, is positive. Let b_d stand for the number of standard tableaux of the form τd. The following result is quite similar to Lemma 3.4.

Lemma 3.8. *Let τd, $\tau' d$ belong to the set Y of standard tableaux. If $\tau d > \tau' d$ then $e_{\tau' d} e_{\tau d} = 0$.*

Proof. Using the proof of Lemma 3.4, we can see that it is enough to find a pair of symbols in the same row of τd and in the same column of $\tau' d$. Let (p, q) be the number of the first box where τd and $\tau' d$ differ. Suppose this box is occupied by s in τd and by t in $\tau' d$. We have $s > t$. Let us find the number (u, v) of the box occupied by t in τd. The tableaux being standard, $p \geq u$, $q \geq v$. Since (p, q) is the first place where the tableaux diverge, we must have $u > p$, $v > q$. In this case the element r in the box with number (u, q) in both tableaux is the same. Thus s, r belong to the same row of τd and of the same column of $\tau' d$. □

Theorem 3.9. *The group algebra kS_n of the symmetric group S_n over the field k of characteristic zero is decomposable as the direct sum of minimal left ideals generated by idempotents $e_{\tau d}$ corresponding to the standard Young tableaux.*

Proof. By Section 3.2.1(ii), it is sufficient to show that $\sum b_d^2 = n! = |S_n|$ and that the sum in equation (3.3) is direct. In fact, using Theorem 3.7, it suffices to prove that the sum

$$kS_n e_{\tau_1 d} + \cdots + kS_n e_{\tau_q d}$$

taken over standard tableaux corresponding to the same Young diagram d is direct. Suppose $\tau_1 d > \cdots > \tau_q d$. If

$$a_1 e_{\tau_1 d} + a_2 e_{\tau_2 d} + \cdots + a_q e_{\tau_q d} = 0$$

then, multiplying by $e_{\tau_q d}$ on the left, we obtain $a_q e_{\tau_q d}^2 = 0$. Proceeding in the same way, we arrive at the desired result. As concerns the equation $\sum_q b_d^2 = n!$, this is left to the reader using hints in Exercise 3.5.4. □

We remark that, by the above, the number b_d of standard Young tableaux corresponding to a given diagram d is equal to the dimension of the unique simple S_n-module T_d associated with d.

3.2.5 Some formulas

This subsection is devoted to giving, without proof, some formulas for the dimension b_d and the character χ_d of an irreducible S_n-module associated with the diagram d and to describing the Richardson–Littlewood algorithm for decomposing tensor products of irreducible S_n-modules. So, let d be a Young diagram corresponding to a partition of n of the form $n = n_1 + n_2 + \cdots + n_q$, where $n_1 \geq n_2 \geq \cdots \geq n_q$. With each box (i,j) on the intersection of the ith row with the jth column, we associate a *hook* as a subdiagram consisting of the part of the ith row to the right of our box and the part of the jth column below our box, including the box itself. The number of boxes in the hook h_{ij}^d is called its length and is denoted by $|h_{ij}^d|$. Now the *dimension formula* is

$$b_d = \frac{n!}{\prod_{(i,j)\in d} |h_{ij}^d|}. \tag{3.4}$$

To write down the *character formula*, we define the *rim* r_{ij}^d of d corresponding to a hook h_{ij}^d as the set of all boxes (p,q) such that $p \geq i$, $q \geq j$. Now the leg l_{ij}^d of the hook h_{ij}^d is the part of the column within h_{ij}^d excluding (i,j). Its length is denoted by $|l_{ij}^d|$. We note that if we discard from a diagram a rim of a hook we arrive at a Young diagram. The recurrence formula for $\chi_d(\pi)$, $\pi \in S_n$, is as follows. Let the decomposition of $\pi \in S_n$ contain a cycle of length t. We denote by π^* an element of S_{n-t} resulting from π by deleting the above mentioned cycle (and renumbering symbols, if necessary). Then

$$\chi_d(\pi) = \sum_{\substack{i,j \\ |h_{ij}^d|=t}} (-1)^{|l_{ij}^d|} \chi^{d\setminus r_{ij}^d}(\pi^*). \tag{3.5}$$

Now let d and e denote two diagrams of degrees m and n, respectively, with M and N being irreducible S_m- and S_n-modules associated with these diagrams. We consider S_m and S_n as subgroups in S_{m+n} of permutations acting on the set of symbols $\{1, \dots, m\}$ and $\{m+1, \dots, m+n\}$. Then $H = S_m S_n$ is a subgroup in S_{m+n} isomorphic to their direct product. We turn the vector space tensor product $P = M \otimes N$ into an H-module by putting

$$(\sigma\tau)(x \otimes y) = \sigma(x) \otimes \tau(y)$$

where $\sigma \in S_m$, $\tau \in S_n$, $x \in M$, $y \in N$. Then we consider the *induced* S_{m+n}-module $Q = kS_{m+n} \otimes_{kH} P$. By Maschke's theorem, Q is decomposable as the direct sum of irreducible S_{n+m}-modules. Their associated Young diagrams can be found using the *Richardson–Littlewood rule* as follows:

(i) Add to d boxes from the first row of e in such a way that the resulting Young diagram does not contain two new boxes in one column.
(ii) Apply the same procedure to the subsequent rows of e so that if a box (i,j) is joined to a certain row of d then the box $(i+1,j)$ should be joined to a row of d with a greater number.

Formally, this procedure can be described as follows. If $(i,j) \to (p,q)$, $(i_1,j_1) \to (p_1,q_1)$ ($\to$ means "is placed to") then we should have
(a) if $i = i_1, j \le j_1$ then $p \le p_1$;
(b) if $i = i_1, j < j_1, p = p_1$ then $q < q_1$;
(c) if $i = i_1, j < j_1, p < p_1$ then $q > q_1$;
(d) if $i < i_1, j = j_1$ then $q < q_1$.

As a result of all possible applications of this procedure, we obtain all the Young diagrams associated with irreducible S_{m+n}-modules entering the decomposition of Q.

3.3 Frobenius algebras and their modules

3.3.1 Definition and examples

A finite-dimensional associative algebra R with 1 over a field k is called *Frobenius* if there exists a bilinear associative nondegenerate function f on R with values in k. The associativity condition means that for all $x, y, z \in R$ we have

$$f(xy, z) = f(x, yz).$$

We remark that this condition is equivalent to the existence of a linear function φ such that its kernel $\ker\varphi$ does not contain nonzero right or left ideals of R. Indeed, if R is Frobenius with respect to f then we define $\varphi : R \to k$ by putting $\varphi(x) = f(x, 1)$, $x \in R$. If for a nonzero left ideal I we have $\varphi(I) = 0$ then for a nonzero $x \in I$ and any $y \in R$ we may write

$$0 = \varphi(yx) = f(yx, 1) = f(y, x),$$

i. e., x belongs to the right kernel of f. Similarly, $\ker\varphi$ contains no right ideals. If, conversely, R possesses a linear form φ with no right or left ideals in its kernel then we may put $f(x,y) = \varphi(xy)$. If, say, x is in the left kernel of f then $\{0\} = f(x, R) = \varphi(xR)$, whence $xR = \{0\}$, that is, $x = 0$.
Examples. (a) Any finite-dimensional algebra with a *monolith*, that is, a nonzero two-sided ideal contained in each nonzero two-sided ideal of R.

(b) The group algebra of a finite group G over a field of characteristic zero. If

$$x = \sum_{g \in G} \alpha_g g, \quad y = \sum_{g \in G} \beta_g g$$

then

$$xy = f(x, y) \cdot 1 + \sum_{g \neq 1} y_g g.$$

Theorem 3.10. *Let M be a free module of finite rank over a Frobenius algebra R. Then M is injective.*[1]

Proof. It is sufficient to show that R is injective as a right R-module. We note first that this module is isomorphic to the R-module R^* of all k-linear functions on the left R-module R. Indeed, let θ be a map from R into R^* given by $\theta(x)(y) = f(x, y)$. If $z \in R$ then $\theta(xz)(y) = f(xz, y) = f(x, zy) = \theta(x)(zy) = (\theta(x)z)(y)$, that is, θ is a homomorphism of right R-modules. Now we show that if R is considered as a left R-module then R^* is a right injective R-module. So we assume that P is a right R-module containing R^* as an R-submodule. We pass to the left R-module P^* and denote by $(R^*)^\perp$ the annihilator of R^* in P^*, that is, the set of all functions φ from P^* into k such that $\varphi(R^*) = \{0\}$. The restriction $\varphi : P \to k$ on R^* defines a homomorphism from P^* into $(R^*)^*$ and an isomorphism

$$P^*/(R^*)^\perp \cong (R^*)^* = R.$$

Since R is a free left R-module, we see that $(R^*)^\perp$ is a direct summand of P^*, that is, for a suitable R-submodule N ($\cong R$) we have $P^* = (R^*)^\perp \oplus N$. Passing to annihilators and using their usual properties, we may write

$$\{0\} = (P^*)^\perp = (R^*)^{\perp\perp} \cap N^\perp = R^* \cap N^\perp,$$
$$P = \{0\}^\perp = (R^* \cap N)^\perp = R^* + N^\perp.$$

Thus, $P = R^* \oplus N$. Now R^* is injective. □

3.3.2 Symmetric Frobenius algebras

A Frobenius algebra R over a field k is called *symmetric* if its associated form f is symmetric. For instance, the group algebra of a finite group over a field is symmetric. The associated form has been exhibited in Example (b) of Section 3.3.1.

Let M be a right R-module and $C = \operatorname{End}_R M$ its endomorphism algebra. Thus, a k-endomorphism $\varphi : M \to M$ is in C if and only if for any $r \in R$ and $x \in M$ we have $\varphi(xr) = \varphi(x)r$. We will be interested in the structure of M as a C-module. Let $M^* = \operatorname{Hom}_k(M, k)$ be the left R-module of k-forms on M with values in k. If $\theta \in M^*$, $r \in R$,

1 The definition of injective module is given in Section 2.5.6. As a matter of fact, it is equivalent to saying that any R-module N containing M splits, i. e., $N = M \oplus Q$, Q being an R-submodule in N.

$x \in M$ then $(r\theta)(x) = \theta(xr)$. We define an R-bilinear map $\tau : M \times M^* \to R$ as follows. Since $f : R \times R \to k$ is nondegenerate, we can find bases $\{x_1, \dots, x_n\}$ and $\{y_1, \dots, y_n\}$ in R such that $f(x_i, y_j) = \delta_{ij}$. We set

$$\tau(x, \psi) = \sum_{i=1}^{n} \psi(xx_i)y_i.$$

The R-bilinearity can be proved in the following way. Choose $a \in R$. From the associativity of f, it follows that if $A' = (\alpha'_{ij})$, $A'' = (\alpha''_{ij})$ are matrices of multiplication by a on the left in $\{x_1, \dots, x_n\}$ and on the right in $\{y_1, \dots, y_n\}$, respectively, then ${}^tA' = A''$. Hence we may write

$$\begin{aligned}\tau(xa, \psi) &= \sum_{i=1}^{n} \psi(xax_i)y_i = \sum_{i=1}^{n}\sum_{m=1}^{n} \psi(\alpha'_{mi}xx_m)y_i \\ &= \sum_{i=1}^{n}\sum_{m=1}^{n} \psi(xx_m)\alpha'_{mi}y_i = \sum_{m=1}^{n}\sum_{i=1}^{n} \psi(xx_m)\alpha''_{im}y_i \\ &= \sum_{m=1}^{n} \psi(xx_m)(y_m a) = \tau(x, \psi)a.\end{aligned}$$

Similarly, one may verify $\tau(x, a\psi) = a\tau(x, \psi)$. Now we prove that τ is nondegenerate. Suppose that $\tau(x, a\psi) = 0$ for all $x \in M$. From the linear independence of $\{y_1, \dots, y_n\}$ it follows that $\psi(xR) = \{0\}$. In particular, $\psi(x) = (0)$ for all x, hence $\psi = 0$. From τ being R-bilinear it follows that the linear span of $\tau(M, M^*)$ is a two-sided ideal in R called the *kernel* R_M of M. A right R-module M is called *τ-regular* if R_M is a ring with unity. We note that if R is the group algebra of a finite group over a field of characteristic zero then any two-sided ideal of this algebra is the direct sum of simple rings with 1 generated by central idempotents (Curtis and Reiner [95]; Skornyakov [345]). Hence in this case any module is τ-regular.

Lemma 3.11 (Double centralizer property). *Let M be a τ-regular module over a symmetric algebra R, $C = \operatorname{End}_R M$, $D = \operatorname{End}_C M$. Then for any $f \in D$ there exists an $r \in R$ such that $f(x) = xr$ for all $x \in M$.*

Proof. We define an endomorphism $\psi\Box x$ ($\psi \in M^*$, $x \in M$) by setting

$$(\psi\Box x)(y) = x\tau(y, \psi),$$

with y being an arbitrary element in M.

Since τ is R-bilinear, we may write ($r \in R$):

$$(\psi\Box x)(yr) = x\tau(yr, \psi) = x\tau(y, \psi)r = ((\psi\Box x)(y))r.$$

Now let

$$\varepsilon = \sum_i \tau(x_i^0, \psi_i^0)$$

be the unity element of R_M. We put

$$r = \sum_i \tau(f(x_i^0), \psi_i^0) \in R_M.$$

Then

$$yr = \sum_i y\tau(f(x_i^0), \psi_i) = f\Big(\sum_i y\tau(x_i^0, \psi_i^0)\Big) = f(y\varepsilon) = f(y).$$

Here we can conclude that $y\varepsilon = y$ is true since τ is nondegenerate. Indeed,

$$\tau(y\varepsilon - y, \psi) = \tau(y\varepsilon, \psi) - \tau(y, \psi) = \tau(y, \psi)\varepsilon - \tau(y, \psi) = 0$$

since ε is the unity in R_M and $\tau(y, \psi) \in R_M$ for all y, ψ. □

3.3.3 Submodules in M and ideals in R_M

If N is a C-submodule in a τ-regular R-module M then, due to R-bilinearity of τ, the set $\tau(N, M^*)$ is a left ideal of R contained in R_M. If I is a left ideal in R_M then, conversely, MI is a left C-submodule in M. We note that using this notation

$$I \supseteq \tau(MI, M^*) \quad \text{and} \quad N \supseteq M\tau(N, M^*).$$

Indeed, $\tau(MI, M^*) \subseteq \tau(M, M^*)I \subseteq I$ since I is a left ideal. Besides, if $x \in M$, $y \in N$ then $x\tau(y, \psi) = (\psi\square x)(y) \subseteq N$ since $\psi\square x \in C$ and N is a C-submodule.

Lemma 3.12. *If N is a direct summand of the C-module M then R_M contains an idempotent e such that $\tau(N, M^*) = Re$ and then $M\tau(N, M^*) = N$. If, conversely, e is an idempotent in R_M, $I = Re$, then $MI = Me$ is a direct summand in the C-module M and $\tau(MI, M^*) = I$.*

Proof. Let $\pi \in \mathrm{End}_C(M)$ denote the corresponding projection of M onto N. According to Lemma 3.11, $\pi(x) = xe$ where $e = \sum \tau(\pi(x_i^0), \psi_i^0) \in \tau(N, M^*)$. If $a = \sum \tau(y_i, \psi_i)$ is any element in $\tau(N, M^*)$, $y_i \in N$ then

$$ae = \sum \tau(y_i e, \psi_i) = a$$

since $y_i e = \pi(y_i) = y_i$. Hence, $\tau(N, M^*) = Re$ and $e^2 = e$. If $x \in N$ then $x = xe \in M\tau(N, M^*)$. By the above remark, $N = M\tau(N, M^*)$. Now let, conversely, $I = Re$ where e is an idempotent. Then $MI = MRe = Me$ and $M = Me + M(1-e)$, where Me and $M(1-e)$ are C-submodules in M. Suppose now $r \in I$. Then

$$r = \varepsilon r = \sum \tau(x_i^0 r, \psi_i^0) \in \tau(MI, M^*)$$

and, as previously, we get $I = \tau(MI, M^*)$. □

Before we formulate the main theorem on regular modules over symmetric algebras, we recall that an idempotent is called *primitive idempotent* if it cannot be written in the form $e = e_1 + e_2$ where e_1, e_2 are idempotents such that $e_1 e_2 = e_2 e_1 = 0$.

Theorem 3.13. *Let M be a right τ-regular R-module over a symmetric Frobenius algebra R. We assume that its kernel R_M is semisimple R_M-module. Denote $C = \mathrm{End}_R(M)$. Then*
(i) *the map $Re \to Me$ ($e^2 = e$, $e \in R_M$) is a bijection of the set of left ideals of R_M and the set of all C-submodules in M;*
(ii) *C-module Me is simple if and only if e is a primitive idempotent in R_M;*
(iii) *C-module M is semisimple.*

Proof. By semisimplicity each left ideal of R_M has the form Re, where e is an idempotent, $e^2 = e$, and an application of Lemma 3.12 gives the bijectivity of the correspondence between direct summands of C-module M and submodules of R_M-module R_M. Now let e be a primitive idempotent, that is, the equation $e = e_1 + e_2$ where e_1, e_2 idempotents with $e_1 e_2 = e_2 e_1 = 0$ is impossible. We consider Me. If N is a nonzero C-submodule in Me then, by Lemma 3.12,

$$Re = \tau(Me, M^*) \supseteq \tau(N, M^*).$$

Since, by primitivity of e, the ideal $Re = R_M e$ is minimal and $\tau(N, M^*) \neq \{0\}$, we have $Re = \tau(N, M^*)$. Applying Lemma 3.12 once more, we find $M\tau(N, M) = Me \subseteq N$. Hence Me is a simple module. Now R_M being semisimple as an R_M-module, we may write $R_M = \sum_i R_M e_i$, where all e_i are primitive idempotents. Whence $M = \sum_i Me_i$, where all the submodules Me_i are minimal. Hence M is semisimple and, in this case, each of its submodules is a direct summand. □

Proposition 3.14. *Under the hypothesis of Theorem 3.13, left ideals $I_1 = Re_1$ and $I_2 = Re_2$ in R_M are R-isomorphic if and only if their associated C-modules Me_1 and Me_2 are C-isomorphic. For any primitive idempotent e of R, either $Me = \{0\}$ or Me is a simple C-module.*

Proof. If $\alpha : Re_1 \to Re_2$ is an R-isomorphism and β is its inverse then $\alpha(e_1) = a$, $\beta(e_2) = b$. We define $\tilde{\alpha} : Me_1 \to Me_2$, $\tilde{\beta} : Me_2 \to Me_1$ by setting

$$\tilde{\alpha}(xe_1) = xe_1 a, \tilde{\beta}(ye_2) = ye_2 b, \quad \text{where } x, y \in M.$$

An immediate verification shows that $\tilde{\alpha}$ and $\tilde{\beta}$ are inverses of each other. It is obvious that they are C-homomorphisms. Now let, conversely, δ be a C-isomorphism from Me_1 into Me_2 and γ its inverse. Then

$$\bar{\gamma} : \sum \tau(x_i, \psi_i) \mapsto \sum \tau(\gamma(x_i), \psi_i) \quad \text{and}$$
$$\bar{\delta} : \sum \tau(x_i, \psi_i) \mapsto \sum \tau(\delta(x_i), \psi_i)$$

are correctly defined R-module isomorphisms inverse to each other. To prove the correctness, we note that if $\sum \tau(x_i, \psi_i) = 0$ then ($x_i \in Me_1$, $\psi_i \in M^*$):

$$0 = \gamma\Big(\sum M\tau(x_i, \psi_i)\Big) = \sum \gamma(\psi_i \square M)(x_i) = \sum(\psi_i \square M)\gamma(x_i)$$
$$= \sum M\tau(\gamma(x_i), \psi_i).$$

Now if $\varepsilon = \sum \tau(x_i^0, \psi_i^0)$ is the unity of R_M then

$$\sum \tau(\gamma(x_i), \psi_i) = \sum \tau(\gamma(x_i), \psi_i)\varepsilon$$
$$= \sum_j \tau\Big(\sum_{\substack{i,\psi \\ j}} \tau(\gamma(x_i), \psi_i)x_j^0, \psi_j^0\Big) = 0.$$

Further verifications are left to the reader.

For the proof of the second part, we take the unity of R_M in the form $\varepsilon = \sum \tau(x_i^0, \psi_i^0)$. As we remarked above (Lemma 3.11), $x\varepsilon = x$ for all $x \in M$. Consider $S = \mathrm{Ann}_R M$. Then for any $s \in S \cap R_M$ we have $s = \varepsilon s = \sum \tau(x_i^0, \psi_i^0)s = \sum \tau(x_i^0 s, \psi_i^0) = 0$. But for any $r \in R$ we have $r\varepsilon - \varepsilon r \in S \cap R_M$. Hence ε is a central idempotent of R. Now since $e = \varepsilon e + (1 - \varepsilon)e$, where εe and $(1 - \varepsilon)e$ are pairwise orthogonal idempotents, then for any primitive idempotent e we have either $e = \varepsilon e$ or $e = (1 - \varepsilon)e$. In the former case, applying Theorem 3.13, we see that Me is a simple C-module. In the latter case $Me = M\varepsilon(1 - \varepsilon)e = \{0\}$ as desired. □

3.3.4 Operators of symmetrization

Let V be an m-dimensional space over a field k, $G = \mathrm{GL}(V)$ the group of all invertible linear operators on V. We consider the nth tensor power $T_n(V)$ of V. Then $T_n(V)$ becomes a natural G-module if we define $g(v_1 \otimes \cdots \otimes v_n) = gv_1 \otimes \cdots \otimes gv_n$. Our aim is to determine all irreducible G-submodules of $T_n(V)$. To do this, we define an action of S_n on $T_n(V)$ by setting ($\sigma \in S_n$, $v_1, \ldots, v_n \in V$):

$$(v_1 \otimes \cdots \otimes v_n)\sigma = v_{\sigma(1)} \otimes \cdots \otimes v_{\sigma(n)}.$$

Then $T_n(V)$ becomes an R-module, where $R = kS_n$. The elements $\sum_{\sigma \in S_n} \alpha_\sigma \sigma$ considered as linear operators of $T_n(V)$ are called *operators of symmetrization* in $T_n(V)$. Now let $\bar{G}$ denote the image of G in $\mathrm{End}_k(T_n(V))$. Then our definition yields $\bar{G} \subseteq C = \mathrm{End}_R(T_n(V))$.

Lemma 3.15. *C is a linear span of $\bar{G}$.*

Proof. We will be using capital letters $I, J, \ldots$ to designate multiindices of the form $I = (i_1, i_2, \ldots, i_n)$, $1 \le i_1, i_2, \ldots, i_n \le m$. Then a basis of $T_n(V)$ consists of tensors $e_1 = e_{i_1} \otimes e_{i_2} \otimes \cdots \otimes e_{i_n}$ where $\{e_1, e_2, \ldots, e_m\}$ is a basis of V. Given $\sigma \in S_n$ and a multiindex I, we define $\sigma I = \sigma(i_1, \ldots, i_n) = (i_{\sigma(1)}, i_{\sigma(2)}, \ldots, i_{\sigma(n)})$. The proof of the lemma will be complete if we show that any linear function on C which is zero on all elements of $\bar{G}$ is zero on all elements of C. So, let $y \in C$. Then $y(e_I) = \sum_J y_{IJ}e_J, y_{IJ} \in k$. If $\sigma \in S_n$ then $\sum y_{\sigma I,\sigma J}e_{\sigma J} = y(e_{\sigma I}) = y(e_I\sigma) = y(e_I)\sigma = \sum y_{IJ}e_{\sigma J}$. Therefore, for any $\sigma \in S$, we have $y_{\sigma I,\sigma J} = y_{IJ}$. Let $g \in G$. Then g defines a linear operator $y^g \in C$ such that

$$y^g_{IJ} = g_{j_1i_1}g_{j_2i_2}\cdots g_{j_ni_n},$$

where (g_{ij}) is the matrix of a nondegenerate operator g in $\{e_1,\dots,e_m\}$. Now let φ denote a linear function on C equal to zero on all elements of the form y^g. Then $\varphi(y) = \sum \alpha_{IJ}y_{IJ}$, $\alpha_{IJ} \in k$, and the sum is expanded over all indices of the form (I,J) such that $(\sigma I,\sigma J) = (I,J)$. We may write

$$0 = \varphi(y^g) = \sum_{I,J} \alpha_{IJ}g_{j_1i_1}g_{j_2i_2}\cdots g_{j_ni_n}, \tag{3.6}$$

where (g_{ij}) is a nondegenerate matrix. The right-hand side of equation (3.6) is a value of a polynomial $P(x_{11},x_{12},\dots,x_{mm})$ at $x_{ij} = g_{ij}$, this polynomial having a form

$$P(x_{11},x_{12},\dots,x_{mm}) = \sum \beta_{k_{11},k_{12},\dots,k_{mm}}x_{11}^{k_{11}}x_{12}^{k_{12}}\cdots x_{mm}^{k_{mm}}.$$

According to our condition $y_{IJ} = y_{\sigma I,\sigma J}$ for all $\sigma \in S_n$, coefficient $\beta_{k_{11},k_{12},\dots,k_{mm}}$ is merely a multiple of any of the coefficients of the form α_{IJ} such that among the pairs $(j_1,i_1),(j_2,i_2),\dots,(j_n,i_n)$ the pair $(1,1)$ occurs k_{11} times, $(1,2)$ occurs k_{12} times, ..., (m,m) occurs k_{mm} times. By our assumption, we have, in the polynomial ring $k[x_{11},x_{12},\dots,x_{mm}]$,

$$P(x_{ij})\det(x_{ij}) = 0. \tag{3.7}$$

Indeed, if (g_{ij}) is a set of values for x_{ij} such that $\det(x_{ij}) \neq 0$ then, by our assumption, $P(g_{ij}) = 0$. Otherwise, $\det(g_{ij}) = 0$ and again the value of equation (3.7) is zero. Now since $\det(x_{ij})$ is a nonzero polynomial and the polynomial ring has no zero divisors, we conclude $P(x_{ij}) = 0$, i. e., for any collection $k_{11},k_{12},\dots,k_{mm}$ we have $\beta_{k_{11},k_{12},\dots,k_{mm}} = 0$. Thus all α_{IJ} are equal to zero. Hence φ is a zero function. □

Theorem 3.16. *Let V be a linear space over a field k of characteristic zero*

$$G = \mathrm{GL}(V), T_n(V) = \underbrace{V\otimes\cdots\otimes V}_{n},$$

the nth tensor power of the G-module V. Then $T_n(V)$ is a semisimple G-module and each simple submodule is isomorphic to a submodule of the form $T_n(V)e_{\tau d}$ where τd is a Young tableau of degree n. Simple submodules of the form $T_n(V)e_{\tau d}$ and $T_n(V)e_{\tau' d'}$ are G-isomorphic if and only if $d = d'$. For any tableau τd, the module $T_n(V)e_{\tau d}$ is nonzero if and only if the number of rows in d is at most $m = \dim V$.

Proof. According to Lemma 3.15, the G-submodules in $T_n(V)$ are precisely the C-submodules where $C = \mathrm{End}_R V$, $R = kS_n$. Since the group algebra $R = kS_n$ for S_n over a field of characteristic zero is symmetric Frobenius (see Example (b) in Section 3.3.1) all conclusions of Theorem 3.16, except that concerning the number of rows, trivially follow from Theorem 3.13 and Proposition 3.14. Now suppose the height of the first column I_1 of d is more than m. Then for any tensor e_1 the tensor $e_1 e_{\tau d}$ is a tensor of degree n alternating in I_1 variables on an m-dimensional space. Clearly, this tensor is zero. If, conversely, $l_1 \leq m$ then we pick out a tensor $p = (e_1 \otimes \cdots \otimes e_{l_1}) \otimes (e_1 \otimes \cdots \otimes e_{l_2}) \otimes \cdots \otimes (e_1 \otimes \cdots \otimes e_{l_t})$, where $l_1, l_2, \ldots, l_t$ are the heights of the columns of the diagram d. Then, obviously,

$$pe_d = cr(e_1 \wedge \cdots \wedge e_{l_1}) \otimes (e_2 \wedge \cdots \wedge e_{l_2}) \otimes \cdots \otimes (e_1 \wedge \cdots \wedge e_{l_t}),$$

where $\wedge$ denotes exterior product, $c = |C_d|$ the order of the subgroup leaving invariant the columns of d, $r = |R_d|$ the order of the subgroup leaving invariant the rows of d. Clearly, $pe_d \neq 0$. Taking $p\tau$ in place of p, we arrive at the same result for $e_{\tau d}$. □

3.3.5 The dimension formula

Now we give (without proof) a formula for the dimension of irreducible GL(V)-modules in $T_n(V)$. According to Theorem 3.16, any simple GL(V)-submodule has the form $V_d = T_n(V)e_d$. Then it can be represented in the form

$$V_d = T_n(V)e_d = T_n(V)kS_n e_d = T_n(V) \otimes_{kS_n} kS_n e_d = T_n(V) \otimes_{kS_n} T_d,$$

where T_d is an S_n-submodule corresponding to the Young diagram d. If $\chi = \chi^d$ is the character of this module then we also write $T_d = T_\chi$ and $V_d = V_\chi$. If $d = \{n_1, n_2 \ldots, n_r\}$, $n_1 + n_2 + \cdots + n_r = n$, $V_d = T_n(V)e_d$, $\dim V = m$ then

$$\dim V_d = \prod_{m \geq i \geq j \geq 1} \frac{n_j - n_i + i - j}{i - j}.$$

3.4 Structure of GL(V)-module $L_n(V)$

3.4.1 Categories, after Cohn [93]

By a category $\mathscr{C}$ we mean any class of objects $A, B, \ldots$ and sets of morphisms $\mathrm{Hom}(A, B)$ for any couple of objects $A, B \in \mathscr{C}$. If $\mu \in \mathrm{Hom}(A, B)$, also written as $\mu : A \to B$, and $\nu \in \mathrm{Hom}(B, C)$ then there exists their product, a morphism $\mu\nu \in \mathrm{Hom}(A, C)$. This (partial) operation is associative (if the factors are well-defined) and in each $\mathrm{Hom}(A, A)$ there is a unity 1_A, a morphism such that $\mu 1_A = \mu$, $1_A \nu = \nu$ if $\mu \in \mathrm{Hom}(A, B)$, $\nu \in \mathrm{Hom}(C, A)$.

In our case the role of $\mathscr{C}$ will be played by classes of modules over a fixed ring R and the role of $\mathrm{Hom}(A, B)$ by the sets of all homomorphism from A into B as R-modules.

A map $\mathscr{F} : \mathscr{C}_1 \to \mathscr{C}_2$ is called a *covariant functor* if, for any object $A \in \mathscr{C}_1$, $\mathscr{F}(A)$ is an object in $\mathscr{C}_2$ and for any morphism $\mu : A \to B$, $\mathscr{F}(\mu)$ is a morphism $\mathscr{F}(\mu) : \mathscr{F}(A) \to \mathscr{F}(B)$ and $\mathscr{F}(1_A) = 1_{\mathscr{F}(A)}$. In addition, if $\mu = \mu_1\mu_2$ then $\mathscr{F}(\mu) = \mathscr{F}(\mu_1)\mathscr{F}(\mu_2)$. If, instead, $\mathscr{F}(\mu) : \mathscr{F}(B) \to \mathscr{F}(A)$ and $\mathscr{F}(\mu_1\mu_2) = \mathscr{F}(\mu_2)\mathscr{F}(\mu_1)$ then we say that $\mathscr{F}$ is a *contravariant functor*.

Now let $\mathscr{F}$ and $\mathscr{G}$ be two covariant functors from $\mathscr{C}_1$ and $\mathscr{C}_2$. Functor *morphism* $f : \mathscr{F} \to \mathscr{G}$ is defined as a family of morphisms $f(A) : \mathscr{F}(A) \to \mathscr{G}(A)$ $A \in \mathscr{C}_1$, with the following property. If $\mu \in \mathrm{Hom}(A, B)$ is a morphism in $\mathscr{C}_1$ then the following diagram is commutative:

$$\begin{array}{ccc} \mathscr{F}(A) & \xrightarrow{f(A)} & \mathscr{G}(A) \\ {\scriptstyle \mathscr{F}(\mu)}\downarrow & & \downarrow{\scriptstyle \mathscr{G}(\mu)} \\ \mathscr{F}(B) & \xrightarrow{f(B)} & \mathscr{G}(B) \end{array}$$

Here commutativity means $\mathscr{G}(\mu)f(A) = f(B)\mathscr{F}(\mu)$. In the case of contravariant functors, the vertical arrows in the diagram should be reversed. Finally, if for each A in $\mathscr{C}_1$, $f(A)$ is an isomorphism, we say that f is an isomorphism of functors $\mathscr{F}$ and $\mathscr{G}$.

3.4.2 Functors L_n and C_n

Let k be a commutative ring with unity, V a k-module, $T(V) = \bigoplus_{n=0}^{\infty} T_n(V)$ the tensor algebra of V, $L(V) = \bigoplus_{n=1}^{\infty} L_n(V)$ the Lie subalgebra in $T(V)$ generated by V with respect to bracketing and graded by degrees of tensors. If V is a free k-module with basis X then $T(V) \cong A(X)$, $L(V) \cong L(X)$, $T_n(V) \cong A_n(X)$, $L_n(V) \cong L_n(X)$. It is obvious that L_n is a functor from the category of k-modules into itself.

Now suppose k contains $1/n$ and a primitive root ε of unity. Let Γ be a cyclic group of order n generated by τ acting in $T_n(V)$ by the rule

$$(x_1 \otimes x_2 \otimes \cdots \otimes x_n)\tau = x_2 \otimes x_3 \otimes \cdots \otimes x_1.$$

We consider k as a Γ-bimodule by putting

$$\tau\lambda = \varepsilon^{-1}\lambda, \quad \lambda\tau = \varepsilon\lambda \quad (\lambda \in k).$$

Let k_ε denote this Γ-bimodule, $k\Gamma$ the group ring of Γ and

$$C_n(V) = T_n(V) \otimes_{k\Gamma} k_\varepsilon.$$

Then, clearly, C_n is also a functor in the category of k-modules. As we know, $T_n(V)$ is a GL(V)-module and $L_n(V)$ its submodule. A natural structure of a G-module exists on $C_n(V)$.

Our aim in this section is the proof of a theorem stating that the two functors L_n and C_n are isomorphic over any ring k containing $1/n$ and a primitive root ε of unity (Theorem 3.20). Such a ring is an algebra over $\mathbb{Z}[1/n, \varepsilon]$ so it is sufficient to construct morphisms of functors $l_n : L_n \to C_n$, $c_n : C_n \to L_n$ defined over $\mathbb{Z}[1/n, \varepsilon]$. We start with a result as follows.

Lemma 3.17. *Let k be the field of characteristic zero containing ε,* $\dim V < \infty$*. Then* $\dim L_n(V) = \dim C_n(V)$.

Proof. According to a classical canonical isomorphism

$$T_n(V) \otimes_{k\Gamma} k_\varepsilon \cong \mathrm{Hom}_{k\Gamma}(k_\varepsilon, T_n(V)),$$

we may identify $C_n(V)$ with the subspace in $T_n(V)$ whose elements v satisfy $v\tau = \varepsilon v$. Then $C_n(V)$ is generated by the elements of the form (in the sequel we replace $\otimes$ by juxtaposition):

$$x_{i_1} x_{i_2} \cdots x_{i_n} + \varepsilon x_{i_n} x_{i_1} \cdots x_{i_{n-1}} + \cdots + \varepsilon^{n-1} x_{i_2} x_{i_3} \cdots x_{i_n} x_{i_1}$$

this element being zero if and only if $(x_{i_1} \cdots x_{i_n})\tau^l = x_{i_1} \cdots x_{i_n}$ for some $1 < l < n$ (x_i is an element of a basis for V). It is easy to see that the set of such nonzero elements is in a bijective correspondence with the set of basic associative words in the sense of Shirshov (Section 2.3.5). Indeed, a cyclic permutation of a word $x_{i_1} \cdots x_{i_n}$ is a basic word if and only if this word is not equal to its proper cyclic permutation. Now since the number of basic associative words of degree n (in a fixed finite set of variables) is $\dim L_n(V)$, our proof is complete. □

3.4.3 Operators l_n and c_n

As we have learnt from Section 3.3.4, the k-module $T_n(V)$ is endowed with a structure of a right S_n-module. We will choose in kS_n two idempotents c_n and l_n with the following property: $c_n l_n = l_n$ and $l_n c_n = c_n$. Then the following will be true.

Lemma 3.18. *If $c_n, l_n \in kS_n$ are such that $c_n^2 = c_n$, $l_n^2 = l_n$, $c_n l_n = l_n$, and $l_n c_n = c_n$ then $T_n(V)c_n$ and $T_n(V)l_n$ are isomorphic* GL(V)*-modules.*

Proof. Let $w \in T_n(V)c_n$. Then $w = uc_n$, $u \in T_n(V)$. We have

$$(wc_n)l_n = ul_n c_n l_n = ul_n^2 = ul_n = w.$$

Hence the restriction of l_n and c_n to $T_n(V)c_n$ and $T_n(V)l_n$ are inverses of each other. Furthermore, if $\dim V < \infty$ then, as we noted in Theorem 3.16, GL(V) belongs to the centralizer of the right action of S_n on $T_n(V)$. Hence $T_n(V)l_n$ and $T_n(V)c_n$ are isomorphic GL(V)-modules. □

We set

$$c_n = \frac{1}{n}\sum_{i=0}^{n-1} \varepsilon^{-i}\tau^i.$$

Then, as we saw in Lemma 3.17, $C_n(V) = T_n(V)c_n$. Easy calculations in the group ring of the cyclic group $\langle\tau\rangle$ prove that c_n is an idempotent. The definition and verification of the above properties of l_n (it will be introduced later) are far less obvious. To start we define two numbers $\operatorname{ind}\sigma$ and $\operatorname{def}\sigma$ (index and defect) for a permutation

$$\sigma = \begin{pmatrix} 1 & 2 & \dots & n \\ i_1 & i_2 & \dots & i_n \end{pmatrix}$$

as follows. We set $\operatorname{ind}\sigma = \sum_{i_t > i_{t+1}} t \pmod{n}$ and

$$\operatorname{def}\sigma = \operatorname{Card}\{t \mid i_t > i_{t+1} (1 \le t \le n)\} \pmod{n}.$$

As usual, $\operatorname{Card} M$ denotes the number of elements in M. In these definitions we assume that $i_{n+1} = i_1$, $i_{n+2} = i_2$, etc. if $\tau = (1, 2, \dots, n)$ then the following relations hold:

(i) $\operatorname{def}\sigma\tau = \operatorname{def}\sigma$;
(ii) $\operatorname{ind}\sigma\tau^i = \operatorname{ind}\sigma - i\operatorname{def}\sigma$;
(iii) $\operatorname{ind}\tau^i\sigma = \operatorname{ind}\sigma - i$;
(iv) $\operatorname{def}\sigma = 1$ if and only if $\sigma = \tau^i$.

(i) is true by definition. To prove (ii), it is sufficient to show that $\operatorname{ind}(\sigma\tau) = \operatorname{ind}\sigma - \operatorname{def}\sigma$. Let us write

$$\sigma\tau = \begin{pmatrix} 1 & 2 & \dots & n \\ j_1 & j_2 & \dots & j_n \end{pmatrix} = \begin{pmatrix} 1 & 2 & \dots & n \\ i_2 & i_3 & \dots & i_1 \end{pmatrix}.$$

Then

$$\operatorname{ind}(\sigma\tau) = \sum_{j_t > j_{t+1}} t = \sum_{i_{t+1} > t_{t+2}} t = \sum_{i_t > t_{t+1}} (t-1) = \operatorname{ind}\sigma - \operatorname{def}\sigma.$$

As previously, it is enough to show $\operatorname{ind}\tau\sigma = \operatorname{ind}\sigma - 1$. We write

$$\sigma\tau = \begin{pmatrix} 1 & 2 & \dots & n \\ j_1 & j_2 & \dots & j_n \end{pmatrix} = \begin{pmatrix} 1 & 2 & \dots & n \\ i_1+1 & i_2+1 & \dots & i_n+1 \end{pmatrix}.$$

Then

$$\operatorname{ind}\tau\sigma = \sum_{j_t > j_{t+1}} t = \sum_{i_t+1 > t_{t+1}+1} t = \sum_{\substack{i_t > t_{t+1} \\ i_t \ne n}} t + (m-1),$$

where $i_m = n$. Now

$$\operatorname{ind}\sigma = \sum_{i_t > i_{t+1}} t = \sum_{\substack{i_t > i_{t+1} \\ i_t \neq n}} t + m.$$

Hence $\operatorname{ind}\tau\sigma = \operatorname{ind}\sigma - 1$ as required. Finally, (iv) is immediate from the definition.

Now we are ready to introduce $l_n \in kS_n$. We set

$$l_n = \frac{1}{n} \sum_{\sigma \in S_n} \varepsilon^{\operatorname{ind}\sigma} \sigma.$$

We verify $l_n^2 = l_n$ and $l_n c_n = c_n$, $c_n l_n = l_n$. In doing so, we will be using relations (i)–(iv). We start with the last equation:

$$\begin{aligned} c_n l_n &= \frac{1}{n^2} \sum_{\sigma,i} \varepsilon^{-i+\operatorname{ind}\sigma} \tau^i \sigma = \frac{1}{n^2} \sum_{\rho,i} \varepsilon^{-1+\operatorname{ind}\tau^{-i}\rho} \rho \\ &= \frac{1}{n^2} \sum_{\rho,i} \varepsilon^{\operatorname{ind}\rho} \rho = l_n. \end{aligned}$$

Passing to the second relation, we have

$$\begin{aligned} l_n c_n &= \frac{1}{n^2} \sum_{\sigma,i} \varepsilon^{-i+\operatorname{ind}\sigma} \sigma \tau^i = \frac{1}{n^2} \sum_{\rho,i} \varepsilon^{-i+\operatorname{ind}\rho\tau^{-i}} \rho \\ &= \frac{1}{n^2} \sum_{\rho,i} \varepsilon^{-i+\operatorname{ind}\rho + i \operatorname{def}\rho} \rho = \frac{1}{n^2} \sum_{\rho} \varepsilon^{\operatorname{ind}\rho} \left(\sum_{i=0}^{n-1} \varepsilon^{i(\operatorname{def}\rho - 1)} \right) \rho. \end{aligned}$$

The inner sum is equal to zero for all ρ except those with $\operatorname{def}\rho = 1$. In the latter case it is equal n. Using (iii) and (v), we arrive at

$$l_n c_n = \frac{1}{n} \sum_{t=1}^{n-1} \varepsilon^{\operatorname{ind}\tau^t} \tau^t = c_n.$$

Finally, $l_n^2 = l_n l_n = l_n(c_n l_n) = (l_n c_n) l_n = c_n l_n = l_n$.

Proposition 3.19. $T_n(V) l_n = L_n(V)$.

Proof. Let X_σ denote the product $x_{\sigma(1)} \cdots x_{\sigma(n)}$. Then the action of l_n on $x_1 \cdots x_n \in T_n(V)$ can be described as follows:

$$X_1 l_n = \frac{1}{n} \sum_{\sigma} \varepsilon^{\operatorname{ind}\sigma} X_\sigma.$$

We extend the action of this operator to arbitrary collections $(u_1, u_2, \ldots, u_n)$ of homogeneous elements $u_i \in T_{m_i}(V)$, $m_i \geq 1$. This should be anticipated by introducing the *generalized index* of $\sigma = \left(\begin{smallmatrix} 1 & 2 & \cdots & n \\ i_1 & i_2 & \cdots & i_n \end{smallmatrix}\right)$ in the following way:

$$\operatorname{ind}\sigma = \sum_{i_t > i_{t+1}} (m_{i_1} + \cdots + m_{i_t}) \pmod{m}, \qquad m = \sum_{i=1}^{n} m_i.$$

If all $m_i = 1$, then this definition coincides with that given above. We set $U_\sigma = u_{\sigma(1)}u_{\sigma(2)}\cdots u_{\sigma(n)}$ and

$$(u_1, u_2, \ldots, u_n)l_n = \sum_{\sigma\in S_n} \frac{\varepsilon^{\operatorname{ind}\sigma} U_\sigma}{\prod_{1\le t<n}(1-\varepsilon^{m_{\sigma(1)}+\cdots+m_{\sigma(t)}})}. \tag{3.8}$$

If all $m_i = 1$ then

$$\prod_{t=1}^{n-1}(1-\varepsilon^{m_{\sigma(1)}+\cdots+m_{\sigma(t)}}) = \prod_{t=1}^{n-1}(1-\varepsilon^t) = n \tag{3.9}$$

and, therefore $(x_1, x_2, \ldots, x_n)l_n = x_1x_2\cdots x_n l_n$.

We note that

$$\begin{aligned}(u_1, \ldots, u_i, u_{i+1}, \ldots, u_n)l_n - (u_1, \ldots, u_{i+1}, u_i, \ldots, u_n)l_n \\ = (u_1, \ldots, [u_i, u_{i+1}], \ldots, u_n)l_{n-1}.\end{aligned} \tag{3.10}$$

Let A_σ and B_σ be coefficients of U_σ in the first and in the second summands of the left-hand side of equation (3.10). If i and $i+1$ are not neighbors in σ then $A_\sigma = B_\sigma$. Hence we may think that in all monomials of the left-hand side of equation (3.10) u_i and u_{i+1} are neighbors. So we must compute $A_\sigma - B_\sigma$ in the case where for some l we have $\sigma(l) = i$, $\sigma(l+1) = i+1$. Then we will have

$$\begin{aligned}A_\sigma - B_\sigma &= \frac{\varepsilon^{\operatorname{ind}\sigma}}{\prod_{1\le t<n}(1-\varepsilon^{m_{\sigma(1)}+\cdots+m_{\sigma(t)}})} - \frac{\varepsilon^{\operatorname{ind}\sigma+(m_{\sigma(1)}+\cdots+m_{\sigma(l)})}}{\prod_{1\le t<n}(1-\varepsilon^{m_{\sigma(1)}+\cdots+m_{\sigma(t)}})} \\ &= \frac{\varepsilon^{\operatorname{ind}\sigma}}{\prod_{1\le t<n}(1-\varepsilon^{m_{\sigma(1)}+\cdots+m_{\sigma(t)}})}(1-\varepsilon^{m_{\sigma(1)}+\cdots+m_{\sigma(l)}}) \\ &= \frac{\varepsilon^{\operatorname{ind}\sigma'}}{\prod_{1\le t<n,\ t\ne l}(1-\varepsilon^{m_{\sigma(1)}+\cdots+m_{\sigma(t)}})}\end{aligned}$$

where $\sigma' = \begin{pmatrix} 1 & \ldots & l-1 & l+1 & \ldots & n \\ \sigma(1) & \ldots & \sigma(l-1) & \sigma(l+1) & \ldots & n\end{pmatrix}$. Thus the coefficient is the same as that of $u_{\sigma(1)}\cdots[u_i, u_{i+1}]\cdots u_{\sigma(n)}$ on the right-hand side of equation (3.10).

The just proved relation enables us to say that given a collection of $u_i \in L_{m_i}(V)$ such that $\sum m_i = m$ then $(u_1, u_2, \ldots, u_t)l_t = (u_1u_2\cdots u_t)l_m$. Indeed, by linearity we may restrict ourselves to the case where u_i is a commutator. If $t = m$ then we arrive at equation (3.9). Otherwise there exists $u_1 = [v, w]$ and

$$(u_1, \ldots, [v, w], \ldots, u_t)l_t = (u_1, \ldots, vw, \ldots, u_t)l_{t+1} - (u_1, \ldots, wv, \ldots, u_t)l_{t+1}.$$

Using induction downwards over t (with basis $t = m$), we get

$$(u_1, \ldots, [v, w], \ldots, u_t)l_t = (u_1\cdots vw\cdots u_t)l_m - (u_1\cdots wv\cdots u_t)l_m$$

$$= (u_1 \cdots [v, w] \cdots u_t) l_m.$$

If $t = 1$, we get $ul_n = u$.

So, $L_n(V) \subseteq T_n(V)l_n$. It remains to verify that any element of the form

$$(x_1 x_2 \cdots x_n) l_n = \frac{1}{n} \sum_{\sigma \in S_n} \varepsilon^{\operatorname{ind} \sigma} X_\sigma \tag{3.11}$$

is in $L_n(V)$. Without loss of generality, we may assume $\dim V = n < \infty$. We note that, by Lemma 3.17 in the case of a field, $\dim C_n(V) = \dim L_n(V)$. Besides, $C_n(V) = T_n(V)c_n \cong T_n(V)l_n \supseteq L_n(V)$. Hence, in the case of fields, we have $T_n(V)l_n = L_n(V)$. Now we consider the field $k' = \mathbb{Q}(\varepsilon)$. Then equation (3.11) is a Lie element over k'. By the Specht–Wever criterion (Section 2.5, Theorem 2.24), we can see that equation (3.11) is invariant under the mapping $\varphi_n : x_1 x_2 \cdots x_n \mapsto (1/n)[x_1, [x_2, [\ldots, [x_{n-1}, x_n] \ldots]]]$. Then

$$(x_1 x_2 \cdots x_n) l_n = \frac{1}{n} \sum_{\sigma} \varepsilon^{\operatorname{ind} \sigma} \varphi_n(X_\sigma).$$

All the coefficients of this equation are in $\mathbb{Z}[1/n, \varepsilon]$ hence it is true over any arbitrary ring k provided it contains $1/n$ and ε. Hence $T_n(V)l_n \subset L_n(V)$. Finally, $T_n(V)l_n = L_n(V)$. □

Theorem 3.20. *Let k be a commutative ring with 1 containing $1/n$ and a primitive root of unity of degree n. Then the functors L_n and C_n from the category of k-modules into itself are isomorphic. If k is a field and $\dim V < \infty$ then $L_n(V)$ and $C_n(V)$ are isomorphic $\mathrm{GL}(V)$-modules.*

Proof. It is noted in Section 3.4.2 that $T_n(V)c_n = C_n(V)$. Now Proposition 3.19 says that $T_n(V)l_n = L_n(V)$. By Lemma 3.17, the application of c_n transforms $T_n(V)l_n$ into $T_n(V)c_n$ and the application of l_n is its inverse, these two maps being $\mathrm{GL}(V)$-isomorphisms in the case where k is a field and V is finite-dimensional. □

3.4.4 Frobenius reciprocity theorem

In the remaining sections we will need some more facts in addition to those given in Section 3.2.1. We will be using the notions of the character, scalar product of characters and some basic theorems concerning them (all this can be found in Kostrikin [178, Chapter 8, Section 4]). One more fact we will be using concerns induced modules. Let H be a subgroup of G, and V a simple H-module. We consider the *induced* G-module

$$V^G = kG \otimes_{kH} V.$$

Let T be a simple G-module, $m(T, V^G)$ denote the multiplicity of T in V^G, and $m(V, T)$ the multiplicity of H-module V in the H-module T.

Theorem 3.21. *Let k be a field of characteristic zero then*

$$m(T, V^G) = m(V, T).$$

Its proof can be found in Lang [199, Chapter 16, Sections 6 and 7].

3.4.5 Functor L_n in the case of a field of characteristic zero

As we remarked in Section 3.3.5, given an irreducible left S_n-module T_χ with character χ, the GL(V)-module $V_\chi = T_n(V) \otimes_{kS_n} T_\chi$ is either zero (if the corresponding Young diagram has more than m rows, $m = \dim V$), or it is irreducible. Moreover, any irreducible submodule over $G = \mathrm{GL}(V)$ can be obtained in this way (in $T_n(V)$). Now let Γ be a subgroup in S_n generated by a cycle τ of length n. Let $m(V_\chi, L_n(V))$ denote the multiplicity of V_χ in the GL(V)-decomposition of the G-module $L_n(V)$ which is a submodule in the semisimple module $T_n(V)$. By $m(k_\varepsilon, T_\chi)$ we denote the multiplicity of the Γ-module k_ε in Γ-module T_χ (the restriction of the action of S_n on its subgroup $\Gamma = \langle\tau\rangle \subset S_n$).

Proposition 3.22.

$$m(V_\chi, L_n(V)) = m(k_\varepsilon, T_\chi) = \frac{1}{n}\sum_{d|n}\mu(d)\chi(\tau^{n/d}).$$

Proof. By Lemma 3.18,

$$\begin{aligned} L_n(V) \cong C_n(V) \cong T_n(V) \otimes_{k\Gamma} k_\varepsilon = T_n(V) \otimes_{kS_n} (kS_n \otimes_{k\Gamma} k_\varepsilon),\\ m(V_\chi, L_n(V)) = m(V_\chi, C_n(V)) = m(T_\chi, kS_n \otimes_{k\Gamma} k_\varepsilon). \end{aligned}$$

By Frobenius reciprocity theorem (Section 3.4.4), the multiplicity of irreducible representation T_χ in the induced module $kS_n \otimes_{k\Gamma} k_\varepsilon$ is equal to the multiplicity of k_ε in the restriction of T to Γ. Hence

$$m(T_\chi, kS_n \otimes_{k\Gamma} k_\varepsilon) = m(k_\varepsilon, T_\chi) = \frac{1}{|\Gamma|}\sum_{i=0}^{n-1}\varepsilon^i\chi(\tau^i). \tag{3.12}$$

If the greatest common divisor (n, i) and (n, j) are the same then τ^i and τ^j have the same cycle decomposition, i. e., they are conjugate in S_n. Then we have $\chi(\tau^i) = \chi(\tau^j)$ and equation (3.12) can be rewritten as follows:

$$\frac{1}{n}\sum_{d|n,(t,\frac{n}{d})=1}\varepsilon^{dt}\chi(\tau^d) = \frac{1}{n}\sum_{d|n}\Big(\sum_{(t,\frac{d}{n})=1}\varepsilon^{dt}\Big)\chi(\tau^d).$$

By the example in Section 3.1.2, the sum of all primitive roots of unity of degree n is equal to $\mu(d)$. Thus we finally get

$$m(V_\chi, L_n(V)) = \frac{1}{n}\sum_{d|n}\mu\Big(\frac{n}{d}\Big)\chi(\tau^d),$$

which is, clearly, equivalent to what we are proving. □

Corollary 3.23. *Let $\varphi(n)$ denote the Euler function of n. Then*

$$m(V_\chi, L_n(V)) \leq (\dim T_\chi)/\varphi(n).$$

In particular, the symmetric and exterior powers of the GL(V)*-module V do not enter the decomposition of the* GL(V)*-module $L_n(V)$ where $n > 2$.*

Proof. Since any irreducible representation is defined over k and has the form $\tau \mapsto \varepsilon^i$, T_χ can be decomposed as the direct sum of the modules of the form k_{ε^i}. According to Proposition 3.22, $m(k_\varepsilon, T_\chi)$ does not depend on the choice of the primitive root ε of unity. Thus

$$\begin{aligned}\dim T_\chi &= \sum_{i=0}^{n-2} m(k_{\varepsilon^i}, T_\chi) \geq \sum_{(i,n)=1} m(k_{\varepsilon^i}, T_\chi)\\ &= \varphi(n) m(V_\chi, L_n(V)),\end{aligned}$$

proving the first part of the corollary. Now we may remark that the symmetric power corresponds to the idempotent $e_1 = \frac{1}{n!}\sum_{\sigma\in S_n}\sigma$ and the exterior one to $e_2 = \frac{1}{n}\sum_{\sigma\in S_n}\varepsilon_\sigma\sigma$, that is, to two one-dimensional representations of S_n. From the above inequalities, we derive that the multiplicity of either of these two representations does not exceed $1/\varphi(n)$. If $n > 2$ then this multiplicity is obviously zero. □

Theorem 3.24. *For $n \neq 4, 6$, any irreducible representation of* GL(V) *that enters the decomposition of $T_n(V)$ and is different from the symmetric and exterior powers of V also enters the decomposition of the* GL(V)*-module $L_n(V)$.*

Proof. According to the equation $m(V_\chi, L_n(V)) = m(k_\varepsilon, T_\chi)$, it is sufficient to prove that any irreducible representation $T = T_\chi$ of the symmetric group S_n contains faithful irreducible representation of $\Gamma = \langle\tau\rangle$ in all cases except

(i) $\dim T = 1$;

(ii) T is associated with the diagram

$d =$ 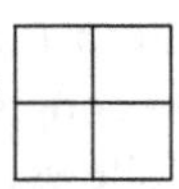, $n = 4$;

(iii) T is associated with the diagram

$d =$ 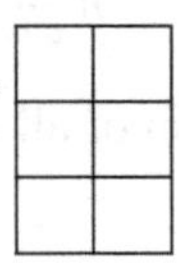, $n = 6$.

For the proof, we start with the case where $n = p^\alpha$.

(a) Consider $n = p^\alpha$, where p is a prime. In this case if T does not contain a faithful irreducible representation of $\Gamma = \langle\tau\rangle$ of order p^α then the image of $\tau^{p^{\alpha-1}}$ is necessarily

1. Hence T is not an faithful representation. But, except for $n = 4$, S_n has no non-trivial normal subgroups different from A_n. Thus, if $n \neq 4$, we arrive at (i). If $n = 4$ then, applying Proposition 3.22 and the fact that Klein's group V_4 belongs to the kernel of the two-dimensional representation associated with the diagram from (ii), we get $m(V_\chi, L_n(V)) = 0$.

(b) Now suppose $n = dq$, where $q = p^\alpha$, $(q, d) = 1$. Then τ^d factors into the product of independent cycles of lengths p^α: $\tau^d = \tau_1\tau_2 \cdots \tau_d$ and these cycles are cyclically permuted by the inner automorphism of S_n induced by τ^d. We set $G = \langle\tau_1\rangle \times \langle\tau_2\rangle \times \cdots \times \langle\tau_d\rangle \subset S_n$. The restriction of T to G decomposes as the direct sum of one-dimensional representations. At least one of them is faithful on $\langle\tau^d\rangle$ (this has the order p^α). Let χ be the character and v a basis of this representation. Then $\chi(\tau_t) = \xi_t$, $\xi_t^q = 1$ and $\chi(\tau^d) = \chi_1 \cdots \chi_t = \chi$ the primitive root of 1 of degree q. Replacing χ by a suitable character $\chi^\sigma : \alpha \mapsto \chi(\sigma^{-1}\alpha\sigma)$ $(\sigma \in S_n)$, we may think that

$$(\xi_1, \ldots, \xi_d) = (\underbrace{\tilde{\xi}_1, \ldots, \tilde{\xi}_1}_{d_1}, \ldots, \underbrace{\tilde{\xi}_m, \ldots, \tilde{\xi}_m}_{d_m}), \tag{3.13}$$

where $\tilde{\xi}_i \neq \tilde{\xi}_j$ for $i \neq j$ and, as before, $\tilde{\xi}_1 \cdots \tilde{\xi}_m$ is the primitive root of unity of degree q. Indeed, for any permutation $(i_1, i_2, \ldots, i_d)$ of indices $(1, 2, \ldots, d)$ there exists a permutation $\sigma_1 \in S_n$ such that $\sigma_1^{-1}\tau_t\sigma_1 = \tau_{i_t}$. If $m \geq 1$ then the characters $\chi^{\tau^{qt}}$, $t = 0, 1, \ldots, d-1$, are pairwise different. Indeed,

$$\chi^{\tau^{qt}}(\tau_i) = \chi(\tau^{-qt}\tau_i\tau^{qt}) = \chi(\tau_{i+1}) = \xi_{i+t}.$$

In this case the collections of eigenvalues of operators $\tau_1, \ldots, \tau_d$ on the subspaces $\langle v\rangle, \langle\tau^q(v)\rangle, \ldots, \langle\tau^{q(d-1)}(v)\rangle$ are pairwise different. Thus the system of elements of the form $\tau^{qt}(v)$ is linearly independent. Let ε be a primitive root of 1 of degree d. Then

$$u = v + \varepsilon^{-q}\tau^{-q}(v) + \cdots + \varepsilon^{-q(d-1)}\tau^{q(d-1)}(v)$$

is a nonzero vector and $\tau^q(u) = \varepsilon u$. Since $\tau^q(u) = \xi u$, where ξ is a primitive root of unity of degree q, we see that u is an eigenvector for τ, the eigenvalue being equal to the primitive root of unity of degree $qd = n$. Now it remains to consider the case where in (3.13) we have $m = 1$, that is, all $\xi_1, \ldots, \xi_d$ are equal to $\tilde{\xi}$. Then $\tilde{\xi}^d = \tilde{\xi}$. We remark that if $r_1, \ldots, r_d$ are numbers coprime to q then one can choose a permutation ρ such that $\rho^{-1}\tau_i\rho = \tau_i^{r_i}$, that is, $\chi^\rho(\tau_i) = \xi_t^{r_t}$, $i = 1, \ldots, d$. Now if $q > 3$ then the numbers r_t can be chosen so that at least two of them are different modulo q and $\chi^\rho(\tau^d) = \chi(\tau^d)$. Indeed, if $q \neq 2^\alpha$, then we may put $r_1 = -2$, $r_2 = 4$, $r_t = 1$ for $t > 2$; if $q \neq 3^\alpha$, then we may put $r_1 = -1$, $r_2 = 3$, $r_t = 1$ for $t > 2$. Then

$$\chi^\rho(\tau^d) = \tilde{\xi}^{r_1} \cdots \tilde{\xi}^{r_d} = \tilde{\xi}^d = \tilde{\xi} = \chi(\tau^d).$$

This brings us back to the case where $m > 1$.

The only remaining case is where $n = 6$. This can be dealt with by direct calculations using Proposition 3.22 and some formulas in Section 3.2.5. □

3.5 Exercises

3.5.1. For what n is the kS_n-module $T_n(V)$ faithful?

3.5.2. Find the multiplicities of irreducible components in $L_6(V)$ where $\dim V = n$ for different values of n.

3.5.3. Let G be a finite group and $\rho : G \to \mathrm{GL}(V)$ an irreducible representation of G. By Schur's lemma,

$$\sum_{g,h \in G} \rho(ghg^{-1}h^{-1}) = \lambda\, \mathrm{id}_V .$$

Find λ.

3.5.4. Let b_d denote the number of standard tableaux of order n corresponding to the Young diagram d. Prove that

$$\sum_d b_d^2 = n!.$$

3.6 Comments

The first three sections of this chapter are classical and can be found in other books (Weyl [369]; Bourbaki [80]; Curtis and Rainer [95]; James [153]). Applications of this theory to identical relations are still rather small in number (Volichenko [360, 363]; Drensky [105, 107]; Mishchenko [230, 232]). There is no doubt that this number will constantly grow. The results of Section 3.4 are due to Klyačko [172]. Formulas for the dimensions of irreducible modules are borrowed from Kerber [164] and Weyl [369].

4 Varieties of Lie algebras

4.1 Identical relations and varieties

4.1.1 Definition of variety

Let k be a commutative ring with 1. A class $\mathbf{V}$ of Lie algebras over k is called a *variety* if there exists a set of identical relations $\{v \equiv 0 \mid v \in V\}$ where V is a collection of nonassociative polynomials such that a Lie algebra G is in $\mathbf{V}$ if and only if G satisfies all the identities of the above set. We note that V can be viewed as a subset of a fixed free Lie algebra $L(X)$ where $X = \{x_1, x_2, \ldots, x_n, \ldots\}$ is some countable set of variables. This is immediate since the left-hand side of each identity depends on finitely many variables. The set of elements λ in k such that $\lambda x \equiv 0$ is an identity of $\mathbf{V}$ is an ideal of k and it is denoted by $\operatorname{Ann}\mathbf{V}$. A variety is called *nontrivial* if $\operatorname{Ann}\mathbf{V} \neq k$. In a trivial variety we have an identity of the form $x \equiv 0$ so it consists of zero algebra only.

If $\mathbf{V}$ is a variety of Lie algebras and V an arbitrary set of nonassociative polynomials defining $\mathbf{V}$ then it is clear that the verbal ideal $V(G)$ (see Section 2.2.6) of a Lie algebra G is not a function of V but solely of $\mathbf{V}$. Thus we may write $\mathbf{V}(G) = V(G)$. This ideal will be called *verbal corresponding to variety* $\mathbf{V}$ (or to the set of nonassociative polynomials V).

4.1.2 Free algebra of a variety

Let $\mathbf{V}$ be a variety defined by a set of nonassociative polynomials V. For a nonempty set Y, let $L(Y)$ denote a free Lie algebra with free generating set Y. We consider the verbal ideal $V(L(Y))$ in $L(Y)$. Then, clearly, $L(Y)/V(L(Y)) \in \mathbf{V}$. Moreover, given a homomorphism π from $L(Y)$ into an algebra in $\mathbf{V}$, we have $V(L(Y)) \subseteq \ker \pi$. Thus an algebra of the form

$$L(Y, \mathbf{V}) = L(Y)/V(L(Y))$$

possesses the following universal property.

Proposition 4.1. *For any map φ from Y into $G \in \mathbf{V}$ there exists a unique homomorphism $\bar{\varphi} : L(Y, \mathbf{V}) \to G$ extending φ.*

We call $L(Y, \mathbf{V})$ the *free algebra of $\mathbf{V}$ with free generating set Y*. It is convenient to identify Y with its image in $L(Y, \mathbf{V})$.

It follows from Proposition 4.1 that if Y and Z are bijective then $L(Y, \mathbf{V}) \cong L(Z, \mathbf{V})$. The converse is also true (see Proposition 4.2). Another obvious corollary of the proposition is that any relation among free generators of a free Lie algebra of a variety is an identical relation in this variety.

https://doi.org/10.1515/9783110566659-004

Proposition 4.2. *If* **V** *is a nontrivial variety of Lie algebras over a ring k, and Y, Z two nonempty sets, then* $L(Y, \mathbf{V}) \cong L(Z, \mathbf{V})$ *if and only if Y, Z have the same cardinality.*

Proof. We noted already that the condition is sufficient. To prove its necessity, we consider a maximal ideal M of k containing $\operatorname{Ann}\mathbf{V}$ and define $U \subseteq L(\{x_1, x_2\})$ as the set of monomials of the form λx_1, $\lambda \in M$, and x_1x_2. Then the quotient algebra $L(Y, \mathbf{V})/U(L(Y, \mathbf{V}))$ is isomorphic to the abelian algebra $(k/M)Y$ and $L(Z, \mathbf{V})/U(L(Z, \mathbf{V})) \cong (k/M)Z$. But $(k/M)Y$ and $(k/M)Z$ are vector spaces over the field k/M with bases Y and Z. From the theory of vector spaces, we know that in such a case Y and Z, the bases of the same (up to isomorphism) vector spaces, have the same cardinality. □

According to this proposition, we may speak, for any cardinal number m, of a free algebra of a variety **V** over k of *rank of free algebra m*. This algebra will be denoted by $L(m, \mathbf{V})$.

There is no difficulty in verifying that any variety **V** over k is a class of Lie algebras which is closed under taking subalgebras, quotient algebras and Cartesian products of algebras from **V**. That the converse is true is the contents of a theorem due to G. Birkhoff. The proof below gives another way of constructing a free algebra of a variety.

Theorem 4.3 (Birkhoff). *A nonempty class* **K** *of Lie algebras over a commutative ring k is a variety if and only if it is closed under forming subalgebras, homomorphic images, and Cartesian products of its algebras.*

Proof. Let V denote the set of all elements $v \in L(\{x_1, x_2, \dots, x_n, \dots\})$ such that $v \equiv 0$ is satisfied by every algebra in **K**. Suppose that G is an algebra such that $V(G) = \{0\}$. The proof will be complete if we show that $G \in \mathbf{K}$.

We represent G as a quotient algebra of a suitable free algebra $L(Y)$. Then G is isomorphic also to a quotient algebra of a free algebra $L(Y, \mathbf{V}) = L(Y)/V(Y)$ of the variety **V** defined by V. Let **P** be an arbitrary subset of **K** determining the same system of identities as **K**. Let **Q** denote the set of all homomorphic images of $L = L(Y)$ in **P** and Ψ the set of all epimorphisms from $L(Y)$ onto algebras in **Q**. We consider the Cartesian product

$$R = \coprod_{\psi \in \Psi} \psi(L).$$

Let f_y, $y \in L$, denote an element in R such that $f_y(\psi) = \psi(y)$. We set $S = \operatorname{alg}\{f_y \mid y \in Y\}$. Then S is a homomorphic image of $L(Y)$ under a homomorphism $\bar{\chi}$ extending the map $\chi : y \mapsto f_y$. We will show that $V(L(Y)) \supseteq \ker\bar{\chi}$. Indeed, if $w \in \ker\bar{\chi}$ then $\psi(w) = 0$ for any homomorphism from $L(Y)$ into any algebra in **P**. Hence $w \equiv 0$ is satisfied by every algebra in **K**. Having written down w in the variables $x_1, x_2, \dots, x_n, \dots$, we will obtain an element $w' \in V$. Coming back to variables in Y shows that $w \in V(L(Y))$. Thus $V(L(Y)) \supseteq \ker(\bar{\chi})$. According to the second part of the homomorphism theorem

(Section 1.1), we will see that $L(Y, \mathbf{V})$ is the homomorphic image of S. By our definition of $\mathbf{K}$, we see that $L(Y, \mathbf{V})$ is in $\mathbf{K}$. Now $\mathbf{K} = \mathbf{V}$. □

If we assume from the very beginning that $\mathbf{K} = \mathbf{V}$ then $S = L(Y, \mathbf{V})$ becomes a construction of a free algebra of the variety $\mathbf{V}$ with free generating set Y. We will introduce one more notion. Let $\mathbf{M}$ be a class of Lie algebras over k, then we define $\mathbf{V} = \mathrm{var}(\mathbf{M})$, the *variety generated by* $\mathbf{M}$, as the least variety of Lie algebras over k containing $\mathbf{M}$. If $\mathbf{M}$ contains precisely one algebra G then we simply write $\mathbf{V} = \mathrm{var}(G)$ (not $\mathrm{var}(\{G\})$). We remark that any variety of Lie algebras is generated by its free Lie algebra of countable rank, as well as by the set of all its free algebras of finite ranks.

Theorem 4.4. *Any variety* $\mathbf{V} = \mathrm{var}(G)$ *generated by a finite Lie algebra* G *is locally finite. In other words, any finitely generated algebra in* V *is finite.*

Proof. We apply the construction of Theorem 4.3 to the case where $\mathbf{P}$ consists of precisely one algebra G. If Y is finite then, G being finite, we see that $\mathbf{Q}$, thus Ψ, is finite. In this case R is finite, hence also finite is its subalgebra S isomorphic to $L(Y, \mathbf{V})$. Since any finitely generated algebra $A \in \mathbf{V}$ can be represented as a quotient algebra of a $L(Y, \mathbf{V})$ with Y finite, we see that A is finite. □

4.1.3 Examples of varieties. Notation

Let the base ring k be fixed.

(i) The variety, defined by a single law $x \equiv 0$, consists of a single zero algebra and its is called the *unity variety* $\mathbf{E}$.
(ii) If M is a proper ideal in k then $\mathbf{O}_M$ denotes the variety defined by the laws of the form $\lambda x \equiv 0$, $\lambda \in M$. If $M = \{0\}$, we write $\mathbf{O}_M = \mathbf{O}$.
(iii) Variety $\mathbf{A}$ consists of all abelian Lie algebras over k; it is defined by the law $x_1x_2 \equiv 0$.
(iv) Variety $\mathbf{N}_c$ consists of all nilpotent Lie algebras whose nilpotent step is at most c, that is, given by the law $x_1x_2 \cdots x_{c+1} \equiv 0$ (see Section 2.1.3).
(v) Variety $\mathbf{S}_l$ consists of all solvable Lie algebras over k whose solvable step is at most l; this is given by the law $\delta_l(x_1, x_2, \ldots, x_{2^l}) \equiv 0$ (see Section 2.1.3).
(vi) Given two varieties $\mathbf{U}$ and $\mathbf{V}$, their intersection $\mathbf{U} \cap \mathbf{V}$ is a variety, too. On the contrary, the set-theoretic union of two varieties, as a rule, is not a variety. We use $\mathbf{U} \cup \mathbf{V}$ to denote the variety generated by all algebras from $\mathbf{U}$ and $\mathbf{V}$. Thus we obtain new examples of varieties, called *intersection* and *union* of two given ones.

4.2 Multihomogeneous varieties. Linearization

4.2.1 Multihomogeneous varieties

A variety **V** is called *multihomogeneous* if its associated ideal of identities (the verbal ideal $\mathbf{V}(L)$ of free Lie algebra L of countable rank) is multihomogeneous. Saying it differently, a variety is multihomogeneous if, given an identity $v \equiv 0$ in **V** such that $v = \sum_\alpha v_\alpha$ is the multihomogeneous decomposition of v, for any α, $v_\alpha \equiv 0$ is also an identity in **V**. We will soon see that any variety of a Lie algebra over an infinite field is multihomogeneous. On the other hand, any variety generated by a finite nonnilpotent algebra is not multihomogeneous. The study of multihomogeneous varieties uses some standard techniques. One of them is the *process of linearization*. It is an essential part of the following.

Theorem 4.5. *Any nontrivial identity has a nontrivial multilinear consequence.*

Proof. We consider a nontrivial identity of the form

$$v(x_1, x_2, \dots, x_s) \equiv 0, \tag{4.1}$$

where $v(x_1, x_2, \dots, x_s)$ is a nonzero element of a subalgebra of a free Lie algebra $L(\{x_1, \dots, x_n, \dots\})$ generated by the set $\{x_1, x_2, \dots, x_s\}$. We know that v is a linear combination of its multihomogeneous components (see Section 2.2.4):

$$v = \sum_\alpha v_\alpha, \tag{4.2}$$

where $m(v_\alpha) = \alpha \in \Phi$. If one has $v_1(x_2, \dots, x_s) = v(0, x_2, \dots, x_s) \neq 0$ then $v_1(x_2, \dots, x_s) \equiv 0$ is a nontrivial consequence of identity (4.1), and we may apply induction by s with obvious basis if $s = 1$. The same argument shows that, given an α such that $v \neq 0$ in equation (4.2), we may think that $\alpha(x_i) \geq 1$, $i = 1, 2, \dots, s$. We choose a function $\beta \in \Phi$ such that, for some i, the number $\beta(x_i)$ is the greatest in the set of all numbers $\alpha(x_j)$, $j = 1, 2, \dots, s$, $\alpha \in \Phi$ is such that $v_\alpha \neq 0$. There will be no loss of generality if we put $i = 1$. Then we consider a polynomial

$$\begin{aligned} w(x_{s+1}, x_{s+2}, x_2, \dots, x_s) &= v(x_{s+1} + x_{s+2}, x_2, \dots, x_s) \\ &= v(x_{s+2}, x_2, \dots, x_s) - v(x_{s+2}, x_2, \dots, x_s). \end{aligned} \tag{4.3}$$

Clearly, $w(x_{s+1}, x_{s+2}, x_2, \dots, x_s) \equiv 0$ is a consequence of identity (4.1). On the other hand, it is obvious that if

$$w = \sum_{\gamma \in \Phi} w_\gamma$$

then $\gamma(x_{s+1}), \gamma(x_{s+2}) < \beta(x_1)$ and $\gamma(x_2), \dots, \gamma(x_s) \leq \beta(x_1)$. This enables us to apply induction over the number of variables x_i for which there exists α such that $\beta(x_1) = \alpha(x_i)$ and over $\beta(x_1)$ itself. It remains to note that any transformation by formula (4.3) brings a nonzero element of the form (4.1) into a nonzero element w. Recall that $\beta(x_1) > 1$ for some $\beta \in \Phi$. We may view $L(X)$ as a Lie subalgebra in $A(X)$ (see Section 2.3.4). For different $\alpha \in \Phi$, the nonzero elements

$$w^\alpha(x_{s+1}, x_{s+2}, x_2, \dots, x_s) = v_\alpha(x_{s+1} + x_{s+2}, x_2, \dots, x_s) \\ - v_\alpha(x_{s+1}, x_2, \dots, x_s) - v_\alpha(x_{s+2}, x_2, \dots, x_s)$$

are linearly independent over k. Thus it is sufficient to prove $w^\beta \neq 0$. Let $\lambda a_0 x_1^{l_1} a_1 x_1^{l_2} \cdots a_{t-1} x_t^{l_t} a_t$ enter the associative expression for v_β, $\lambda \in k$, $\lambda \neq 0$, $l_1 + l_2 + \cdots + l_t = \beta(x_1)$. This and only this monomial can give rise to a monomial of the form

$$a_0 x_{s+1} x_{s+2}^{l_1 - 1} a_1 x_{s+2}^{l_2} \cdots a_{t-1} x_{s+2}^{l_t} a_t$$

which enters with nonzero coefficient λ in the associative expression for the element $w(x_{s+1}, x_{s+2}, x_2, \dots, x_s)$. Since different associative monomials form a basis of $A(X)$, we finally see that w is a nonzero monomial. □

Theorem 4.6. *Let k be an infinite field. Then any identity of the form $v \equiv 0$ is equivalent to a system of $\{w \equiv 0 \mid w \in W\}$ of multihomogeneous identities.*

Proof. We represent v in the form $v = \sum_{\alpha \in \Phi} v_\alpha$ and set $v_i = \sum_{\alpha(x_1)=i} v_\alpha$, $i = 1, 2, \dots, d$. Then $v = v_0 + v_1 + \cdots + v_d$. We choose $d + 1$ pairwise different elements $\lambda_0, \lambda_1, \dots, \lambda_d$ in k and consider the elements of the form

$$v(\lambda_i x_1, x_2, \dots, x_d) = v_0 + \lambda_i v_1 + \lambda_i^2 v_2 + \cdots + \lambda_i^d v_d, \quad i = 0, 1, \dots, d. \tag{4.4}$$

All elements of the form of equation (4.4) are in the linear subspace $v(L(X))$. Now the system of equations (4.4) can be uniquely solved with respect to $v_0, v_1, \dots, v_d$ since its determinant is the Vandermonde determinant $\det(\lambda_i^j)$, $0 \le i, j \le d$. Hence $v_0, v_1, \dots, v_d \in v(L(X))$, the converse being obviously true. The polynomials $v_0, v_1, \dots, v_d$ are homogeneous with respect to x_1. Proceeding by induction over the number of variables involved in v, we arrive at the conclusion of the theorem. □

Corollary 4.7. *Any variety of Lie algebras over an infinite field is multihomogeneous.*

Proof. In the course of our proof, we showed that, for each element v such that $v \equiv 0$ is an identity, we have $v_\alpha \equiv 0$, v_α being an arbitrary multihomogeneous component of v. □

Theorem 4.8. *Over any field of characteristic zero, any identity is equivalent to a system of multilinear identities.*

Proof. Using Theorem 4.6, we may think that the original identity is multihomogeneous. Let this identity have the form $v(x_1, x_2, \dots, x_s) \equiv 0$, $m(v) = \alpha$. The process of linearization is an algorithm of deriving a multilinear consequence of $v \equiv 0$. If we manage to show that each step of this algorithm can be reversed then our proof will be complete. But if in equation (4.3) we replace x_{s+1} and x_{s+2} by x_1 then we will see that the following is true:

$$w(x_1, x_1, x_2, \dots, x_s) = v(2x_1, x_2, \dots, x_s) - 2v(x_1, x_2, \dots, x_s).$$

Since v is multihomogeneous, we may write

$$v(2x_1, x_2, \dots, x_s) = d \cdot v(x_1, x_2, \dots, x_s), \quad \text{where } d = \alpha(x_1) > 1.$$

Hence

$$v(x_1, x_2, \dots, x_s) = \frac{1}{2^d - 2} w(x_1, x_1, x_2, \dots, x_s). \qquad \square$$

In the case of infinite fields of characteristic $p > 0$, we have a weakened version of this theorem as follows.

Theorem 4.9. *Over any field k which is infinite of characteristic $p > 0$, any identity of the form $v \equiv 0$ is equivalent to a system of multihomogeneous identities of the form $w \equiv 0$, $w \in L_\alpha(\{x_1, x_2, \dots, x_n, \dots\})$, such that either $\alpha(x_i) = 0$ or $\alpha(x_i) = p^{t_i}$, $i = 1, 2, \dots$*

Proof. By Theorem 4.6, we may think that $v = v(x_1, x_2, \dots, x_s)$ is a multihomogeneous polynomial. If $\alpha = m(v)$ satisfies our claim, the proof is complete. Otherwise, without loss of generality, we may assume that $\alpha(q) = r$ where r is not a power of p. Since k is infinite there exists a $\lambda \in k$ such that $(1+\lambda)^r \neq 1 + \lambda^r$. Indeed, the polynomial

$$f(t) = (1+t)^r - 1 - t^r = \sum_{i=1}^{r-1} \binom{r}{i} t^i$$

is nonzero. For instance, if $r = p^m q$ where $(p, q) = 1$, then $\binom{r}{p^m}$ is a nonzero coefficient of this polynomial. What remains is to take λ which is not a root of this polynomial. Now we may apply a transformation similar to the one applied in Theorem 4.5:

$$\begin{aligned} w(y_1, y_2, x_2, \dots, x_s) = {} & v(y_1 + \lambda y_2, x_2, \dots, x_s) - v(y_1, x_2, \dots, x_s) \\ & - v(\lambda y_2, x_2, \dots, x_s). \end{aligned} \tag{4.5}$$

Replacing $v \equiv 0$ by $w \equiv 0$ is invertible since

$$w(x_1, x_1, x_2, \dots, x_s) = ((1+\lambda)^r - 1 - \lambda^r) v(x_1, x_2, \dots, x_s).$$

Replacing w by its homogeneous components with respect to y_1, y_2 gives us an opportunity to apply induction on the number of indices i with $\alpha(i) = r$ and then on r. $\square$

It should be noted that varieties over infinite fields that cannot be given by a set of multilinear laws do exist. For instance, this is the case for the variety given by $(\operatorname{ad} x)^{p+1} \equiv 0$, $p = \operatorname{char} k$.

4.2.2 Partial and complete linearization

Let G be a Lie algebra over a field k, $E = (e_\alpha)_{\alpha\in I}$ a basis of G. The verification of the validity of a multilinear identity $v = v(x_1, x_2, \dots, x_s) \equiv 0$ in G reduces to computing v in the elements of E. If we want to simplify verification similarly in the case where v is arbitrary, we must substitute scalar multiples of elements in E into all possible consequences of $v \equiv 0$ which have the form

$$v(y_{11} + \dots + y_{1a}, y_{21} + \dots + y_{2b}, \dots, y_{s1} + \dots + y_{sc}),$$

where all y_{ij} are pairwise different variables. Up to equivalence, any identity of this form can be obtained from $v \equiv 0$ by a finite number of reductions of the form of equation (4.3). A substitution of elements into $v(y_1 + y_2, x_2, \dots, x_s)$ is equivalent to a substitution of elements into $v(x_1, x_2, \dots, x_s)$ and $w(y_1, y_2, \dots, x_s)$. It should be remarked that the "new" polynomial $w(y_1, y_2, x_2, \dots, x_n)$ will be obtained as a result of such an "elementary" step if and only if v is expressed through basic monomials with nonzero coefficients such that one of these monomials involves x_1 at least twice. Thus, to verify the validity of an identity $v \equiv 0$ in G by substituting scalar multiples of basic elements, it is sufficient to replace v by the least (finite) set $\mathscr{L}(v)$ of nonzero polynomials which is closed under equation (4.3) and contains v. This passage is called the linearization of identity $v \equiv 0$. Any identity of the form $w \equiv 0$ where w is an element of $\mathscr{L}(v)$ is called a linearization of $v \equiv 0$. If, in addition, w is a multilinear polynomial, we say that $w \equiv 0$ is a *complete linearization*. Otherwise $w \equiv 0$ is a *partial linearization*. It is not hard to see that the complete linearization is unique up to equivalence, provided v is multihomogeneous.

In the case of an infinite field, the above argument and the procedure of deriving multihomogeneous components (Theorem 4.6), together with Theorem 4.9, give the following.

Theorem 4.10. *Let $L = L(x_1, x_2, \dots, x_n, \dots)$ be a free Lie algebra of countable rank and* $\mathbf{V}$ *the variety of Lie algebras over an infinite field of characteristic $p > 0$ given by $v \equiv 0$. Then there exists a finite multihomogeneous set $W = \{w_1, \dots, w_m\} \subset L$ such that:*
(i) *$\{w_1 \equiv 0, \dots, w_m \equiv 0\}$ is equivalent to $v \equiv 0$;*
(ii) *a Lie algebra G with a basis E (see Section 4.2.2) is in* $\mathbf{V}$ *if and only if W vanishes in the elements of E;*
(iii) *the degree of each $w_i \in W$ in every x_j is a p-power.*

4.2.3 Some additional definitions

A subalgebra H of a Lie algebra G over a ring k is called *characteristic* if this is stable under all derivations of G, that is, for any $\delta \in \operatorname{Der} G$ and any $h \in H$ we have $\delta(h) \in H$. For instance, this property holds valid for verbal subalgebras V in the case where k is a field of characteristic zero. Indeed, in this case it is sufficient to consider the action of δ on a multilinear polynomial $v(x_1,\dots,x_n) \in \mathbf{V}$:

$$\delta(v(x_1,\dots,x_i,\dots,x_n)) = \sum_{i=1}^{n} v(x_1,\dots,\delta(x_i),\dots,x_n).$$

It is easily seen that each summand of the right-hand side is in V, hence V is stable under δ.

Now a subalgebra H in G is called *invariant* if this is stable under all automorphisms of G. It is called *fully invariant* if "automorphisms" are replaced by "endomorphisms." Obviously, any verbal subalgebra is invariant, even fully invariant. The converse is not always true. For instance, a one-dimensional subalgebra $H = \langle xy \rangle$ in a free Lie algebra $L = L(x, y)$ is invariant but, obviously, it is not verbal. Our nearest purpose (Theorem 4.12) is to show that, in the case of an infinite field and of infinite rank, any invariant subalgebra of a free Lie algebra is always verbal. We start with an auxiliary result.

Lemma 4.11. *Let L be a free Lie algebra over a ring k with an infinite free generating set X. For an arbitrary $x \in X$, let φ_x denote an endomorphism of L extending the map $x \mapsto 0$, $y \mapsto y$ $(y \neq x)$ of X. If $M \subseteq L$ is an invariant subalgebra which is stable under all φ_x then it is fully invariant.*

Proof. Let $w = w(x_1, x_2, \dots, x_m)$ be an element of M and φ an endomorphism of L transforming x_i into $v_i(x_1,\dots,x_n)$, $i = 1, 2, \dots, m$. We consider an automorphism ω of L given by $\omega(x_i) = x_{n+i} + v_i(x_1,\dots,x_n)$, $1 \le i \le m$, $\omega(x_{m+j}) = x_j$, $1 \le j \le n$; $\omega(y) = y$, $y \in X \setminus \{x_1, x_2, \dots, x_{n+m}\}$. The bijectivity of ω follows since it is easy to construct its inverse. We compute $\omega(w)$:

$$\begin{aligned}\omega(w) = w(&x_{n+1} + v_1(x_1,\dots,x_n), x_{n+2} + v_2(x_1,\dots,x_n),\dots,\\ &x_{n+m} + v_m(x_1,\dots,x_n)).\end{aligned}$$

By invariance, $\omega(w) \in M$. Applying a sequence of endomorphisms as in the statement of Lemma 4.11, we see that

$$\begin{aligned}&w(v_1(x_1,\dots,x_n), v_2(x_1,\dots,x_n),\dots, v_m(x_1,\dots,x_n))\\ &= \varphi_{x_{n+1}}\varphi_{x_{n+2}}\cdots\varphi_{x_{n+m}}(\omega(w))\end{aligned}$$

is in M. □

Theorem 4.12. *Let $L = L(X)$ be a free Lie algebra over a ring k with an infinite set X of free generators and M an invariant subalgebra in L. Then M is a fully invariant subalgebra. If, in addition, M is a multihomogeneous subalgebra then M is a verbal ideal.*

Proof. We will show first that M is a fully invariant subalgebra. We achieve this by applying Lemma 4.11. Let x be in X and φ_x denote the same as in Lemma 4.11. We fix an element $u \in M$ and represent it in the form $u = u_0 + u_1 + \cdots + u_s$, $u_s \neq 0$, where u_i is homogeneous of degree i in x. Arguing by induction over s with trivial basis $s = 0$, we will prove that $\varphi_x(u) \in M$. Now suppose $s > 0$ and assume that u does not depend on a $z \in X$. Let θ_1 and θ_2 be automorphisms of L induced by the maps $x \mapsto z$, $z \mapsto x$ and $x \mapsto x+z$, $z \mapsto z$, respectively (here $\theta_2(y) = y$ if $y \neq z, x$, $i = 1, 2$). By our hypothesis, $v = u+\theta_1(u)-\theta_2(u)$ is in M. As in Theorem 4.5, the degree of this element in x is strictly less than s. Since $\varphi_x(\theta_2(u)) = \theta_1(u)$, we see that $\varphi_x(v) = \varphi_x(u) + \theta_1(u) - \theta_1(u) = \varphi_x(u) \in M$. Since x is an arbitrary element of X, by Lemma 4.11 M is a fully invariant subalgebra.

Now suppose M is a multihomogeneous subalgebra. We fix a multihomogeneous polynomial $v(x_1, x_2, \ldots, x_n) \in M$ and let $x_{n+1} \in X \setminus \{x_1, \ldots, x_n\}$. It is enough to show that $x_{n+1}v(x_1, \ldots, x_n) \in M$. To do this, we consider an element of the form

$$w = v(x_1 + x_{n+1}x_1, x_2 + x_{n+1}x_2, \ldots, x_n + x_{n+1}x_n) \in M. \tag{4.6}$$

It is easy to see that the sum of all monomials in the expression for (4.6) depending on x_{n+1} linearly is equal to $x_{n+1}v(x_1, \ldots, x_n)$. Since M is multihomogeneous, we conclude that $x_{n+1}v(x_1, \ldots, x_n) \in M$. Now since M is fully invariant, we see that it is closed with respect to multiplying by arbitrary elements in L. Let V be equal to the set of all nonassociative polynomials comprising M. Then our result can be written as $M = V(L)$. □

4.2.4 Free algebra of a multihomogeneous variety

We recall that a free algebra $L(X, \mathbf{V})$ of a variety of Lie algebras $\mathbf{V}$ with free generating set X can be defined as the quotient algebra of a free Lie algebra $L = L(X)$ by its verbal ideal $V = \mathbf{V}(L)$ (see Section 4.1.1). Now the following is obvious.

Proposition 4.13. *Let $\mathbf{V}$ be a multihomogeneous variety of Lie algebras over a ring k, $G = L(X, \mathbf{V})$ its free algebra with free generating set X. Then G is a Lie algebra graded by the subspaces $G_n = L_n + V/V$, $n = 1, 2, \ldots$ In particular,*

$$\bigcap_{n=1}^{\infty} G^n = \{0\}.$$

An important application of this is as follows.

Theorem 4.14. *Let $G = L(Y, \mathbf{V})$ be a free algebra of a multihomogeneous variety $\mathbf{V}$ of Lie algebras over a field k, Z a subset of G which is linearly independent modulo the derived algebra G^2 and H a subalgebra of G generated by Z. Then $H \cong L(Z, \mathbf{V})$.*

Proof. Let X be a set of the same cardinality as Z, φ a bijection of X onto Z and $\bar{\varphi}$ a homomorphism from $L(X, \mathbf{V})$ onto H that extends this map. If $\ker \bar{\varphi} \neq \{0\}$ then there exist $x_1, \ldots, x_m \in X$ and $z_1 = \varphi(x_1), \ldots, z_m = \varphi(x_m)$ such that the restriction $\bar{\varphi}$: $L(x_1, \ldots, x_m, \mathbf{V}) \to \operatorname{alg}(z_1, \ldots, z_m)$ is not injective. This enables us to restrict ourselves to the case where $Z = \{z_1, \ldots, z_m\}$ is a finite set. It is obvious here that we may think that the set $Y = \{y_1, \ldots, y_n\}$ is finite, too (replacing L by its subalgebra lessens the derived algebra, leaving the hypotheses unchanged). Now $n = \dim G/G^2 \geq m$. Interchanging, if necessary, the elements of Y, we may assume that $Z' = \{z_1, \ldots, z_m, y_{m+1}, \ldots, y_n\}$ is linearly independent modulo G^2. We set $H' = \operatorname{alg} Z'$. Now the map $\chi : y_1 \mapsto z_1, \ldots, y_m \mapsto z_m, y_{m+1} \mapsto y_{m+1}, \ldots, y_n \mapsto y_n$ extends to a homomorphism $\bar{\chi} : G \to G$. The proof will be complete if we manage to show $\ker \bar{\chi} = 0$. Suppose, conversely, that there exists $0 \neq v \in \ker \bar{\chi}$. Since $\bigcap_{n=1}^{\infty} G^n = \{0\}$ (see Section 4.2.4) there exists $n \geq 2$ such that $v \notin G^n$. Furthermore, since $\bar{\chi}(G^n) \subseteq G^n$, it then induces a homomorphism $\tilde{\chi} : G/G^n \to G/G^n$. Now we remark that G/G^n is a nilpotent algebra and the image of Z' in this algebra generates it modulo its derived algebra. Hence (Section 1.7.2) $\tilde{\chi}$ is a surjection of a finite-dimensional space onto itself. Thus $\tilde{\chi}$ is an isomorphism and $\tilde{\chi}(v + G^n) = \{0\}$. Therefore $v \in G^n$, contradicting our assumption. Thus we may conclude that $\bar{\chi}$ is an injective homomorphism so that

$$H = \operatorname{alg} Z \cong L(y_1, \ldots, y_m, \mathbf{V}). \qquad \square$$

Corollary 4.15. *Any free Lie algebra of a variety of Lie algebras over an infinite field is freely generated by any generating set which is linearly independent modulo its derived algebra.*

We remark that Theorem 4.14 with certain changes holds valid also in the case of varieties over rings (see Exercise 4.10.11). However, it is not sufficient to require the k-freeness of Z modulo G^2.

4.2.5 Independent system of identities

A system of identities of the form $\{v \equiv 0 \mid i \in I\}$ is called *independent* if any of its members is not a consequence of the system of all the remaining identities. Any variety that is given by a finite system of identities can also be given by an independent system. For instance, if the system considered of $v_1(x_1, \ldots, x_n) \equiv 0, \ldots, v_m(x_1, \ldots, x_n) \equiv 0$ then it would be equivalent to a single identity $u(x_1, \ldots, x_{mn}) \equiv 0$ of the form

$$u(x_1, \ldots, x_{mn}) = v_1(x_1, \ldots, x_n) + v_2(x_{n+1}, \ldots, x_{2n}) + \cdots + v_m(x_{(m-1)n+1}, \ldots, x_{mn}).$$

Obviously, a system of one identity is independent. The question whether any variety of Lie algebras over a commutative ring with unity can be given by an independent system of identical relations is open. A theorem below gives a positive answer in the case of multihomogeneous varieties over fields. Before we formulate it, we define the notion of a *normal* polynomial. By this we mean a linear combination of nonassociative monomials, each of which depending on the same set of variables. It is obvious that any system of identities is equivalent to a system of *normal* identities, i. e., identities of the form $v \equiv 0$, v being a normal polynomial.

Theorem 4.16. *Any multihomogeneous variety* $\mathbf{V}$ *over a field k can be given by an independent system of identical relations.*

Proof. By our hypothesis, $V = \mathbf{V}(L)$, where L is a free Lie algebra of countable rank, is multihomogeneous. In particular, if we put $V_n = V \cap L_n$ then we may write $V = V_1 \oplus V_2 \oplus \cdots \oplus V_n \oplus \cdots$. We note that, by a remark preceding our theorem, the system $\{v \equiv 0 \mid v \in V_n\}$ is equivalent to a subsystem depending entirely on $x_1, \ldots, x_n$. But the set of all left-hand sides of this subsystem is a finite-dimensional subspace V_n' in V_n. Choosing a basis in V_n' of the form $\{v_1, \ldots, v_m\}$ and applying the same argument as in Section 4.2.5, we will see that the system $\{v \equiv 0 \mid v \in V_n\}$ is equivalent to a single identity $w_n \equiv 0$. Now we will build a sequence of natural numbers $t_1, t_2, \ldots, t_m, \ldots$ such that $\{w_{t_i} \equiv 0 \mid i = 1, 2, \ldots\}$ is a system of identities which is independent and yet equivalent to our original system. We use induction over m. For t_1 we take the least natural number such that $V_{t_1} \neq \{0\}$. Assuming $t_1, t_2, \ldots, t_{m-1}$ already selected, we chose t_m as the least natural number such that V_{t_m} does not belong to the verbal ideal generated by $w_{t_1}, w_{t_2}, \ldots, w_{t_{m-1}}$. If such a t_m does not exist then our construction is over. Otherwise we apply induction. Now we must show that the system of identities of the form $W = \{w_{t_m} \equiv 0 \mid m = 1, 2, \ldots\}$ is independent. Again we proceed by induction over m. Indeed, $w_{t_1} \equiv 0$ is an identity whose degree is strictly less than that of the other identities. Hence the basis of induction is obvious. Now suppose $w_{t_m} \equiv 0$ is a consequence of the system of all remaining identities in W. By the same argument as above $w_{t_m} \equiv 0$ is in fact a consequence of a subsystem $w_{t_1} \equiv 0, \ldots, w_{t_{m-1}} \equiv 0$. Hence w_{t_m} belongs to the verbal ideal generated by polynomials $w_{t_1}, \ldots, w_{t_{m-1}}$. Thus all w such that $w \equiv 0$ is a consequence of $w \equiv 0$ are in that ideal. By the definition of identities of the form $w_n \equiv 0$, the whole subspace V_{t_m} belongs to this verbal ideal, contradicting the construction of the sequence $t_1, t_2, \ldots, t_m, \ldots$ By our construction, we also see that the system $\{w_{t_m} \equiv 0 \mid m = 1, 2, \ldots\}$ is equivalent to the original system of identities. □

4.3 Product varieties

4.3.1 Definition and description of identities in the product variety

Given two classes $\mathbf{U}, \mathbf{V}$ of algebras over a ring k, their product $\mathbf{UV}$ can be defined as follows. An algebra G is in $\mathbf{UV}$ if and only if G has an ideal $H \in \mathbf{U}$ such that G/H is in $\mathbf{V}$. If $\mathbf{U}$ and $\mathbf{V}$ are varieties then it follows from Theorem 4.3 (Birkhoff's) that $\mathbf{UV}$ is again a variety. We verify, e. g., that any subalgebra F of G has similar structure to that of G. Indeed, $F \cap H$ is an ideal of F and $F \cap H \subseteq H$, hence $F \cap H \in \mathbf{U}$. Now by the isomorphism theorem (Section 1.1) $F/F \cap H \cong F + H/H \subset G/H \in \mathbf{V}$. The verification of the invariance under quotient algebras and Cartesian products is even simpler.

To study the identities of the product variety, we consider a free Lie algebra L of countable rank. Set $V = \mathbf{V}(L)$, $P = \mathbf{U}(V)$. It should be remarked that P is not necessarily an ideal of L. So we put $\bar{P} = \mathrm{id}_L(P)$. Then $\bar{P} = (\mathbf{UV})(L)$. Indeed, $L/\bar{P}$ has an ideal $V/\bar{P}$ which is, obviously, in $\mathbf{U}$ and $L/V \in \mathbf{V}$. On the other hand, it is clear that $\bar{P}$ belongs to the kernel of any homomorphism of L into an algebra in $\mathbf{UV}$. Now if we form products $(\mathbf{UV})\mathbf{W}$ and $\mathbf{U}(\mathbf{VW})$ then the above remark indicates (and this is confirmed by examples of varieties over $\mathbb{Z}_2$, see Bahturin [22]) that the operation of multiplication on the set of varieties over a fixed commutative ring k is not necessarily associative. However, if k is an infinite field then the associativity takes place. Let $v(k)$ denote the groupoid of all varieties of Lie algebras over a ring k with respect to the above multiplication.

Proposition 4.17. *Let k be an infinite field. Then $v(k)$ is a semigroup with 0 and 1.*

Proof. Let L be a free Lie algebra of countable rank. We compute the ideals of identities of both $(\mathbf{UV})\mathbf{W}$ and $\mathbf{U}(\mathbf{VW})$. We note that $P = \mathbf{U}(\mathbf{V}(L))$ (see the beginning of this subsection) is obviously stable under all automorphisms of L. By Theorem 4.12, P is a verbal subalgebra in L, that is, $\bar{P} = P$ and P is the ideal of identities of $\mathbf{UV}$. It is obvious from here now that the ideal of identities for both $(\mathbf{UV})\mathbf{W}$ and $\mathbf{U}(\mathbf{VW})$ is the subalgebra $\mathbf{U}(\mathbf{V}(\mathbf{W}(L)))$. It remains to notice that the role of 0 and 1 in $v(k)$ is respectively played by the variety $\mathbf{O}$ of all Lie algebras over k and the variety $\mathbf{E}$ consisting of zero algebra only. □

The main result of this section is Theorem 4.23 saying that, in the case of an infinite field k, $v(k)$ is a free semigroup with 0 and 1. In the proof we will need some auxiliary facts which are true under less restrictive conditions. Now a number $N = N(\mathbf{V})$ is called the *norm* of a nonzero variety $\mathbf{V}$ if N is the least degree of nonzero homogeneous components of a polynomial $v \in L(x_1, x_2, \dots, x_n, \dots)$ such that $v \equiv 0$ is an identity in $\mathbf{V}$. It is clear that the norm of any nontrivial (that is distinct from $\mathbf{O}$ and $\mathbf{E}$) variety is strictly greater than 1. Besides, the explicit form of the ideal $\bar{P}$ of identities of $\mathbf{UV}$ shows that $N(\mathbf{UV}) \gneqq N(\mathbf{U}), N(\mathbf{V})$. Thus we have the following.

Proposition 4.18. *Any proper variety over an arbitrary field k is decomposable as the product of indecomposable varieties.*

Now we present the so-called *monotonicity theorem* which is important not only in the context of the present section.

Theorem 4.19. *Let $L = L(X)$ be a nonabelian free Lie algebra over a field k, $\mathbf{V}$ a nonzero variety, H, K ideals in L. If $\mathbf{V}(K) \subseteq \mathbf{V}(H)$ then $K \subseteq H$.*

Proof. Suppose, for a pair of ideals H, K in L our claim is not true, that is, $\mathbf{V}(K) \subseteq \mathbf{V}(H)$ but $K \nsubseteq H$. Then we can find a $q \in K \backslash H$. We set $K_1 = K \cap H + kq$, $L_1 = K_1 + H$. Then, H being nonzero, we see that L_1 is a nonabelian free Lie algebra (see Section 2.4, Theorem 2.18). Further, $\mathbf{V}(K_1) \subseteq \mathbf{V}(K) \subseteq \mathbf{V}(H)$ and, as before, $K_1 \nsubseteq H$. Thus we may think from the very beginning that $L = K + H$ and $\dim L/H = 1$. We can choose in L a free generating set of the form $Z = Z' \cup \{x\}$, where $Z' \subseteq H$. Indeed, let x be a generator outside H. Then for any $y \in X$ there exists an $\alpha_y \in k$ such that $y - \alpha_y x \in H$. The map $x \mapsto x$, $y \mapsto y - \alpha_y x$ is invertible ($y \neq x$). Hence it extends to an automorphism of L and Z' can be chosen in the form $\{y - \alpha_y x \mid y \in X \setminus \{x\}\}$. Now we pick a $y \in Z'$ and we set $\varphi(x) = x$, $\varphi(y) = y$, $\varphi(z) = 0\,(z \in Z\setminus\{x,y\})$. This map extends to a homomorphism $\bar{\varphi} : L \to L(\{x,y\})$. Now we have $\bar{\varphi}(L) = \bar{\varphi}(K)+\bar{\varphi}(H)$, $\mathbf{V}(\bar{\varphi}(K)) = \bar{\varphi}(\mathbf{V}(K)) \subseteq \bar{\varphi}(\mathbf{V}(H)) = \mathbf{V}(\bar{\varphi}(H))$, $\dim \bar{\varphi}(L)/\bar{\varphi}(H) = 1$. This argument shows that we may restrict ourselves to the case where L is a free Lie algebra freely generated by x, y with two ideals K, H such that $L = K + H$, $H = \mathrm{id}_L(\{y\})$. As previously $\mathbf{V}(K) \subseteq \mathbf{V}(H)$.

Applying the argument from the proof of Theorem 4.5, we may find an element $v \in \mathbf{V}(L_\infty)$ $[L_\infty = L(x_1, x_2, \ldots, x_n, \ldots)]$ whose lowest term is multilinear and is of degree $d = N(\mathbf{V})$. We will denote this lowest term by $w(x_1, x_2, \ldots, x_d)$. Using Section 2.4, the corollary of Proposition 2.15, we choose a free generating set T for H in the form $\{t_r = (\operatorname{ad} x)^r(y), r = 0, 1, 2, \ldots\}$. Since $y \in H$, there exists an element $q_1 \in L^2$ such that $q = x + q_1 \in K$. We consider the value of v in elements of ideal K having the form

$$v(q, qt_1, qt_3, \ldots, qt_{2d-3}). \tag{4.7}$$

The lowest term of this element as a polynomial both in x, y and in T has the form

$$w(x, xt_1, xt_3, \ldots, xt_{2d-3}) = w(x, t_2, t_4, \ldots, t_{2d-2}), \tag{4.8}$$

provided (4.8) is nonzero. Now since element (4.7) is in $\mathbf{V}(H) = \mathbf{V}(L(T))$, this element is a linear combination of elements of the form $u(t_1, t_2, \ldots, t_p)$ $u \equiv 0$, being an identity in $\mathbf{V}$. By the definition of the norm of a variety (Section 4.3.1) the degree of the lowest term of element (4.7) with respect to T is strictly less than d since any homogeneous element whose degree in T is at the least d must have degree in y at least d.

It remains to show that (4.8) is a nonzero element of L. For this we represent $w(x_1, x_2, \ldots, x_d)$ as a linear combination of linearly independent monomials of the form $x_{i_1} \cdots x_{i_{d-1}} x_1$ (see Section 4.8.1). Then (4.8) is a linear combination of monomials of the form

$$t_{2i_1-2} \cdots t_{2i_d-2} x = -t_{2i_1-2} \cdots t_{2i_d-2} t_{2i_d-1}.$$

The monomials we obtain in this way are multilinear with respect to T and they can be linearly dependent only if they have the same i_d. But an argument similar to the above shows that the monomials with the same i_d are linearly dependent only if $i_1, \ldots, i_{d-1}$ are the same. Now since $w(x_1, \ldots, x_d)$ is nonzero, the same is true for the element in (4.8). Thus we have found an element (4.7) in $\mathbf{V}(K)$ outside $\mathbf{V}(H)$, a contradiction. □

Corollary 4.20. *Let k be an infinite field, H an ideal of a free Lie algebra L of countable rank, and let $\mathbf{U}, \mathbf{V}$ be two nonzero varieties of Lie algebras over k. If $\mathbf{U}(L) = \mathbf{V}(H)$ then there exists a variety $\mathbf{W}$ such that $H = \mathbf{W}(L)$ and $\mathbf{U} = \mathbf{VW}$.*

Proof. Let α be an automorphism of L. Then

$$\mathbf{V}(\alpha(H)) = \alpha(\mathbf{V}(H)) = \alpha(\mathbf{U}(L)) = \mathbf{U}(\alpha(L)) = \mathbf{U}(L) = \mathbf{V}(H).$$

By Theorem 4.19, $\alpha(H) = H$. Applying Theorem 4.12, we see that H is a verbal ideal of the form $H = \mathbf{W}(L)$ where $\mathbf{W}$ is a suitable variety. Thus we obtain $\mathbf{U}(L) = \mathbf{V}(\mathbf{W}(L))$. Since L is a free Lie algebra of countable rank this yields $\mathbf{U} = \mathbf{VW}$. □

Proposition 4.21. *Let H, K be ideals of a free Lie algebra of countable rank over an arbitrary field k, and let $\mathbf{U}$, $\mathbf{V}$ be two varieties. If $\mathbf{U}(K) \subseteq \mathbf{V}(H)$ then either $\mathbf{U} \supseteq \mathbf{V}$ or $K \subseteq H$.*

Proof. Suppose $K \not\subseteq H$. As in the proof of Theorem 4.11, one may think that $L = H + K = L(Z)$, $\dim L/H = 1$, $Z = \{x\} \cup Z'$, $H = \mathrm{id}_L(Z')$. Since any nontrivial ideal of a free Lie algebra is not finitely generated, Z is infinite. There is an element q in K which has the form $q = x + h$, $h \in H$. Let Z_1 denote a finite subset in Z' such that $q \in \mathrm{alg}(\{x\} \cup Z_1)$ and $Z_2 = Z' \setminus Z_1$. We want to show that a free generating set for H can be chosen in the form

$$P = \{x^t z, q^u z' \mid z \in Z_1, z' \in Z_2, t, u = 0, 1, 2, \ldots\}.$$

Indeed, by Section 2.4, the corollary of Proposition 2.15, H is freely generated by $Q = \{x^t z \mid x \in Z', t = 0, 1, 2, \ldots\}$. We note that elements of P and Q are the same modulo H^2. Since $\bigcap_{n=1}^{\infty} L^n = \{0\}$, we may apply Theorem 4.12, according to which P is a free generating set for subalgebra $H_1 = \mathrm{alg}(P)$. It remains to verify $H_1 = H$. Using induction on u, we will show that $x^u z' \in H_1$, where $z' \in Z_2$. The basis of induction $u = 0$ is obvious. Now if $u > 0$ then

$$q^u z' = (x + h)^u z' = x^u z' + \sum_{m_1 + \cdots + m_p > 0} x^{l_1} h^{m_1} \cdots x^{l_p} h^{m_p} z'. \tag{4.9}$$

Since $h \in \mathrm{alg}(\{x\} \cup Z_1)$, we see that h is in a subalgebra generated by the elements of the form $x^t z$, $z \in Z_1$, $t = 0, 1, 2, \ldots$ Since multiplication by x is a derivation, we immediately see that the sum in equation (4.9) is, in fact, a linear combination of products of the form $x^r z$, $z \in Z_1$, $r = 0, 1, \ldots$, and $x^s z'$, $z' \in Z$, $s = 0, 1, \ldots, u - 1$. By the induction hypothesis, we get $x^u z' \in H_1$. Thus P is indeed a free generating set for H. Now let $H' = \mathrm{alg}\{q^u z' \mid z' \in Z_2, u = 1, 2, \ldots\} \subseteq H$. Since H' is generated by a part of a free generating set P for H, there exists a projection $\pi : H \to H'$. Here if $x \in \mathbf{V}(H) \cap H'$ then $\pi(x) = x \in H$ and $\pi(x) \in \pi(\mathbf{V}(H)) = \mathbf{V}(\pi(H)) = \mathbf{V}(H')$. Thus $x \in \mathbf{V}(H')$, that is, $\mathbf{V}(H') = \mathbf{V}(H) \cap H'$. In this case

$$\mathbf{U}(H') \subseteq \mathbf{U}(K) \cap H' \subseteq \mathbf{V}(H) \cap H' = \mathbf{V}(H').$$

Since H' is a free Lie algebra of countable rank, $\mathbf{U} \supseteq \mathbf{V}$. □

Corollary 4.22. *Let H and K be nonzero ideals of a free Lie algebra of countable rank over an infinite field k, and let $\mathbf{U}$ and $\mathbf{V}$ be two varieties over k such that $\mathbf{U}(K) = \mathbf{V}(H)$. Then there exists a variety $\mathbf{W}$ such that either $H = \mathbf{W}(K)$ and $\mathbf{U} = \mathbf{VW}$ or $K = \mathbf{W}(H)$ and $\mathbf{V} = \mathbf{UW}$.*

Proof. By Proposition 4.21, if none of the ideals H, K is contained in the other then $\mathbf{U} = \mathbf{V}$. By Theorem 4.19, then also $H = K$ so that we can put $\mathbf{W} = \mathbf{E}$. Otherwise, say, if $H \subseteq K$, then, using the corollary of Theorem 4.19 and the fact that a nonzero ideal of free Lie algebra of countable rank is of countable rank, too, we get $H = \mathbf{W}(K)$ and $\mathbf{U} = \mathbf{V}\mathbf{W}$ for a suitable variety $\mathbf{W}$. □

Theorem 4.23. *Let $v(k)$ be the semigroup of varieties of Lie algebras over an infinite field k. Then $v(k)$ is a free semigroup with 0 and 1, whose free generators are indecomposable varieties.*

Proof. By Proposition 4.18, every nontrivial variety is equal to the product of a set of indecomposable varieties. From Section 4.3.1, $v(k)$ is a semigroup. It remains to show that if $\mathbf{U}_1 \cdots \mathbf{U}_r = \mathbf{V}_1 \cdots \mathbf{V}_s$ where $\mathbf{U}_1, \ldots, \mathbf{U}_r, \mathbf{V}_1, \ldots, \mathbf{V}_s$ are indecomposable, then $r = s$ and $\mathbf{U}_1 = \mathbf{V}_1, \ldots, \mathbf{U}_r = \mathbf{V}_s$. So we assume that

$$\mathbf{U}_1\mathbf{U}_2 \cdots \mathbf{U}_r = \mathbf{V}_1\mathbf{V}_2 \cdots \mathbf{V}_s.$$

By indecomposability, we may assume also $r, s \geq 2$. We denote $\mathbf{U}_1' = \mathbf{U}_2 \cdots \mathbf{U}_r$, $\mathbf{V}_1' = \mathbf{V}_2 \cdots \mathbf{V}_s$. If L is of countable rank, we may rewrite $\mathbf{U}_1(\mathbf{U}_1'(L)) = \mathbf{V}_1(\mathbf{V}_1'(L))$. By the above corollary, there exists a variety $\mathbf{W}$ such that, for example, $\mathbf{U}'(L) = \mathbf{W}(\mathbf{V}_1(L))$ and $\mathbf{U} = \mathbf{V}_1\mathbf{W}$. From the first equation, we derive $\mathbf{U}_1' = \mathbf{W}\mathbf{V}_1'$. Since $\mathbf{U}$ is indecomposable, we have $\mathbf{W} = \mathbf{E}$, that is, $\mathbf{U}_1 = \mathbf{V}_1$ and $\mathbf{U}_1' = \mathbf{V}_1'$. Induction on r completes the proof. □

4.3.2 Indecomposable varieties

The complete description of indecomposable varieties is a very difficult problem which is still open. Since the number of varieties over any field of characteristic $p > 0$ (see Section 5.4, the corollary of Theorem 5.24), as well as over any uncountable field of characteristic zero (see Section 4.8.3), is uncountable, we see that this question is essentially equivalent to the question of describing all varieties. Obvious examples of indecomposable varieties are nilpotent varieties, as well as any variety generated by one simple algebra. Now let the *commutator* $\mathbf{W} = [\mathbf{U}, \mathbf{V}]$ of two varieties over a ring k be defined as a variety such that $\mathbf{W}(L) = \mathbf{U}(L)\mathbf{V}(L)$ (if $L = L(X)$ is considered as embedded in $A(X)$ then $\mathbf{W}(L) = [\mathbf{U}(L), \mathbf{V}(L)]$ which explains our terminology). Then the commutator of two indecomposable varieties, as a general rule, is itself indecomposable (see Zaicev [375] as well as Exercise 4.10.4).

4.4 Embedding theorems

4.4.1 Wreath product

Let $\mathbf{V}$ be a variety of Lie algebras over a commutative ring k with 1, A a Lie algebra in $\mathbf{V}$ and G an arbitrary Lie algebra over k. We call a Lie algebra W the $\mathbf{V}$-*wreath product*

of algebras A and *G* if the following conditions hold:
(i) $W = \mathrm{alg}(A, G)$.
(ii) Let $K = \mathrm{id}_W(A)$ be an ideal generated by A in W, then $K \in \mathbf{V}$.
(iii) Let $\varphi : A \to C$, $\psi : G \to C$ be homomorphisms from A and G into a k-Lie algebra C such that
 (a) $C = \mathrm{alg}(\varphi(A), \psi(G))$,
 (b) $S = \mathrm{id}_C(\varphi(A)) \in \mathbf{V}$,
 then there exists a unique homomorphism $\chi : W \to C$ such that $\chi|_A = \varphi, \chi|_g = \psi$.

The construction of the $\mathbf{V}$-wreath product of A by G will be denoted by $A \,\mathrm{wr}_{\mathbf{V}}\, G$.

To give an explicit form of the wreath product (we omit $\mathbf{V}$ if there is no risk of ambiguity), we start with the free product $P = A * G$ (see Section 2.6). Let Q be an ideal generated by A, V the least ideal of P containing the verbal ideal $\mathbf{V}(Q)$. Then $W \cong P/V$. The validity of all conditions as above is immediate from the universality of the free product and from the definition of the verbal subalgebra.

Proposition 4.24. *Let $W = A \,\mathrm{wr}_{\mathbf{V}}\, G$. Then W is a semidirect product of a subalgebra G by an ideal $K = \mathrm{id}_W(A)$. If $\mathbf{V}$ is a multihomogeneous variety and G is free as k-module and A is a $\mathbf{V}$-free algebra then K is also a $\mathbf{V}$-free algebra.*

Proof. In the construction of the wreath product given just above, K is the image of the ideal $Q = \mathrm{id}_P(A)$ where $P = A * G$. Hence Q is the kernel of the natural homomorphism from P onto G, that is, $Q \cap G = \{0\}$. Now if $\mathbf{V}$ is a multihomogeneous variety and $A = L(X, \mathbf{V})$ is a $\mathbf{V}$-free algebra then $W = A \,\mathrm{wr}_{\mathbf{V}}\, G$ can be constructed, starting with $A = L(X) = L$ where L is a free Lie algebra with the same generating set as A. We set $F = L * G$ and $J = \mathrm{id}_F L$. By homogeneity, $\widetilde{V} = \mathbf{V}(J)$ is an ideal of F (Theorem 4.12). Put $\widetilde{W} = F/\widetilde{V}$. Since any homomorphism φ of $A = L/\mathbf{V}(L)$ lifts back to a homomorphism ψ of L such that $\varphi = \psi\varepsilon$, ε being a natural homomorphism from L onto A, and any homomorphism ψ from L into an algebra in $\mathbf{V}$ factors through $A = L/\mathbf{V}(L)$, there is no problem in proving all the properties (i)–(iii) of the definition of the wreath product $W = A \,\mathrm{wr}_{\mathbf{V}}\, G$ for the algebra $\widetilde{W}$. Now $K = \mathrm{id}_W A \cong J/\mathbf{V}(J)$. By Section 2.6, Proposition 2.32, K is a $\mathbf{V}$-free algebra and a set of free generators for K can be chosen in the form

$$e_1 e_2 \cdots e_s x, \quad e_1 \geq e_2 \geq \cdots \geq e_s,\ e_i \in E,\ x \in X,\ s \geq 0,$$

where E is a well-ordered basis for G. □

Lemma 4.25. *Let A be an abelian ideal of a Lie algebra B over a ring k, $G = B/A$, G being k-free. Then B is isomorphic to a subalgebra (also denoted by B) of a semidirect product $S = G \ltimes C$ where C is an abelian Lie and $C \cap B = A$.*

Proof. By Section 1.6.5, A naturally becomes a G-module. We represent B in the form (G, A, f) (see Section 1.6.6). If U is the universal enveloping algebra for G then A is a left U-module. Let C be an injective hull for the U-module A (Section 2.5.6). We may construct the extension $S = (G, C, f)$. This is possible since the values of f are in the subalgebra A, embedded in C. Clearly, $B \subseteq S$ and $B \cap C = A$. With C being injective, the extension S splits (Section 2.5.6). Hence S can be considered as a semidirect product $G \rtimes C$, as required. □

Now we are ready to pass to the main results of this section. We will be denoting the **A**-product simply by wr (not $\mathrm{wr}_{\mathbf{A}}$).

Theorem 4.26. *Let M be an ideal of an absolutely free Lie algebra $L = L(\{x_i \mid i \in I\})$ over a ring k, A a free abelian algebra freely generated by $\{a_i \mid i \in I\}$. We denote the image of a g in L/M by $\bar{g}$. If L/M is k-free then the map $\varphi(x_i) = \bar{x}_i + a_i$, $i \in I$, induces an isomorphic imbedding $\tilde{\varphi} : L/M^2 \to A\,\mathrm{wr}(L/M)$.*

Proof. We denote the images of the generators x_i in L/M^2 by the same letters. Now φ extends to a homomorphism $\bar{\varphi} : L \to W = A\,\mathrm{wr}(L/M)$. It is obvious that $\ker \bar{\varphi} \supset M^2$. Hence there exists a homomorphism $\tilde{\varphi} : L/M^2 \to W$ mapping x_i in the same way as φ. Now suppose $\ker \tilde{\varphi} \neq \{0\}$. We apply Lemma 4.25 to L/M^2. So we set $A = M/M^2$ and $B = L/M^2$. Then $G \cong L/M$. By Lemma 4.25, L/M^2 is isomorphic to a subalgebra in an appropriate semidirect product $S = L/M \rtimes C$. We denote this isomorphism by μ. We write $\mu(x_i) = \bar{x}_i + c_i$ $(i \in I)$. By the definition of the wreath product (Section 4.4.1), the identity map from L/M into L/M and the map $a_i \mapsto c_i$ $(i \in I)$ extends to a homomorphism $\varphi : A\,\mathrm{wr}(L/M) \to S = L/M \rtimes C$. Thus we have a commutative diagram

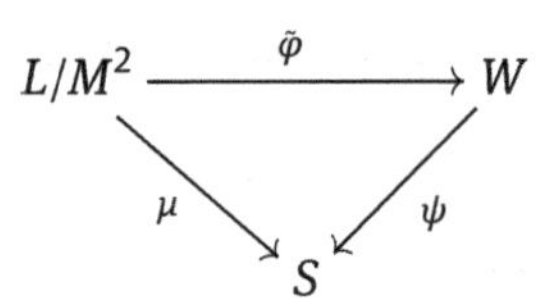

Since μ is a monomorphism and $\mu = \mu\tilde{\varphi}$, we see that $\tilde{\varphi}$ is a monomorphism too. □

Corollary 4.27. *Let k be an arbitrary ring, and suppose $M = L(\{x_i \mid i \in I\}, \mathbf{A}^2)$ is a free algebra in the variety of all metabelian Lie algebras over k with free generating set $\{x_i \mid i \in I\}$. We denote by A the free abelian algebra with basis $\{y_i \mid i \in I\}$, $R = k[y_i \mid i \in I]$ the polynomial ring in $\{y_i \mid i \in I\}$, and P a free R-module with a basis $\{z_i \mid i \in I\}$. Then the map $x_i \mapsto y_i + z_i$ extends to a monomorphism from M to the semidirect product $S = A \rtimes P$.*

Theorem 4.28 (A. L. Shmelkin). *Let $\mathbf{V}$ be a multihomogeneous variety of Lie algebras over a field k, $L = L(\{x_i \mid i \in I\})$ a free Lie algebra over k with free generating set $\{x_i \mid i \in I\}$, M an ideal in L, and $A = L(\{y_i \mid i \in I\}, \mathbf{V})$. Let $\bar{g}$ denote the image of $g \in L$ in L/M. Then the map $x_i \mapsto \bar{x}_i + y_i$ $(i \in I)$ extends to a monomorphism*

$$L/\mathbf{V}(M) \to A\,\mathrm{wr}_{\mathbf{V}}(L/M).$$

Proof. As in Theorem 4.26, there exists a homomorphism $\psi : L/\mathbf{V}(M) \to A\,\mathrm{wr}_{\mathbf{V}}(L/M)$ extending the map $\varphi : x_i \to \bar{x}_i + y_i$. We want to show that ψ is a monomorphism. By our hypothesis, $\mathbf{V}(M) \subseteq M^2$ and therefore, adopting the notation in the proof of Theorem 4.26, we arrive at a commutative diagram of the form

$$\begin{array}{ccc} L/\mathbf{V}(M) & \xrightarrow{\psi} & A\,\mathrm{wr}_{\mathbf{V}}(L/M) \\ \Big\downarrow{\scriptstyle \varepsilon} & & \Big\downarrow{\scriptstyle \chi} \\ L/M^2 & \xrightarrow[\tilde{\varphi}]{} & (A/A^2)\,\mathrm{wr}(L/M) \end{array}$$

According to Theorem 4.26, the map in the lower row is injective. If we assume, for a moment, that ψ is not injective, there must exist an element $g \in M/\mathbf{V}(M)$ such that $\psi(g) = 0$. But $M/\mathbf{V}(M)$ is a free algebra of a multihomogeneous variety $\mathbf{V}$. Under ψ, it is mapped into a free algebra K of $\mathbf{V}$ (see Proposition 4.24), where $K = \mathrm{id}_W A$, $W = A\,\mathrm{wr}_{\mathbf{V}}(L/M)$. From the explicit form of a free generating set for K (see the quotation above), we derive that under χ the ideal K is mapped onto $K_1 = \mathrm{id}_{W_1}(A/A^2)$ where $W_1 = (A/A^2)\,\mathrm{wr}(L/M)$. But, under $\tilde{\varphi}\varepsilon$, the quotient algebra M/M^2 is mapped into K_1 injectively. Hence ψ is an injective map of $M/\mathbf{V}(M)$ into K modulo the derived subalgebras of these algebras. By Theorem 4.14, $\psi(M/\mathbf{V}(M))$ is a free $\mathbf{V}$-free algebra freely generated by a set of its elements linearly independent modulo the derived subalgebra of K. Therefore ψ is a monomorphism. □

Corollary 4.29. *Let* $\mathbf{U}$ *and* $\mathbf{V}$ *be two varieties of Lie algebras over a field* k, $\mathbf{U}$ *being multihomogeneous. Then the free algebra* $L(\{x_i \mid i \in I\}, \mathbf{UV})$ *of the product variety* $\mathbf{UV}$ *is isomorphic to a subalgebra in a* $\mathbf{U}$*-wreath product of the form*

$$L(\{y_i \mid i \in I\}, \mathbf{U})\,\mathrm{wr}_{\mathbf{U}}\,L(\{z_i \mid i \in I\}, \mathbf{V})$$

generated by the elements $y_i + z_i$, $i \in I$.

4.5 Free algebras of varieties. Identities

4.5.1 Basic rank of variety

Let $\mathbf{V}$ be a variety over a ring k. It is an easy consequence of Birkhoff's theorem (Theorem 4.3) that $\mathbf{V}$ is generated by its free algebra $L_\infty(\mathbf{V})$ of countable rank. In many cases, $\mathbf{V}$ can also be generated by a free algebra of some finite rank r. A minimal positive number r with this property is then called the *basic rank* of $\mathbf{V}$ and it is denoted by $r_b(\mathbf{V})$. For instance, $r_b(\mathbf{A}) = 1$, $r_b(\mathbf{O}) = 2$. The former of these two being trivial, the latter follows since a free Lie algebra of rank 2 possesses free subalgebras of countable rank. If $\mathbf{V}$ is not generated by a free algebra of some finite rank (hence by any collection of algebras with bounded number of generators) then we say that $\mathbf{V}$ is a variety of *infinite basic rank* and we write $r_v(\mathbf{V}) = \infty$.

Theorem 4.30. *Let* $\mathbf{V}$ *be a nilpotent variety of nilpotent step at most* c *(that is,* $\mathbf{V} \subseteq \mathbf{N}_c$*) over an arbitrary ring* k*. Then* $r_b(\mathbf{V}) \le c$.

Proof. Since any variety is generated by a set of all its free algebras of finite rank, it is sufficient to show that $G_n = L(n, \mathbf{V})$, $n > c$, is a subalgebra of a direct power of $G_c = L(c, \mathbf{V})$. Now let J denote the set of all increasing sequences of the form $\alpha = (m_1, m_2, \dots, m_q)$, $1 \le m_i \le n$, $i = 1, \dots, c$, $1 \le q \le c$. We consider G_c^J. Let $x_1, \dots, x_c$ be free generators of G_c and suppose the elements $t_1, \dots, t_n \in G_c^J$ have the form $t_{m_i}(\alpha) = x_i$ if $m_i \in \alpha$ and $t_s(\alpha) = 0$ if $s \notin \alpha$.

We will show that $G_n \cong \operatorname{alg}(t_1, \dots, t_n)$. Indeed, let $f(y_1, \dots, y_n)$ be a nonassociative polynomial such that $f(t_1, \dots, t_n) = 0$. We have to show that $f(y_1, \dots, y_n) \equiv 0$ is an identity in $\mathbf{V}$. Clearly, we may suppose from the very beginning that f is of degree at most c. If $\alpha \in J$ then we define π_α as a projection of $L(y_1, \dots, y_n)$ such that $\pi_\alpha(y_{m_i}) = y_{m_i}$ if $m_i \in \alpha$ and $\pi_\alpha(y_s) = 0$ otherwise. Then, clearly,

$$f(y_1, \dots, y_n) = \sum_{\alpha \in J} \lambda_\alpha \pi_\alpha f(y_1, \dots, y_n), \quad \lambda_\alpha \in k,$$

so that the validity of $f(y_1, \dots, y_n) \equiv 0$ is equivalent to the validity of the system of identities of the form $\pi_\alpha f(y_1, \dots, y_n) \equiv 0$ for all $\alpha \in J$. But if $f(t_1, \dots, t_n) = 0$ then $f(t_1(\alpha), \dots, t_n(\alpha)) = 0$ for all α. Replacing in the equation thus obtained elements x_i by y_i, $i = 1, \dots, c$, we will get precisely $\pi_\alpha f(y_1, \dots, y_n) \equiv 0$. This replacement is correct since $x_1, \dots, x_c$ are free generators of G_c. Hence $f(y_1, \dots, y_n) \equiv 0$ is an identity in $\mathbf{V}$, and $\operatorname{alg}(t_1, \dots, t_n) \cong L(n, \mathbf{V})$. □

Now we pass to the varieties of infinite basic rank.

Proposition 4.31. *Over an arbitrary commutative ring* k *with unity the basic rank of a variety of the form* $\mathbf{V} = \mathbf{N}_c\mathbf{N}_d\mathbf{N}_e$, $c, d, e \ge 1$ *is infinite.*

Proof. Let n be an arbitrary natural number, $n \ge 2$. We want to show that $G = L(n, \mathbf{V})$ satisfies an identity which is nontrivial in $\mathbf{V}$. Let $p = \dim L(n, \mathbf{V})$, $q = p + c + 1$. We set

$$f(x_1, \dots, x_q, y_1, \dots, y_q, z) = ((z^e y_1)^d \bar{x}_1) \cdots ((z^e y_q)^d \bar{x}_q), \tag{4.10}$$

where the bars mean that f is obtained by alternation over the set of variables $\{x_1, \dots, x_q\}$. We choose a basis in G_n so that all its elements, except some p of them, are in G_n^{e+1}. Since $f(x_1, \dots, z)$ is multilinear in $x_1, \dots, x_n$, when we compute a value of this polynomial in G it is sufficient to substitute for $x_1, \dots, x_n$ an arbitrary collection of pairwise distinct elements of the chosen basis. While doing this, in each of the $q!$ summands at least $c + 1$ factors take value in $(G_n^{e+1})^{d+1}$. Since $((G_n^{e+1})^{d+1})^{c+1} = \{0\}$, we see that $f(x_1, \dots, x_q, y_1, \dots, y_q, z) \equiv 0$ is an identity of G_n. Now we will show that equation (4.10) is a nontrivial element in $G = L(\{x_1, \dots, x_q, y_1, \dots, y_q, z\}, \mathbf{V})$. For this we note, as it follows by Section 2.4, Proposition 2.15, that G^{e+1} is a free algebra in $\mathbf{N}_c\mathbf{N}_d$ whose free generators are all right-normed basic monomials of length at least $e + 1$.

Selecting a usual Hall's basis (Section 2.3, Theorem 2.10) on the set $\{x_1, \ldots, x_q, y_1, \ldots, y_q, z\}$ with the ordering $x_1 > \cdots > x_q > y_1 > \cdots > y_q > z$, we will see that each monomial f_σ of equation (4.10) up to a sign is a product of generators of G^{e+1}: if we put $u_i = z^e y_i$, $v_{ij} = x_j z^e y_i$, $i, j = 1, \ldots, q$, then

$$\begin{aligned} f_\sigma &= ((z^e y_1)^d x_{\sigma(1)}) \cdots ((z^e y_q)^d x_{\sigma(q)}) \\ &= \pm(u^{d-1} v_{1\sigma(1)}) \cdots (u_q^d v_{q\sigma(q)}), \end{aligned} \tag{4.11}$$

with σ being a permutation from S_q. Any relation among free generators, as we know, is an identity in the corresponding variety. Since for distinct σ the monomials f_σ turn out to depend on distinct sets of free generators of the free algebra G^{e+1}, we see that $\mathbf{N}_c\mathbf{N}_d$ satisfies an identity of the form

$$g = (u_1^{d-1} v_{11}) \cdots (u_q^{d-1} v_{qq}) \equiv 0. \tag{4.12}$$

As a matter of fact, this identity does not hold even in $\mathbf{AN}_d$. For the verification, it is enough to take a semidirect product of the form $S = P \rtimes M$, where P is a free nilpotent algebra of nilpotent step d with free generating set $a_1, \ldots, a_q$ and M is a *free cyclic* P-module freely generated by an element b (Section 1.6.7). We substitute a_1 for v_{11}, $\ldots$, a_{q-1} for $v_{q-1,q-1}$, a_q for $u_1, \ldots, u_q$, and b for v_{qq}. Then we will get

$$(a_q^{d-1} a_1) \cdots (a_q^{d-1} a_{q-1})(a_q^{d-1} b). \tag{4.13}$$

Since the elements $a_q^{d-1} a_1, \ldots, a_q^{d-1} a_{q-1}, a_q$ are linearly independent in P, then by the PBW theorem (Section 2.5) the element $(a_q^{d-1} a_1) \cdots (a_q^{d-1} a_{q-1}) a_q^{d-1}$ is nonzero in $U(P)$ and then monomial (4.13) is nonzero in S. □

Proposition 4.32. *Over an arbitrary commutative ring k with unity the basic rank of any variety of the form* $\mathbf{V} = \mathbf{N}_c\mathbf{N}_d$, $c \geq 1$, $d \geq 2$ *is infinite.*

Proof. Let n be an arbitrary natural number $n \geq 2$. We want to indicate an identity that is true in $G_n = L(n, \mathbf{V})$ but is not true in the whole of $\mathbf{V}$. Now let $p = \dim L(n, \mathbf{N}_{d-1})$, $q = p + c + 1$. We set

$$f(x_1, \ldots, x_q, y_1, \ldots, y_q) = (\bar{x}_1 y_1) \cdots (\bar{x}_q y_q), \tag{4.14}$$

the bars having the same meaning as previously. Then, by choosing a basis in G_n in the same way as is done in Proposition 4.31, we will see that, when substituting for $x_1, \ldots, x_n$ some basic elements of G_n, at least $c+1$ factor in each term of the sum in equation (4.14) will take value in G_n^d. Hence $f(x_1, \ldots, x_q, y_q, \ldots, y_q) \equiv 0$ is an identity of G_n. To verify its nontriviality, we consider a semidirect product of the form $S = G \rtimes M$ of a free d-step nilpotent algebra G freely generated by a set of the form $a_1, \ldots, a_q, b_1, \ldots, b_{q-1}$ with a free cycling G-module M freely generated by an element m. We consider a value of $f(x_1, \ldots, y_q)$ in S of the form

$$g = (\bar{a}_1 b_1)(\bar{a}_2 b_2) \cdots (\bar{a}_{q-1} b_{q-1})(\bar{a}_q m). \tag{4.15}$$

It is enough to show that $u = (\bar{a}_1 b_1)(\bar{a}_2 b_2)\cdots(\bar{a}_{q-1} b_{q-1})\bar{a}_q$ is a nonzero element in $U(G)$. But since $d \geq 2$, the set of elements of the form $\{a_i b_j \mid i,j = 1,\ldots,q-1\}$ is linearly independent, and we can endow it with an ordering such that $a_i b_j > a_m b_l$ if $j > l$. Hence, by the PBW theorem (Section 2.5), we can say that u is a linear combination with coefficients ± 1 of distinct elements of a basis for $U(G)$. Therefore, monomial (4.15) is a nonzero element in S. Since $S \in \mathbf{A}\mathbf{N}_d \subseteq \mathbf{N}_c\mathbf{N}_d = \mathbf{V}$, we see that $f \equiv 0$ is not an identity in $\mathbf{V}$, so that the basic rank of $\mathbf{V} = \mathbf{N}_c\mathbf{N}_d$ is infinite. □

To formulate the main theorem of this section, we introduce the *variety of polynilpotent Lie algebras* associated with a sequence of natural numbers $(c_1,\ldots,c_t)$, $c_i \geq 1$, as the class of all Lie algebras over k such that if G is such an algebra then it possesses a finite series of the form

$$G = G_1 \supset G_2 \supset \cdots \supset G_t \supset G_{t+1} = \{0\},$$

where, for each $i = 1,2,\ldots,t$, the quotient algebra G_i/G_{i+1} is nilpotent of the nilpotent step at most c_i. It is clear that this class, denoted by $\mathbf{P}(c_1,\ldots,c_t)$, is a variety and it is equal to the product variety of the form $\mathbf{N}_{c_t}\cdots\mathbf{N}_{c_1}$.

Theorem 4.33. *Over an arbitrary commutative ring k with unity, any variety $\mathbf{P}(c_1,\ldots,c_t)$ of all polynilpotent Lie algebras corresponding to a sequence $(c_1,\ldots,c_t)$ of natural numbers has infinite basic rank unless, possibly, $t = 1$ or $t = 2$, $c_1 = 1$.*

Proof. We remark that if $\mathbf{V}$ is a variety of infinite basic rank then, for any $c \geq 1$, the variety $\mathbf{N}_c\mathbf{V}$ is of infinite basic rank, too. Indeed, given a nontrivial identity of $L(n,\mathbf{V})$, the identity of the form

$$f(x_1^1,\ldots,x_m^1)\cdots f(x_1^{c+1},\ldots,x_m^{c+1})x_{m+1} \equiv 0 \tag{4.16}$$

is a nontrivial identity of $L(n,\mathbf{N}_c\mathbf{V})$. Its validity in $L(n,\mathbf{N}_c\mathbf{V})$ is obvious. The verification that this identity is nontrivial in $\mathbf{N}_c\mathbf{V}$ can be achieved as in Proposition 4.31 and 4.32, by considering a semidirect product of the form $S = G \rtimes M$ where G is a free algebra in $\mathbf{V}$ of rank $(c+1)m$ and M a free cyclic G-module with a free generator x_{m+1}. Here we choose that a free generating set for G is $\{x_1^1,\ldots,x_m^{c+1}\}$. In this case the left-hand side of equation (4.16) is an element of S. This element is nontrivial since, obviously, $f(x_1^1,\ldots,x_m^1)\cdots f(x_1^{c+1},\ldots,x_m^{c+1})$ is a nonzero element of the universal enveloping algebra $U(G)$ for G.

This remark reduces our proof to already proved Propositions 4.31 and 4.32. □

4.5.2 Exceptional cases

As has been shown previously (Theorem 4.30), the basic rank of $\mathbf{N}_c$ is finite and is at most c. Concerning $\mathbf{N}_c\mathbf{A}$, the basic rank of such a variety can be both finite or infinite depending on the base ring k. An exhaustive answer to this question has not been found, yet. Thus we restrict ourselves to two examples.

Example 4.34. Let k be a finite field, $|k| = p^m$, where p is a prime number. Then $r_b(\mathbf{A}^2) = \infty$.

Indeed, in this case $G = L(n, \mathbf{A}^2)$ satisfies an identity of the form

$$\tilde{x}_1^q \tilde{x}_2^{q^2} \cdots \tilde{x}_s^{q^s} y_1 y_2 \equiv 0, \tag{4.17}$$

where tildes over letters mean symmetrization and $s = n(p-1) + 1$. This identity is nontrivial since the commutative polynomial $f = \tilde{x}_1^q \tilde{x}_2^{q^2} \cdots \tilde{x}_s^{q^s}$ is a nonzero polynomial in $k[x_1, x_2, \ldots, x_s]$. Now the validity of equation (4.17) on G follows since f becomes zero each time when $x_1, \ldots, x_s$ are replaced by elements in the linear span of $x_1, \ldots, x_n$. Indeed, a standard argument shows that, over k, our polynomial f is multilinear. Now if we substitute elements of the set $\{x_1, \ldots, x_n\}$ for $x_1, \ldots, x_s$ then at least one of these elements will occur $t \geq p$ times. Having assembled like terms, we will see that each monomial has coefficient divisible by $t!$ and this is equal to zero in k.

Example 4.35. Let k be an infinite field, c a natural number, $c \geq 1$. Then $r_b(\mathbf{N}_c\mathbf{A}) \leq c+1$.

Since the algebra of all triangular matrices of order $c + 1$, $G = \mathfrak{t}(c + 1, k)$, can be generated by $c + 1$ elements it is sufficient to prove that $\mathbf{N}_c\mathbf{A}$ is generated by G. This will, at the same time, show that the identities of the algebra of all upper triangular matrices are consequences of a single one of them

$$(x_1 x_2) \cdots (x_{2c+1} x_{2c+2}) \equiv 0.$$

For the proof we consider the field $\bar{k}$ of rational functions in the variables $x_{\alpha i}, y_{\alpha i}$, $\alpha = 1, 2, \ldots, i = 1, 2, \ldots, c$. Now let $\bar{G} = \mathfrak{t}(c+1, \bar{k})$. Since the base field is infinite, the identities of G have multihomogeneous basis. Thus $\bar{G} = G \otimes_k \bar{k}$ belongs to the variety generated by G. We will show that the free algebra $L_\infty(\mathbf{N}_c\mathbf{A})$ is isomorphic to a subalgebra of $\bar{G}$. For this we consider matrices of the form

$$u_\alpha = \begin{pmatrix} x_{\alpha 1} & y_{\alpha 1} & 0 & \ldots & 0 & 0 \\ 0 & x_{\alpha 2} & y_{\alpha 2} & & 0 & 0 \\ \vdots & \vdots & \vdots & \ddots & \vdots & \vdots \\ 0 & 0 & 0 & \ldots & x_{\alpha c} & y_{\alpha c} \\ 0 & 0 & 0 & \ldots & 0 & 0 \end{pmatrix}, \quad \alpha = 1, 2, \ldots$$

We want to show $L_\infty(\mathbf{N}_c\mathbf{A}) \cong \mathrm{alg}(\{u_\alpha \mid \alpha = 1, 2, \ldots\})$. Let R be the polynomial ring $R = k[x_{\alpha i}, y_{\beta i} \mid \alpha, \beta = 1, 2, \ldots, n,\ i = 1, 2, \ldots, c]$, G the Lie algebra of upper triangular matrices over R of order $c + 1$. We consider the $\mathbf{N}_c$-wreath product $W = A \operatorname{wr}_{\mathbf{N}_c} B$ where A is a free nilpotent Lie algebra of nilpotent step c and free generators $y_1, \ldots, y_n$ and B

a free abelian Lie algebra with free generating set $z_1, \dots, z_n$. We map

$$z_\alpha \mapsto \begin{pmatrix} x_{\alpha 1} & & & & \\ & x_{\alpha 2} & & & \\ & & \ddots & & \\ & & & x_{\alpha c} & \\ & & & & 0 \end{pmatrix} \quad (\alpha = 1, \dots, n),$$

$$y_\beta \mapsto \begin{pmatrix} 0 & y_{\beta 1} & 0 & \dots & 0 \\ 0 & 0 & y_{\beta 2} & \dots & 0 \\ \vdots & \vdots & \vdots & \ddots & \vdots \\ 0 & 0 & 0 & \dots & y_{\beta c} \\ 0 & 0 & 0 & \dots & 0 \end{pmatrix} \quad (\beta = 1, \dots, n).$$

By the definition of the wreath product, this map extends to a homomorphism ψ from W into $\bar{G}$. It is sufficient to verify that the images of elements of the form $z_{\alpha_1} z_{\alpha 2} \cdots z_{\alpha i} y_i$, $\alpha_1 \geq \alpha_2 \geq \dots \geq \alpha_t$, $\alpha = 1, \dots, n$, generate a free nilpotent algebra of nilpotent step c. A matrix associated with this element has the form

$$\begin{pmatrix} 0 & (x_{\alpha_1 1} - x_{\alpha_1 2}) \cdots (x_{\alpha_t 1} - x_{\alpha_t 2}) y_{\beta 1} & 0 & \dots & 0 \\ 0 & 0 & (x_{\alpha_1 2} - x_{\alpha_1 3}) \cdots (x_{\alpha_t 2} - x_{\alpha_t 3}) y_{\beta_2} & \dots & 0 \\ \vdots & \vdots & \vdots & \ddots & \vdots \\ 0 & 0 & 0 & \dots & x_{\alpha_1 c} \cdots x_{\alpha_t c} y_{\beta c} \\ 0 & 0 & 0 & \dots & 0 \end{pmatrix}.$$

By a change of variables of the form $x'_{\alpha i} = x_{\alpha i} - x_{\alpha, i+1}$ $(i = 1, \dots, c-1)$, $x'_{\alpha c} = x_{\alpha c}$, we may assume that we are dealing with the set of matrices of the form

$$\begin{pmatrix} 0 & x_{\alpha_1 1} \cdots x_{\alpha_t 1} y_{\beta 1} & 0 & \dots & 0 \\ 0 & 0 & x_{\alpha_1 2} \cdots x_{\alpha_t 2} y_{\beta 2} & \dots & 0 \\ \vdots & \vdots & \vdots & \ddots & \vdots \\ 0 & 0 & 0 & \dots & x_{\alpha_1 c} \cdots x_{\alpha_t c} y_{\beta c} \\ 0 & 0 & 0 & \dots & 0 \end{pmatrix}. \tag{4.18}$$

It is sufficient to show that these matrices generate a free nilpotent algebra of nilpotent step c in the associative sense. Applying induction over c proves that we may restrict ourselves to considering the monomials in the matrices of the form (4.18) of degree precisely c. Now it is easy to compute

$$\begin{pmatrix} 0 & a_{11} & 0 & \dots & 0 \\ 0 & 0 & a_{12} & \dots & 0 \\ \vdots & \vdots & \vdots & \ddots & \vdots \\ 0 & 0 & 0 & \dots & a_{1c} \\ 0 & 0 & 0 & \dots & 0 \end{pmatrix} \cdots \begin{pmatrix} 0 & a_{c1} & 0 & \dots & 0 \\ 0 & 0 & a_{c2} & \dots & 0 \\ \vdots & \vdots & \vdots & \ddots & \vdots \\ 0 & 0 & 0 & \dots & a_{cc} \\ 0 & 0 & 0 & \dots & 0 \end{pmatrix}$$

$$= \begin{pmatrix} 0 & 0 & \dots & a_{11} \dots a_{cc} \\ 0 & 0 & \dots & 0 \\ \dots & \dots & \dots & \dots \\ 0 & 0 & \dots & 0 \\ 0 & 0 & \dots & 0 \end{pmatrix}. \tag{4.19}$$

Denoting $(\alpha_1, \dots, \alpha_t) = A$, we may rewrite equation (4.18) in the form

$$\begin{pmatrix} 0 & x_{A1}y_{\beta 1} & 0 & \dots & 0 \\ 0 & 0 & x_{A2}y_{\beta 2} & \dots & 0 \\ \vdots & \vdots & \vdots & \ddots & \vdots \\ 0 & 0 & 0 & \dots & x_{Ac}y_{\beta c} \\ 0 & 0 & 0 & \dots & 0 \end{pmatrix}.$$

Any product of such matrices which are the images of elements $z_{A1}y_{\beta 1}, \dots, z_{Ac}y_{\beta c}$ is a matrix of the type (4.19), in the upper-right corner of which we find the element

$$x_{A_1 1}y_{\beta_1 1}x_{A_2 2}y_{\beta_2 2} \cdots x_{A_c c}y_{\beta_c c}.$$

Clearly, all such monomials are linearly independent. Hence, applying induction over c, we see that $\psi : W \to G$ is an imbedding.

To complete the proof, we remark that $\bar{G} = \mathfrak{t}(c+1,k) \otimes_k R$ is in $\operatorname{var}(\mathfrak{t}(c+1,k))$. By Theorem 4.28, $L(n, \mathbf{N}_c\mathbf{A})$ can be embedded in W as above for any natural number n, hence $\mathbf{N}_c\mathbf{A} \subseteq \operatorname{var}(\mathfrak{t}(c+1,k))$. Now $\mathfrak{t}(c+1,k)$ can be generated by $c+1$ elements. Hence $r_b(\mathbf{N}_c\mathbf{A}) \le c+1$.

4.6 Free algebras of varieties. Subalgebras

4.6.1 Definition of Schreier variety

A variety of Lie algebras over k is called *Schreier* if any subalgebra H of a free algebra $L(X, \mathbf{V})$ is free in this variety, that is, for a suitable Y, we have $H \cong L(Y, \mathbf{V})$. This terminology comes down to us from group theory where there is a theorem of Nielsen–Schreier on the freeness of subgroups in free groups. By Section 4.2, Theorem 4.6, if k is a field then any subalgebra of an absolutely free Lie algebra is itself (absolutely) free. Thus any variety of the form $\mathbf{O}_I$ determined by all identities of the form $\lambda x \equiv 0$, $\lambda \in I$, is Schreier, provided that I is maximal. On the other hand, if I is such that k/I is the principal ideal domain (that is, principal ideal ring without zero divisors) then $\mathbf{A}_I$, the variety given by $xy \equiv 0$ and $\lambda x \equiv 0$ for all $\lambda \in I$, is obviously Schreier. Indeed, in this case, a subalgebra of a free algebra is merely a submodule of a free k/I-module. But over any principal ideal domain any submodule of a free module is itself free (see,

e. g., Lang [199, Chapter XV]), proving our claim. A trivial example of Schreier varieties is that of **E**, the variety consisting of zero algebra only. Our principal result in this section is the proof that these examples exhaust the list of Schreier varieties.

It should be noted that, given a Schreier variety **V** and its annihilator Ann **V** (see Section 4.1.1), we always have $\bar{k} = k/\operatorname{Ann}\mathbf{V}$ integral, i. e., without zero divisors. Indeed, the free algebra of rank 1 in **V** is just $\bar{k}$ and its subalgebras are left ideals of $\bar{k}$. By our hypothesis, all nonzero ideals of $\bar{k}$ must be isomorphic to $\bar{k}$ itself. Hence, trivially, $\bar{k}$ has no zero divisors. It is clear that, for our purposes, it is sufficient to consider **V** just over $\bar{k}$. Hence, in what follows, we assume Ann $\mathbf{V} = \{0\}$ and consider k having no zero divisors. We also recall that $N(\mathbf{V})$ denotes the norm of a variety **V** (Section 4.3.1).

Lemma 4.36. *If* $\mathbf{V} \neq \{\mathbf{O}\}$ *is a Schreier variety then* $N(\mathbf{V}) \leq 3$.

Proof. Let L denote a free Lie algebra of countable rank. If **V** is a verbal ideal in L corresponding to **V** such that $N(\mathbf{V}) = r$ then there exists an element $g(x_1, x_2, \dots, x_r)$ whose lowest term is of degree r. Using linearization (Theorem 4.5), we may assume that this lowest term is multilinear. Now let **V** be a Schreier variety and $N(\mathbf{V}) = r$. Then there is a polynomial $v = v(x_1, \dots, x_{2r-1}) \in V \cap (L^2)^{r-1}$ with lowest term of degree $2r-1$. Indeed, we choose $g(x_1, x_2, \dots, x_r)$ as above and replace in it x_1 by x_1x_r, x_2 by $x_2x_{r+1}, \dots, x_{r-1}$ by $x_{r-1}x_{2r-2}$, x_r by x_{2r-1}. Using bases (see, e. g., Section 4.7) consisting of monomials of the form

$$x_{\sigma(1)}x_{\sigma(2)}\cdots x_{\sigma(r-1)}x_r \quad \text{and} \quad x_{\tau(1)}x_{\tau(2)}\cdots x_{\tau(2r-2)}x_{2r-1}$$

in the spaces of multilinear polynomials in $x_1, \dots, x_r$ and $x_1, \dots, x_{2r-1}$, we will see the following. After a substitution as above, $x_{\sigma(1)}x_{\sigma(2)}\cdots x_{\sigma(r-1)}x_r$ takes the form

$$(x_{\sigma(1)}x_{\sigma(1)+r-1})(x_{\sigma(2)}x_{\sigma(2)+r-1})\cdots(x_{\sigma(r-1)}x_{\sigma(r-1)+r-1})x_{2r-1}.$$

Removing the brackets, we will see that the above product at distinct σ gives rise to different nonzero elements of the second of the bases indicated above. Hence the lowest term of the polynomial obtained in this way is indeed of degree $2r-1$. Obviously, it belongs to $(L^2)^{r-1}$.

Now suppose $N(\mathbf{V}) > 3$. We choose an element $v \in V \cap (L^2)^{r-1}$ with lowest term of degree $2r-1$. Then, for a suitable $u(x_1, \dots, x_m) \in L^{r-1}$ there exist $y_1, \dots, y_m \in L^2$ such that $v(x_1, \dots, x_{2r-1}) = u(y_1, \dots, y_m)$. Let $\{h_1, h_2, \dots, h_s, \dots\}$ be a set of elements in L^2 generating L^2 modulo V. Then $y_i = f_i(h_1, \dots, h_s) + v_i$, $v_i \in V$, $i = 1, \dots, m$, and, further,

$$\begin{aligned} v(x_1, \dots, x_{2r-1}) &= u(f_1(h_1, \dots, h_s) + v_1, \dots, f_m(h_1, \dots, h_s) + v_m) \\ &= u(f_1(h_1, \dots, h_s), \dots, f_m(h_1, \dots, h_s)) + v'(x_1, \dots, x_p) \\ &= w(h_1, \dots, h_s) + v'(x_1, \dots, x_p). \end{aligned}$$

Here the degree of the lowest term of $v'(x_1, \dots, x_p)$ is at least $r+(r-2)2 = 3r-4 > 2r-1$ when $r > 3$. Since $h_1, \dots, h_s \in L^2$, the degree of the lowest term of $w(x_1, \dots, x_s)$ must be less than r. Now we consider a free algebra L of countable rank in $\mathbf{V}$. Then L^2/V is a free algebra in $\mathbf{V}$. If $\{h_1, h_2, \ldots, h_s, \ldots\}$ freely generate L^2 modulo V then there exists a polynomial $w(x_1, \ldots, x_s)$ whose lowest term has degree strictly less than r and such that $w(h_1, \ldots, h_s) = 0$. Thus $w(x_1, \ldots, x_s) = 0$, as soon as we substitute free generators of L^2/V for $x_1, \ldots, x_s$. As a result we have obtained an identity of $\mathbf{V}$ whose lowest term has degree strictly less than r, a contradiction. □

Proposition 4.37. *If k is an infinite ring and $\mathbf{V}$ is a Schreier variety with nontrivial identity of degree greater than 1, then $\mathbf{V}$ is an abelian variety.*

Proof. Suppose $v = v_1 + v_2 + \cdots + v_n$ is a nonassociative polynomial such that $v \equiv 0$ is a nontrivial identity of $\mathbf{V}$ mentioned in the statement and $\deg v_q = q$. Since k is infinite, a usual trick with Vandermonde's determinant (Theorem 4.6) shows that there exist $\lambda_1, \lambda_2, \ldots, \lambda_n \in k$ such that $\lambda_1 v_1 \equiv 0, \ldots, \lambda_n v_n \equiv 0$ are all identities of $\mathbf{V}$.

By Lemma 4.36, we may assume that $N(\mathbf{V}) \leq 3$. If $N(\mathbf{V}) = 3$ then $\mathbf{V}$ satisfies an identity of the form

$$\lambda x_2 x_1 x_3 + \mu x_3 x_1 x_2 \equiv 0.$$

If $\mu \neq 0$ then, replacing x_3 by x_1, we will get $\mu x_1^2 x_2 \equiv 0$. If $x_1^2 x_2 \equiv 0$ is not an identity of $\mathbf{V}$ then $k x_1^2 x_2$ is a $\mathbf{V}$-free algebra of rank 1, hence $\mathbf{V}$ satisfies $\mu x \equiv 0$, contradicting our assumption that the annihilator of a variety must be trivial (see the final part of Section 4.6.1). Thus $\mathbf{V}$ satisfies $x_1^2 x_2 = 0$. In this case the subalgebra generated by elements x_1 and $x_1 x_2$ must be abelian, hence of rank 1. If $w(x_1, x_2)$ is a generator of this subalgebra then $x_1 - \lambda' w(x_1, x_2) \in V$ and $x_1 x_2 - \mu' w(x_1, x_2) \in V$ so that $N(\mathbf{V}) \leq 2$, a contradiction.

Now if $N(\mathbf{V}) = 2$ then separating homogeneous components as previously shows that $\mathbf{V}$ is indeed abelian. □

Lemma 4.38. *Let $\mathbf{V}$ be a variety of Lie algebras over a finite field k and $N(\mathbf{V}) \leq 3$. If H is a $\mathbf{V}$-free algebra of finite rank then $H/H^{(2)}$ is a finite-dimensional algebra.*

Proof. We pass to $\mathbf{V} \cap \mathbf{A}^2$. Since its norm is 3, we may think that $H^{(2)} = \{0\}$. Proceeding as in the previous proof, we see that V contains an element of the form

$$x_1^2 x_2 + f(x_1, x_2),$$

where the degrees of monomials in f are strictly greater than 3. Replacing x_1 by $x_1 x_2$, we will get

$$x_1^3 x_2 + \lambda x_1^4 x_2 + \cdots + \mu x_1^m x_2 \in V + H^{(2)}. \tag{4.20}$$

This shows that each element in H modulo $H^{(2)}$ is a linear combination of right-normed monomials in free generators, every generator occurring to a power at most m, i. e., bounded by a fixed number. □

Proposition 4.39. *The norm of a Schreier variety over a finite field k is at most 2.*

Proof. Suppose $N(\mathbf{V}) > 2$. Let F be a subalgebra in L generated by two elements $x = x_1$ and $y = x_2$. Let S denote the ideal of F generated by y. As a Lie algebra, S is generated by all elements of the form $z_0 = y$ and $z_j = x^j y$, $j \geq 1$. By the hypothesis, $H = S + V/V$ is free in $\mathbf{V}$. The rank of this algebra is equal to $\dim(H/H^2)$. But $H/H^2 \cong S + V/S^2 + V$. Since $S^2 \supset F^{(2)}$ and by Lemma 4.38, $\dim H/H^2 \leq \dim(F + V/F^2 + V) < \infty$. Let n be a maximal number such that $z_0, z_1, \dots, z_n$ are linearly independent modulo $S^2 + V$. Then the rank of H is precisely $n+1$. Let $\bar{z}_i$ denote the element $z_i + V$, $i = 0, 1, \dots, n$, in H/V. We designate also by $h_0, h_1, \dots, h_n$ some free generators of H such that $h_i \in \bar{z}_i + H^2$. Since $N(\mathbf{V}) = 3$, we have $y^2 x + f(x, y) \in V$ where $f(x, y) \in L^4$. Considering, if necessary, the polynomial $(x+y)^2 y + f(y, x+y) - x^2 y - f(y, x)$, we may think that each monomial in $f(x, y)$ has degree in y at least 2. Since $S \supseteq F^2$, we may write $f(x, y) = f'(z_0, z_1, \dots, z_m)$ where m is a natural number. The latter polynomial f' has no linear terms and is representable in the form

$$f'(z_0, z_1, \dots, z_m) = g_1(z_0, z_1, \dots, z_m) + g_2(z_0, z_1, \dots, z_m),$$

where $g_1(z_0, z_1, \dots, z_m)$ is a linear combination of monomials of degree 2 in $z_0, z_1, \dots, z_m$ and $g_2(z_0, z_1, \dots, z_m) = f'(z_0, z_1, \dots, z_m) - g_1(z_0, z_1, \dots, z_m)$. We note that the expression for $g_1(z_0, z_1, \dots, z_m)$ does not contain $z_0 z_1$. Since $N(\mathbf{V}) = 3$, modulo $S^2 + V$, the elements z_i, $i > n$, are linear combinations of elements of the form z_i where $i = 2, 3, \dots, n$. This yields a relation of the form

$$z_0 z_1 + f_1(z_0, \dots, z_n) + f_2(z_0, \dots, z_n) \in V,$$

where f_1 is a linear combination of monomials of degree 2 in $z_0, z_1, \dots, z_n$ excluding $z_0 z_1$ and f_2 is a linear combination of monomials of degree at least 3. Passing to H, we get an equation of the form

$$\bar{z}_0 \bar{z}_1 + f_1(\bar{z}_0, \dots, \bar{z}_n) + f_2(\bar{z}_1, \dots, \bar{z}_n) = 0.$$

Replacing every z_i by its expression through h_i, $i = 1, \dots, n$, we will get

$$h_0 h_1 + f_1(h_0, \dots, h_n) + f_3(h_0, \dots, h_n) = 0, \tag{4.21}$$

where f_3 is a linear combination of monomials of degree at least three in $h_0, \dots, h_n$. Now equation (4.21) is an identity of $\mathbf{V}$, contradicting the fact that $N(\mathbf{V}) = 3$. □

Proposition 4.40. *Let* $\mathbf{V}$ *be a Schreier variety of Lie algebras over a finite field k such that* $N(\mathbf{V}) = 2$. *Then* $\mathbf{V} = \mathbf{A}$.

Proof. Let L denote a free Lie algebra with free generators x, y, z. We denote by S the subalgebra generated by $\{x, y, xz, yz\}$. Suppose first that this set is linearly dependent modulo $S^2 + V$. Then, for an appropriate nonassociative polynomial $u(x_1, x_2, x_3, x_4)$, we may write

$$\lambda xz + \mu yz + u(x, y, xz, xy) \in V,$$

where at least one of the coefficients λ, μ, say λ, is distinct from zero. Replacing y by zero, we will get

$$xz + g(x, xz) \in V,$$

where $g(x_1, x_2)$ is a suitable polynomial in the derived algebra of a free Lie algebra of countable rank. Replacing x by xy, we will get $g(xy, (xy)z) \in L^{(2)}$, and then $(xy)z \in V + L^{(2)}$. Thus, $L^3 \subseteq V + L^{(2)}$. Since $N(\mathbf{V}) = 2$, we also have $L^2 + V = L^3 + V$. Finally, $L^2 = V + L^{(2)}$, that is, either $L^2 \subseteq V$ and then $\mathbf{V} = \mathbf{A}$, or $L^2 + V/V$ is a free algebra of $\mathbf{V}$ equal to its derived algebra $L^{(2)} + V/V$ which is impossible. Hence $\mathbf{V} = \mathbf{A}$.

Now suppose x, y, xy, xz are linearly independent modulo $S^2 + V$. Then the rank of $S + V/V$ is equal to four. But $xy + f(x, y) \in V$, where $f(x_1, x_2)$ is a polynomial in the cube of a free Lie algebra of countable rank. Hence, multiplying by z on the right, we may write

$$(xz)y + x(yz) + f(x, y)z \in V.$$

In other words, we can find a polynomial $g(x_1, x_2, x_3, x_4)$ in the cube of a free Lie algebra of countable rank such that

$$(xz)y + x(yz) + g(x, y, xz, xy) \in V, \tag{4.22}$$

with $g(x, y, xz, yz)$ being linear in z. Mapping $x_1 \mapsto xz$, $x_2 \mapsto y$, $x_3 \mapsto x$, $x_4 \mapsto yz$, we arrive at an epimorphism from $L(x_1, x_2, x_3, x_4)$ onto $S + V/V$. By Lemma 4.38, this is an isomorphism modulo $V + L^{(2)}$. Then, using equation (4.22), we will get

$$x_1x_2 + x_3x_4 + g(x_3, x_2, x_1, x_4) \in V + L^{(2)}.$$

Substituting 0 for x_3 and x_4, we arrive at a relation of the form

$$x_1x_2 + g'(x_1, x_2) \in V + L^{(2)},$$

where g' is linear in x_1. Hence $g'(x, yz) \in L^{(2)}$, that is, $x(yz) \in V + L^{(2)}$. This yields $L^3 \subset V + L^{(2)}$ and, as previously, $L^2 = V + L^{(2)}$, that is, $\mathbf{V} = \mathbf{A}$. □

Theorem 4.41. *Any Schreier variety of Lie algebras over a ring k is given either by* $\{\lambda x \equiv 0, \lambda \in I\}$, *I being a maximal ideal of k, or by* $\{xy \equiv 0, \lambda x \equiv 0, \lambda \in J\}$ *where J is an ideal such that k/J is a principal ideal domain, or else by* $x \equiv 0$.

Proof. It was remarked in the final part of Section 4.6.1 that we may restrict ourselves to the case where Ann $\mathbf{V} = 0$. In this case we have to prove that k is a principal ideal domain and, if, in addition, $\mathbf{V}$ is not abelian, then k is a field. Now by Lemma 4.36 for a Schreier variety, we always have $N(\mathbf{V}) \leq 3$. If k is finite then, having no zero divisors, it turns out to be a field. In this case we may use Propositions 4.39 and 4.40, provided that $N(\mathbf{V})$ is either 2 or 3. If $N(\mathbf{V}) = 1$ then, trivially, $\mathbf{V}$ is the one element variety.

If k is infinite then, by Proposition 4.37, we know that either $\mathbf{V}$ is the variety of all Lie algebras over k or $\mathbf{V}$ is abelian. Suppose, in the former case, that k contains a non-invertible element λ. Then, in the free algebra $L(x, y)$, we may consider a subalgebra H generated by λx, y, and xy. Similar to an example in Section 2.4.1, we can prove that H is not free. This contradiction shows that any nonzero element in k is invertible, i. e., k is a field, as required. In the latter case as we saw in the final part of Section 4.6.1, any ideal of k must be isomorphic to k itself, proving that k is a principal ideal ring.

Considering the argument in Section 4.6.1, we come to the conclusion that our proof is complete. □

4.7 Metabelian varieties

In this small section we turn reader's attention to one of the simplest classes of varieties of Lie algebras consisting of solvable Lie algebras of solvable step at most 2, called *metabelian* varieties. They admit a satisfactory description in the case where the base ring is an infinite field. The situation becomes much harder if k is a finite field.

4.7.1 Bases of free metabelian algebras

Let $M(X) = L(X, \mathbf{A}^2)$ be a free metabelian Lie algebra over a commutative ring k with 1. Since $M(X) \cong L(X)/L(X)^{(2)}$, a basis for this algebra can be chosen by using Section 2.4.2 where we have found a free generating set for the derived algebra of a free Lie algebra. Thus a basis can be chosen in the form of the union of X and the set of all monomials

$$x_1 x_2 \cdots x_{t-1} x_t, \tag{4.23}$$

where $x_1 \geq x_2 \geq \cdots \geq x_{t-1} < x_t$, $t \geq 2$.

Theorem 4.42. *Let* $\mathbf{V}$ *be a solvable variety over an infinite field* k. *If* $\mathbf{A}^2 \not\subseteq \mathbf{V}$ *then* $\mathbf{V}$ *is a locally nilpotent variety.*

Proof. It is sufficient to show that $\mathbf{V}$ is an Engel variety, that is, with an identity of the form $(\operatorname{ad} x)^m = 0$ for a suitable natural number m. Indeed, in this case our proof may be finished as follows. Let G be an algebra in $\mathbf{V}$ with generators $a_1, \ldots, a_n$. The derived algebra, as a subalgebra, is generated by all elements of the form

$$a_{i_1}^{t_1} a_{i_2}^{t_2} \cdots a_{i_s}^{t_s} a_{i_{s+1}} a_{i_{s+2}},$$

where $i_1 > i_2 > \cdots > i_s > i_{s+1} < i_{s+2}$. Since $(\operatorname{ad} x)^m = 0$, we see that all exponents t_i may be uniformly bounded by m. Hence the number of nonzero elements of the form as in (4.21) is finite. Therefore, G^2 is a finitely generated algebra. Since the solvable step of G^2 is strictly less than that of G, the induction hypothesis applies proving that G^2 is nilpotent, hence finite-dimensional. As a result, we obtain that G is finite-dimensional, hence by the corollary of Engel's theorem (Section 1.7) it is nilpotent, as required.

Now, for the proof that for some m we have $(\operatorname{ad} x)^m \equiv 0$, we consider a free Lie algebra of countable rank in $\mathbf{A}^2 \cap \mathbf{V}$ with free generators $\{x_1, x_2, \ldots, x_l, \ldots\}$. Since L is a proper quotient algebra of $M(x_1, x_2, \ldots, x_l)$, L satisfies a (normal) identity of the form

$$\sum_{i=1}^{s-1} \alpha_i x_1^{t_1} \cdots x_i^{t_i - 1} \cdots x_s^{t_s} x_i \equiv 0.$$

The degree of each monomial in every fixed variable is the same since in the case of an infinite field we may pass from an identity to its multihomogeneous components. If $\alpha_1 \neq 0$ then we substitute $x_1 = x$, $x_2 = x_3 = \cdots = x_s = y$. This will bring us to an identity of the form $x^u y^v x \equiv 0$. Replacing y by $x + y$ and equating to zero the component of degree 1 in y will come to $-x^{u+v-1} yx = x^{u+v} y \equiv 0$. It remains to put $m = u + v$ and then $(\operatorname{ad} x)^m = 0$ in $\mathbf{V} \cap \mathbf{A}^2$. Now in the free algebra $L = L(\{x, y\}, \mathbf{V})$ we have $x^m y \in L^{(2)}$. But any element in $L^{(2)}$ may be homogeneous of degree 1 in one of the two variables (here y) only if it is zero. Hence $(\operatorname{ad} x)^m \equiv 0$ is an identity of $\mathbf{V}$. □

Corollary 4.43. *Let k be a field of characteristic zero and $\mathbf{V}$ a solvable variety over k such that $\mathbf{V} \not\supseteq \mathbf{A}^2$. Then $\mathbf{V}$ is a nilpotent variety. If, in addition, $\mathbf{V} \subseteq \mathbf{A}^2$ then there exists a natural c such that $\mathbf{V} = \mathbf{N}_c \cap \mathbf{A}^2$. In particular, the lattice $\mathscr{L}(\mathbf{A}^2)$ of subvarieties of $\mathbf{A}^2$ is a chain.*

Proof. As we learned from the proof of Theorem 4.42, an identity of degree $d + 1$ in $\mathbf{V} \cap \mathbf{A}^2$ implies $(\operatorname{ad} x)^d \equiv 0$. A simple argument involving linearization and considering commutativity of $\operatorname{ad} L|_{L^2}$ shows that any $L \in \mathbf{A}^2$ with $(\operatorname{ad} x)^d \equiv 0$ is at most d-step nilpotent. Hence $\mathbf{V} \cap \mathbf{A}^2 \subseteq \mathbf{N}_d \cap \mathbf{A}^2$. Taking c equal to the minimal degree of identities of $\mathbf{V} \cap \mathbf{A}^2$ which are not true in $\mathbf{A}^2$, we will get $\mathbf{V} \cap \mathbf{A}^2 = \mathbf{N}_c \cap \mathbf{A}^2$. If $\mathbf{V} \subseteq \mathbf{A}^2$ then the proof is finished. Otherwise we may apply induction over the solvable step of $\mathbf{V}$. If $G \in \mathbf{V}$ then G^2 has a lesser solvable step, that is, G^2 is nilpotent. Now $G/(G^2)^2 \in \mathbf{V} \cap \mathbf{A}^2$ is nilpotent. By Theorem 1.19 (Section 1.7), G is a nilpotent Lie algebra. □

To simplify the statement of the last result of this section, we want to give meaning to an expression of the form

$$a = x_1^{r_1} x_2^{r_2} \cdots x_{n-1}^{r_{n-1}} x_n^{r_n}$$

as an element of a free metabelian algebra $M(x_1, x_2, \dots, x_n)$. This is done by setting

$$a = x_n^{r_n-1} x_1^{r_1} \cdots x_{n-1}^{r_{n-1}} x_n.$$

Theorem 4.44. *Let k be an infinite field of characteristic $p > 0$. Then every proper metabelian variety $\mathbf{V}$ can be given within $\mathbf{A}^2$ by a finite system of identities of the form*

$$x_1^{p^{r_1}} x_2^{p^{r_2}} \cdots x_n^{p^{r_n}} x_{n+1}^{p^{r_{n+1}}} \equiv 0 \quad (r_1 \geq r_2 \geq \cdots \geq r_n = r_{n+1} \geq 0) \tag{4.24}$$

and

$$\sum_{i=m+1}^{n} x_1^{p^{r_1}} \cdots x_m^{p^{r_m}} x_{m+1}^{p^r} \cdots x_n^{p^r} x_i^{p^r} x_{n+1}^{p^r} \equiv 0 \quad (r_1 \geq \cdots \geq r_m \geq r \geq 0, p|(n-m+1)). \tag{4.25}$$

Proof. We apply Theorem 4.10. According to this theorem, any identity is equivalent to a system of multihomogeneous identities depending on each variable to the degree which is a p-power. Applying Hall's basis (Section 2.3, Theorem 2.10) with respect to the ordering $x_1 > x_2 > \cdots > x_{n+1}$, we may write any identity in the form

$$\sum_{i=1}^{n} \alpha_i A_i \equiv 0 \quad \text{where } A_i = x_1^{p^{r_1}} \cdots x_i^{p^{r_{i-n}}} \cdots x_n^{p^{r_n}} x_i x_{n+1}^{p^{r_{n+1}}}, \tag{4.26}$$

where we may assume that $r_1 \geq r_2 \geq \cdots \geq r_n \geq r_{n+1}$. We reduce first to the case where for all i such that $\alpha_i \neq 0$ we have $r_i = r_{n+1}$. Indeed, suppose $r = r_n > r_{n+1} = l$ and $\alpha_n \neq 0$. We represent p^r in the form

$$p^r = p^{r-1}(p-1) + p^{r-2}(p-1) + \cdots + p^{l+1}(p-1) + p^l p.$$

Then we replace x_n in equation (4.26) by an expression of the form

$$\sum_{i=1}^{p} y_{r-l,i} + \sum_{i=1}^{r-l-1} \sum_{j=1}^{p-1} y_{ij}.$$

Next we consider a multihomogeneous component of the resulting identity which is homogeneous of degree p^{r-1} in $y_{11}, \dots, y_{1,p-1}$, of degree p^{r-2} in $y_{21}, \dots, y_{2,p-1}$, and so on, of degree p^l in $y_{r-l,1}, \dots, y_{r-l,p}$. We want to see what happens then to the components A_i of equation (4.26). If $i \neq n$ then the coefficient of the part originating from A_i has the form

$$S = p^r / (p^{r-1}!)^{p-1} (p^{r-2}!)^{p-1} \cdots (p^l!)^p.$$

We recall that the exponent N of a prime p such that p^N is the maximal p-power dividing m is given by the formula

$$N = \left[\frac{m}{p}\right] + \left[\frac{m}{p^2}\right] + \cdots,$$

where $[a]$ is the integral part of a. Hence the numerator of S is divisible by p^N where $N = p^{r-1} + p^{r-2} + \cdots + 1$ and the denominator by p^M where $M = (p^{r-2} + \cdots + 1)(p-1) + \cdots + (p^{l-1} + \cdots + 1)p = N - (r-1) < N$. Therefore, $S = 0$ in k. Now if $i = n$ then we have a consequence of the form

$$\sum_{j=1}^{r-1} c_j \sum_s B_{js} \quad \text{where } B_{sj} = x_1^{p^{r_1}} \cdots x_{n-1}^{p^{r_{n-1}}} y_{11}^{p^{r-1}} \cdots y_{r-l,p}^{p^l} y_{js}^{p^{r-j}} x_{n+1}^{p^l}. \tag{4.27}$$

If $j \neq r - l$ then the coefficient c_j is equal to zero, otherwise $c_{r-1} \neq 0$. For

$$c_j = (p^r - 1)!/(p^{r-1}!)^{p-1} \cdots (p^{l+j-1}!)^{p-2} \cdots (p^l!)^p,$$

the numerator of this fraction is divisible by p^{N-r}, since the denominator this power is $p^{N-r-j+1}$, proving the required. Finally, we find that equation (4.27) takes the form

$$\sum_{s=1}^{p} B_{r-l,s} \equiv 0, \tag{4.28}$$

where each monomial $B_{r-l,s}$ is of the form $\cdots y_{r-1,s}^{p^l} x_{n+1}^{p^l}$. On the other hand, it is obvious that $A_n \equiv 0$ is a consequence of equation (4.28) (it is sufficient to equate all y_{js}). Hence equation (4.26) is equivalent to equation (4.28), along with $\sum_{i=1}^{n-1} A_i \equiv 0$. Applying the induction on the number of nonzero coefficients in equation (4.26), we may, indeed, assume that in equation (4.26) we have $r_i = r_{n+1}$ as soon as $\alpha_i \neq 0$.

In this case, after possible renumbering of variables, we may write the left-hand side of equation (4.26) in the form

$$f = \sum_{i=m+1}^{n} \alpha_i A_i.$$

Let τ_i denote an automorphism of $M(X)$ permuting x_{n+1} and x_i and leaving all the other variables invariant, $i = m+1, \ldots, n$. Then $\tau_i(A_i) = -A_i$ and $\tau_i(A_j) = -A_i + A_j$ $(i \neq j)$. Thus

$$\tau_i(f) = \alpha_1 A_1 + \cdots + (-\alpha_1 - \alpha_2 - \cdots - \alpha_n) A_i + \cdots + \alpha_n A_n.$$

Using equation (4.26) itself, we will obtain in $\mathbf{V}$ a system of identities of the form

$$(\alpha_1 + \cdots + 2\alpha_i + \cdots + \alpha_n) A_i \equiv 0, \quad i = m+1, \ldots, n. \tag{4.29}$$

If for some i a coefficient in equation (4.29) is nonzero then $A_i \equiv 0$ is an identity in $\mathbf{V}$ which is equivalent to equation (4.26) and has the form of equation (4.24). Otherwise equation (4.29) is a system of equations for the coefficients and an immediate computation shows that its determinant is equal to $n - m + 1$. Thus we obtain a contradiction, unless $p | (n - m + 1)$. In this remaining case, this system has rank $n - m$, with $(1, 1, \ldots, 1)$ being a basis for the space of solutions. This brings us to an identity of the form of equation (4.25). The finiteness of a system of identities of the form of equations (4.24) and (4.25) determining $\mathbf{V}$ follows from the finiteness of a basis of identities of a metabelian variety over any noetherian ring, which will be proved in Section 5.3, Lemma 5.15. □

4.8 Young diagrams and identities

4.8.1 Module P_n

Let $X = \{x_1, \dots, x_n\}$ be a free generating set of a free Lie algebra $L = L(X)$. Let P_n denote the k-submodule of multilinear polynomials of degree n in L.

Lemma 4.45. *A basis of a free k-module P_n can be chosen in the form*

$$x_{\sigma(1)}x_{\sigma(2)} \cdots x_{\sigma(n-1)}x_n, \tag{4.30}$$

where σ ranges over the set of all permutations on the set of symbols $\{1, 2, \dots, n-1\}$.

Proof. Let w be a monomial in P_n. Using induction on n with obvious basis $n = 1$, we may assume that $w = uv$ where v is a right-normed monomial of the form (4.30). If $\deg(u) = 1$ then w is of the form (4.30). Otherwise $u = u_1u_2$. Then $w = u_1u_2v - u_2u_1v$. Applying induction by n to u_2v and u_1v, we lower the degree of u in $w = uv$. Applying induction by $\deg(u)$, we prove that $w = uv$ is a linear combination of monomials of the form (4.30). As for the uniqueness of such a representation, this can be proved by the same argument as that used in Section 2.1.3 in the case of the standard identity. □

So we have $\dim P_n = (n-1)!$. Now we turn P_n to an S_n-module by setting

$$\sigma \circ f(x_1, \dots, x_n) = f(x_{\sigma(1)}, \dots, x_{\sigma(n)}), \quad \sigma \in S_n,\ f(x_1, \dots, x_n) \in P_n.$$

If $f, g \in P_n$, then, obviously, $g \equiv 0$ is a consequence of $f \equiv 0$ if and only if $g \in kS_nf$. Thus we have as many pairwise inequivalent systems of (multilinear) identities of degree n as the number of distinct kS_n-submodules in P_n. By Lemma 4.45, any element in P_n is a linear combination of monomials of the form (4.30). Hence any one of them, say, $x_1x_2 \cdots x_n$, can be chosen as a generator of P_n which is a cyclic kS_n-module. Therefore P_n is a quotient module of the regular S_n-module kS_n under a homomorphism π transforming 1 into $x_1x_2 \cdots x_n$. An intrinsic characterization of P_n can be obtained if we use the results in Chapter 3.

Adopting our notation in Section 3.4 with $m = n$, we set $D_n(X) = kS_n \otimes_{k\Gamma} k_\varepsilon$. In fact, $D_n(X)$ is the restriction of $C_n(X)$ to the multilinear component of $T_n(X)$. Arguing along the lines of the proof of Section 3.4, Lemma 3.17, we immediately see that P_n is a k-module isomorphic to $D_n(X)$. Of course, $D_n(X)$ is not invariant with respect to the action of $\mathrm{GL}(V)$, but this is the case for S_n as a subgroup of $\mathrm{GL}(V)$. Hence

$$P_n \cong kS_n \otimes_{k\Gamma} k_\varepsilon, \tag{4.31}$$

and this is an isomorphism of S_n-modules. Now relation (4.31) enables us to apply to P_n all formulas for multiplicities derived in Section 3.4 concerning $L_n(X)$.

A particularly clear case is that of $\operatorname{char} k = 0$. Here (see Section 3.2.4) $1 = \sum_{\tau,d} y_{\tau d} e_{\tau d}$, where $y_{\tau d} \in kS_n$ and the elements $e_{\tau d}$ are defined as in Section 3.2.2.

Here $e_{\tau d}e_{\tau' d'} = 0$ if $d \neq d'$. The group ring $A = kS_n$ as a left S_n-module is representable in the form

$$A = \sum_{\tau,d} Ae_{\tau d},$$

where $Ae_{\tau d} \cong Ae_{\tau' d'}$ if and only if $d = d'$. Applying the above homomorphism π, we may write

$$P_n = \sum_{\tau,d} Af_{\tau d}, \quad f_{\tau d} = e_{\tau d} \circ (x_1 x_2 \cdots x_n). \tag{4.32}$$

With $Ae_{\tau d}$ being a simple S_n-module, we see that $Af_{\tau d}$ is either zero or simple. It should be noted that some of the summands in equation (4.32) can be discarded in such a way that the sum becomes direct. Summing up all the above, we arrive at the following result.

Theorem 4.46. *Let k be a field of characteristic zero. Then P_n as S_n-module decomposes as the direct sum of simple S_n-modules of the form $kS_n f_{\tau d} \cong kS_n e_{\tau d}$, corresponding to Young tableaux of the form τd. If $d \neq d'$ then $kS_n f_{\tau d} \not\cong kS_n f_{\tau' d'}$. Any system of multilinear identities of degree n is equivalent to a system of identities of the form $f_{\tau d} \equiv 0$, $d \in T$, where T is a set of standard Young tableaux of degree n. If $n \neq 4, 6$ then P_n contains a simple S_n-submodule corresponding to an arbitrary Young diagram except those associated with the partitions of the form $n = 1 + \cdots + 1$ and $n = n$ (in fact, the latter diagrams should be discarded for all $n > 2$).*

4.8.2 Polynomials $f_{\tau d}$ and $g_{\tau d}$

We recall that, given a Young tableau τd, the subgroup $R_{\tau d}$ consists of all permutations in S_n leaving invariant the sets of symbols in the rows of τd, whereas $C_{\tau d}$ is the set of permutations with the same property for columns. Now $e_{\tau d}$ and $f_{\tau d}$ have the form

$$e_{\tau d} = \sum_{\substack{\rho \in R_{\tau d} \\ \sigma \in C_{\tau d}}} \varepsilon_\sigma \rho\sigma, \quad f_{\tau d} = \sum_{\substack{\rho \in R_{\tau d} \\ \sigma \in C_{\tau d}}} \varepsilon_\sigma x_{\rho\sigma(1)} \cdots x_{\rho\sigma(n)}.$$

If we identify, in $f_{\tau d}$, the variables that belong to each fixed row of τd and then divide the resulting polynomial by the order of $R_{\tau d}$ then we will arrive at a polynomial $g_{\tau d}$ depending on m variables where m is the number of rows in d. It is clear that $g_{\tau d} \equiv 0$ is a consequence of $f_{\tau d} \equiv 0$. The converse is also true since, obviously, $f_{\tau d} \equiv 0$ is a full linearization of $g_{\tau d} \equiv 0$. A particularly simple form for $g_{\tau d}$ exists in the case of the trivial tableau d (see Section 3.2.2). Here, if $l_1, l_2, \ldots, l_t$ are the heights of the columns with the numbers $1, 2, \ldots, t = n_1$ then $g_{\tau d}$ results by the right-normed arrangement of brackets on the product of associative standard polynomials of the form

$$s_{l_1}(x_1, \ldots, x_{l_1}) s_{l_2}(x_1, \ldots, x_{l_2}) \cdots s_{l_t}(x_1, \ldots, x_{l_t}),$$

where, as previously,

$$s_m(x_1,\dots,x_m)=\sum_{\sigma\in S_m}\varepsilon_\sigma x_{\sigma(1)}\cdots x_{\sigma(n)}.$$

4.8.3 Structure of P_n where n is small

(i) Suppose $n = 3$. Here $\dim P_3 = 2$ and only one diagram is admissible:

$d =$ 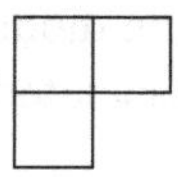 .

By Section 3.2, formula (3.4), the dimension b_d of the corresponding S_3-modules equal to 2 and $g_d = (x_1x_2-x_2x_1)x_1 = 2(x_1x_2)x_1 \neq 0$. Thus $x_1^2x_2 \equiv 0$ is the only up-to-equivalence identity of degree 3. It is obvious that $x_1x_2x_3 \equiv 0$ is an identity equivalent to this latter.

(ii) Let $n = 4$. Here $\dim P_4 = 6$. Admissible diagrams have the form

$d_1 =$ 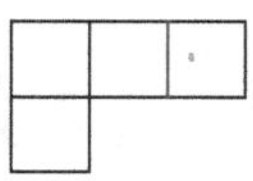 , $d_2 =$ 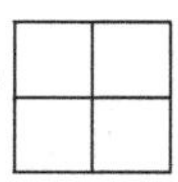 , $d_3 =$ 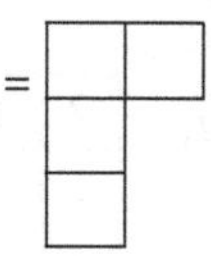 .

The dimensions of the corresponding modules are $b_{d_1} = b_{d_3} = 3$, $b_{d_2} = 2$. Taking the tableaux

$\tau d_1 =$ 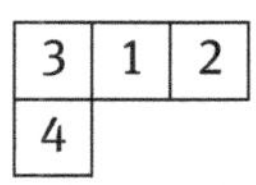

and $d_3 =$ 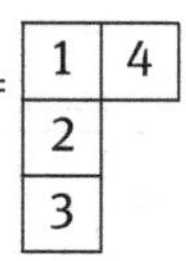

,

we will obtain the identities of the form

$$g_{\tau d_1} = 2x_1^3x_2 \neq 0 \quad \text{and} \quad g_{d_3} = s_3(\operatorname{ad} x_1, \operatorname{ad} x_2, \operatorname{ad} x_3)(x_1) \neq 0.$$

Hence d_2 does not correspond to a nonzero simple component in P_4 (this is equivalent to a part of the proof of Section 3.4, Theorem 3.24). So we may write $P_4 = Af_{\tau d_1} \oplus Af_{d_3}$ and $Af_{\tau d_1} \not\cong Af_{d_3}$. It should be noted that $Af_{d_3} = Af$ where $f = (x_1x_2)x_3x_4$ is the left-hand side of the metabelian identity. This follows since Af is a nonzero submodule, $Af \neq P_4$ since A^2 is nonnilpotent. Besides, an explicit form of the basis of a free metabelian algebra (Section 4.6.1) shows that $g_{\tau d_1} \equiv 0$ is not an identity of A^2. Hence $Af = Af_{d_3}$.

(iii) Let $n = 5$. Here $\dim P_5 = 24$. Admissible types of diagrams have the form

$d_1 =$ 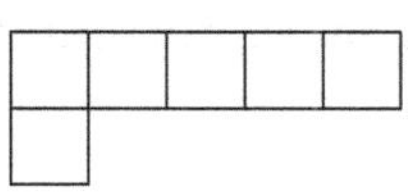 , $d_2 =$ 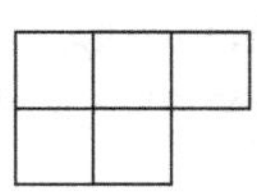 , $d_3 =$ 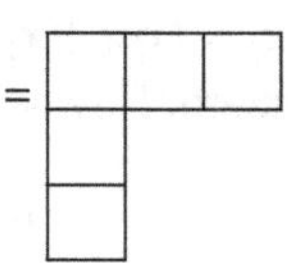

$$d_4 = \begin{array}{|c|c|}\hline \ & \ \\ \hline \ & \ \\ \hline \ \\ \cline{1-1}\end{array}, \quad d_5 = \begin{array}{|c|c|}\hline \ & \ \\ \hline \ \\ \cline{1-1} \ \\ \cline{1-1} \ \\ \cline{1-1}\end{array}.$$

The dimensions of the corresponding modules have the form

$$b_{d_1} = b_{d_5} = 4, \quad b_{d_2} = b_{d_4} = 5, \quad b_{d_3} = 6.$$

By Theorem 4.46, all the corresponding modules enter the decomposition of P_5. Now we have $24 = 4 + 5 + 6 + 5 + 4$. The associated identities result by using the following tableaux:

$$\tau_1 d_1 = \begin{array}{|c|c|c|c|}\hline 4 & 1 & 2 & 3 \\ \hline 5 \\ \cline{1-1}\end{array}, \quad g_{\tau_1 d_1} = 2x_1^4 x_2,$$

$$\tau_2 d_2 = \begin{array}{|c|c|c|}\hline 1 & 4 & 3 \\ \hline 2 & 5 \\ \cline{1-2}\end{array}, \quad g_{\tau_2 d_2} = 2(x_1 x_2)x_1^2 x_2,$$

$$\tau_3 d_3 = \begin{array}{|c|c|c|}\hline 2 & 1 & 5 \\ \hline 3 \\ \cline{1-1} 4 \\ \cline{1-1}\end{array}, \quad g_{\tau_3 d_3} = x_1 s_3(\operatorname{ad} x_1, \operatorname{ad} x_2, \operatorname{ad} x_3)x_1$$

$$\tau_4 d_4 = d_4 = \begin{array}{|c|c|}\hline 1 & 4 \\ \hline 2 & 5 \\ \hline 3 \\ \cline{1-1}\end{array}, \quad g_{d_4} = 2s_3(\operatorname{ad} x_1, \operatorname{ad} x_2, \operatorname{ad} x_3)x_1 x_2,$$

$$\tau_5 d_5 = d_5 = \begin{array}{|c|c|}\hline 1 & 5 \\ \hline 2 \\ \cline{1-1} 3 \\ \cline{1-1} 4 \\ \cline{1-1}\end{array}, \quad g_{d_5} = s_4(\operatorname{ad} x_1, \operatorname{ad} x_2, \operatorname{ad} x_3, \operatorname{ad} x_4)x_1.$$

That each of these identities is nontrivial follows by using the usual Hall's basis of a free Lie algebra. For instance,

$$\begin{aligned} g_{d_5} &= (x_1x_2)(x_3x_4)x_1 - (x_1x_3)(x_2x_4)x_1 + (x_1x_4)(x_2x_3)x_1 + (x_3x_4)(x_1x_2)x_1 \\ &= -(x_2x_4)(x_1x_3)x_1 + (x_2x_3)(x_1x_4)x_1. \end{aligned}$$

If we assume that x_1 is the greatest variable and the other variables are ordered in their natural order then any monomial in this decomposition is equal to a basic one. Thus, $(x_2x_3)(x_1x_4)x_1 = (x_2x_3)(x_1x_4x_1)$. Hence $g_{d_5} \neq 0$. Similarly, we can verify that all the other $g_{\tau_i d_i}$ are nonzero. Now we may write

$$P_5 = Af_{\tau_1 d_1} \oplus Af_{\tau_2 d_2} \oplus Af_{\tau_3 d_3} \oplus Af_{d_4} \oplus Af_{d_5}.$$

All these modules being pairwise nonisomorphic, any submodule is equal to a direct sum of some of these submodules. Hence any system of homogeneous identities of degree 5 is equivalent to precisely one system of the form

$$\{g_{\tau d} \equiv 0 \mid d \in T\},$$

where $T \subseteq \{\tau_1 d_1, \tau_2 d_2, \tau_3 d_3, d_4, d_5\}$. The total number of such identities (varieties) is equal to 31. Interpretations of some of the above identities are given in Section 4.10 (exercises).

(iv) Consider now $n = 6$. In this case $\dim P_6 = 120$. Admissible decompositions of 6 corresponding to irreducible components of P_6 as S_6-submodule by Section 3.4, Theorem 3.24, have the form

$$d_1 = \{5,1\}, \quad d_2 = \{4,2\}, \quad d_3 = \{4,1,1\}, \quad d_4 = \{3,3\}, d_5 = \{3,2,1\},$$
$$d_6 = \{3,1,1,1\}, \quad d_7 = \{2,2,1,1\}, \quad d_8 = \{2,1,1,1,1\}.$$

The dimensions of the corresponding S_6-modules have the form: $b_{d_1} = b_{d_8} = 5$, $b_{d_2} = b_{d_7} = 9$, $b_{d_3} = b_{d_6} = 10$, $b_{d_4} = 5$, and $b_{d_5} = 16$. Since the sum of these numbers is strictly less than 120, the decomposition of S_6-module P_6 into the direct sum of irreducible components has isomorphic terms (in fact, the multiplicity is 3 in the case of d_5 and 2 in the case of d_3 and d_7). This observation leads to a result as follows.

Proposition 4.47. *The cardinality of the set of varieties of Lie algebras over a field k is not less than the cardinality of k.*

Proof. Let V_1 and V_2 denote two distinct isomorphic S_n-submodules in P_n, $0 \neq f_1 \in V_1$ and f_2 the image of f_1 under an isomorphism of V_1 onto V_2. Then for distinct $\lambda, \mu \in k$ the identities $f_1 + \lambda f_2 \equiv 0$ and $f_1 + \mu f_2 \equiv 0$ are inequivalent. For, if this is not the case, there exists an element $a \in A = kS_n$ such that, say, $f_1 + \mu f_2 = a(f_1 + \lambda f_2)$. Since V_1 and V_2 form a direct sum, we have $0 = af_1 - f_1 = -a\lambda f_2 + \mu f_2$. Hence $(a-1)f_1 = 0$. By the isomorphism then $(a-1)f_2 = 0$, that is, $0 = -a\lambda f_2 + \mu f_2 = (-\lambda + \mu)f_2$. Hence $\lambda = \mu$. Taking $n = 6$, we see that such V_1 and V_2 exist, making our argument meaningful. □

4.8.4 Lattices of varieties

Let **V** be a variety. The set $\mathscr{L}(\mathbf{V})$ of its subvarieties is a lattice with respect to the union and intersection of varieties. A complete description of the lattice of varieties is known in a very limited number of cases. In the corollary of Theorem 4.42, we have seen that the lattice $\mathscr{L}(\mathbf{A}^2)$ of metabelian varieties over any field of characteristic zero is a chain, hence distributive. It should be stressed out that the property of distributivity is rather rare. In the case of fields of characteristic zero, this question is closely related to the question on the lattice of S_n-submodules in S_n-module $P_n(\mathbf{V})$ whose definition is the following. By Section 4.8.1, for any $n \geq 6$, P_n is a semisimple S_n-module. It decomposes as the direct sum $P_n = P_n(\mathbf{V}) \oplus P'_n(\mathbf{V})$ where $f \in P'_n(\mathbf{V})$ if and only if $f \equiv 0$ is an identity satisfied by $\mathbf{V}$. It should be noticed that $P_n(\mathbf{V})$, defined in this way, is unique up to isomorphism.

Proposition 4.48. *The lattice $\mathscr{L}(\mathbf{V})$ is distributive if and only if, for any n, $P_n(\mathbf{V})$ contains no distinct isomorphic submodules.*

Proof. Indeed, if, for some n, $P_n(\mathbf{V})$ contains two distinct isomorphic S_n-modules V_1 and V_2 then, following the notation in the proof of Proposition 4.47, we take for $\mathbf{X}$ a subvariety of $\mathbf{V}$ given by $f_1 \equiv 0$, for $\mathbf{Y}$ the subvariety given by $f_2 \equiv 0$, and for $\mathbf{Z}$ the subvariety given by $f_1 + f_2 \equiv 0$. Then $\mathbf{X} \cap \mathbf{Y}$ will be given by a system of two identities $f_1 \equiv 0, f_2 \equiv 0$ so that $\mathbf{X} \cap \mathbf{Y} \subseteq \mathbf{Z}$ and $(\mathbf{X} \cap \mathbf{Y}) \cup \mathbf{Z} = \mathbf{Z}$. On the other had, the identities of $\mathbf{X} \cup \mathbf{Z}$ of degree n have the form $g \equiv 0$ where $g \in Af_1 \cap A(f_1 + f_2) = \{0\}$. Similarly, $\mathbf{Y} \cup \mathbf{Z}$ has no identities of degree n. Hence $(\mathbf{X} \cup \mathbf{Z}) \cap (\mathbf{Y} \cup \mathbf{Z}) \neq (\mathbf{X} \cap \mathbf{Y}) \cup \mathbf{Z}$, proving that $\mathscr{L}(\mathbf{V})$ is not distributive.

Let conversely, for any n, $P_n(\mathbf{V})$ be the direct sum of pairwise nonisomorphic S_n-modules. The question of the distributivity of $\mathscr{L}(\mathbf{V})$ reduces of the question of the coincidence of multilinear identities of any given degree holding in $(\mathbf{X} \cap \mathbf{Y}) \cup \mathbf{Z}$, and $(\mathbf{X} \cup \mathbf{Z}) \cap (\mathbf{Y} \cup \mathbf{Z})$. If X_n, Y_n, Z_n are the sets of left-hand sides of multilinear identities of degree n for $\mathbf{X}$, $\mathbf{Y}$, $\mathbf{Z}$, respectively, then we must answer the question on the coincidence of S_n-submodules of the form $(X_n \cap Y_n) + (Y_n \cap Z_n)$ and $(X_n + Y_n) \cap Z_n$. By our hypothesis about $P_n(\mathbf{V})$ each S_n-submodule is merely the direct sum of all simple submodules contained in it. Hence all reduces to the set theoretic union and intersection in the lattice of subsets of the set of all simple submodules of $P_n(\mathbf{V})$. Since the lattice of subsets of a set is distributive, the same is true as concerns $\mathscr{L}(\mathbf{V})$. □

Corollary 4.49. *$\mathscr{L}(\mathbf{N}_5)$ is distributive, and $\mathscr{L}(\mathbf{N}_6)$ is not distributive.*

Proof. In the case of $\mathbf{N}_5$, we have $P_n(\mathbf{N}_5) = P_n$ if $n \leq 5$ and $P_n(\mathbf{N}_5) = \{0\}$ if $n \geq 6$. Applying the above proposition and Section 4.8.3, we see that $\mathscr{L}(\mathbf{N}_5)$ is indeed distributive. In the case of $\mathbf{N}_6$ we have $P_6(\mathbf{N}_6) = P_6$. Using once more Proposition 4.48 and Section 4.8.3, we conclude that $\mathscr{L}(\mathbf{N}_6)$ is not distributive. □

4.8.5 Center-by-metabelian varieties of Lie algebras

By now a complete description (and distributivity) has been achieved for two non-nilpotent varieties $\mathbf{V}_1$ and $\mathbf{V}_2$ given by identities of degree 6. The former $\mathbf{V}_1 = \mathbf{AN}_2 \cap \mathbf{N}_2\mathbf{A}$ is determined by $(x_1x_2)(x_3x_4)x_5x_6 \equiv 0$ and $(x_1x_2x_3)x_4x_5x_6 \equiv 0$. The latter, $\mathbf{V}_2 = [\mathbf{E}, [\mathbf{E}, \mathbf{A}^2]]$, is determined by a single identity of the form $x_1x_2(x_3x_4)x_5x_6 \equiv 0$. As an illustration of the preceding theory, we will show that the lattice of the variety $\mathbf{C} = [\mathbf{E}, \mathbf{A}^2] \subseteq \mathbf{V}_2$ is distributive. The algebras in $\mathbf{C}$ are called *center-by-metabelian*. If $G \in \mathbf{C}$

then the second derived algebra $G^{(2)}$ is in the center of G, $G^{(2)}G = \{0\}$. We want to estimate $\dim P_n(\mathbf{C})$. For this, note that $\mathbf{C}$ is determined by $x_1(x_2x_3)x_4x_5 \equiv 0$. A consequence of the latter identity has the form

$$(x_1x_2x_3)(x_4x_5) = -(x_2x_3)(x_1x_4x_5). \tag{4.33}$$

Now let $L = L(X, \mathbf{C})$. Applying equation (4.33) enables us to write any multilinear element in $L^{(2)}$ in the form of a linear combination of the products of the form

$$(x_{i_1}x_{i_2})x_{i_3}\cdots x_{i_{n-1}}x_n, \quad \text{where } n > i_1 > i_2,\ i_3 > \cdots > i_{n-1}.$$

The number of such monomials is obviously equal to $\binom{n-1}{2} = [(n-1)(n-2)]/2$. Since $P_n(\mathbf{C}) \cong P_n(\mathbf{A}^2) \oplus (P_n(\mathbf{C}) \cap L^{(2)})$ and considering a basis of $P_n(\mathbf{A}^2)$ (see Section 4.7.1), we can write

$$\dim P_n(\mathbf{C}) \le n - 1 + \frac{(n-1)(n-2)}{2} = \frac{n(n-1)}{2}.$$

Now let d_1, d_2, d_3 be the diagrams corresponding to decompositions of the forms $n = (n-1)+1$, $n = (n-2)+2$, $n = (n-2)+1+1$. The dimensions of their associated modules are respectively $n-1$, $[n(n-3)]/2$, and $[(n-1)(n-2)]/2$. Their corresponding identities have the forms $g_1 = (\operatorname{ad} x)^{n-1}(y)$, $g_2 = (xy)(x^{n-3}y)$, and $g_3 = (x_1x_2)(x_1^{n-3}x_3) - (x_1x_3)(x_1^{n-3}x_2)$. Using equation (4.33) and a linearization of $g_2 \equiv 0$ by y, we come to a conclusion that both identities $g_2 \equiv 0$ when n is odd and $g_3 \equiv 0$ when n is even are equivalent to the same identity $(xy)(x^{n-3}z) = 0$. Using an example in Section 6.6.4, we may notice that $g_1 \equiv 0$ and $(xy)(x^{n-3}z) \equiv 0$ are not identities of $\mathbf{C}$. By comparison of dimensions, we derive that $P_n(\mathbf{C}) = Af_{\tau_1 d_1} \oplus Af_{\tau_2 d_2}$ for suitable τ_1, τ_2 and even n, and $P_n(\mathbf{C}) = Af_{\tau_1, d_1} \oplus Af_{\tau_3 d_3}$ for suitable τ_1, τ_3 and odd n. Using Proposition 4.48, we come to the following.

Theorem 4.50. *Let k be a field of characteristic zero and $\mathbf{C}$ the variety of center-by-metabelian Lie algebras over k. Then $\mathcal{L}(\mathbf{C})$ is distributive.*

A difficult open question concerning lattices of subvarieties is to describe all the varieties of Lie algebras over fields of characteristic zero with distributive lattice of subvarieties.

4.8.6 Numerical series related to varieties

Let $\mathbf{V}$ be a multihomogeneous variety over a field k. We set $c_n = \dim P_n(\mathbf{V})$ (see Section 4.8.4). A numerical series $c(\mathbf{V}, t) = \sum_{n=1}^{\infty} c_n t^n$ is called the *codimension series* of $\mathbf{V}$. The rate of its growth is an important invariant of $\mathbf{V}$ and of its lattice $\mathcal{L}(\mathbf{V})$. If there exists a real number q such that for all sufficiently large n we have $c_n \le n^q$ then we say that the growth of $\mathbf{V}$ is *polynomial*. If no such q exists and there exists a real number d such that for all sufficiently large n we have $c_n \le d^n$ then we say that the growth of

V is at most *exponential*. According to our considerations in Section 4.8.5, the growth of the variety of all center-by-metabelian Lie algebras is polynomial, $c_n \le n^2$. In general, it is known (see Mishchenko [232]) that any variety with polynomial growth is in $\mathbf{N}_c\mathbf{A}$ for a suitable c. Using Section 5.5, Theorem 5.32, we can derive that any variety with polynomial growth has finite basis for its identities. On the other hand, there exist varieties with superexponential growth. An example of such a variety is given in Volichenko [362]. We sketch some relevant calculations. Using Section 2.4.2, the reader may see that the following system of monomials is linearly independent:

$$(x_{i_1}x_{j_1})\cdots(x_{i_q}x_{j_q})(x_{l_1}x_{l_2}x_{l_3}),$$

where

$$j_1 > j_2 > \cdots > j_q,\ l_1 > l_2 < l_3,\ i_1 < j_1, \ldots, i_q < j_q,$$

and

$$\{i_1, \ldots, i_q, j_1, \ldots, j_q, l_1, l_2, l_3\} = \{1, \ldots, n\}.$$

It is easy to see that the number of these monomials equals to

$$\frac{2}{q!}\binom{2q+3}{3}\binom{2q}{2}\binom{2q-2}{2}\cdots\binom{2}{2} = \frac{1}{3}(2q+3)(2q+3)(2q+1)\frac{(2q)!}{2^q q!}.$$

It is easy to see that the growth of this number is superexponential.

Another invariant of a variety or, to be precise, of its free algebra of finite rank m, where m is arbitrary natural, is its *mth Hilbert series*, that is, a series of the form

$$H_m(\mathbf{V}, t) = \sum_{n=1}^{\infty} b_m^{(n)} t^n,$$

where $b_m^{(n)} = \dim L_n(m, \mathbf{V})$. A more explicit invariant is a series of the form

$$H(\mathbf{V}, t_1, \ldots, t_m) = \sum_{\alpha} b_\alpha t_1^{\alpha(1)} \cdots t_m^{\alpha(m)},$$

where $b_\alpha = \dim L_\alpha$ is the dimension of the multihomogeneous component of a free algebra in **V** of rank m corresponding to the multidegree α. Here one of the intriguing questions is that of rationality of such series, that is, whether or not such a series is equal to a fraction whose numerator and denominator are polynomials. From Witt's formula (Section 3.1.3), one can easily observe that $H_m(\mathbf{O}, t)$ is not rational for all $m \ge 2$.

As an example, we compute the Hilbert series $H_m(t) = H(\mathbf{A}^2, t)$ in the case of the variety of all metabelian Lie algebras. For this, we remark that $P_n(\mathbf{A}^2)$ is an irreducible S_n-module associated with the diagram

$d =$ 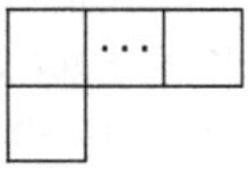

Indeed, by formulas in Section 3.2.5, we know that $b_d = n-1$. By Section 4.6.1, we also have $\dim P_n(\mathbf{A}^2) = n-1$. Besides, if

$$\tau d = \begin{array}{|c|c|c|c|}\hline n-1 & n-2 & \dots & 1 \\ \hline n & \multicolumn{3}{c}{} \\ \cline{1-1} \end{array}$$

then $(1/2)g_{\tau d} = x_1^{n-1}x_2$, which is an element of the canonical basis of $M(x_1, x_2)$. Using this and the coincidence of multiplicities of irreducible modules corresponding to the same Young diagrams in P_n and $L_n(m, \mathbf{A}^2)$, it is easy to derive that $L_n(m, \mathbf{A}^2)$ is also an irreducible $\mathrm{GL}(m, k)$-module. Its dimension can be computed by a formula in Section 3.3.5. In this formula the sequence of numbers $(n_1, n_2, \dots, n_m)$ takes the form $(n-1, 1, 0, \dots, 0)$. If $m \geq 2$, $n \geq 2$ then we get

$$b_m^{(n)} = b_{m-1}^{(n)} \frac{n+m-2}{m-1} \cdot \frac{m-1}{m-2} = \frac{n+m-2}{m-2} b_{m-1}^{(n)}.$$

Assuming that $b_m^{(1)}$ and $b_m^{(0)}$ are bound by the same formula, we arrive at a recurrence formula for an auxiliary series of the form $\widetilde{H}_m(t)$ where $H_m(t) = 1 + t + \widetilde{H}_m(t)$:

$$\begin{aligned} \widetilde{H}_m(t) &= \frac{1}{m-2}\left(\sum_{n=0}^{\infty}(n+1)b_{m-1}^{(n)}t^n + (m-3)\widetilde{H}_{m-1}(t)\right) \\ &= \frac{1}{m-2}\left(\left(\sum_{n=0}^{\infty} b_{m-1}^{(n)}t^{n+1}\right)' + (m-3)\widetilde{H}_{m-1}(t)\right) \\ &= \frac{1}{m-2}((t\widetilde{H}_{m-1}(t))' + (m-3)\widetilde{H}_{m-1}(t)) \\ &= \widetilde{H}_{m-1}(t) + \frac{t}{m-2}\widetilde{H}'_{m-1}(t). \end{aligned} \tag{4.34}$$

If $m = 2$ then we have $b_2^{(n)} = n-1$; we may also set $b_2^{(1)} = 0$ and $b_2^{(0)} = -1$. Thus we obtain

$$\begin{aligned} H_2(t) &= \sum_{n=0}^{\infty}(n-1)t^n = \sum_{n=0}^{\infty}(n+1)t^n - 2\sum_{n=0}^{\infty}t^n \\ &= \left(t\sum_{n=0}^{\infty}t^n\right)' - \frac{2}{1-t} = \left(\frac{t}{1-t}\right)' - \frac{2}{1-t} = \frac{2t-1}{(1-t)^2}. \end{aligned}$$

Now an immediate computation by induction using equation (4.34) gives us

$$\widetilde{H}_m(t) = \frac{mt-1}{(1-t)^m}, \quad \text{and} \quad H_m(t) = 1 + t + \frac{mt-1}{(1-t)^m}, \quad m = 2, 3, \dots$$

It has recently been shown by Drensky (unpublished) that, given a variety $\mathbf{V}$ with polynomial growth, for any m the series $H_m(\mathbf{V}, t)$ is rational. Some interesting computations related to Hilbert's series for the variety generated by Lie algebra $\mathfrak{sl}(2, k)$ are given in Drensky [105].

4.8.7 Passing from P_n to P_{n+1}

Suppose that a variety **V** is determined by an identity of the form $f_{\tau d} \equiv 0$ where d is a Young diagram of degree n. Then $P_s(\mathbf{V}) = P_s$ for $s < n$, $P_n(\mathbf{V}) = P_n/Af_{\tau d}$. The structure of $P_s(\mathbf{V})$ where $s > n$ is an open question. Still, using Richardson–Littlewood's rule, a certain piece of information can be obtained in the case $s = n + 1$. The thing is that the S_{n+1}-module U of consequences of $f_{\tau d} \equiv 0$ of degree $n + 1$ is generated by the elements of the following form. If $f_{\tau d} = f(x_1, \dots, x_{n-1}, x_n)$ then, up to the action of S_n, we may restrict ourselves to a consequence of the form $f' = f(x_1, \dots, x_{n-1}, x_{n+1}x_n)$.

We write

$$f_{\tau d} = \sum_{\sigma \in S_{n-1}} \alpha_\sigma x_{\sigma(1)} \cdots x_{\sigma(n-1)} x_n.$$

Then

$$f' = \sum_{\sigma \in S_{n-1}} \alpha_\sigma x_{\sigma(1)} \cdots x_{\sigma(n-1)} x_{n+1} x_n.$$

The S_{n-1}-submodules generated by these elements in P_n and P_{n+1} are isomorphic. Now let Q be a module to which the two mentioned above are isomorphic. Since $T = Af_{\tau d}$ and U as S_n- and S_{n+1}-modules, respectively, are generated by the submodules isomorphic to Q we may conclude that T (respectively, U) is isomorphic to a quotient module of the induced module $kS_n \otimes_{kS_{n-1}} Q$ (respectively, $kS_{n+1} \otimes_{kS_{n-1}} Q$). Moreover, it is easy to see that the subgroup $S_2 \subset S_{n+1}$ generated by the cycle $(n, n + 1)$ acts on U in the alternating way, that is, according to the diagram

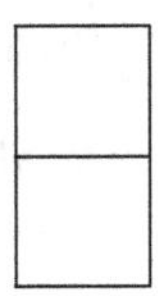

Thus we derive a procedure for the determination which Young diagrams may correspond to irreducible components of the module of consequences U. This can be done by application of Richardson–Littlewood's rule. We must first determine which simple S_{n-1}-modules may enter the decomposition of the S_n-module T and then determine what is induced by them in the degree $n + 1$.

For instance, the Engel identity of degree n, $n \geq 5$, does not imply the standard identity of degree $n + 1$. Indeed, the former identity corresponds to the diagram d_1 and

the latter to d_2 where

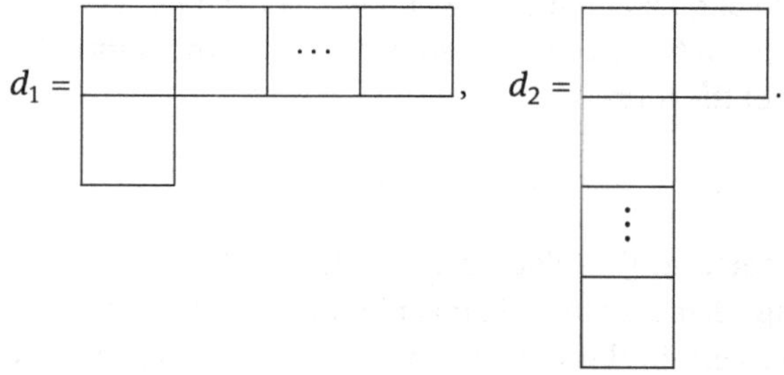

It is clear that in the process of applying the above algorithm the length of the first row cannot become shorter than $n-2$. Thus d_2 does not correspond to a simple component in the decomposition of the module of consequences of the former identity in the degree $n+1$.

4.9 Quasivarieties

4.9.1 Filtered product

Let I be a set and $\mathscr{P}(I)$ the lattice of all its subsets. A subset $\mathscr{F} \subseteq \mathscr{P}(I)$ is called a *filter* if this is a sublattice in $\mathscr{P}(I)$, i. e., for any $A, B \in \mathscr{F}$ we have $A \cap B \in \mathscr{F}$ and $A \cup B \in \mathscr{F}$ and if, in addition, given $C \in \mathscr{F}$ and $D \supseteq C$, we always have $D \in \mathscr{F}$. A filter $\mathscr{F}$ is called an *ultrafilter* if this is a proper filter, that is, $\mathscr{F} \neq \mathscr{P}(I)$ and if $\mathscr{F}$ is not a proper subset of another proper filter. Now we consider a family of algebras $(G_\alpha)_{\alpha\in I}$. Let $G = \prod_{\alpha\in I} G_\alpha$ be the Cartesian product of this family. Let $H_{\mathscr{F}}$ be a set of functions f such that the set $\{\alpha \mid f(\alpha) = 0\}$ belongs to $\mathscr{F}$. The verification of $H_{\mathscr{F}}$ being an ideal is straightforward and we leave it to the reader. The quotient algebra $G_{\mathscr{F}} = G/H_{\mathscr{F}}$ is called the *filtered product* of the family $(G_\alpha)_{\alpha\in I}$ with respect to the filter $\mathscr{F}$. If $\mathscr{F}$ is an ultrafilter we arrive at the concept of the ultraproduct. A filter is called *principal* if it consists of all subsets containing a fixed one, J. Such a filter is denoted by $\mathscr{F} = (J)$. It is clear that $G_{\mathscr{F}} \cong \coprod_{\alpha\in I\setminus J} G_\alpha$. In the case of finite sets any filter is principal, thus an essential case for filtered product is that of infinite families of algebras.

4.9.2 Quasiidentities and quasivarieties

Any formula of the form

$$f_1(x_1,\dots,x_n) = 0\,\&\,\cdots\,\&f_m(x_1,\dots,x_n) = 0 \Rightarrow f_{m+1}(x_1,\dots,x_n) = 0, \tag{4.35}$$

where $f_1,\dots,f_{m+1}$ are some nonassociative polynomials, is called a *quasiidentity*. We say that a quasiidentity holds in an algebra G (not necessarily a Lie algebra) if this is true every time when $x_1, \dots, x_n$ are replaced by elements of G. Obviously, an identity is a particular case of quasiidentities (when $m = 0$). To give a nontrivial example, we notice that in a nonabelian free Lie algebra (which has no nontrivial identities) there is a quasiidentity of the form

$$x(xy) \equiv 0 \Rightarrow xy \equiv 0.$$

A *quasivariety* of algebras (Lie algebras, associative algebras, and so on) is a class of algebras satisfying a fixed system of quasiidentities. A quasivariety determined by all quasiidentities of a certain class S of algebras is denoted by qvar(S) (or simply qvar(G) if $S = \{G\}$).

An important characteristic property of quasivarieties is given by a theorem due to Malcev [219] which is similar to Birkhoff's theorem (Theorem 4.3).

Theorem 4.51. *A nonempty abstract*[1] *class* $\mathbf{K}$ *of algebras (over a commutative ring* k*) is a quasivariety if and only if* $\mathbf{K}$ *is closed under forming subalgebras and filtered products of its algebras.*

The theory of quasivarieties of Lie algebras is quite rudimentary, although some results can be proven using the analogy with the group theory.

4.10 Exercises

4.10.1. If in a variety $\mathbf{V}$ all algebras are solvable then for some l we have $\mathbf{V} \subseteq \mathbf{S}_l$.

4.10.2. Let $\mathbf{V}$ be a variety over an infinite field k in which the nilpotent steps of finite-dimensional algebras are totally bounded. Then there exists a natural number c such that $\mathbf{V} \subseteq \mathbf{N}_c$.

4.10.3. Prove that any identity of degree 5 in W_1 over a field of characteristic zero is a consequence of $s_4(\operatorname{ad} x_1, \operatorname{ad} x_2, \operatorname{ad} x_3, \operatorname{ad} x_4)x_5 \equiv 0$. It is not known if this is true for identities of arbitrary degree.

4.10.4. Prove that the free metabelian algebra $M(x,y) = L(x,y)/(L(x,y))^{(2)}$ is not finitely related.

4.10.5. Prove that the center of a $\mathbf{V}$-free algebra of countable rank is a verbal ideal. Give an example of a Lie algebra in which the center is not a verbal ideal.

4.10.6. Let L be a free Lie algebra and M its ideal. Prove that any nilpotent subalgebra in L/M^2 is either one-dimensional or a subalgebra in M/M^2.

4.10.7. Prove that the center of a free algebra of a nontrivial product variety over a field of characteristic zero is equal to zero.

4.10.8. Let $A = \{a_1, \dots, a_n\}$, $B = \{b_1, \dots, b_n\}$ be abelian Lie algebras, $W = A \operatorname{wr} B$, M a subalgebra in W generated by elements $m_i = a_i + b_i$, $i = 1, \dots, n$. Prove that

1 A class is called *abstract* if it is closed under isomorphism.

an element c of the form

$$c = b + \sum_{i=1}^{n} f_i(b_1, \ldots, b_n)a_i$$

is in M if and only if

$$\sum_{i=1}^{n} f_i(b_1, \ldots, b_n)b_i = b.$$

Hints: (1) Reduce to $b = 0$; (2) rove that the map $\psi : a_i \to b_i$ extends to a B-module homomorphism from $K = \mathrm{id}_W(A)$ into $U(B)$, our claim being equivalent to $\ker \psi = M \cap K$.

4.10.9. Let X be a nonempty set, $\mathbf{V}$ a variety over an infinite field k, $L = L(X, \mathbf{V})$, $Y \subseteq L$ be linearly independent modulo L^2 and L/L^2 be generated by the image of Y. Is it true that Y is a free generating set for L? Describe varieties $\mathbf{V}$ for which this is true.

4.10.10. Let $\mathbf{V} \subseteq \mathbf{N}_2$ be the variety of Lie rings determined within $\mathbf{N}_2$ by $2xy = 0$, $L = L(\{a, b\}, \mathbf{V})$. Prove that the elements $a, 2b$ are linearly independent modulo L^2 but they do not generate a free algebra of $\mathbf{V}$.

4.10.11. Let $L = L(X, \mathbf{V})$ be a free ring of a multihomogeneous variety of Lie rings $\mathbf{V}$ and $Y \subseteq L$ is such that the image of Y generates a $\mathbf{V} \cap \mathbf{A}$-free subring which is a direct summand of L/L^2. Then $\mathrm{alg}(Y)$ is free with free generating set Y.

4.10.12. Produce a counterexample to the monotonicity theorem (Theorem 4.19) in the case of algebras over a ring.

4.10.13. Prove claims in Section 4.3.2.

4.10.14. Let $\mathbf{V}$ be a variety of Lie algebras over a field of characteristic zero. Then $[\mathbf{V}, \mathbf{E}]$ is an indecomposable variety.

4.10.15. Prove the following equations (the base ring is a field):
(a) $(\mathbf{U}_1 \cup \mathbf{U}_2)\mathbf{V} = \mathbf{U}_1\mathbf{V} \cup \mathbf{U}_2\mathbf{V}$,
(b) $(\mathbf{U}_1 \cap \mathbf{U}_2)\mathbf{V} = \mathbf{U}_1\mathbf{V} \cap \mathbf{U}_2\mathbf{V}$,
(c) $[\mathbf{U}_1, \mathbf{U}_2]\mathbf{V} = [\mathbf{U}_1\mathbf{V}, \mathbf{U}_2\mathbf{V}]$.

4.10.16. Prove
(a) $\mathbf{U}(\mathbf{V}_1 \cup \mathbf{V}_2) \supseteq \mathbf{U}\mathbf{V}_1 \cup \mathbf{U}\mathbf{V}_2$,
(b) $\mathbf{U}(\mathbf{V}_1 \cap \mathbf{V}_2) \subseteq \mathbf{U}\mathbf{V}_1 \cap \mathbf{U}\mathbf{V}_2$,
where the inclusions may be proper.

4.10.17. Show that the varieties $\mathbf{U}[\mathbf{V}_1, \mathbf{V}_2]$ and $[\mathbf{U}\mathbf{V}_1, \mathbf{U}\mathbf{V}_2]$ may turn out to be incomparable.

4.10.18. Let $\mathbf{U}, \mathbf{V}$ be varieties of rings with $\mathbf{U} \cap \mathbf{V} = \mathbf{E}$. Then $\mathbf{U}\mathbf{V} \cap \mathbf{V}\mathbf{U} = \mathbf{U} \cup \mathbf{V}$.

4.10.19. Let $\mathbf{U}_1 \supset \mathbf{U}_2 \supset \cdots$ be a descending chain of varieties. Then $\mathbf{V}(\bigcap_{i=1}^{\infty} \mathbf{U}_i) = \bigcap_{i=1}^{\infty} \mathbf{V}\mathbf{U}_i$. Give a correct claim and its proof in the case of an infinite union.

4.10.20. Which of the two identities $x(yz)uv = 0$ and $(xyz)uv \equiv 0$ implies the other? Prove that these are not equivalent.

4.10.21. Prove that a free algebra of finite rank in a variety over an infinite field cannot be mapped into itself with nonzero kernel (i. e., it is a hopfian algebra).

4.10.22. Prove that any algebra over a field of characteristic zero with identity of the form

$$x_1 x_2 \cdots x_c x_1 \equiv 0$$

is nilpotent.

4.10.23. Prove the claim following Theorem 4.9 concerning the Engel identity.

4.10.24. Prove that any variety generated by a finite nonnilpotent Lie algebra is not multihomogeneous.

4.10.25. Construct bases for wreath products of one-dimensional algebras of the form

$$W_1 = \langle e_1 \rangle \operatorname{wr} \langle e_2 \rangle, \quad W_2 = \langle e_1 \rangle \operatorname{wr}_{\mathbf{N}_2} \langle e_2 \rangle.$$

4.10.26. Compute the basic rank of $\mathbf{N}_c \cap \mathbf{A}^2$.

4.10.27. Find implications between the identities of metabelian varieties over an infinite field of characteristic $p > 0$:
(a) $v_1 = x^{p^2} y^{p^2} \equiv 0$, $v_2 = x_1^p \cdots x_p^p y_1^p \cdots y_p^p \equiv 0$,
(b) $v = x^p \sum_{i=1}^{p-1} x_1 \cdots \hat{x}_i \cdots x_{p-1} x_i x_p \equiv 0$, $v_2 = x^p x_1 \cdots x_p \equiv 0$.

4.10.28. (See notation in Section 4.8.3.) Which of the canonical identities $g_{\tau_i d_i} \equiv 0$ of degree five imply nilpotence of some step? Solubility?

4.10.29. (Notation in Section 4.8.3.) Prove that $g_{d_5} \equiv 0$ is equivalent to the standard identity of the same degree (see Exercise 4.10.3), while $g_{\tau_3 d_3} \equiv 0$ is equivalent to the following property: the cube of an inner derivation is a derivation.

4.10.30. If the cardinalities of all factors of an ultraproduct do not exceed a fixed natural number n then the same is true of the ultraproduct itself.

4.10.31. A finite algebra G belongs to $\operatorname{qvar}(H)$, where H is a finite algebra, if and only if G is a subalgebra of a finite direct power of H.

4.10.32. Let G be a three-dimensional nilpotent nonabelian Lie algebra over a finite field k. Prove that $\operatorname{var}(G) \neq \operatorname{qvar}(G)$.

4.11 Comments

First treatments of the theory of varieties of Lie algebras in a book form appeared in Amayo and Stewart [5] and Bahturin [26]. Many preliminary results are general for the varieties of arbitrary algebras. In such a general form, varieties are treated in Malcev [219]. It is a pleasure for us to acknowledge the influence of H. Neumann's monograph [247]. Processes of linearization in Section 4.2 are classical; their application to the identities has appeared in Malcev [218]. Theorem 4.10 has been formulated in Drensky [103]. Some results in Section 4.2 are similar to group-theoretic results in H. Neumann [247]. Theorem 4.12 is based on Andreev and Shabel'nikova [9] and Bahturin [26]. Further results concerning the question on the coincidence of invariant and verbal subalgebras can be found in Bahturin and Olshanskii [51]. In connection with Theorem 4.14, we remark that the question on hopficity of a finitely generated **V**-free algebra, **V** being a variety of Lie algebras over a finite field, is still open. A theorem on the possibility of determining a multihomogeneous variety by an independent system of identities is due to Olshanskii [252]. In the case of varieties over a finite field, this is an open question. Section 4.3 is based on the papers of the author (Bahturin [19, 21, 22]), Parfenov [255, 256], and Genov [118]. In the case of the two-element field $\mathbb{Z}_2$, the groupoid $v(\mathbb{Z}_2)$ is not associative (see Bahturin [22]). A thorough study of indecomposable varieties is given in Zaĭcev [375]. Section 4.4 is built on Shmelkin [342]. Section 4.5 stems from Bahturin [29], the present proofs being new. In the same source one can find the precise value of $r_b(\mathbf{N}_c)$ over an arbitrary commutative ring. A basis for identical relations of the group of nondegenerate triangular matrices of order n has been found by Romanovsky. Theorem 4.41, in the case of an infinite field, has been obtained in Bahturin [19, 21]. The final solution has been given by Zaĭcev [374]. Results of Section 4.7 in the zero characteristic case are due to Shmelkin. The case of prime characteristic has been studied by Vaughan-Lee [353] and Bahturin [24]. The case of finite fields is studied by Sheina [332]. Section 4.8 is introductory. Complete description of the lattice of center-by-metabelian varieties is due to Mishchenko. He has also given a description of the lattice $\mathscr{L}(\mathbf{V}_1)$ of hypercenter-by-metabelian varieties over a field of characteristic zero [228, 230]. The lattice of subvarieties of $\mathbf{V}_2 = \mathbf{N}_2\mathbf{A} \cap \mathbf{A}\mathbf{N}_2$ has been described by Drensky [107]. Numerical series associated with varieties are virtually unknown. An exception is the Hilbert series in the variety generated by $\mathfrak{sl}(2, k)$ where k is of characteristic 0 (see Drensky [105]). A remark due to Mishchenko is that the series of codimensions of a variety over zero characteristic fields cannot have intermediate growth (that is, subexponential but not polynomial). Deep applications of the theory of representations of the symmetric group to varieties of Lie algebras have been given by Volichenko. In particular he has found (Volichenko [360]) the least variety **G** over a field of characteristic zero that does not satisfy a standard identity. This variety is generated by a Lie algebra of the form

$$\begin{pmatrix} G & G \\ 0 & 0 \end{pmatrix},$$

where G is an infinite dimensional Grassmann's algebra (see Section 6.4.3). The basic rank of **G** is infinite but every proper subvariety in **G** has finite basic rank.

Artamonov studied free metabelian Lie algebras in [14, 16] and varieties of restricted Lie algebras in [12]. Homomorphisms and approximation of solvable Lie algebras were studied in [20, 28].

5 Finite basis problem

5.1 Statement of the problem

5.1.1 Dependence on the base ring

One of the principal questions of the theory of identities of any class of algebras is the question of a possibility of reducing any system of identities to its finite subsystem equivalent to the whole system. It is called the *finite basis problem*. This question is answered in the negative if the base ring is not noetherian. In the case of Lie algebras, a counterexample is provided by the variety $\mathbf{A} \cap \mathbf{O}_M$ (see Section 4.1.3) where M is an ideal of the base ring which has no finite generating system. A system of identities determining this variety has the form $\{xy \equiv 0\} \cup \{\lambda x \equiv 0 \quad \text{where } \lambda \in M\}$. Suppose this system is equivalent to a finite subsystem $\{v_1 \equiv 0, v_2 \equiv 0, \ldots, v_m \equiv 0\}$. Since any identity is a consequence of a finite number of identities of the former system, the latter system can be replaced by a system of the form $\{xy \equiv 0, \lambda_1 x \equiv 0, \ldots, \lambda_m x \equiv 0\}$. Let N denote the ideal of k generated by $\lambda_1, \ldots, \lambda_m$. Pick out a $\lambda \in M \setminus N$. Then the abelian Lie algebra k/N satisfies all the identities $xy \equiv 0$ and $\lambda_i x \equiv 0$, $i = 1, \ldots, m$, but $\lambda x \equiv 0$ is not an identity since $\lambda(1 + N) = \lambda + N \neq 0$.

Thus, in our further study of the finite basis problem, it will be assumed that the base ring k is noetherian.

5.1.2 Axiomatic rank of variety

If $\mathbf{V}$ is a finitely based variety, that is, if $\mathbf{V}$ is determined by a finite system of identities $\{v_1 \equiv 0, v_2 \equiv 0, \ldots, v_m \equiv 0\}$ then, obviously, the number of variables occurring in the expression of all polynomials $v_1, v_2, \ldots, v_m$ is finite. To generalize this situation, we define the *axiomatic rank* of a variety $\mathbf{V}$ of Lie algebras as the minimal number of variables used for expressing a basis of identities for $\mathbf{V}$. This value is denoted by $r_a(\mathbf{V})$. Thus $r_a(\mathbf{V})$ is either a natural number or $+\infty$ depending on the possibility of determining $\mathbf{V}$ by a system of identities written in finitely many variables.

Proposition 5.1. *The axiomatic rank of any nilpotent variety* $\mathbf{V}$ *is finite. If* $\mathbf{V} \subseteq \mathbf{N}_c$ *then* $r_a(\mathbf{V}) \leq c + 1$.

Proof. As remarked in Section 4.2.5, any identity is equivalent to a system of normal identities, that is, having the form $f(x_1, x_2, \ldots, x_n) \equiv 0$ where each monomial in f has as its factors all the variables $x_1, x_2, \ldots, x_n$. If $\mathbf{V} \subseteq \mathbf{N}_c$ and $n \geq c+1$ then $f \equiv 0$ turns out to be a consequence of $x_1 x_2 \cdots x_{c+1} \equiv 0$ satisfied by $\mathbf{V}$. Therefore the system of all identities holding in $\mathbf{V}$ is equivalent to its subsystem formed by normal identities depending on at most c variables and by the nilpotent identity $x_1 x_2 \cdots x_{c+1} \equiv 0$. □

https://doi.org/10.1515/9783110566659-C05

5.1.3 Specht variety

A variety of Lie algebras $\mathbf{V}$ is called *Specht* (in honor of German mathematician W. Specht) if any subvariety $\mathbf{U}$ of $\mathbf{V}$, including $\mathbf{V}$ itself, is a finitely based variety. If $L = L(x_1, x_2, \dots, x_n, \dots)$ is a free Lie algebra on countably many variables then the verbal ideals determining subvarieties of $\mathbf{V}$ are precisely the verbal ideals of L containing $V = \mathbf{V}(L)$. If $L(m) = L(x_1, x_2, \dots, x_m)$ then the subvarieties of $\mathbf{V}$ which can be determined modulo $\mathbf{V}$ by systems of identities depending on at most m variables are fully determined by the verbal ideals of $L(m)$ containing $V_m = \mathbf{V}(L(m))$. A Lie algebra G is called *verbally noetherian* if G satisfies the maximality condition for verbal ideals.

Proposition 5.2. *For a finitely based variety* $\mathbf{V}$*, the following conditions are equivalent:*
(i) $\mathbf{V}$ *is a Specht variety;*
(ii) $\mathscr{L}(\mathbf{V})$ *is an artinian lattice;*
(iii) $L/\mathbf{V}(L)$ *is verbally noetherian.*

If, in addition, $r_a(\mathbf{U}) \le m$ *for any subvariety* $\mathbf{U} \subseteq \mathbf{V}$ *then these conditions are equivalent to:*
(iv) $L(m)/\mathbf{V}(L(m))$ *is verbally noetherian.*

Proof. Since the lattice of subvarieties of any variety is antiisomorphic to the lattice of verbal ideals containing $\mathbf{V}(L)$, $L = L(x_1, x_2, \dots, x_n, \dots)$, then (i) and (ii) are obviously equivalent. Now if (i) holds then any verbal ideal containing $\mathbf{V}(L)$ is finitely generated, hence it cannot be equal to the union of the properly increasing chain of verbal ideals, containing $\mathbf{V}(L)$. Let, conversely, $U = \mathbf{U}(L)$ be a verbal ideal of L containing $\mathbf{V}(L)$. Then, by the noetherian property, there exists a finite system of elements $\{v_1, v_2, \dots, v_s\}$ in L such that

$$V = \mathbf{V}(L) \subset \{V, v_1\}(L) \subset \{V, v_1, v_2\}(L) \subset \cdots \subset \{V, v_1, \dots, v_s\}(L) = \mathbf{U}(L) = U.$$

Since V is a finitely generated verbal ideal, the union of its generating system with $\{v_1, v_2, \dots, v_s\}$ is a finite generating system for U. This proves the equivalence of (i), (ii), and (iii). Now we consider $L(m)$ as a subalgebra in L generated by $\{x_1, \dots, x_m\}$. Then $V_m = V(L(m)) = \mathbf{V}(L) \cap L(m)$ (see the proof of Proposition 4.18 in Section 4.3). Under our hypothesis, we also have $\mathbf{V}(L) = V_m(L)$. These relations show that $L/\mathbf{V}(L)$ is verbally noetherian if and only if $L(m)/\mathbf{V}(L(m))$ is a verbally noetherian algebra. □

Theorem 5.3. *Identical relations of a nilpotent variety over a noetherian ring are finitely based.*

Proof. According to Sections 5.1.2 and 5.1.3, it is sufficient to prove the verbal noetherian property for $L(c+1)/L(c+1)^{c+1}$. But this algebra is a finitely generated k-module whose generators are all monomials in $x_1, \dots, x_{c+1}$ of degree at most c. Since k is noetherian, this module is noetherian, too. □

5.1.4 Finite basis property and operations on the varieties

It is obvious that the intersection of two finitely based varieties is itself finitely based. It is determined by the union of bases for the varieties being intersected. However, the question on the finite basis for the union of two finitely based varieties is an example of open questions in the theory under consideration. There is no final answer to the question on the finite basis property of the product and the commutator of two finitely based varieties. There are only some partial results on this subject and these are due to Zaicev [373].

Before we pass to these results, we need a concept of a strictly finitely based variety. Namely, we say that $\mathbf{V}$ is a *strictly finitely based variety* if the associated verbal ideal $\mathbf{V}(L)$ is the k-span of values of finitely many elements in $\mathbf{V}(L)$.

Proposition 5.4.
(i) *Any nilpotent variety over a noetherian ring is strictly finitely based.*
(ii) *Any finitely based multihomogeneous variety is strictly finitely based.*

Proof. If L is a free Lie algebra of countable rank and $u_1 \equiv 0, \dots, u_p \equiv 0$ is a basis of identities of a nilpotent variety $\mathbf{V}$ of nilpotent step c including the nilpotent identity of nilpotent step c then $\mathbf{V}(L)$ is strictly generated by finitely many polynomials of the form $x_1 \cdots x_t u_i(y_1, \dots, y_s)$ of total degree at most c where all $x_1, \dots, x_t$ are pairwise distinct and have no common letters with $y_1, \dots, y_s, t \le c$. Now if $\mathbf{V}$ is multihomogeneous and $v_1 \equiv 0, \dots, v_p \equiv 0$ is its multihomogeneous basis of identities then the k-span of the values of polynomials $v_1, \dots, v_p$ is automorphism invariant and multihomogeneous, hence, by Section 4.2, Theorem 4.14, it is a verbal ideal, which is, of course, $\mathbf{V}(L)$ itself. □

Theorem 5.5. *Let $\mathbf{U}$ be a finitely based variety and $\mathbf{V}$ a strictly finitely based variety. Then $\mathbf{UV}$ and $[\mathbf{U}, \mathbf{V}]$ are finitely based varieties.*

Proof. If $\mathbf{V}$ is a strictly finitely based variety then in $V = \mathbf{V}(L_\infty)$, where L_∞ is a free Lie algebra of countable rank, one can find a finite set of polynomials $v_1, v_2, \dots, v_s$ such that any $v \in V$ is a k-combination of the values of these polynomials in L_∞. If we write down the values of $v_1, v_2, \dots, v_s$ in pairwise disjoint sets of variables, we will come to the elements $v_1', v_2', \dots, v_s'$ whose sum is a strict generator $v = v(x_1, \dots, x_n)$ for V. Now we choose in $U = \mathbf{U}(L_\infty)$ a finite subsystem of normal polynomials which is closed under partial linearizations (see Section 4.2).

(I) We concentrate first on the product variety $\mathbf{W} = \mathbf{UV}$. In this case $\mathbf{W}(L_\infty)$ is an ideal of L_∞ generated by all elements of the form $u_i(y_1, \dots, y_{t_i})$, $i = 1, \dots, m$, where $y_i \in V = \mathbf{V}(L)$. Now each y_j is a linear combination of the values of v in L_∞ with coefficients in k. Since $\{u_1, \dots, u_s\}$ is closed under linearizations, it is sufficient to consider the case where $y_j = \lambda_j v(\dots)$ (see Section 4.2.2). Thus we may assume that our ideal is generated by a system of the elements of the form

$$u_i(\lambda_0 v^0, \lambda_1 v^1, \ldots, \lambda_s v^s), \quad i = 1, 2, \ldots, m,\ \lambda_1, \lambda_2, \ldots, \lambda_s \in k \tag{5.1}$$

(s depending on i), $v^q(x_{qn+1}, \ldots, x_{qn+n})$, $q = 0, 1, \ldots$ The finite basis property will be proved if we manage to bound the number of s-tuples $(\lambda_1, \ldots, \lambda_s)$. We write $u_i(x_1, \ldots, x_s)$ in the form of a linear combination $\sum_{j=1}^{r} \mu_j w_j$ of basic monomials. Then for any selection $(\lambda_1, \ldots, \lambda_s) \in k^s$ we will be having $u_i(\lambda_1 x_1, \ldots, \lambda_s x_s) = \sum_{j=1}^{r} \gamma_j \mu_j w_j$. This gives rise to a map $\psi : k^s \to k^r$ such that $\psi(\lambda_1, \ldots, \lambda_s) = (\gamma_1, \ldots, \gamma_r)$. Now k^r is a noetherian k-module, hence the k-module $\psi(k^s)$ is finitely generated and it is possible to choose a finite generating set for it in the form $\psi(\bar{\lambda}^1), \ldots, \psi(\bar{\lambda}^p) \in \psi(k^s)$, $\bar{\lambda}^i = (\lambda_1^i, \ldots, \lambda_s^i)$, $i = 1, \ldots, p$. For any $\bar{\lambda} = (\lambda_1, \ldots, \lambda_s)$ there exists $\alpha_1, \ldots, \alpha_p \in k$ such that

$$\psi(\bar{\lambda}) = \alpha_1 \psi(\bar{\lambda}^1) + \cdots + \alpha_p \psi(\bar{\lambda}^p).$$

We set $\psi(\bar{\lambda}) = (\gamma_1, \ldots, \gamma_r)$, $\psi(\bar{\lambda}^i) = (\gamma_1^i, \ldots, \gamma_r^i)$, $i = 1, \ldots, p$. Then

$$\begin{aligned} u_1(\lambda_1 x_1, \ldots, \lambda_s x_s) &= \gamma_1 \mu_1 w_1 + \cdots + \gamma_r \mu_r w_r \\ &= (\alpha_1 \gamma_1^1 + \cdots + \alpha_p \gamma_1^p)\mu_1 w_1 + \cdots + (\alpha_1 \gamma_r^1 + \cdots + \alpha_p \gamma_r^p)\mu_r w_r \\ &= \alpha_1(\gamma_1^1 \mu_1 w_1 + \cdots + \gamma_r^1 \mu_r w_r) + \cdots + \alpha_p(\gamma_1^p \mu_1 w_1 + \cdots + \gamma_r^p \mu_r w_r) \\ &= \alpha_1 u_i(\lambda_1^1 x_1, \ldots, \lambda_s^1 x_s) + \cdots + \alpha_p u_i(\lambda_1^p x_1, \ldots, \lambda_s^p x_s). \end{aligned}$$

Thus, a generating set for $\mathbf{W}(L_\infty)$ can be chosen in the form (5.1) where the set of s-tuples is finite and it is chosen in the way described just above as $(\lambda_1^1, \ldots, \lambda_s^1), \ldots, (\lambda_1^p, \ldots, \lambda_s^p)$. Therefore we conclude that $\mathbf{W} = \mathbf{UV}$ is finitely based variety.

(II) Passing to $\mathbf{W} = [\mathbf{U}, \mathbf{V}]$, we notice that $W = \mathbf{W}(L) = \mathbf{U}(L)\mathbf{V}(L)$. It is convenient to assume that both $\mathbf{U}$ and $\mathbf{V}$ are generated by one element each, say by u and v, respectively. Moreover, as above, it will be assumed that v is a strict generator for $\mathbf{V}(L)$. Then, writing $u = u(x_1, \ldots, x_m)$ and $v = v(x_{m+1}, \ldots, x_n)$, we obtain a generating set for W in the form

$$(x_{m+1} x_{m+2} \cdots x_t u)(x_{t+1} \cdots x_s v). \tag{5.2}$$

Applying formula $(ab)(cd) = a(bcd) - b(acd)$, we come to the conclusion that it is sufficient to consider only the products (5.2) of the form $u x_{t+1} \cdots x_s v$. But $x_{t+1} \cdots x_s v \in \mathbf{V}(L)$; hence, by our choice of v, this is a linear combination of the values of the latter. Hence, in fact, $\mathbf{W}(L)$ is generated by a single element $u(x_1, \ldots, x_m)v(x_{m+1}, \ldots, x_n)$, proving that $\mathbf{W}$ is indeed a finitely based variety. □

Corollary 5.6. *Let* $\mathbf{U}$ *and* $\mathbf{V}$ *be finitely based varieties over a noetherian ring* k *such that* $\mathbf{V}$ *is either nilpotent or multihomogeneous. Then* $\mathbf{UV}$ *and* $[\mathbf{U}, \mathbf{V}]$ *are finitely based varieties.*

Corollary 5.7. *Let* k *be an infinite field. Then* $\mathbf{W} = \mathbf{UV}$ *is finitely based if and only if this holds for* $\mathbf{U}$ *and* $\mathbf{V}$.

Proof. By the above, all that remains is to prove that if $\mathbf{W}$ is finitely based then the same is true for $\mathbf{U}$ and $\mathbf{V}$. Suppose, by contrast, that $\mathbf{V}$ is not a finitely based variety. Then $V = \mathbf{V}(L)$ is equal to the union of a strictly increasing chain of verbal ideals of L of the form $V = \bigcup_{i=1}^{\infty} \mathbf{V}_i$. Then $W = \mathbf{W}(L) = \bigcup_{i=1}^{\infty} \mathbf{U}(V_i)$. By the monotonicity theorem (Section 4.3, Theorem 4.19), all the ideals $\mathbf{U}(V_i)$ are pairwise distinct. Hence $\mathbf{W}$ is not finitely based, a contradiction. Now if we admit $\mathbf{U}$ not being finitely based then $U = \mathbf{U}(L) = \bigcup_{i=1}^{\infty} U_i$ where $U_i \subset U_{i+1}$ and $W = \mathbf{W}(L) = \bigcup_{i=1}^{\infty} U_i(V)$. Since V is a free Lie algebra of countable rank (Theorem 2.18), all ideals $U_i(V)$ are pairwise distinct and again $\mathbf{W}$ is a variety which cannot have finite basis for its identities. □

An essential part of the proof of Theorem 5.9 about the union of finitely based varieties is the following.

Lemma 5.8. *Let U be a verbal ideal of a free Lie algebra L of countable rank. Then, for any $n \geq 1$, the ideal $U \cap L^n$ is a finitely generated verbal ideal if and only if this is true of $U \cap L^{n+1}$.*

Proof. We consider $L(n) = L(x_1,\dots,x_n) \subset L$. If we pass to the normal polynomials generating $U \cap L^n$, we will see that, as a verbal ideal, U is generated by $U \cap (L(n))^n + U \cap L^{n+1}$. Now the following are the relations between k-modules:

$$U \cap (L(n))^n + U \cap L^{n+1}/U \cap L^{n+1} \cong U \cap (L(n))^n/U \cap (L(n))^{n+1}$$
$$\cong U \cap (L(n))^n + (L(n))^{n+1}/(L(n))^{n+1} \subseteq (L(n))^n/(L(n))^{n+1}.$$

Since the last term of this chain is a finitely generated, hence noetherian, k-module, we conclude that $U \cap L^n$ is generated as a verbal ideal by generators of $U \cap L^{n+1}$ and by finitely many inverse images of k-generators of the k-module $U \cap (L(n))^n + U \cap L^{n+1}/U \cap L^{n+1}$, proving the "if" part of Lemma 5.8.

Now we assume that $U \cap L^n$ is a verbally finitely generated ideal. Suppose $\{u_1,\dots,u_s,v_1,\dots,v_r\}$ is its finite normal generating system such that $u_1,\dots,u_s \notin U \cap L^{n+1}$, $v_1,\dots,v_r \in U \cap L^{n-1}$ and $\{v_1,\dots,v_r\}$ is closed under forming partial linearizations. Pick out a variable z which is not used in the expression of the above basis. We form a system P of nonassociative polynomials in $U \cap L^{n+1}$ including all $v_1,\dots,v_r$, as well as all polynomials of the form $zu_i(x_1,\dots,x_l)$ and $u_i(x_1,\dots,zx_j,\dots,x_l)$, $i = 1,2,\dots,s$, $j = 1,2,\dots,l$, l being the number of variables occurring in the normal polynomial u_i. Now let w be an arbitrary normal element in $U \cap L^{n+1}$. Since $w \in U \cap L^n$, this element is representable in the form $w = w_1 + w_2$ where w_1 belongs to the ideal generated by $\{v_1,\dots,v_r\}$ and w_2 is an element of the ideal generated by $\{u_1,\dots,u_s\}$. We further decompose w_2 as $w_2 = w_2' + w_2'' + w_2'''$ where the summands are as follows: w_2' is a linear combination of values of the form $u_i(g_1,\dots,g_l)$ where at least one of g_j is in L^2; w_2'' is a linear combination of elements of the form $g_{l+1}u_i(g_1,\dots,g_l)$; w_2''' is a linear combination of the values of $u_1,\dots,u_s$ in $x_1,x_2,\dots,x_n,\dots$ It is clear that w_2' and w_2'' belong to the verbal ideal generated by P. Now replacing w_2''' by its equivalent normal polynomial, we will get $e(w) \in U \cap L^{n+1}$. This element depends on $x_1, \dots, x_t$, $t \le n$. Hence the set E of all $e(w)$, $w \in L^{n+1} \cap U$, belongs to a finitely generated k-submodule whose elements are all possible linear combinations of values of $u_1, \dots, u_s$ in $x_1, \dots, x_q$. Thus if $e_1 = e(w_1), \dots, e_q = e(w_q)$ is a generating set for kE then $U \cap L^{n+1}$ is generated as a verbal ideal by $P \cup \{e_1, \dots, e_q\}$. □

Theorem 5.9. *Let* $\mathbf{U}$ *and* $\mathbf{V}$ *be varieties of Lie algebras over a noetherian ring* k*,* $\mathbf{V}$ *being nilpotent. Then their union* $\mathbf{W} = \mathbf{U} \cup \mathbf{V}$ *is finitely based if and only if* $\mathbf{U}$ *is a finitely based variety.*

Proof. Since, for an appropriate c, we have $\mathbf{V} \subseteq \mathbf{N}_c$, we may write $(\mathbf{U} \cup \mathbf{V}) \cup \mathbf{N}_c = \mathbf{U} \cup \mathbf{N}_c$. Therefore it is sufficient to prove our theorem in the case where $\mathbf{V} = \mathbf{N}_c$. In this case we are dealing with an ideal of the form $U \cap L^{c+1}$ where $U = \mathbf{U}(L)$. It is clear that our claim is true for $c = 1$. Hence the induction over c applies, the induction step being supplied by Lemma 5.8. □

The question on the finite basis property of the union of two finitely based varieties without imposing restrictions on the components is open even in the case where the base field is of characteristic zero.

5.2 Partially well-ordered sets

5.2.1 Definition

A partially ordered set P with order $\le$ is called a *partially well-ordered* set if, given an arbitrary nonempty subset Q in P, there exist finitely minimal elements $q_1, \dots, q_r$ in Q such that for any $q \in Q$ there exists a number i such that $q_i \le q$. For instance, if A and B are two well-ordered sets then $P = A \times B$ with direct product of orders (i. e., $(a, b) \le (a', b')$ if and only if $a \le a'$, $b \le b'$) is a partially well-ordered set. Indeed, it is obvious that P is artinian, that is, satisfies the minimality condition. Suppose now that a nonempty set Q has infinitely many minimal elements $(u_1, v_1), (u_2, v_2), \dots$ Rewriting these pairs so that $u_1 < u_2 < \cdots$ is an increasing sequence, we are forced to conclude that $v_1 > v_2 > \cdots$ is a decreasing sequence. But B is a well-ordered set, and we are easily led to a contradiction.

Proposition 5.10. *A partially ordered set* $(P, \le)$ *is partially well-ordered if and only if, given an infinite sequence* $p_1, p_2, \dots$ *of its elements there exist numbers* $i < j$ *such that* $p_i \le p_j$.

Proof. It is obvious that this condition is necessary. To prove its sufficiency, we consider a nonempty subset Q in P. Let Q_0 be the set of minimal elements in Q. This is a finite set since any two elements in Q_0 are either incomparable or equal. This is a nonempty set. Indeed, let a_1 be an element which is not minimal. Then there exists an

a_2 such that $a_1 > a_2$. Repeating this we easily derive that our hypothesis $Q_0 = \varnothing$ gives rise to an infinite strictly decreasing sequence $a_1 > a_2 > \cdots$, a contradiction. The same very argument shows that each element of Q is comparable with some minimal. □

Corollary 5.11. *Let A, B be two partially well-ordered sets, $P = A \times B$ their direct product endowed with the direct product of orders (as in Section 5.2.1). Then P is a partially well-ordered set, too.*

Proof. We notice that any infinite subset Q of elements of a partially well-ordered set A contains an infinite chain. Indeed, let $u_1, \ldots, u_s$ be all minimal elements of Q. Denote by $a^\Delta = a_Q^\Delta$ be the set (cone) of elements x in Q such that $a \leq x$. Then $Q = u_1^\Delta \cup u_2^\Delta \cup \cdots \cup u_s^\Delta$. One of the cones, say u_1^Δ, is infinite. Now let $Q_1 = u_1^\Delta \setminus \{u_1\}$ and repeat the same argument to Q_1. Then $Q_1 = v_1^\Delta \cup \cdots \cup v_r^\Delta$. If v_1^Δ is infinite then we set $Q_2 = v_1^\Delta \setminus \{v_1\}$. Continuing this procedure indefinitely will eventually yield an infinite increasing sequence $u_1 < v_1 < \cdots$, as required. Now let $(a_1, b_1), (a_2, b_2), \ldots, (a_n, b_n), \ldots$ be an infinite sequence of elements in P. From what was just proved, we may assume that $a_1 \leq a_2 \leq \cdots \leq a_n \leq \cdots$. Then our proposition produces $i < j$ such that $b_i < b_j$ so that $(a_1, b_i) \leq (a_j, b_j)$. Applying our proposition in the opposite direction, we derive that P is a partially well-ordered set, as required. □

5.2.2 Ordering by divisibility

Let P be a partially ordered set. Let $D(P)$ denote the set of all finite sequences of the form $(p_1, \ldots, p_r)$ where the components are in P and the order is given by $(p_1, \ldots, p_r) \leq (q_1, \ldots, q_s)$ if and only if there exists an order preserving injection ψ of the set of natural numbers $\mathbb{N}$ into itself such that $\psi(r) \leq s$ and $p_i \leq q_{\psi(i)}$ for any $i = 1, \ldots, s$.

Theorem 5.12. *If P is a partially well-ordered set then $D(P)$ is also partially well-ordered.*

Proof. We adopt the notation of the corollary of Proposition 5.10 and write $D = D(P)$. An important remark is that if, in a partially ordered set D, any subset of the form $D \setminus a^\Delta$ is partially well-ordered then the same is true of D itself. Indeed, if $E \neq \varnothing$, $E \subseteq D$, and $a \in E$ then we may consider $E \setminus a^\Delta$. This is a subset of a partially well-ordered set $D \setminus a^\Delta$. Hence either $E \subseteq a^\Delta$ or there exist $b_1, \ldots, b_s$ in E such that for any $b \in E \setminus a^\Delta$ there exists i with $b_i \leq b$. In either case E possesses a finite subset of minimal elements which is contained in $\{a, b_1, \ldots, b_s\}$, and D is indeed partially well-ordered.

Thus, in proving our theorem, we may assume that D does not satisfy the above condition. To arrive at a contradiction, we will find in D a subset D' which is not partially well-ordered but in which the above condition holds valid in a particular case: for any $p \in P$, we have $D' \setminus (p)^\Delta$ partially well-ordered. For this we consider the descending chain of subsets S_i which are not partially well-ordered in the following way. We set $S_1 = D$ and, admitting S_i constructed, we set $S_{i+1} = S_i$ in the case where for each $p \in P$ the subset $S_i \setminus (p)^\Delta$ is partially well-ordered. If this is not the case, we set $S_{i+1} = S_i \setminus (p_i)^\Delta$

where $p_i \in P$ is an arbitrary element such that $S_i \setminus (p_i)^\Delta$ is not partially well-ordered. This chain stabilizes at a finite step since a sequence $p_1, p_2, \ldots, p_i, \ldots$ associated with it is a sequence of elements in a partially well-ordered set in which for no pair $i < j$ we have $p_i \leq p_j$ (namely, by construction, $p_j \notin p_i^\Delta$). If $S_i = S_{i+1} = \cdots$ then we set $D' = S_i$.

With D' being a set which is not partially well-ordered, there is a sequence u of minimal length in D' such that $D' \setminus u^\Delta$ is not a partially well-ordered set. If $u = (p_1, \ldots, p_r)$ then $r > 1$ and for $v = (p_1, \ldots, p_{r-1})$ and $p = p_r$ the subsets $D' \setminus v^\Delta$ and $D' \setminus (p)^\Delta$ are partially well-ordered. Now any element w in $D' \setminus u^\Delta$ can be written as a triple (any of the components can be empty) $w = (a, b, c)$ where $a \in D' \setminus v^\Delta$, $b \in P$, $c \in D' \setminus (p)^\Delta$. Indeed, if $w \in D' \setminus v^\Delta$ then we simply put $a = w$, $b = c = \varnothing$. If $w \in v^\Delta$ then $w = (a, b, c)$ where $b \in P$, $(a, b) \in v^\Delta$ and $a \notin v^\Delta$. It is clear that $c \notin (p)^\Delta$ since otherwise $w \in u^\Delta$. Now any infinite sequence of elements in $D' \setminus u^\Delta$ can be considered as an increasing sequence in $(D' \setminus v^\Delta) \times P \times (D' \setminus (p)^\Delta)$ (possibly in the direct product of two or one of these sets). This set with componentwise order is shown in the corollary of Proposition 5.10 to be partially well-ordered. The same is true then for $D' \setminus u^\Delta$ with its original order. This is a contradiction showing that our hypothesis of D not being partially well-ordered is wrong. □

To generalize Theorem 5.12, we consider the set $D_n = D_n(P)$ of finite sequences of the form $(i_1, \ldots, i_n; p_1, \ldots, p_s)$ where n is fixed, $i_1, \ldots, i_s \in \mathbb{N}$, $p_1, \ldots, p_s \in P$. If P is partially ordered then an order on $D_n(P)$ can be defined by setting $(i_1, \ldots, i_n; p_1, \ldots, p_s) \leq (j_1, \ldots, j_n; q_1, \ldots, q_r)$ if there exists an order preserving injection $\xi : \mathbb{N} \to \mathbb{N}$ such that $\xi(i_1) = j_1, \ldots, \xi(i_n) = j_n$, $\xi(s) \leq r$, $p_1 \leq q_{\xi(1)}, \ldots, p_s \leq q_{\xi(s)}$.

Theorem 5.13. *Let P be a partially well-ordered set and let n be a natural number. Then $D_n(P)$ is partially well-ordered.*

Proof. Let $a_1, a_2, \ldots$ be a sequence of elements in D_n. Since $\mathbb{N}^n$ with the above order is partially well-ordered (cf. corollary of Proposition 5.10), we may assume that in our sequence the sequence of n-tuples formed by the first components is increasing. So if $a_i = (i_1, \ldots, i_n; p_1, \ldots, p_s)$, $a_{i+1} = (j_1, \ldots, j_n; q_1, \ldots, q_r)$ then there exists an injection ξ_1 such that $\xi_i(i_1) = j_1, \ldots, \xi(i_n) = j_n$. Now interchanging simultaneously the first n components in $a_1, a_2, \ldots$ in such a way that they form an increasing sequence in a_1, we easily observe that the same property will be inherited by the remaining terms $a_2, a_3, \ldots$ (this is true since each ξ_i is an order-preserving injection). In this case the second part $(p_1, \ldots, p_s)$ of each a_i decomposes into $n + 1$ parts (some of them can be empty) of the form

$$c^1 = (p_1, \ldots, p_{i_1-1}), \quad c_1^2 = (p_{i_1+1}, \ldots, p_{i_2-1}), \quad \ldots, \quad c_i^n = (p_{i_n+1}, \ldots, p_s).$$

All $c_r^i \in D(P) \cup \{\varnothing\}$. We assume, for convenience, that $\varnothing$ is the least element in $D(P) \cup \{\varnothing\}$.

Now an infinite increasing subsequence in $a_1, a_2, \ldots$ can be chosen in the following way. We first choose an infinite subsequence (this is identified with the original sequence) in which the sequence $c_1^1, c_2^1, \dots$ is increasing (Theorem 5.12). This will give rise to an injection ψ such that $\psi(i_1 - 1) \le j_1 - 1$ and $p_1 \le q_{\psi(q)}, \dots, p_{i_1-1} \le q_{\psi(i_1-1)}$. It is clear that ψ can be modified in such a way that all the above inequalities hold and in addition we have $\psi(i_1) = \xi_i(i_1), \dots, \psi(i_n) = \xi(i_n)$. Having done this, we select from this new sequence a subsequence in which the components p_{i_1} increase. Then we pass to an infinite subsequence such that $c_1^2, c_2^2, \dots$ is infinitely increasing.

In this case for each i we will find an injection χ such that $j_1 + 1 \le \chi(i_1 + 1), \chi(i_2 - 1) \le j_2 = 1$, and $p_{i_1+1} \le q_{\chi(i_1+1)}, \dots, p_{i_2-1} \le q_{\chi(i_2-1)}$. As previously, χ can be improved to be commensurable with ψ above. Proceeding similarly, we will arrive at an infinite sequence $a_1', a_2', \dots$ and, for any i to an order-preserving injection $\tau : \mathbb{N} \to \mathbb{N}$ such that $\tau(i_1) = \xi_i(i_1) = j_1, \dots, \tau(i_n) = \xi_i(i_n) = j_n$; $\tau(s) \le r$, $p_1 \le q_{\tau(1)}, \dots, p_s \le q_{\tau(s)}$. Hence $D_n(P)$ is a partially well-ordered set. □

5.2.3 A generalization of Hilbert's basis theorem

Let R be a polynomial ring in variables $x_1, x_2, \dots, x_n, \dots$ over a noetherian ring k, and M a free R-module with countable free generating set $\{t_1, t_2, \dots, t_n, \dots\}$. Let also Ψ denote the set of all order-preserving injections from $\mathbb{N}$ in $\mathbb{N}$. A submodule T of M is called a Ψ-*submodule* if for any $p = \sum f_i(x_1, x_2, \dots)t_i \in T$ and any $\psi \in \Psi$ any element of the form

$$\psi(p) = \sum f_i(x_{\psi(1)}, x_{\psi(2)}, \dots)t_{\psi(i)}$$

is in T.

Theorem 5.14. *Module M satisfies the maximality condition for Ψ-submodules.*

Proof. Let P denote the set of all finite sequences of nonnegative integers. We give P a well-ordering by setting $(i_1, \dots, i_m) \preccurlyeq (j_1, \dots, j_n)$ if either $m < n$ or $m = n$ and then there exists an r such that $i_r < j_r$ and $i_s = j_s$ for all $s > r$. We will also be considering P with the order $\le$ of the set $D(\mathbb{N} \cup \{0\})$. Now to each monomial of the form $x_1^{i_1} x_2^{i_2} \cdots x_n^{i_n}$ we attach its weight $\mathrm{wt}(u)$ as a sequence of the form $(i_1, i_2, \dots, i_n) \in P$. If f is a polynomial then its weight $\mathrm{wt}(f)$ is the weight of its nonzero monomial which is maximal in the sense of order $\le$. Now, given an element y in M of the form

$$y = \sum_{i=1}^{n} f_i t_i, \quad f_n \ne 0,$$

we define its weight as $\mathrm{wt}(y) = (n, \mathrm{wt}(f_n))$. Finally, we set $\theta(y) = (\lambda, \mathrm{wt}(y))$ where λ is the leading coefficient in f_n, and $\theta(0) = (0, 1, 0)$. By the definition, we have $\theta(y) \in Q = k \times \mathbb{N} \times P$. Let $\mathbb{N} \times P$ be ordered as $D_1(\mathbb{N})$. Now a subset $Q' \subseteq Q$ is called closed if from $(\lambda_1, n_1, v_1), \dots, (\lambda_s, n_s, v_s) \in Q'$ it follows that any element (λ, n, v) such that $\lambda \in k\lambda_1 + \cdots + k\lambda_s$ and $(n_1, v_1), \dots, (n_s, v_s) \le (n, v)$ is in Q', too. We notice that for any Ψ-submodule $M' \subseteq M$ the subset $Q' = \theta(M')$ is closed. Indeed, it is obvious that $\mathrm{wt}(\psi(y)) = \psi(\mathrm{wt}(y))$.

By our hypothesis, there exist $y_1, \dots, y_s \in M'$ such that $\mathrm{wt}(y_1) = (n_1, v_1), \dots, \mathrm{wt}(y_s) = (n_s, v_s)$. Hence there exist $\psi_1, \dots, \psi_s \in \Psi$ such that $\mathrm{wt}(y) = \psi_1(\mathrm{wt}(y_1)), \dots, \mathrm{wt}(y_s) = \psi_s(\mathrm{wt}(y_s))$. If $\lambda = \mu_1\lambda_1 + \cdots + \mu_s\lambda_s$ then $z = \mu_1 y_1 + \cdots + \mu_s y_s \in M'$ and $\theta(z) = (\lambda, n, v) \in Q'$.

Now we prove that each closed subset of $Q' \subseteq Q$ is a closure of a finite subset. Let $Q'(n,p)$, $(n,p) \in \mathbb{N} \times P$, be the set of elements $\lambda \in k$ such that $(\lambda, n, p) \in Q'$. Then $Q'(n,p)$ is an ideal of k; moreover, if $(n,p) \le (n_1,p_1)$ then $Q'(n,p) \subseteq Q'(n_1,p_1)$. We introduce an order on the set $\{(Q'(n,p), n, p) \mid (n,p) \in \mathbb{N} \times P\}$ by setting $(Q'(n,p), n, p) \le (Q'(n_1,p_1), n_1, p_1)$ if $(n,p) \le (n_1,p_1)$ and $Q'(n,p) = Q'(n_1,p_1)$. By the above monotonicity and the noetherian property of k, the set of such triples is partially well-ordered (cf. Proposition 5.10). Now let $Q'(n_1,p_1), \dots, Q'(n_s,p_s)$ be such that for any $(n,p) \in \mathbb{N} \times P$ there exist an i with $(Q'(n_i,p_i), n_i, p_i) \le (Q'(n,p), n, p)$. Then $(n_i,p_i) \le (n,p)$ and $Q'(n_i,p_i) = Q'(n,p)$. Due to k being noetherian, there exist $\kappa_{ij} \in k$ such that $Q'(n_i,p_i) = k\kappa_{i1} + \cdots + k\kappa_{it_i}$, $i = 1, \dots, s$. Hence Q' is the closure of a finite set of triples of the form (κ_{ij}, n_i, p_i), $j = 1, \dots, t_i$, $i = 1, \dots, s$.

To complete the proof, it suffices to verify that any Ψ-submodule M' with $Q' = \theta(M')$ being a closure of finite set is itself finitely generated. A standard argument will then give the maximality condition. So let $y_1, \dots, y_s \in M'$ be such that Q' is the closure of $\theta(y_1) = (\lambda_1, n_1, p_1), \dots, \theta(y_s) = (\lambda_s, n_s, p_s)$. We want to show that then M' is generated by $y_1, \dots, y_s$. We consider $y \in M'$ with $\theta(y) = (\lambda, n, p)$. Since Q' is closed the Ψ-submodule generated by $y_1, \dots, y_s$ contains an element z with $\theta(z) = (\lambda, n, p)$. We consider then $y' = y - z$ with $\theta(y') = (\lambda', n', p')$. The leading terms of y and z being the same, we conclude $(n',p') \preccurlyeq (n,p)$, $(n',p') \ne (n,p)$. Since $\preccurlyeq$ is a well-order the process of replacing y by y' after a finite number of applications reaches zero which means that y is expressible through $y_1, \dots, y_s$. □

5.3 Partially well-ordered sets. Applications

The techniques of the preceding section will be applied here first to proving finite basis property of metabelian varieties over an arbitrary noethering ring (Theorem 5.16). Then the property of being Specht will be shown for $\mathbb{N}_2\mathbb{A}$ (Theorem 5.18) and, more generally, for $\mathbf{N}_c\mathbf{A} \cap \mathbf{N}_2\mathbf{N}_c$ (Theorem 5.21) over an arbitrary field k of characteristic different from 2. This restriction is essential since it will be shown in Section 5.4 that over any field of characteristic $p > 0$ any variety of the form $\mathbf{N}_p\mathbf{A}$ is not Specht.

We start with a simple result.

Lemma 5.15. *Let G be a Lie algebra and H its verbal ideal. Then G is verbally noetherian if and only if G/H is verbally noetherian and H satisfies the maximality condition for verbal ideals of G.*

Proof. It is obvious that this condition is necessary. Now suppose G/H and H satisfy the above maximality conditions. Let $V_1 \subseteq \cdots \subseteq V_n \subseteq V_{n+1} \subseteq \cdots$ be an ascending chain of verbal ideals in G. Then $V_1 + H/H \subseteq \cdots \subseteq V_n + H/H \subseteq V_{n+1} + H/H \subseteq \cdots$ is an ascending chain of verbal ideals in G/H and $V_1 \cap H \subseteq \cdots \subseteq V_n \cap H \subseteq \cdots$ is an ascending chain of verbal ideals of G contained in H. By our hypothesis, these chains stabilize, say, at the nth step. Then $V_n + H/H = V_{n+1} + H/H$ and $V_n \cap H = V_{n+1} \cap H$. Now choose $x \in V_{n+1}$. Then $x \in V_n + H$, i. e., $x = v_n + h$, $v_n \in V_n$, $h \in H$. Hence $h = x - v_n \in V_{n+1} \cap H = V_n \cap H$. Then $x \in V_n + V_n \cap H$ and $V_{n+1} \subseteq V_n$, showing that the original chain stabilizes at the nth step, too. □

Theorem 5.16. *Let k be a noetherian ring with 1 and $\mathbf{V} \subseteq \mathbf{A}^2$ be an arbitrary metabelian variety. Then $\mathbf{V}$ enjoys the finite basis property.*

Proof. It has been remarked in Section 5.1 that a variety $\mathbf{U}$ of Lie algebras is Specht if and only if the maximality condition holds for the verbal ideals of the free Lie algebra L_∞ which contain $U = \mathbf{U}(L)$ and U itself is finitely generated as a verbal ideal. In our particular case this is equivalent to the verbal noetherian property for the free metabelian algebra $\bar{L} = L/(L^2)^2$ of countable rank. By Lemma 5.15, this amounts to proving the verbal noetherian property for $\bar{L}/\bar{L}^2$ and the maximality condition for verbal ideals contained in $\bar{L}^2$. Since $\bar{L}/\bar{L}^2$ is a free abelian (that is, nilpotent) algebra, the first part follows from Theorem 5.3. Thus the core of the proof is the maximality condition for the verbal ideals of $\bar{L}$ contained in $\bar{L}^2$. In fact, a stronger theorem holds: the maximality condition is satisfied by the set of Ψ-ideals of $\bar{L}$, that is, the ideals of $\bar{L}$ which are contained in $\bar{L}^2$ and which are stable under all endomorphisms of $\bar{L}$ which are extensions of the maps of the form $y_i \mapsto y_{\psi(i)}$, $i = 1, 2, \ldots$, $\psi : \mathbb{N} \to \mathbb{N}$ being an order-preserving injection and $\{y_i \mid i = 1, 2, \ldots\}$ a free generating set for $\bar{L}$.

We will be using the corollary of the embedding theorem (Section 4.4, Theorem 4.26). By this corollary, $\bar{L}$ is isomorphic to a subalgebra of the semidirect product $S = A \rtimes M$ generated by all elements $y_i = x_i + t_i$, $i = 1, 2, \ldots$, where $\{x_1, x_2, \ldots\}$ is a basis of a free abelian algebra A and $\{t_1, t_2, \ldots\}$ is a basis of a free A-module M. It has been remarked in Section 4.4, the corollary of Theorem 4.26, that M is a free R-module where $R = U(A) = k[x_1, x_2, \ldots]$ with basis $\{t_1, t_2, \ldots\}$. It is important that the image of any Ψ-ideal of L belongs to M and it is a Ψ-submodule in this latter. By Theorem 5.14, M satisfies the maximality condition for Ψ-submodules. Hence any ascending chain of Ψ-ideals of $\bar{L}$ stabilizes and $\bar{L}$ is verbally noetherian. Since $\mathbf{A}^2$ itself can be determined by a single identity, $(xy)(zu) \equiv 0$, we conclude that any subvariety $\mathbf{V} \subseteq \mathbf{A}^2$ is finitely based. □

5.3.1 Variety $\mathbf{N}_2\mathbf{A}$

By definition, $\mathbf{N}_2\mathbf{A}$ consists of Lie algebras which are extensions of a 2-step nilpotent ideal by an abelian algebra. In other words, $G \in \mathbb{N}_2\mathbf{A}$ if and only if $(G^2)^3 = \{0\}$. Now let F be a free algebra in $\mathbf{N}_2\mathbf{A}$ freely generated by a set $\{y_1, y_2, \ldots\}$. As we know, the Specht property is equivalent to the maximality condition for the verbal ideals of F. According to Theorem 5.16, such a condition is satisfied by the set of verbal ideals containing

$(F^2)^2$. Now if we prove the validity of this condition for the set of verbal ideals in $(F^2)^2$ then by Lemma 5.15 the maximality condition will hold for arbitrary verbal ideals of F. The following is important for the determination of the structure of $(F^2)^2$.

Lemma 5.17. *$(F^2)^2$ is equal to the k-span of the set of all elements of the form*

$$y_1^{\alpha(1)} \cdots y_m^{\alpha(m)} (y_i y_j) y_1^{\beta(1)} y_2^{\beta(2)} \cdots y_n^{\beta(n)} y_t y_l \tag{5.3}$$

where $\alpha(r)$ and $\beta(r)$ are nonnegative integers, $r = 1, 2, \ldots$

Proof. By Section 2.4.2, a system of free generators for the derived algebra of a free algebra $L(x_1, x_2, \ldots)$ of countable rank can be chosen in the form $x_{i_1} x_{i_2} \cdots x_{i_s} x_{i_{s+1}}$, $i_1 \geq i_2 \geq \cdots \geq i_s < i_{s+1}$, where $s \geq 1$. Now $F^2/(F^2)^3$ is a free 2-step nilpotent Lie algebra whose free generating set is equal to the set of images of these monomials under the map $x_i \mapsto y_i, \ldots, i = 1, 2, \ldots$ Thus a basis for $(F^2)^2$ can be taken in the form of all possible ordered products of generators as just above, that is,

$$(y_1^{\alpha(1)} \cdots y_m^{\alpha(m)} y_i y_j) y_1^{\beta(1)} \cdots y_m^{\beta(m)} y_t y_l.$$

Now an immediate application of the Jacobi identity leads to monomial (5.3). □

Theorem 5.18. *Let k be an arbitrary field of characteristic not 2. Then $\mathbf{N}_2\mathbf{A}$ is a Specht variety.*

Proof. We associate to each monomial of the form (5.3) a vector $s = (i, j, t, l, \alpha, \beta)$ where $\alpha = (\alpha(1), \ldots, \alpha(m))$, $\beta = (\beta(1), \ldots, \beta(n))$ (of course, m and n are not fixed). The set S of all such vectors will be ordered in two ways. The first order, $\leq$, is lexicographic: $s \leq s'$ if the first (from the left) component where s and s' differ is smaller in s. Here the order of the first four components is the usual order on the set of nonnegative integers whereas the elements of the remaining two components are ordered according to their weights (see the beginning of the proof of Theorem 5.14).

The second order, $\preccurlyeq$, is introduced in two steps. If $s = (i, j, t, l, \alpha, \beta)$ and $s' = (i', j', t', l', \alpha', \beta')$ then we write $s \leq_1 s'$ if there exists an injection $\psi \in \Psi$ such that $\psi(i) = i'$, $\psi(j) = j'$, $\psi(t) = t'$, $\psi(l) = l'$, $\alpha(r) \leq \alpha'(\psi(r))$, $\beta(r) \leq \beta'(\psi(r))$ for all possible r (we may consider α and β as the functions of natural argument equal to zero almost everywhere). The set P of pairs (a, b) with componentwise order $(a, b) \leq_1 (a', b')$ is known to be partially well-ordered (corollary of Proposition 5.10). Now any pair (α, β) can be considered as a finite set of pairs in P: $p_1 = (\alpha(1), \beta(1))$, $p_2 = (\alpha(2), \beta(2)), \ldots, p_m = (\alpha(m), \beta(m))$ where m is such that $\alpha(n) = \beta(n) = 0$ for all $n > m$. Hence S consists of the sequences of the form $(i, i, t, l; p_1, p_2, \ldots, p_m)$ and the order $\leq_1$ is identical to that considered in Theorem 5.13. Hence $(S, \leq_1)$ is a partially well-ordered set. Now, as the second step of our definition for $\preccurlyeq$, we write $s \preccurlyeq s'$ if $s \leq_1 s'$ and, in addition, $\sum_{r=1}^{\infty} \beta(r) \equiv \sum_{r=1}^{\infty} \beta'(r) \pmod 2$. Applying Proposition 5.10, we easily derive that $(S, \preccurlyeq)$ is a partially well-ordered set.

Coming back to $(F^2)^2$, we represent each element u of this subspace in the form $u = \lambda_1 u_1 + \cdots + \lambda_n u_n$, where u_i are as in (5.3), $\lambda_i \in k$, $\lambda_i \neq 0$, $i = 1, \ldots, n$. Let $s_1, \ldots, s_n$ be the vectors as above associated with $u_1, \ldots, u_n$ and i be such a number that $s = s_i$ is the greatest vector in the sense of the lexicographic order. We say that s is the weight of u. By our previous argument, the set S of weights with order $\preccurlyeq$ is partially well-ordered. Then any closed (relative to $\preccurlyeq$) subset of S (cf. the proof of Theorem 5.13) is a closure of its finite subset. Hence our proof will be complete as soon as we manage to prove that the set of weights of an arbitrary verbal ideal $I \subseteq (F^2)^2$ is closed.

So let h be an element in $(F^2)^2$ with weight s and s' an element in S with $s \preccurlyeq s'$. We are going to show that the verbal ideal generated by h contains an element h' whose weight is s'. Let ψ be the injection in the definition of the relation $s \leq_1 s'$. We write $\xi(y_r) = y_{\psi(r)}$, $r \in \mathbb{N}$. Then the weight of $h'' = \xi(h)$ is equal to $(i', j', t', l', \alpha'', \beta'')$ where, for all r, $\alpha''(r) \leq \alpha'(r)$, $\beta''(r) \leq \beta'(r)$ and $\sum(\beta'(r) - \beta''(r)) \equiv 0 \pmod 2$. If $m \in \mathbb{N}$ is such that $\alpha'(r) = \alpha''(r) = 0$ for all $r > m$ then we consider

$$g = y_1^{\alpha'(1)-\alpha''(1)} y_2^{\alpha'(2)-\alpha''(2)} \cdots y_m^{\alpha'(m)-\alpha''(m)} h''. \tag{5.4}$$

Since h'' is in the verbal ideal generated by h, the same is obviously true of g. Now the weight of g is equal to $(i', j', t', l', \alpha', \beta'')$. Proceeding by induction over $\sum(\beta'(r) - \beta''(r))$, it is sufficient to restrict ourselves to the case where $\sum(\beta'(r) - \beta''(r)) = 2$. So let p, q be natural numbers (not necessarily distinct) such that $\beta'(r) = \beta''(r)$ for $r \notin \{p, q\}$ and, if $p = q$, then $\beta'(p) - \beta''(p) = 2$, if $p \neq q$, then $\beta'(p) - \beta''(p) = \beta'(q) - \beta''(q) = 1$. Let n be a natural number such that y_n does not occur in the expression of (5.4). We define an endomorphism θ by setting $\theta(y_r) = y_r + y_n y_r$, $r = 1, 2 \ldots$ Then to obtain an element of the desired weight in the verbal closure of g it is sufficient, first, to write $g_1 = \theta(g) - y_n g - g$, and, second, pass from $g_1 = g_1(y_n)$ to $g_2 = g_1(y_p + y_q) - g(y_p) - g_1(y_q)$ (that is, to consider the linearization with respect to y_n). Now g_2 is the desired element since using $(F^2)^3 = \{0\}$ and the Jacobi identity we conclude that the leading term of g_1 is equal to the leading term of

$$-y_1^{\alpha'(1)} \cdots y_m^{\alpha'(m)} (y_{i'} y_{j'}) y_n^2 y_1^{\beta''(1)} \cdots y_m^{\beta''(m)} y_{t'} y_{l'}$$
$$+ y_n y_1^{\alpha'(1)} \cdots y_m^{\alpha'(m)} (y_{i'} y_{j'}) y_n y_1^{\beta''(1)} \cdots y_m^{\beta''(m)} y_{t'} y_{l'}.$$

Applying linearization, it is easy to see that the leading term of g_2 is equal to the leading term of

$$-2 y_1^{\alpha'(1)} \cdots y_m^{\alpha'(m)} (y_{i'} y_{j'}) y_p y_q y_1^{\beta''(1)} \cdots y_m^{\beta''(m)} y_{t'} y_{l'}$$
$$+ y_1^{\alpha'(1)} \cdots y_m^{\alpha'(m)} y_p (y_{i'} y_{j'}) y_q y_1^{\beta''(1)} \cdots y_m^{\beta''(m)} y_{t'} y_{l'}$$
$$+ y_1^{\alpha'(1)} \cdots y_m^{\alpha'(m)} y_q (y_{i'} y_{j'}) y_p y_1^{\beta''(1)} \cdots y_m^{\beta''(m)} y_{t'} y_{l'}.$$

Since the characteristic of the base field is different from 2, it is evident that the weight of g_2 is equal to $(i', j', t', l', \alpha', \beta')$ as required. □

5.3.2 Variety $\mathbf{N}_c\mathbf{A} \cap \mathbf{N}_2\mathbf{N}_c$

Let L denote a free Lie algebra of countable rank over a field k, $X = \{x_1, x_2, \ldots\}$ a free generating set for L. We denote by $L(n_1, n_2, \ldots, n_t)$ the product of ideals of the form $L^{n_1}L^{n_2}\cdots L^{n_t}$ and by $Q_{p,q}$ $(1 \le p < q \le t)$ an ideal of the form

$$\begin{aligned} Q_{p,q}(n_1, \ldots, n_t) = & \sum_{i \neq p,q} L(n_1, \ldots, n_i + 1, \ldots, n_t) \\ & + L(n_1, \ldots, n_q, 2, n_{q+1}, \ldots, n_t) + L(2, n_1, \ldots, n_t). \end{aligned} \tag{5.5}$$

Using this notation, the verbal ideal V of identities for $\mathbf{N}_c\mathbf{A} \cap \mathbf{N}_2\mathbf{N}_c$ takes the form

$$V = L(\underbrace{2, 2, \ldots, 2}_{c+1}) + L(c+1, c+1, c+1).$$

In order to prove the maximality conditions for the verbal ideals of L containing V, it is sufficient to restrict ourselves only to those which are contained in $(L^2)^3$, that is, in $L(2, 2, 2)$ (this is due to Theorem 5.18, according to which $\mathbf{N}_2\mathbf{A}$ is a Specht variety). Thus we arrive at the problem of constructing a finite chain of verbal ideals from $L(2, 2, 2)$ to V, each factor of which satisfies the maximality condition for the images of verbal ideals of L.

Proposition 5.19. *There exists a finite descending chain of ideals of L joining $L(2, 2, 2)$ with V, each term of which has the form*

$$M = \sum_{j=1}^{l} L(n_1^{(j)}, \ldots, n_{t(j)}^{(j)})$$

and such that the next term M' of this chain has the following property:

$$M' \supseteq \sum_{j=1}^{l} Q_{p(j),q(j)}(n_1^{(j)}, \ldots, n_{t(j)}^{(j)}) \quad (1 \le p(j) < q(j) \le t(j),\ j = 1, \ldots, l).$$

Proof. The inductive process of building such a chain is as follows. For each $L(n_1, \ldots, n_t)$ occurring in the above decomposition of M, we choose a pair of indices p, q such that $1 \le p < q \le t$ and n_p, n_q are the greatest possible among $n_1, \ldots, n_t$. Now we consider $Q_{p,q}(n_1, \ldots, n_t)$. This is a sum of certain $L(n_1', \ldots, n_t')$. The sum of all such ideals is the desired ideal M'. The proof will be complete if we prove that after finitely many steps of this sort all $L(n_1', \ldots, n_t')$ arising are in V. We remark that in any $L(n_1, \ldots, n_t)$ which is obtained in applying the above procedure the values of three maximal indices differ from each other by at most 1. Indeed, this is the case for the first term of our chain, i. e., for $L(2, 2, 2)$. Now since all $n_1, \ldots, n_t$ are at least 2, while passing to subsequence terms, it is sufficient to consider those terms in equation (5.5) which occur as summands of the first sum. Now our claim is straightforward. To conclude, we consider the ideals $L(n_1, \ldots, n_t)$ arising after $s = c^2 + c - 6$ steps of our procedure. If $t \ge c+1$

then such an $L(n_1,\dots,n_t)$ is in $L(\underbrace{2,\dots,2}_{c+1})$ since the number of indices grows by adjoining new indices equal to 2. Now if $t < c+1$ then the value of the greatest index is at least $c+2$ (if not then the sum of indices would be less than $c(c+1)$ whereas equation (5.5) shows that it should be at least $s+6$). In this case, recalling our claim on three maximal indices gives three indices $n_i, n_j, n_l \geq c+1$ so that $L(n_1,n_2,\dots,n_t) \subseteq L(c+1,c+1,c+1)$. □

We notice that, by our argument, the chain constructed above can be refined by considering elementary steps of passing from $L(n_1,\dots,n_t)$ to $Q_{p,q} = Q_{p,q}(n_1,\dots,n_t)$.

Lemma 5.20. *$L(n_1,\dots,n_t)/Q_{p,q}$ is a linear span of monomials of the form*

$$y_1^{\alpha(1)}\cdots y_m^{\alpha(m)}a_1\cdots a_q y_1^{\beta(1)}\cdots y_m^{\beta(m)}a_{q+1}\cdots a_t \tag{5.6}$$

where each a_j is a right-normed monomial in $y_1, y_2, \dots$, $j = 1,\dots,t$, and $y_1, y_2, \dots$ are the images of $x_1, x_2, \dots$ in $L/Q_{p,q}$, $\alpha(i), \beta(i) \geq 0$, $i = 1,\dots,m$.

Proof. The following formulas easily follow by applying formula (5.5) and the Jacobi identity (g_i being an element of the image of L^n, $i = 1,\dots,t$ and $b \in L/Q_{p,q}$):

$$\begin{aligned} &g_1\cdots g_{j-1}(bg_j)\cdots g_t = 0 \quad (j \neq p,q),\\ &g_1\cdots g_{q-1}(bg_q)g_{q+1}\cdots g_t = g_1\dots g_{q-1}bg_q\cdots g_t,\\ &g_1\cdots g_{p-1}(bg_p)g_{p+1}\cdots g_t = bg_1\cdots g_t - g_1\cdots bg_q\cdots g_t. \end{aligned} \tag{5.7}$$

For instance, to prove the third of these relations, it is sufficient to write

$$\begin{aligned} bg_1\cdots g_p\cdots g_q\cdots g_t = &\sum_{i\neq p,q} g_1\cdots(bg_i)\cdots g_t + g_1\cdots(bg_p)\cdots g_q\cdots g_t\\ &+ g_1\cdots g_p\cdots(bg_q)\cdots g_t. \end{aligned}$$

By equation (5.5), all the summands under the summation sign are equal to zero. It remains to shift the rightmost term to the left-hand side and to apply the second of the relations in (5.7).

Now we recall that $L(n_1,n_2,\dots,n_t) = L^{n_1}L^{n_2}\cdots L^{n_t}$ is a verbal ideal equal to the linear span of products of the form $w_1w_1\cdots w_t$ where w_i is a right-normed product in L^{n_i}, $i = 1,\dots,t$. Due to the first two relations in (5.7), each such product is equal to a linear combination of similar products in which each w_i, except possibly for w_p and w_q, is of degree precisely n_i. Now applying the latter two relations in (5.7) shows that w_p and w_q can also be taken of degrees n_p and n_q, respectively, but then additional factors as in (5.6) arise. □

Now we can formulate and prove the main result of this section. Its proof is very close to that of Theorem 5.18, the main idea being the translation into the language of ordered sets.

Theorem 5.21. *Over any field of characteristic not 2, any variety of Lie algebras of the form $\mathbf{N}_c\mathbf{A} \cap \mathbf{N}_2\mathbf{N}_c$ is Specht.*

Proof. We start by proving the maximality condition for the set of images of verbal ideals of L contained in $L(n_1, \dots, n_t)$ modulo $Q_{p,q}(n_1, \dots, n_t)$. For convenience, we call these images the *verbal subspaces*. Now we define the weight of a monomial of the form (5.6) as a vector of the form $(i_1, i_2, \dots, i_n; \alpha, \beta)$ where $a_1 \cdots a_t = y_{i_1} y_{i_2} \cdots y_{i_n}$, $n = n_1 + \cdots + n_t$. We order the set S of weights as in the proof of Theorem 5.18, namely lexicographically (well-order) and partially, by setting $(i_1, \dots, i_n; \alpha, \beta) \preccurlyeq (i'_1, \dots, i'_n; \alpha', \beta')$ if there exists an order-preserving injection $\psi : \mathbb{N} \to \mathbb{N}$ such that $\psi(i_d) = i'_d$, $d = 1, 2, \dots, n$, $\alpha(i) \le \alpha'(\psi(i))$, $\beta(i) \le \beta'(\psi(i))$, $i = 1, 2, \dots$, and finally, $\sum_r \beta(r) \equiv \sum_r \beta'(r) \pmod 2$. As in the proof of Theorem 5.18, we see that S is a partially well-ordered set. Now we define the weight of an element in $L(n_1, \dots, n_t)/Q_{p,q}$ written as a linear combination of monomials of the form (5.6) as the maximum of the weights of those monomials which enter this combination with nonzero coefficients. As in Theorem 5.18, it is sufficient to prove that if s is the weight of $h \in L(n_1, \dots, n_t)/Q_{p,q}$ and $s' \in S$ is such that $s \preccurlyeq s'$ then the verbal closure of h contains an element h' whose weight is equal to s'. If $s' = (i'_1, \dots, i'_n; \alpha', \beta')$ and $\psi : \mathbb{N} \to \mathbb{N}$ as above then, applying an endomorphism ξ such that $\xi(y_i) = y_{\xi(i)}$ and multiplying by appropriate powers of $y_1, \dots, y_m$, we will obtain an element

$$g = y_1^{\alpha'(1)} \cdots y_{m'}^{\alpha'(m')} a'_1 \cdots a'_p y_1^{\beta''(1)} \cdots y_{m'}^{\beta''(m')} a'_{p+1} \cdots a'_t$$

in the verbal closure whose weight is equal to $s'' = (i'_1, \dots, i'_n; \alpha', \beta'') \le s'$. Now arguing by induction over $\sum_r (\beta'(r) - \beta''(r))$, we may assume that this latter value is equal to 2. In this case sequences β' and β'' differ not more than by two entries. Proceeding as in Theorem 5.18, we choose u and u', $1 \le u, u' \le m'$, such that if $u = u'$ then $\beta'(u) = \beta''(u) = 2$ and if $u \ne u'$ then $\beta'(u) - \beta''(u) = \beta'(u') - \beta''(u') = 1$. Suppose y_l is not used in expressing g. We define an endomorphism θ_l by setting $\theta_l(y_r) = y_r + y_l y_r$ for all $r = 1, 2, \dots$ As in Theorem 5.18, we first put $g_1 = g_1(y_l) = \theta_l(g) - y_l g - g$. Then the desired element g_2 (of weight s) in the verbal closure of h can be taken in the form

$$g_2 = g_1(y_u + y_{u'}) - g_1(y_u) - g_1(y_{u'}).$$

This can be verified as follows. We notice first that for any monomial a modulo $(L^2)^2$ we always have $\theta_l(a) = a + y_l a$. Assembling together equation (5.5), relations (5.7), and the equation just above, we may write the following equation modulo $Q_{p,q}$ in which w is a monomial of the form (5.6) which is the leading term for g:

$$\begin{aligned}\theta_l(w) = {} & (y_1 + y_l y_1)^{\alpha'(1)} \cdots (y_{m'} + y_l y_{m'})^{\alpha'(m')} a'_1 \cdots a'_{p-1} (a'_p + y_l a'_p) a'_{p+1} \\ & \cdots a'_{q-1} (a'_q + y_l a'_q)(y_1 + y_l y_q)^{\beta''(1)} \cdots (y_{m'} + y_l y_{m'})^{\beta''(m')} a'_{q+1} \cdots a'_t.\end{aligned}$$

Recalling that $L(n_1,\dots,n_p,2,n_{p+1},\dots,n_t)$ and $L(2,n_1,\dots,n_t)$ are in $Q_{p,q}$, we rewrite

$$\theta_l(w) = y_1^{\alpha'(1)}\cdots y_{m'}^{\alpha'(m')}a_1'\cdots a_{p-1}'(a_p'+y_la_p')a_{p+1}' \\ \cdots a_{q-1}'(a_q'+y_la_q')y_1^{\beta''(1)}\cdots y_{m'}^{\beta''(m')}a_{q+1}'\cdots a_t'.$$

By the same reason, y_l commutes with $y_1,\dots,y_{m'}$. Recalling that the multiplication by y_l is a derivation and using the relations in (5.7), we see that the leading term of g_1 is equal to that of

$$\theta_l(w)-y_lw-w = -y_1^{\alpha'(1)}\cdots y_{m'}^{\alpha'(m')}a_1'\cdots a_q'y_l^2y_1^{\beta''(1)}\cdots y_{m'}^{\beta''(m')}a_{q+1}'\cdots a_t' \\ + y_1^{\alpha'(1)}\cdots y_{m'}^{\alpha'(m')}y_la_1'\cdots a_q'y_ly_1^{\beta''(1)}\cdots y_{m'}^{\beta''(m')}a_{q+1}'\cdots a_t'.$$

Passing to w_2, the leading term of g_2, we will find that this is equal to the leading term of the following expression:

$$-2y_1^{\alpha'(1)}\cdots y_{m'}^{\alpha'(m')}a_1'\cdots a_q'y_uy_{u'}y_1^{\beta''(1)}\cdots y_{m'}^{\beta''(m')}a_{q+1}'\cdots a_t' \\ -y_1^{\alpha'(1)}\cdots y_{m'}^{\alpha'(m')}y_ua_1'\cdots a_q'y_{u'}y_1^{\beta''(1)}\cdots y_{m'}^{\beta''(m')}a_{q+1}'\cdots a_t' \\ -y_1^{\alpha'(1)}\cdots y_{m'}^{\beta'(m')}y_{u'}a_1'\cdots a_q'y_uy_1^{\beta''(1)}\cdots y_{m'}^{\beta''(m')}a_{q+1}'\cdots a_t'.$$

It is clear that the weight of this element is equal to $s' = (i',\dots,i';\alpha',\beta')$. This also proves that any verbal subspace U in $L(n_1,\dots,n_t)/Q_{p,q}$ is generated by finitely many elements, namely by those elements whose weights form the set of minimal elements in the set of weights for U in S.

To finish the proof, we recall that, according to Proposition 5.19, $L(2,2,2)$ has a finite chain of ideals of the form $M = \sum_{j=1}^{l} L(n_1^{(j)},\dots,n_{t(j)}^{(j)})$ such that each subsequent term contains a sum of the form $\sum_{j=1}^{l} Q_{p(j),q(j)}(n_1^{(j)},\dots,n_{t(j)}^{(j)})$ for pairs $(p(j),q(j)), 1\le p(j) < q(j)\le t(j)$. Then there is a chain of ideals M_s, $s=1,2,\dots,t-1$ between M and M' of the form

$$M_s = \sum_{j=1}^{s} Q_{p(j),q(j)} + \sum_{j=s+1}^{l} L(n_1^{(j)},\dots,n_{t(j)}^{(j)}).$$

All factors of this chain are homomorphic images of the spaces of the form $L(n_1,\dots,n_t)/Q_{p,q}$ (with the above verbal structure). Since these spaces are shown to be verbally noetherian, it follows that the same is true of the set of verbal ideals in $L(2,2,2)/V$ (see Lemma 5.15). □

5.4 Examples of varieties which are not finitely based

The main aim of this section is two-fold. First, we are going to construct, over an arbitrary field of characteristic $p > 0$, an example of a variety which does not admit a finite basis for its identities (Theorem 5.24). Second, we will give an example of a finite-dimensional Lie algebra over an arbitrary infinite field of characteristic $p > 0$ whose identities are not finitely based (Theorem 5.28).

5.4.1 Lie algebras A_n and their derivations

Let k be a field of characteristic $p > 0$. For any natural number n, let T_n denote the set of all subsets in $\{1, 2, \dots, 2n-1\}$. If $\tau \in T_n$ then $|\tau|$ denotes the number of elements in τ. We consider a Lie algebra A_n over k whose basis is formed by the set

$$\{a_\sigma^{(1)}, a_\tau^{(s)}, b \mid \sigma, \tau \subseteq T_n, |\tau| \text{ odd}, s = 2, 3, \dots, p-1\}$$

and the multiplication table (in which, for convenience, we set $a_\tau^{(p)} = 0$ for $|\tau| < 2n-1$ and $a_{T_n}^{(p)} = b$) is as follows:

$$a_\sigma^{(s)} a_\tau^{(r)} = \begin{cases} (-1)^{|\sigma|} a_{\sigma\cup\tau}^{(s+r)} & \text{if } \sigma \cap \tau = \varnothing, s + r \leq p, |\sigma| + |\tau| \equiv 1 \pmod 2, \\ 0 & \text{otherwise.} \end{cases} \tag{5.8}$$

In particular, b is a central element in A_n.

We also introduce a set of linear operators on A_n of the form

$$\{g_\lambda, h_\mu \mid \lambda, \mu \in T_n, |\lambda| \text{ odd}, |\mu| \text{ even}\} \tag{5.9}$$

whose action on the basis of A_n is as follows:

$$g_\lambda(a_\sigma^{(s)}) = \begin{cases} a_{\lambda\cup\sigma}^{(s)} & \text{if } \lambda \cap \sigma = \varnothing \text{ and } |\sigma| \text{ is even,} \\ 0 & \text{otherwise,} \end{cases}$$

$$h_\mu(a_\sigma^{(s)}) = \begin{cases} s a_{\mu\cup\sigma}^{(s)} & \text{if } \mu \cap \sigma = \varnothing, \\ 0 & \text{otherwise.} \end{cases}$$

As for b, this is annihilated by each of the above operators.

Proposition 5.22. *For any natural number n, each A_n is a metabelian nilpotent algebra of nilpotent step p and (5.9) is the set of pairwise commuting derivations of this algebra.*

Proof. To verify the Jacobi identity in A_n, it is sufficient to take pairwise different $a_\rho^{(r)}, a_\sigma^{(s)}, a_\tau^{(t)}$. By equation (5.8), one of the numbers, say $|\rho|$, is odd and the rest are even. Besides, $\rho \cap \sigma = \rho \cap \tau = \sigma \cap \tau = \varnothing$ and $r + s + t \leq p$. In this case we may write

$$\begin{aligned} a_\rho^{(r)} a_\sigma^{(s)} a_\tau^{(t)} + a_\sigma^{(s)} a_\tau^{(t)} a_\rho^{(r)} + a_\tau^{(t)} a_\rho^{(r)} a_\sigma^{(s)} &= 0 + a_\sigma^{(s)} a_{\tau\cup\rho}^{(t+r)} - a_\tau^{(t)} a_{\rho\cup\sigma}^{(r+s)} \\ &= a_{\sigma\cup\tau\cup\rho}^{(s+t+r)} - a_{\tau\cup\rho\cup\sigma}^{(t+r+s)} = 0. \end{aligned}$$

The anticommutativity follows from equation (5.8). The property of being metabelian is also a consequence of equation (5.8) since A_n^2 is the span of those $a_\sigma^{(s)}$ for which $|\sigma|$

is an odd number. Furthermore, any product of at least $p+1$ elements is zero in A_n and $\underbrace{a_{\varnothing}^{(1)}a_{\varnothing}^{(1)}\cdots a_{\varnothing}^{(1)}}_{p-1}a_{T_n}^{(1)} = b \neq 0$, proving the claim on the nilpotent class. Now we want to prove $g_\lambda(a_\sigma^{(s)}a_\tau^{(t)}) = g_\lambda(a_\sigma^{(s)})a_\tau^{(t)} + a_\sigma^{(s)}g_\lambda(a_\tau^{(t)})$. Here $|\lambda|$ is odd and $\lambda\cap\sigma = \lambda\cap\tau = \sigma\cap\tau = \varnothing$. The left-hand side is certainly zero according to our claim on A_n^2. Now if $|\sigma|$ and $|\tau|$ are of different parity then $|\lambda\cup\sigma|$ and $|\tau|$, as well as $|\sigma|$ and $|\lambda\cup\tau|$, are of the same parity. In this case, the right-hand side is obviously zero. Now if $|\sigma|$ and $|\tau|$ have the same parity then, by the definition of g_λ, it is sufficient to consider the case of even $|\sigma|$ and $|\tau|$. Then $g_\lambda(a_\sigma^{(s)})a_\tau^{(t)} + a_\sigma^{(s)}g_\lambda(a_\tau^{(t)}) = a_{\lambda\cup\sigma}^{(s)}a_\tau^{(t)} + a_\sigma^{(s)}a_{\lambda\cup\tau}^{(t)} = (-1)^{|\lambda|+|\sigma|}a_{\lambda\cup\sigma\cup\tau}^{(s+t)} + (-1)^{|\sigma|}a_{\lambda\cup\sigma\cup\tau}^{(s+t)} = 0$. Finally, the fact that h_μ, for $|\mu|$ even, is a derivation and the commutativity of the set (5.9) are immediate from the definitions. □

We denote by D_n the abelian Lie subalgebra in $\mathrm{Der}(A_n)$ spanned by the set (5.9). We consider the semidirect product (Section 1.4.4) $B_n = D_n \rtimes A_n$, $n = 1, 2, \ldots$ It is obvious that $B_n \in \mathbf{A}^3 \cap \mathbf{N}_p\mathbf{A}$ and that B_n satisfies an identical relation of the form

$$x_1(x_2x_3)(x_4x_5)\cdots(x_{2p}x_{2p+1}) \equiv 0. \tag{5.10}$$

5.4.2 Lie algebras C_n

Our main tool in what follows is the set of Lie algebras C_n, $n = 1, 2, \ldots$, each defined as a subalgebra in B_n generated by the set

$$a_{\varnothing}^{(1)}, g_{\{1\}}, h_{\varnothing}, h_{\{2,3\}}, h_{\{4,5\}}, \ldots, h_{\{2n-2,2n-1\}}. \tag{5.11}$$

Lemma 5.23. *If $a_1, a_2 \in A_n$, $c_1, c_2, \ldots, c_s \in C_n$, $\alpha_1, \alpha_2 \in k$ then*

$$\begin{aligned}&(\mathrm{ad}(\alpha_1a_1 + \alpha_2a_2))^{p-1}(c_1c_2\cdots c_s(\alpha_1a_1 + \alpha_2a_2))\\ &= \alpha_1^p(\mathrm{ad}\,a_1)^{p-1}(c_1c_2\cdots c_sa_1) + \alpha_2^p(\mathrm{ad}\,a_2)^{p-1}(c_1c_2\cdots c_sa_2).\end{aligned} \tag{5.12}$$

Proof. Since both sides of equation (5.12) are linear in $c_1, c_2, \ldots, c_s$ and due to $A_n^{p+1} = \{0\}$, it is sufficient to restrict ourselves to the case where all $c_1, c_2, \ldots, c_s$ are the elements from (5.11) and a_1, a_2 are the linear combinations of elements in (5.8) with superscript (5.1). Furthermore, since $g^2(A_n) = \{0\}$ and since D_n is commutative, it remains to consider the following two cases:

(i) $c_1 = h_{\mu_1}, \ldots, c_s = h_{\mu_s}$ and then $c_1c_2\cdots c_sa = h_\mu(a)$ (or zero), $\mu = \mu_1\cup\cdots\cup\mu_s$.

(ii) $c_1 = h_{\mu_1}, c_2 = h_{\mu_2}, \ldots, c_{s-1} = h_{\mu_{s-1}}, c_s = g_{\{1\}}$ and then $c_1c_2\cdots c_sa = \pm g_\mu(a)$ (or zero) where $\mu = \mu_1\cup\mu_2\cup\cdots\cup\mu_{s-1}\cup\{1\}$.

In the former case it suffices to show $(\mathrm{ad}\,a)^p(h_\mu) = 0$, while in the latter we have to prove

$$(\mathrm{ad}(\alpha_1a_1 + \alpha_2a_2))^p(g_\lambda) = \alpha_1^p(\mathrm{ad}\,a_1)^p(g_\lambda) + \alpha_2^p(\mathrm{ad}\,a_2)^p(g_\lambda). \tag{5.13}$$

In the former case we can prove even $(\mathrm{ad}\, a)^2(h_\mu) = 0$. Indeed, applying linearization, we may consider the following two relations:

$$a_\sigma^{(s)} a_\sigma^{(s)} h_\mu = 0 \quad \text{and} \quad a_\sigma^{(s)} a_\tau^{(t)} h_\mu + a_\tau^{(t)} a_\sigma^{(s)} h_\mu = 0.$$

The first of these is obvious since $|\mu|$ is even. As for the second, its left-hand side is either zero or takes the form

$$-(-1)^{|\sigma|} sta_{\sigma\cup\tau\cup\mu}^{(s+t)} - (-1)^{|\tau|} sta_{\tau\cup\sigma\cup\mu}^{(s+t)}.$$

Now if $|\sigma|$ and $|\tau|$ are of the same parity then $|\sigma| + |\tau\cup\mu| \equiv 0 \pmod 2$ and $|\tau| + |\sigma\cup\mu| \equiv$ (mod 2), i. e., both products are zero. If $|\sigma|$ and $|\tau|$ have different parities then one of the numbers $(-1)^{|\sigma|}$ and $(-1)^{|\tau|}$ is 1 and the other is -1. Again we conclude that the left-hand side is zero, and (i) is finished.

The proof of equation (5.13) in case (ii) will be complete if we prove that for any $a_1, a_2, \ldots, a_p$ of the form $a_\sigma^{(1)}$ we have

$$a_1 a_2 \cdots a_p g_\lambda = a_{i_1} a_{i_2} \cdots a_{i_p} g_\lambda \tag{5.14}$$

where $(i_1, i_2, \ldots, i_p)$ is a permutation of $(1, 2, \ldots, p)$. Since A_n is metabelian, we certainly have $xyzu = yxzu$, hence $a_1, a_2, \ldots, a_{p-2}$ can be arbitrarily permuted in equation (5.14). Now we remark that for $s > 1$ we always have $g_\lambda(a_\sigma^{(s)}) = 0$, hence $g_\lambda(A^2) = \{0\}$. Thus $a_{p-1}a_p g_\lambda = a_p a_{p-1} g_\lambda + (a_{p-1}a_p) g_\lambda = a_p a_{p-1} g_\lambda$. Finally,

$$\begin{aligned} a_{p-2}a_{p-1}a_p g_\lambda &= a_{p-1}a_{p-2}a_p g_\lambda + (a_{p-2}a_{p-1})a_p g_\lambda \\ &= a_{p-1}a_{p-2}a_p g_\lambda + a_p(a_{p-2}a_{p-1}) g_\lambda = a_{p-1}a_{p-2}a_p g_\lambda, \end{aligned}$$

proving equation (5.14). The validity of equation (5.14) enables us to apply the binomial decomposition to write

$$\begin{aligned} (\mathrm{ad}(\alpha_1 a_1 + \alpha_2 a_2))^p (g_\lambda) &= \sum_{i=0}^{p} \alpha_1^i \alpha_2^{p-i} \binom{p}{i} (\mathrm{ad}\, a_1)^i (\mathrm{ad}\, a_2)^{p-i} (g_\lambda) \\ &= \alpha_1^p (\mathrm{ad}\, a_1)^p (g_\lambda) + \alpha_2^p (\mathrm{ad}\, a_2)^p (g_\lambda). \end{aligned}$$

This is true since the coefficients of the remaining terms of the decomposition are zero. This completes the proof of equation (5.13) and then the whole of Lemma 5.23. □

We introduce the nonassociative monomials of the form

$$u_s = (\mathrm{ad}(x_1 x_2))^{p-1} (x_3 x_4 \cdots x_{s+2} x_1 x_2), \quad s = 1, 2, \ldots \tag{5.15}$$

Theorem 5.24. *Let k be an arbitrary field of characteristic $p > 0$. Then the system of identities $U = \{u_s \equiv 0 \mid s = 1, 2, \ldots\}$ is independent, so that the variety of Lie algebras determined by this system admits no finite basis of identities.*

Proof. We will show that, for $n = 1, 2, \dots$, $u_n \equiv 0$ is not satisfied by C_n in which all $u_s \equiv 0$, $s < n$, are identically true. Indeed,

$$u_n(h_\varnothing, a_\varnothing^{(1)}, h_{\{2,3\}}, \dots, h_{\{2n-2,2n-1\}}, g_{\{1\}}) = (\operatorname{ad} a_\varnothing^{(1)})^{p-1} g_{T_n} a^{(1)} = b \neq 0,$$

i. e., $u_s(C_n) \neq \{0\}$. To verify the validity of $u_s \equiv 0$, $s < n$, it is sufficient, by Lemma 5.23, to show that

$$(\operatorname{ad} a_\sigma^{(1)})^{p-1}(c_1 c_2 \cdots c_s a_\sigma^{(1)}) = 0,$$

where c_i is a derivation in (5.11) chosen according to one of the rules (i) or (ii), $i = 1, 2, \dots, s$. This is obvious if $\sigma \neq \varnothing$. Now if $\sigma = \varnothing$ and $s < n$ then on the left-hand side we will find an element of the form $\pm a_\pi^{(p)}$ where $|\pi| < 2n - 1$. By our convention (see the beginning of Section 5.4.1), this element is equal to zero. It follows, by the above argument, that $u_n \equiv 0$ is not a consequence of the set of identities of the form $u_s \equiv 0$, $s < n$. But U being homogeneous, it follows that if $u_n \equiv 0$ is a consequence of a subsystem $W \subseteq U$ then $u_n \equiv 0$ is a consequence of a subsystem of W whose elements are all of degree at most n. Thus U is an independent system. □

Corollary 5.25. *Let k be an arbitrary field of characteristic $p > 0$. Then $\mathbf{V} = \mathbf{A}^3 \cap \mathbf{N}_p\mathbf{A}$ is not a Specht variety. Moreover, this variety has uncountably many subvarieties.*

Proof. We adjoin to U (the system of identities in Theorem 5.24) the following multilinear identities determining $\mathbf{A}^3 \cap \mathbf{N}_p\mathbf{A}$:

$$\delta_3(x_1, \dots, x_8) \equiv 0, \quad (x_1x_2)(x_3x_4)\cdots(x_{2p+1}x_{2p+2}) \equiv 0.$$

Since all algebras C_n are in this variety (see Section 5.4.1), the proof of Theorem 5.24 shows that the system thus obtained is again independent. But now we have a system of identities determining a subvariety in $\mathbf{A}^3 \cap \mathbf{N}_p\mathbf{A}$ so that the proof of the first part is complete.

Now let $\mathscr{T}$ be the set of subsets in $\mathbb{N}$. Then for any $S \in \mathscr{T}$, $S \neq \varnothing$, we can consider a variety $\mathbf{U}_S$ which is a subvariety of $\mathbf{V}$ given by a system of identities of the form $\{u_s \equiv 0 \mid s \in S\}$. Then $\mathbf{U}_S \neq \mathbf{U}_T$ if $S \neq T$. Indeed, if s is the least number in $(S \setminus T) \cup (T \setminus S)$, say $s \in S \setminus T$, then, according to the proof of Theorem 5.24, $u_s \equiv 0$ is not a consequence of the system $\{u_n \equiv 0 \mid n \in T\}$. □

5.4.3 Lie algebras C_n (infinite field case)

For any pair (l, n), $l < n$, we define the map $\psi : C_n \to C_l$ by setting $\psi(g_{\{1\}}) = g_{\{1\}}$, $\psi(a_\sigma^{(1)}) = a_{\sigma'}^{(1)}$, $\psi(h_\mu) = h_{\mu'}$ where $\sigma' = \sigma \setminus \{2l, \dots, 2n-1\}$. We remark that a basis for C_n can be chosen in the form

$$\{a_\sigma^{(1)} (\sigma \in T_n), g_{\{1\}}, h_\varnothing, h_{\{2,3\}}, \dots, h_{\{2n-2,2n-1\}}\}. \tag{5.16}$$

Lemma 5.26. *Let k be an infinite field of characteristic $p > 0$. Then any identity of a Lie algebra C_l, $l \le n$, is satisfied by C_n.*

Proof. The base field being infinite, it suffices to consider the multihomogeneous identities of the form $v(x_1, x_2, \dots, x_m) \equiv 0$, the degree of v with respect to x_i being equal to p^{t_i} for some t_i (Section 4.2, Theorem 4.9). Moreover, if some $v \equiv 0$ is not an identity in C_n then some consequence of like form, $v' = 0$, does not vanish after substituting elements of a basis, say (5.16), of C_n (Section 4.2, Theorem 4.10). Without any loss of generality, we may assume that this is the case already for v. Thus, there exist $c_1, c_2, \dots, c_m$ in (5.16) such that $v(c_1, c_2, \dots, c_m) \ne 0$. An immediate verification using induction over the degree of v shows that if $v(c_1, c_2, \dots, c_m) \ne 0$ then

$$v(\psi(c_1), \psi(c_2), \dots, \psi(c_m)) = \psi v(c_1, c_2, \dots, c_m). \tag{5.17}$$

Now from the definition of operations in B_n it follows that a basis element with subscript σ can occur in $v(c_1, c_2, \dots, c_m)$ more than once only in the case $\sigma = \varnothing$. Hence the coincidence of subscripts in equation (5.17) is immediate since $\rho \cap \sigma = \varnothing$ implies $(\rho \cup \sigma) \setminus \tau = (\rho \setminus \tau) \cup (\sigma \setminus \tau)$. Concerning superscripts, they do not depend on the algebra (i. e., C_n or C_l) where we perform our computation. Now to complete the proof of Lemma 5.26, it remains to notice that the value of each $v(c_1, \dots, c_m)$ is an element of the form $\lambda a_\sigma^{(s)}$, where σ and s are fixed and $\lambda \ne 0$. In this case $v(\psi(c_1), \dots, \psi(c_m)) = \lambda a_{\sigma'}^{(s)} \ne 0$, where $\sigma' = \sigma \setminus \{2t, \dots, 2n-1\}$. □

5.4.4 Lie algebra L with no finite basis for its identities

Let L denote a subalgebra of C_2 generated by the set

$$\{a_\varnothing^{(1)}, a_{\{2,3\}}^{(1)}, g_{\{1\}}, h_\varnothing\}.$$

This is a $(2p+3)$-dimensional algebra whose basis is of the form

$$\{a_\varnothing^{(1)}, a_{\{1\}}^{(1)}, a_{\{1,2,3\}}^{(1)}, \dots, a_{\{1\}}^{(p-1)}, a_{\{1,2,3\}}^{(p-1)}, a_{\{1,2,3\}}^{(p)}, g_{\{1\}}, h_\varnothing\}.$$

Lemma 5.27. *If $v(x_1, x_2, \dots, x_m) \equiv 0$ is an identity in Lie algebra L then there exists a natural number n such that $v(x_1, x_2, \dots, x_m) \equiv 0$ is identically true in C_n.*

Proof. Admitting the opposite, we may assume that we are given an identity $v \ne 0$ where v is a multihomogeneous polynomial of degree d whose degree in each x_i is of the form p^{t_i}, $t_i \ge 0$, $v(x_1, x_2, \dots, x_m)$ being nonzero after substituting some $c_1, c_2, \dots, c_m$ of the basis of (5.16) for C_n for an appropriate n. By Section 5.4.3, we may also assume $n \ge d$. We notice that if an element with subscript σ is substituted for a variable x_i of degree p^{t_i}, $t_i \ge 1$, then $\sigma = \varnothing$. In other words, there are only two options: first, where elements of the form $a_\sigma^{(s)}$ are substituted for variables of degree p^{t_i}, $t_i \ge 1$; second, where an element of the form $a_\sigma^{(s)}$ is substituted for a linear argument x_i.

In the former case, $\sigma = \varnothing$, $s = 1$, $t_i = 1$, and $v(c_1, c_2, \ldots, c_m) = \lambda a_\pi^{(p)}$. Moreover, the subscripts of the elements c_j may be different from $\varnothing$ only in the case if these are derivations and the number of derivations is less than $d-1$. Since the subscripts of derivations are at most two-element sets, we have $|\pi| \leq 2(d-1) < 2n-1$. Then $a_\pi^{(p)} = 0$, that is, the first of the two possibilities does not take place.

In the latter case we will be using the map $\psi : C_n \to C_2$. In this case the relation $\psi(v(c_1, \ldots, c_m)) = v(\psi(c_1), \ldots, \psi(c_m))$ enables us to reduce to $n = 2$. Since the bases for C_2 and for L differ only in $h_{\{2,3\}}$, at least one of the elements $c_1, \ldots, c_m$, say, c_m, must be equal to $h_{\{2,3\}}$. It is clear that then $t_m = 0$. Thus $v(a_\sigma^{(s)}, c_2, \ldots, c_{m-1}, h_{\{2,3\}}) \neq 0$. At the same times it is immediate from our hypotheses that

$$v(a_\sigma^{(s)}, c_2, \ldots, c_{m-1}, h_{\{2,3\}}) = v(a_{\sigma\cup\{2,3\}}^{(s)}, c_2, \ldots, c_{m-1}, h_\varnothing).$$

Now the right-hand side of this equation is already a value of v in the elements of L, hence it is zero, a contradiction. Thus, the second possibility cannot be realized either. Therefore, there exists a natural n such that $v(x_1, \ldots, x_m) \equiv 0$ is satisfied by C_n. □

Theorem 5.28. *Over an arbitrary infinite field k of characteristic $p > 0$, there exists a Lie algebra of dimension $2p+3$ whose identities do not admit a finite basis.*

Proof. Suppose the opposite holds and $v_1 \equiv 0, \ldots, v_t \equiv 0$ is a basis for identical relations of L. Then, according to Lemmas 5.26 and 5.27, there exists an $n > 2$ such that all these identities are satisfied by a single C_n. Now we remark that by Theorem 5.24 $u_n \equiv 0$ is not an identity of C_n. An argument in the proof of Theorem 5.24 shows that to verify the validity of $u_n \equiv 0$ in C_2 it suffices to consider an element of the form

$$(\operatorname{ad} a_\sigma^{(1)})^{p-1}(c_1 c_2 \cdots c_s a_\sigma^{(1)}) = \pm a_\pi^{(p)},$$

where $c_1, c_2, \ldots, c_s$ are derivations. Here $\sigma = \varnothing$ and the subscripts of derivations $c_1, c_2, \ldots, c_s$ are equal to either $\{1\}$ or $\varnothing$. Clearly, $\pi \neq \{1,2,3\}$, hence $a_\pi^{(p)} = 0$. Therefore $u_n \equiv 0$ is an identity of L, a consequence of $v_1 \equiv 0, \ldots, v_t \equiv 0$, holding both in L and C_n. This is a contradiction showing that our former hypothesis was wrong. □

5.4.5 Remark

Over an arbitrary infinite field of characteristic 2, the identities of the matrix algebra of order 2 admit no finite basis (see Vaughan-Lee [353]; Drensky [104]).

5.5 Finite basis problem over zero characteristic fields

5.5.1 Capelli identities

Let k be a field of characteristic zero. For any natural number l, we denote by $\mathbf{C}_l$ the variety of Lie algebras over k determined by all identical relations of the form $f_{\tau d} \equiv 0$,

d being an arbitrary Young diagram with at least l rows (see Section 4.7). The ideal $C_l = \mathbf{C}_l(L_\infty)$ in the free Lie algebra L_∞ of countable rank is called the set of *Capelli polynomials* while the set of all identities of the form $f \equiv 0, f \in C_l$, is called the *system of Capelli identities of order l*. Since any identity of the form $f_{\tau d} = 0$ determined by the Young tableau τd is equivalent to an identity with as many variables as the number of rows in d (Section 4.7) then the following is true.

Proposition 5.29. *For any $l = 2, 3, \ldots$, each subvariety* $\mathbf{V}$ *in* $\mathbf{C}_l$ *is given modulo* $\mathbf{C}_l$ *by a system of identical relations depending on at most l variables.*

Theorem 5.30. *Over an arbitrary field k of characteristic zero, the system of all Capelli identities of order l is finitely based.*

Proof. We remark that alternatively the system of Capelli identities of order l can be described as the set of consequences of all multilinear identities of the form $f(x_1, x_2, \ldots, x_l, x_{l+1}, \ldots, x_m) \equiv 0$ which are skew-symmetric over the set of variables of the form $\{x_1, x_2, \ldots, x_l\}$. This enables us to give sense to the systems of Capelli identities even in the case where the base field is not necessarily of characteristic zero, our theorem holding valid in this new setting, provided the characteristic of k is greater than l. That such a description is true can be proved as follows. Clearly, the identities as in Section 5.5.1 follow from this new system. Let, conversely, n be the degree of the polynomial $f(x_1, x_2, \ldots, x_n)$. Denoting by u the sum of idempotents of the group algebra which are multiples of elements in kS_n equal to $e_{\tau d}$ such that d has at least l rows and by v the sum of idempotents associated with all remaining $e_{\tau\sigma}$, we will be able to write $1 = u + v$ (Section 3.2.4) and $f = uf + vf = uf$ since $vf \equiv 0$ is a consequence of $f \equiv 0$ depending on a number of variables which is less than l.

So we may restrict ourselves to considering identical relations of the form

$$\sum_\sigma \varepsilon_\sigma a_{11} \cdots a_{1p_1} x_{\sigma(1)} a_{21} \cdots a_{2p_2} x_{\sigma(2)} \cdots a_{l1} \cdots a_{lp_l} x_{\sigma(l)} a_{l+1} \equiv 0.$$

It is a simple matter, however, using the Jacobi identity and induction over l with obvious basis, to verify that the system of such identities is equivalent to the system of identities of the form

$$\sum \varepsilon_\sigma (a_{11} \cdots a_{1p_1} x_{\sigma(1)})(a_{21} \cdots a_{2p_2} x_{\sigma(2)}) \cdots (a_{l1} \cdots a_{lp_l} x_{\sigma(l)}) \equiv 0.$$

This will be written, for short, as

$$\sum \varepsilon_\sigma (a_1 x_{\sigma(1)})(a_2 x_{\sigma(2)}) \cdots (a_l x_{\sigma(l)}) \equiv 0, \tag{5.18}$$

where each a_i is an associative word $a_{i1} \cdots a_{ip_i}$, $i = 1, \ldots, l$. It is sufficient to prove that any identity of the form of equation (5.18) is a consequence of those ones for which the length of any of the words $a_1, a_2, \ldots, a_l$ is at most $l - 1$.

So let $\xi(a_1, a_2, \ldots, a_l)$ denote the left-hand side of equation (5.18). If y is a variable in $X = \{x_1, x_2, \ldots\}$ which does not occur in equation (5.18) and $r = r(t_1, \ldots, t_l)$ is a commutative polynomial in $t_1, \ldots, t_l$ over k then we define a nonassociative polynomial $r\xi(a_1, \ldots, a_l)$ by setting

$$t_1^{q_1} t_2^{q_2} \cdots t_l^{q_l} \circ \xi(a_1, \ldots, a_l) = \xi(a_1 y^{q_1}, \ldots, a_l y^{q_l}).$$

If $s_1, s_2, \ldots, s_l$ are the elementary symmetric polynomials in $t_1, t_2, \ldots, t_l$ then, for any $m = 1, 2, \ldots, l$, $s_m \xi(a_1, \ldots, a_l) \equiv 0$ is a consequence of $\xi(a_1, \ldots, a_l) \equiv 0$. To verify this, it is sufficient to choose any m indices $i_1, \ldots, i_m$ from $\{1, \ldots, l\}$ and to replace x_{i_1} by $yx_{i_1}, \ldots, x_{i_m}$ by yx_{i_m} and then take the sum of all expressions obtained. Using the Vieta formula

$$(y - t_1)(y - t_2) \cdots (y - t_l) = y^l - s_1 y^{l-1} + \cdots \pm s_l,$$

we may write $t_i^l = s_1 t_i^{l-1} - s_2 t_i^{l-2} + \cdots \pm s_l$. Thus, for any i, $1 < i \le l$, the identity $\xi(a_1, \ldots, a_i' y^l, \ldots, a_l) = 0$ is a consequence of the set of identities of the form $\xi(a_1, \ldots, a_i' y^q, \ldots, a_l) \equiv 0$ with $q = 0, 1, \ldots, l-1$. Now let $\xi(b_1, \ldots, b_i, \ldots, b_l) \equiv 0$ be such that the length of b_i is at least l, that is, $b_i = x_1 x_2 \cdots x_r$, $r \ge l$. Modulo the consequences of identities of the form $\xi(c_1, c_2, \ldots, c_l) \equiv 0$ of lower degrees, the variables $x_1, x_2, \ldots, x_r$ commute. Now we set $b' = x_1 \cdots x_{r-1}$. Then $\xi(b_1, \ldots, b' y^l, \ldots, b_l) \equiv 0$ is a consequence of identities $\xi(b_1, \ldots, b' y^q, \ldots, b_l) \equiv 0$ of lower degrees. Using linearization and the above commutativity of the variables $x_1, \ldots, x_r$ (here we need a restriction on the characteristic of the base field, if this is nonzero), we finally conclude that this is a consequence of Capelli identities of order l and of lower degrees. □

Using Proposition 5.29 and Theorem 5.30, we have the following.

Corollary 5.31. *Let* $\mathbf{V}$ *be a variety of Lie algebras over a field of characteristic zero. If* $\mathbf{V} \subseteq \mathbf{C}_l$ *then* $\mathbf{V}$ *has finite axiomatic rank, namely* $r_a(\mathbf{V}) < l^2$.

5.5.2 Variety $\mathbf{N}_c\mathbf{A}$ over a field of characteristic zero

We consider the identity (5.18). If for some l the associative word a_i on the left-hand side of this identity is not empty then $a_i x_{\sigma(i)}$ is an element of the derived subalgebra. Moreover, for any $i = 0, \ldots, [l/2] - 2$, we may write

$$\sum_{\sigma} \varepsilon_\sigma U x_{\sigma(2i+1)} x_{\sigma(2i+2)} V = \sum_{\substack{\sigma, \\ \sigma(2i+1) < \sigma(2i+2)}} \varepsilon_\sigma U (x_{\sigma(2i+1)} x_{\sigma(2i+2)}) V.$$

Hence if $l \ge 2c + 2$ then identity (5.18) is satisfied by any Lie algebra whose derived subalgebra is nilpotent of step at most c. In other words, $\mathbf{N}_c\mathbf{A} \subseteq \mathbf{C}_l$. By the corollary of Theorem 5.30, $\mathbf{N}_c\mathbf{A}$ and all its subvarieties have finite axiomatic rank r bounded by

$4(c+1)^2$. Now, by Section 5.1.3, the proof of this variety being Specht reduces to proving the verbal noetherian property for the free algebra F in $\mathbf{N}_c\mathbf{A}$ whose rank is r. By the embedding theorem (Section 4.4, Theorem 4.28) $F = L(\{x_1,\dots,x_r\},\mathbf{N}_c\mathbf{A})$ is embedded in the c-step nilpotent wreath product $W = A \operatorname{wr}_{\mathbf{N}_c} B$ of a free c-step nilpotent Lie algebra A with free generating set $y_1,\dots,y_r$ and a free abelian Lie algebra B with basis $z_1,\dots,z_r$. The map $x_i \mapsto y_i + z_i$, $i = 1,2,\dots,r$, effects this embedding of F in W. Now the main result of this section is as follows.

Theorem 5.32. *Over any field k of characteristic zero and for any $c \geq 1$, the variety of Lie algebras $\mathbf{N}_c\mathbf{A}$ is Specht.*

Proof. We proceed by induction over c with obvious basis $c = 1$ (Theorem 5.16). Now let H denote the last nonzero term of the lower central series for F^2. It is sufficient to prove that H satisfies the maximality condition for the set of ideals of L admitting all endomorphisms $\psi_{\lambda i} : x_j \to x_j + \lambda x_i x_j$, $1 \leq i \leq r$, where $\lambda \in k$, $i = 1,2,\dots,r$. As a vector space in W our ideal H is generated by the set of elements of the form

$$(z_1^{m_{11}}\cdots z_r^{m_{1r}} e_{i_1j_1})(z_1^{m_{21}}\cdots z_r^{m_{2r}} e_{i_2j_2})\cdots(z_1^{m_{c1}}\cdots z_r^{m_{cr}} e_{i_cj_c}), \tag{5.19}$$

where $e_{i_lj_l} = x_{i_l}x_{j_l} = y_{i_l}y_{j_l} + y_{i_l}z_{j_l} - y_{j_l}z_{i_l}$, $1 \leq i,j \leq r$, $l = 1,\dots,c$. It is easy to verify that the action of $\psi_{\lambda i}$ on an element of the form of equation (5.19) amounts to multiplying each of its factors by $1+\lambda_i z_i$. We introduce the polynomial ring R over k in the variables $u_1,\dots,u_r; v_1,\dots,v_r;\dots,w_1,\dots,w_r$ (altogether c types of variables, in each group being r variables). Also let M denote a free R-module freely generated by the elements of the form $a(i_1j_1i_2j_2\cdots i_cj_c)$, $1 \leq i_l,j_l \leq r$, $l = 1,2,\dots,c$. Let $\theta : M \to H$ be a k-linear map determined by mapping an element of the form

$$u_1^{m_{11}}\cdots u_r^{m_{1r}} v_1^{m_{21}}\cdots v_r^{m_{2r}}\cdots w_1^{m_{c1}}\cdots w_r^{m_{cr}} a(i_1j_1\cdots i_cj_c)$$

into the element (5.19) in H. Now we set

$$q_{\lambda i} = (1+\lambda u_i)(1+\lambda v_i)\cdots(1+\lambda w_i).$$

We notice that, for any $a \in M$, we have $\psi_{\lambda i}(\theta(a)) = \theta(q_{\lambda i}a)$. Let R_1 denote the subalgebra in R generated by all elements $q_{\lambda i}$ ($\lambda \in k$, $i = 1,2,\dots,r$). We have seen that the inverse image under θ of a subalgebra invariant under all $\psi_{\lambda i}$ ($\lambda \in k$, $i = 1,2,\dots,r$) is an R_1-submodule in M. Hence it is sufficient to prove that M is a noetherian R_1-module. We fix an l, $1 \leq l \leq r$. Then $q_{\lambda l} = 1 + \lambda(u_l + v_l + \cdots + w_l) + \lambda^2(u_lv_l + \cdots + u_lw_l + \cdots) + \cdots + \lambda^c u_lv_l\cdots w_l$. An argument with Vandermonde's determinant shows that, in fact, all symmetric polynomials $\sigma_1^{(l)} = u_l + v_l + \cdots + w_l$, $\dots$, $\sigma_c^{(l)} = u_lv_l\cdots w_l$ are in R_1 for any $l = 1,2,\dots,r$. Hence R_1 is generated by the set of polynomials $\sigma_m^{(l)}$, $1 \leq m \leq c$, $1 \leq l \leq r$, proving that R_1 is a noetherian algebra. Since M was chosen finitely generated over R, the only thing that remains is to prove that R is finitely generated as a natural R_1-module. This, however, can be proved using the same trick as in Theorem 5.30 with Vieta's formulas. As in Theorem 5.30, R is generated over R_1 by those monomials which include powers of each of the variables $u_1, \ldots, w_r$ whose exponents are strictly less than c. □

5.6 Identities of Lie algebra $\mathfrak{sl}(2,k)$

In this section we will concentrate on proving the Specht property for the variety **V** generated by Lie algebra $\mathfrak{gl}(2,k) = [k_2]$ of all second order matrices over a field k of characteristic zero. This theorem fails to be true in the case where the base field is infinite of characteristic 2 (Section 5.4.5). Since in the case of characteristic different from 2, we always have $\mathrm{var}(\mathfrak{gl}(2,k)) = \mathrm{var}(\mathfrak{sl}(2,k))$, in what follows we deal with $\mathfrak{sl}(2,k)$, the Lie algebra of trace-zero matrices of order 2 over a field k of characteristic zero.

5.6.1 Weak identities

Let (R,G) be a pair whose components are respectively an associative algebra R and its Lie subalgebra $G \subseteq [R]$, over a field k. Any relation of the form $f(x_1, x_2, \ldots, x_n) = 0$, f being an associative polynomial which is a nonzero element in $A(x_1, x_2, \ldots, x_n)$, the free associative algebra over k freely generated by $x_1, x_2, \ldots, x_n$, is called a *weak identity* of the pair (R,G) if for any choice of $g_1, g_2, \ldots, g_n \in G$ we have $f(g_1, g_2, \ldots, g_n) = 0$. Let us set $a \circ b = ab + ba$. Then an immediate verification shows that the pair $(k_2, \mathfrak{sl}(2,k))$ satisfies the following weak identity:

$$[x \circ y, z] \equiv 0. \tag{5.20}$$

Lemma 5.33. *The following are consequences of identity* (5.20) *(that is, they hold in any pair satisfying* (5.20)*):*

$$[x,y,z] \equiv 2(x \circ y)z - 2(x \circ z)y, \tag{5.21}$$

$$x(y \circ [z,u]) \equiv (x \circ y)[z,u] - (x \circ z)[y,u] + (x \circ u)[y,z], \tag{5.22}$$

$$4(x \circ y)[z,u] \equiv [z,x,y,u] + [z,y,x,u] - [x,u,y,z] - [y,u,x,z]. \tag{5.23}$$

Proof. The first of these follows by removing brackets with the use of equation (5.20). Next, if we multiply both sides of equation (5.22) by 2 and shift the first term of the right-hand side to the left, then, applying equations (5.20) and (5.21), we will be having on either side the following:

$$\begin{aligned}2(y \circ [z,u])x - 2(y \circ x)[z,u] &= [y,[z,u],x] \\ &= [y,x,u,z] - 2[y,(x \circ z)u] + 2[y,(x \circ u)z] \\ &= [y,x,u,z].\end{aligned}$$

This proves equation (5.22). To verify equation (5.23), we rewrite each term on the right-hand side in the form

$$\begin{aligned}[z,x,y,u] &= 2[z,(x\circ y)u] - 2[z,(x\circ u)y],\\ [z,y,x,u] &= 2[z,(y\circ x)u] - 2[z,(y\circ u)x],\\ -[x,u,y,z] &= -2[x,(u\circ y)z] + 2[x,(u\circ z)y],\\ -[y,u,x,z] &= -2[y,(u\circ x)z] + 2[y,(u\circ z)x].\end{aligned}$$

Summing up all these expressions and having collected like terms, we will arrive, considering the centrality of $a\circ b$, at $4(x\circ y)[z,u]$. □

Theorem 5.34. *Any weak identity of the pair* $(k_2, \mathfrak{sl}(2,k))$ *over a field k of characteristic zero is a consequence of equation* (5.20) *which is itself a weak identity of this pair.*

Proof. Let $f(x_1,x_2,x_3,\dots,x_l)$ be a multilinear polynomial in the free associative algebra $A(x_1,x_2,\dots,x_l)$ and $f' = f(x_2,x_1,x_3,\dots,x_l)$. Using induction over l, we show first the existence of a multilinear polynomial $v(y_0,y_1,\dots,y_{l-2})$ such that

$$f - f' \equiv v([x_1,x_2],x_3,\dots,x_l) \tag{5.24}$$

is a consequence of equation (5.20). We notice that applying equations (5.20)–(5.22) makes it possible to write f as a C-linear combination of commutators of degree one and/or two in $x_1,x_2,\dots,x_l$, C being a subalgebra in $A(x_1,x_2,\dots,x_l)$ generated by the elements $x_i\circ x_j$ and $x_i\circ[x_j,x_m]$, $1\le i,j,m\le l$, whose images in k_2 are central. Indeed, according to equations (5.20) and (5.21), any commutator is equal to a linear combination of commutators of degree 1 and 2 (modulo the identity (5.20)). Thus it remains to show that the same holds for the product of the form $x_i[x_j,x_m]$. We write

$$\begin{aligned}x_i[x_j,x_m] &= \frac{1}{2}\,x_i\circ[x_j,x_m] + \frac{1}{2}\,[x_i,[x_j,x_m]]\\ &= \frac{1}{2}\,x_i\circ[x_j,x_m] + (x_i\circ x_j)x_m - (x_i\circ x_m)x_j.\end{aligned}$$

This argument and equation (5.22) together show that to prove the existence of the consequence of the form of equation (5.24) it is sufficient to consider f having one of the forms: (1) $f=[x_1,x_2]$; (2) $f=(x_1\circ x_2)$; (3) $f=(x_1\circ x_3)x_2$; and (4) $f=[x_1,x_3]\circ x_2$.

In case (1) it is sufficient to set $v(y_0)=y_0$; in case (2) $v=0$; in case (3) $(x_1\circ x_3)x_2 - (x_2\circ x_3)x_1 = \frac{1}{2}[x_3,x_1,x_2]$, and we may set $v=\frac{1}{2}[y_1,y_2]$. In case (4) we remark that $0=[x,y\circ z]=[x,y]\circ z + x\circ[y,z]$. Hence $f=[x_1,x_3]\circ x_2 = x_3\circ[x_1,x_2]$, and case (3) applies.

Now suppose $f\equiv 0$ is a weak identity of our pair. Then the same is true of $f'\equiv 0$, hence $v([x_1,x_2],x_3,\dots,x_l)\equiv 0$ is a weak identity of $(k_2,\mathfrak{sl}(2,k))$. Since $\mathfrak{sl}(2,k)$ is equal to its derived subalgebra, $v(y_1,y_2,\dots,y_{l-1})\equiv 0$ is a weak identity, too. By induction with an obvious basis, this is a consequence of identity (5.20). So we find that $f-f'\equiv 0$ is a consequence of identity (5.20). The choice of the pair (x_1, x_2) being arbitrary, it can be assumed that for any permutation σ of indices $1, 2, \dots, l$ the following is a consequence of identity (5.20):

$$f(x_1, x_2, \dots, x_l) - f(x_{\sigma(1)}, x_{\sigma(2)}, \dots, x_{\sigma(l)}) \equiv 0.$$

Summing up all such equations, we observe that

$$l!f(x_1, x_2, \dots, x_l) - \lambda \sum_{\sigma \in S_l} x_{\sigma(1)} x_{\sigma(2)} \dots x_{\sigma(l)} \equiv 0$$

is a consequence of identity (5.20), λ being a suitable element of k. Substituting

$$x_1 = x_2 = \dots = x_l = h = \begin{pmatrix} 1 & 0 \\ 0 & -1 \end{pmatrix},$$

we will have $\lambda l! h^l = 0$ so that $\lambda = 0$. Hence $f(x_1, x_2, \dots, x_l) \equiv 0$ is a consequence of identity (5.20). □

5.6.2 Some identities of Lie algebra $\mathfrak{sl}(2, k)$

Since $G = \mathfrak{sl}(2, k)$ is 3-dimensional, it satisfies all standard identities

$$[\bar{x}_1, \bar{x}_2, \dots, \bar{x}_{n-1}, x_n] \equiv 0, \quad n = 5, 6, \dots \tag{5.25}$$

The bars over $x_1, x_2, \dots, x_{n-1}$ mean that the left-hand side of equation (5.25) is an alternating sum of such commutators. Now we will prove that for any natural s, $\mathfrak{sl}(2, k)$ satisfies

$$(\operatorname{ad} x)^{2s+1}([y, z]) = [(\operatorname{ad} x)^{2s+1}(y), z] + [y, (\operatorname{ad} x)^{2s+1}(z)]. \tag{5.26}$$

Indeed, by equation (5.23), we have

$$4x^2[x, y] = (\operatorname{ad} x)^3(y). \tag{5.27}$$

Hence

$$\begin{aligned}(\operatorname{ad} x)^{2s+1}([y, z]) &= 2^{2s} x^{2s} [x, y, z] \\ &= [2^{2s} x^{2s} [x, y], z] + [y, 2^{2s} x^{2s} [x, z]] \\ &= [(\operatorname{ad} x)^{2s+1}(y), z] + [y, (\operatorname{ad} x)^{2s+1}(z)].\end{aligned}$$

Turning our attention to identities of the lowest, fifth, degree (the identities of lesser degrees imply solubility), we remark that, according to Section 4.8.3, the space $P_5(X)$

of all multilinear commutator polynomials (i. e., Lie polynomials with respect to bracketing) is equal to the direct sum of five irreducible S_5-modules:

$$P_5 = Af_{\tau_1 d_1} \oplus Af_{\tau_2 d_2} \oplus Af_{\tau_3 d_3} \oplus Af_{d_4} \oplus Af_{d_5}.$$

Denoting by $P_5'(\mathbf{V})$, $\mathbf{V} = \operatorname{var}(G)$, the set of all multilinear identities of degree 5 holding in $\mathbf{V}$, we will find

$$P_5'(\mathbf{V}) = Af_{\tau_3 d_3} \oplus Af_{d_5}.$$

Indeed, according to Section 4.8.3, each S_5-submodule in P_5 is equal to a sum of some of $Af_{\tau d}$, $\tau d \in T$. In this case it is sufficient to verify the validity of identities of the form $g_{\tau_i d_i} \equiv 0$ in G. Now $g_{d_5} = [\bar{y}_1, \bar{y}_3, \bar{y}_4, y_1] = 0$ is a consequence of identity (5.25) with $n = 5$. Furthermore,

$$\begin{aligned} g_{\tau_3 d_3} &= 2[y_1, [y_1, y_3], y_1, y_2] \\ &= \frac{1}{3}\,(2(\operatorname{ad} y_1)^3([y_3, y_2]) - 2[(\operatorname{ad} y_1)^3(y_3), y_2] - 2[y_3, (\operatorname{ad} y_1)^3(y_2)]). \end{aligned}$$

Hence $g_{\tau_3 d_3} \equiv 0$ is a consequence of identity (5.26) with $s = 1$. Now $g_{\tau_1 d_1} = 2(\operatorname{ad} x)^4(y)$, $g_{\tau_4 d_4} = 2[\bar{y}_1, \bar{y}_2, \bar{y}_3, y_1, y_2]$, $g_{\tau_2 d_2} = 2[[y_1, y_2], y_1, y_1, y_2]$. Choosing in $\mathfrak{sl}(2, k)$ a basis of the form

$$e = \begin{pmatrix} 0 & 1 \\ 0 & 0 \end{pmatrix}, \quad f = \begin{pmatrix} 0 & 0 \\ 1 & 0 \end{pmatrix}, \quad h = \begin{pmatrix} 1 & 0 \\ 0 & -1 \end{pmatrix}$$

with the multiplication table $[h, e] = 2e$, $[h, f] = -2f$, $[e, f] = h$, we easily see that $g_{\tau_1 d_1}$ is nonzero for $x = h$, $y = e$, $g_{\tau_4 d_4}$ takes a nonzero value if $y_1 = e$, $y_2 = f$, $y_3 = h$; in the case of $g_{\tau_2 d_2}$ it is sufficient to put $y_1 = e$, $y_2 = f$.

We will also need the following two identities of degree 6:

$$[x, x, x, y, z, u] = [x, y, z, x, x, u] - [x, y, x, u, x, z], \tag{5.28}$$

$$[z, u, x, x, x, y] = [z, x, x, u, x, y] - [x, [u, x, y], x, z]. \tag{5.29}$$

These follow using (5.27) and $4x^2[y, z] = [y, x, x, z] - [x, z, x, y]$, which are the consequences of the weak identity (5.23).

Indeed,

$$[x, x, x, y, z, u] = 4x^2[x, y, z, u] = [x, y, 4x^2[z, u]] = [x, y, z, x, x, y] - [x, y, x, u, x, z],$$

$$[z, u, x, x, x, y] = 4x^2[z, u, x, y] = [z, x, x, u, x, y] - [x, [u, x, y], x, z].$$

Our main result in this section is Theorem 5.36 stating that identity (5.25) with $n = 5$ and identity (5.26) with $s = 2$ form a basis for identical relations of $\mathfrak{sl}(2, k)$, k being an arbitrary field of characteristic zero. Its proof uses the same method as applied while proving Theorem 5.34. An important part of the present proof where the above-mentioned method is used will be formulated as a separate lemma. In stating this lemma, we are using the notation $[\dots \tilde{x}_{i_1} \dots \tilde{x}_{i_2} \dots \tilde{x}_{i_t} \dots]$ to abbreviate the symmetric sum

$$\sum_{\sigma \in S_t} [\dots x_{i_{\sigma(1)}} \dots x_{i_{\sigma(2)}} \dots x_{i_{\sigma(t)}} \dots].$$

We will also write $\sigma_{ij} f(x_1, \dots, x_i, \dots, x_j, \dots, x_l) = f(x_1, \dots, x_j, \dots, x_i, \dots, x_l)$. Finally, we use ^ to mean that the variable under this sign should be omitted.

Lemma 5.35. *In the variety of Lie algebras determined by the identities*

$$[\tilde{x}_1, \tilde{x}_2, \tilde{x}_3, \tilde{x}_4, x_5] = 0 \quad \text{and} \quad (\operatorname{ad} x)^3([y,z]) = [(\operatorname{ad} x)^3(y), z] + [y, (\operatorname{ad} x)^3(z)],$$

given a multilinear commutator polynomial f of degree $l > 4$ and two indices i, j, $1 \le i \le j \le l$, there exists a multilinear commutator polynomial v of degree $l-1$ such that

$$\begin{aligned} f - \sigma_{ij} f = c([x_i, \tilde{x}_1, \dots, \hat{x}_i, \dots, \hat{x}_j, \dots, \tilde{x}_l, x_j] \\ - [x_j, \tilde{x}_1, \dots, \hat{x}_i, \dots, \hat{x}_j, \dots, \tilde{x}_l, x_i]) + v(x_1, \dots, [x_i, x_j], \dots, \hat{x}_j, \dots, x_l). \end{aligned} \tag{5.30}$$

In the case of odd l, we can even put $c = 0$.

Proof. Without loss of generality, we may assume $i = 1$, $j = l$. Since a basis in $P_l(x_1, \dots, x_l)$ is formed by all monomials of the form $[x_{\sigma(1)}, x_{\sigma(2)}, \dots, x_{\sigma(l-1)}, x_l]$ (Section 4.8.1), we may restrict ourselves to the case where f is a monomial of the form $f = [x_{\sigma(1)}, \dots, x_1, \dots, x_{\sigma(l-1)}, x_l]$.

If $l = 5$, we may consider f in one of the forms:

$$[x_2, x_3, x_4, x_1, x_5], [x_2, x_3, x_1, x_4, x_5], [x_2, x_1, x_3, x_4, x_5], [x_1, x_2, x_3, x_4, x_5].$$

In the first and second cases, the desired representation with $c = 0$ results by applying the anticommutativity and the Jacobi identity. Before we pass to the remaining cases, we remark that any commutator identity of degree 5 which is a consequence of identity (5.20) is an identity of $\mathfrak{sl}(2,k)$ of degree 5, hence, by the decomposition for $P'(\mathbf{V})$ in Section 5.6.2, it is implied by the identities in the statement of Lemma 5.35. This means that, in finding a representation as in equation (5.30) in the case of degree 5, we are allowed to use the weak identity (5.20) and its consequences. So we write

$$\begin{aligned} &[x_2, x_1, x_3, x_4, x_5] - [x_2, x_5, x_3, x_4, x_1] \\ &\quad = 2[x_2, x_1, (x_3 \circ x_4) x_5] - [2x_2, x_1, (x_3 \circ x_5) x_4] \\ &\qquad - 2[x_2, x_5, (x_3 \circ x_4) x_1] + 2[x_2, x_5, (x_3 \circ x_1) x_4] \\ &\quad = 4(x_3 \circ x_4)[x_2, x_1, x_5] - [x_2, 2(x_3 \circ x_5) x_1 - 2(x_3 \circ x_1) x_5, x_4] \\ &\quad = 4(x_3 \circ x_4)[x_2, x_1, x_5] - [x_2, [x_3, x_5, x_1], x_4]. \end{aligned}$$

Applying equation (5.23) with $x = x_3$, $y = x_4$, $z = x_2$, $u = [x_1, x_5]$, we obtain the sum of commutators each of which contains $[x_1, x_5]$ as a factor, yielding the desired decomposition. Finally, applying equation (5.21) we can write

$$\begin{aligned}
&[x_1, x_2, x_3, x_4, x_5] - [x_5, x_2, x_3, x_4, x_1] \\
&\qquad = 2(x_3 \circ x_4)([x_1, x_2, x_5] - [x_5, x_2, x_1]) \\
&\qquad\quad - 2(x_3 \circ x_5)[x_1, x_2, x_4] + 2(x_3 \circ x_1)[x_5, x_2, x_4] \\
&\qquad = 2(x_3 \circ x_4)[x_2, x_1, x_5] + [2(x_3 \circ x_1)x_5 - 2(x_3 \circ x_5)x_1, x_2, x_4] \\
&\qquad = 2(x_3 \circ x_4)[x_2, x_1, x_5] + [[x_3, x_1, x_5], [x_2, x_4]].
\end{aligned}$$

Applying equation (5.23) as above, we arrive at a representation of the form (5.30), with $c = 0$.

Now we pass to the case $l \geq 6$. We notice first that a commutator of the form $f = [x_1, x_2, \ldots, x_{l-1}, x_l]$ is representable as the sum of $f_0 = [x_1, \tilde{x}_2, \ldots, \tilde{x}_{l-1}, x_l]$ and a sum of commutators of the form $f_i = [x_1, x_2, \ldots, [x_i, x_{i+1}], \ldots, x_{l-1}, x_l]$ (up to a renumbering of $x_2, \ldots, x_{l-1}$). If l is odd then for $f = f_0$ we obviously have equation (5.30) with $c = 1$ and $v = 0$. For $f = f_i$, the induction hypothesis applies, and then $c = 0$.

Anticipating Theorem 5.36, we remark that it will be shown in the course of its proof that any identity of $\mathfrak{sl}(2, k)$ of the form $f \equiv 0$ with a representation of the form (5.30) is a consequence of identities of lower degree. Thus, in finding a representation of the form (5.30) for the identities of degree $l \geq 7$, we are allowed to apply identities (5.28) and (5.29) of degree 6.

Coming back to our proof, we recall that we are considering the case where l is an odd number with $l \geq 7$. Here the induction hypothesis with $c = 0$ applies to the commutators of the form $[x_2, x_3, [\ldots, x_1, \ldots, x_{l-1}, x_l]]$ (it is sufficient to consider a factor of the form $[\ldots, x_1, \ldots, x_{l-1}, x_l]$ of degree $l - 2$). In the case of $[x_2, x_1, x_3, \ldots, x_{l-1}, x_l]$, by the induction hypothesis, we write

$$\begin{aligned}
f - \sigma_{1l} f = {} & c([x_2, x_1, \tilde{x}_3, \ldots, \tilde{x}_{l-1}, x_l] - [x_2, x_l, \tilde{x}_3, \ldots, \tilde{x}_{l-1}, x_1]) \\
& + [x_1, v'([x_1, x_l], x_3, \ldots, x_{l-1})].
\end{aligned}$$

Now the latter summand is of the desired form. To deal with the former, it is useful to remark that this is a linearization of the polynomial of the form

$$c([x_2, x_1, \underbrace{y, \ldots, y}_{l-3}, x_l] - [x_2, x_l, \underbrace{y, \ldots, y}_{l-3}, x_1]).$$

Both terms can be transformed using the identity (a consequence of (5.29))

$$[z, u, x, \underbrace{x, \ldots, x}_{2s}, y] = [z, x, x, u, \underbrace{x, \ldots, x}_{2s-1}, y] - [x, [u, \underbrace{x, \ldots, x}_{2s-1}, y], x, z]$$

to yield

$$\begin{aligned}
&[x_2, x_1, \tilde{x}_3, \ldots, \tilde{x}_{l-1}, x_l] \\
&\quad = [x_2, \tilde{x}_3, \tilde{x}_4, x_1, \tilde{x}_5, \ldots, \tilde{x}_{l-1}, x_1] - [\tilde{x}_4, [x_1, \tilde{x}_5, \ldots, \tilde{x}_{l-2}, x_l], \tilde{x}_{l-1}, x_2].
\end{aligned}$$

Now the induction hypothesis applies to $[\tilde{x}_3, \tilde{x}_4, x_1, \tilde{x}_5, \dots, \tilde{x}_{l-1}, x_l]$, as well as to appropriate factors of degree $l-1$ in the latter symmetric sum of commutators. What remains is to consider the following cases:

(1) $f = [x_1, \tilde{x}_2, \dots, \tilde{x}_{l-1}, x_l]$. Here we may apply the complete linearization of identity (5.26) with $2s+1 = l-2$ (which is a consequence of (5.26) with $s = 1$).

(2) $f = [x_1, x_2, \dots, [x_i, x_{i+1}], \dots, x_{l-1}, x_l]$. In this case by the induction hypothesis, we write

$$\begin{aligned} f - \sigma_{1l} f = c'([x_1, \tilde{x}_2, \dots, \widetilde{[x_i, x_{i+1}]}, \dots, \tilde{x}_{l-1}, x_l] - [x_1, \tilde{x}_2, \dots, \widetilde{[x_i, x_{i+1}]}, \dots, \tilde{x}_{l-1}, x_1]) \\ + v'([x_1, x_l], x_2, \dots, [x_i, x_{i+1}], \dots, x_{l-1}). \end{aligned} \tag{5.31}$$

The latter summand is as required. To transform the former, we will use a consequence of (5.28) which has the form

$$[\underbrace{x, \dots, x}_{t}, y, z, u] = [\underbrace{x, \dots, x}_{t-2}, z, x, x, u] - [\underbrace{x, \dots, x}_{t-2}, y, x, u, x, z].$$

Applying this identity repeatedly, we arrive at the following:

$$[\underbrace{x, \dots, x}_{t}, y, z, u] = \sum_i [x, x, y, w_i'] + \sum_j [x, y, u_j', v_j'], \tag{5.32}$$

where u_j', v_j', w_j' are some commutators in x, z, u. We rewrite

$$g = [x_1, \tilde{x}_2, \dots, \widetilde{[x_i, x_{i+1}]}, \dots, \tilde{x}_{l-1}, x_l]$$

as a linear combination of commutators of the form $[\dots, [x_i, x_{i+1}]]$. Now g is symmetric in $x_2, \dots, x_{i-1}, x_{i+2}, \dots, x_{l-1}$. Applying a linearization of equation (5.32), we will obtain

$$g = \sum_i \pm[x_1, x_p, x_q, x_l, w_i] + \sum_j \pm[x_1, x_p, x_l, u_j, v_j] + \sum_s \pm[x_1, x_l, a_s, b_s, c_s],$$

with a_s, b_s, c_s, u_j, w_i being commutators in the variables distinct from x_1, x_l. Hence g can be handled as a commutator polynomial of degree r, proving the existence of representation of the form of equation (5.30) for the former summand, too. □

Theorem 5.36. *Identical relations of Lie algebra $G = \mathfrak{sl}(2,k)$ over a field k of characteristic zero admit a basis of the form*

$$[\bar{x}_1, \bar{x}_2, \bar{x}_3, \bar{x}_4, x_5] \equiv 0, \tag{5.33}$$

$$(\operatorname{ad} x)^3([y, z]) = [(\operatorname{ad} x)^3(y), z] + [y, (\operatorname{ad} x)^3(z)]. \tag{5.34}$$

Proof. We start with showing that if $f \equiv 0$ is a multilinear identity of degree l in G then in equation (5.30) we may always put $c = 0$. Indeed, it suffices to substitute $x_i = e$,

$x_j = f$, $x_t = h$ $(t \neq i,j)$. Since l is even and $f \equiv 0$ is an identity, the left-hand side vanishes and we get

$$0 = c(l-2)!2^{l-1}h + v(h, \dots, [e,f], \dots, h).$$

Now $[e,f] = h$. Since v is a commutator polynomial, we obtain $c(l-2)!2^{l-1} = 0$, hence $c = 0$. Therefore for any polynomial identity $f \equiv 0$ of degree $l \geq 6$ and for any pair i,j, $1 \leq i \leq j \leq l$, there exists a multilinear commutator polynomial v of degree $l-1$ such that

$$f - \sigma_{ij}f \equiv v(x_1, \dots, [x_i, x_j], \dots, \hat{x}_j, \dots, x_l)$$

is a consequence of identities (5.33) and (5.34). Since $f - \sigma_{ij}f = 0$ is an identity of G and $G = G^2$, then $v \equiv 0$ is an identity of G of degree $l-1$. By the induction hypothesis, $v \equiv 0$ follows from equations (5.33) and (5.34). Hence also $f - \sigma_{ij}f = 0$ is a consequence of equations (5.33) and (5.34). It is a simple matter then to conclude that, for any permutation $\tau \in S_l$, (5.33) and (5.34) imply $f - \tau f = 0$. Taking the sum of all these identities, we derive that

$$n!f \equiv \sum_{\tau \in S_l} \tau f$$

is a consequence of identities (5.33) and (5.34). Now the right-hand side is a multiple of the sum $\sum_{\tau \in S_l} [x_{\tau(1)}, x_{\tau(2)}, \dots, x_{\tau(l)}]$, that is, an identical zero in G. □

5.6.3 Free pair

Let $X = \{x_1, x_2, \dots, x_n, \dots\}$ be a countable set, $A = A(X)$, $L = L(X)$ free associative and free Lie algebra over X, respectively, and I the ideal of A whose elements are all polynomials $f(x_1, \dots, x_m)$ such that $f(x_1, \dots, x_m) \equiv 0$ is a weak identity of the pair $(k_2, \mathfrak{sl}(2,k))$. Denote by (A_1, L_1) the pair where $A_1 = A/I$, $L_1 = L + I/I$. The set of weak identities of (A_1, L_1) coincides with the set of weak identities of $(k_2, \mathfrak{sl}(2,k))$.

Denoting the images of $x_1, x_2, \dots, x_n, \dots$ in A_1 by the same letters, we will see, as in the proof of Theorem 5.34, that any element a in A_1 is representable in the form $a = c + v$ where c is a central element of A_1 and v is a linear combination of commutators of degree 1 and 2 in X with coefficients in the center C. We will show that, in fact, this representation is unique. Indeed, if $c + v(x_1, \dots, x_l) = 0$ then $[x_{l+1}, v(x_1, \dots, x_l)] = 0$. Hence $[x_{l+1}, v] = 0$ is a weak identity of $(k_2, \mathfrak{sl}(2,k))$. Applying equation (5.22) and recalling that $\mathfrak{sl}(2,k)$ is equal to its derived subalgebra enables us to rewrite any value of v in $\mathfrak{sl}(2,k)$ as a linear combination of the elements of the form

$$(v_{i_1} \circ v_{j_1}) \cdots (v_{i_s} \circ v_{j_s})[v_p, v_q] \quad \text{where } v_{i_1}, \dots, v_q \in \mathfrak{sl}(2,k).$$

By formula (5.23), this element is in $\mathfrak{sl}(2,k)$. Since $\mathfrak{sl}(2,k)$ over any field k of characteristic different from 2 has trivial center, we see that v vanishes under any substitution of elements in $\mathfrak{sl}(2, k)$, hence $v \equiv 0$ is a weak identity of $(k_2, \mathfrak{sl}(2, k))$. Thus $v(x_1, \ldots, x_l) = 0$ in A_1, proving the desired uniqueness. □

Lemma 5.37. *Consider an ideal S of L generated by all elements of the form $[v_0, u(v_1, \ldots, v_l)]$ where $u(x_1, \ldots, x_l)$, $v_0, v_1, \ldots, v_l \in L_1$. If T is an ideal of A_1 generated by S then $S = L_1 \cap T$.*

Proof. It is obvious that $S \subseteq T \cap L_1$. To prove the converse, we note that any element in T is representable as a linear combination of the commutator $[v_0, u]$ and the polynomials in $x_1, \ldots, x_n, \ldots$ Using the weak identity (5.20), we reduce this to the form $c + v$ where $c \in C'$ is a polynomial in the generators of the form $x_i \circ x_j$, $x_i \circ [v_0, u]$ and $x_i \circ [x_j, [v_0, u]]$. These elements are in the center of A_1. Now v is representable as a linear combination of the commutators of degree 1 and 2 with coefficients in C'. Thus, in the expression of monomials forming v, the commutator $[u, v]$ occurs in one of the factors of the form $[v_0, u]$, $[x_i, v_0, u]$, $(x_i \circ [v_0, u])x_s$, $(x_i \circ [x_j, v_0, u])x_s$. Since $[x_j, v_0, u]$ is a linear combination of elements of the form $[v_0', u] = [v_0', u(v_1', \ldots, v_l')]$ and since $(x_i \circ [v_0, u])x_s$ using formulas (5.22) and (5.23) can easily be reduced to a linear combination of the elements of the form $[v_0'', u''] \in S$, the desired representation follows.

Now suppose $a \in T \cap L_1$. In this case a can be written in the form $a = c + v$, where c is in the center of A_1 and $v \in S$. In this case, $c + v = z \in L_1$. By formulas (5.21)–(5.23), any commutator is a linear combination of the commutators of degree 1 and 2. By our remarks just before Lemma 5.37, we have $c = 0$ and $v = z$. Hence $a \in S$, proving the inclusion $S \supseteq T \cap L_1$. □

5.6.4 Nilpotent elements in the free pair

The just proved Lemma 5.37 and the one that follows (Lemma 5.38) form intermediate stages in proving a theorem stating that each proper subvariety in $\mathbf{V} = \mathrm{var}(\mathfrak{sl}(2, k))$ is in $\mathbf{N}_c\mathbf{A}$ for an appropriate c. In the course of proving this theorem, we will encounter the situation where in a free pair (A, L) with identity (5.20) the element $\bar{x}_1\bar{x}_2\bar{x}_3$ is nilpotent, $\{x_1, x_2, \ldots, x_n, \ldots\}$ being a free generating set for the pair. Let us denote by A' the subalgebra in A generated by $\bar{v}_1\bar{v}_2\bar{v}_3$ $(v_1, v_2, v_3 \in L)$.

Lemma 5.38. *A' is a nilpotent subalgebra of A.*

Proof. Let $f_0 = (\bar{x}\bar{y}\bar{z})^m \equiv 0$ be a weak identity of the pair under consideration. We are going to prove that $f = (\bar{x}_1\bar{y}_1\bar{z}_1) \cdots (\bar{x}_m\bar{y}_m\bar{z}_m) \equiv 0$ is a weak identity, too. We consider the group algebra $R = kS_{3m}$. We represent 1 in R (see Section 3.2.4) as the sum of idempotents associated with standard Young tableaux and write $f = e_1f + e_2f$, where e_1 is a linear combination of elements $e_{\tau d}$ associated with those diagrams in which the number of rows is at least four. Then e_1f is a linear combination of multilinear polynomials each alternating in a group of elements with at least four variables in each such group. Since $\mathfrak{sl}(2, k)$ is three-dimensional then $e_1f = 0$ is a weak identity. Now we write $e_2 = e_1' + e_2''$, where e_1' is a linear combination of the elements $e_{\tau d}$ associated with those diagrams in which the number of columns is greater than m. Then $e_2'f$ is a linear combination of multilinear polynomials each being symmetric in a group of elements with at least $m+1$ elements in such a group. Since all the variables in f split into m triples of alternating variables we deduce $e_2'f = 0$. Finally, e_2'' is a linear combination of elements as $e_{\tau d}$ associated with the diagram of the form

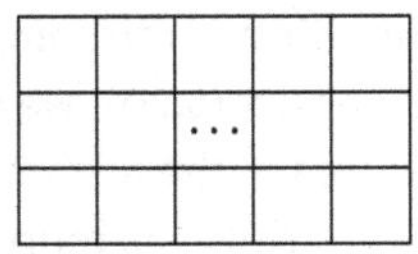

Now for any tableau τd of such a form the identification of the variables in the rows of $e_{\tau d}f$ leads to a multiple of the original polynomial f_0, hence $e_{\tau d}f \equiv 0$ is a weak identity in (A, L). Hence $e_2''f = 0$ and then also $f \equiv 0$ is a weak identity in (A, L). □

Theorem 5.39. *Every proper subvariety* $\mathbf{U}$ *of* $\mathbf{V} = \mathrm{var}(\mathfrak{sl}(2,k))$ *over any field* k *of characteristic zero is in* $\mathbf{N}_c\mathbf{A}$ *for a suitable* c.

Proof. By our hypothesis, there exists an identity $u(x_1, \ldots, x_l) \equiv 0$ satisfied by $\mathbf{U}$ and nontrivial in $\mathbf{V}$. Let (A, L) denote the free pair determined by the weak identity (5.20) and with A_1, S, T being the same as in Lemma 5.37 (we remark that $[u(x_1, \ldots, x_l), x_{l+1}] = 0$ is not an identity in $\mathfrak{sl}(2,k)$ since this is a centerless algebra). If $\varphi : A_1 \to A/T$ is a natural homomorphism then the pair $(\varphi(A_1), \varphi(L_1))$ is free and it is clear that $\varphi(L_1) \cong L_1/S$. It is obvious that the free algebra of countable rank in $\mathbf{U}$ is a homomorphic image of $\varphi(L_1)$. So it is sufficient to verify that $\varphi(L_1)$ is in $\mathbf{N}_c\mathbf{A}$ for a suitable c. Considering a subpair (A_2, L_2) in $(\varphi(A_1), \varphi(L_1))$ generated by three elements x, y, z, we will see that, in the terminology of Section 6.1.1, A_2 is a finitely generated associative PI-algebra. Now the Jacobson radical J of such an algebra is a nil-ideal (Section 6.1, Theorem 6.3). The quotient algebra A_2/J is a subdirect product of matrix algebras of bounded order over suitable skew fields. Furthermore, the image of L_2 in each of these algebras is an irreducible Lie algebra of linear operators of a finite-dimensional vector space. Such algebra is either abelian or it contains a subalgebra isomorphic to $\mathfrak{sl}(2,k)$ (Section 6.4.2). Since $\mathbf{U}$ is proper the latter is impossible. Hence the commutator of L_2 is in J. Now $\bar{x}\bar{y}\bar{z}$ is easily seen to belong to the ideal of A_2 generated by L_2^2, hence, for some m, we have $(\bar{x}\bar{y}\bar{z})^m = 0$. By Lemma 5.38, the subalgebra A' in $\varphi(A_1)$ generated by $\bar{a}_1\bar{a}_2\bar{a}_3$, $a_i \in v(L_1)$, is nilpotent. Now let $v_1, v_2 \in \varphi(L_1^2)$. Then an immediate application of equation (5.20) gives $v_1 \circ v_2 \in A'$. Applying weak identity (5.21) then gives $[v_1, v_2, \ldots, v_{2m+1}] = 0$ for any $v_1, v_2, \ldots, v_{2m+1} \in \varphi(L_1)^2$. Thus $\varphi(L_1^2)$ is a nilpotent algebra. □

Corollary 5.40. *Variety of Lie algebras* $\mathbf{V} = \mathrm{var}(\mathfrak{sl}(2,k))$ *over any field* k *of characteristic zero is Specht.*

Proof. This is immediate from Theorems 5.32, 5.36 and 5.39. □

5.7 Exercises

5.7.1. Let **U** and **V** satisfy the minimality condition for subvarieties. Then the same is true as concerns $\mathbf{U} \cup \mathbf{V}$ and $\mathbf{U} \cap \mathbf{V}$.

5.7.2. Prove that the Capelli identities of order 3 follows from the standard identity of degree 4.

5.7.3. Prove the finite basis property for the varieties of finite axiomatic rank which are subvarieties in $\mathbf{AN}_c$.
Hint: use Section 6.6, Theorem 6.41.

5.7.4. Any variety generated by a finite-dimensional algebra is of finite axiomatic rank.

5.7.5. Let $\prod$ denote the set of decreasing sequences of natural numbers $(m_1, \dots, m_r)$, with an ordering of the form $(m_1, \dots, m_r) \le (m'_1, \dots, m'_t)$ if and only if $r \le t$ and $m_i \le m'_i$ $(i = 1, \dots, r)$. Is this set partially well-ordered?

5.7.6. Prove that $(k_2, \mathfrak{sl}(2, k))$ satisfies a weak identity of the form $[\bar{x}\bar{y}\bar{z}, u] = 0$ (bars mean alternation).

5.7.7. Prove that k_2 satisfies $\bar{x}\bar{y}\bar{z}\bar{u} = 0$.

5.7.8. Prove that $\mathfrak{sl}(2, k)$ satisfies the following Lie identity:

$$\bar{x}_1\bar{x}_2\bar{x}_3yz = y\bar{x}_1\bar{x}_2\bar{x}_3z.$$

5.8 Comments

Theorems 5.5 and 5.9 are due to Zaĭcev [373]. In the case of an infinite field, the main Corollary 5.7 was previously proved by the author (Bahturin [21]). The technique of partially well-ordered sets goes back to Higman [142]. It was first used for the finite basis problem by Cohen [89] who proved that $\mathbf{A}^2$ in the case of groups is a Specht variety. Theorem 5.18 is due to Bryant and Vaughan-Lee [88], see also [355, 356]. Its generalization to the case of $\mathbf{N}_c\mathbf{A} \cap \mathbf{N}_2\mathbf{N}_c$ has been achieved by Sheina [330]. It should be noted that $\mathbf{N}_c\mathbf{A} \cap \mathbf{AN}_c$ is a Specht variety over any noetherian ring (Vaughan-Lee [354]). The first example of an infinitely based variety of Lie algebras over a field was given by Vaughan-Lee [354] in the case of characteristic 2. Section 5.4 here is based on a paper by Drensky [103] who extended the results of Vaughan-Lee's paper to the case of fields of arbitrary prime characteristic. A refinement of these results led Drensky to exhibiting a locally finite infinitely based variety (Drensky [104]). Volichenko [361] has given an example of a just-nonfinitely based variety of Lie algebras. Examples of infinitely based finite-dimensional linear algebras (which are not Lie algebras) over an arbitrary field are given by Polin [292], Malcev and Parfenov [220], and L'vov [208]. Capelli identities were first introduced into the theory of varieties by Razmyslov [304]. Theorem 5.30 is due to Stovba [347]. The validity of this result has been pointed out in the case of fields of characteristic zero by Razmyslov [306]. The Specht property for $\mathbf{N}_c\mathbf{A}$ was first shown by Krasil'nikov [184]. This result can now be derived from a theorem of Stovba, which can now be derived from a theorem due to Stovba [348], which we discuss in more detail in Chapter 14: if $\mathbf{V}$ is a variety of finite axiomatic rank which is a subvariety in $\mathbf{N}_c\mathbf{N}_d$ over a field k such that $|k| \geq c$ then $\mathbf{V}$ is a finitely based variety. An important recent result is due to Volichenko who has shown that over an arbitrary field of characteristic zero $\mathbf{AN}_2$ is a Specht variety (Volichenko [363]).

The results in Section 5.6 are due to Razmyslov [301, 303]. Two-variable identities of $\mathfrak{sl}(n, \mathbb{C})$ have been studied in [30]. There is a good reason to conjecture that the identities of any of $\mathfrak{sl}(n, k)$, k being a field of characteristic zero, have finite basis for an arbitrary $n > 2$. The principal open question in this area is the existence of a variety admitting no finite basis for its identities in the case of zero characteristic base field. This problem remains unsolved even in the case of identities of finite-dimensional algebras. It is plausible that the subvarieties in $\mathbf{N}_{p-1}\mathbf{A}$ are finitely based in the case of characteristic $p > 0$. It is interesting to study the finite basis property for the products of finitely based varieties even in the case where the first factor is nilpotent. It is not known if there exists a finitely based variety whose identities depending on two variables are infinitely based. There is no answer to the question on the finite basis property in the case of identities of Lie algebra W_n even in the case $n = 1$.

6 Special Lie algebras: structure, identities, and applications

In this chapter we will be occupied with a variety of questions relating the connection between Lie algebras satisfying identical relations and associative algebras with polynomial identities (PI-algebras). Here, by necessity, we deviate from our principle of closedness. However, all results we are using without proofs can be easily recovered in well-known sources. These results are accumulated in preliminary Sections 6.1 and 6.2. The former is a collection of some rather common results on associative algebras. The details can be found in such books as Herstein's [140], Jacobson's [152], Rowen's [322]. In the latter we write down some classical results on finite-dimensional Lie algebras. The references are given in Chapter 1. The next three sections are devoted to special Lie algebras, a fairly wide class of infinite-dimensional Lie algebras, defined in the language of identities and inheriting many usual properties of the class of finite-dimensional Lie algebras. In the second part of the chapter, we come back to the universal enveloping algebras and prove a theorem on the necessary and sufficient conditions on a Lie algebra under which its universal enveloping algebra is a PI-algebra. This theorem finds its application in the theory of representations of Lie algebras. Section 6.9 provides a description of Lie superalgebras over a field of characteristic zero whose universal enveloping algebra is a PI-algebra.

6.1 Some results on associative PI-algebras

6.1.1 Definitions

An associative algebra R over a commutative ring with unity 1 is called a *PI-algebra* if there exists an admissible polynomial $f(x_1,\dots,x_n)$ without constant term in a free associative algebra $A(x_1,x_2,\dots,x_n,\dots)$ of countable rank such that $f(r_1,r_2,\dots,r_n)=0$ for any $r_1,r_2,\dots,r_n \in R$. Here a polynomial $f(x_1,\dots,x_n)$ is called *admissible* if at least one of its coefficients is equal to 1. Any PI-algebra satisfies a power of a standard identity (see Jacobson [151, 152]). Using this and linearization, we conclude that if k^* is an extension of the base ring k then a k-algebra is a PI-algebra if and only if this is true for R_{k^*} (see Section 1.1.7).

An algebra is called *primitive* if it possesses a faithful simple module. In this case R can be considered as an irreducible subalgebra in the endomorphism algebra of a k-module M.

Theorem 6.1 (Density theorem; Jacobson [151, Chapter 2]; Herstein [140, Chapter 2]). *Let R be an irreducible algebra of linear operators of a linear space V over a field k. Then the centralizer of R in $\operatorname{End}_k V$ is a skew field D which is an extension of k. If $\dim_D V = n < \infty$ then $R \cong D_n$. If $\dim_D V = \infty$ then for any n there is a subalgebra in R with a homomorphic image isomorphic to D_n.*

https://doi.org/10.1515/9783110566659-006

Theorem 6.2 (Herstein [140, Chapter 6]). *Let R be a primitive algebra over a field k satisfying an identity of degree d. Then the center P of R is a field and $\dim_P R \leq [d/2]^2$.*

Here $[a]$ denotes the integral part of a.

We define the *Jacobson radical* $J(R)$ of R as the intersection of annihilators of simple left R-modules. Equivalently, $J(R)$ is the intersections of kernels of all irreducible representations. Any one-sided nil-ideal of R (i. e., an ideal in which any element is nilpotent) is in $J(R)$.

Theorem 6.3 (Amitsur [6]). *The Jacobson radical $J(R)$ of a finitely generated PI-algebra R over a field k is a nil-ideal.*

Theorem 6.4 (Jacobson [151, Chapter 1]). *Let G be a Lie algebra over a field k and $U = U(G)$ its universal enveloping algebra. Then $J(U) = \{0\}$.*

Let R be an associative algebra over a field k. An algebra S with 1 is called a *classical (right) ring of quotients for* R if R is a subalgebra in S, each regular element of R is invertible in S and each element of S can be written in the form xy^{-1}, where $x, y \in R$ and y is regular (that is, y is not a zero divisor in R). If R has no zero divisors then S as above is called a *classical field of quotients*. We will further say that R is *prime* if the product IJ of any two nonzero two-sided ideals in R is itself nonzero.

Theorem 6.5 (Herstein [140, Chapter 6]). *Let R be a prime PI-algebra. Then its classical ring of quotients S exists and it is isomorphic to the matrix ring D_n over a skew field D which is finite-dimensional over its center. If R has no zero divisors then $n = 1$.*

An important result due to P. M. Cohn [91] is this: For any Lie algebra G, the universal enveloping algebra $U(G)$ can be isomorphically embedded in a division algebra. For finite-dimensional G, $U(G)$ has a classical ring of fractions.

6.1.2 The notion of the height

Let u be an associative monomial in $x_1, x_2, \ldots, x_n$ of the form

$$u = x_{i_1}^{s_1} x_{i_2}^{s_2} \cdots x_{i_h}^{s_h} \quad \text{where } i_r \neq i_{r+1},\ r = 1, \ldots, h-1.$$

Then h is called the height of u with respect to the given set of variables. If R is an associative algebra generated by a finite set $\{r_1, \ldots, r_n\}$, $A = A(x_1, \ldots, x_n)$ the free associative algebra with free generating set $\{x_1, \ldots, x_n\}$, $F = \{f_1, \ldots, f_l\}$ a finite set of homogeneous polynomials in A, and $v = v(x_1, \ldots, x_n)$, then the *height of the monomial* $\bar{v} = v(r_1, \ldots, r_n)$ *relative to*

$$\bar{F} = \{f_1(r_1, \ldots, r_n), \ldots, f_l(r_1, \ldots, r_n)\}$$

is defined as the least height of monomials $u_1, \ldots, u_t$ in all possible representations of the form

$$\bar{v} = \lambda_1 u_1(f_1, \ldots, f_l) + \cdots + \lambda_t u_t(f_1, \ldots, f_l).$$

If the heights of all monomials relative to $\bar{F}$ are uniformly bounded by a natural number h and h is the least possible with this property then we say that h is the *height of R relative to $\bar{F}$*.

Theorem 6.6 (Zhevlakov et al. [386, Chapter 5]). *Let R be an associative PI-algebra with a nontrivial admissible identity of degree d and a finite generating set $\{r_1, \dots, r_n\}$. Then R is an algebra of finite height relative to the set of words in $r_1, \dots, r_n$ whose lengths are less than d.*

An associative algebra R over k is called *algebraic* if each of its elements r is a root of a polynomial $f_r(t) \in k[t]$ with leading coefficient 1. If $f_r(t) = t^{s(r)}$, for all $r \in R$, then we say that R is a *nil-algebra*. In this latter case each element of R is called *nilpotent*.

Corollary 6.7.
(1) *Any finitely generated associative algebraic PI-algebra over a ring is finite-dimensional (see Section* 1.1.8*).*
(2) *Any finitely generated associative nil-algebra with nontrivial admissible identity over a commutative ring is nilpotent.*

6.1.3 Growth of varieties of associative algebras

Let $\mathbf{V}$ be a variety of associative algebras over a field k. Proceeding as in Section 4.8.5, we may define $P_n(\mathbf{V})$ as the quotient module of the module P_n of all associative multilinear polynomials in the set of fixed n variables over the submodule $P'_n(\mathbf{V})$ whose terms are polynomials f such that $f \equiv 0$ is an identity of $\mathbf{V}$. Let $c_n = \dim P_n(\mathbf{V})$. As it was mentioned in Section 4.8.5, the growth of a variety of Lie algebras may be superexponential. In the case of associative algebras, this is impossible. We give this comparatively new result by A. Regev [315] with an elegant proof due to Latyshev [202].

Lemma 6.8. *The number of permutations $s = (i_1, i_2, \dots, i_n)$ of $1, 2, \dots, n$, in which any selection of d elements, d being fixed, $2 \le d \le n$, contains at least one order, is bounded by $(d-1)^{2n}$.*

Proof. We consider $\mathbb{Z}_n = \{1, 2, \dots, n\}$ and fix a permutation s of the form as in the statement of Lemma 6.8. We introduce a partial order on $\mathbb{Z}_n$ by putting $i < j$ if (i, j) is an order in s. Then, in $\mathbb{Z}_n$, any ordered d-element set contains at least one pair of comparable elements. In this case $\mathbb{Z}_n$ splits as the disjoint union of at most $d-1$ chains (see Hall [138, Lemma 7.2.1]). Now any permutation as in the lemma is recovered as follows. Take a partition of $\mathbb{Z}_n$ as the disjoint union of $d-1$ chains and form a permutation, using these chains while keeping the order within any such chain. Clearly, the number of partitions into $d-1$ chains is bounded by $(d-1)^n$ (this is the number of maps from $\mathbb{Z}_n$ into $\mathbb{Z}_{d-1}$). Now on each of n places of s we find an element of one of the $d-1$

chains. Thus the number of ways of making a permutation from the given union is at most $(d-1)^{2n}$. □

Theorem 6.9. *For any variety* $\mathbf{V}$ *of associative algebras over a field k there exists a constant d (the degree of any nontrivial identity holding in* $\mathbf{V}$*) such that*

$$c_n = \dim P_n(\mathbf{V}) \le (d-1)^{2n}.$$

Proof. Applying linearization for associative algebras (cf. Section 4.2, Theorem 4.5), we easily derive that $\mathbf{V}$ satisfies an identity of the form

$$y_1 y_2 \cdots y_d = \sum_{\sigma \ne 1} \lambda_\sigma y_{\sigma(1)} y_{\sigma(2)} \cdots y_{\sigma(d)}, \quad \lambda_\sigma \in k. \tag{6.1}$$

Our proof will be complete as soon as we manage to prove that, modulo $P_n'(\mathbf{V})$, the space P_n is spanned by the monomials of the form $x_{\sigma(1)} x_{\sigma(2)} \cdots x_{\sigma(n)}$ in which the permutation σ satisfies the conditions of Lemma 6.8 (with $s = \sigma$).

Suppose not. We choose the lexicographically least monomial u which cannot be expressed as conjectured. Then u contains a subword of the form

$$u' = (x_{\alpha_1} \cdots x_{\beta_1})(x_{\alpha_2} \cdots x_{\beta_2}) \cdots (x_{\alpha_{d-1}} \cdots x_{\beta_{d-1}}) x_{\alpha_d}$$

where $\alpha_1 > \alpha_2 > \cdots > \alpha_d$. Substituting into equation (6.1) $y_i = x_{\alpha_i} \cdots x_{\beta_i}$, $i = 1, \ldots, d-1$, $y_d = x_{\alpha_d}$, we will immediately see that, modulo $P_n'(\mathbf{V})$, our monomial u can be expressed in the form

$$u = \sum_{\sigma \ne 1} \lambda_\sigma v_\sigma, \quad \lambda_\sigma \in k,$$

where $v_\sigma \in P_n$, $\sigma \ne 1$, is a monomial which is lexicographically lower than u. This is a contradiction. □

6.1.4 Tensor product of k-algebras

As usual, we define the *tensor product* $T = A \otimes_k B$ of two algebras A and B over a field k as the tensor product of corresponding vector spaces with the multiplication correctly defined by $(r \otimes s)(r' \otimes s') = rr' \otimes ss'$. Here $r, r' \in A$, $s, s' \in B$. If C is any associative k-algebra generated by the subalgebras A', B', $A \cong A'$, $B \cong B'$, and if the elements of A' commute with the elements of B' then the isomorphisms of A onto A' and B onto B' extend to a homomorphism from T onto C.

Theorem 6.10. *Let A and B be PI-algebras over a field k. Then* $T = A \otimes_k B$ *is also a PI-algebra.*

Proof. Let V be a verbal ideal of the free associative algebra $A_\infty(X)$ of countable rank such that $V(A) = V(B) = \{0\}$. Suppose V contains a nontrivial polynomial of some degree $d \geq 2$. Let $m_1(x), \dots, m_q(x)$ be monomials forming a basis of $P_n(X)$ modulo V, $q \leq (d-1)^{2n}$. For any permutation $\sigma \in S_n$, we can find $\lambda_1(\sigma), \dots, \lambda_q(\sigma) \in k$ such that

$$x_{\sigma(1)} \cdots x_{\sigma(n)} = \sum_{i=1}^{q} \lambda_i(\sigma) m_i(x)$$

is an identity in both A and B. We will construct a polynomial

$$f(x_1, \dots, x_n) = \sum_{\sigma \in S_n} \gamma_\sigma x_{\sigma(1)} \cdots x_{\sigma(n)},$$

where $\gamma_\sigma \in k$, such that $f \equiv 0$ is an identity in $A \otimes_k B$. We write

$$\begin{aligned} f(a_1 \otimes b, \dots, a_n \otimes b_n) &= \sum_{\sigma \in S_n} \gamma_\sigma a_{\sigma(1)} \cdots a_{\sigma(n)} \otimes b_{\sigma(1)} \cdots b_{\sigma(n)} \\ &= \sum_{i=1}^{n} \sum_{j=1}^{n} \sum_{\sigma \in S_n} \lambda_i(\sigma) \lambda_j(\sigma) \gamma_\sigma m_i(a) \otimes m_j(b). \end{aligned}$$

For sufficiently large values of n, we have $q^2 \leq (d-1)^{2n} \leq n!$, so that the system of linear equations of the form

$$\sum_{\sigma \in S_n} \lambda_i(\sigma) \lambda_j(\sigma) \gamma_\sigma = 0 \quad (1 \leq i, j \leq q)$$

has a nontrivial solution $\gamma_\sigma = \gamma_\sigma^0$ and then the expression of the form

$$f(x_1, \dots, x_n) = \sum_{\sigma \in S_n} \gamma_\sigma^0 x_{\sigma(1)} \cdots x_{\sigma(n)} \equiv 0$$

is an identity in $A \otimes_k B$. □

6.2 Some further results on Lie algebras

6.2.1 Definitions

A Lie algebra L over a field k is called *semisimple* if L has no nonzero abelian ideals. From the definition of the solvable radical $S(L)$ of a Lie algebra L (see Section 1.9.2), it follows that the quotient algebra $L/S(L)$ of an arbitrary Lie algebra over its solvable radical is semisimple.

Theorem 6.11. *Let L be a finite-dimensional Lie algebra over a field k of characteristic zero and $S = S(L)$ its solvable radical. Then there exists a semisimple subalgebra G such that $L = G \oplus S$. Furthermore, any semisimple algebra G is representable in the form*

$$G = G_1 \oplus G_2 \oplus \cdots \oplus G_t,$$

where $G_1, G_2, \ldots, G_t$ are ideals which are simple Lie algebras. The ideal $N = [G, S]$ is nilpotent.

Any subalgebra G as in the first claim of the above theorem is called a *Levi subalgebra* of L and the claim itself is called *Levi's theorem* (Bourbaki [81, Section 6]).

6.2.2 Structure of a finite-dimensional semisimple Lie algebra over an algebraically closed field of characteristic zero

Let L be a Lie algebra and H its arbitrary subalgebra. We consider L as an adjoint H-module. For any linear function $\alpha \in H^*$, we set

$$L_\alpha = \{x \mid [h, x] = \alpha(h)(x), \forall h \in H\}.$$

Theorem 6.12. *Let L be a semisimple Lie algebra over an algebraically closed field of characteristic zero. Then there are an abelian subalgebra H in L and a finite subset $R \subseteq H^*$, $0 \notin R$, such that $H = L_0$ and*

$$L = H \oplus \sum_{\alpha \in R} L_\alpha.$$

Moreover, $\dim L_\alpha = 1$ *for any $\alpha \in R$, and $\alpha \in R$ implies $-\alpha \in R$.*

The subalgebra H from the statement of this theorem is called the *Cartan subalgebra*. The set R is called the *root system* of L with respect to H. Given an $\alpha \in R$, the subspace L_α is called the *root subspace* corresponding to the root α. For any $\alpha, \beta \in R$ the product $[L_\alpha, L_\beta]$ is either zero or it is equal to $L_{\alpha+\beta}$ if $\alpha + \beta \in R$. For any $\alpha \in R$, the subspace $L_\alpha \oplus [L_\alpha, L_{-\alpha}] \oplus L_{-\alpha}$ is a subalgebra of L isomorphic to $\mathfrak{sl}(2, k)$. Choosing a basis $\{e_\alpha\}$ in each L_α and setting $[e_\alpha, e_{-\alpha}] = h_\alpha$, we will have $\alpha(h_\alpha) = 2$. Any subset $\beta = \{\alpha_1, \ldots, \alpha_l\} \subset R$ such that B is a basis in H^* and each element $\alpha \in R$ is representable in the form $\alpha = \lambda_1\alpha_1 + \cdots + \lambda_l\alpha_l$ where all coefficients $\lambda_1, \ldots, \lambda_l$ are simultaneously nonnegative or simultaneously nonpositive integers is called a *basis of the root system R*. Those roots for which $\lambda_1, \ldots, \lambda_l \geq 0$ are called *positive*, all the remaining roots are called *negative*. The set of positive roots is denoted by R_+ and the set of negative roots is denoted by R_-. We define a *triangular decomposition* of any semisimple algebra as that of the form

$$L = N_- \oplus H \oplus N_+,$$

where H is a Cartan subalgebra and

$$N_+ = \sum_{\alpha \in R_+} L_\alpha, \quad N_- = \sum_{\alpha \in R_-} L_\alpha.$$

It is clear that N_+ and N_- are nilpotent subalgebras and the subalgebras $H \oplus N_+$ and $N_- \oplus H$ are solvable (Bourbaki [82]).

6.2.3 Finite-dimensional representations

As we easily conclude from Section 2.5, PBW theorem, any Lie algebra L over a field is isomorphic to a subalgebra of a Lie algebra of the form $[A]$ where A is an associative algebra. In the case of a finite-dimensional algebra over a field, A can be chosen finite-dimensional. An exact form of this reads as follows.

Theorem 6.13 (Ado–Iwasawa). *Any finite-dimensional Lie algebra over a field k has a faithful finite-dimensional representation.*

It is clear that if V is the underlying space of the representation in the statement of the theorem then A can be chosen as the algebra $\operatorname{End}_k V$ of all linear operators of V over k.

We give a proof of this theorem in the case where the field k is of positive characteristic p. The way we prove it will be useful in our further argument. We start with an important auxiliary result.

Lemma 6.14. *Let k be a field, $\operatorname{char} k = p > 0$, and let G be a Lie algebra such that $\dim G = n$. Then the universal enveloping algebra $U(G)$ for G is a finitely generated free module over an appropriate central subalgebra which is isomorphic to the polynomial ring in n variables.*

Proof. We choose a basis $\{e_1, e_2, \dots, e_n\}$ in G. By Section 2.5, PBW theorem, a basis in $U(G)$ can be chosen in the form

$$e_1^{l_1} e_2^{l_2} \cdots e_n^{l_n}, \quad l_i = 0, 1, 2, \dots, \; i = 1, 2, \dots, n. \tag{6.2}$$

We consider the operators $\varphi_i = \operatorname{ad} e_i \in E = \operatorname{End}_k G$. Since $\dim E = n^2$, for an arbitrary $r \geq n^2$, the operators $\varphi_i, \varphi_i^p, \dots, \varphi_i^{p^r}$ are linearly dependent, that is, there exist not all zero $\lambda_0, \lambda_1, \dots, \lambda_r$ such that

$$\lambda_0 \varphi_i + \lambda_1 \varphi_i^p + \cdots + \lambda_r \varphi_i^{p_r} = 0.$$

We consider a polynomial $f = f(t) = \lambda_0 t + \lambda_1 t^p + \cdots + \lambda_r t^{p^r}$. Then $f_i(\operatorname{ad} e_i) = 0$, $i = 1, 2, \dots, n$. According to Section 1.11.1, we have $(\operatorname{ad}_U e_i)^p = \operatorname{ad}_u e_i^p$. Hence

$$f_i(\operatorname{ad} e_i) = \sum_{j=0}^{r} \lambda_j (\operatorname{ad} e_i)^{p^j} = \sum_{j=0}^{r} \lambda_j \operatorname{ad} e_i^{p_j} = \operatorname{ad} f_i(e_i).$$

If $x \in G$ then $[f_i(e_i), x] = \operatorname{ad} f_i(e_i)(x) = f_i(\operatorname{ad} e_i)(x) = 0$. Since G generates $U(G)$, we conclude that $f_i(e_i)$ are in the center of $U(G)$. We put $y_i = f_i(e_i)$, $i = 1, 2, \dots, n$. If d_i is the degree of f_i then a basis of $U(G)$ can be formed by the monomials

$$e_1^{l_1} e_2^{l_2} \cdots e_n^{l_n} y_1^{s_1} y_2^{s_2} \cdots y_n^{s_n}, \quad 0 \leq l_i < d_i, s_i \geq 0, \; i = 1, 2, \dots, n. \tag{6.3}$$

Indeed, it is easy to understand that the elements of the form (6.3) generate $U(G)$. In fact, $e_i^{d_i}$ is a linear combination of y_i and of elements of the form e_i^t, $t < d_i$. Let $V_{q_1q_2\dots q_n}$ denote the subspace in $U(G)$ spanned by the monomials of the form (6.3) in which $s_1 \le q_1, \dots, s_n \le q_n$. Expressing each y_i as a polynomial in e_i shows that $V_{q_1q_2\dots q_n}$ is generated over k by the monomials of the form (6.2) such that $l_i \le (q_i + 1)d_i - 1$, $i = 1, 2, \dots, 2n$. It is easy to compute that the number of the monomials of both types generating $V_{q_1q_2\dots q_n}$ is the same. Thus, the system of monomials of the type (6.3) with restrictions $s_1 \le q_1, \dots, s_n \le q_n$ is linearly independent for any choice of $(q_1, \dots, q_n)$. Since any finite system of elements of the form (6.3) is in a suitable $V_{q_1q_2\dots q_n}$, the desired linear independence follows.

In particular, a subsystem of system (6.3) consisting of polynomials with $l_1 = l_2 = \dots = l_n = 0$ is a basis of a central subalgebra R which is equal to the polynomial ring $k[y_1, \dots, y_n]$. A basis of U as a right (left) R-module is formed by the monomials of the form (6.3) with $0 = s_1 = \dots = s_n$. Since the number of these monomials is equal to $d_1 d_2 \cdots d_n$, U is a free R-module of finite rank $d_1 d_2 \cdots d_n$. □

6.2.4 Proof of Ado–Iwasawa's theorem in the case where *k* is a field of characteristic *p* > 0

We adopt the notation from Lemma 6.14. Let I denote the subspace in $U(G)$ spanned by the monomials (6.3) in which, for some $i = 1, 2, \dots, n$, we have $s_i \ne 0$. It is obvious that I is a two-sided ideal in $U(G)$. Besides, $U(G) = E \oplus I$, E being a subspace of $U(G)$ spanned by the monomials of the form (6.3) in which $s_1 = s_2 = \dots = s_n = 0$. The choice of the polynomials f_i being to a considerable degree optional, we may think that $d_i > 1$ for all $i = 1, \dots, n$. In this case $G \subseteq E$ and, under the composition of the sequence of maps

$$G \to U(G) \to U(G)/I = A,$$

G can be isomorphically embedded in an associative algebra A such that $\dim A \le d_1 d_2 \cdots d_n < \infty$. Therefore G is isomorphically embedded in a Lie algebra of the form $[A]$ where A is a finite-dimensional associative algebra with unity. Now the desired representation is the left regular representation of A (see the final paragraph of Section 1.6.2).

6.2.5 Simple modules in the zero characteristic case

Theorem 6.15 (Bourbaki [81, Chapter 1]). *Let G be a semisimple Lie algebra over a field k of characteristic 0. Then any finite-dimensional G-module V is decomposable as the direct sum of simple G-modules.*

If k is an algebraically closed field and G a semisimple Lie algebra with a Cartan subalgebra H, a basis B of the root system R associated with H, then any finite-dimensional simple G-module is completely determined by a linear function $\omega \in H^*$ such that $\omega(h_\alpha)$ is a nonnegative integer for all $\alpha \in B$. If $p = (1/2)\sum_{\alpha\in R}\alpha$ then the dimension of such a module is given by the formula (Bourbaki [82, Résumé])

$$\dim M = \prod_{\alpha\in R_+}\left(1 + \frac{\omega(h_\alpha)}{\rho(h_\alpha)}\right).$$

It is clear that the number of simple modules of any fixed dimension is finite. Now ω is a weight of M, called the *highest weight* and if v is a nonzero vector of weight ω in M then M is a linear span of the set of vectors of the form

$$e^{p_1}_{-\alpha_1}\cdots e^{p_s}_{-\alpha_s}v, \quad \alpha_1,\dots,\alpha \in R_+, p_1,\dots,p_s \geq 0.$$

An immediate consequence of this is that, for any natural number l, the number of simple G-modules M with $e^l_\alpha x = 0$ for all $\alpha \in R$ and $x \in M$ is finite.

6.3 General results on special Lie algebras

6.3.1 Definition and examples

A Lie algebra L over a finite-dimensional commutative ring k with 1 is called a *special Lie algebra* if it is isomorphic to a Lie subalgebra of a Lie algebra of the form $[A]$ where A is an associative PI-algebra over k.

It is obvious that an abelian Lie algebra over an arbitrary ring is special. The following example (Shirshov [335]) shows the existence of finite-dimensional nilpotent Lie algebras over a suitable k which cannot be embedded in any associative algebras (not mentioning PI-algebras). So let k be the quotient ring of the polynomial ring $\mathbb{Z}_2[t_1, t_2, t_3]$ in three variables t_1, t_2, t_3 over the two-element field $\mathbb{Z}_2$ modulo the ideal generated by t_1^2, t_2^2, t_3^2. Let L be a free nilpotent Lie algebra of nilpotent step 2 with a free generating set x_1, x_2, x_3. Also let $\lambda_1, \lambda_2, \lambda_3$ designate the images of t_1, t_2, t_3 in k. We denote by M the ideal of L generated by $u = \lambda_1 x_2 + \lambda_2 x_2 + \lambda_3 x_3$. This ideal is the k-span of the elements u, $x_1 u = \lambda_2 x_1 x_2 + \lambda_3 x_1 x_3$, $x_2 u = \lambda_1 x_2 x_1 + \lambda_3 x_2 x_3$, $x_3 u = \lambda_1 x_3 x_1 + \lambda_2 x_3 x_2$. From this it is obvious that $v = \lambda_1\lambda_2 x_1 x_2 + \lambda_1\lambda_3 x_1 x_3 + \lambda_2\lambda_3 x_2 x_3$ is not an element of M. Now let φ be an arbitrary map from L into an associative k-algebra A such that $\varphi(ab) = [\varphi(a), \varphi(b)]$. Then

$$\begin{aligned}(\varphi(u))^2 &= \lambda_1\lambda_2(\varphi(x_1)\varphi(x_2) + \varphi(x_2)\varphi(x_1)) + \lambda_1\lambda_3(\varphi(x_1)\varphi(x_3) + \varphi(x_3)\varphi(x_1)) \\ &\quad + \lambda_2\lambda_3(\varphi(x_2)\varphi(x_3) + \varphi(x_3)\varphi(x_2)) \\ &= \lambda_1\lambda_2[\varphi(x_1), \varphi(x_2)] + \lambda_1\lambda_3[\varphi(x_1), \varphi(x_3)] + \lambda_2\lambda_3[\varphi(x_2), \varphi(x_3)] \\ &= \varphi(v).\end{aligned}$$

This shows that under any homomorphism of L/M into a Lie algebra of the form $[A]$, A being an associative algebra, the image of a nonzero element $v + M$ is necessarily zero.

Still, in the case where k is a field, any finite-dimensional Lie algebra is special. Indeed, by Ado–Iwasawa's theorem (see Section 6.2.3), such an L possesses an exact finite-dimensional representation, i. e., it is embeddable into a Lie algebra of the form $[\mathrm{End}_k V]$, where $\dim V < \infty$. Since any finite-dimensional associative algebra satisfies a standard identity of a suitable degree, we conclude that L is special. We will see below that a nilpotent Lie algebra L over a ring k for which the canonical map $\varepsilon : L \to U(L)$ is injective is necessarily special (corollary of Theorem 5.30).

6.3.2 Heredity

The final remark in Section 6.3.1 provides an easy example of a special Lie algebra whose homomorphic image is not special. For this, we choose as a base ring k' any commutative ring whose homomorphic image is equal to k from this example. Now we consider a free 2-step nilpotent Lie algebra N over k' of sufficiently large rank (so that L is a homomorphic image of N). By the PBW theorem, Section 2.5, N, being a free k'-module, is embeddable in $U(N)$. Applying the corollary of Theorem 6.20, N is a special Lie algebra. However, using the argument in the above section, its homomorphic image L/M is not a special Lie algebra. As an example of k', one may take the integral factorial ring $\mathbb{Z}[t_1, t_2, t_3]$ of integral polynomials in three variables. We remark that in the case where k is a field the question on the homomorphic image of a special Lie algebra is still open.

The heredity question for Cartesian products and subalgebras is being solved in what follows.

Proposition 6.16. *Let G be a special Lie algebra over a ring k and $\mathbf{Q}$ the quasivariety generated by G (see Section 4.9). Then any algebra L in $\mathbf{Q}$ is a special Lie algebra. In particular, any free algebra of $\mathbf{V} = \mathrm{var}(G)$ is special.*

Proof. By Section 4.8.2, any quasivariety is a nonempty class which is closed under passing to subalgebras and filtered products. In our case it is sufficient to consider subalgebras of filtered powers of G. But if G is a subalgebra of an associative PI-algebra A then L is a subalgebra of a filtered power of A with respect to the same filter, that is, by Birkhoff's theorem on varieties, it is a subalgebra of an associative PI-algebra. □

6.3.3 Adjoint associative algebra

Let G be a Lie algebra over a ring k. The associative subalgebra $\mathrm{Ad}(G)$ in the algebra $\mathrm{End}_k(G)$ of all endomorphisms of the k-module G generated by all endomorphisms of the form $\mathrm{ad}\, g$, $g \in G$, is called the *adjoint associative algebra* of the Lie algebra G.

Proposition 6.17. *Let G be a Lie algebra such that* $\mathrm{Ad}(G)$ *is a PI-algebra and* $H \in \mathrm{var}(G)$. *Then* $\mathrm{Ad}(H)$ *is a PI-algebra, too.*

Proof. We write H in the form $H = L/M$ where L is a free algebra of $\mathbf{V} = \mathrm{var}(G)$ and M is an ideal in L. We know that L is isomorphic to a subalgebra in a Cartesian power G^I, I being a set of indices (Section 4.1.2). Let $a \in \mathrm{Ad}(L)$. Then

$$a = \sum_{i_1,\dots,i_t} \mathrm{ad}\, x_{i_1} \cdots \mathrm{ad}\, x_{i_t},$$

where $x_{i_j} \in L$. Considering the values of this expression on indices, we write, for any $\alpha \in I$,

$$a(\alpha) = \sum \mathrm{ad}\, x_{i_1}(\alpha) \cdots \mathrm{ad}\, x_{i_t}(\alpha) \in \mathrm{Ad}(G).$$

Now for any polynomial $f(y_1, \dots, y_p)$, we obtain

$$[f(a_1, \dots, a_p)(y)](\alpha) = f(a_1(\alpha), \dots, a_p(\alpha))(y(\alpha)).$$

If $f(y_1, \dots, y_p) \equiv 0$ is an identity in $\mathrm{Ad}(G)$ then the latter equation shows that this is also an identity of $\mathrm{Ad}(L)$. Now $\mathrm{Ad}(H)$ being a homomorphic image of $\mathrm{Ad}(L)$ proves that $\mathrm{Ad}(H)$ is a PI-algebra, too. □

Theorem 6.18. *Let G be a special Lie algebra over an arbitrary field k. Then its adjoint associative algebra* $\mathrm{Ad}(G)$ *is a PI-algebra.*

Proof. Let A be an associative PI-algebra (with unity) such that $G \subseteq [A]$. We consider subalgebra A_r and A_l in $\mathrm{End}\, A$ generated respectively by all the operators of the form r_a and l_a such that $r_a(x) = xa$ and $l_a(x) = ax$, $x \in A$. Then A is isomorphic to A_r and anti-isomorphic to A_l. The subalgebra $\mathrm{Ad}([A])$ is a subalgebra in A_rA_l generated by all operators of the form $\mathrm{ad}\, a = l_a - r_a$. Since the elements of the subalgebras A_r and A_l are pairwise commutative, we may conclude, using the universal property of the tensor product, that A_rA_l is a homomorphic image of $A_r \otimes A_l$ (Section 6.1.4). Applying another result in Section 6.1.4, we conclude that A_rA_l is a homomorphic image of a PI-algebra. Now we consider a subalgebra S in A_rA_l generated by operators of the form $\mathrm{ad}_A\, x$, $x \in G$. This is an associative PI-algebra and G is invariant with respect to S. Restricting the operators of S to G, we derive that $\mathrm{Ad}(G)$, a homomorphic image of S, is a PI-algebra. □

Corollary 6.19.
(1) *Any algebra with trivial center in a variety generated by a special Lie algebra is itself special.*
(2) *Let G be a special Lie algebra. Then G satisfies a nontrivial identity as a Lie algebra.*

Proof. (1) Let G be as in this claim. Then $\mathrm{Ad}(G)$ is a PI-algebra. Since $\mathrm{ad} : G \to \mathrm{Ad}(G)$ is a homomorphism whose kernel is equal to the center of G, we obtain that $G = G/Z(G)$ is a special Lie algebra.

(2) By Proposition 6.17, if G does not satisfy a nontrivial identity, we may conclude that a free Lie algebra in arbitrarily many variables is special. But $\mathrm{Ad}(L(x_1, \ldots, x_n, \ldots))$ is not a PI-algebra: indeed, if $f(y_1, \ldots, y_p) \equiv 0$ is an identity in $\mathrm{Ad}(L(x_1, \ldots, x_n, \ldots))$ then $f(\mathrm{ad}\, x_1, \ldots, \mathrm{ad}\, x_p)(x_{p+1}) = 0$. But according to Section 4.8.1, the monomials of the form $x_{i_1} \cdots x_{i_t} x_{p+1}$, $x_{i_j} \neq x_{p+1}$ are linearly independent in L, proving $f = 0$. □

Our next aim is to prove a sufficient condition for a Lie algebra to be special. For simplicity, we impose a restriction on the ring of coefficients although the proof works in a more general situation.

Theorem 6.20. *Let L be a Lie algebra which is a free module over the base ring k which is commutative integral. Suppose that N is a nilpotent ideal in L and the quotient algebra $G = L/N$ is such that $U(G)$ is a PI-algebra. Then L is a special Lie algebra.*

Proof. Passing to the field of quotients for k enables us to restrict ourselves to the case where k is a field. So we put $N^0 = L$ and choose a basis E for L as follows. Let c be such that $N^c \neq \{0\}$ and $N^{c+1} = \{0\}$. We choose a basis E_c for N^c and let E_i, $i = 0, 1, \ldots, c-1$, be such that $E_i \cup E_{i+1} \cup \cdots \cup E_c$ is a basis for N^i. We define the weight $\mathrm{wt}(x)$ of an element $x \in E$ by setting $\mathrm{wt}(x) = i$, if $x \in E_i$. Now let $\preceq$ be a well order on E such that $\mathrm{wt}(x) \leq \mathrm{wt}(y)$ implies $x \preceq y$. By the PBW theorem (Section 2.5), a basis of $U(L)$ is formed by 1 and the set of all monomials of the form

$$e_1 e_2 \cdots e_n, \quad e_1 \preceq e_2 \preceq \cdots \preceq e_n, \ n \geq 1. \tag{6.4}$$

If m is an element of the form (6.4) then we put $\mathrm{wt}(m) = \mathrm{wt}(e_1) + \cdots + \mathrm{wt}(e_n)$. If $u = \sum_s \alpha_s m_s$, where $0 \neq \alpha_s \in k$ and m_s is of the form (6.4) then we put $\mathrm{wt}(u) = \min_s\{\mathrm{wt}(m_1)\}$. We put, for convenience, $\mathrm{wt}(0) = +\infty$ and $\mathrm{wt}(1) = 0$. Since in L we have $[N^i, N^j] \subseteq N^{i+j}$ for all $i, j = 0, 1, \ldots, c$, we may write $\mathrm{wt}([x, y]) \geq \mathrm{wt}(x) + \mathrm{wt}(y)$ where $x, y \in L$. This immediately give for all, possibly unordered, $f_1, \ldots, f_q \in E$ the inequality

$$\mathrm{wt}(f_1 \cdots f_q) \geq \mathrm{wt}(f_1) + \cdots + \mathrm{wt}(f_q)$$

hence, for all $u, v \in U(L)$, we have

$$\mathrm{wt}(uv) \geq \mathrm{wt}(u) + \mathrm{wt}(v). \tag{6.5}$$

Hence for each $s = 0, 1, 2, \ldots$, the set U_s of all elements u in $U(L)$ such that $\mathrm{wt}(u) \geq s$ is an ideal in $U(L)$. Moreover, $U_i^s \subseteq U_i$ and $U_s \cap L = \{0\}$ for all $s \geq c+1$. Indeed, the former relation is a consequence of equation (6.5). As for the latter, this follows since the weights of nonzero elements of L are bounded by c. To complete the proof, we notice that from $L \cap U_{c+1} = \{0\}$ it follows that L is isomorphically embedded in $A = U/U_{c+1}$. Now A is a PI-algebra since it possesses a nilpotent ideal $U_1 + U_{c+1}/U_{c+1}$ which is at the same time the kernel of the natural homomorphism from $U(L)$ onto $U(G) = U(G) = U(L/N)$. Therefore, if $w \equiv 0$ is a nontrivial identity holding in $U(G)$ then $w^{c+1} \equiv 0$ is a nontrivial identity holding in A, proving that L is indeed a special Lie algebra. □

In the case $G = L/N$ is abelian $U(G)$ is commutative. Hence we have a corollary.

Corollary 6.21. *Let L be a Lie algebra with nilpotent derived subalgebra, in particular, a nilpotent Lie algebra, which is a free module over the base ring k. If k is an integral ring then L is a special Lie algebra.*

A proposition we are going to prove is important in studying extensions of special Lie algebras.

Proposition 6.22. *Let L be a Lie algebra over a ring k with a series of ideals of the form*

$$L = L_1 \rhd L_2 \rhd \cdots \rhd L_{n+1} = \{0\}$$

and consider $A = \mathrm{Ad}(L)$, with A_i being the associative algebra of endomorphisms of L_i/L_{i+1}, $i = 1, \dots, n$, induced by A. Then A is a PI-algebra if and only if, for each $i = 1, \dots, n$, A_i is a PI-algebra.

Proof. Passing from A to A_i is obvious since the latter is a homomorphic image of the former. The converse follows since the extension of PI-algebras is a PI-algebra. □

6.3.4 Examples

(1) Let L be a centerless algebra with an ideal M such that L/M is a special Lie algebra and $[L^2, M] = \{0\}$. Then L is a special Lie algebra. Indeed, in this case we consider the series of ideals of the form $L \rhd M \rhd \{0\}$. We have $(\mathrm{Ad}(L))_{L/M} \cong \mathrm{Ad}(L/M)$ which is special by Theorem 6.18 and $(\mathrm{Ad}(L))_M$ is a commutative algebra (for the details see Section 6.5.2). By Proposition 6.22, then $\mathrm{Ad}(L)$ is a PI-algebra and $L \subseteq \mathrm{Ad}(L)$.

(2) Let G be a finite-dimensional nonabelian Lie algebra and $\rho : G \to \mathrm{End}(M)$ its infinite-dimensional absolutely irreducible representation. Then the corresponding semidirect product $L = G \rtimes M$ is not a special Lie algebra. Indeed, since the extension of the base field leaves a special Lie algebra special and since M is absolutely irreducible, we may think that the base field is algebraically closed. By Theorem 6.37, any G-endomorphism of M is scalar. Now $(\mathrm{Ad}(L))_M$ is primitive with module M and its centralizer k. By the density theorem (Section 6.1), Theorem 6.2, and our previous argument, the dimension of M over k must be finite, which is not the case.

(3) As an illustration of the general rule of the preceding example, we exhibit a Lie algebra $L \in \mathbf{AN}_2$ which is not special. Consider $G = \langle x, y, z \mid xy = z, xz = yz = 0 \rangle$, $M = k[t]$. We give M a G-module structure by setting

$$x \circ f(t) = f'(t), \quad y \circ f(t) = tf(t), \quad z \circ f(t) = f(t).$$

If $\operatorname{char} k = 0$, then M is an absolutely irreducible G-module and $L = G \rtimes M$ is a Lie algebra as desired.

(4) For the future, we need an example of irreducible G-module for G which is two-dimensional nonabelian, $G = \langle x, y \mid xy = y \rangle$. Then again we choose $M = k[t]$, but in this case we set $x \circ f(t) = tf(t)$, $y \circ f(t) = f(t-1)$. As before $L = G \rtimes M$ is a nonspecial Lie algebra (this time $L \in \mathbf{A}^3$).

(5) Let G be a Lie algebra and $A = \mathrm{Ad}(G)$ its adjoint algebra. The centralizer $C(G)$ of A in $\mathfrak{gl}(G)$ is called the *centroid* of G. It is clear that if G is a simple Lie algebra then, by Schur's lemma, $C(G)$ is a field containing k (which is then a field). From what we have proved above and the density theorem (Section 6.1), it easily follows that *any special simple Lie algebra is finite-dimensional over its centroid.*

6.4 Special Lie algebras over zero characteristic fields

6.4.1 Ado's theorem for special Lie algebras

Classical Ado's theorem (Section 6.2) saying that any finite-dimensional Lie algebra over a field of characteristic zero has a faithful finite-dimensional representation can also be put in the following way. Each finite-dimensional Lie algebra L is isomorphic to a subalgebra of a Lie algebra of the form $[A]$ where A is a finite-dimensional associative algebra. This theorem, however, is trivial in the case of Lie algebras with trivial center: it suffices to consider the canonical homomorphism $\mathrm{ad} : L \to \mathrm{End}\, L$. Passing to the general case in characteristic zero reduces, with the help of Theorem 6.11 and Lie's theorem (Section 1.8), to considering Lie algebras of the form $L = G \oplus M$, where G is a semisimple subalgebra (for which the theorem is true as above) and M is a solvable ideal such that $N = [L, M]$ is nilpotent. A direct generalization of this is as follows.

Theorem 6.23. *Let L be a Lie algebra over a field, $L = G \oplus M$, where G is a special Lie algebra and $N = [L, M]$ is a nilpotent ideal. Let V denote the annihilator of the adjoint G-module N in $U(G)$. If $U(G)/V$ is a PI-algebra then L is a special Lie algebra.*

For the proof we will need two lemmas. The proof of the first of them is left to the reader: this is an immediate verification of definitions.

Lemma 6.24. *Let L be a Lie algebra of the form $L = G \oplus M$ where G is a subalgebra and M is an ideal. Let he maps $\pi : M \to [\mathrm{End}_k\, U(M)]$ and $\varphi : G \to [\mathrm{End}_k\, U(M)]$ be given by $\pi(m)(u) = mu$ and $\varphi(g)(m) = [g, m]$. Then the map $\psi : L \to [\mathrm{End}_k\, U(M)]$ such that $\psi(g + m) = \varphi(g) + \pi(m)$ is a homomorphism of Lie algebras.*

Lemma 6.25. *Let G be a Lie algebra and N be a submodule of a G-module M such that $GM \subseteq N$. Let T_c denote the subspace of the tensor algebra $T = T(M) = \bigoplus_{t=0}^{\infty} M^{\otimes t}$, where $M^{\otimes 0} = k \cdot 1$, generated by all tensors of the form $m_1 \otimes m_2 \otimes \cdots \otimes m_t$, where $m_i \in M$ and where at least $c + 1$ of the elements m_i are in N. We assume that $U(G)/\mathrm{Ann}\, N$ is a PI-algebra. Then also $U(G)/\mathrm{Ann}(T/T_c)$ is a PI-algebra.*

Proof. We denote $U(G)$ by Q. By definition, the G-module structure on

$$M^{\otimes t} = \underbrace{M \otimes \cdots \otimes M}_{t}$$

is a composition of the natural $U(G)^{\otimes t}$-module structure on $M^{\otimes t}$, that is, for any $u_1, \dots, u_t \in U(G)$ and $m_1, \dots, m_t \in M$, we have

$$(u_1 \otimes \cdots \otimes u_t)(m_1 \otimes \cdots \otimes m_t) = u_1 m_1 \otimes \cdots \otimes u_t m_t, \tag{6.6}$$

and the diagonal map $\Delta_t : G \to U(G)^{\otimes t}$ given by

$$\Delta_t(g) = g \otimes 1 \otimes \cdots \otimes 1 + 1 \otimes g \otimes \cdots \otimes 1 + 1 \otimes 1 \otimes \cdots \otimes g.$$

Hence, it is sufficient to prove that there exists a polynomial identity satisfied by the homomorphic image of each $U(G)^{\otimes t}$ in $\mathrm{End}_k(M^{\otimes t}/T_c \cap M^{\otimes t})$. Now since $Q/\mathrm{Ann}_Q(N)$ is a PI-algebra, we easily conclude that $Q/\mathrm{Ann}_Q(M)$ is a PI-algebra. It is clear that $Q^{\otimes t}/\mathrm{Ann}_{Q^{\otimes t}}(M^{\otimes t})$ is a homomorphic image of $Q/\mathrm{Ann}_Q(M)^{\otimes t}$. Indeed, if $(\bar{q}_1, \dots, \bar{q}_t) \in (Q/\mathrm{Ann}_Q(M), \dots, Q/\mathrm{Ann}_Q(M))$ then we can define a map

$$(\bar{q}_1, \dots, \bar{q}_t) \to q_1 \otimes \cdots \otimes q_t + \mathrm{Ann}_{Q^{\otimes t}}(M^{\otimes t}). \tag{6.7}$$

If $q_1 = q_1' + p$, where $p \in \mathrm{Ann}_Q(M)$, then $q_1 \otimes \cdots \otimes q_t - q_1' \otimes \cdots \otimes q_t = p \otimes \cdots \otimes q_t$ and this element annihilates $M^{\otimes t}$. By the universal property of the tensor product, the map (6.7) extends to an epimorphism. Since $Q/\mathrm{Ann}_Q(M)$ is a PI-algebra then, by Theorem 6.10, the same is true of each tensor product $Q/\mathrm{Ann}_{Q^{\otimes t}}(M)^{\otimes t}$. Hence each algebra of the form $Q^{\otimes t}/\mathrm{Ann}_{Q^{\otimes t}}(M^{\otimes t})$ is a PI-algebra. Let $w(x_1, \dots, x_n) \equiv 0$ be a multilinear identity holding in all such algebras, with $t = 1, \dots, c$.

We want to prove that $w(x_1, \dots, x_n) \equiv 0$ holds valid in the images of all $Q^{\otimes t}$ in $\mathrm{End}_k(M^{\otimes t}/T_c \cap M^{\otimes t})$. Of course, we may assume that $t \geq c+1$. It is clear from equation (6.6) that the image of $Q^{\otimes t}$ is equal to a sum of the images of subalgebras $Q_{i_1,\dots,i_c}$ which are the linear span of tensors $u_1 \otimes \cdots \otimes u_t$ in which all vectors are equal to 1 with the possible exception of those with numbers $i_1, \dots, i_c$, $1 \leq i_1 \leq \cdots \leq i_c \leq r$. Each of these algebras is isomorphic to $Q^{\otimes t}$ and is acting on $M^{\otimes t}$ in the same way as $Q^{\otimes c}$ is acting on $M^{\otimes c}$. If, further, $(i_1, \dots, i_c) \neq (j_1, \dots, j_c)$ and $u \in Q_{i_1,\dots,i_c} \setminus Q_{j_1,\dots,j_c}$, $v \in Q_{j_1,\dots,j_c} \setminus Q_{i_1,\dots,i_c}$ then uv is acting on $M^{\otimes t}/T_c \cap M^{\otimes t}$ trivially. Hence, while computing a value of $w(x_1, \dots, x_n)$ in the image of $Q^{\otimes t}$, one may restrict oneself to substituting elements from some particular subalgebra of the form $Q_{i_1,\dots,i_c}$. Therefore $w(x_1, \dots, x_n) \equiv 0$ is indeed an identity in the images of all $Q^{\otimes t}$. Finally, since the G-module T is the direct sum of the G-modules of the form $M^{\otimes t}$ it follows that $Q/\mathrm{Ann}_Q(T/T_c)$ is a PI-algebra. □

Proof of Theorem 6.23. Since G is a special Lie algebra, there exists a homomorphism α from L into a PI-algebra A such that $\mathrm{Ker}\,\alpha = M$. Suppose we have constructed a homomorphism β into another PI-algebra B such that $\mathrm{Ker}\,\beta \cap M = \{0\}$. We define γ :

$L \to A \times B$ by setting $\gamma(l) = (\alpha(l), \beta(l))$. Then γ is the desired monomorphism of L into a PI-algebra. Now for the construction of β we will be proceeding as in Theorem 6.20. We denote by W the intersection of U_{c+1} with $U = U(M)$ (c being the nilpotent step of N). Let π, φ, ψ be as in Lemma 6.24. Since W is invariant with respect to π, φ, the same is true for ψ. Hence we derive a homomorphism $\bar{\psi} : L \to \mathrm{End}_k(U/W)$. We will prove that $\psi(U(L))$ is a PI-algebra. From the definition of the universal enveloping algebra (Section 2.5.1), it is easy to see that $U(M)$ is isomorphic to a quotient algebra of $T(M)$ modulo the ideal generated by all tensors of the form $a \otimes b - b \otimes a - [a, b]$, $a, b \in M$. Furthermore, as a G-module, $U(M)$ is a quotient module of the G-module $T(M)$ and the images of the elements of T_c are of weight $c + 1$ in $U(L)$ (we again exploit the notation of Theorem 6.20). Hence, in fact, U/W is a quotient module of T/T_c. If S is the annihilator of U/W in $Q = U(G)$ then Lemma 6.25 shows that Q/S is a PI-algebra. Now let $w(x_1, \dots, x_n) \equiv 0$ be a multilinear identity holding in Q/S. We will show that

$$w(x_1, \dots, x_n)^{c+1} \equiv 0 \tag{6.8}$$

is identically true in $\bar{\psi}(U(L))$. First of all, we notice that if P is an ideal of $U(L)$ generated by N then $\bar{\psi}(P^{c+1}) = \{0\}$. Indeed, each element in P^{c+1} has weight at least $c + 1$ and is equal to a linear combination of monomials of the form $e_{i_1} \cdots e_{i_t} e_{j_1} \cdots e_{j_r}$ where $e_{i_1}, \dots, e_{i_t} \in E$ and $e_{j_1}, \dots, e_{j_r} \in E_1 \cup \cdots \cup E_c$. We can write $E_0 = E_0' \cup E_0''$ where E_0' is a basis for G and $E_0'' \cup E_1 \cup \cdots \cup E_c$ is a basis for M, so that for some s we have $e_{i_1}, \dots, e_{i_s} \in E_0'$ and $e_{i_{s+1}}, \dots, e_{i_t} \in E_0''$. Applying $\bar{\psi}$ to the above monomial, we can write

$$\bar{\psi}(e_{i_1} \dots e_{j_r})(m) = [e_{i_1}, \dots, e_{i_s}, e_{i_{s+1}} \dots e_{j_r} m].$$

Now the right-hand side has weight at least $c + 1$, hence it is in W.

Next, $\bar{\psi}(U(L))$ is an extension of $\bar{\psi}(P)$ by $\bar{\psi}(U(L))/\bar{\psi}(P)$. The latter algebra is an image of $U(L/P) = U(L/N)$, the enveloping algebra for $G \times M/[L, M]$. The universal enveloping algebra for the direct product is equal to the tensor product of universal envelopes of the factors, hence we arrive at $U(G) \otimes U(M/[L, M])$. Since $\bar{\psi}(U(G))$ satisfies a multilinear identity and $w(x_1, \dots, x_n) \equiv 0$, and since $U(M/[L, M])$ is commutative, the same identity is satisfied by $\bar{\psi}(U(L))/\bar{\psi}(P)$. Thus identity (6.8) is satisfied by $\psi(U(L))$. □

6.4.2 Solvable ideals in special Lie algebras over fields of characteristic zero

Lemma 6.26. *Let L be a Lie algebra over a field k of characteristic zero with a finite-dimensional faithful module V which is simple over an extension $\bar{k} \supset k$. Then any solvable ideal M of L is central in L. If L is not commutative and k is algebraically closed then it contains a subalgebra isomorphic to $\mathfrak{sl}(2, k)$.*

Proof. Applying the argument in Lie's theorem, it is easy to see that the elements of $N = [L, M]$ act on V by nilpotent transformations. Now Engel's theorem (Section 1.7) supplies us with a maximal nonzero subspace W such that $NW = \{0\}$. This subspace is invariant relative to L, hence $W = V$. Since V is a faithful L-module, we conclude $N = \{0\}$. Thus, the solvable radical of L coincides with its center. By Levi's theorem (Theorem 6.11), $L = G \oplus Z(L)$. If L is not commutative then $G \neq \{0\}$. Now according to Section 6.2.2, G possesses a subalgebra $\mathfrak{sl}(2, k)$ (if the base field is algebraically closed). □

Theorem 6.27. *Let k be a field of characteristic zero and M a locally solvable ideal of a special Lie algebra L. Then $N = [L, M]$ is a locally nilpotent ideal.*

Proof. There is no loss of generality in assuming L finitely generated. Then, by speciality, L is a subalgebra of a finitely generated associative PI-algebra A. The Jacobson radical J of A then is a nil-algebra (Theorem 6.3). The quotient algebra A/J is a subdirect product of primitive algebras A_α, $\alpha \in I$, which are finite-dimensional over their centers k_α. Each A_α is generated by the image of L under the canonical homomorphism $\gamma_\alpha : A \to A_\alpha$. Hence $L_\alpha = \gamma_\alpha(L)$ is a Lie algebra with faithful simple module (one may take such module for A_α) over k_α. By Lemma 6.26, $M_\alpha = \gamma_\alpha(M)$ is central in L_α, that is $[L_\alpha, M_\alpha] \subseteq J$ for all $\alpha \in I$. Since the intersection of all $\ker \gamma_\alpha$ is in J, we derive $N = [L, M] \subseteq J$. Now each finitely generated subalgebra of N belongs to a finitely generated subalgebra S of J. Since J is a nil-algebra, applying the corollary of Theorem 6.6 shows that B is nilpotent. A standard argument (cf. Section 1.7.3) proves that B being c-step nilpotent as an associative algebra implies B being nilpotent of nilpotent step at most $2c - 1$ as a Lie algebra. □

Corollary 6.28.

(i) *Let R, N denote, respectively, the locally solvable and the locally nilpotent radicals of a special Lie algebra L over a field k of characteristic zero. Then $[L, R] \subseteq N$.*

(ii) *Let k be a field of characteristic zero, L a special Lie algebra over k, G a solvable subalgebra in L, and x an element in G^2. Then the inner derivation* $\operatorname{ad} x$ *is nilpotent.*

6.4.3 Main theorem on finitely generated solvable special Lie algebras

A Lie algebra L is called *virtually solvable* if it has a solvable ideal of finite codimension. To obtain a description of special virtually solvable algebras, we will need a lemma.

Lemma 6.29. *Let A be a finitely generated associative PI-algebra. If J is a nil-ideal in A finitely generated as an ideal of A then it is nilpotent.*

Proof. Let $\{a_1, \dots, a_n\}$ be a generating set for A and $\{b_1, \dots, b_m\}$ a generating set for J as an ideal. We consider a free algebra F of the variety $\mathbf{V} = \operatorname{var}(A)$ of associative algebras generated by A, with a free generating set of the form $\{x_1, \dots, x_n, y_1, \dots, y_m\}$. According to Shirshov's height theorem (Theorem 6.6), if d is the degree of a nontrivial identity of A then F has finite height, say, h, with respect to a finite set W of monomials in the alphabet $\{x_1, \dots, y_n\}$ whose degree does not exceed $d-1$. Hence any element of F is equal to a linear combination of monomials of the form

$$w_{i_1}^{p_1} w_{i_2}^{p_2} \cdots w_{i_s}^{p_s}, \quad w_{i_j} \in W, s \le h. \tag{6.9}$$

We consider a homomorphism $\varphi : F \to A$ determined by $\varphi(x_i) = a_i$, $\varphi(y_j) = b_j$, $i = 1, \dots, n$, $j = 1, \dots, m$. Let T be a subalgebra of J generated by a finite set of monomials $c_1 \cdots c_r$, $r \le d-1$, where $c_1, \dots, c_r$ are in $\{a_1, \dots, b_m\}$ and at least one of them is in $\{b_1, \dots, b_m\}$. Then T is a finitely generated nil-algebra, hence by the corollary of Theorem 6.6, T is a nilpotent algebra of nilpotent step, say, l. Now let q be any natural number with $q > hl$. Our proof will be complete as soon as we prove that any product of the form

$$t = u_1 b_{i_1} v_1 \cdots u_q b_{i_q} v_q, \tag{6.10}$$

where $u_1, v_1, \dots, u_q, v_q$ are arbitrary products of elements in $\{a_1, \dots, a_n\}$, is equal to zero. Now equation (6.10) is the image of an element t' of the form $u_1' y_{i_1} v_1' \cdots u_q' y_{i_q} v_q'$ where $u_1', v_1', \dots, u_q', v_q'$ are words in the alphabet $x_1, \dots, x_n$. Since any element of F is a linear combination of the elements of equation (6.9), t is a linear combination of the products of the form $\bar{w}_{i_1}^{p_1} \bar{w}_{i_2}^{p_2} \cdots \bar{w}_{i_s}^{p_s}$. The degree of monomial (6.10) with respect to $\{y_1, \dots, y_m\}$ is equal to $q > hl$. Hence the degree of some w_{i_j} in $\{y_1, \dots, y_m\}$ is equal to $p_{i_j} > l$. In this case $\bar{w}^{p_{i_j}} = 0$, and therefore monomial (6.10) is equal to zero. □

Theorem 6.30. *Let k be a field of characteristic zero and L a special algebra over k. If L is solvable then L^2 is locally nilpotent. If L is finitely generated (virtually) solvable then it is (virtually) an algebra with nilpotent derived subalgebra. There exists a solvable special Lie algebra L such that L^2 is not nilpotent.*

Proof. The first claim is immediate from Theorem 6.27. Let, further, L be a subalgebra of an associative PI-algebra A. With L being finitely generated, the same can be assumed for A. By the hypothesis, L possesses a solvable ideal M such that $\dim(L/M) < \infty$. Applying the same argument as in the proof of Theorem 6.27, we conclude that $N = [L, M]$ is in the Jacobson radical of A. Now L/N is an extension of the ideal $M/[L, M]$ by L/M. Both algebras being finite-dimensional, we conclude that L/N is finite-dimensional and N is a finitely generated ideal of L (see Section 6.6.3). Now let I be two-sided ideal of A generated by N. Since L generates A, it follows that I is a finitely generated two-sided ideal in A. Now J/I is an ideal of A/I which is a homomorphic image of the universal enveloping algebra U for L/N. It is well known (and it will be proved in Theorem 6.41) that U is a noetherian algebra, hence J/I is a finitely generated ideal of A/I. Consequently, J is a finitely generated ideal of A. Applying Lemma 6.29, we derive that, as desired, $N = [L, M]$ is a nilpotent ideal. If L is solvable then $M = L$, hence L^2 is nilpotent.

Now to prove the final claim, we choose any field k of characteristic not 2 and consider L in the form $L = G \rtimes M$ where G is a Grassmann algebra with generating set $\{x_1, \dots, x_n, \dots\}$ and M is its left regular module (Section 1.6.2), that is, the same vector space relative multiplication by elements of G on the left. Now G is a Lie algebra with respect to commutator $[x, y] = xy - yx$ and M is an abelian Lie algebra. A basis for G is formed by the elements $x_{i_1} x_{i_2} \cdots x_{i_n}$, $i_1 < i_2 < \cdots < i_n$, $n \geq 1$. To avoid ambiguity, we denote the corresponding elements of M by the letters y_i, i as above. Thus, the commutator of two elements $x_{i_1} \cdots x_{i_m}$ in G and $y_{j_1} \cdots y_{j_n}$ in M is given by the formula

$$[x_{i_1} \cdots x_{i_m}, y_{j_1} \cdots y_{j_n}] = y_{i_1} \cdots y_{i_m} y_{j_1} \cdots y_{j_n}.$$

It is clear that M is a G-module if G is considered as a Lie algebra. Consequently, L is a Lie algebra. Now it is an easy exercise to verify that G satisfies the identity $[[x, y], z] \equiv 0$, that is, it is a 2-step nilpotent Lie algebra. Hence L is a solvable Lie algebra. The derived algebra L^2 for L is spanned by elements of the form $x_{i_1} \cdots x_{i_{2t}}$ and $y_{j_1} \cdots y_{j_r}$ where $t, r \geq 1$. Then for any $n > 1$, the following commutator of elements in L^2 is different from zero:

$$[x_1 x_2, x_3 x_4, \dots, x_{2n-1} x_{2n}, y_{2n+1}] = y_1 \cdots y_{2n+1}.$$

Hence L is not an algebra with nilpotent derived algebra. That L is a special Lie algebra is immediate from Section 6.4.1. Now the proof of all claims of our theorem is complete. □

It was remarked by Volichenko [362] that L as above can be presented as

$$L = \begin{pmatrix} G & G \\ 0 & 0 \end{pmatrix},$$

where G is as above with respect to the usual commutator of two matrices. This yields another proof that L is special.

Corollary 6.31. *A homomorphic image of a finitely generated solvable special Lie algebra over a field of characteristic zero is a special Lie algebra.*

The proof of the main result of the present section will be anticipated by an auxiliary lemma.

Lemma 6.32. *Let G be a finite-dimensional semisimple Lie algebra over a field k of characteristic zero and V a G-module. Suppose that the annihilators of all chief factors of V belong to a finite set $\{P_1, \dots, P_s\}$ of ideals of $U(G)$ and $\dim U(G)/P_i < \infty$, $i = 1, \dots, s$. Then V is a locally finite-dimensional module.*

Proof. It is sufficient to consider the case where V is a cyclic module. Since $U(G)$ is a noetherian algebra (Theorem 6.41), V is noetherian. Let T be an arbitrary submodule in V such that $\dim V/T < \infty$. Since G is semisimple, V/T is the direct sum of the modules whose annihilators are in the set $\{P_1, \dots, P_s\}$ (see Section 6.2.5). Consequently, if $I = \mathrm{Ann}_{U(G)}(V)$ then, for some natural m, we have $\dim U(G)/I \leq m$. This shows that $\dim V/T$ is always bounded by m. Now let T_0 be a submodule such that $\dim(V/T_0)$ is the greatest possible. Since V is noetherian, $T_0 \neq V$ and T_0 is finitely generated. If T_0 is nontrivial then the same argument as above can be applied to a cyclic quotient module of T_0, yielding a submodule T_1 of T_0 for which $\dim(V/T_0) < \dim(V/T_1) < \infty$. Thus T_0 is trivial and the proof is complete. □

Theorem 6.33. *Let k be a field of characteristic zero and L a finitely generated virtually solvable special Lie algebra. Then L a split extension of a finite-dimensional semisimple subalgebra G with solvable kernel M such that each chief factor of adjoint G-module M is of finite bounded dimension.*

Proof. Let M be a solvable ideal of L such that $G = L/M$ is a finite-dimensional semisimple Lie algebra. By Theorem 6.30, $N = [L, M]$ is a nilpotent Lie algebra and L/N is a finite-dimensional algebra. Applying Levi's theorem (Theorem 6.11), we may write

$$L/N = \bar{G}/N \times M/N,$$

where $G = \bar{G}/N$ is a semisimple subalgebra and M/N is a central ideal. Now G is naturally represented by the endomorphisms of the spaces N^t/N^{t+1}, $t = 1, \ldots, c$, where c is such that $N^c \neq \{0\}$ and $N^{c+1} = \{0\}$. To put this explicitly, we define δ_t, $t = 1, \ldots, c$, as follows. Let $x \in G$ and $g + N$ be a coset in $\bar{G}/N$ ($g \in \bar{G}$) corresponding to x and $y = n + N^{t+1}$ an element in N^t/N^{t+1} ($n \in N^t$). We put

$$\delta_t(x)(y) = [g, n] + N^{t+1}. \tag{6.11}$$

This definition is correct since N centralizes each factor N^t/N^{t+1}.

Notice that passing to a finite extension of the base field enables us to assume that G itself, as well as its finite-dimensional representations, are such as in the case of an algebraically closed field. In this case G is called *split*. We start our proof with this particular case, i. e., where G is split. Now let T be a finitely generated subalgebra of L such that $T + N/N = H \oplus N_-$ (we adopt the notation from Section 6.2). Since T is special solvable, Theorem 6.30 applies. Hence, there exists a natural number d such that any associative product of at least d elements in T^2 is equal to zero. Similarly, there exists an n such that any product of at least n elements of the associative ideal C generated by N in A is equal to zero, A being the associative PI-algebra generated by L. Now let $e_\alpha = x_\alpha + N$, $h_\alpha = y_\alpha + N$, $x_\alpha, y_\alpha \in T$. Then $x_\alpha + N = e_\alpha = \frac{1}{2}[x_\alpha, y_\alpha] + N$, that is, $x_\alpha = \frac{1}{2}[x_\alpha, y_\alpha] + p$, $p \in N$. We want to compute $x_\alpha^{d+n-1} = (\frac{1}{2}[x_\alpha, y_\alpha] + p)^{d+n-1}$. The right-hand side is a linear combination of the monomials of the form

$$[x_\alpha, y_\alpha]^s p_1 p_2 \cdots p_{d+n-1-s},$$

where $0 \leq s \leq d + n - 1$ and $p_1, p_2, \ldots, p_{d+n-1-s} \in C$. It is obvious that this gives the desired result. Since G is finite-dimensional and arguing in the same way, we arrive at a system of elements x_α, $\alpha \in R$, and a natural number l' such that $e_\alpha + x_\alpha + N$ and $x_\alpha^{l'} = 0$. Recalling definition (6.11) of δ_t, we derive the existence of a natural number l such that $\delta_t(x)^l = 0$ for all $x \in N_-$. By Section 6.2.5, there exists only finitely many irreducible representations $\rho_1, \dots, \rho_q$ such that each chief factor of N^t/N^{t+1} is isomorphic to a module of one of $\rho_1, \dots, \rho_q$. Thus only finitely many ideals $P_1, \dots, P_s$ of $U(G)$ can serve as kernels of irreducible representations occurring in δ_t and $\dim U(G)/P_i < \infty$, $i = 1, \dots, s$, $t = 1, \dots, c$.

If we abandon the hypothesis that $G = L/M$ is a split semisimple Lie algebra then as noticed above there exists a finite extension $\bar{k}$ of k such that $G_{\bar{k}} \cong L_{\bar{k}}/M_{\bar{k}}$. Since $[X, Y]_{\bar{k}} = [X_{\bar{k}}, Y_{\bar{k}}]$, $X, Y \subseteq L$, all the above applies to $L_{\bar{k}}$. In particular, only finitely many ideals $Q_1, \dots, Q_r$ of $U(G_{\bar{k}}) = U(G)_{\bar{k}}$ may be annihilators of the chief factors of $G_{\bar{k}}$-modules of the form $N_{\bar{k}}^t/N_{\bar{k}}^{t+1} = (N^t/N^{t+1})_{\bar{k}}$. Now let P be the annihilator in $U(G)$ of a chief factor S of N^t/N^{t+1}. Then $P_{\bar{k}}$ is the annihilator of $S_{\bar{k}}$ in $U(G_{\bar{k}})$. It is known (see the footnote in Section 1.8, corollary of Lie's theorem (Theorem 1.8)) that $S_{\bar{k}}$ is the direct sum of finitely many irreducible $G_{\bar{k}}$-modules. Hence $P_{\bar{k}} = Q_{i_1} \cap \dots \cap Q_{i_s}$, where $1 \le i_1, \dots, i_s \le r$. Hence it follows that the number of annihilators of the form P is also finite. This proves our claim on the boundedness of dimensions of chief factors in the general case.

To complete the proof, we need to show that L is a split extension of a subalgebra $G' \cong G$ and M. We apply induction over c, the nilpotent step of N. Since L/N is a split extension, the basis of induction is at hand. For the induction step, we consider a subalgebra $\bar{G}'$ in L such that $\bar{G}' \supset N^c$ and $\bar{G}'/N \cong G$. Let $\{e_1, \dots, e_n\}$ be a basis in $\bar{G}'$ modulo N^c. Then $[e_i, e_j] = \sum_t \gamma_{ij}^t e_t + u_{ij}$ where $\gamma_{ij}^y \in k$ and $u_{ij} \in N^c$. Let V be a submodule of the G-module N^c generated by all u_{ij}, $1 \le i, j \le n$. By Lemma 6.32, V is finite-dimensional. Now let $\widetilde{G}$ denote a subalgebra in L generated by elements $e_1, \dots, e_n$. It is clear that $\widetilde{G} \cap N^c = V$ and $\widetilde{G}/V = \widetilde{G} + N^c/N^c \subseteq \bar{G}'/N^c = G$. Hence $\widetilde{G}$ is finite-dimensional. By Levi's theorem there exists a subalgebra $\bar{G}$ in $\widetilde{G}$ isomorphic to G. Since $\bar{G}$ is semisimple, $\bar{G} \cap M = \{0\}$ and hence L is a split extension of M by $\bar{G}$. Now each chief factor S of M as G-module is isomorphic to a chief factor of a G-module of one of the forms M/N and N^t/N^{t+1}, $t = 1, 2, \dots, c$. It follows from Section 6.2.5 and Lemma 6.32 that $\dim_k S$ is a finite bounded number. □

Assembling together Theorems 6.23 and 6.33, we arrive at the main result of this section.

Theorem 6.34. *Let k be a field of characteristic zero. A finitely generated virtually solvable Lie algebra is special if and only if it is isomorphic to a semidirect product of a finite-dimensional semisimple Lie algebra G by a solvable ideal M such that $N = [L, M]$ is a nilpotent ideal and all the chief factors of M as a G-module are of finite bounded dimension. Moreover, M has finitely many generators as a Lie algebra.*

Proof. The necessity has just been proved (see Theorem 6.33). To cope with sufficiency, one should remark that, in our case, $U(G)/\mathrm{Ann}_{U(G)} N$ is a PI-algebra, so that Theorem 6.23 applies. The proof of the latter claim using the PBW theorem (Section 2.5) and Lemma 6.32 is left to the reader as an exercise. □

6.5 Identities of special Lie algebras over the fields of characteristic zero

6.5.1 Special varieties of Lie algebras

A variety of Lie algebras over commutative ring k is called *special* if it is generated by a special algebra. In this section we discuss special varieties over a field of characteristic zero. An important role in the study of special varieties is played by the following result.

Proposition 6.35. *Let L be a free algebra of a variety $\mathbf{V}$ of Lie algebras over a field k of characteristic zero and $Z = Z(L)$ is the center of this algebra. Suppose L/Z is a special Lie algebra. Then L is a special Lie algebra, too.*

Proof. Let X denote a free generating set for L. Since $L \cong F/\mathbf{V}(F)$ where F is a free Lie algebra and $\mathbf{V}(F)$ is its verbal ideal, and since $\mathbf{V}(F)$ is stable under derivations (see Section 4.2, Theorem 4.12), we conclude that any map of X onto L extends to a derivation of L (see Section 2.5.5). So the identity map of X extends to a derivation $d : L \to L$ under which any homogeneous element a of degree n is transformed into na. Hence $\ker d = \{0\}$. Taking $M = kd \ltimes L$, we will have a Lie algebra M with trivial center and such that L is embedded as an ideal in M. We consider $\operatorname{Ad} M$ and its subalgebra S generated by all inner derivations of the form $\operatorname{ad}_M x$, $x \in L$. Then from $Z(M) = 0$ it follows that $x \mapsto \operatorname{ad}_M x$ is a monomorphism from L into $[S]$. But S is a PI-algebra. Indeed, let $f(x_1, \dots, x_n) \equiv 0$ be a nontrivial identity of $\operatorname{Ad}(L/Z)$ (see Theorem 6.18). Then $x_0 f(x_1, \dots, x_n) \equiv 0$ is a nontrivial identity of $\operatorname{Ad}(L)$. Since $L \triangleleft M$, it is obvious that $x_0 f(x_1, \dots, x_r)x_{r+1} \equiv 0$ is an identity in the associative algebra S. Hence L is a special Lie algebra. □

We recall that a subset S of a lattice R is called its ideal if for any $x, y \in S$ and $a \in R$ we have $x \vee y \in S$ and $x \wedge a \in S$ where $\vee$ is the operation of union and $\wedge$ that of intersection.

Proposition 6.36. *The set $s(k)$ of all special varieties over a field k of characteristic 0 is an ideal of the lattice $v(k)$ of all varieties of Lie algebras over k.*

Proof. Suppose $\mathbf{U}, \mathbf{V}$ are in $s(k)$, that is, there exist special Lie algebras $G \in \mathbf{U}$, $H \in \mathbf{V}$ generating these varieties. In this case the direct product $K = G \times H$ generates $\mathbf{U} \cup \mathbf{V}$, and it is obvious that K is a special Lie algebra. Now suppose that $\mathbf{U}$ is a subvariety of a special variety $\mathbf{V}$. We consider a free algebra of countable rank L in $\mathbf{U}$. Let G be a special Lie algebra generating $\mathbf{V}$. By Proposition 6.17, $\operatorname{Ad}(G)$ is a PI-algebra. Hence L/Z

is a special Lie algebra. By Proposition 6.35, L is itself special. Since L generates $\mathbf{U}$, we conclude that $\mathbf{U}$ is a special variety. □

The main result of this section is finding necessary and sufficient conditions under which the product and the commutator of two varieties are special varieties.

Theorem 6.37. *Let* $\mathbf{U}$ *and* $\mathbf{V}$ *be nontrivial varieties of Lie algebras over a field of characteristic zero. Then their product* $\mathbf{UV}$ *is a special variety if and only if* $\mathbf{U}$ *is nilpotent and* $\mathbf{V}$ *is abelian.*

Proof. Any nontrivial variety of Lie algebras contains $\mathbf{A}$, the variety of all abelian Lie algebras, and any nonabelian variety over an infinite field contains $\mathbf{N}_2$, the variety of all nilpotent Lie algebras of nilpotent step at most 2. Thus, if $\mathbf{V}$ is not abelian then $\mathbf{N}_2 \subseteq \mathbf{V}$ and $\mathbf{AN}_2 \subseteq \mathbf{UV}$. But Lie algebra L in Example 3 of Section 6.3.4 is in $\mathbf{AN}_2$ and has trivial center. According to the corollary of Theorem 6.18, $\mathbf{AN}_2$ is not a special variety. In this case, by Proposition 6.36, $\mathbf{UV}$ is not a special variety either. This shows that, in fact, $\mathbf{V}$ must be assumed abelian.

Next we consider the verbal $\mathbf{U}$-product of two one-dimensional Lie algebras $L = (ka)\,\mathrm{wr}_{\mathbf{U}}(kb)$ (see Section 4.4.1). Then $L \in \mathbf{UA}$ and it is an extension of a $\mathbf{U}$-free algebra of countable rank by a one-dimensional quotient algebra. The number of generators of L is equal to 2. Suppose $\mathbf{UA}$ is a special variety. Then $A = \mathrm{Ad}(L)$ is a PI-algebra. Now A is a finitely generated PI-algebra. Applying Theorem 6.3 and Lemma 6.29 shows that any finitely generated ideal in the Jacobson radical is nilpotent. If all primitive factors of A are commutative then L is solvable. Indeed, in this case the derived algebra of L is in a finitely generated ideal of A which is in $J(A)$. Hence we conclude that the $\mathbf{U}$-free algebra of countable rank, which is a subalgebra of L, is solvable. Then $\mathbf{U}$ is a solvable variety. By Section 4.6, Lemma 4.36, either $\mathbf{U}$ is nilpotent or $\mathbf{U}$ contains $\mathbf{A}^2$, the variety of all metabelian algebras. In the latter case, $\mathbf{UA} \supseteq \mathbf{A}^3$ and this variety is not special (see Section 6.3.4).

Now we may assume that some primitive factor of A is not commutative. In this case L has a homomorphic image which is isomorphic to a noncommutative Lie algebra M which is an irreducible Lie algebra of operators of a finite-dimensional vector space over a suitable extension k' of the base field k. Tensoring by the algebraic closure k'' of k', it is easy to see (Section 6.4.1) that $\mathbf{UA}$ contains an algebra G of the form $G = \mathfrak{sl}(2, k'')$ of matrices of order 2 with trace zero. From the definition of the product variety, the derived subalgebra of $G = \mathfrak{sl}(2, k'')$ must be in $\mathbf{U}$. Since $G = G^2$, we derive $G \in \mathbf{U}$. Now G contains a two-dimensional nonabelian subalgebra. This subalgebra is not nilpotent, hence it generates $\mathbf{A}^2$. Again we have $\mathbf{UA} \supseteq \mathbf{A}^3$ so that $\mathbf{UA}$ is not a special variety as before. Summing up all the above, we conclude that if $\mathbf{UA}$ is special then $\mathbf{U}$ is nilpotent, proving the necessity of the conditions of our theorem. The sufficiency follows from Theorem 6.20. Alternatively, one may just use the fact that $\mathbf{N}_c\mathbf{A}$ is generated by Lie algebra of upper triangular matrices of order $c + 1$ (see Section 4.5.2, Example 2). □

6.5.2 Commutators of special varieties

Theorem 6.37 gives no new examples of special varieties. With this in mind, of certain interest is a theorem on the commutators of varieties which we are going to prove. In its proof, we will need the following.

Lemma 6.38. *Let M be an ideal of a Lie algebra L such that $C_L(M) \supseteq L^2$. If $\mathrm{Ad}(L/M)$ is a PI-algebra then $R = \mathrm{Ad}(L)$ is a PI-algebra, too.*

Proof. Considering the proof of Proposition 6.22, it is sufficient to verify that $(\mathrm{Ad}(L))_M$ is a commutative algebra. But if $a = \mathrm{ad}\, e_1 \cdots \mathrm{ad}\, e_s$, $b = \mathrm{ad}\, f_1 \cdots \mathrm{ad}\, f_t$ are two arbitrary monomials in $\mathrm{Ad}(L)$ then

$$[a, b] = \sum_{i,j} \mathrm{ad}\, e_1 \cdots \mathrm{ad}\, e_{i-1}\, \mathrm{ad}\, f_1 \cdots \mathrm{ad}\, f_{i-1}\, \mathrm{ad}[e_i, f_j]\, \mathrm{ad}\, e_{i+1} \cdots \mathrm{ad}\, e_s\, \mathrm{ad}\, f_{j+1} \cdots \mathrm{ad}\, f_t.$$

Obviously, $[a, b] \in \mathrm{Ann}_R(M)$, as required. □

Theorem 6.39. *The commutator $[\mathbf{U}, \mathbf{V}]$ of two special varieties over a field k of characteristic zero is special if and only if one of these varieties is abelian.*

Proof. If we admit that neither $\mathbf{U}$ nor $\mathbf{V}$ is abelian then $[\mathbf{U}, \mathbf{V}] \supseteq [\mathbf{N}_2, \mathbf{N}_2]$. The latter variety is, in fact, another expression for $\mathbf{AN}_2$, a nonspecial variety, a contradiction. To prove the converse, we consider the free algebra of countable rank L in the commutator variety $[\mathbf{U}, \mathbf{A}]$. By definition, L possesses an ideal M such that $L/M \in \mathbf{U}$ and $[M, L^2] = \{0\}$. As previously, then $\mathrm{Ad}(L/M)$ is a PI-algebra. By Lemma 6.38, also $\mathrm{Ad}(L)$ is a PI-algebra. According to Section 6.5.1, then L is a special Lie algebra. Since $[\mathbf{U}, \mathbf{A}]$ is generated by L, this is a special variety. □

6.5.3 The codimension series of a special variety

Let $\mathbf{V}$ be a special variety of Lie algebras over a field k and $L = L(X, \mathbf{V})$ a $\mathbf{V}$-free algebra of countable rank. Then $L \subseteq A$ where A is a PI-algebra generated by X. Now let $F = F(X, \mathbf{W})$ be a free algebra of a variety $\mathbf{W}$ of associative algebras generated by A. Then the identity map of X extends to a homomorphism φ from F into A. Clearly, then $P_n(\mathbf{V}) \subseteq \varphi(P_n(\mathbf{W}))$. This argument and Theorem 6.9 together show that the following is true.

Proposition 6.40. *The growth of any special variety of Lie algebras over a field is at most exponential.*

6.6 The universal enveloping algebra of a finite-dimensional Lie algebra. Some results and applications

In this section we are proving some auxiliary (for the present context) results about the structure and representations of universal envelopes of finite-dimensional Lie algebras.

Theorem 6.41. *Let G be a finite-dimensional Lie algebra over a field k and $U(G)$ its universal enveloping algebra. Then $U(G)$ is a noetherian algebra.*

Proof. We consider $\bar{U} = \operatorname{gr} U$ (see Section 2.5.3). As we know, if $\{e_1, e_2, \dots, e_n\}$ is a basis for G then $\bar{U}$ is isomorphic to the polynomial ring $k[e_1, e_2, \dots, e_n]$ under the isomorphism transforming each e_i into the coset $e_i + U_0$, $i = 1, \dots, n$. Let us show that any, say, left, ideal I of U can be generated by a finite number of elements. For this, we consider the ideal $\bar{I} = \operatorname{gr} I$ (see Section 1.10.4). By Hilbert's basis theorem, $\bar{I}$ is a finitely generated ideal of the polynomial ring. Now let $u_1, u_2, \dots, u_m$ in I be such that $\operatorname{gr} u_1, \dots, \operatorname{gr} u_m$ are the generators of $\bar{I}$. Let us show that any $u \in I$ is in the ideal generated by $u_1, \dots, u_m$. Suppose $u \in U_s$. We consider $a_1, \dots, a_m$ such that

$$\operatorname{gr} u = \operatorname{gr} a_1 \cdot \operatorname{gr} u_1 + \cdots + \operatorname{gr} a_m \cdot \operatorname{gr} u_m = \operatorname{gr}(a_1 u_1 + \cdots + a_m u_m).$$

Now $u - a_1 u_1 - \cdots - a_m u_m \in U_{s-1} \cap I$. Applying induction with obvious basis when $s = 0$ completes the proof. □

6.6.1 Endomorphisms of simple *G*-modules

By Schur's lemma, the endomorphism ring D of a simple G-module M is a skew field, which is an algebra over the base field k. If M is finite-dimensional over k and k is algebraically closed then, of course, $D = L$. But if M is infinite-dimensional then for arbitrary Lie algebras such a conclusion fails to be true (cf. Section 6.8). Still, if $\dim G < \infty$ then this is true, and we are going to prove it in the final part of the next subsection. But first we need an auxiliary result. In its statement, A_f, with A being a commutative k-algebra and f its regular element, stands for the subring of the ring of quotients for A which is generated over k by A and f^{-1}.

Lemma 6.42. *Let A be a commutative ring without zero divisors, B a commutative A-algebra with a finite number of generators and with unity 1, and M a finitely generated B-module. Then we can find a nonzero $f \in A$ such that $A_f \otimes M$ is a free A_f-module.*

Proof. We proceed by induction on the number of the generators for M. Applying induction, we may assume that M is a cyclic module, i. e., the quotient module of B. Since in our case the quotient module is just a finitely generated commutative algebra, there will be no loss of generality if we consider B itself. Let $\{x_1, \dots, x_n\}$ be a generating set for the A-algebra B. For $\nu = (\nu_1, \dots, \nu_n) \in \mathbb{N}^n$, we put $x^\nu = x_1^{\nu_1} \cdots x_n^{\nu_n}$. We give $\mathbb{N}^n$ the structure of an ordered set by considering the direct power of the usual order on $\mathbb{N}$ (Section 5.2.1). We set

$$B_\nu = \sum_{\nu' \le \nu} Ax^{\nu'}, \quad B_\nu^- = \sum_{\nu' < \nu} Ax^{\nu'}.$$

Let x_*^ν denote the image of x^ν in B_ν / B_ν^-. We denote by I_ν the annihilator of x_*^ν in A. It is clear that $\nu_1 < \nu_2$ implies $I_{\nu_1} \subseteq I_{\nu_2}$. We consider the subset S in $\mathbb{N}^n$ consisting of those ν for which $I_\nu \neq \{0\}$. By Section 5.2, the corollary of Proposition 5.10, S has only finitely many minimal elements, each element majorizing a minimal one. If $\nu_1, \dots, \nu_r$ are such minimal elements then the ideals $I_{\nu_1}, \dots, I_{\nu_r}$ are minimal. Since A has no zero divisors, the ideal $I = I_{\nu_1} \cap \cdots \cap I_{\nu_r}$ is nonzero. Pick an f in I. Then for each ν the A_f-module $A_f \otimes (B_\nu / B_\nu^-)$ is either zero or free. Since $A_f \otimes B$ is a successive extension of free A_f-modules, it must be free. □

Theorem 6.43 (J. Dixmier). *Let C be a k-algebra endowed with an ascending filtration $\{C_0, C_1, \dots, C_n, \dots\}$. Suppose the associated graded algebra $\bar{C} = \operatorname{gr} C$ is finitely generated and commutative. We consider a simple C-module M and its skew field D of all C-endomorphisms. If $d \in D$ then d is algebraic over k.*

Proof. Let A be a k-subalgebra in D generated by 1 and d. If d is transcendental then $A = k[d]$. We denote $A \otimes C$ by V. Then M naturally becomes a V-module if we set $(d \otimes c) \cdot m = d(cm)$, where $d \in A, c \in C, m \in M$. Now V has a filtration by subspaces of the form $V_r = A \otimes C_r$. If m_0 is a nonzero element in M then M becomes a filtered V-module if we define $M_r = V_r m_0$. Thus $V_r M_s \subseteq M_{r+s}$ for all $r, s = 0, 1, \dots$ It is clear that $\operatorname{gr} M$ is a cyclic $\operatorname{gr} V$-module. Now $\operatorname{gr} V$ is a finitely generated A-module. By Lemma 6.42, there exists $f \neq 0$ such that $A_f \otimes \operatorname{gr} M$ is a free A_f-module. Since A_f is a principal ideal ring, the direct summands of a free module are free, i. e., for each $r = 1, 2, \dots, M_r / M_{r-1}$ is a free A_f-module. This forces $A_f \otimes M$ to be a free A_f-module. Now let $g \neq 0$ be an element in A which is not a divisor of any power of f. Then the map $a \mapsto ga$ of A_f into itself is not surjective. Hence also the multiplication by g in $A_f \otimes M$ is a nonsurjective map of this module. But $g(a \otimes m) = ga \otimes m = a \otimes gm$. Since $g \in D$, a skew field, the map $g \mapsto gm$ is an isomorphism, a contradiction. □

A G-module is called *absolutely simple* (absolutely irreducible) if it remains simple under any extensions of the base field.

Corollary 6.44. *Let G be a finite-dimensional Lie algebra over a field k and M an absolutely simple G-module. Then any G-endomorphism of M is scalar.*

Proof. To apply Theorem 6.43, we put $C = U(G)$ keeping the natural filtration of the latter. Now any endomorphism of M as G-module is its endomorphism as $U(G)$-module. Let d be such an endomorphism. Passing to the algebraic closure $\bar{k}$ of k, we may consider the $G_{\bar{k}}$-module $\bar{k} \otimes M$. The operator $1 \otimes d$ belongs to the algebra of $G_{\bar{k}}$-endomorphism of this module. Since M is absolutely simple, our new module $\bar{k} \otimes M$ is a simple $U(G)_{\bar{k}}$-module. Now Theorem 6.43 gives $1 \otimes d \in \bar{k}$. But $1 \otimes d$ transforms $1 \otimes M$ into itself. Hence $d \in k$ is as required. □

Theorem 6.45 (Hilbert's Nullstellensatz). *Let k' be a field which is an extension of the base field k such that k' is a finitely generated k-algebra. Then k' is algebraic over k.*

Proof. By our condition $k' = k[x_1, \dots, x_n]/I$ where I is a maximal ideal. Hence k' is a simple G-module where G is an abelian Lie algebra with basis $\{x_1, \dots, x_n\}$. Besides, $k' \subseteq \operatorname{End}_G k'$. Now from Theorem 6.43 we immediately derive that k' is algebraic over k. □

An immediate corollary of the above is the fact that if G is a finite-dimensional abelian Lie algebra then each irreducible G-module is finite-dimensional.

6.6.2 Simple G-module in the case where the base field is of prime characteristic

As we saw in Section 6.3.4, over zero characteristic fields finite-dimensional Lie algebras may have infinite-dimensional simple modules. Our nearest aim is to show that in the case of prime characteristic any simple G-module, G being a finite-dimensional Lie algebra, is finite-dimensional. Such a module is isomorphic to a quotient module of the universal enveloping algebra $U(G)$ by a maximal left ideal I. By Lemma 6.14, $U(G)$ has a central subalgebra R isomorphic to the polynomial ring k in n variables, $n = \dim G$, and such that $U(G)$ is a finitely generated R-module. A basis of $U(G)$ can be chosen in the form

$$e_1^{l_1} \cdots e_n^{l_n} y_1^{s_1} \cdots y_n^{s_n} \quad \text{where } 0 \le l_i < d_i,\ s_i \ge 0.$$

The elements $y_1, \dots, y_n$ are central and $R = k[y_1, \dots, y_n]$. So given a nonzero element m_0 in M, then M is a linear span of the set of elements of the form

$$e_1^{l_1} \cdots e_n^{l_n} y_1^{s_1} \cdots y_n^{s_n} m_0, \quad 0 \le l_i < d,\ s_i \ge 0. \tag{6.12}$$

We recall that all y_i commute with all e_j, hence, as elements of $\operatorname{End}_k M$ they are in $\operatorname{End}_G M$. By Theorem 6.43, these elements must be algebraic, that is, for any $i = 1, \dots, n$, there exists a polynomial $g_i \in k[t]$ such that $g_i(y_i)m = 0$. If q_i is the degree of g_i then it is clear that M is, in fact, the span of those elements of the form (6.12) for which $s_i < q_i$, $i = 1, \dots, n$. Hence M is finite-dimensional.

Theorem 6.46. *Let G be a finite-dimensional Lie algebra over a field k of characteristic $p > 0$ and M a simple G-module. Then $\dim M < \infty$. Moreover, if k is algebraically closed then $\dim M$ is bounded by a constant depending entirely on G.*

Proof. The former claim having been proved just above, we pass to the latter one. It is a simple matter to notice that if the base field is algebraically closed then the numbers q_i of the preceding proof are all equal to 1, that is, M is a linear span of the elements of the form (6.12) with $s_1 = s_2 = \cdots = s_n = 0$. In this case $\dim M \le d_1 d_2 \cdots d_n$, which does not depend on M but entirely on G, as required. □

6.6.3 Residually finite Lie algebras

We apply the above theory to the study of one of the finiteness conditions for infinite-dimensional Lie algebra, called residual finiteness. A Lie algebra G is called *residually finite* if for any $g \neq 0$ in G there exists a homomorphism φ_g of G into a finite-dimensional algebra F_g such that $\varphi_g(g) \neq 0$ in F_g. As in the case of an arbitrary finiteness condition (in fact, this can be viewed as a definition of such a condition) this condition holds in any finite-dimensional Lie algebra, but there exists an infinite-dimensional algebra in which it fails to be satisfied. Obviously, any abelian Lie algebra over a field is residually finite. But it is easy to give an example of a nilpotent Lie algebra which is not residually finite. For instance, such an example is supplied by a Lie algebra H with basis $\{x_i, y_i, z \mid i \in \mathbb{Z}\}$ and a multiplication table in which any product of two basic elements is zero except the products of the form $x_i y_i = -y_i x_i = z$, $i \in \mathbb{Z}$. In the case of this algebra, given a homomorphism $\varphi : H \to F$, F a finite-dimensional algebra, we always have $\varphi(z) = 0$. For the proof, we remark that there must exist not all zero coefficients $\lambda_1, \dots, \lambda_t$ such that $\lambda_1\varphi(x_{i_1}) + \cdots + \lambda_t\varphi(x_{i_t}) = 0$. If $\lambda_1 \neq 0$ then we can write

$$\begin{aligned} 0 &= (\lambda_1\varphi(x_{i_1}) + \cdots + \lambda_t\varphi(x_{i_t}))\varphi(y_{i_1}) \\ &= \varphi((\lambda_1 x_{i_1} + \cdots + \lambda_t x_{i_t})y_{i_1}) = \lambda_1\varphi(z), \end{aligned}$$

as asserted. This example shows that the most interesting case in the study of residually finite Lie algebras is that of finitely generated algebras.

Theorem 6.47. *Let S be a finitely generated Lie algebra over a field k with an abelian ideal A and the quotient algebra $G = S/A$ such that G is a finite-dimensional algebra, which is abelian if* $\operatorname{char} k = 0$. *Then S is a residually finite Lie algebra.*

Proof. As noted in Section 1.6.5, the abelian ideal A becomes a G-module if, for $a \in A$ and $g = x + A$, we set $ga = [x, a]$. Let $s_1, \dots, s_n$ be generators for S. We remark that the G-module A is a finitely generated module. Indeed, we may represent S as a quotient algebra of a finitely generated free Lie algebra $L(x_1, \dots, x_n)$. Let R be the inverse image of A in this representation. Then L/R is a finitely generated algebra. If $\{e_1, \dots, e_n\}$ is a basis for L modulo R then for any $1 \leq i, j \leq m$ we have

$$r_{ij} = [e_i, e_j] - \sum_{t=1}^{m} c_{ij}^t e_t \in R,$$

where the coefficients c_{ij}^t are so-called *structure constants* of L/R in the basis $\{e_1 + R, \dots, e_m + R\}$. We want to show that r_{ij}, as polynomials in $x_1, \dots, x_n$, are the generators of the ideal R. Indeed, let J be an ideal generated by these elements. Then, on the one hand, $J \subset R$. Hence the map $x_1 \mapsto s_1, \dots, x_n \mapsto s_n$ extends to a homomorphism of L/J onto L/I. On the other hand, if all r_{ij} are trivial in a quotient algebra of L then this algebra is of dimension less than or equal to n. Hence $J = I$ as required. Now it is clear that A, being abelian, is a finitely generated G-module.

To prove the residual finiteness of S, it is sufficient, for any $s \neq 0$, to find an ideal I of finite codimension such that $s \notin I$. If $s \notin A$ then we simply put $I = A$. What remains is to find for each $a \in A$ a G-submodule T in A such that $\dim A/T < \infty$ and $a \in T$. Let Q be a maximal G-submodule in A with $a \notin Q$. Such a submodule exists by Zorn's lemma. Then S/Q satisfies all our hypotheses but, additionally, its ideal M generated by $\bar{a} = a + Q$ is nonzero and it is contained in any nonzero ideal of S/Q. We call M a *monolith* of S/Q. Applying such a reduction, we may assume from the very beginning that S is a monolithic algebra, that is, an algebra with a nonzero monolith. In this case it is sufficient (and necessary) to prove that A is a finite-dimensional G-module. We notice that M is a simple G-module, hence, by Theorems 6.45 and 6.6.2, it is finite-dimensional.

Now let $R = k[y_1, \ldots, y_n]$ be a central subalgebra in $U(G)$ such that $U(G)$ is a finitely generated R-module. Such a subalgebra in the case where $\operatorname{char} k = p > 0$ is supplied by Lemma 6.14. If $\operatorname{char} k = 0$, then, by our hypothesis, G is abelian and we may put $R = U(G)$. Now let $r \in R \cap \operatorname{Ann}_U M$. We consider a descending chain of U-submodules of the form $A_0 = A$, $A_i = r^i A$, $i = 1, 2, \ldots$ Either there exists a natural number s such that $A_s = \{0\}$ or, for all i, we have $A_i \supseteq M$. Suppose the latter holds. For each $i = 0, 1, 2, \ldots$, we consider a U-submodule of the form $B_i = \ker r^i$. Since U is a noetherian algebra and A is a finitely generated U-module, it follows that A is a noetherian U-module. Then one can find a natural number p such that $B_p = B_{p+1}$. By the choice of r, we have also $B_p \supseteq M$. Let $m \neq 0$ be an element in M. Then, for some $a \in A$, we have $m = r^p a \neq 0$. But $0 = rm = r^{p+1} a = 0$. Hence $a \in B_{p+1} \setminus B_p$, a contradiction. Thus $r^p A = \{0\}$. Now it follows that each element $r \in R \cap \operatorname{Ann}_U M$ is acting in A as a nilpotent operator. We remark that M being finite-dimensional, implies $\dim_k R/(R \cap \operatorname{Ann}_U M) < \infty$. This proves the existence, for each y_i, $i = 1, \ldots, n$, of a polynomial $g_i \in k[t]$ such that $g_i(y_i) \in R \cap \operatorname{Ann}_U M$. As in the proof of Theorem 6.46, we derive that A is the linear span of elements of the form

$$e_1^{l_1} \cdots e_n^{l_n} y_1^{s_1} \cdots y_n^{s_n} z_1^{t_1} \cdots z_n^{t_n} a_i, \quad i = 1, \ldots, q,$$

where $a_1, \ldots, a_q$ are the generators of the U-module A, $z_i = g_i(y_i)$, $i = 1, \ldots, n$, and $0 \le l_i < d_i$, $0 \le s_i < \deg g_i$, $0 \le t_i < n_i$, with n_i being a natural number such that $z_i^{n_i} = \{0\}$. Now A is shown to be a finite-dimensional vector space over k, and the proof is complete. □

6.6.4 Examples

Here we give two examples of finitely generated Lie algebras which are sufficiently close to metabelian but, in contrast to the metabelian case, they are not residually finite.

(1) Let char $k = 0$ and let L be one of the algebras in Section 6.3.4, Examples 3 and 4. Then $L = G \rtimes M$, where $\dim G = 2, 3$, and M is an infinite-dimensional simple G-module. Under any homomorphism of L of such form onto a finite-dimensional Lie algebra, M is mapped trivially, showing that L is not residually finite. We remark that in the case $\dim G = 3$ we have $L \in \mathbf{AN}_2$.

(2) Over an arbitrary field, one may construct a center-by-metabelian Lie algebra which is finitely generated but not residually finite. For this, we take a Lie algebra H as in the first paragraph in Section 6.6.3 and consider a derivation $d : H \to H$ given by $d(x_i) = x_{i-1} - x_{i+1}$, $d(y_i) = y_{i-1} - y_{i+1}$, $d(z) = 0$, $i \in \mathbb{Z}$. The verification of d being a derivation is left to the reader. Then we consider the semidirect product d, $L = kd \rtimes H$ (see Section 1.4.4). Then L can be generated by d, x_0, x_1, y_0, y_1. Besides, it contains a subalgebra H with a property that under any homomorphism onto a finite-dimensional Lie algebra its element z is mapped into zero. Hence L is not a residually finite algebra.

These examples lead us to the following problems:

(1) Describe all varieties of Lie algebra with *all* algebras residually finite. Clearly, this is of interest in the case of finite fields only.
(2) Describe all varieties of Lie algebras in which all *finitely generated* algebras are residually finite. Both problems are still open.

6.6.5 Residual properties and wreath products

A Lie algebra G is called *residually nilpotent* if for any $0 \neq g \in G$ there exists a homomorphism φ_g from G into a nilpotent algebra N_g such that $\varphi_g(g) \neq 0$. An equivalent condition is the following: the intersection of the terms of the lower central series of G is equal to zero algebra. Indeed, if $\bigcap_{n=1}^{\infty} G^n = \{0\}$ then for any $0 \neq g \in G$ we can find n such that $g \notin G^n$. Under the homomorphism $\varepsilon_n : G \to G/G^n$, we have $\varepsilon_n(g) \neq 0$ and G/G^n is a nilpotent algebra. Let, conversely, G be residually nilpotent and $g \in \bigcap_{n=1}^{\infty} G^n$. If $g \neq 0$ then for some $\varphi : G \to N$, where N is a nilpotent Lie algebra of nilpotent step, say, c, we have $\varphi(g) \neq 0$. But $\varphi(G^{c+1}) = (\varphi(G))^{c+1} = N^{c+1} = \{0\}$. Since $g \in G^{c+1}$, we may write $\varphi(g) \in \varphi(G^{c+1}) = \{0\}$, a contradiction, proving the equivalence of both definitions given above.

The following result will be used in Chapter 8 for the group theory purposes.

Proposition 6.48. *Consider $W = A \operatorname{wr}_{\mathbf{U}} L$, the $\mathbf{U}$-wreath product of a nilpotent Lie algebra L and a free algebra A of a multihomogeneous variety $\mathbf{U}$ over a field k. Then W is a residually nilpotent Lie algebra.*

Proof. We adopt the notation of the proof of Theorem 6.20, L being equal to its nilpotent radical N. Let U_s, $s = 0, 1, 2, \ldots$, be an ideal of $U(L)$ which is the set of all elements of this associative algebra whose weight is at least s. By Section 4.4, Proposition 4.24,

$W = L \rtimes K$ where K is a **U**-free algebra with free generating set Y of the form

$$ux, \quad \text{with } u = e_1 \cdots e_s, \quad e_1 \geq \cdots \geq e_s, \ e_i \in E, \ x \in X, \ s \geq 0, \tag{6.13}$$

where X is a free generating set for A. We define the weight of ux in (6.13) as $\mathrm{wt}(u)+1$. If a is a monomial in the generators of the form (6.13) then its weight is defined as the sum of the weights of its factors. Now let K_n be the linear span of the monomials of weight at least n. Then K_n is an ideal in K. Moreover, if $e_i \in E$ and u is a monomial in U_{n-1} then $(\mathrm{ad}\, e_i)(ux) = (e_i u)x$. Since $e_i u \in U_n$, $(\mathrm{ad}\, e_i)(ux) \in K_{n+1}$ hence K_n is an ideal of W. The weight of each element in K being a well-defined number, it is clear that $\bigcap_{s=1}^{\infty} K_s = \{0\}$. Furthermore, for each s, W/K_s is a nilpotent Lie algebra. This follows by applying Section 1.7, Theorem 1.19, to W/K_s with its ideal $K + K_s/K_s$. By the above property, $eK_n \subseteq eK_{n+1}$, and we conclude that $(W/K_s)(K + K_s/K_s)^2$ is nilpotent. Since, obviously, K/K_s is nilpotent, we finally conclude that W/K_s is a nilpotent algebra. Thus, W is residually an algebra in the family of nilpotent algebras of the form W/K_s, $s = 1, 2, \ldots$ □

6.7 Identities of the universal enveloping algebra

A particular class of special Lie algebras is formed by those Lie algebras G for which already the universal enveloping algebra $U(G)$ is a PI-algebra. We will give a complete description of this class which is not merely interesting by itself but has applications to the representation theory of Lie algebras in characteristic $p > 0$.

6.7.1 Description of the universal envelopes with identity in the case char $k = 0$

It will be shown here that if $U(G)$ is a PI-algebra then G is an abelian Lie algebra.

Lemma 6.49. *Any finite-dimensional nonabelian Lie algebra G over an algebraically closed field either contains a two-dimensional nonabelian Lie algebra or a three-dimensional nonabelian nilpotent subalgebra.*

Proof. This is left to the reader as an easy exercise based on considering eigenvalues of operators of the form $\mathrm{ad}\, x$, $x \in G$. □

Now we remark that if G has an infinite-dimensional absolutely simple module M then $U = U(G)$ cannot be a PI-algebra. Indeed, by the density theorem (Section 6.1), if $I = \mathrm{Ann}_U M$ then, for any n, U/I has a subalgebra whose homomorphic image is isomorphic to k_n. This follows since, by Theorem 6.43, we must have $\mathrm{End}_G V = k$. Now by Theorem 6.2, U cannot satisfy a nontrivial identity. Hence it follows that, over an algebraically closed field of characteristic zero, $U(G)$, G being a finite-dimensional nonabelian, is not a PI-algebra. To complete our argument, it remains to pass to an arbitrary field of characteristic zero and to any arbitrary dimension of G. The former is simple; indeed, if $U(G)$ is a PI-algebra then it satisfies a nontrivial multilinear identity. Now if $\bar{k}$ is the algebraic closure of the base field k then $U(\bar{k}\otimes G) \cong \bar{k}\otimes U(G)$. Hence, $\bar{k}\otimes G$ is an abelian Lie algebra, implying the same for G. Now suppose G is a Lie algebra of arbitrary dimension over a field k and $U(G)$ is a PI-algebra. By Theorems 6.4 and 6.5, $U(G)$ is embedded in a field D which is finite-dimensional over its center Z. Hence $\widetilde{G} = GZ \subseteq D$ is finite-dimensional over Z. Its basis can be chosen to consist of the elements of G, namely $e_1, \dots, e_n$. We may consider the universal enveloping algebra V for a finite-dimensional Z-algebra G. Its associative k-subalgebra W generated by the k-subalgebra G is a PI-algebra (as a homomorphic image of $U(G)$). On the other hand, $V = WZ$ since a basis of V over Z consists of ordered monomials in $e_1, \dots, e_n$. Thus V is a PI-algebra and, by the first part of the proof, $\widetilde{G}$ is abelian. Of course, then G is also abelian. So we have the following.

Theorem 6.50. *Let k be a field of characteristic zero. The universal enveloping algebra $U(G)$ for a Lie algebra G over k is a PI-algebra if and only if G is an abelian Lie algebra.*

Our main theme in the remaining part of this section will be the case of a field of prime characteristic. In the next subsection we are proving an important auxiliary result. In its statement the *breadth* of an element x of a Lie algebra G is a number $b(x) = \dim G/C_G(x)$ (see Section 1.5.4).

Theorem 6.51. *Let G be a Lie algebra over a field k such that for a natural number b we have $b(x) < b$ for any $x \in G$. Then $\dim G^2 < b^2$.*

Proof. It is convenient to translate this statement to the language of bilinear maps. So we consider a more general problem: given a bilinear map $\varphi : U \times V \to W$, we denote $b_\varphi(u) = \dim(V/\mathrm{Ann}_\varphi u)$ and $b_\varphi(v) = \dim(V/\mathrm{Ann}_\varphi v)$, where $u \in U$, $v \in V$. It is clear that $b_\varphi(u) = \dim \varphi(u, V)$ and $b_\varphi(v) = \dim \varphi(U, v)$. Since $\varphi(U, V)$ is not necessarily a subspace in W, we denote by W' the linear span of $\varphi(U, V)$. Suppose there exist natural numbers r, s such that for all $u \in U$, $v \in V$, we have $b_\varphi(u) \le r$, $b_\varphi(v) \le s$. It is clear that the proof of Theorem 6.51 will be complete as soon as we prove $\dim W' \le rs$. Indeed, it is possible, in this case, to identify each of U, V, W with G and to consider $\varphi : G \times G \to G$ such that $\varphi(g_1, g_2) = g_1 g_2$. It is clear that the linear span $k\varphi(G, G)$ will then be equal to G^2. So we concentrate on the problem of bilinear maps. It should be remarked that in Section 6.9 we will be using the result on the bilinear maps in its full form. Now, adopting the above notation, we assume that $r \le s$. We will proceed by induction on r, the base being obvious in the case $r = 0$. So if we had $b_\varphi(u) = r - 1$ then we would be able to write $\dim W' \le (r - 1)s$. Now we pick a $u_1 \in U$ with $b_\varphi(u_1) = r$. Denote $X = \mathrm{Ann}_\varphi(u_1) \subseteq V$, $Y = \mathrm{Ann}_\varphi(X) \subseteq U$. We consider the natural map $\psi_1 : W \to W/\varphi(u_1, V)$ and let φ_1 be the bilinear map $\varphi_1 = \psi_1\varphi : U \times V \to W_1$. It is clear that $\mathrm{Ann}_{\varphi_1}(u_1) = V$. We notice that for any $u \in U \setminus Y$ we have $b_{\varphi_1}(u) \le r - 1$. Indeed, in this case there exists an element $x \in X$ such that $\varphi(u, x) \ne 0$. Then also $\varphi(u + u_1, x) \ne 0$. We set $U^* = ku + ku_1 = k(u + u_1) + ku$ and $T = k\varphi(U^*, V)$. Then

$$T = \varphi(u, V) + \varphi(u_1, V) = \varphi(u, V) + \varphi(u + u_1, V).$$

Since $\varphi(u, V) \cap \varphi(u + u_1, V) \ni \varphi(u, x) = \varphi(u + u_1, x) \neq 0$, we have $\dim T \le 2r - 1$. But $\dim \varphi(u_1, V) = r$ so that

$$b_{\varphi_1}(u) = \dim \varphi_1(u, V) \le \dim T/\varphi(u_1, V) \le r - 1.$$

Now if also for all $y \in Y$ we have $b_{\varphi_1}(y) \le r - 1$ then, by the induction hypothesis, $\dim k_{\varphi_1}(U, V) \le (r - 1)s$. In this case the proof is complete since

$$\dim k\varphi(U, V) = \dim k\varphi_1(U, V) + \dim \varphi(u_1, V) \le (r - 1)s + r \le (r - 1)s + s \le rs.$$

In the remaining case, there exists an element $u_2 \in Y$ such that $b_{\varphi_2}(u_2) = r$. Let ψ_2 denote the natural map from W_1 into $W_2 = W_1/\varphi_1(u_2, V)$. Under the bilinear map $\varphi_2 = \psi_2\varphi_1 : U \times V \to W_2$ we have $\mathrm{Ann}_{\varphi_2}(u_2) = \mathrm{Ann}_{\varphi_2}(u_1) = V$. If u is not in Y then $\varphi_2(u, V) = \varphi_1(u, V) + \varphi_1(u_2, V)/\varphi_1(u_2, V)$. As before, we can derive $b_{\varphi_2}(u) \le r - 2$ for all $u \in U \setminus Y$. Proceeding in a similar way, we list a sequence $\varphi_1, \varphi_2, \dots, \varphi_t$ of bilinear maps and a sequence $\psi_1, \psi_2, \dots, \psi_t$ of linear maps, $t \le r$, such that:

(i) $\varphi_{i+1} = \psi_{i+1}\varphi_i$, $\dim(\ker \psi_i) = r$,
(ii) $b_{\varphi_i}(u) \le r - 1$ for all $u \notin Y$,
(iii) $b_{\varphi_t}(y) \le r - t$ for all $y \in Y$,

or $r = t$. In the case $r = t$, we have $k\varphi_r(U, V) = k\varphi_r(Y, V)$ and $k\varphi_r(U, V) = k\varphi_r(Y, V)$ since $\ker(\psi_r \cdots \psi_1) \subseteq k\varphi(Y, V)$. But if $\{v_1, \dots, v_r\}$ is a basis of V modulo X then

$$k\varphi(y, V) = \varphi(Y, v_1) + \cdots + \varphi(Y, v_r) + \varphi(Y, X).$$

Since $\varphi(X, Y) = 0$ and $\varphi(Y, v_i) \le s$, we finally have $\dim W' \le rs$. Now suppose $t < r$. In this case $b_\varphi(u) \le r - t$ for all u which, by the induction hypothesis, gives $\dim k\varphi_t(U, V) \le (r - t)s$. But $\dim k\varphi(U, V) = \dim k\varphi_t(U, V) + \dim \ker(\psi_t \cdots \psi_1) \le (r - t)s + \mathrm{tr} \le (r - t)s + ts = rs$, completing the proof of Theorem 6.51 in all cases. □

Theorem 6.52. *Let G be a Lie algebra over a field k of characteristic p > 0. Its universal enveloping algebra is a PI-algebra if and only if the following two conditions hold:*

(1) *G possesses an abelian ideal H such that* $\dim(G/H) < \infty$;
(2) *For any* $x \in G$ *the inner derivation* $\mathrm{ad}\, x$ *is algebraic of bounded degree depending entirely on G.*

Proof. The necessity of these conditions will be proved in Proposition 6.55 and Theorem 6.58. We show first that these are sufficient.

We write G as a direct sum of H and a complementing subspace F: $G = H \oplus F$ and choose a well-ordered basis E in H and a basis $\{f_1, f_2, \ldots, f_n\}$ in F where $n = \dim(G/H)$. Then by the PBW theorem (Section 2.5), a basis in $U(G)$ can be chosen in the form

$$e_1 e_2 \cdots e_t f_1^{r_1} f_2^{r_2} \cdots f_n^{r_n}, \quad e_1 \geq e_2 \geq \cdots \geq e_t, e_i \in E, r_j \geq 0.$$

By Section 6.2.3, we can find p-polynomials $\varphi_1, \varphi_2, \ldots, \varphi_n$ such that $\varphi_i(\operatorname{ad} f_i) = 0$ and $z_i = \varphi_i(f_i) \in Z(U)$, $i = 1, 2, \ldots, n$, and then a basis for U can be chosen in the form

$$e_1 e_2 \cdots e_t z_1^{s_1} z_2^{s_2} \cdots z_n^{s_n} f_1^{r_1} f_2^{r_2} \cdots f_n^{r_n}, \quad e_1 \geq \cdots \geq e_t,\ e_i \in E,$$
$$s_j \geq 0, 0 \leq r_j < \deg \varphi_j. \tag{6.14}$$

It is easy to see that the elements of the form (6.14) with $r_1 = r_2 = \cdots = r_n = 0$ form a commutative subring S in U, the polynomial ring $k[E; z_1, z_2, \ldots, z_n]$, and that U is a free left S-module with a finite number d of generators, $d \leq (\deg \varphi_1) \cdots (\deg \varphi_n)$. We consider $\operatorname{End}_S(U)$ and view U as a subalgebra in $\operatorname{End}_S(U)$ by identifying each $x \in U$ with operator R_x in $\operatorname{End}_S(U)$. This is correct since U is a left S-module and R_x is a multiplication on the right. But $\operatorname{End}_S(U) \cong S_d$ satisfies a standard identity of any degree at least $d^2 + 1$ (in fact, $2d$) and thus it is a PI-algebra. Hence we conclude that U is a PI-algebra, proving the sufficiency part of the main theorem. □

Now we pass to a much tougher part, the necessity, in the main theorem. Let G be a Lie algebra over a field k and $\Delta_n = \Delta_n(G)$ the set of all $x \in G$ such that $b(x) \leq n$ (Theorem 6.51). Although Δ_n is not a subspace, we say that the set of elements $x_1, x_2, \ldots, x_n$ is linearly dependent modulo Δ_n if there exist not all zero coefficients $\alpha_1, \alpha_2, \ldots, \alpha_m \in k$ such that $\alpha_1 x_1 + \alpha_2 x_2 + \cdots + \alpha_m x_m \in \Delta_n$.

Proposition 6.53. *Suppose there exist, in a Lie algebra G, two natural numbers m, n such that any m elements are linearly dependent modulo Δ_n. Then G has a subalgebra H such that* $\dim(G/H)$ *and* $\dim(H^2)$ *are finite.*

Proof. Let m be the least with this property, and $e_1, e_2, \ldots, e_{m-1}$ linearly independent modulo Δ_n. Let H denote the subalgebra generated by Δ_n. Now $H = \Delta_n + H \cap E$ where E is the linear span of $\{e_1, \ldots, e_{m-1}\}$. By our hypothesis, then for each $h \in H$ we have $h + \alpha_1 e_1 + \cdots + \alpha_{m-1} e_{m-1} \in \Delta_n$. Let $\{f_1, f_2, \ldots, f_q\}$ be a basis of $H \cap E$. We notice that if $x, y \in \Delta_n(G)$ then $x + y$ and $[x, y] \in \Delta_{2n}(G)$ because, say, $C_G(x + y) \supset C_G(x) \cap C_G(y)$ and $\dim(G/C_G(x) \cap C_G(y)) \leq 2n$. Using this remark, it is now a simple matter to find a natural number N such that $b_G(h) \leq N$ for any $h \in H$. Indeed, each element in H can be written in the form of a linear combination of at most m elements of one of the form $x \in \Delta_n$ and $\beta_i f_i$, $i = 1, \ldots, q$; each of these elements is in $H = \operatorname{alg}(\Delta_n)$, hence it is representable in the form of a linear combination of finitely many monomials of finite degree in Δ_n. Having noticed this, it is clear that $b_H(h) \leq b_G(h) \leq N$. By Theorem 6.51, then $\dim(H^2) \leq N^2$. Now $\dim(G/H) < \infty$ follows since $H \supseteq \Delta_n$. □

Proposition 6.54. *Let G be a Lie algebra over a field k and A its abelian subalgebra of finite codimension. Then G has an abelian ideal of finite codimension in G.*

Proof. Let $[nG, A] = [\underbrace{G, \dots, G}_{n}, A]$ be the subspace of G generated by all monomials of the form

$$[g_1, g_2, \dots, g_n, a], \quad g_i \in G, i = 1, 2, \dots, n, a \in A, n \geq 1,$$

and let $\{B_i, i = 0, 1, 2, \dots\}$ be a nondecreasing chain of subspaces of the form $B_0 = A$, $B_{s+1} = B_s + [(s+1)G, A]$. By our condition, for some r we have $B_{r+1} = B_r$ and therefore B_r is an ideal of G. Now suppose $m = \dim(G/A)$ and let $g_1, g_2, \dots, g_m$ form a basis of G modulo A. We define a descending chain of abelian subalgebras A_j, $j = 0, 1, 2, \dots$, by setting $A_0 = A$ and choosing A_l in the form of the set of all elements in A_{l-1} such that $[x, g_j] \in A_{l-1}$, $j = 1, \dots, m$. It is clear that $[G, A_l] \subseteq A_{l-1}$. Moreover, using induction over l, we can prove the following:
(1) $\dim(G/A_l) < \infty$, $l = 0, 1, 2, \dots$,
(2) $[A_l, B_l] = 0$, $l = 0, 1, 2, \dots$

The basis of induction, $l = 0$, is trivial. So we consider the bilinear map $A_{l-1} \times G \to G \to G/A_{l-1}$ induced by the bracket operation in G. Since A_l is the left kernel of this map, the induction hypothesis, $\dim(G/A_{l-1}) < \infty$, immediately gives $\dim(A_{l-1}/A_l) < \infty$, hence the same is true for $\dim(G/A_{l-1})$. The second claim is simple as well. Indeed, if already $[A_{l-1}, B_{l-1}] = \{0\}$ then, by the Jacobi identity, we may write

$$\begin{aligned}
[A_l, B_l] &= [A_l, B_{l-1} + [lG, A]] = [A_l, B_{l-1}] + [A_l, lG, A] \\
&= [[A_l, G], [(l-1)G, A]] + [G, A_l, (l-1)G, A] \subseteq [A_{l-1}, B_{l-1}] = 0.
\end{aligned}$$

Now we consider the ideal B_r of G and let H be its center. Then, by (6.1) and (6.2), we have

$$\dim(B_r/H) \leq \dim(B_r/A_r) \leq \dim(G/A_r) < \infty.$$

Since the center of an ideal is itself an ideal, our proof is complete. □

Proposition 6.55. *Let G be a Lie algebra over a field k, $U(G)$ its universal enveloping algebra satisfying a nontrivial polynomial identity. Then there exist natural numbers m and n such that any m elements of G are linearly dependent modulo $\Delta_n(G)$.*

Proof. Since $U(G)$ has no zero divisors, Theorem 6.5 applies. Let $m-1$ be the dimension of the right ring of quotients Q of $U = U(G)$ over its center. Then, given an m-element subset $x_1, \dots, x_m$ in G, there exist not all zero central elements $a_1b_1^{-1}, a_2b_2^{-1}, \dots a_mb_m^{-1}$, $a_ib_i \in U$, $1 \leq i \leq m$, such that

$$x_1a_1b_1^{-1} + x_2a_2b_2^{-1} + \dots + x_ma_mb_m^{-1} = 0. \tag{6.15}$$

Let x be an arbitrary element in G. Taking commutators of both sides of equation (6.15) with x and then multiplying both sides by $b_1 b_2 \cdots b_m$, we will write

$$[x_1, x]c_1 + [x_2, x]c_2 + \cdots + [x_m, x]c_m = 0,$$

where each value c_i is an element of U of the form $c_i = a_i b_i^{-1} b_1 \cdots b_m = b_1 \cdots b_{i-1} a_i b_i^{-1} b_i \cdots b_m$, $i = 1, \ldots, m$. Not all c_i are zero. Now we consider $R = \mathrm{gr}(U) = k[G]$, where G is identified with the subspace of homogeneous polynomials of degree 1 in $k[G]$. Let $f_1, f_2, \ldots, f_m$ be the images of $c_1, c_2, \ldots, c_m$ in R. Then we may write

$$[x_1, x]f_1 + [x_2, x]f_2 + \cdots + [x_m, x]f_m = 0 \tag{6.16}$$

an equation in $R = k[G]$, which is identically true over $x \in G$. Now we arrive at a system

$$a_{x1}t_1 + a_{x2}t_2 + \cdots + a_{xm}t_m = 0, \quad x \in G, a_{xi} = [x, x_i] \tag{6.17}$$

of linear equations over the polynomial ring $R = k[G]$ which has a nontrivial solution $(f_1, f_2, \ldots, f_m)$ as in equation (6.16). Since R has no zero divisors, there exists an l, $0 \leq l < m$, such that all the minors of the matrix of coefficients of system (6.17) of order $> l$ are zero. We choose l minimal with this property. If $l = 0$ then $a_{xi} = [x, x_i] = 0$ for all $x \in G$, hence $x_i \in Z(G) \subseteq \Delta_n$, $n \geq 0$. If $l > 0$ then there exists a nontrivial minor P of order l with elements $a_{y_s i_r}$, $y_1, y_2, \ldots, y_l \in G$, $\{i_1, i_2, \ldots, i_l\} \subset \{1, 2, \ldots, m\}$. Let $i_{l+1} \in \{1, 2, \ldots, m\} \setminus \{i_1, i_2, \ldots, i_l\}$, and let x be an arbitrary element in G. Then expanding the determinant of order $l + 1$ of the form

$$0 = \begin{vmatrix} & & & a_{y_1 i_{l+1}} \\ & P & & \vdots \\ & & & a_{y_l i_{l+1}} \\ a_{xi_1} & \cdots & a_{xi_l} & a_{xi_{l+1}} \end{vmatrix} = \begin{vmatrix} [y_1, x_{i_1}] & \cdots & [y_1, x_{i_l}] & [y_1, x_{i_{l+1}}] \\ \vdots & \ddots & \vdots & \vdots \\ [y_l, x_{i_1}] & \cdots & [y_l, x_{i_l}] & [y_l, x_{i_{l+1}}] \\ [x, x_{i_1}] & \cdots & [x, x_{i_l}] & [x, x_{i_{l+1}}] \end{vmatrix}$$

along the last row, we obtain a relation of the form

$$[x_1, x]g_1 + [x_2, x]g_2 + \cdots + [x_m, x]g_m = 0 \tag{6.18}$$

where not all $g_1, g_2, \ldots, g_m$ are equal to zero (for instance, $g_{i+1} = P \neq 0$). Moreover, if V is a subspace of G spanned by the commutators $[y_s, x_{i_r}]$, $1 \leq s \leq l$, then $\dim(V) \leq l(l + 1) \leq m^2$ and $g_1, g_2, \ldots, g_m \in k[V]$.

Now we put $n = m^2$ and, assuming with no loss of generality that $g_m \neq 0$, we will prove that $x_1, x_2, \ldots, x_m$ are linearly independent modulo $\Delta_n(G)$. In doing this, we will use induction over the number of nonzero coefficients in equation (6.18). If $m = 1$ then $[x_m, x]g_m = 0$, which in R, free of zero divisors, implies $[x_m, x] = 0$ for any $x \in G$, hence $x_m \in Z(G) \subseteq \Delta_n$, $n \geq 1$. Now suppose $m > 1$ and, say, $g_1 \neq 0$. We assume first that there exists an i such that $[x_i, x] \in V$ for all $x \in G$. Then $[x_i, G] \subseteq V$ and $b(x_i) = \dim(G/C(x_i)) \leq \dim V < m^2$ so that $x_i \in \Delta_n$ and we are finished. If that hypothesis does not hold then we choose an element $y_0 \in G$ such that $[x_1, y_0] = e \notin V$. We choose a linear basis in G including a basis of V and e and write $[x_i, y_0] = \alpha_i e + h_i$, $i = 2, \ldots, m$; the elements h_i are written through the elements of this basis without e. Substituting y_0 into equation (6.18), we will get

$$e(g_1 + \alpha_2 g_2 + \cdots + \alpha_m g_m) + h_2 g_2 + \cdots + h_m g_m = 0.$$

Using a standard basis of the polynomial ring, we derive from this equation the following:

$$g_1 + \alpha_2 g_2 + \cdots + \alpha_m g_m = 0,$$

which turns equation (6.18) into the equation of the form

$$[x_2 - \alpha_2 x_1, x]g_2 + [x_3 - \alpha_3 x_1, x]g_3 + \cdots + [x_m - \alpha_m x_1, x]g_m = 0.$$

Now the induction hypothesis applies. But it is obvious that a nontrivial dependence of $x_2 - \alpha_2 x_1, \ldots, x_m - \alpha_m x_1$ modulo Δ_n implies such a dependence for $x_1, x_2, \ldots, x_m$. □

Corollary 6.56. *A Lie algebra G with $U(G)$ a PI-algebra has a subalgebra H with* $\dim H^2 < \infty$ *and* $\dim(G/H) < \infty$.

Proof. This is immediate from the just-proved Proposition 6.55 and Proposition 6.53. □

Our nearest aim is the following.

Theorem 6.57. *Let G be a Lie algebra over a field k and suppose its universal enveloping algebra $U(G)$ is a PI-algebra. Then G has an abelian ideal of finite codimension depending solely on the degree of the nontrivial identity holding in $U(G)$.*

We start with an example of a Lie algebra whose universal enveloping algebra is not a PI-algebra.

Example. Let H_{d+1} be a two-step nilpotent Lie algebra with a basis

$$\{x_1, x_2, \ldots, x_{d+1}, y_1, y_2, \ldots, y_{d+1}, z\},$$

and the multiplication table for H_{d+1} being

$$[x_i, x_j] = [y_i, y_j] = [x_i, z] = [y_i, z] = 0, \quad 1 \le i, j \le d+1, \; [x_i, y_j] = \delta_{ij} z,$$

where δ_{ij} is Kronecker's delta. We will show that $U(H_{d+1})$ satisfies no polynomial identity of degree at most d. Indeed, let

$$X_1 X_2 \cdots X_d + \sum_{\sigma \neq 1} \alpha_\sigma X_{\sigma(1)} X_{\sigma(2)} \cdots X_{\sigma(d)} = 0 \tag{6.19}$$

be a multilinear identity of degree d satisfied by $U(H_{d+1})$. We order the above basis for H_{d+1} in the following way

$$x_1 > x_2 > \cdots > x_{d+1} > y_1 > y_2 > \cdots > y_{d+1} > z.$$

Then by the PBW theorem (Section 2.5), a basis for $U(H_{d+1})$ is formed by monomials of the form

$$x_1^{s_1} x_2^{s_2} \cdots x_{d+1}^{s_{d+1}} y_1^{r_1} y_2^{r_2} \cdots y_{d+1}^{r_{d+1}} z^t \quad (s_i, r_j, t \geq 0). \tag{6.20}$$

To compute the value of the left-hand side of equation (6.19) in $X_1 = x_1 y_2$, $X_2 = x_2 y_3$, $\ldots, X_d = x_d y_{d+1}$, we need the following result.

Let $Q = (q_{ij})$ be an upper triangular $d \times d$ matrix each column of which has precisely one nonzero element and this is equal to 1. Then for each permutation $\sigma \in S_n$ the element

$$x_{\sigma(1)} y_{\sigma(1)+1} x_{\sigma(2)} y_{\sigma(2)+1} \cdots x_{\sigma(d)} y_{\sigma(d)+1}$$

can be rewritten in the form

$$\sum_Q \pm x_{\sigma(1)}^{q_{11}} x_{\sigma(2)}^{q_{12}} \cdots x_{\sigma(d)}^{q_{1d}} ((\operatorname{ad} x_{\sigma(d2)})^{q_{2d}}) y_{\sigma(1)+1}$$
$$\cdots ((\operatorname{ad} x_{\sigma(d)})^{q_{dd}})(y_{\sigma(d-1)+1}) y_{\sigma(d)+1}, \tag{6.21}$$

where the summation is taken over the matrices of the above form.

This follows by a direct calculation using induction on d and considering the fact that $(\operatorname{ad} x_i)(y_j)$ is central. Each summand in (6.21) is either zero or equals to a basic element in $U(H_{d+1})$. We consider the set of those summands in (6.21) which have the form as in (6.20) with $t = d - 1$. It is obvious that such a monomial arises only in the case where Q is a unity matrix, that is, $q_{ii} = 1$ for all $1 \leq i \leq d$. This unique monomial has the form

$$\pm x_{\sigma(1)} (\operatorname{ad} x_{\sigma(2)})(y_{\sigma(1)+1}) \cdots (\operatorname{ad} x_{\sigma(d)})(y_{\sigma(d-1)+1}) y_{\sigma(d)+1} z^{d-1}. \tag{6.22}$$

If $\sigma = 1$ then monomial (6.22) takes the form $x_1 y_{d+1} z^{d-1}$. If this is not the case then, for some j, we have $\sigma(j) \neq \sigma(j-1) + 1$ so that

$$(\operatorname{ad} x_{\sigma(j)})(y_{\sigma(j-1)+1}) = [x_{\sigma(j)}, y_{\sigma(j-1)+1}] = 0.$$

This argument shows that the above-mentioned value of the left-hand side in equation (6.19) is different from zero, contradicting our former hypothesis. Thus, an example of a 2-step nilpotent Lie algebra whose universal enveloping algebra is not a PI-algebra is given by a Lie algebra of the form H_∞, that is, with the same multiplication table as for H_{d+1} but with infinitely many basic elements x_i, y_j.

Proof of Theorem 6.57. By the corollary of Proposition 6.55, it is sufficient to restrict to the case of a Lie algebra H with $\dim(H^2) < \infty$. Suppose we have proven our theorem for H so that K is an abelian ideal of finite codimension in H. Then K an abelian subalgebra of finite codimension in G and Proposition 6.54 gives the desired result. So, coming back to H, we consider $C = C_H(H^2)$. By Section 1.5.4, C is an ideal of finite codimension in H. It is clear that C is a nilpotent Lie algebra of nilpotent step at most 2. By the same reason as previously, it is sufficient to prove that the center $Z = Z(C)$ of C is of finite codimension in C. Moreover, we can reduce even further in the following way. Since $\dim(C^2) = t < \infty$ and $C^2 \subseteq Z(D)$, we can represent $\{0\}$ as an intersection of ideals D_i, $i = 1, 2, \dots, t$, $\dim C^2/D_i = 1$. Then C is a subdirect product of Lie algebras $C_i = C/D_i$ with $\dim C_i^2 = 1$. In other words, the map $\varphi : C \to \prod_i C_i$ given by $\varphi(c) = (c + D_1, \dots, c + D_t)$ is a monomorphism such that the projection of $\varphi(C)$ onto C_i is equal to the whole of C_i. Then, clearly, $U(C_i)$ is a PI-algebra (it is a homomorphic image of $U(C)$), so it suffices to verify that the center of C_i is of finite codimension in C_i. Since C_i^2 is one-dimensional and it is generated by an element, say, e, the multiplication in C_i determines an alternating bilinear form $(x, y) = \alpha$ where $[x, y] = \alpha e$. It is well known that such a form has a canonical basis $\{x_1, \dots, x_s, \dots; y_1, \dots, y_r; z_1, z_2, \dots\}$ such that $(y_r, x_s) = -(x_s, y_r) = \delta_{rs}$, $(x_i, x_j) = (y_i, y_j) = (y_i, z_j) = (z_i, z_j) = 0$. If the center of some C is not of finite codimension, then, for any $d > 0$, C_i cannot satisfy a nontrivial identity, a contradiction. □

In the remaining part of the section, we treat inner derivations in Lie algebras whose universal envelopes are PI-algebras.

Theorem 6.58. *Let G be a Lie algebra over a field k of characteristic $p > 0$ such that $U(G)$ is a PI-algebra. Then for any $x \in G$, $\operatorname{ad} x$ is algebraic of some degree D where D depends entirely on G.*

Proof. Suppose $U(G)$ satisfies a multilinear identity of the form

$$P(X_1, X_2, \dots, X_d) = \sum_{\sigma \in S_d} \alpha_\sigma X_{\sigma(1)} X_{\sigma(2)} \cdots X_{\sigma(d)} \equiv 0, \tag{6.23}$$

where S_d is the permutation group of $\{1, 2, \dots, d\}$, $\alpha_1 \neq 0$. We assume for a moment that the elements $v_0 = v$, $v_1 = (\operatorname{ad} g)(v), \dots, v_{N+1} = (\operatorname{ad} g)^{N+1}(v)$ are linearly independent where $N = 1 + p + p^2 + \cdots + p^{2d}$, $g, v \in G$. Obviously, then $g, v_0, v_1, \dots, v_N$ form a linearly independent system. We include this system as an initial segment into a well-ordered basis for $U(G)$. Then a basis of $U(G)$ is formed by monomials of the form

$$g^n v_0^{n_0} v_1^{n_1} \cdots v_N^{n_N} e_1^{l_1} e_2^{l_2} \cdots e_t^{l_t}, \tag{6.24}$$

where $e_1 \geq e_2 \geq \cdots \geq e_t$ and almost all nonnegative integers $n, n_0, n_1, \dots, n_N$, $l_1, l_2, \dots, l_t, \dots$ are equal to zero. We are going to show that

$$P(g^p v_{p^{d+1}}, g^{p^2} v_{p^{d+2}}, \dots, g^{p^d} v_{p^{2d}}) \tag{6.25}$$

is a nonzero element in $U(G)$, in contradiction with equation (6.23). Indeed, the left-hand side of (6.25) is a linear combination of the monomials of the form

$$B_\sigma = g^{p^{\sigma(1)}} v_{p^{d+\sigma(1)}} g^{p^{\sigma(2)}} v_{p^{d+\sigma(2)}} \cdots g^{p^{\sigma(d)}} v_{p^{d+\sigma(d)}}. \tag{6.26}$$

Lemma 6.59. *Let Q be an upper triangular $d \times d$ matrix as in Proposition 6.55. Then B_σ in equation (6.26) is representable in the form*

$$\sum_Q \pm g^{p^{\sigma(1)}+q_{12}p^{\sigma(2)}+\cdots+q_{1d}p^{\sigma(d)}} v_{p^{d+\sigma(1)}+q_{22}p^{\sigma(2)}+\cdots+q_{2d}p^{\sigma(d)}} \cdots v_{p^{d+\sigma(d-1)}+q_{dd}p^{\sigma(d)}} v_p^{d+\sigma(d)}, \tag{6.27}$$

where the summation is taken over all matrices as above.

We proceed by induction over d, our lemma being trivial if $d = 1$. The general case is dealt with by transforming the initial segment of equation (6.26) (i. e., without $g^{p^{\sigma(d)}} v_{p^{d+\sigma(d)}}$) using the induction hypothesis, and permuting the "v-tail" of equation (6.26) with $g^{p^{\sigma(d)}}$, considering the fact that commutation is a derivation such that $(\operatorname{ad} g^p)(v_r) = (\operatorname{ad} g)^p(v_r) = v_{r+p}$. Standard calculations using Section 1.11.1 are left to the reader.

Now it is clear that by our hypothesis, for different sequences $(s, \varepsilon_1, \varepsilon_2, \ldots, \varepsilon_d)$, the elements of the form

$$v_{p^{d+s}+\varepsilon_1 p+\varepsilon_2 p^2+\cdots+\varepsilon_d p^d}, \quad 1 \le s \le d, \varepsilon_r = 0, 1,\ r = 1, 2, \ldots, d \tag{6.28}$$

are pairwise different elements of the above basis of G. Denoting by $\bar{M}$ the element of the highest degree in the representation of a monomial M in (6.28) in the form (6.24), we easily see that $\bar{M}$ is different from M only by the permutation of "v-factors", that is, factors of the form (6.28).

So we choose from (6.28) the unique monomial C_σ in which only one "v-factor" has the form as in (6.28) with $\varepsilon_1 = \varepsilon_2 = \cdots = \varepsilon_d = 0$. It is clear that this monomial can be written as

$$C_\sigma = g^{p^{\sigma(1)}} v_{p^{d+\sigma(1)}+p^{\sigma(2)}} v_{p^{d+\sigma(2)}+p^{\sigma(3)}} \cdots v_{p^{d+\sigma(d)}}, \tag{6.29}$$

and it is linearly independent of the set of all like monomials for the remaining values of σ. Suppose now that $C_1 = C_\sigma$ for some σ. Then, clearly, $p^{\sigma(1)} = p^1$, hence $\sigma(1) = 1$. Therefore, $v_{p^{d+\sigma(1)}+p^{\sigma(2)}} = v_{p^{d+1}+p^2}$, hence $\sigma(2) = 2$. Proceeding in a similar way, we eventually get $\sigma = 1$. Thus C_1 is linearly independent of all the remaining summands in (6.25) and it has coefficients $\alpha_1 \neq 0$.

This contradiction proves that, given $g \in G$, for any $v \in G$ there exists a polynomial $\varphi_v(t) \in k[t]$ such that $(\varphi_v(\operatorname{ad} g))(v) = 0$ and $\deg \varphi \le N+1$. In this case, we consider G as a periodic $k[t]$-module relative to the action $tv = (\operatorname{ad} g)(v)$ and decompose a the direct sum of primary components $G = \bigoplus_\mu G_\mu$, μ being an arbitrary irreducible polynomial in $k[t]$. It is obvious that at most $N+1$ components in this decomposition can be zero. Now since each G_{μ_i} is annihilated by $\mu_i(\operatorname{ad} g)^{N+1}$, $\operatorname{ad} g$ is annihilated by $(\mu_1\mu_2\cdots\mu_{N+1})^{N+1}$, hence is algebraic of bounded degree D. This completes the proof of Theorem 6.58. It is also clear that the combination of all the above theorems gives the proof of the main Theorem 6.52. □

As an application of Theorem 6.52 to the study of special Lie algebras, we have the following sufficient conditions of speciality of Lie algebras over a field of positive characteristic.

Theorem 6.60. *Let* $\operatorname{char} k = p > 0$. *Suppose L is a Lie algebra with a solvable ideal M such that M^2 is nilpotent and* $\dim L/M < \infty$. *If the natural action of L on M/M^2 is algebraic of bounded degree then L is a special Lie algebra. In particular, if G has an ideal M of finite codimension such that $[L, M]$ is a nilpotent algebra then L is a special Lie algebra.*

6.8 Application: description of Lie algebras all of whose irreducible representations have finite bounded degree

Theorem 6.61. *Let k be an algebraically closed field of characteristic $p > 0$ and G a Lie algebra over k. Then all irreducible representations of G have finite bounded dimension if and only if:*
(1) *G has an abelian ideal H of finite codimension;*
(2) *all inner derivations* $\operatorname{ad} x$, $x \in G$, *are algebraic;*
(3) $\dim_k(H) < |k|$.

Corollary 6.62. *A Lie algebra as in the above theorem is finite-dimensional if any of the following conditions hold:*
(1) *G is finitely generated;*
(2) *k is countable.*

In the zero characteristic case we have the following.

Proposition 6.63. *Let G be a Lie algebra over an algebraically closed field k of characteristic zero. Then all irreducible representations of G are of finite bounded degree if and only if G is abelian and* $\dim(G) < |k|$.

We will need the following auxiliary result.

Lemma 6.64. *Let k be an algebraically closed field, X a set of variables and $A = k[X]$ is the polynomial ring over k. Then A has an infinite-dimensional irreducible representation if and only if* $|X| \geq |k|$.

Proof. We remark first that if M is a simple A-module then $M \cong A/I$ where I is a maximal ideal in A, that is, M is an extension of k. Hence if X is finite then M is a field which is finitely generated over k as a k-algebra. In this case by Hilbert's Nullstellensatz (Section 6.6), M is algebraic over k, hence $M = k$. Therefore, we may consider X infinite. Moreover,

$$\dim(M) = \dim(A/I) \leq \dim(A) = |X|.$$

Now we consider an element $u \in A/I$ and let $B = k(u)$ be the subfield in A/I generated by u. If u is transcendental over k then $\dim(B) = |k|$. Indeed, $k(u)$ contains a subset $\{(u-\alpha)^{-1} \mid \alpha \in k\}$ which is linearly independent and of cardinality $|k|$. Hence $\dim(B) \geq |k|$. The converse is also true since B as a ring is generated by

$$R = \{u, (u-\alpha)^{-1}\},$$

the cardinality of R being equal to that of k. So, if A has an irreducible representation in $M = A/I$ and, at the same time, we have $|X| < |k|$ then any element $u \in A/I$ is algebraic over k, hence $\dim(A/I) = 1$ which is not the case. If, otherwise, $|X| \geq |k|$ then any map from X onto R extends to an epimorphism of $A = k[X]$ onto $B = k(u)$. Then A acquires an infinite-dimensional irreducible representation. □

Lemma 6.65. *Let k be an algebraically closed field of characteristic $p > 0$. If G is a Lie algebra over k satisfying (1) and (2) in Theorem 6.61 then G has an infinite-dimensional irreducible representation if and only if the same is true of H.*

Proof. Let $\{e_1, e_2, \ldots, e_n\}$ be a basis of G modulo H. By Section 6.2.3, there exist polynomials $f_1, f_2, \ldots, f_n$ such that $z_1 = f_1(e_1)$, $z_2 = f_2(e_2)$, $\ldots, z_n = f_n(e_n)$ are central in $U = U(G)$. We consider a subalgebra R in $U(G)$ generated by $z_1, z_2, \ldots, z_n$ together with $U(H)$. Since $z_1, z_2, \ldots, z_n$ are central, R is abelian. Moreover, $U(G)$ being a free left $U(H)$-module whose basis consists of the monomials of the form $e_1^{m_1} \cdots e_n^{m_n}$ ($\sum m_i \geq 0$), it is immediate that R is isomorphic to the polynomial ring $k[z_1, z_2, \ldots, z_n, u_\alpha \mid \alpha \in B]$ where $\{u_\alpha \mid \alpha \in B\}$ is a basis for H over k. Since, obviously, it is sufficient to view B as infinite, we derive $R \cong U(H) = k[u_\alpha \mid \alpha \in B]$. This shows that R has an infinite-dimensional irreducible representation if and only if this holds for H.

Suppose now H has no infinite-dimensional irreducible representation. By Lemma 6.64, then $\dim_k(H) < |k|$. In this case $U(G)$ is a finitely generated R-module, the dimension of R being equal to that of H. Hence the dimension of $U(G)$ over k is equal to $\dim(H)$ and, by the above, $\dim(U(G)) < |k|$.

Any simple $U(G)$-module is isomorphic to a certain $U(G)/I$, I being a maximal left ideal in $U(G)$. Let B be the greatest subalgebra in $U(G)$ containing I and such that I is a two-sided ideal in B. It is known (see Jacobson [151, Chapter 2]) that then $K = B/I$ is a skew field and then, by the density theorem (Section 6.1), either $\dim_K(U(G)/I)$ is finite or for any natural n there is a subalgebra in $U(G)$ whose homomorphic image is isomorphic to the matrix ring K_n. Now by Theorem 6.52, $U(G)$ satisfies a nontrivial identity. This identity then should be satisfied by every K_n, which is forbidden by Theorem 6.3.

Thus we may concentrate on the case where $\dim_K(U(G)/I) < \infty$. Here, K being algebraic over k immediately gives $K = k$. The property of being algebraic can be verified as follows. Suppose that some u is transcendental. As in Lemma 6.64 then $\dim(k(u)) = |k|$. This contradicts our former hypothesis:

$$\dim(k(u)) \leq \dim(K) \leq \dim(B/I) \leq \dim(U(G)/I) \leq \dim(U(G)) < |k|.$$

The "if" part of the lemma is even simpler. Let V be an infinite-dimensional simple left module for R. Considering that $U(G)$ is a right R-module, we construct a left $U(G)$-module by putting $W = U(G) \otimes_R V$. We identify V with a submodule $1 \otimes V$ in W. Taking a nonzero element $v \in V$, we consider a cyclic $U(G)$-submodule V' generated by v. Let M be a maximal $U(G)$-submodule of W which does not contain v (it exists by Zorn's lemma). Since V is R-irreducible we have $M \cap V = \{0\}$. Now we set $V'' = V' + M$. It is obvious that $T = V''/M$ is a simple $U(G)$-module. On the other hand, considering R-module, we may write

$$V \cong V/V \cap M \cong V + M/M \subset V' + M/M = V''/M = T,$$

which is an evidence of T being infinite-dimensional over k. □

6.8.1 Proofs of Theorem 6.41 and Proposition 6.63

First, let G be a Lie algebra all of whose irreducible representations have finite bounded degree. Since $U(G)$ is a Jacobson semisimple algebra, it follows that $U(G)$ is a subdirect product of primitive algebras having, by our hypothesis, finite bounded dimensions. Then $U(G)$ is a PI-algebra, so that we may apply Theorem 6.52 deriving the validity of conditions (1) and (2) in Theorem 6.61 in the case where the characteristic of the base field is prime. If the characteristic of k is zero then, by Section 6.7.1, it follows that G is abelian. Now (3) in Theorem 6.61 follows since by Lemma 6.64 in this case $\operatorname{char} k = p > 0$. If $\operatorname{char} k = 0$ then Lemma 6.64 yields $\dim(G) < |k|$.

These very lemmas work in proving the "if" part of our theorem and proposition. Indeed, if (1), (2) and (3) hold then G cannot have an infinite-dimensional irreducible representation. Now if M is a finite-dimensional irreducible G-module with $\dim_k(M) = n$ then by the density theorem, k_n is a homomorphic image of $U(G)$. Since $U(G)$ satisfies a nontrivial identity, this bounds n and completes the proof of our results. □

6.9 Description of Lie superalgebras whose universal enveloping algebra is a PI-algebra

6.9.1 Main definitions

The techniques of preceding sections are also efficient in finding necessary and sufficient conditions on a Lie superalgebra $L = L_0 \oplus L_1$ (see the definition in Section 1.10.2 (iii)) under which its universal enveloping algebra $U(L)$ is a PI-algebra. The notion of the universal enveloping algebra for a Lie superalgebra is quite similar to that for an ordinary Lie algebra. This is natural since, given a $\mathbb{Z}_2$-graded associative algebra $A = A_0 \oplus A_1$, setting

$$[x, y] = xy - (-1)^{\alpha\beta} yx, \quad \text{where } x \in A_\alpha, y \in A_\beta,$$

gives A a structure of a $\mathbb{Z}_2$-graded algebra, this time a Lie superalgebra. So, a $\mathbb{Z}_2$-graded associative algebra $U(L)$ with unity 1 is called the universal enveloping algebra for a Lie superalgebra L if there exists a graded homomorphism of Lie superalgebras $\varepsilon : L \to [U(L)]$ such that for any graded homomorphism $\varphi : L \to [A]$, A begin an associative $\mathbb{Z}_2$-algebra with 1, there exists unique graded homomorphism $\psi : U(L) \to A$ of associative algebras with unity such that $\psi\varepsilon = \varphi$. Similarly to our argument in the PBW theorem (Section 2.5), it is easy to prove that $U(L)$ and ε always exist and that the following analogue of the PBW theorem holds.

Theorem 6.66. *Let k be a field of characteristic different from 2, $L = L_0 \oplus L_1$ a Lie superalgebra over k, $E = (e_i)_{i\in I}$ a basis in L_0, $F = (f_i)_{i\in J}$ a basis in L_1, I, J being linearly ordered. Then a basis of $U(L)$ is formed by 1 and the set of all monomials of the form*

$$\varepsilon(e_{i_1})\varepsilon(e_{i_2})\cdots\varepsilon(e_{i_s})\varepsilon(f_{j_1})\varepsilon(f_{j_2})\cdots\varepsilon(f_{j_t}) \tag{6.30}$$

where $i_1 \le i_2 \le \cdots \le i_s$, $j_1 < j_2 < \cdots < j_t$.

It is immediate from this theorem that ε is injective. Identifying L with its canonical image in $U(L)$ shows that a basis of $U(L)$ can be written in the form

$$e_{i_1} e_{i_2} \cdots e_{i_s} f_{j_1} f_{j_2} \cdots f_{j_t}, \tag{6.31}$$

the ordering of indices being the same as in (6.30).

As in Section 6.8, a connection can be established of the question of being a nontrivial identity in $U(L)$ and the question of the boundedness of degrees of irreducible representations of a Lie superalgebra L. Unfortunately, this connection is not as strong as it is in the case of ordinary Lie algebras. Accordingly, we leave out the definition of a representation for a Lie superalgebra. Now the principal result of this section is as follows.

Theorem 6.67. *Let k be a field of characteristic zero, $L = L_0 \oplus L_1$ a Lie superalgebra over k, and $U = U(L)$ its universal enveloping associative algebra. Then U is a PI-algebra if and only if:*
(1) *L_0 is abelian;*
(2) *the L_0-module L_1 has a submodule M of finite codimension such that $[M, M] = \{0\}$;*
(3) $\dim[L_0, M] < \infty$.

6.9.2 Proving the sufficiency part in Theorem 6.67

We set $N = [L_0, M]$. Then $L \triangleright L' = L_0 \oplus M \triangleright N$. If U' stands for the universal enveloping algebra of L' and U'' is the universal algebra of the abelian superalgebra $L'' = L_0 \oplus (M/N)$ then

$$U'' \cong U(L_0) \otimes E(M/N),$$

where $E(M/N)$ is the Grassmann's algebra of the vector space M/N. It was mentioned in the final part of Section 6.4.3 that $E(M/N)$ satisfies an identity of the form $[x, [y, z]] \equiv 0$. From its multilinearity it follows that the same identity holds in U''. Now if $p = \dim N$ then (6.31) shows that any product of more than p elements in K, the kernel of the natural map of U' onto U'', is equal to zero. Hence U' satisfies an identity of the form

$$[x_1, [y_1, z_1]] \cdots [x_{p+1}, [y_{p+1}, z_{p+1}]] \equiv 0.$$

Now U is a free left module of finite type over U' (the number of generators in U is equal to $2^{\dim(L_1/M)}$). As in Theorem 6.52, it follows that U can be embedded in the endomorphism algebra of a finitely generated module over a PI-algebra U'. It is well known that in this case U is itself a PI-algebra (see Procesi and Small [296]).

We finally remark that here the base field is arbitrary.

6.9.3 Proving the necessity in the case where $[L_1, L_1] = \{0\}$

Using the techniques developed in Theorems 6.57 and 6.58, it is possible to prove the following.

Lemma 6.68. *In each of the cases indicated below the algebra $U = U(L)$, where $L = L_0 \oplus L_1$, $L_0 = kx$, and $[L_1, L_1] = 0$, is not a PI-algebra:*
(i) *k is arbitrary, $L_1 = \langle y_0, y_1, \dots, y_m, \dots \rangle$, $[x, y_{i-1}] = y_i$, $i = 1, 2, \dots$;*
(ii) *k is infinite, $[x, y] = y$ for any $y \in L_1$, $\dim L_1 = \infty$;*
(iii) char $k = 0$, $L_1 = \langle y_1, y_2, \dots, y_m, \dots \rangle \oplus \langle z_1, z_2, \dots, z_m, \dots \rangle$, $[x, y_i] = z_i$, $[x, z_i] = 0$, $i = 1, 2, \dots$

Proof. Suppose, on the contrary, that

$$\sum_{\sigma\in S_n} \alpha_\sigma X_{\sigma(1)}X_{\sigma(2)}\cdots X_{\sigma(n)} \equiv 0$$

is a multilinear identity satisfied by $U(L)$ with $\alpha_1 \neq 0$. We will fairly easily arrive at a contradiction in either of the above cases if we substitute for X_i the following elements of $U(L)$:

Case (i) $X_i \mapsto x^{i-1}y_{in}$, $i = 1, 2, \dots, n$;

Case (ii) $X_i \mapsto x^i y_1^{(i)} y_2^{(i)} \cdots y_{s_i}^{(i)}$ where the set of all elements $y_j^{(i)}$ for pairwise different i, j is linearly independent and $(s_1, \dots, s_n)$ is an arbitrary sequence of non-negative integers, $i = 1, 2, \dots, n$;

Case (iii) $X \mapsto x^{i-1} y_{q_i} \cdots y_{q_i+i-1}$ where $q_1 = 1$ and $q_i = q_{i-1} + i - 1$, $i = 2, \dots, n-1$.

In Case (i) one can see that a basic monomial of the form

$$y_{n+1}y_{2n+2}y_{3n+3}\cdots y_{n^2}$$

enters the expression of the value of the left-hand side of our identity through the basis with coefficient α_1. In Case (ii) it is possible to choose an n-tuple $(s_1, s_2, \dots, s_n)$ so that the coefficient of

$$xy_1^{(1)}y_2^{(1)}\cdots y_{s_1}^{(1)}\cdots y_1^{(n)}\cdots y_{s_n}^{(n)}$$

is again α_1. In Case (iii) a nonzero integral multiple of α_1 will be a coefficient of the basic monomial

$$z_1 z_2 \cdots z_{s-1} y_s \cdots y_r.$$

Easy details are left to the reader. □

An argument similar to that used in the proof of part (iii) of Lemma 6.68 is sufficient for proving the following.

Lemma 6.69. *Let k be a field, $L = \langle x_1, x_2, \dots, x_m, \dots\rangle$, and consider $L_1 = \langle y_1, y_2, \dots, y_m, \dots\rangle \oplus \langle z_1, z_2, \dots, z_m, \dots\rangle$, $[L_1, L_1] = 0$, $[x_i, y_i] = z_i$, $[x_i, z_j] = 0$ for all i, j, $[x_i, y_j] = 0$ for $i < j$. Then $U(L)$ is not a PI-algebra.*

Now if x is a fixed element in L_0 then the operator $\operatorname{ad} x|_L$ is algebraic of bounded degree (follows from (i)). Moreover, in the Fitting decomposition $L_1 = L_1^0 \oplus L_1^1$ of L_1 relative to $\operatorname{ad} x$ the codimension of L_1^1 is bounded by a number independent of the particular choice of x (follows from (ii)). Finally, the codimension of the centralizer for x in L_1^1 is also bounded (follows from (iii)). Now choosing an $x_0 \in L_0$ with $\dim L_1/L_1^0$ the greatest possible, we can prove in a standard way that L_1^0 for this element belongs to the zero component of any arbitrary $x \in L_0$. This enables us to restrict ourselves to the case where the action of any $x \in L_0$ on L_1 is nilpotent of bounded index.

The following definition is similar to that in Theorem 6.52:

$$\Delta_p(L_1) = \{y \in L_1 \mid \dim L_1/C_{L_0}(y) \le p\},$$

where $C_{L_0}(y)$ is the centralizer of y in L_0. Proceeding along the lines of the proof of Proposition 6.53, we derive that, given natural numbers p, r, s such that $\dim L/C_{L_1}(x) \le r$ for any $x \in L_0$ and any s elements in L_1 are linearly dependent modulo $\Delta_p(L_1)$, there exists an L_0-submodule M such that $\dim L/M \le s$ and $\dim[L_0, M] \le psr$. It should be emphasized that Theorem 6.51 is used here in its full volume, that is, in the setting of bilinear maps.

Now let q be a number such that $(\operatorname{ad} x|_{L_1})^q = 0$ for all $x \in L_0$. Since L_0 is commutative, an application of Section 1.8, Lemma 1.24, shows that for any $x_1, \dots, x_q \in L_0$ the operator

$$\operatorname{ad} x_1 \cdots \operatorname{ad} x_q$$

acts on L_1 trivially. We consider in L_1 a series of L_0-modules of the form

$$L_1 \supseteq [L_0, L_1] \supseteq \cdots \supseteq [\underbrace{L_0, \dots, L_0}_{q}, L_1] \supseteq \{0\}.$$

Using induction over q with obvious basis $q = 1$ allows us to reduce the proof of the necessity of our theorem in our particular case to the proof of the following.

Lemma 6.70. *Let P, Q be L_0-submodules in L_1 such that $[L_0, L_1] \subseteq P$, $[L_0, Q] = \{0\}$, and $\dim P/Q < \infty$. Then L_1 contains L_0-submodules M, N such that $\dim L/M, \dim N < \infty$ and $[L_0, M] \subseteq N$.*

Proof. Let m be a maximal natural number such that there exist $y_1, \dots, y_m \in L_1$, $x_1, \dots, x_m \in L_0$ for which the elements $z_i = [x_i, y_i]$, $i = 1, \dots, m$, are in Q, they are linearly independent, and $[x_i, y_j] = 0$ for $i < j$. By Lemma 6.69, m is bounded from above. Now we choose a subspace

$$R = C_{L_1}(x_1) \cap \cdots \cap C_{L_m}(x_m).$$

By the above, R is of bounded codimension. We select a $y_{m+1} \in R$ and consider a subspace of the form $[L_0, y_{m+1}]$. Suppose $d = \dim P/Q$. If $\dim[L_0, y_{m+1}] \le m + d$ then $y_{m+1} \in \Delta_{m+d}$. If this is not the case, we put $S_m = \langle z_1, \dots, z_m \rangle$. Then there exists an $x_{m+1} \in L_0$ such that $z_{m+1} = [x_{m+1}, y_{m+1}] \in Q \setminus S_m$. This contradicts maximality in our choice of m. Therefore $R \subseteq \Delta_{m+d}$. Now we may apply our remark just after the definition of $\Delta_p(L_1)$ to conclude that there exists a submodule M in L_1 such that $\dim L/M$ and $\dim[L_0, M]$ are finite. □

6.9.4 Proof of necessity in the case where L_1 is a trivial L_0-module

If $[L_0, L_1] = \{0\}$, the associative subalgebra $U_0 = U(L_0)$ generated in $U(L)$ by even elements is integral and central. Thus we put $Z = U_0 \setminus \{0\}$ and construct the ring of quotients U_Z of elements in U with denominators in Z. Let also K denote the subfield of U_Z generated by U_0. It is easy to verify that U_Z satisfies all the multilinear identities of U. If $F = (f_j)_{j\in J}$ is a basis in L_1 then a basis of U_Z as a vector space over K can be chosen to consist of 1 and all monomials of the form

$$f_{j_1} f_{j_2} \cdots f_{j_t} \quad (j_1 < j_2 < \cdots < j_t).$$

Besides, $f_i \circ f_j = [f_i, f_j]$ where $a \circ b = ab + ba$. We remark that the bilinear symmetric map from $L_1 \times L_1$ into L_0 can be considered as a symmetric K-bilinear map from $KL_1 \times KL_1$ into K. Here $V = KL_1$ is viewed as a K-subspace of U_Z spanned by F. Now we consider the Clifford algebra $C(\varphi)$ (see Bourbaki [80, Chapter IX]). It is easy to observe that $U_Z \cong C(\varphi)$. Let $W = \ker \varphi$ and let $\bar{\varphi}$ be a bilinear function induced by φ on V/W. There exists a canonical map $\alpha : C(\varphi) \to C(\bar{\varphi})$. Since $C(\varphi) \cong U$ is a PI-algebra, the same is true for $C(\bar{\varphi})$.

Now we select in V/W a subspace of arbitrarily high even dimension $p < \dim V/W$ such that the restriction $\tilde{\varphi}$ of $\bar{\varphi}$ to this subspace is not degenerate. We have $C(\tilde{\varphi}) \subset C(\varphi)$. Again by Bourbaki [80, Chapter IX], it is known that $C(\tilde{\varphi})$ is a central simple algebra over K of some finite dimension 2^p. By Theorem 6.2, then $C(\tilde{\varphi})$, and hence $C(\bar{\varphi})$, cannot satisfy any identity of degree less than 2^{p+1}.

Thus, in $V = KL_1$ there is a subspace W of finite codimension such that $[W, KL_1] = \{0\}$. In order to get consequences about the structure of L_1 as a vector space over k, it is convenient to introduce the following subset:

$$\delta_p(L_1) = \{y \in L_1 \mid \dim L_1/C_{L_1}(y) \le p\}.$$

Lemma 6.71. *Under our general hypotheses, there exist p and q such that any q elements of L_1 are linearly dependent modulo $\delta_p(L_1)$.*

Proof. The proof of this lemma is another application of the techniques of Proposition 6.55. Now equation (6.16) in the proof of Proposition 6.55 is replaced by the equation

$$r_1(y_1 \circ y) + r_2(y_2 \circ y) + \cdots + r_q(y_q \circ y) = 0,$$

where $q - 1 = \dim_K V/\ker \varphi$, $y_1, \ldots, y_q$ are arbitrary fixed elements in L_1, $r_1, r_2, \ldots, r_q$ the elements in U_0 such that $r_1 y_1 + r_2 y_2 + \cdots + r_q y_q \in \ker \varphi$, and y is an arbitrary element in L. □

Using an argument similar to that applied in Proposition 6.53, we will find in L_1 a subspace M of finite codimension which is in $\delta_{(q+1)p}$. Applying P. M. Neumann's theorem (Theorem 6.51) once more, we derive $\dim[M, M] \le (q+1)^2 p^2$. Hence, in the case where L_1 is a trivial L_0-module, it contains a subspace M of finite codimension such that $\dim[M, M] < \infty$.

6.9.5 Proof of necessity in the general case

Let $L = L_0 \oplus L_1$ be an arbitrary Lie superalgebra over a field k. We associate with it a Lie superalgebra $\tilde{L}$ on the same $\mathbb{Z}_2$-graded space $L_0 \oplus L_1$ and the multiplication in which differs only in that $[\tilde{L}_1, \tilde{L}_1] = 0$. It is easy to verify that such an $\tilde{L}$ is indeed a Lie superalgebra (Section 1.10.2). If $\widetilde{U} = U(\tilde{L})$ and $U = U(L)$ then a connection between $\widetilde{U}$ and U is the following. We define a chain of subspaces in U by setting

$$U_{-1} = \{0\}, \quad U_0 = U(L_0), \quad U_1 = U_0 \oplus U_0 L_1, \dots, U_{n+1} = U_n \oplus U_n L_1, \dots \tag{6.32}$$

It is easy to verify that (6.32) is an ascending filtration in $U(L)$. Applying Theorem 6.66, it is possible to prove that the graded algebra $G = \mathrm{gr}(U)$ associated with U by means of filtration (6.32) is isomorphic to $\widetilde{U} = U(\tilde{L})$. Hence U being a PI-algebra immediately implies the same property of $\widetilde{U}$. Now the results of Section 6.9.3 apply: since the structure of L_1 as an L_0-module is the same in L and $\tilde{L}$, we conclude that there is an L_0-submodule M in L_1 such that $\dim L_1/M < \infty$, $\dim[L_0, M] < \infty$.

Without loss of generality, we identify L_1 with M. Then we are dealing with a Lie superalgebra $L = L_0 \oplus L_1$ such that $\dim[L_0, L_1] < \infty$. We set $P_1 = [L_0, L_1]$ and consider an ideal $P = P_0 \oplus P_1$ of L generated by P_1. Since $[L_0, L_0] = \{0\}$ it is easy to verify that $\dim P_0 \le (\dim P_1)^2 < \infty$. Now we consider the quotient Lie superalgebra $L/P = L_0/P_0 \oplus L_1/P_1$. Then L/P satisfies the hypothesis of Section 6.9.4. Hence there is an L_0-submodule Q/P_1 of finite codimension in L_1/P_1 such that $R/P_0 = [Q/P_1, Q/P_1]$ is a finite-dimensional space. Passing to the inverse images, we find in L_1 an L_0-submodules Q of finite codimension such that $R = [Q, Q]$ is finite-dimensional. As previously, we have $\dim[L, Q] < \infty$.

This reduces our problem to the case of a Lie superalgebra $L = L_0 \oplus L_1$ such that $L^2 = [L_1, L_1] \oplus [L_0, L_1]$ is finite-dimensional. We choose a basis $\{x_1, \dots, x_n\}$ in $[L_1, L_1]$. Let T_i denote the centralizer of x_i in L_1. If $z \in T_i$ and $x \in L_0$ then

$$[[x, z], x_i] = -[x_i, [x, z]] = -[x, [x_i, z]] + [[x, x_i], z] = 0.$$

Hence T_i is an L_0-submodule in L. By Section 6.9.2, its codimension is a bounded number. We set $T = \bigcap_{i=1}^{r} T_i$. Then T is an L_0-submodule of finite codimension in L_1. Replacing L_1 by T, we may assume that $[L_1, L_1]$ centralizes L_1 and, in addition, $\dim[L_1, L_1] < \infty$.

We consider the subalgebra $D = [L_1, L_1] \oplus L_1$. It is a simple matter to verify that D is an ideal in L. Let C denote the center of D. It is clear that C is a homogeneous ideal in L, $C = C_0 \oplus C_1$, $[L_0, C_1] \subset C_1$ and $[C_1, C_1] = \{0\}$. Hence C_1 is an L_0-submodule in L_1 such that $[C_1, C_1] = \{0\}$ and $\dim[L_0, C_1] \le \dim[L_0, L_1] < \infty$. If we prove that $\dim L_1/C_1 < \infty$ then setting $M = C_1$ will complete the whole proof.

Lemma 6.72. *Let $L = kx \oplus L_1$ be a Lie superalgebra such that $[x, L_1] = \{0\}$. If $U = U(L)$ is a PI-algebra then the center C of L is of finite codimension in L.*

Proof. The proof of Lemma 6.72 is a slightly modified version of our argument in Section 6.9.4. In this case the kernel of the bilinear map is of finite codimension $s - 1$ over $k(x)$, the rational function field in one variable. Choosing $y_1, y_2, \ldots, y_s$ in L_1, there exist polynomials $f_1(x), f_2(x), \ldots, f_s(x) \in k[x]$ such that

$$f_1(x)y_1 + f_2(x)y_2 + \cdots + f_s(x)y_s \in \operatorname{Ann}\varphi. \tag{6.33}$$

For any $y \in L_1$, we have $\varphi(y, y_i) = \pi_i(y)x$ where π_i is a linear function $\pi_i : L_1 \to k$. Applying equation (6.33) to an arbitrary $y \in L_1$ and canceling out x gives

$$f_1(x)\pi_1(y) + f_2(x)\pi_2(y) + \cdots + f_s(x)\pi_s(y) = 0. \tag{6.34}$$

Since at least some of the polynomials are nontrivial, collecting coefficients at an appropriate x^t, we obtain a nontrivial relation of the form

$$\lambda_1\pi_1(y) + \lambda_2\pi_2(y) + \cdots + \lambda_s\pi_s(y) = 0.$$

In this case, given an arbitrary $y \in L_1$, we always have

$$0 = \varphi(y, \lambda_1 y_1 + \lambda_2 y_2 + \cdots + \lambda_s y_s) = [y, \lambda_1 y_1 + \lambda_2 y_2 + \cdots + \lambda_s y_s].$$

Thus $\lambda_1 y_1 + \lambda_2 y_2 + \cdots + \lambda_s y_s \in C_1$. □

Now we are close to the finish. So we are given a Lie superalgebra $D = [L_1, L_1] \oplus L_1$ where $\dim[L_1, L_1] = r$ and $[[L_1, L_1], L_1] = \{0\}$. We present the zero subspace in the form $\{0\} = P_1 \cap P_2 \cap \cdots \cap P_r$, where $P_1, P_2, \ldots, P_r$ are some $(r-1)$-dimensional subspaces in $[L_1, L_1]$. It is clear that P_i is an ideal in D, $i = 1, 2, \ldots, r$. Now each D/P_i satisfies the hypotheses of Lemma 6.72. Then $C = D \cap C'$, where C' is the center of $D/P_1 \oplus \cdots \oplus D/P_s$, has finite codimension in D. The proof of Theorem 6.67 is now complete. □

6.10 Exercises

6.10.1. Let L be a metabelian Lie algebra and $\operatorname{Ad} L$ its adjoint algebra. Find an identity of degree 3 holding in $\operatorname{Ad} L$. If L is a free metabelian algebra then $\operatorname{Ad} L$ satisfies no identity of degree less than 3.

6.10.2. Let A be a relatively free associative algebra with the free countable set X of generators and with identity $[x, y]z \equiv 0$ and L a Lie superalgebra in A generated by X, with respect to bracketing. Prove that L is relatively free and find the identities of the corresponding variety.

6.10.3. For what n can it be guaranteed that a Lie algebra satisfying a standard identity of degree n is special?

6.10.4. Give an example of Lie algebra satisfying a system of Capelli identities of some fixed order, but which is not special.

6.10.5. Given an example of a locally nilpotent nonnilpotent special Lie algebra over a field of characteristic zero.

6.10.6. Give an example of a locally nilpotent Lie algebra which is not special.

6.10.7. Find an exact upper bound for the dimensions of irreducible modules for 2-dimensional metabelian algebra over an algebraically closed field of characteristic 2.

6.10.8. According to Section 6.7.1, each nonabelian finite-dimensional algebra over an algebraically closed field contains either a metabelian subalgebra of dimension 2 or a nilpotent subalgebra of dimension 3. Is it true in the case of the fields which are not algebraically closed?

6.10.9. Prove that any finitely generated residually finite Lie algebra is hopfian; that is, cannot be mapped onto itself with a nonzero kernel.

6.10.10. Let $\mathscr{M}$ be a class of Lie algebras over a ring k. We say that a Lie algebra L is residually an algebra in $\mathscr{M}$ if for each $x \neq 0$ there exists an ideal $H \lhd L$ such that $x \notin H$ and $L/H \in \mathscr{M}$. Prove that for any $n \geq 3$ the free algebra of rank n is residually a free algebra of rank $n-1$.

6.10.11. A Lie algebra G is called *finitely separable* if for any finitely generated subalgebra H and any $g \notin H$ there exists a homomorphism $\varphi : G \to G'$, where $\dim G' < \infty$, such that $\varphi(g) \notin (H)$. Now let $G = A \rtimes P$ where A is an abelian Lie algebra with a basis $\{x, y\}$, $P = k\langle z\rangle$ is the group ring of the infinite cyclic group $\langle z\rangle$. We give P the structure of an A-module by setting $xa = za$, $ya = z^{-1}a$, $a \in P$. Prove that G is a finitely generated Lie algebra which is not finitely separable. Is it true that G is residually finite?

6.11 Comments

A much sharper form of Amitsur's Theorem 6.3 was proven by Braun [86]: The Jacobson radical of a finitely generated PI-algebra is nilpotent. Razmyslov [304] and later Kemer [162] proved the same result in the case where the base field is of characteristic 0. Theorem 6.6 is a famous Shirshov's Height Theorem [337], its consequences (1) and (2) in Corollary 6.7 are earlier results of Kaplansky [158] and Levitzki [204], respectively. Theorem 6.9 and Section 6.1.4 are due to Regev [315], their present elegant proof is due to Latyshev [202].

A considerable portion of this chapter is based on the author's results (Bahturin [23, 27, 31–33]). Sections 6.3 and 6.5 contain some results of Pihtil'kov [289], namely Theorem 6.18 and Section 6.5.1. Section 6.6 contains some known results, say, those in Theorem 6.41, Lemma 6.42, and Theorem 6.43, can be found in Dixmier [101]. Theorem 6.46 is due to Curtis [94]. The results in Section 6.6.3, which are due to the author, can also be found in Bahturin [26]. The description of Lie algebras G over zero characteristic fields such that $U(G)$ is a PI-algebra is due to Latyshev [201]. In the case of prime characteristic this has been done by the author (Bahturin [23]). This work was inspired by the paper of Passman [257] about the description of groups whose group algebra is a PI-algebra. Theorem 6.51 is due to P. M. Neumann [248]. Section 6.8 consists of the author's results (Bahturin [27]). The results on superalgebras in Section 6.9 are also due to the author. A motivation of the study of superalgebras can be found *inter alia* in the books of Berezin [70] and of Scheunert [324]. The following question remains open: Is a central extension of a special Lie algebra over a field itself special? This question is shown to be equivalent to the question on the homomorphic image of a special Lie algebra (Pihtil'kov [289]). It would be interesting to describe solvable special varieties of Lie algebras. It does not seem altogether meaningless to try to describe the structure of special Lie algebras if some reasonable restrictions are imposed (cf. Bahturin [34]). It seems plausible that special varieties over zero characteristic field are finitely based. We conclude by mentioning a corollary of a recent result due to Mishchenko: Engel special varieties are nilpotent.

7 Identities in finite Lie algebras

The main aim of this chapter is the proof of the finite basis property for the identities of a finite Lie algebra over a finite commutative ring, in particular, for the identities of a finite Lie ring. In some of its parts, this proof is close to the proof due to Kovács and Newman [183] of Oates–Powell's theorem [249] on the finite basis of identities for a finite group. However, in many places, especially where we deal with limiting the number of variables in the basis of identities of a finite Lie algebra, we are forced to develop new techniques to overcome some difficulties specific to the situation of Lie algebras. These techniques turn out to be efficient in solving some other problems in the theory of locally finite varieties of Lie algebras.

7.1 Some finiteness conditions

Let k be a finite commutative ring with unity and G a Lie algebra over k. Suppose for each $g \in G$ there exists a polynomial $f_g(t)$ with leading coefficient 1 such that $f_g(\operatorname{ad} g) = 0$. Then we call G an *algebraic* Lie algebra. Any finite-dimensional Lie algebra over a field is algebraic. In this section we are going to prove two results on algebraic Lie algebras which are similar to some group theoretic theorems.

Theorem 7.1. *Let k be a field of prime characteristic $p > 0$, G a finitely generated Lie algebra over k, H a subalgebra of finite codimension in G, $n = \dim G/H$. Suppose there exist $e_1, e_2, \dots, e_n \in G$ which are linearly independent modulo H, and nonzero $f_1, f_2, \dots, f_n \in k[t]$ such that $f_i(\operatorname{ad} e_i) = 0$, $i = 1, 2, \dots, n$. Then H is a finitely generated Lie algebra.*

Proof. Suppose first that k is algebraically closed. We denote by $U = U(G)$ the universal enveloping algebra for G. By Section 6.2.3, we can adjust f_i in such a way that the elements $z_i = f_i(e_i)$ become central in U and the elements

$$e_1^{l_1} e_2^{l_2} \cdots e_n^{l_n} z_1^{s_1} z_2^{s_2} \cdots z_n^{s_n} u, \qquad (7.1)$$
$$0 \le l_i < \deg f_i,\ s_j \ge 0,\ u \in U(H)$$

form a basis of U. Now let I be a linear subspace of U spanned by the monomials of the form (7.1) in which $s_1 + s_2 + \cdots + s_n > 0$. It is easy to verify that I is an ideal of U such that $U(H) \cap I = G \cap I = \{0\}$. Consequently, G and $U(H)$ can be isomorphically embedded in $Q = U(G)/I$.

Now we consider a p-subalgebra R generated in the p-algebra $[U]$ by G and also let S be such a p-subalgebra for H. Then, clearly, R is a finitely generated p-algebra such that $\dim(R/S) < \infty$. Indeed, it is a simple matter to show that R is generated modulo S over k by a system of elements of the form

$$e_i^{p^{s_i}}, \quad i = 1, 2, \dots, m; 0 \le s_i < \deg f_i.$$

https://doi.org/10.1515/9783110566659-007

Indeed, according to Section 1.11, Proposition 1.32, R is generated over k by p-powers of the elements $e_1, e_2, \dots, e_n$ and by all $b \in H$. Since $f_i(e_i) = 0$, $i = 1, 2, \dots, m$, and $b^p \in S$ by the definition of S, our claim follows. Applying Section 2.7, the corollary of Theorem 2.37, we conclude that S is a finitely generated Lie p-algebra. Notice, by Section 1.11, Proposition 1.32, that each element in S is a linear combination of p-powers of elements in H. Thus we may assume that S possesses a finite system of p-generators of the form $b_1, b_2, \dots, b_r$, $b_i \in H$. Let us prove that the same system of elements generate H as a Lie algebra. Let b be an arbitrary element of H and suppose

$$b = \sum_i \alpha_{i0} w_{i0} + \sum_i \alpha_{i1} w_{i1}^p + \dots + \sum_i \alpha_{it} w_{it}^{p^t}, \tag{7.2}$$

where w_{ij} are some values of basic monomials of a free Lie algebra L of rank r in $b_1, b_2, \dots, b_r$. Since k is an algebraically closed field there exist $\beta_{ij} \in k$ such that $\alpha_{ij} = \beta_{ij}^p$ for all i, j. Then by Section 1.11.1 (see equation (1.3)), there exists an element b' in the ordinary subalgebra of H generated by $b_1, b_2, \dots, b_r$, i. e., $b' \in H$, such that

$$b' + \sum_i \alpha_{i1} w_{i1}^p + \dots + \sum_i \alpha_{it} w_{it}^{p^t} = \Big(\sum_i \beta_{i1} w_{i1} + \dots + \sum_i \beta_{it} w_{it}^{p^{t-1}}\Big)^p. \tag{7.3}$$

By equation (7.2), the left-hand side of equation (7.3) is in H and it is a p-power of an element in $U(H)$. But it is immediate from the PBW theorem (Section 2.5) that no element in H can be written as a p-power of an element of $U(H)$. Hence the left-hand side of equation (7.3) is equal to zero and $b = \sum_i \alpha_{i0} w_{i0} - b' \in H$, completing our proof in the case where the base field is algebraically closed.

Now let k be an arbitrary field of characteristic $p > 0$ and k^* its algebraic closure. We embed G canonically as a k-subspace in $G_{k^*} = k^* \otimes_k G$ and let H_{k^*} be a subalgebra of G_{k^*} generated by H and equal to $k^* \otimes H$. It is obvious that G_{k^*} and H_{k^*} satisfy the conditions of our theorem, so that by the first part of the proof there exist $b_1^*, b_2^*, \dots, b_s^*$ generating H_{k^*} as a Lie algebra over k^*. Each of these elements has the form

$$b_1^* = \alpha_{i1} \otimes b_{i1} + \dots + \alpha_{is_i} \otimes b_{is_i}, \quad i = 1, 2, \dots, s; \alpha_{ij} \in k^*, b_{ij} \in H.$$

Now it is clear that H can be generated by the system of elements of the form b_{ij}, $i = 1, 2, \dots, s$; $j = 1, 2, \dots, s_i$, completing the proof in the general case. □

It should be noticed that the theorem fails to be true if we abandon algebraicity. For instance, let $G = M(x_1, x_2)$ be a free metabelian algebra. Then $\dim(G/G^2) = 2$ whereas G^2 is a free abelian algebra of infinite rank (see Section 4.7.1). An important corollary of the just proved theorem is the following.

Theorem 7.2. *Let k be a finite commutative ring with unity and G a finitely generated algebraic Lie algebra over k. If H is a subalgebra in G of finite index as a subgroup of an abelian group then it is a finitely generated Lie algebra.*

Proof. Let J denote the Jacobson radical of k, that is, the greatest nilpotent ideal of k. Applying an argument similar to that used in the theorem on generators of a nilpotent Lie algebra (Section 1.7, Theorem 1.13), it is a simple matter to conclude that H can be generated by any set of its elements that generate it modulo JH. Moreover, since $H \cap JG/JH$ is a finite module, it suffices to show that $H/H \cap JG$ is a finitely generated algebra. Furthermore, since

$$H/H \cap JG \cong H + JG/JG = \bar{H} \subseteq \bar{G} = G/JG,$$

we may restrict ourselves to the case where k is a semisimple ring (i. e., with $J = \{0\}$). Now let $\varepsilon_1, \varepsilon_2, \dots, \varepsilon_n$ be a complete system of pairwise orthogonal idempotents of k and

$$k = k\varepsilon_1 + k\varepsilon_2 + \cdots + k\varepsilon_n,$$

where $k\varepsilon_i$ is a field, $i = 1, 2, \dots, n$. It is clear that

$$G = \sum_{i=1}^{n} \varepsilon_i G \supset \sum_{i=1}^{n} \varepsilon_i H = H,$$

so that each $\varepsilon_i H$ is a subalgebra of finite index in $\varepsilon_i G$, the latter being already a Lie algebra over $k\varepsilon_i$ which is a field. Hence Theorem 7.1 applies. □

The last result of this section need heavier restrictions.

Theorem 7.3. *Let k be a finite commutative ring with 1 and G a finitely generated Lie algebra over k, f a polynomial with leading coefficient 1 in $k[t]$ such that $f(\operatorname{ad} g) = 0$ for any $g \in G$. Then, for any natural number n, G contains only finitely many subalgebras of index n.*

Proof. We will show first that if k is a field then there exists a p-polynomial annihilating all $\operatorname{ad} g$, $g \in G$. Indeed, if f is of degree m then

$$t^{p^i} = f(t)q_i(t) + r_i(t), \quad i = 0, 1, \dots, m, \ \deg(r_i) < m.$$

Since the space of all polynomials of degree less than m is m-dimensional, there exist $\alpha_0, \alpha_1, \dots, \alpha_m \in k$ not all zero such that

$$\sum_{i=0}^{m} \alpha_i r_i = 0.$$

But then f is a divisor of the p-polynomial

$$\sum_{i=0}^{m} \alpha_i t^{p^i}.$$

Thus we may assume that $f(t)$ is itself a p-polynomial, moreover, that $f(t)$ is of degree $d \geq p$ and has no constant term.

To prove our theorem in the case of a finite field, we associate with each subalgebra H of index n a linear representation of degree $(n+1)d^n$, for distinct H these representations being pairwise inequivalent. Since G is finitely generated and $k_{(n+1)d^n}$ is finite, this will imply our claim. To do this, we fix a basis $\{e_1, e_2, \ldots, e_n\}$ of G modulo H. Denoting by z_i an element of the form $f(e_i)$ we will get, as previously, that a basis of $U(G)$ can be chosen in the form

$$e_1^{s_1} e_2^{s_2} \cdots e_n^{s_n} z_1^{l_1} z_2^{l_2} \cdots z_n^{l_n} b, \tag{7.4}$$

where $0 \leq s_i < d$, $l_i \geq 0$, $i = 1, 2, \ldots, n$, b is a basic element of $U(H)$. Now let V_H denote a subspace of $U = U(G)$ generated by those monomials of the form (7.4) for which either $b \neq 1$ or $l_1 + l_2 + \cdots + l_n \geq 2$. Then, as is easy to see, V_H is a submodule of the left G-module U. Let ρ_H be the corresponding representation of G in U/V_H. This is the desired representation. Indeed, we are going to show that $\rho_H(f(x)) = 0$ if $x \in H$ and $\rho_H(f(x)) \neq 0$ if $x \notin H$. The first claim is obvious since $f(x)$ is central in U and, at the same time, it is an element of $U(H)$. To prove the second, we consider a filtration $(U_m)_{m \geq 0}$ of U introduced in Section 2.5.3. Then, given an element $x = \alpha_1 e_1 + \alpha_2 e_2 + \cdots + \alpha_n e_n + h$, $h \in H$, $\alpha_i \in k$, $i = 1, 2, \ldots, n$, using the fact that f is a p-polynomial, $p = \mathrm{char}(k)$, we get

$$f(x) = f(\alpha_1 e_1) + f(\alpha_2 e_2) + \cdots + f(\alpha_n e_n) + u, \quad u \in U_{d-1}. \tag{7.5}$$

Recalling $f(e_i) = z_i$, $i = 1, 2, \ldots, n$, we may rewrite equation (7.5) in the form

$$f(x) = \alpha_1^t z_1 + \alpha_2^t z_2 + \cdots + \alpha_n^t z_n + f(h) + v, \quad v \in U_{d-1}. \tag{7.6}$$

Using equation (7.6), we immediately get

$$\rho_H(f(x))(1 + V_H) = f(x) + V_h \neq 0,$$

as soon as for some i we have $\alpha_i \neq 0$. This proves our theorem in the case where k is a finite field.

Suppose now that k is a finite ring with Jacobson's radical J, $J^l = \{0\}$. If $J = \{0\}$ and $\varepsilon_1, \varepsilon_2, \ldots, \varepsilon_s$ is a complete system of pairwise orthogonal idempotents of k then $G = \varepsilon_1 G \oplus \varepsilon_2 G \oplus \cdots \oplus \varepsilon_s G$, where $\varepsilon_i G$ is a finitely generated Lie algebra over $\varepsilon_i k$. Since in our case H is the direct sum of its projections onto $\varepsilon_1 G, \varepsilon_2 G, \ldots, \varepsilon_s G$, the case $J = \{0\}$ easily follows using the above argument for fields. If $J \neq \{0\}$ then we have shown, in fact, that there exist finitely many subalgebras of the form $H + JG$, H being a subalgebra of index n in G. Viewing this as a basis of induction, we will prove that for any $s > 0$ there exist only finitely many subalgebras of the form $H + J^s G$, $|G : H| = n$. Indeed, let $C = H + J^s G$ be one such subalgebra. Then $C = H + JD$, where $D = H + J^{s-1} G$. Being of finite index in G, D is a finitely generated algebra, hence from $|D/H| \leq n$ we derive the finiteness of the number of subalgebras of the form C in D. By the induction hypothesis, the number of subalgebras like D is itself finite, proving the step of induction. Putting $s = l$, we arrive at the validity of our claim in the general case. □

As previously, it is easy to give an example showing that the conditions of the theorem are indispensable.

7.2 A characterization of the variety generated by a finite Lie algebra

7.2.1 Class $\mathbf{C}(f, m, c)$

Let k be a finite commutative ring with 1 and G a finite Lie algebra over k. Throughout Sections 7.2–7.6, $\mathbf{V}$ denotes the variety of Lie algebras generated by G, $\mathbf{V} = \mathrm{var}(G)$. By Section 4.1, Theorem 4.4, $\mathbf{V}$ is a locally finite variety. Our nearest aim is a more precise characterization of $\mathbf{V}$. This will be done in terms of classes of the form $\mathbf{C}(f, m, c)$ where m, c are two natural numbers and f is a polynomial in one variable with leading coefficient 1. We say that Q is a *factor* of a Lie algebra L if $Q = R/S$ where R, S are subalgebras in L and S is an ideal of R. A factor Q is called *chief factor* if this is a chief factor of the adjoint L-module L (see Section 1.6.2). Now we say that a Lie algebra L over k belongs to $\mathbf{C}(f, m, c)$ if:

(1) for any $x \in L$ we have $f(\mathrm{ad}\, x) = 0$;
(2) for any chief factor M of L we have $|M| \leq m$;
(3) for any nilpotent factor N of L we have $N^{c+1} = \{0\}$.

We will see in the sequel that any such class is a locally finite variety of Lie algebras. Still, in this section, we are able only to prove the following.

Proposition 7.4. *Let G be a finite Lie algebra in a class $\mathbf{C}(f, m, c)$ and $\mathbf{V} = \mathrm{var}(G)$. Then $\mathbf{V} \subseteq \mathbf{C}(f, m, c)$.*

Proof. Pick an algebra L in $\mathbf{V}$ and prove that L satisfies the above conditions (1), (2), and (3).

(1) This is trivial since this condition is, in fact, equivalent to the identical relation $f(\mathrm{ad}\, x)(y) = 0$ holding in G, hence in any algebra of $\mathbf{V}$.

(2) We assume first that L is finite. We apply the construction in the proof of Birkhoff's theorem (Section 4.1, Theorem 4.3):

$$L \cong B/C, \quad B \subseteq P = \prod_{i=1}^{n} G_i, \quad G_i \cong G,\ i = 1, 2, \ldots, n, \tag{7.7}$$

where n is a natural number. Any chief factor of P is isomorphic as an abelian group to a chief factor of some $G_i = G$, hence its order is bounded by m. Now according to Section 1.6.4, chief factors of subalgebras and quotient algebras are isomorphic to subfactors of chief factors of the original algebra (again as abelian groups) proving (2) in the case where L is finite. Suppose next that L is infinite and that M is a minimal ideal of L of order $> m$. We may consider a finite k-submodule $N \subseteq M$ such that $|N| > m$. For any $0 \neq x \in N$, there exist finitely many elements $a_1, a_2, \ldots, a_s$ in L such that $\operatorname{alg}(\{a_1, a_2, \ldots, a_s, x\}) \supseteq N$. Now let D be a finitely generated (hence finite) subalgebra of L which contains N and such that any ideal of D having nontrivial intersection with N contains this subalgebra. Then, choosing a maximal ideal T of D with $N \cap T = \{0\}$ and a minimal ideal S of D containing T, we obtain a chief factor S/T of a finite algebra $D \in \mathbf{V}$ whose order is greater than $|N| > m$ in contradiction with what we have proved just above.

(3) Since any factor of an algebra in $\mathbf{V}$ is itself in $\mathbf{V}$ and since a Lie algebra has nilpotent step c if and only if this is true for its $(c+1)$-generated subalgebras (Section 5.1.2), we may restrict ourselves to the case where L is a $(c+1)$-generator, hence finite, Lie algebra L represented, as in (2), in the form (7.7). Suppose first $C \subseteq \Phi(B)$, $\Phi(B)$ being the Frattini subalgebra of B. Then by Section 1.7, Theorem 1.17, B is nilpotent, hence nilpotent are its projections onto each G_i. By our hypotheses, these projections then have nilpotent step at most c. This implies that B, hence L, are nilpotent of nilpotent steps at most c. So we may assume $C \not\subseteq \Phi(B)$ and that B is a minimal subalgebra for which equation (7.7) holds. Now since C is not in the Frattini subalgebra, there exists a maximal subalgebra M in B such that $B = M + C$. In this case,

$$L \cong (M + C)/C \cong M/M \cap C,$$

contradicting the minimality of B and proving (3). □

Another piece of information on $\mathbf{C}(f, m, c)$ is given in what follows.

Proposition 7.5. *$\mathbf{C}(f, m, c)$ is locally finite.*

Proof. We choose a finitely generated Lie algebra L in $\mathbf{C}$. Let F denote the intersection of all *chief centralizers* of L, that is, all centralizers of chief factors. Given a chief factor M, we have $|L/C_L(M)| < m^m$ (Section 1.5.4). By Theorem 7.3, there exists only a finite number of such centralizers, hence $|L/F| < \infty$. Now applying Theorem 7.2, we see that F is an ideal of L which is a finitely generated subalgebra. Let Q be the intersection of all ideals of finite index in L. We are going to show: (i) $|L/Q| < \infty$, (ii) $Q = \{0\}$. So let N be an ideal of finite index in L and $N_1 = F \cap N$. We consider a chief series of L/N_1 which is a refinement of the series $L/N_1 \rhd F/N_1 \rhd N_1/N_1$. By the definition of F, all chief factors which are between F/N_1 and N_1/N_1 are central, hence F/N_1 is nilpotent of nilpotent step at most c. Since k is finite, we may conclude that the order of F/N_1 is bounded in terms of $|k|$, c, and the number of generators in F only. Applying Theorem 7.2 once more, we see that there exist only finitely many subalgebras of the form N_1 in N and, consequently, $|L/N|$ is also bounded, proving (i). Now (ii) is immediate. Indeed, by Theorem 7.2, Q is finitely generated and if $Q \neq \{0\}$ then there must exist a maximal ideal R of L contained in Q so that Q/R is a chief factor of L. By our choice of Q, $|L/R|$ and hence $|Q/R|$ are infinite, contradicting the definition of $\mathbf{C}$. Finally, $L = L/Q$ is finite. □

7.3 Critical algebras

7.3.1 Definition

Let G be a Lie algebra and Q its factor R/S (see Section 7.2.1). We say that Q is a *proper factor* if either $R \neq G$ or $S \neq \{0\}$. A finite algebra G is called *critical* if it does not belong to the variety generated by all its proper factors. Examples of critical algebras are given in the exercises. Their significance is clear from the following.

Lemma 7.6. *Any locally finite variety can be generated by its critical algebras.*

Proof. As we know, any variety can be generated by the set of its free algebras of finite ranks. In our case this yields $\mathbf{V}$ being generated by the set of its finite algebras. We are going to apply induction over the order of a finite algebra to prove that such an algebra G belongs to the variety generated by its critical factors. Indeed, if G does not belong to the variety generated by its proper factors then it is critical and we are finished. Otherwise G is in the variety generated by its proper factors $Q_1, \dots, Q_r$. Since $|Q_i| < |G|$, $i = 1, \dots, r$, we may apply to each Q_i the induction hypothesis. But the factors of Q_i are isomorphic to the factors of G, $i = 1, \dots, r$. □

An important result of the next section is that any class $\mathbf{C}(f, m, c)$ contains only finitely many pairwise nonisomorphic critical algebras. To achieve this, we need a sufficient condition for an algebra to be noncritical.

Proposition 7.7. *Let G be a finite Lie algebra with a set of ideals $M_1, M_2, \dots, M_s$ and a subalgebra L such that:*

(1) $G = \operatorname{alg}(L, M_1, M_2, \dots, M_s)$;
(2) $G \neq \operatorname{alg}(L, M_1, \dots, \widehat{M_i}, \dots, M_s)$ *where* ^ *means that the ideal under this hat must be omitted,* $i = 1, \dots, s$;
(3) $M_{\sigma(1)} M_{\sigma(2)} \cdots M_{\sigma(s)} = \{0\}$ *for any permutation σ of indices* $1, 2, \dots, s$.

Then G is a noncritical algebra.

Proof. We set $B_0 = L$, $B_i = M_i$, $i = 1, 2, \dots, s$, and $P = B_0 * B_1 * \cdots * B_s$ (see Section 2.6). Let θ_j denote an endomorphism of P induced by the identity map of every B_i, $i \neq j$, and by the zero map of B_j. We set $K = \bigcap_{j=1}^{s} \ker \theta_j$. It is clear that any element in K is representable in the form of a linear combination of commutators each of which includes factors from each of $B_1, B_2, \dots, B_s$. Therefore, given a homomorphism α of P onto G induced by the isomorphisms $B_0 \to L$, $B_i \to M_i$, $i = 1, 2, \dots, s$, we have, by (3), $K \subseteq \ker \alpha$.

Now we consider an element of the form $w = (1 - \theta_1)(1 - \theta_2) \cdots (1 - \theta_s)u$ where u is an element in P. Then, clearly, $w \in K$ and we may write

$$u = w + v_1 + v_2 + \cdots + v_t, \tag{7.8}$$

where $v_i = \pm\beta_i(u)$, β_i being a nonempty product of endomorphisms of the form θ_j, $i = 1,\dots,t$; $w \in \ker\alpha$. Now we introduce the algebras $G_i = \beta_i(P)$. According to (2), all G_i, $i = 1,\dots,t$, are proper subalgebras of G. To complete the proof of the proposition, it is sufficient to show that G is isomorphic to a homomorphic image of the direct product $S = G_1 \times G_2 \times \cdots \times G_t$. But S is representable as a homomorphic image of P by putting $\varphi(u) = (\alpha(v_1), \alpha(v_2), \dots, \alpha(v_t))$ for any $u \in P$ written in the form of equation (7.8). It is obvious from the definitions that $\ker\varphi \subseteq \ker\alpha$, proving that G is indeed a homomorphic image of S. □

Another auxiliary result which will be useful in proving the theorem mentioned above is the following.

Lemma 7.8. *Let $M_1, M_2, \dots, M_s$ be nonnilpotent ideals of a finite Lie algebra G such that $\operatorname{alg}(M_1, M_2, \dots, M_s) = M_1 \oplus M_2 \oplus \cdots \oplus M_s$. Then G possesses a subalgebra L for which $G = \operatorname{alg}(L, M_1, M_2, \dots, M_s)$ but*

$$G \neq \operatorname{alg}(L, M_1, M_2, \dots, \widehat{M_i}, \dots, M_s), \quad i = 1, 2, \dots, s.$$

Proof. Let L be a minimal subalgebra in G for which $G = \operatorname{alg}(L, M_1, \dots, M_s)$. Proving by contradiction, without loss of generality, we may assume that $G = \operatorname{alg}(L, M_2, \dots, M_s)$. We set $M = M_1 \oplus M_2 \oplus \cdots \oplus M_s$ and $M^* = M_2 \oplus \cdots \oplus M_s$. Then $G = L + M = L + M^*$. Using the modular law, we may write $M = M^* + M \cap L$. Now if $N = M^* \cap L$ then $M_1 = M/M^* = M \cap L + M^*/M^* \cong M \cap L/M^* \cap L = M \cap L/N$ so that $M \cap L/N$ is a nonnilpotent ideal of L/N. By Section 1.7.5, $M \cap L/N \not\subseteq \Phi(L/N)$, hence, for a maximal proper subalgebra L_1/N of L/N, we may write $M \cap L/N + L_1/N = L/N$. Using this latter equation, we write $L = L_1 + M \cap L$, hence $G = L + M = L_1 + M \cap L + M = L_1 \oplus M$, contradicting the minimality of our choice of L. □

7.3.2 Minimal representation

A set $\mathscr{S}$ of algebras is called *factor closed* if all factors of algebras in $\mathscr{S}$ are itself in $\mathscr{S}$. We consider $\mathbf{U} = \operatorname{var}(\mathscr{S})$ where $\mathscr{S}$ is a finite factor closed set of finite Lie algebras over k. Then, following the proof of Birkhoff's theorem (Section 4.1, Theorem 4.3), we may represent any finite $G \in \mathbf{U}$ in the form of a factor of a finite direct product P of algebras in $\mathscr{S}$:

$$G = B/C, \quad B \subseteq P = \prod_{i=1}^{n} S_i, \quad S_i \in \mathscr{S},\ i = 1, \dots, n. \tag{7.9}$$

Without any loss of generality, we may assume $|S_1| \geq |S_2| \geq \cdots \geq |S_n|$. Then a representation of G in the form of equation (7.9) is called *minimal* if the n-tuple of orders $(|S_1|, |S_2|, \dots, |S_n|)$ is lexicographically the least possible. Some properties of minimal representations are listed below.

Proposition 7.9. *Let (7.9) be a minimal representation of a finite Lie algebra G in a variety* **U** *generated by a finite factor closed set $\mathscr{S}$ of finite algebras. Then we have:*
1. *all algebras S_i, $i = 1, 2, \ldots, n$, are critical;*
2. *B is a subdirect product in P;*
3. *a subalgebra $D \subseteq S_i$ is an ideal of S_i if and only if $BD \subseteq D$;*
4. *any nontrivial ideal $D \subseteq S_i$ has nontrivial intersection with B;*
5. *$C \cap S_i = \{0\}$ for all $i = 1, 2, \ldots, n$.*

Proof. (1) Suppose a factor S_i is not a critical algebra. Then S_i is isomorphic to a factor of a finite direct product of its proper factors which have smaller orders. This obviously leads to a contradiction with the minimality of the representation (7.9).

(2) If π_i is a projection of P onto S_i, $i = 1, 2, \ldots, n$, then, in fact, B is isomorphic to a subalgebra in $\pi_1(B) \times \pi_2(B) \times \cdots \times \pi_n(B)$. By minimality we obviously come to $\pi_i(B) = S_i$, $i = 1, 2, \ldots, n$.

(3) This is immediate from (2).

(4) If this is not the case then for some nontrivial ideal D of S, say, S_1, we have $D \cap B = \{0\}$ so that B is isomorphic to a subalgebra of $(S_1/D) \times S_2 \times \cdots \times S_n$, contradicting the minimality of (7.9).

(5) Let $\{0\} \neq D \lhd S_1$ be an ideal of S_1 such that $D \cap C \neq \{0\}$. Then G is isomorphic to a factor of $S_1/D \cap C \times S_2 \times \cdots \times S_n$. □

Proposition 7.10. *Let G be a finite Lie algebra in the variety* **U** *generated by a finite factor closed set $\mathscr{S}$ of finite Lie algebras in which the orders of critical algebras are bounded by a natural number q. Then G has a semisimple series (see Section* 1.6.4*) whose length is at most q.*

Proof. Let (7.9) be a minimal representation for G in **U**. We use Proposition 7.9. By (1), all S_i are critical algebras with monoliths M_i. According to (3) and (4), each M_i is a minimal ideal in B, $i = 1, 2, \ldots, n$. Let Q_1 denote the direct product $Q_1 = M_1 \times M_2 \times \cdots \times M_n$. It is obvious that B/Q_1 is isomorphic to a subalgebra of $S_1/M_1 \times S_2/M_2 \times \cdots \times S_n/M_n$, so that we may apply the induction hypothesis over q. Then B/Q_1 has a semisimple series of length at most $q - 1$. The inverse images of the terms of this series together with Q_1 form a semisimple series

$$B = Q_q \supseteq Q_{q-1} \supseteq \cdots \supseteq Q_1 \supseteq Q_0 = \{0\}$$

of B. A semisimple series for B/C of length at most c is now immediate as soon as we apply Section 1.6.4. □

7.4 Critical algebras in C(f, m, c)

7.4.1 Nilpotent radical of finite Lie algebra

In this section we will be using the structure theory of finite Lie algebras developed in Section 1.9. We recall that, given a finite Lie algebra G, its nilpotent radical N and Frattini ideal φ, one can find ideals $M_1, M_2, \ldots, M_s$ of G such that $N/\varphi = M_1/\varphi \oplus M_2/\varphi \oplus \cdots \oplus M_s/\varphi$, M_i/φ being minimal abelian in G/φ, and a subalgebra L in G such that $G/\varphi = N/\varphi \oplus L/\varphi$. Let C denote the centralizer of N/φ in G and F the intersection of all chief centralizers.

Lemma 7.11. *$F = N$. Moreover, if G is a solvable Lie algebra then $C = N$.*

Proof. Since F centralizes its G-chief series, it has a central series, that is, it is nilpotent. Hence $F \subsetneq N$. Now we consider the lower central series of N; its terms are G-invariant. We consider a chief series of G refining the G-series $G \rhd N \rhd N^2 \rhd \cdots \rhd N^{d+1} = \{0\}$. Any factor of this series is either "inside" or "above" N. In either case it is centralized by N. But any chief factor of a finite algebra by Section 1.6, Jordan–Hölder theorem, is isomorphic to a factor of any fixed chief series. In our case we see that $N \subseteq F$, completing the proof of the first claim.

Now suppose G solvable. If $C \neq N$ then we may consider an abelian chief factor D/N which is contained in C/N. Since D centralizes N/φ and N/φ is nilpotent, we conclude that D/φ is nilpotent, too. However, in this case, by Section 1.7.5, D is itself nilpotent, contradicting the definition of N as the greatest nilpotent ideal of G. □

Lemma 7.12. *Let G be a critical algebra in* **C**(f, m, c). *Then $|N/\varphi| \le m^c$ and $|G/C| \le m^{mc}$. If G is solvable then its order is bounded in terms of* **C** *only. Thus* **C** *contains only finitely many pairwise nonisomorphic solvable critical algebras.*

Proof. If $|N/\varphi| > m^c$ then Proposition 7.7 applies to $M_1, M_2, \ldots, M_s, L$. The conditions of that proposition are satisfied since $|M_i/\varphi| \le m$, hence $s > c$, so that $M_{\pi(1)}M_{\pi(2)} \cdots M_{\pi(s)} = \{0\}$. The verification of the remaining two conditions being also simple, we immediately derive that if G is critical then $|N/\varphi| \le m^c$ and $s \le c$. But C is the intersection of chief centralizers, namely the centralizers of chief factors M_i/φ, $i = 1, 2, \ldots, s$. Using Section 1.5, Proposition 1.4, we get $|G/C| \le (m^m)^c$.

Now if G is solvable then, by Lemma 7.11, we may write $|G/\varphi| \le m^{(m+1)c}$, hence G can be generated by a set of cardinality at most $m^{(m+1)c}$. We then consider a free algebra of the variety determined by $f(\operatorname{ad} x)(y) \equiv 0$ of rank at most $m^{(m+1)c}$. By Theorem 7.2, any subalgebra of index $d \le m^{(m+1)c}$ has at most $a = a(f, m, c)$ generators. Clearly, the same value bounds the number of generators in $\varphi(G)$. Since $\varphi(G)$ is nilpotent of nilpotent step at most c and since the base ring k is finite, the order of $\varphi(G)$ is bounded by a certain number $b = b(f, m, c)$ depending entirely on **C**. Thus the order of G, $|G| = |G/\varphi| \cdot |\varphi|$, is bounded. □

Now we are in a position to formulate and prove the main result of this section.

Theorem 7.13. *The number of critical algebras in any class having the form* $\mathbf{C}(f, m, c)$ *is finite.*

Proof. In view of Lemma 7.12 and its proof, it is enough to prove that, given a critical algebra G in $\mathbf{C}$, N its nilpotent radical, C the centralizer of N/φ, then the order C/N is bounded by a number depending solely on $\mathbf{C}$. By Lemma 7.11, N is representable in the form $N = \bigcap_{i=1}^{t}(C \cap C_i)$ where C_i is the centralizer of a chief factor of G. Suppose t is minimal in such a representation of N. Then for each $i = 1, 2, \dots, t$, N is properly contained in the ideal $H_i = \bigcap_{j \neq i}(C \cap C_j)$ and

$$\operatorname{alg}(H_1/N, H_2/N, \dots, H_t/N) = H_1/N \oplus H_2/N \oplus \cdots \oplus H_t/N.$$

We choose in each H_i/N a minimal ideal N_i/N of G/N, $i = 1, 2, \dots, t$. By the same reason as in the proof of Lemma 7.11, we conclude that each N_i/N is non-nilpotent. By Lemma 7.8, there exists a subalgebra L/N that generates G/N together with the whole of $\{N_1/N, \dots, N_t/N\}$ but not with any of its proper subsets. In order to apply Proposition 7.7, it is sufficient to verify that for some t depending on $\mathbf{C}$ only we always have

$$N_{\pi(1)}N_{\pi(2)} \cdots N_{\pi(t)} = \{0\}.$$

For this we choose some elements $b_1, b_2, \dots, b_t$ in $N_1, N_2, \dots, N_t$, respectively, and consider the subalgebra B generated by all these elements along with N. Now B is a solvable algebra, and if $S_1, S_2, \dots, S_t$ are its critical factors then their orders by Lemma 7.12 are bounded by a number depending on $\mathbf{C}$ only. Applying Proposition 7.10 to $B \in \operatorname{var}(\mathscr{S})$, $\mathscr{S}$ being the set of the factors of B, we conclude that B possesses a semisimple series of length at most q. Clearly, then a semisimple series (with respect to B) of length at most q can be chosen also in the B-ideal N:

$$N = Z_q \rhd Z_{q-1} \rhd \cdots \rhd Z_1 \rhd Z_0 = \{0\}. \tag{7.10}$$

Assuming $t = 1 + qm^m$, we choose a subset $\{b_1, b_2, \dots, b_{m^m}\}$ in $\{b_1, b_2, \dots, b_t\}$ and a chief factor T_λ/Z_r in the B-series (7.10). We want to prove that, given an $x \in T_\lambda$, we always have

$$b_1 b_2 \cdots b_{m^m} x \in Z_r. \tag{7.11}$$

Indeed, any B-factor is isomorphic, as a B-module, to a B-subfactor of a chief G-factor (Section 1.6.4). Hence T_λ/Z_r is B-isomorphic to a B-subfactor of a chief G-factor X_λ/Y_λ. Let C_λ denote the centralizer of X_λ/Y_λ and $U = C_\lambda/N \cap (N_1/N \oplus \cdots \oplus N_{m^m}/N)$. We will show that, in fact, U is the product of some of N_i/N, $1 \le i \le m^m$.

Indeed, if $0 \neq u + N \in U$ then $u = \sum_i d_i + N$, $d_i \in N_i$; if, say, $d_1 \notin N$ then, N_1/N being minimal nonabelian, we can find $d_1' \in N_1$ such that

$$0 \neq (u + N)(d_1' + N) = d_1 d_1' + N \in N_1/N.$$

By the minimality of N_1/N, we conclude that this ideal is in U, proving our claim for U. Now since $|G/C_\lambda| < m^m$, the same inequality holds for the order of $(N_1/N \oplus \cdots \oplus N_{m^m}/N)/U$. This enables us to conclude that at least one of $b_1, b_2, \ldots, b_{m^m}$, say, b_1, is in C_λ. Hence $b_1 T_\lambda \subseteq Z_r$. Since T_λ and Z_r are B-invariant, relation (7.11) is true. Recalling the definition of series (7.10), each factor of this series is a direct sum of chief factors, each of which is of the form of T_λ/Z_r. Thus our preceding argument gives $b_1 b_2 \cdots b_{m^m} Z_r \subseteq Z_{r+1}$. Since B/N is abelian, we finally get

$$b_1 b_2 \cdots b_{qm^m} b = b_1 b_2 \cdots b_t = 0.$$

If $t > qm^m$ then we can apply Proposition 7.10 and conclude that G is not critical. Hence, for G critical, we must have $t \leq 1 + qm^m$, which bounds the order of C/N in terms of $\mathbf{C}$. □

Theorem 7.14. *For any parameters f, m, c as in Section 7.2.1, $\mathbf{C}(f, m, c)$ is a variety.*

Proof. Let G denote the direct product of all critical algebras and $\mathbf{V}$ the variety of Lie algebras generated by $\mathbf{G}$. To prove that $\mathbf{V} = \mathbf{C}$, it is sufficient to verify that every finitely generated subalgebra B of an algebra C in $\mathbf{C}$ belongs to $\mathbf{V}$. If C is itself finitely generated then, by Proposition 7.5, C is finite and B belongs to the variety generated by its critical factors which, by Proposition 7.4, are critical algebras of $\mathbf{C}$. Now we pass to the case where C is an arbitrary algebra of $\mathbf{C}$. Suppose that $\mathbf{V}(B)$ is a nonzero ideal of B. By Proposition 7.4, $B/\mathbf{V}(B) \in \mathrm{var}(G) \subseteq \mathbf{C}$. In particular, $B/\mathbf{V}(B)$ is a finite algebra and $\mathbf{V}(B)$ is finitely generated as an ideal of B. Now, similar to the proof of Proposition 7.4, there exists a maximal ideal N of C (hence of B) such that $N \subsetneq \mathbf{V}(B)$. Let M be the intersection of all C-ideals properly containing N. By the choice of N, we have $M \supseteq \mathbf{V}(B)$; moreover, M/N is a chief factor of C. Hence $|M/N| \leq m$. Now we put $L = \mathbf{V}(B) \cap N$ and consider B/L. Since $\mathbf{V}(B)/L \cong \mathbf{V}(B) + N/N \subseteq M/N$ and since any chief factor of B/L is isomorphic either to a chief factor of $B/\mathbf{V}(B)$ or to a chief factor of $\mathbf{V}(B)/L$, we can see that all chief factors of B/L have order at most m. Furthermore, since B/L is a factor of C it is bounded by the parameters c and f. Consequently, $B/L \in \mathbf{C}$ and, by the first part of the proof, $B/L \in \mathbf{V}$. On the other hand, $\mathbf{V}(B/L) = \mathbf{V}(B)/L \neq \{0\}$, which is impossible. This contradiction shows that $\mathbf{V}(B) = \{0\}$. □

7.5 Variety $\mathbf{V}^{(n)}$

7.5.1 Definition

Let $\mathbf{V}$ be a variety and n a natural number. Let V_n denote the set of identities holding in $\mathbf{V}$ and depending on at most n variables. Then V_n determines a variety of Lie algebras over the base ring k which we denote by $\mathbf{V}^{(n)}$. It is equivalent to saying that $\mathbf{V}^{(n)}$ is a class of all Lie algebras whose r-generator subalgebras are in $\mathbf{V}$, $0 \leq r \leq n$. It is clear that $\mathbf{V} \subseteq \mathbf{V}^{(n)}$ for all $n = 1, 2, \ldots$ If $\mathbf{V}$ is a finitely based variety then it can be determined by a single identity (Section 4.2.5) and, for an appropriate n, we have $\mathbf{V} = \mathbf{V}^{(n)}$. A partial converse of this assertion is the following proposition which also shows the significance of such varieties for the study of identities in finite algebras.

Proposition 7.15. *Let $\mathbf{V}$ be a locally finite variety over a ring k and n a natural number. Then $\mathbf{V}^{(n)}$ is a finitely based variety.*

Proof. Indeed, if $L = L(X)$ is a free Lie algebra of countable rank and $L' = L(\{x_1, x_2, \dots, x_n\})$, $x_i \in L$, its subalgebra, then the verbal ideal V_n determining $\mathbf{V}^{(n)}$ is equivalent to $V \cap L' = V(L')$, where $V = \mathbf{V}(L)$ is the verbal ideal determining $\mathbf{V}$. By our hypothesis, $L'/V(L')$ is a finite algebra. Hence $V(L')$ is finitely generated as an ideal of L', hence as a verbal ideal of this algebra. □

Proposition 7.16. *Suppose $\mathbf{U}$ is a finitely based subvariety of $\mathbf{C}(f, m, c)$ and $\mathbf{W}$ is a subvariety in $\mathbf{U}$, $\mathbf{W} \subseteq \mathbf{U}$. Then $\mathbf{W}$ is a finitely based variety.*

Proof. Indeed, according to Section 7.3.1, any locally finite variety, in particular, $\mathbf{C}$, $\mathbf{W}$, and $\mathbf{U}$, is generated by their critical algebras. Theorem 7.13 then shows that $\mathbf{C}(f, m, c)$ has only finitely many subvarieties. Thus there exists a chain

$$\mathbf{W} = \mathbf{W}_0 \subset \mathbf{W}_1 \subset \cdots \subset \mathbf{W}_{s-1} \subset \mathbf{W}_s = \mathbf{U}$$

which cannot be refined, that is, such that there exists no variety $\mathbf{V}$ with $\mathbf{W}_i \subsetneq \mathbf{V} \subseteq \mathbf{W}_{i+1}$, $i = 0, 1, 2, \dots, s-1$. We will proceed by induction over s, the case $s = 0$ being trivial. Now let $u_1 \equiv 0$, $u_2 \equiv 0$, $\dots$, $u_t \equiv 0$ be a finite basis of identities holding in $\mathbf{U}$ and $u_{t+1} \equiv 0$ an identity of $\mathbf{W}_{s-1}$ which does not hold in $\mathbf{U} = \mathbf{W}_s$. Let $\mathbf{V}$ be the variety determined by $u_1 \equiv 0$, $u_2 \equiv 0$, $\dots$, $u_t \equiv 0$, $u_{t+1} \equiv 0$. Then $\mathbf{W}_{s-1} \subseteq \mathbf{V} \subsetneq \mathbf{W}_s$ hence $\mathbf{V} = \mathbf{W}_{s-1}$. Thus $\mathbf{W}_{s-1}$ is a finitely based variety and we may apply induction hypothesis to $\mathbf{W}$ as a subvariety of $\mathbf{W}_{s-1}$. □

Corollary 7.17. *Let G be a finite algebra and $\mathbf{V} = \operatorname{var}(G)$. Then $\mathbf{V}$ is finitely based if and only if for some n, f, m, c we have $\mathbf{V}^{(n)} \subseteq \mathbf{C}(f, m, c)$.*

The proof of the existence of such an n occupies this and the next sections. In fact, it will be proved that such n exists for any f, m, c such that $G \in \mathbf{C}(f, m, c)$. This will finally enable us to state our main result (Theorem 7.25), the finiteness of the basis for identities of a finite Lie algebra. In the present section we will only be able to prove a partial result (Proposition 7.21). This will be anticipated by three lemmas.

Lemma 7.18. *Let G be a finite algebra and $\mathbf{V} = \operatorname{var}(G) \subseteq \mathbf{C}(f, m, c)$. Suppose that n is a natural number, $n \geq m + 1$. Then there exists a natural $q = q(\mathbf{V})$ such that for any monomial $x_{i_1} x_{i_2} \cdots x_{i_s} y$, $i_j \in \{1, 2, \dots, m\}$, there exists a monomial $x_{j_1} x_{j_2} \cdots x_{j_r} y$, $j_r \in \{1, 2, \dots, m\}$, with $r \leq q$, and such that*

$$x_{i_1} x_{i_2} \cdots x_{i_s} y \equiv x_{j_1} x_{j_2} \cdots x_{j_r} y \tag{7.12}$$

is an identical relation in $\mathbf{V}^{(n)}$.

Proof. By Proposition 7.5, the free algebra L of $\mathbf{V}$ generated by $x_1, x_2, \dots, x_m, y$ is finite, so that the number of distinct monomials of the form $x_{j_1}x_{j_2}\cdots x_{j_r}y$ in L is finite. Choosing q as the maximum of degrees of such monomials, we arrive at the desired result. □

Lemma 7.19. *Let G be a finite algebra and $\mathbf{V} = \mathrm{var}(G) \subseteq \mathbf{C}(f, m, c)$. Suppose $n \geq m + 1$. Then there exists an $r = r(\mathbf{V})$ such that if:*
(1) *$P \in \mathbf{V}^{(n)}$ and M is a minimal ideal of P;*
(2) *Q is a subalgebra of P such that $|P/Q| \leq m$;*
(3) *N is an ideal of Q such that $\{0\} \neq N \subseteq M$*

then $|M| \leq |N|^r$.

Proof. We assume that $\{x_1 + Q, x_2 + Q, \dots, x_m + Q\}$ is a generating set of the k-module P/Q and q is the number found in Lemma 7.18. We consider a k-module $S = \sum_{t=0}^{q} S_t$, where

$$S_t = \sum_{i_1,i_2,\dots,i_t=1}^{m} x_{i_1}x_{i_2}\dots x_{i_t}N.$$

We are going to prove that S is a P-module. According to identities of the form (7.12), we have $x_iS \subseteq S$. To prove its Q-invariance, we set $T_l = \sum_{t=0}^{l} S_t$. It is easy to see that $T_0 = N$ and $T_q = S$. Using induction on l, we will prove that $aT_l \subseteq T_l$ for any $a \in Q$. By (3), the case $l = 0$ is trivial. Now we write

$$\begin{aligned} aT_l = aT_{l-1} + aS_l &\subseteq T_{l-1} + \sum_{i_1,i_2,\dots,i_l=1}^{m} (ax_{i_1})\cdots x_{i_l}N \\ &\subseteq T_{l-1} + \sum_{i_1,\dots,i_l=1}^{m} x_{i_1}ax_{i_2}\cdots x_{i_l}N + \sum_{i_1,\dots,i_l=1}^{m} ax_{i_1}\cdots x_{i_l}N \\ &\subseteq T_{l-1} + \sum_{i_1=1}^{m} x_{i_1}T_{l-1} + T_l \subseteq T_l. \end{aligned}$$

We have used here the Jacobi identity and the induction hypothesis. Since $|x_{i_1}x_{i_2}\cdots x_{i_l}N| \leq N$, we see that

$$|S| \leq |N|^{\sum_{t=0}^{q} m^t} = |N|^r,$$

where r depends entirely on $\mathbf{V}$. To complete the proof, we notice that, since M is minimal, we must have $M = S$. □

Lemma 7.20. *Let G be a finite algebra and $\mathbf{V} = \mathrm{var}(G) \subseteq \mathbf{C}(f, m, c)$. Suppose $n \geq m + 1$. Then there exists $s = s(\mathbf{V})$ such that if:*
(1) *P is a finite algebra in $\mathbf{V}^{(n)}$;*

(2) *M is a minimal ideal of P such that $P/M \in \mathbf{C}(f, m, c)$;*
(3) $|M| > m$

then there exists a subalgebra $Q \subseteq P$ generated by at most s elements and possessing a minimal ideal N such that $|N| > m$.

Proof. Let Q be a minimal subalgebra of P possessing a minimal ideal of order more than m. Then Q is equal to an extension of an ideal $Q \cap M \in \mathbf{C}$ by a subalgebra $Q + M/M \in \mathbf{C}$. Since the orders of chief factors in algebras of $\mathbf{C}$ are bounded by m, there exists a subalgebra (even an ideal) J in Q with $|Q/J| \leq m$. Since a minimal J-ideal has order at most m, by Lemma 7.19 we have $|N| \leq m^r$. Since N is minimal, for each proper k-submodule V of N there exists an $x_V \in Q$ such that $x_V V \subseteq V$. Since the number of submodules in N does not exceed 2^{m^r}, it is sufficient to put $s = 2^{m^r}$: being minimal, Q is generated by all elements of the form x_V and a nonzero element in N. □

Proposition 7.21. *Let G be a finite Lie algebra and $\mathbf{V} = \operatorname{var}(G) \subseteq \mathbf{C}(f, m, c)$. Then there exists $t > 0$ such that any finite algebra $P \in \mathbf{V}^{(n)} \cap \mathbf{C}^2$ is in $\mathbf{C}$.*

Proof. By definition, all n-generator algebras of $\mathbf{V}^{(n)}$ are in $\mathbf{V}$. Hence if $n \geq \max\{2, c+1\}$ then, by Proposition 7.4, we conclude that P satisfies $f(\operatorname{ad} x)y \equiv 0$ and that the nilpotent steps of nilpotent factors are bounded by c. What remains is to bound the orders of chief factors of finite algebras in $\mathbf{V}^{(t)} \cap \mathbf{C}^2$ by m, where $t \geq \max\{n, s\}$ (s is as in Lemma 7.20). Suppose that P is a finite algebra in such a variety of minimal order such that $P \notin \mathbf{C}$. By minimality, P possesses a minimal ideal $M \neq \{0\}$ such that $|M| \geq m$. By our condition, also $M \neq P$ and therefore $P/M, M \in \mathbf{C}$. But in this case Lemma 7.20 provides an s-generator subalgebra $Q \subseteq P$ with a minimal ideal N of order $|N| > m$. This, however, is impossible by Proposition 7.4 since $Q \in \mathbf{V} \subseteq \mathbf{C}(f, m, c)$. □

7.6 Chief centralizers and maximal subvarieties

We continue our study of the variety $\mathbf{V} = \operatorname{var}(G)$ generated by a finite Lie algebra G over a finite ring k. We have learned in Section 7.4 that $\mathbf{V}$ has a finite number of subvarieties. We pick one of the maximal proper subvarieties and denote it $\mathbf{V}_0$.

Lemma 7.22. *Let B be a finite algebra in $\mathbf{V}$ and $\mathbf{V}_0(B)$ the verbal ideal of B corresponding to $\mathbf{V}_0$. Then either $\mathbf{V}_0(B) = \{0\}$ or $\mathbf{V}_0(B)$ is a direct sum of a family of minimal ideals of B.*

Proof. There is no difficulty to verify that a family of critical algebras in $\mathbf{V}$ can be chosen in the form $G_0, G_1, G_2, \dots, G_s \in \mathbf{V}$, where $G_1, G_2, \dots, G_s$ are all critical algebras of $\mathbf{V}_0$, $G_0 \notin \mathbf{V}_0$, and the variety generated by all proper factors of G_0 belongs to $\mathbf{V}_0$. Hence $\mathbf{V}$ is generated by the set $\mathscr{S}$ of all factors of $G_0, G_1, \dots, G_s$, which is a factor closed set. If M_0 is a monolith of G_0 then, by our choice, we have $\mathbf{V}_0(G_0) = M_0$. We consider a minimal representation

$$B = C/D, \quad C \subseteq P = \prod_{i=1}^{n} S_i, \quad S_i \in \mathscr{S},\ i = 1, 2, \ldots, n,$$

for B in $\operatorname{var}(\mathscr{S}) = \mathbf{V}$. By Proposition 7.9, equation (7.9), all S_i are critical algebras. By our choice, all but a certain number of isomorphic copies $S_1, S_2, \ldots, S_t$ of G_0 are in $\mathbf{V}_0$. Hence $\mathbf{V}_0(P) \subseteq \mathbf{V}_0(S_1 \times S_2 \times \cdots \times S_t) = M_1 \times \cdots \times M_t$ where $M_i \cong M$ is a minimal ideal of C. Thus we may write

$$\mathbf{V}_0(C) \subseteq T = M_1 \times M_2 \times \cdots \times M_t \subseteq C$$

so that $\mathbf{V}_0(C)$ turns out to be an ideal of the socle (the sum of minimal ideals) of C. As noted in Section 1.6.4, we may conclude that $\mathbf{V}_0(C)$ is the direct sum of a family of minimal ideals. Now we consider $\mathbf{V}_0(C/D)$. This is equal to the image of $\mathbf{V}_0(C)$ under the canonical C-homomorphism $C \to C/D$ and is equal to $\mathbf{V}_0(C) + D/D$. We have a C-isomorphism

$$\mathbf{V}_0(C) + D/D \cong \mathbf{V}_0(C)/\mathbf{V}_0(C) \cap D.$$

Hence, being a quotient module of a semisimple module, $\mathbf{V}_0(C) + D/D$ is itself a semisimple C-, hence C/D-module. In other words, $\mathbf{V}_0(C/D)$ is the direct sum of minimal ideals of C/D, as needed. □

7.6.1 Formula $\mathscr{F}^v_{d,s}$

All the remaining arguments of this section are connected with some formulas which enable us to restrict ourselves the indices of certain chief centralizers in the algebras where they are true. These formulas depend on two natural numbers d, s and a nonassociative polynomial v. Each of these formulas is denoted by $\mathscr{F}^v_{d,s}$. We define

$$\begin{aligned}\mathscr{F}^v_{d,s} &= (\forall a, z_1, \ldots, z_l)[(\forall x_1, x_2, \ldots, x_d)(av(z_1, \ldots, z_l)) \\ &= ax_1 v(z_1, \ldots, z_l) = \cdots = ax_1 \cdots x_d v(z_1, \ldots, z_l) = 0 \\ &\Rightarrow (\forall y_1, \ldots, y_s)(ay_1 \cdots y_s v(z_1, \ldots, z_l) = 0)].\end{aligned}$$

It is clear that the formulas $\mathscr{F}^v_{d,s}$ with $s < d$ are identically true in any algebra. The significance of these formulas in the study of chief centralizers is demonstrated by the following.

Lemma 7.23. *Suppose $\mathscr{F}^v_{s,s+1}$ is true in a Lie algebra G and $v = v(z_1^0, \ldots, z_l^0) \in G$. An element a is in $C_G(I)$, where $I = \operatorname{id}_G(v)$ if and only if*

$$av = ax_1 v = \cdots = ax_1 \cdots x_s v = 0 \tag{7.13}$$

for all $x_1, x_2, \ldots, x_s \in G$.

Proof. We first remark that $\mathcal{F}^v_{s,s+1}$ is a consequence of $\mathcal{F}^v_{s-1,s}$. Indeed, suppose that $\mathcal{F}^v_{s-1,s}$ is true in G. In order to derive $\mathcal{F}^v_{s,s+1}$, it suffices to prove that if equation (7.13) is true in G then for all $y_1, y_2, \dots, y_{s+1} \in G$ we have $ay_1 \cdots y_{s+1}v = 0$. But it is immediate from the Jacobi identity that

$$\begin{aligned} ay_1y_2\cdots y_{s+1}v &= y_1ay_2\cdots y_{s+1}v + (ay_1)y_2\cdots y_{s+1}v \\ &= by_2\cdots y_{s+1}v, \end{aligned} \tag{7.14}$$

where $b = ay_1$. Since $\mathcal{F}^v_{s-1,s}$ is true in G, to prove that the last term of equation (7.14) is zero, it is sufficient to verify that $bx_1x_2\cdots x_rv = 0$ for all $x_1, x_2, \dots, x_r \in G$, $r < s$. Again applying the Jacobi identity and equation (7.13), we may write

$$bx_1x_2\cdots x_rv = (ay_1)x_1\cdots x_rv = ay_1x_1\cdots x_rv - y_1ax_1\cdots x_rv = 0$$

since $r + 1 \le s$.

Now we come back to the proof of Lemma 7.23. The necessity of equation (7.13) being obvious, the sufficiency can be proved by applying formulas $\mathcal{F}^v_{s,s+1}$, $\mathcal{F}^v_{s+1,s+2}$, $\dots$, $\mathcal{F}^v_{r-1,r}$ for any $r > s$ since any element in I can be represented as a linear combination of monomials of the form $y_1y_2\cdots y_rv$ for some $y_1, y_2, \dots, y_r$ in G. □

Now we return to the variety $\mathbf{V}$ generated by a finite Lie algebra G and suppose that $G \in \mathbf{C}(f, m, c)$. Also let $\mathbf{V}_0$ be a maximal subvariety of $\mathbf{V}$ and $v(z_1, z_2, \dots, z_l) \equiv 0$ be an identical relation satisfied by $\mathbf{V}_0$. Then the following is true.

Lemma 7.24. *$\mathcal{F}^v_{m-1,s}$ is true in any algebra of* $\mathbf{V}$.

Proof. It is clear that we may restrict ourselves to considering finitely generated, hence finite, algebras (since $\mathbf{V}$ is locally finite). So we choose elements $a, z_1, \dots, z_l$ in a finite algebra $H \in \mathbf{V}$ such that for all $x_1, x_2, \dots, x_{m-1}$ we have

$$av(z_1, \dots, z_l) = ax_1v(z_1, \dots, z_l) = \cdots = ax_1\cdots x_{m-1}v(z_1, \dots, z_l) = 0.$$

If $v(z_1, \dots, z_l) = 0$ then there is nothing to prove. Assume $v(z_1, z_2, \dots, z_l) \ne 0$. Using Lemma 7.22, we represent v in the form $v = \sum_{i=1}^t v_i$ where v_i belongs to some minimal ideal N_i of H. To prove that $\mathcal{F}^v_{m-1,s}$ is true in H, it is sufficient to verify that $ay_1\cdots y_sv_i = 0$ for all $y_1, \dots, y_s \in H$. By Proposition 7.4, $|N_i| \le m$, $i = 1, 2, \dots, r$, and consequently, the chain

$$0 \subseteq kv_i \subseteq H(kv_i) + kv_i \subseteq HH(kv_i) + H(kv_i) + kv_i \subseteq \cdots$$

stabilizes no later than on the mth term. This means that any product of the form $ay_1\cdots y_sv_i$ can be represented as a linear combination of products of the form $ax_1x_2\cdots x_rv$ where $r \le m-1$. By our hypotheses, all the latter products equal to zero. □

Now we are in a position to prove the main result of this chapter.

Theorem 7.25. *Identical relations of a finite Lie algebra over a finite ring admit a finite basis.*

Proof. This theorem can be proved using induction on the number of proper subvarieties of a variety $\mathbf{V}$ generated by a finite Lie algebra G. This is correct since, on the one hand, the number of such subvarieties is finite and, on the other hand, each of such subvarieties can be generated by the direct product of its critical algebras, that is, by a finite Lie algebra. An important role in the proof is played by the following lemma in which $v_0(z_1, z_2, \dots, z_l) \equiv 0$ stands for an identical relation determining a maximal subvariety $\mathbf{V}_0$ which is finitely based by the induction hypothesis. The order of G is denoted by $g = |G|$.

Lemma 7.26. *Let $P \in \mathbf{V}^{(n)}$ where*

$$n \geq l + g + \frac{(m-1)g(g+1)}{2}$$

and Q/I a principal factor of P. If C is the centralizer of Q/I in P and $Q/I \notin \mathbf{V}_0$ then $|P/C| \leq g$.

Proof. Without loss of generality, we may assume $I = \{0\}$. Now let on the contrary $|P/C| > g$. The desired contradiction will be obtained as soon as we manage to construct an identity depending on at most n variables holding in $\mathbf{V}$ but not in P.

Since $P \notin \mathbf{V}_0$, $\mathbf{V}_0$ is defined by $v_0 \equiv 0$, and Q is a minimal ideal of P, there exist $z_1^0, z_2^0, \dots, z_l^0$ such that this ideal is generated by an element of the form $v_0(z_1^0, z_2^0, \dots, z_l^0) \neq 0$. Let $a_1^0, a_2^0, \dots, a_g^0$ be the elements of P belonging to pairwise distinct cosets of C. Now $\mathscr{F}_{m-1,m}^{v_0}$ is true in $\mathbf{V}$ and it depends on at most n variables. Hence it is true in $\mathbf{V}^{(n)}$ and then, by Lemma 7.23, we can find $x_1^0, \dots, x_r^0 \in P$, $0 \leq r \leq m-1$, such that $0 \neq a_1^0 x_1^0 \cdots x_r^0 v_0(z_1^0, \dots, z_l^0)$. Thus an identity $v_1(a_1, x_1, \dots, x_r, z_1, \dots, z_l) = a_1 x_1 \cdots x_r v_0 \equiv 0$ holding in $\mathbf{V}$ is not satisfied by P. Now using $\mathscr{F}_{m-1,m}^{v_1}$, we can find $x_{r+1}^0, \dots, x_{r+s}^0$ in P, $0 \leq s \leq m-1$ such that

$$(a_2^0 - a_1^0)x_{r+1}^0 \cdots x_{r+s}^0 v_1(a_1^0; x_1^0, \dots, x_r^0; z_1^0, \dots, z_l^0) \neq 0.$$

We set

$$\begin{aligned} &v_2(a_1, a_2; x_1, \dots, x_{r+1}; z_1, \dots, z_l) \\ &\quad = (a_2 - a_1)x_{r+1} \cdots x_{r+s} v_1(a_1; x_1, \dots, x_r; z_1, \dots, z_l). \end{aligned}$$

Proceeding in the same way, we will eventually come to a polynomial

$$\bar{v} = v_{\frac{g(g+1)}{2}}(a_1, \dots, a_g; x_1, \dots, x_d; z_1, \dots, z_l)$$

depending on at most n variables a_i, x_j, z_s which is not identically zero on P and which has as its factors all variables $a_1, \dots, a_g$ and their pairwise differences $a_i - a_j$, $i > j$. At each step of our construction, we have been applying formulas $\mathscr{F}^{V_s}_{m-1,m}$, $0 \le s \le \frac{g(g+1)}{2}$ which are true in **V**. Since these formulas depend on at most n variables, we must conclude that they are true in $G \in \mathbf{V}^{(n)}$. On the other hand, $\bar{v} \equiv 0$ is an identity in G since it is immediate from $|G| = g$ that either a value of a variable a_i is equal to that of a_j, $i \neq j$, or a value of some a_i is equal to zero. □

Now we assemble all heretofore done to complete the proof of the main theorem. Let s be an integer found in Lemma 7.20, n as just above, $t = \max\{s, n\}$. Using the corollary of Proposition 7.16, it is sufficient to verify that $\mathbf{V}^{(n)} \subseteq \mathbf{C}(f, m, c)$, as soon as $\mathbf{V} \subseteq \mathbf{C}(f, m, c)$. Since, by Theorem 7.14, **C** is a variety, it remains to prove that all finitely generated algebras in $\mathbf{V}^{(t)}$ are in **C**. Let $\mathbf{C}(P)$ be the verbal ideal of such an algebra P corresponding to **C**. Suppose $\mathbf{C}(P) \neq \{0\}$. By our hypothesis and Proposition 7.5, $P/\mathbf{C}(P)$ is a finite algebra, hence $Q = \mathbf{C}(P)$ is a finitely generated ideal in P. Then there exists a maximal ideal I of P which is contained in Q and such that $I \neq Q$. To arrive at a contradiction, it is sufficient to show that $P/I \in \mathbf{C}$. Thus we may assume that $I = \{0\}$, that is, Q is a minimal ideal in G. Let us prove that Q is a finite algebra in **C**. Indeed, by Theorem 7.2, Q is finitely generated as a subalgebra and if $Q \in \mathbf{V}_0 \subseteq \mathbf{C}$ then Q is finite. If R is the centralizer of Q in P then either $R \supseteq Q$ and then $Q \in \mathbf{C}$ (since $Q \in \mathbf{V}^{(t)} \subseteq \mathbf{V}^{(1)}$ implies $Q \in \mathbf{V} \subseteq \mathbf{C}$) or $R \cap Q = \{0\}$ and then $|Q| \le |P/R|$. If $Q \notin \mathbf{V}_0$ then, by Lemma 7.26, $|P/R| \le g$, which gives $Q \in \mathbf{V}$ (for $Q \in \mathbf{V}^{(t)} \subseteq \mathbf{V}^{(g)}$).

Thus we may conclude that P is a finite algebra in $\mathbf{V}^{(t)} \cap \mathbf{C}^2$, $t \ge s$. Applying Proposition 7.21, we get $P \in \mathbf{C}$ which is in contradiction with our hypothesis $\mathbf{C}(P) = Q \neq \{0\}$. Now the proof of the main theorem is complete. □

Corollary 7.27. *For any values of parameters f, m, c, $\mathbf{C}(f, m, c)$ is a finitely based variety.*

7.7 Locally finite varieties of Lie algebras

7.7.1 Cross varieties

Due to a number of reasons (see H. Neumann [247]), any variety of Lie algebras **V** over a finite ring k generated by a finite Lie algebra is called a *Cross variety*. It can also be characterized by three properties:

(a) **V** is a locally finite variety;
(b) **V** is a finitely based variety;
(c) **V** contains only finitely many pairwise nonisomorphic critical algebras.

In Sections 7.1–7.6 we proved that any variety of the form $\mathbf{V} = \mathrm{var}(G)$, $|G| < \infty$, satisfies (a), (b), and (c). If, conversely, **V** satisfies (a), (b), and (c) then it is clear that it is generated by a single finite algebra, namely, by the direct product of all critical algebras contained in **V**. An important property of a Cross variety is its embeddability in a class of the form $\mathbf{C}(f, m, c)$ (see Section 7.2). By Theorem 7.14, $\mathbf{C}(f, m, c)$ is itself a Cross variety and, by Section 7.5, any of its subvarieties is a Cross variety. All these facts enable us to come to the conclusion that Cross varieties form a class of varieties having a nice internal structure.

The next step in describing locally finite varieties of Lie algebras is describing varieties which are not Cross but all of whose proper subvarieties are already Cross varieties. Such varieties will be called *just-non-Cross*. The importance of having a list of such varieties is well illustrated by the following.

Proposition 7.28. *Any variety which is not a Cross variety contains a subvariety which is just-non-Cross.*

Proof. We pass to considering verbal ideal of a free Lie algebra of countable rank L_∞. Let V the verbal ideal of identities of a non-Cross variety $\mathbf{V}$ and $\mathscr{P}$ the set of verbal ideals corresponding to non-Cross subvarieties of $\mathbf{V}$. Let

$$V = V_0 \subset V_1 \subset \cdots \subset V_n \subset \cdots \subset U = \bigcup_{n=0}^{\infty} V_n$$

be an infinite increasing chain of elements of P. Its union U is a verbal ideal of L_∞ which cannot be generated by finitely many elements as a verbal ideal. Hence U cannot correspond to a Cross subvariety, that is, $U \in \mathscr{P}$. Thus we may apply to $\mathscr{P}$ Zorn's lemma. A result of its application is a verbal ideal W corresponding to a non-Cross subvariety $\mathbf{W}$ of $\mathbf{V}$ such that each verbal ideal properly containing W is a verbal ideal of a Cross variety. Finally, $\mathbf{W}$ is a just-non-Cross variety which is contained in a non-Cross variety $\mathbf{V}$. □

We notice that having a list of just-non-Cross varieties enables us to solve the following problem: Given an identity $w(x_1, x_2, \dots, x_n) \equiv 0$, is it possible to determine whether this identity determines a variety generated by a finite algebra? In view of the above, it is sufficient to verify the validity of this identity in a free algebra of any variety in the list of just-non-Cross varieties of rank n.

7.7.2 An approach to the classification of just-non-Cross varieties of Lie algebras

In this book we restrict ourselves to studying solvable just-non-Cross varieties of Lie algebras. Our final result will be the complete description of solvable just-non-Cross varieties over an arbitrary commutative ring (Theorem 7.45). As for nonsolvable varieties, their existence has been proved by Razmyslov [306], but their *description* in our opinion should be based on some entirely different ideas.

As it will be shown in Proposition 7.31, any solvable just-non-Cross variety is locally finite and satisfies an identity of the form $f(\operatorname{ad} x)(y) \equiv 0$. Moreover, we will prove, in the final part of this subsection, that, in a locally finite variety $\mathbf{V}$ the collection of all locally nilpotent algebras is a subvariety $\mathbf{V}_{LN}$. In this case either $\mathbf{V}_{LN} = \mathbf{V}$ or $\mathbf{V}_{LN} \subseteq \mathbf{N}_c$

for some c. Indeed, if $\mathbf{V}_{LN} \neq \mathbf{V}$ then $\mathbf{V}_{LN}$ is a Cross variety and it is generated by a single locally nilpotent finite algebra, that is, $\mathbf{V}_{LN}$ is nilpotent. In the former case it is obvious that the orders of the chief factors in the algebras of $\mathbf{V}$ are bounded by $q = |k|$, and the nilpotent steps of nilpotent factors are unbounded. In the latter case there must exist a sequence of algebras $L_n \in \mathbf{V}$, $n = 1, 2, \ldots$, with chief factors M_n (they can be viewed as minimal ideals) whose orders are not bounded. Indeed, in either case if we assume to the contrary we will immediately see that, for some values of f, m, c, we have $\mathbf{V} \subseteq \mathbf{C}(f, m, c)$. By the results of the previous sections, we may conclude that $\mathbf{V}$ is a Cross variety. Hence the study of solvable just-non-Cross varieties splits into two essentially different parts:

(1) the study of locally nilpotent solvable just-non-Cross varieties;
(2) the study of solvable just-non-Cross varieties which are not locally nilpotent.

The first case will be dealt with in the present section. It will be shown that if k is a field then such a variety exists and it is unique. The second possibility will be discussed in Section 7.8. The main result of that section is the proof that, over a finite field, there exist no solvable just-non-Cross varieties which are not locally nilpotent. This will bring us to the classification theorems and to some results of algorithmic nature.

Now we prove an auxiliary result concerning locally nilpotent algebras.

Lemma 7.29. *Let $\mathbf{V}$ be a locally finite variety of Lie algebras over a finite ring k. Then the set $\mathbf{V}_{LN}$ of locally nilpotent algebras in $\mathbf{V}$ is a variety.*

Proof. We apply Birkhoff's theorem (Section 4.1, Theorem 4.3). Clearly, $\mathbf{V}_{LN}$ is closed with respect to subalgebras and quotient algebras. Now let $\{G_\alpha \mid \alpha \in I\}$ be a family of locally nilpotent algebras in $\mathbf{V}$. We consider a subalgebra H in $G = \prod_{\alpha\in I} G_\alpha$ generated by $h_1, \ldots, h_s$. If H_α is the kernel of projection of H_α onto G_α then $\bigcap_{\alpha\in I} H_\alpha = \{0\}$. Since H is finite, for some $\alpha_1, \ldots, \alpha_t$ we may write $H_{\alpha_1} \cap \cdots \cap H_{\alpha_t} = \{0\}$. Hence $H \subseteq H/H_{\alpha_1} \times \cdots \times H/H_{\alpha_t}$. But, for any α, H/H_α is isomorphic to a finitely generated subalgebra in G_α, that is, H/H_α is nilpotent. It is immediate now that H is a nilpotent Lie algebra. □

7.7.3 An example of a solvable just-non-Cross variety of Lie algebras over a finite field of characteristic $p > 0$

We consider a variety $\mathbf{M}$ determined by a system of identical relations of the form (for the notation see Section 4.7, the corollary of Theorem 4.26):

(i) $(x_1x_2)x_3x_4 \equiv 0$;
(ii) $x^p yz \equiv 0$;
(iii) $x^p y^p \equiv 0$.

As we see, $\mathbf{M}$ consists of metabelian algebras. It is locally nilpotent: indeed, according to (ii), $\mathbf{M}$ is Engel, hence, by Section 4.7, Theorem 4.42, it is locally nilpotent. We also notice that an element as on the left-hand side of (iii) is in the center of a Lie algebra satisfying (i) and (ii):

$$zx^{p-1}y^px = x^{p-1}y^{p-1}zyx = x^{p-1}y^pzx - x^py^{p-1}zy = 0.$$

Hence, if a Lie algebra satisfying (i) and (ii) is not nilpotent then its quotient algebra over the verbal ideal determined by (iii) is nonnilpotent, too. To verify that M is nonnilpotent it is sufficient to consider a Lie algebra in Section 1.8.3. Now we want to show that any proper subvariety in **M** is nilpotent. Let V be the verbal ideal of a free metabelian algebra $M(X)$ of countable rank determined by x^pyz. Since $(u+v)^pyz = u^pyz + v^pyz$, we can see that V is a multihomogeneous ideal. If $W \supseteq V$ is the verbal ideal in $M(X)$ determined modulo V by (iii) then

$$x^{p-1}(u+v)^{p-1}x(u+v) = x^{p-1}u^{p-1}xu + x^{p-1}v^{p-1}xv + P,$$

where

$$\begin{aligned} P &= x^{p-1}\sum_{s=1}^{p-1}\binom{p-1}{s}u^sv^{p-1-s}xv + x^{p-1}\sum_{s=0}^{p-2}\binom{p-1}{s}u^sv^{p-1-s}xu \\ &= x^p\sum_{s=1}^{p-1}\binom{p-1}{s}u^{s-1}v^{p-1-s}uv + x^{p-1}\sum_{s=1}^{p-1}\binom{p-1}{s}u^{s-1}v^{p-s}xu \\ &+ x^{p-1}\sum_{s=0}^{p-2}\binom{p-1}{s}u^sv^{p-1-s}xu = x^{p-1}\sum_{s=0}^{p-1}\binom{p}{s}u^sv^{p-1-s}xu = 0. \end{aligned}$$

Hence the free algebra $F(X) = M(X)/W$ is multihomogeneous. Thus a polynomial in X is equal to zero in $F(X)$ if and only if its multihomogeneous components are zero. Now let $f(x_1,\dots,x_n) \equiv 0$ be a nontrivial identity in **M**. If this is multilinear then it has the form

$$\sum_{i=1}^{n}\alpha_ix_1\cdots\hat{x}_i\cdots x_nx_ix_{n+1} = 0.$$

Now substituting x_jx_{n+2} for an x_j with $\alpha_j \neq 0$, we immediately obtain nilpotency.

Suppose next that the degree of our identity in, say, x_1 is equal to $t > 1$. Then $t \leq p$. We apply linearization over x_1. By our remark concerning multihomogeneity, it is sufficient to consider a multihomogeneous identity. It can be written in the form

$$P = P(x_1) = \sum_{i=1}^{n}g_ix_1^tx_i, \quad g_i = g_i(x_2,\dots,x_n).$$

We will pass to $P(x_1 + x_{n+1}) - P(x_1) - P(x_{n+1})$ and pick the component of degree $t-1$ in x_1 and 1 in x_{n+1}. This looks like the following:

$$\begin{aligned} Q &= \sum_i g_ix_1^{t-1}x_{n+1}x_i + (t-1)\sum_i g_ix_1^{t-2}x_{n+1}x_1x_i \\ &= t\sum_i g_ix_1^{t-1}x_{n+1}x_i - (t-1)\sum_i g_ix_ix_1^{t-2}x_{n+1}x_1. \end{aligned}$$

If $t \neq p$ then a consequence of $Q \equiv 0$ at $x_{n+1} = x_1$ is that $P \equiv 0$. Hence Q is nontrivial in $F(X)$. Its degree in x_1 has diminished, so that an induction applies. Now if $t = p$ then it should be remarked that, in **M**, we have an identity of the form $xy^pz \equiv zy^px$ which can conventionally be written as $xzy^p = zxy^p$. Hence $P \equiv 0$ takes the form $P = P'x_1^p \equiv 0$ where P' is a polynomial in $x_2, \dots, x_n$. Then Q can be rewritten as $Q = P'x^{p-1}x_{n+1}$. Hence, using an example in Section 1.8.3 (or Section 7.7.4), we can find an index i, $2 \leq i \leq n$, such that P is divisible by x_i^p. In this case $P = P''x_i^px_1^p = 0$ by (iii) which is satisfied in **M**. Therefore, if P is nontrivial then Q is nontrivial, too. Again this leads to diminishing the degree in x_1. This reduction enables us to reach the conclusion that $f(x_1, \dots, x_n) \equiv 0$ is a multilinear identity. By the above, this implies nilpotency, proving our claim about **M**. Thus **M** provides an example of a solvable locally finite just-non-Cross variety over a finite field of characteristic $p > 0$. Our main goal now is to prove that this is the only solvable just-non-Cross variety over a finite field. We will need a detailed knowledge of free algebras of **M**.

7.7.4 Structure of free algebra of M

Let k be a field of characteristic $p > 0$ and B an abelian algebra over k with a basis of the form $\{z_i \mid i \in I\}$. We consider an abelian algebra with the basis of the form ft_i where f is a monomial in $k[z_i \mid i \in I]$ which is not divisible by z_j^p for any j. It will be convenient to write $ft_i = 0$ if f is divisible by z_j^p for some j. We endow A with a B-module structure by setting $z_j \circ (ft_i) = (z_jf)t_i$. Obviously, then A becomes a module over $k[z_i \mid i \in I]$ (again the module action will be denoted by the circle $\circ$). We form a semidirect product $K = B \rtimes A$ corresponding to the given structure of a B-module on A (Section 1.4.4).

Lemma 7.30.
(1) *K is an algebra in* **M**.
(2) *The elements $y_i = z_i + t_i$ freely generate an* **M***-free subalgebra of K.*

Proof. (1) It is obvious that K is metabelian. Let $x = b_1 + a_1$, $y = b_2 + a_2$, $z = b_3 + a_3$, $b_i \in B$, $a_i \in A$, $i = 1, 2, 3$. Then

$$x^pyz = b_1^p(b_2a_3 - b_3a_2) = \left(\sum_j \alpha_jz_j\right)^p \circ \left(\sum_i f_i \circ t_i\right)$$
$$= \sum_{i,j} \alpha_j^pz_j^pf_i \circ t_i = 0.$$

Besides,

$$x^{p-1}y^px = b_1^{p-1}b_2^{p-1}(b_2a_1 - b_1a_2) = b_1^{p-1}b_2^pa_1 - b_2^{p-1}b_1^pa_2 = 0$$

as in the preceding calculation.

(2) We want to show that any relation among the elements y_i, $i \in I$, is an identity of K. This will complete the proof since in the case of infinite I it is easy to see that K is not nilpotent, hence $\operatorname{var} K = \mathbf{M}$, for the contrary would contradict the fact that any proper subvariety in $\mathbf{M}$ by Section 7.7.3 is nilpotent. So we consider a relation as mentioned above, $F(y_1, \dots, y_n) = 0$. Since we are in the semidirect product K, we may rewrite $F(b_1 + a_1, \dots, b_n + a_n) = F(b_1, \dots, b_n) + \sum f_i(b_1, \dots, b_n)a_i$. Since $F(z_1, \dots, z_n) + \sum f_i(z_1, \dots, z_n) \circ t_i = 0$, we derive $F(z_1, \dots, z_n) = 0$ and $\sum f_i(z_1, \dots, z_n) \circ t_i = 0$. By the definition of A, then each monomial f_{im} of each f_i, $i \in I$, is divisible by $z^p_{j_{im}}$, that is, $f_{im} = z^p_{j_{im}} f'_{im}$. But, in this case, if $b_{j_{im}} = \sum_s \alpha_s z_s$, then we have

$$f_{im}(b_1, \dots, b_n) = \sum_s \alpha_s^p z_s^p f'_{im}(b_1, \dots, b_n)$$

so that $f_{im}(b_1, \dots, b_n) \circ a_i = 0$. Now it becomes clear that $F(b_1, \dots, b_n) = 0$. □

Proposition 7.31. *Let* $\mathbf{V}$ *be a locally solvable variety of Lie algebras over a finite field of characteristic* $p > 0$ *in which no identity of the form* $f(\operatorname{ad} x)(y) \equiv 0$ *holds valid, where* f *is a polynomial with leading coefficient* 1. *Then* $\mathbf{V} \supset \mathbf{M}$.

Proof. We consider the variety $\mathbf{W}$ of all metabelian algebras belonging to $\mathbf{V}$. Suppose $\mathbf{W}$ satisfies an identity of the form $f(\operatorname{ad} x)(y) \equiv 0$, f being a polynomial with leading coefficient 1. Now if all (2-generator) algebras in $\mathbf{V}$ have solvable steps at most l, then $f^{l-1}(\operatorname{ad} x)(y) \equiv 0$ in $\mathbf{V}$. Thus we may assume that $\mathbf{V}$ is a metabelian variety. Now suppose G denotes the free algebra of $\mathbf{V}$ with free generators x, y. It follows from our proposition that the elements $xy, x^2y, \dots, x^ny, \dots$ are linearly independent. We set $t = xy$ and consider a subalgebra L generated by x and t. We recall that in a metabelian algebra we always have $xyzt = yxzt$. As in Section 7.7.3, this gives sense to an expression of the form $f(x_1, \dots, x_n)xy$ where $f \in k[t_1, \dots, t_n]$. Any nontrivial identity of a metabelian algebra is equivalent to a system of identities of the form

$$\sum_{i=2}^{n} \lambda_i f_i(x_1, x_2, \dots, x_n)x_i x_1 \equiv 0. \tag{7.15}$$

Suppose L satisfies an identity (7.15) where all λ_i are nonzero. We replace in this identity the elements x_i by the elements of the form $\alpha_i x + h_i t$ where $\alpha_i \in k$, $h_i \in k[x]$, $i = 1, 2, \dots, n$. We will obtain

$$\sum_i \lambda_i f_i(\alpha_1 x, \alpha_2 x, \dots, \alpha_n x)(\alpha_i x + h_i t)(\alpha_1 x + h_1 t) = 0.$$

Let $f_i = f_{i1} + f_{i2} + \dots + f_{is}$ be the decomposition of f_i as the sum of homogeneous components. Then we may rewrite

$$\sum_i \lambda_i \sum_{j=1}^{s} f_{ij}(\alpha_1, \alpha_2, \dots, \alpha_n)x^{j+1}(\alpha_i h_1 - \alpha_1 h_i)t = 0.$$

Since t is a free generator of a free module over $k[x]$, the above equation splits into several equations, one for each degree of x. We first put $h_i = 1$, $h_j = 0$ if $j \neq i \neq 1$. Then we get

$$\alpha_1 f_{ij}(\alpha_1, \dots, \alpha_n) = 0 \quad \text{for all } \alpha_1, \dots, \alpha_n \in k. \tag{7.16}$$

If we set $h_1 = 1$, $h_j = 0$ for all $j \neq 1$ then we will have

$$\sum_{i=2}^{n} \lambda_i \alpha_i f_{ij}(\alpha_1, \dots, \alpha_n) = 0 \tag{7.17}$$

for any $\alpha_1, \alpha_2, \dots, \alpha_n \in k$.

Now we digress to make a remark on homogeneous polynomials. Let F be a homogeneous polynomial in $t_1, \dots, t_n$ such that $F(\alpha_1, \dots, \alpha_n) = 0$ for all $\alpha_1, \alpha_2, \dots, \alpha_n \in k$. We write F in the form $F = \sum_{s=1}^{t} \mu_s M_s$ where M_s is a monomial (of the same degree as F). Also let $M_1, \dots, M_l$ be the polynomials whose degrees in every x_i are less than q, where q is the number of elements in k, and $M_{l+1}, \dots, M_t$ are the remaining monomials in the above decomposition. We denote by I the ideal in $k[t_1, \dots, t_n]$ generated by all monomials of the form $t_i^q - t_i$, $i = 1, 2, \dots, n$. Since $F \in I$, we can replace $M_1, \dots, M_t$ by the $\widetilde{M}_1, \dots, \widetilde{M}_t$ which are already reduced modulo I, and still $\widetilde{F} \in I$. Thus, on the one hand, $\widetilde{F} = 0$ and, on the other hand, $\widetilde{F}$ is equal to the sum of two reduced polynomials of essentially distinct degrees: $\widetilde{P}$ with $\deg \widetilde{P} = \deg F$ and $\widetilde{Q}$ with $\deg \widetilde{Q} < \deg F$, where $\widetilde{P}$ and $\widetilde{Q}$ have the form

$$\widetilde{P} = \sum_{i=1}^{l} \mu_i \widetilde{M}_i = \sum_{i=1}^{l} \mu_i M_i \quad \text{and} \quad Q = \sum_{j=l+1}^{t} \mu_j M_j.$$

Hence $P = 0$. Now we compute the value of the left-hand side of equation (7.15) at $x_i = y_i = z_i + t_i$ (see Lemma 7.29):

$$\begin{aligned} \sum_{i=2}^{n} \lambda_i f_i(y_1, \dots, y_n) y_i y_1 &= \sum_{i=2}^{n} \lambda_i f_i(z_1, \dots, z_n)(z_i t_1 - z_1 t_i) \\ &= \sum_{j} f_{ij}(z_1, \dots, z_n) \circ \sum_{i=2}^{n} \lambda_i (z_i t_1 - z_1 t_i) \\ &= -\sum_{j} f_{ij}(z_1, \dots, z_n) z_1 \circ \sum_{i} \lambda_i t_i \\ &\quad + \sum_{j} \sum_{i} \lambda_i z_i f_{ij}(z_1, \dots, z_n) \circ t_1. \end{aligned}$$

Using equations (7.16), (7.17), and our remark concerning homogeneous polynomials, we conclude that, in polynomials $f_{ij} z_1$ and $\sum \lambda_i f_{ij} z_i$, each monomial is divisible by z_r^p for some r. Now from the definition of K and Lemma 7.30, it follows that (7.15) is an identity in $\mathbf{M}$. Hence $\mathbf{V} \supseteq \operatorname{var} L \supset \mathbf{M}$. □

Corollary 7.32. *Let* $\mathbf{V}$ *be a locally solvable just-non-Cross variety. Then an identity of the form*

$$f(\operatorname{ad} x)(y) = 0 \tag{7.18}$$

holds in $\mathbf{V}$*, where* f *is a polynomial in* $k[t]$ *with leading coefficient* 1*. Moreover,* $\mathbf{V}$ *is locally finite.*

Proof. The first claim follows by assembling the facts that $\mathbf{M}$ satisfies $x^{p+1}y \equiv 0$, that two different just-non-Cross varieties cannot be subvarieties of each other and by applying Section 7.7.3 and Proposition 7.31. To prove the second claim, we apply Theorem 7.1, according to which a subalgebra of finite index of a finitely generated algebra, in which any two elements are bound with a relation of the form of equation (7.18), is itself finitely generated. Having this in mind, it remains to apply the induction over the solvable step of a finitely generated algebra in $\mathbf{V}$ (cf. Section 4.6, Lemma 4.36). □

After these preliminaries we pass to the main result of the present section.

Proposition 7.33. *Let* $\mathbf{V} = \mathbf{V}_{LN}$ *be a solvable locally nilpotent just-non-Cross variety of Lie algebras over a finite field* k *of characteristic* $p > 0$*. Then* $\mathbf{V} = \mathbf{M}$*.*

Proof. We remark that $\mathbf{V}$ is a nonnilpotent variety. Indeed, any nilpotent variety is Cross since it belongs to a class of the form $\mathbf{C}(t^c, q, c)$ where $q = |k|$, c being the nilpotent step of this variety, and from Theorem 7.25. Now suppose $\mathbf{V}$ is not a metabelian variety. Then, if $L \in \mathbf{V}$, $L/(L^2)^2$ is nilpotent and L^2 is nilpotent since these algebras have smaller solvable step, hence belong to a proper subvariety of $\mathbf{V}$ which is necessarily Cross. By Section 1.7, Theorem 1.19, any algebra L in $\mathbf{V}$ is nilpotent, contradicting our assumption.

So $\mathbf{V}$ is a metabelian variety. Let L denote a free algebra of $\mathbf{V}$ with a free generating system $\{x_1, x_2, \ldots, x_n, \ldots\}$. Let $\mathbf{M}(L)$ denote the verbal ideal of L corresponding to $\mathbf{M}$. Suppose $\mathbf{V} \not\supseteq \mathbf{M}$. Then $L/\mathbf{M}(L)$ is in $\mathbf{V} \cap \mathbf{M}$. Since $\mathbf{M}$ is just-non-Cross, we see that $L/\mathbf{M}(L)$ is a nilpotent algebra of nilpotent step, say, c. Therefore $x_1x_2 \cdots x_{c+1} \in \mathbf{M}(L)$. Now, by our definition, any element in $\mathbf{M}(L)$ is a linear combination of products of one of the two forms:

$$w_1 \cdots w_s w_{s+1}^p w_{s+2} w_{s+3} \quad \text{and} \quad w_1 \cdots w_s w_{s+1}^{p-1} w_{s+2}^p w_{s+1}. \tag{7.19}$$

Multiplying by x_{c+2} and renumbering, if necessary, we arrive at an equation of the form (cf. Section 7.7.3):

$$x_1x_2 \cdots x_{c+2} = \sum_i w_{i1} \cdots w_{i,n_i} w_{i,n_i+1}^p w_{i,n_i+2} w_{i,n_i+3}. \tag{7.20}$$

Since $\mathbf{V}$ is metabelian and the expression in equation (7.20) is multilinear in all arguments w_{is}, we may assume that each summand in equation (7.20) has the form

$$y x_{i_1} x_{i_2} \cdots x_{i_s} x_{i_{s+1}}^p x_{i_{s+2}} \cdots x_{i_{t_i}}, \tag{7.21}$$

where $s \geq 2$, $\gamma \in k$, and the monomials of the form (7.21) are such that the variables $x_{i_1}, \ldots, x_{i_{t_i}}$ are the same as $x_1, \ldots, x_{c+2}$. However, in this case the right-hand side of equation (7.20) will be equal to a linear combination of monomials of degree strictly greater than the degree of the left-hand side. Since L is a **V**-free algebra, this equation is an identity in **V**. Thus we are forced to conclude $L^{c+2} = L^{c+3} = (= L^{c+4} = \cdots)$. But **V** is locally nilpotent so that the intersection of the terms of the lower central series in L must be trivial. Therefore $L^{c+2} = \{0\}$. This contradicts our remark at the beginning of our proof. Hence $\mathbf{V} \supseteq \mathbf{M}$ so that, in fact, $\mathbf{V} = \mathbf{M}$ and we are finished. □

7.8 Description of solvable just-non-Cross varieties

7.8.1 Generating sequence of a just-non-Cross variety

Throughout five first subsections of this section **V** will stand for a solvable just-non-Cross variety of Lie algebras over a finite field which is different from **M** (see Section 7.7.3). It was noted in Section 7.7.2 that, in such a variety, a sequence of algebras L_n must exist, with minimal ideals M_n such that the sequence $\dim M_n$, $n = 1, 2, \ldots$, is unbounded from above. Having factored each L_n by a maximal ideal whose intersection with M_n is trivial, we may assume that all M_n are monoliths in the corresponding L_n. Our nearest aim is a thorough study of L_n, $n = 1, 2, \ldots$, which is, in fact, a generating sequence for **V**. We will soon see that our assumption that **V** exists leads to this sequence lying in $\mathbf{AN}_c$ for a suitable choice of c. Then it will turn out that $c = 1$. This will be in contradiction with unbounded growth of the sequence $\dim M_n$, $n = 1, 2, \ldots$ Our final conclusion will be that there is no solvable just-non-Cross variety different from **M**. This will make it possible to complete our classification of solvable just-non-Cross varieties, first in the case of finite fields and, second, in the case of arbitrary finite rings. So we may say that, in the subsections under consideration, we are dealing with objects which do not exist. However, in every particular case it is not difficult to transform a statement of each particular lemma or proposition to make it meaningful.

We start with a useful remark. Let G be a Lie algebra over a finite ring k and A its abelian ideal. If $C = C_G(A)$ is the centralizer of S in G then, as in Section 1.6.5, it is easy to make A into a G/A- and G/C-module and to consider the semidirect products $B = (G/A) \rtimes A$ and $D = (G/C) \rtimes A$.

Lemma 7.34. $B, D \in \operatorname{var} G$.

Proof. We consider two isomorphic copies G_1, G_2 of G. Suppose, under a fixed isomorphism of G_1 onto G_2, the ideals A_1 and A_2 isomorphic to A correspond to each other. We consider the diagonal $\bar{G}$ of the direct product $E = G_1 \times G_2$ corresponding to the same isomorphism as above. Then $\bar{G} \cong G$, $\bar{A} \cong A$. We set $H = A_1 + \bar{G}$. It is obvious that $\bar{A} \lhd H$ and that $K = H/\bar{A}$ is isomorphic to the semidirect product of the ideal $(A_1 + \bar{A})/\bar{A}$ and the subalgebra $\bar{G}/A$. It is easy to see that K is isomorphic to B. By Section 4.1, Theorem 4.3, $B \in \operatorname{var} G$. Furthermore, D is a quotient algebra of B, hence $B \in \operatorname{var} G$. □

By Lemma 7.34, it follows that **V** possesses a sequence of the form $K_n = G_n \rtimes M_n$, $n = 1, 2, \ldots$, of finite semidirect products in which M_n is an irreducible G_n-module and $\dim M_n \to \infty$ as soon as $n \to \infty$. Since **V** is a minimal non-Cross variety, we derive $\operatorname{var}\{K_n \mid n = 1, 2, \ldots\} = \mathbf{V}$. This argument will be repeatedly applied in the subsequent subsections. The two forthcoming lemmas are aimed to improving generating sequences.

Lemma 7.35. *Let the subalgebras H_n in G_n, $n = 1, 2, \ldots$, be such that the dimensions of G_n/H_n are totally bounded. Then the sequence of algebras of the form $K' = H_n \rtimes W_n$, where W_n is an arbitrary irreducible H_n-submodule of M_n, $n = 1, 2, \ldots$, generates* **V**.

Proof. Let W_n be an arbitrary irreducible submodule of M_n as H_n-module. We omit index n and choose a basis $e_1, e_2, \ldots, e_m$ for G modulo H. If d is the degree of the polynomial f and $f(\operatorname{ad} x) = 0$ for any $x \in K$, then the subspace

$$V = \sum_{(t_1, t_2, \ldots, t_m)} e_1^{t_1} e_2^{t_2} \cdots e_m^{t_m} W, \quad 0 \le t_i < d, i = 1, 2, \ldots, m,$$

is a G-submodule of M whose dimension is bounded by a number depending entirely on m, d and $\dim W$. Since $W \neq \{0\}$, $W = M$ and, therefore, $\dim W_n \to \infty$ along with $\dim W_n$ (considering a subsequence it is obvious that we may think $\dim M_n \to \infty$, not merely that the set of dimensions is unbounded). □

Lemma 7.36. *Let L be a Lie algebra, $V = V_1 \oplus \cdots \oplus V_n$ a module such that all direct summands are pairwise isomorphic irreducible L-modules. Let U be a finite subset in V. Then U is contained in an L-submodule W whose order is bounded in terms of $|U|$ and $|V_1|$.*

Proof. We denote by π_i the projection of V onto V_i, $i = 1, \ldots, n$, and by $\{\varphi_{ij}\}$ a system of consistent isomorphisms $\varphi_{ij} : V_i \to V_j$, that is, such that $\varphi_{ij} = \varphi_{it}\varphi_{tj}$ and $1_{V_i} = \varphi_{ij}\varphi_{ji}$, $1 \le i, j, t \le n$. We say that the submodules V_i and V_j are equivalent if the following diagram is commutative:

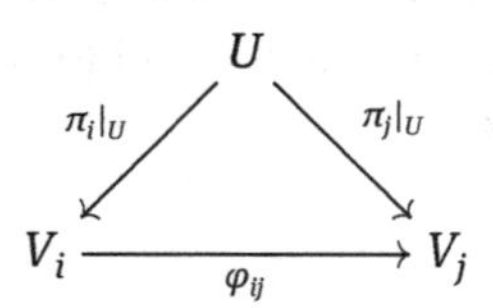

An obvious verification shows that, indeed, this relation is an equivalence on the set $\{V_1, V_2, \ldots, V_n\}$ and the number of equivalence classes is bounded in terms of numbers $|U|$ and $|V_1|$. Let $M_1, M_2, \ldots, M_s$ be the direct sums of submodules forming each of the s equivalence classes and $D_1, D_2, \ldots, D_s$ the diagonals of these direct sums with respect to the isomorphisms φ_{ij}. Then $|D_1| = |D_2| = \cdots = |D_s| = |V_1|$. Besides, $U \subseteq D_1 \oplus D_2 \oplus \cdots \oplus D_s$. Setting $W = D_1 \oplus \cdots \oplus D_s$ completes our proof. □

7.8.2 First principal reduction of the generating sequence of a solvable just-non-Cross variety

In this subsection we are going to prove that a generating sequence $K_n = G_n \rtimes M_n$, $n = 1, 2, \ldots$, for a solvable just-non-Cross variety $\mathbf{V} \neq \mathbf{M}$ can be so chosen that all the subalgebras G_n are nilpotent. We start with a simple result.

Lemma 7.37. *Let G be a finite l-step solvable Lie algebra. Then $G = N + H$ where N is a solvable ideal of G whose solvable step is strictly less than l and H is a nilpotent subalgebra.*

Proof. It is sufficient to choose N equal to G^2 and H equal to a subalgebra of minimal order with $G^2 + H = G$. From the minimality, it easily follows that $H \cap G^2 \subseteq \Phi(H)$ where $\Phi(H)$ is the Frattini subalgebra of H. Since $H/H \cap G^2$ is abelian, H is nilpotent by Section 1.7, Theorem 1.17. □

One more remark concerns the so-called *socle series* of an algebra. Let $\mathbf{U}$ be a Cross variety, P a finite algebra, and M a finite p-module such that $P \rtimes M \in \mathbf{U}$. Let M_0 denote the zero submodule in M and let an ascending chain of submodules $M_0, M_1, \ldots, M_i, \ldots$, be defined by induction according to the rule: M_{i+1} is a submodule of M with $M_{i+1} \supset M_i$ and M_{i+1}/M_i is equal to the sum of all minimal submodules of M/M_i. This chain is called the *socle series* of the P-module M. Using Proposition 7.10 and the results in Section 1.6, it is easy to derive that the length of the socle series as above is bounded by a constant depending solely on $\mathbf{U}$.

Proposition 7.38. *Let $K_n = G_n \rtimes M_n$, $n = 1, 2, \ldots$, be a generating sequence of a solvable just-non-Cross variety $\mathbf{V}$ where M_n is an abelian ideal in K_n which is an irreducible G_n-module and where the solvable step of any of G_n is strictly less than that of $\mathbf{V}$. Suppose $G_n = N_n + H_n$ (as in Lemma 7.37). Then there exists a chief factor L_n in the H_n-module M_n such that the sequence of algebras $\bar{K}_n = H_n \rtimes L_n$, $n = 1, 2, \ldots$, generates $\mathbf{V}$.*

Proof. We consider the sequence of algebras $T_n = N_n \rtimes M_n$, $n = 1, 2, \ldots$ Since the solvable step of each of the algebras T_n is strictly less than that of $\mathbf{V}$, it follows that all T_n belong to a proper subvariety of $\mathbf{V}$ and then the lengths of semisimple series of N_n-submodules of M_n, $n = 1, 2, \ldots$, are bounded by a certain number s. We will be proving our proposition by contradiction, while assuming that s is minimal. As previously, we will be omitting index n. So, let

$$\{0\} = M_0 \subset M_1 \subset \cdots \subset M_{s-1} \subset M_s = M$$

be the socle series of the N-module M. By Section 1.6, Theorem 1.7, each of its factors M_i/M_{i-1}, $i = 1, 2, \ldots, s$, is a direct sum of pairwise isomorphic irreducible N-submodules.

Suppose first that M_s/M_{s-1} is the direct sum of unboundedly many (in n) such N-submodules and chose an arbitrary irreducible H-submodule L in M. We consider a subset of M_s/M_{s-1} which has the form $L + M_{s-1}/M_{s-1}$. The order of this subset is bounded by $|L|$, hence, according to Lemma 7.36, one can find in M_s/M_{s-1} an N-submodule $D/M_{s-1} \supseteq L + M_{s-1}/M_{s-1}$ such that $|D/M_{s-1}|$ is a bounded value, that is, for some n, D_n is distinct from M_n. But $D \supseteq L$, hence the least G-submodule of M containing L belongs to the N-submodule D which is impossible since M is an irreducible G-module. So we conclude that M_s/M_{s-1} is a direct sum of a bounded number of irreducible N-submodules.

Since these latter have the order which is bounded, $|M_s/M_{s-1}|$ is bounded in terms of n and $s > 1$. Using a H-chief series of M, we may choose in M a submodule U whose order is bounded but it is greater than $|M_s/M_{s-1}|$. This can be done since by our hypothesis the chief H-factors of M have bounded order.

Now let C denote the centralizer of U in H. Then $|H/C|$ is bounded and, by Lemma 7.35, if $R = N+C$ and W is an irreducible R-submodule in M, the sequence of algebras of the form $\{R_n \rtimes W_n \mid n = 1, 2, \ldots\}$, generates $\mathbf{V}$. Moreover, since $|U| > |M_s/M_{s-1}|$ and $M_{s-1} \cap U$ is a nonzero C-submodule of the N-module M_{s-1} (C is acting trivially on U), we have $|U \cap M_{s-1}| = |U|\,|U+M_{s-1}/M_{s-1}| > 1$. It is obvious that the least R-submodule W of M containing $U \cap M_{s-1}$ is in M_{s-1}. Therefore, the sequence $\{R_n \rtimes W_n \mid n = 1, 2, \ldots\}$ generating $\mathbf{V}$ consists of algebras in which W_n is an N_n-module possessing a semisimple series whose length is strictly less than s, following because $W \subseteq M_{s-1}$. This contradicts the minimality of s. □

To accomplish our reduction we remark that, given a solvable variety $\mathbf{V}$ of solvable step l and $K = G \rtimes M$ an algebra in $\mathbf{V}$, hence with $K^{(l+1)} = \{0\}$, such that M is an irreducible G-module, we then have $G^{(l)}M \subset M \cap K^{(l)}$. This implies $G^{(l)}(G^{(l)}M) \subseteq G^{(l)}(M \cap K^{(l)}) = \{0\}$. Since $G^{(l)}M$ is a G-submodule of M, we have $G^{(l)}M = \{0\}$, that is, $G^{(l)} \subseteq C_G(M)$. Hence Proposition 7.38 applies, and we may replace our generating sequence by a such $\{G_n \rtimes M_n \mid n = 1, 2, \ldots\}$ in which all the algebras G_n are nilpotent.

7.8.3 The case of subvarieties in $\mathbf{AN}_c$

So we are considering a generating sequence of a solvable just-non-Cross variety $\mathbf{V}$ consisting of algebras $K_n = G_n \rtimes M_n$, $n = 1, 2, \ldots$, in which all G_n are nilpotent and all M_n are irreducible G_n-module. As follows from our remarks in Section 7.7.2, $\mathbf{V}$ is bounded with respect to the parameter c and, therefore, there exists a natural c, which can be chosen the least possible, such that the nilpotent step of each G_n is at most c. Applying factorization over $C_{G_n}(M_n)$ enables us to restrict ourselves to the case where all M_n

are faithful G_n-modules. As previously, we consider boundedness with respect to n even in the case where this parameter is omitted. Since **V** is generated by any infinite subsequence of algebras in $\{K_n \mid n = 1, 2, \dots\}$ for which the sequence of dimensions of monoliths is unbounded, the unboundedness of some sequence $\{a_n\}$ will in fact mean $a_n \to \infty$ as soon as $n \to \infty$. We will need an auxiliary lemma.

Lemma 7.39. *The center of G as above is of bounded order. If $c > 1$, Y/Z is the center of G/Z and $C = C_G(Y)$ is the centralizer of Y in G then $|G/C|$ is unbounded whereas $|G/C_G(y)|$ is a bounded value for any $y \in Y$.*

Proof. It is obvious that the center Z of G belongs to the endomorphism ring D of the G-module M. By Schur's lemma, D is a skew field, hence a commutative field by Wedderburn's theorem. According to equation (7.18), any element z in Z satisfies the equation $f(z) = 0$ which yields $|Z| \leq \deg f$. Now since $C_G(y)$ is the kernel of the linear map $x \mapsto xy$ from G into Z, $|Z|$ being bounded, it follows that $|G/C_G(y)|$ is bounded. We note that the nilpotent step of C is strictly less than c. Indeed, $C^{c-1} \subseteq G^{c-1} \subseteq Y$, hence $C^{c-1} \subseteq C \cap Y$ so that $C^c = \{0\}$. But $|G/C|$ is bounded which, by Lemma 7.35, enables us to write $\mathbf{V} = \operatorname{var}\{C_n \rtimes W_n \mid n = 1, 2, \dots\}$ where W_n is an irreducible C_n-submodule in M_n. This contradicts the minimality of c. □

To prove the main result of this section we will need a well-known number-theoretic theorem due to Warning.

Theorem 7.40. *Let k be a finite field, $f_1, \dots, f_r$ polynomials with zero constant terms in $k[t_1, \dots, t_s]$ of degrees $d_1, \dots, d_r$, respectively. If $s > \sum_{i=1}^{r} d_i$ then these polynomials have a nontrivial common zero.*

The proof of this theorem can be found, e. g., in Lang [199, Chapter 5].

Lemma 7.41. *If $c > 2$ then one can choose the elements $g_1, \dots, g_s \in G$, $y_1, \dots, y_s \in Y$, $z \in Z$, $z \neq 0$ such that:*

(1) *$g_i y_j = 0$ if $i > j$;*
(2) *$g_i y_i = z$;*
(3) *the subalgebra H generated by the elements $g_1, \dots, g_s, y_1, \dots, y_s$ has nilpotent step strictly less than c;*
(4) *$s = s(n)$ is unbounded by n.*

Proof. We keep our notations from Lemma 7.39. The argument below works in the case where $G = G_n$ with sufficiently large n. We note first that the product $x_1 x_2 \cdots x_c$ is multilinear and its value is determined by the cosets of G^2 containing $x_1, \dots, x_c$. It is also easy to see that $G^2 C \subseteq C$. We choose an arbitrary element g_1 in $G \setminus C$. Now let y_1 be an element in Y such that $g_1 y_1$ is a nonzero element in Z. By Section 7.8.3, the dimension of $C_G(y_1)/C$ is unbounded. Hence we may conclude that the same is true for the dimension of A_1/G^2, a direct complement to $C_G(y_1)/G^2$ in C/G^2. We consider the products of the form $f_\alpha = x_1 x_2 \cdots x_c$ where each x_i is either g_1 or $x_i = x \in A_1/G^2$. The values of these products depend on x, belong to Z, and if we introduce the coordinates in both A_1/G^2 and Z then the coordinates of the values of a polynomial f_α are homogeneous polynomials of degree less than c in the coordinates of x. Since the degree and the number of these polynomials are bounded in terms of c and $|Z|$, and the number of variables, which is equal to $\dim A_1/G^2$, is unbounded, by Warning's theorem there exists $g_2 \in A_1 \setminus G^2$ such that any product of degree c whose factors are g_1 and g_2 is equal to zero. Since $g_2 \notin C$ there exists a $y_2 \in Y$ such that $g_2y_2 \neq 0$. Applying Lemma 7.39, we remark that the dimension of $C(y_1) \cap C(y_2)/C$ is unbounded.

Now we choose in $C(y_1) \cap C(y_2)/G^2$ a complement A_2/G^2 to C/G^2 and consider all products of the form $x_1x_2 \cdots x_c$ where either $x_i = g_1$ or $x_i = g_2$ or else $x_i = x \in A_2/G^2$. As above, we will find a $g_3 \in A_2 \setminus G^2$, hence $g_3 \notin C$, such that all the products $x_1x_2 \cdots x_c$ are zero if each x_i is replaced by any one of the three elements g_1, g_2, g_3. Again one can find an element $y_3 \in Y$ such that $g_3y_3 \neq 0$. Thus, if n is sufficiently large, one can choose arbitrarily many pairs g_i, y_i, $i = 1, 2, \ldots, t = t(n)$, satisfying (1) since $g \in \bigcap_{t=1}^{i-1} C(y_t)$.

Recalling that the order of Z is bounded, we conclude that there exist unboundedly many products g_iy_i equal to one and the same element $z \in Z$, where $z \neq 0$. Hence, after renumbering, if necessary, we arrive at elements $g_1, \ldots, g_s, y_1, \ldots, y_s, z$ satisfying the conditions (1), (2), and (4). Now (3) is also satisfied. Indeed, having replaced the factors of the product $x_1x_2 \cdots x_c$ by the elements in the set $g_1, \ldots, g_s, y_1, \ldots, y_s$, we arrive either at $g_{i_2}g_{i_1} \cdots g_{i_c}$ which is equal to zero by construction or at the product of the form $g_{i_1} \cdots y_j \cdots g_{i_c}$. The latter product is equal to zero since $c > 2$ and y_j is an element of Y, the second term of the upper central series of G. □

7.8.4 Second principal reduction of the generating sequence of a solvable just-non-Cross variety

In this section we want to show that the generating sequence $\{L_n = G_n \rtimes M_n\}$ of a solvable just-non-Cross variety $\mathbf{V} \neq \mathbf{M}$ can be so chose so that all G_n are abelian Lie algebras. We start with a simple result.

Lemma 7.42. *Let H be a finite-dimensional Lie algebra which is 2-step nilpotent and H^2 one-dimensional. Then H is equal to the sum of two abelian ideals.*

Proof. Since H^2 is one-dimensional, the product $(x, y) \mapsto xy$ can be considered as a bilinear map on H/H^2. We choose a hyperbolic basis $\{a_1 + H^2, b_1 + H^2, \ldots, a_r + H^2, b_r + H^2, c_1 + H^2, \ldots, c_l + H^2\}$, i. e., $a_ib_j = \delta_{ij}z$, where $\{z\}$ is a basis in H^2, $a_ia_j = b_ib_j = a_ic_j = b_ic_j = 0$. Then the required ideals can be chosen, e. g., as linear spans of $\{a_1, \ldots, a_r, z\}$ and $\{b_1, \ldots, b_r, c_1, \ldots, c_l, z\}$. □

Proposition 7.43. *Let $L_n = G_n \rtimes M_n$, $n = 1, 2, \ldots$, be a generating sequence of a solvable just-non-Cross variety $\mathbf{V}$ such that all M_n are irreducible G_n-modules and all G_n are nilpotent with nilpotent step at most c, over a finite field k. Then $c \leq 1$.*

Proof. We will show, for the beginning, that $c \leq 2$. Assume, on the contrary, that $c > 2$. We consider subalgebras of the form $H = H_n$ of G_n constructed in Lemma 7.41. Let

$$0 \subset N_1 \subset N_2 \subset \cdots \subset N_t = M$$

be a chief series of the H-module M. We remark that there is a number r such that the linear operator induced by z in N_r/N_{r-1} is nonzero. Indeed, if this is not the case, we have $Mz^t = \{0\}$. Then the maximal subspace Q such that $Qz = \{0\}$ is nonzero, G-invariant, hence $Q = M$. Then z is acting trivially in M which contradicts the faithfulness of M as a G-module.

So let z act nontrivially in $W = N_r/N_{r-1}$. Then, for any pair of indices $i > j$, the element $y_i - y_j$ acts nontrivially in W since otherwise the action of $g_i y_i = z$ is the same as that of $g_i y_j = 0$. Now from the unboundedness of $s = s(n)$ of the number of elements $y_1, \ldots, y_s$ we immediately derive the unboundedness of the order of W, which is a chief factor the action of H. As usual, this yields

$$\mathbf{V} = \operatorname{var}\{H_n \rtimes W_n \mid n = 1, 2, \ldots\},$$

in contradiction with the minimality of c and Lemma 7.41. Thus we conclude $c \leq 2$.

Now we want to show that, in fact, $c = 1$. Admitting the opposite, we derive from Lemma 7.35 that the indices of all abelian ideals in G_n are arbitrarily high if n is sufficiently large. Having noticed this, we proceed as follows. Let $x_1, y_1 \in G$ and $x_1 y_1 \neq 0$. Since G coincides with its second center Y, by Lemma 7.39 the ideal $B_1 = C(x_1) \cap C(y_1)$ is of bounded index in G, so for sufficiently large n one can find $x_2, y_2 \in B_1$ such that $x_2 y_2 \neq 0$. We put $B_2 = B_1 \cap C(x_2) \cap C(y_2)$ and then we repeat our procedure. With $|Z|$ being bounded (Lemma 7.39), we argue as in Lemma 7.41 to get unboundedly many pairs $\{x_i, y_i\}$, $i = 1, \ldots, s$, with the condition $x_i y_i = z$ where z is a fixed nonzero element of the center. It is clear that $\{z\}$ is a basis for H^2 where H is a subalgebra generated by all elements x_i, y_i, $i = 1, 2, \ldots, s$, since $x_i y_j = 0$ for $i \neq j$. Now as in the first part of the proof we can find irreducible H_n-modules W_n such that $\mathbf{V} = \operatorname{var}\{H_n \rtimes W_n \mid n = 1, 2, \ldots\}$. By Lemma 7.41, H is representable in the form $H = A + B$, where A and B are abelian ideals. Since W_n is an irreducible H_n-module and $W_n z_n \neq \{0\}$ it follows that the solvable step of $H \rtimes W$ is equal to three. Hence, and from Proposition 7.38, we derive that $\mathbf{V} = \operatorname{var}\{B_n \rtimes U_n\}$ where U is a factor of the B-module W. With B being abelian, we get $c = 1$, which is a contradiction. This forces us to conclude that in a generating sequence with minimal c we must have all G_n abelian. □

7.8.5 Main theorem on solvable just-non-Cross varieties of Lie algebras over a finite field

Theorem 7.44. *Variety* $\mathbf{M}$ *defined in Section 7.7.3 is the only solvable just-non-Cross variety of Lie algebras over a finite field of characteristic* $p > 0$.

Proof. By Section 7.7.3, **M** is a just-non-Cross metabelian variety. Let, conversely, $\mathbf{V} \neq \mathbf{M}$ be a solvable just-non-Cross variety. Then, according to Section 7.7.2, Proposition 7.33, the principal reductions of Section 7.8.2 and Lemma 7.41, we may write $\mathbf{V} = \operatorname{var}\{G_n \rtimes M_n \mid n = 1, 2 \ldots\}$, where G_n is an abelian algebra acting irreducibly and faithfully on M_n, the dimensions of M_n being unbounded. In such circumstances, Lemma 7.39 bounds the orders of G_n, $n = 1, 2, \ldots$, and then Lemma 7.35 bounds the dimensions of M_n, n as above, since in this case $\{0\}$ is an ideal of bounded index in G_n. This is a contradiction. □

Now we are in a position to formulate and prove the main result of the present section.

Theorem 7.45. *Let k be a finite commutative ring with 1. Then any solvable just-non-Cross variety **V** over k has the form $\mathbf{M}_I$ where I is a maximal ideal in k and a basis of identities for $\mathbf{M}_I$ has the form of the union of (1), (2), and (3) in Section 7.7.3, $p = \operatorname{char} k/I$, and the set of identities of the form $rx = 0$, $r \in I$.*

Proof. Let **V** be a solvable just-non-Cross variety of Lie algebras over k. We denote by J the nilpotent radical of k and by **W** the subvariety in **V** selected by a system of identities of the form $jx = 0$, $j \in J$. If $\mathbf{W} \neq \mathbf{V}$ then **W** is a Cross variety, that is, a subvariety in a variety of the form $\mathbf{C}(f, m, c)$. Choose a $G \in \mathbf{V}$. Then for any $x, y \in G$ we have $f(\operatorname{ad} x)(y) \in JG, f^2(\operatorname{ad} x)(y) \in J^2G$, and so on. Since $J^t = \{0\}$ for some t, we conclude $f^t(\operatorname{ad} x)(y) = 0$. A similar argument shows that the nilpotent step of a nilpotent algebra in **V** is bounded by ct. Now let H/K be a chief factor of G. It is isomorphic either to a chief factor of $G/JG \in \mathbf{W}$ or to a factor of the form H_1/K_1, where $H_1 \subseteq JG$, which is abelian since JG is a nilpotent ideal. In the former case the order of H/K does not exceed m. In the latter case it follows from $J^tH = \{0\}$ that $JH \subseteq K$ and $JG \subseteq C$ where C is the centralizer of H/K. Now the semidirect product $D = (G/C) \rtimes (H/K)$ is in **V** (by Lemma 7.34), hence in **W**. To finish, it remains to note that H/K is a minimal ideal of D and then $|H/K| \leq m$ and $\mathbf{V} \subseteq \mathbf{C}(f^t, m, ct)$. Here we have a contradiction since a subvariety of a Cross variety is itself Cross.

Now $\mathbf{W} = \mathbf{V}$, and we may assume that all algebras of **V** are, in fact, S-algebras where $S = k/J$. It is well known that S is the direct sum of some fields $k_1, k_2, \ldots, k_n$. Hence any S-algebra is the direct product of k_i-algebras and **V** is generated by n varieties of the form $\mathbf{M}_{J_r}$ (see the statement of Theorem 7.45). Finally, **V** being just-non-Cross implies $\mathbf{V} = \mathbf{M}_{J_r}$ for an appropriate r, as required. □

Corollary 7.46. *Let **V** be a solvable just-non-Cross variety of Lie rings. Then either $\mathbf{V} = \mathbf{A}$, the variety of all abelian Lie rings, or $\mathbf{V} = \mathbf{M}_p$, the variety given by $px \equiv 0$ (p a prime number) and a system of identities (1), (2), and (3) in Section 7.7.3.*

Proof. Let **V** be such a variety as in the statement. It is an easy exercise that each proper subvariety in **A** is determined by an identity of the form $nx \equiv 0$, hence is generated by a ring with zero multiplication whose additive group is cyclic of order n. Hence **A** is just-non-Cross. Now if $\mathbf{V} \neq \mathbf{A}$ then $\mathbf{V}$ satisfies an identity of the form $nx \equiv 0$, n being a natural number. Thus $\mathbf{V}$ consists of $\mathbb{Z}/n\mathbb{Z}$-algebras so that Theorem 7.45 applies. □

7.8.6 Class $\mathbf{S}_l(f, c)$

We introduce a class of Lie algebras $\mathbf{S}_l(f, c)$ whose members are solvable k-algebras of solvable step at most l with identity (7.18) and with nilpotent factors of nilpotent step at most c. From the results of this chapter it follows easily that $\mathbf{S}_l(f, c)$ is a variety. By Theorem 7.45, this is a Cross variety since the nilpotent algebras in $\mathbf{M}$ have unbounded nilpotent steps. It follows that a variety $\mathbf{V}$ of solvable Lie algebras over k is Cross if and only if $\mathbf{V}$ is contained in a class of the form $\mathbf{S}_l(f, c)$ for appropriate l, f, c. Besides, it follows that a solvable variety $\mathbf{W}$ is a non-Cross variety if and only if it contains nilpotent algebras of arbitrarily high nilpotent step.

Inside any variety $\mathbf{S}_l$ of solvable algebras of solvable step at most l, we can effectively solve the following question: Does a given identity $v(x_1, \ldots, x_n) \equiv 0$ determine all identities of a suitable finite Lie algebra? If the answer is "yes" then there is no difficulty to construct the finite algebra itself whose identities have $v(x_1, \ldots, x_n) \equiv 0$ as its basis. It is worth noting that, in general, this question of writing down a basis for identities of a given finite Lie algebra (not necessarily solvable) is effectively decidable. The explicit form of these algorithms is left to the reader as an exercise.

7.9 Skeletons of locally finite varieties

In this section we discuss in detail the question on the generating classes of locally finite solvable varieties.

7.9.1 Definition

The *skeleton* $\mathscr{S}(\mathbf{V})$ of a variety $\mathbf{V}$ over a finite ring k is the intersection of all factor closed classes of algebras generating this variety. Also let $\mathscr{F}(\mathbf{V})$ denote the class of all finite algebras in $\mathbf{V}$. Let $\mathbf{M}$ denote the unique solvable just-non-Cross variety over a finite field k of characteristic $p > 0$ (Section 7.8.5). The main aim of this section is the proof of Theorem 7.48 stating $\mathscr{S}(\mathbf{M}) = \mathscr{F}(\mathbf{M})$. An important corollary is a necessary and sufficient condition under which the variety generated by a locally finite solvable Lie algebra can be generated by a single finite algebra, that is, it is a Cross variety. This condition says that there exists a finite algebra in $\mathbf{M}$ which is not isomorphic to any given factor of the generating algebra.

7.9.2 Variety $\mathbf{M}_n$. Class $\mathscr{G}(n, r, s)$

Let k be a finite field of characteristic p and n a natural number divisible by p. We denote by $\widetilde{\mathbf{M}}_n$ the subvariety of $\mathbf{A}^2$ determined by an additional identity of the form $x^n yz \equiv 0$ and by $\mathbf{M}_n$ the subvariety of $\widetilde{\mathbf{M}}_n$ given additionally by $x^n y^y \equiv 0$ (see the notation in Section 4.7). As in Section 7.7.3, it is easy to see that if $G \in \widetilde{\mathbf{M}}_n$ then $G/Z(G) \in \mathbf{M}_n$. If $n = p$ then we arrive at $\mathbf{M} = \mathbf{M}_p$. Now let r, s be natural numbers. We define a Lie algebra $G(n, r, s)$ (cf. Section 7.7.4) as a semidirect product of an abelian algebra $B = B(r)$ with a basis $\{z_1, \dots, z_r\}$ by a B-module $A = A(n, r, s)$ defined in the following way. We consider a commutative ring $R = R(n, r)$ which is isomorphic to the quotient ring of the polynomial ring $k[z_1, \dots, z_r]$ over the ideal generated by the monomials $z_1^n, \dots, z_r^n$. Then A is defined as the free R-module freely generated by $\{a_1, \dots, a_s\}$. Now B is the Lie subalgebra in R generated by $\{z_1, \dots, z_r\}$ with respect to the operation of bracketing and A is the left regular B-module (see the final paragraph of Section 1.6.2). As in Section 7.7.4, it is easy to verify that G is in $\mathbf{M}_n$. A useful remark is that, for any $s > 1$, $G(n, r, s)$ is isomorphic to a subalgebra in $G(n, r+1, s-1)$ generated by the elements $z_1, \dots, z_r, a_1, \dots, a_{s-1}, z_{r+1}a_1$. It is also clear that if $A^s = Ra_s$ then $B + A^s \cong G(n, r, 1)$ and $G(n, r, s)/A^s \cong G(n, r, s-1)$.

Let $\mathscr{A} = \mathscr{A}(n, r)$ be the category of all R-modules or, which is the same, of B-modules A with the condition $z_i^n a = 0$, $a \in A$, $i = 1, \dots, r$. We remark that R is a monolithic algebra with the monolith $kz_1^{n-1} \cdots z_r^{n-1}$. By Section 3.3.1, then R is a Frobenius algebra. Since all $A(n, r, s)$ are free R-modules of finite rank, it follows, using Section 3.3, Theorem 3.10, that all these modules are injective. In other words, all modules $A(n, r, s)$ are injective objects in the category $\mathscr{A}$. Applying once more Section 3.3, Theorem 3.10, and the commutativity of R, we write $A \cong A^* = \operatorname{Hom}(A, k)$. Now if M is an arbitrary R-module with $\dim M = s$ then M^* is a homomorphic image of A^*, hence M is an R-submodule in A.

Finally, we denote by $\mathscr{G}(n, r, s)$ the class of all metabelian Lie algebras G over k which are isomorphic to an extension of the $R(r)$-module $A = A(n, r, s)$ by a subalgebra which is isomorphic to $B(r)$. Split extensions of this form are precisely the algebras of the form $G(n, r, s)$. Summing up all the facts in the present subsection, we arrive at the following.

Proposition 7.47. *We consider $G \in F(\mathbf{M}_n)$. If $\dim G^2 = s$, $\dim G/G^s = r$ then G can be embedded in $G(n, r, s)$.*

Proof. It is immediate from the above that G is isomorphic to a subalgebra of an algebra in the class $\mathscr{G}(n, r, s)$. Indeed, $G/G^2 \cong B(r)$. Now $B(r)$-module G^2 is embeddable in $A = A(n, r, s)$. Hence G can be embedded in Lie algebra G_1 which is an extension of the $B(r)$-module $A(n, r, s)$ by $B(r)$. It remains to verify that $G_1 \in \mathbf{M}_n$. It is sufficient to prove $x^n y^n \equiv 0$. Indeed, $G_1 = G + A$, and G satisfies this identity. Now we set $x = g_1 + a_1$,

$y = g_2 + a_2, g_2 \in G, a_1, a_2 \in A$. Then

$$(g_1 + a_1)^{n-1}(g_2 + a_2)^{n-1}(g_1 + a_1)(g_2 + a_2) = g_1^{n-1}g_2^{n-1}(g_1a_2 - g_2a_1)$$
$$= g_2^{n-1}g_1^n a_2 - g_1^{n-1}g_2^n a_1 = 0.$$

For the completion of the proof, it remains to use an auxiliary assertion whose proof, full of technical details, is postponed till the end of this section and which reads as follows:

Any algebra in the class $\mathscr{G}(n, r, s)$ in which $x^n y^n \equiv 0$ is identically true is a split extension (that is, isomorphic to $G(n, r, s)$).

Modulo this assertion, our proof is complete. □

Theorem 7.48. *Let k be a finite field of characteristic $p > 0$ and n a natural number divisible by p. Then the skeleton $\mathscr{S}(\mathbf{M}_n)$ of $\mathbf{M}_n$ is equal to the set $\mathscr{F}(\mathbf{M}_n)$ of its finite-dimensional algebras.*

Proof. Let $\mathbf{X}$ be a factor closed class generating $\mathbf{M}_n$. It was shown in Section 7.9.2 that any algebra in $\mathscr{F}(\mathbf{M}_n)$ can be embedded in an algebra of the form $G(n, r, s)$ and that this latter algebra is embeddable in an algebra $G(n, r, 1)$ for a suitable r. Thus our proof will be complete if we manage to verify that any monolithic algebra $G(n, r, 1)$ belongs to $\mathbf{X}$. Since $G(n, r, 1)$ does not satisfy $x_1^{n-1} \cdots x_r^{n-1} yz \equiv 0$, this identity does not hold valid in a certain algebra G in $\mathbf{X}$. Let $g_1, \dots, g_r \in G$ and $a \in A = G^2$ be elements such that $g_1^{n-1} \cdots g_r^{n-1} a \neq 0$. We set $G' = \mathrm{alg}(g_1, \dots, g_r, A)$, $B' = G'/A$. Let A' be a B'-submodule in A generated by a. We will show that A' is a direct summand of the B'-module A. Indeed, let a linear map $\varphi : B(r) \to B'$ be given by $\varphi(z_i) = g_i + A$, $i = 1, \dots, r$. Then A' becomes a $B(r)$-module. Let ψ be a homomorphism of the free $B(r)$-module $A = A(n, r, 1)$ to A' such that $\psi(a_1) = a$. Since $\psi(z_1^{n-1} \cdots z_r^{n-1} a_1) = g_1^{n-1} \cdots g_r^{n-1} a \neq 0$, under ψ the monolith of A_0 is mapped nontrivially and ψ is an isomorphism of modules. Now, using the faithfulness of the action of B on A, we derive that φ is an isomorphism. It was noted in Section 7.8.2 that A is an injective B-module. Therefore we conclude that also A' is an injective B'-module. Hence A' is a direct summand of a B'-module A, that is, there exists a B'-submodule A'' in A such that $A = A' \oplus A''$. In this case, on the one hand, G'/A'' is an algebra in $\mathbf{X}$ and, on the other hand, this is an algebra in $\mathscr{G}(n, r, 1)$. Applying once more the splitting property from Section 7.9.4, we can see that $G'/A'' \cong G(n, r, 1)$. □

7.9.3 Identities $w_{c,l}$

Let c, l be nonnegative integers. We set

$$w_{c,l}(x_1, \dots, x_c, y, z) = x_1^l \cdots x_c^l yz.$$

Let $W_{c,l}(G)$ stand for the verbal ideal of a Lie algebra G corresponding to this identity. It is obvious that $w_{c,l} \equiv 0$ is a generalization of the Engel identity. We now write down some properties of these identities generalizing those in Chapters 1 and 4.

Proposition 7.49.
(i) *Let* **U** *be a nonnilpotent metabelian variety over a field* k, char $k = p > 0$; *if* $w_{1,m} \equiv 0$ *is an identity in* **U** *then* $m \geq p$ *and there exist natural numbers* $l, c > 0$ *with* $p|l$ *such that* $w_{c,l} \equiv 0$ *is an identity in* **U** *and* $w_{r,l-1} \equiv 0$ *is not an identity in* **U** *for any natural* r.
(ii) *Let* G *be a metabelian algebra. If* $W_{c,n}(G) \subseteq W_{1,l}(G)$ *then for any natural* d *there exists a natural* e *such that* $W_{e,n}(G) \subseteq W_{d,l}(G)$.

Proof. (i) Using the complete linearization of $x^m yz \equiv 0$ with respect to x in a metabelian variety gives $m!x_1 \cdots x_m yz \equiv 0$ (see Section 4.7). If $m < p$ then we deduce nilpotency. Now let l be the minimal of the numbers t for which $w_{c,t} \equiv 0$ is an identity of **U** for some $c > 0$. Then the numbers c, l are as required. That $p|l$ is immediate from the linearization process in a metabelian variety. Indeed, if $x^l yz \equiv 0$ then also $(x_1 + x_2)^l x_2^{l-2} yz \equiv 0$. Multiplying out and discarding zero terms, we get $lx_1^{l-1}x_1^{l-1}yz \equiv 0$. Since l is minimal, we immediately get $p|l$.

(ii) Using induction over $j = 0, 1, \ldots$, we are going to show that for all $g_1, \ldots, g_c \in G$ we have

$$g_1^n \cdots g_c^n W_{j,l}(G) \subseteq W_{j+1,l}(G).$$

If $j = 0$ then this is equivalent to our hypothesis. If $j = 0$ then by the metabelian property we have

$$\begin{aligned} g_1 \cdots g_c\Big(\sum_{g\in G} gW_{j-1,l}(G)\Big) &= \sum_{g\in G} g(g_1 \cdots g_c W_{j-1,l}(G)) \\ &\subseteq \sum_{g\in G} g^l W_{j,l}(G) = W_{j+1,l}(G). \end{aligned}$$

Now it is immediate that putting $e = cd$ completes our proof. □

Theorem 7.50. *Let* **V** *be a solvable locally finite variety of Lie algebras over a finite field* k. *If* **V** *is not a Cross variety then* $\mathscr{S}(\mathbf{V}) \supseteq \mathscr{F}(\mathbf{M})$ *where* **M** *is the unique solvable just-non-Cross variety over* k.

Proof. Let $\mathbf{X}_0$ be a factor closed set of finite algebras generating $\mathbf{V} = \mathbf{V}_0$. Since **V** is not Cross we have $\mathbf{V} \supseteq \mathbf{M}$ (Section 7.8.5). Let $\mathbf{X}_1$ denote the subclass of locally nilpotent algebras in $\mathbf{X}_0$. Since the nilpotency steps of nilpotent algebras in **M** are unbounded, we have $\mathbf{V}_1 = \mathrm{var}(\mathbf{X}_1) \supseteq \mathbf{M}$. Let $\mathbf{X}_2$ be the subclass of metabelian algebras in $\mathbf{X}_1$. Applying Section 1.7, Theorem 1.19, we see that $\mathbf{V}_2$ is a nonnilpotent locally nilpotent variety. By local nilpotency, $\mathbf{V}_2$ satisfies an identity of the form $x^m yz \equiv 0$. Applying Proposition 7.49, we find that $\mathbf{X}_2$ satisfies an identity of the form $w_{c,l} \equiv 0$ with c, l natural, $p|l$; at the same time, $w_{r,l-1} \equiv 0$ is not an identity for any $r = 0, 1, \ldots$ Let $\mathbf{X}_3$ be the subclass of $\mathbf{X}_2$ whose algebras satisfy $w_{1,l} \equiv 0$. Then, by Proposition 7.49 (ii), $\mathbf{X}_3$ does not satisfy an identity of the form $w_{r,l-1} \equiv 0$ for any $r > 0$. Now we have $\mathbf{V}_3 = \mathrm{var}(\mathbf{X}_3) \subseteq \mathbf{M}_l$ and $\mathbf{V}_3$ does not satisfy any identity of the form $w_{r,l-1} \equiv 0$, r being arbitrary. Passing to the quotient algebras over their centers (see Section 7.9.2), we arrive at the factor closed set $\mathbf{X}_4 = \mathbf{X}_3 \cap \mathbf{M}_l$ satisfying no identity of the form $x_1^{l-1} \cdots x_r^{l-1} yz \equiv 0$. But $\mathbf{X}_4$ is already a class to which the proof of Theorem 7.48 applies, giving $\mathbf{X}_4 \supseteq \mathscr{F}(\mathbf{M}_l)$. Since $p|l$, it is obvious that $\mathbf{M} \subseteq \mathbf{M}_l$, that is, $\mathbf{X}_4 \supseteq \mathscr{F}(\mathbf{M})$. □

Corollary 7.51. *Let G be a solvable locally finite Lie algebra over a finite field k of characteristic $p > 0$ and B a finite algebra in $\mathbf{M}$. If B is not a factor of G then* var(G) *is a Cross variety.*

7.9.4 Splitting algebras in $\mathscr{G}(n, r, s) \cap \mathbf{M}_n$

Let G be an extension in $\mathscr{G}(n, r, s) \cap \mathbf{M}_n$, $G = B \oplus A$, where $B = B(n, r)$, $A = A(n, r, s)$ (the vector space direct sum). By Section 1.6, Theorem 1.8, the multiplication in G is determined by an element $f \in H^2(B, A)$. This extension splits if and only if $f = \partial h$ where $h \in C^1(B, A)$, that is, when there exists a linear function $h : B \to A$ such that

$$f(x, y) = xh(y) - yh(x). \tag{7.22}$$

Since $G \in \mathbf{M}_n$, it follows that f satisfies an additional condition

$$x^{n-1} y^{n-1} f(x, y) = 0. \tag{7.23}$$

Indeed, we know that $\mathbf{M}_n$ satisfies $x^n y^n \equiv 0$. Taking $x, y \in B$ and applying equation (7.11) in Section 1.6.6 immediately gives equation (7.23). Since $\partial f = 0$ and considering formula (1.9) in Section 1.6.5, we get

$$xf(y, z) + yf(z, x) + zf(x, y) = 0. \tag{7.24}$$

Thus to prove that any extension $G \in \mathscr{G}(n, r, s) \cap \mathbf{M}_n$ splits it is sufficient, for a bilinear skew-symmetric function $f : B \times B \to A$ satisfying equations (7.23) and (7.24), to construct a linear function $h : B \to A$ for which equation (7.22) is satisfied.

Since $A(n, r, s)$ is the direct sum of the copies of $A(n, r, 1)$, using projections on the direct summands shows that it is enough to restrict to the case $s = 1$, that is, where A is the free $R(n, r)$-module $R(n, r)$ (see Section 7.9.2). So we identify A with $R(n, r)$. Then any cocycle f takes the form of a polynomial in $z_1, \ldots, z_r$. We set $f_{ij} = f(z_i, z_j)$. Since f is skew-symmetric and using equation (7.23) it follows immediately that $f_{ij} = z_j g_i^j - z_i g_j^i$, $i < j$. We write

$$f_{ij} = z_j \tilde{g}_i^j - z_i \tilde{g}_j^i + z_i z_j g_{ij}, \tag{7.25}$$

where $\tilde{g}_i^j$ is not divisible by z_i. Putting $\tilde{g}_i^i = g_{ii} = 0$, we may assume that equation (7.25) holds for all i, j, $1 \le i, j \le r$. From equation (7.24) and from $x^n yz \equiv 0$ in G_0, we have $z_i^{n-1} z_t f_{ij} = z_i^{n-1} z_j f_{it}$. Using equation (7.25) and canceling by z_i^{n-1} (which is admissible), we get $z_t z_j \tilde{g}_i^j = z_j z_t \tilde{g}_i^t$. This shows that there exists a polynomial $\tilde{g}_i$ such that $z_j \tilde{g}_i^j = z_j \tilde{g}_i$ for all $j \neq i$. We write down this assertion in the form of a lemma, as follows.

Lemma 7.52. *Let $q_1, \dots, q_m \in R(n,r)$, $m \le r$. If $z_t z_l q_t = z_l z_t q_t$ for all t, l, $1 \le t, l \le m$, then there exists a polynomial $q \in R(n,r)$ such that for any t, $1 \le t \le m$, we have $z_t q_t = z_t q$.*

Proof. We apply induction over m with obvious basis $m = 1$. Suppose $m > 1$ and let $\tilde{q}$ be such that $z_t \tilde{q} = z_t q_t$ for all $1 \le t < m$. We write

$$\tilde{q} = p_1 z_m^{n-1} p_1^1 + z_1^{n-1} \cdots z_{m-1}^{n-1} p_2^1 + p^1, \quad q_m = z_m^{n-1} p_1^2 + z_1^{n-1} \cdots z_{m-1}^{n-1} p_2^2 + p^2,$$

where p^1 and p^2 do not belong to the ideal generated by z_m^{n-1} and $z_1^{n-1} \cdots z_{m-1}^{n-1}$. By the hypothesis, $z_m z_t q_m = z_m z_t q_t = z_m z_t \tilde{q}$ $(t < m)$. Hence $z_m z_t p^1 = z_m z_t p^2$ $(t < m)$. By the definition of p^1 and p^2, we have $p^1 = p^2 = p^0$. Now q can be taken in the form $q = p_1^1 z_m^{n-1} + p_2^2 z_1^{n-1} \cdots z_{m-1}^{n-1} + p^0$. □

Coming back to the argument preceding Lemma 7.52, we can see that for some polynomials $\tilde{g}_1, \dots, \tilde{g}_r$ we have $f_{ij} = z_j \tilde{g}_i - z_i \tilde{g}_j + z_i z_j g_{ij}$. Since $z_i \tilde{g}_j - z_j \tilde{g}_i$ is a coboundary, $z_i z_j g_{ij}$ is a cocycle and equation (7.23) is satisfied by it. To prove that f is a coboundary, therefore, it is sufficient to show that a cocycle of the form $z_i z_j g_{ij}$ with equation (7.23) is a coboundary. Using equation (7.24), we write

$$\begin{aligned} z_i z_j z_t g_{ijt} &= z_i (z_j z_t g_{jt}) + z_j (z_i z_t g_{ti}) + z_t (z_i z_j g_{ij}) \\ &= z_i f_{jt} + z_j f_{ti} + z_t f_{ij} = 0. \end{aligned}$$

We will need another technical result.

Lemma 7.53. *Let $\{g_{ijt}\}$ be a family of polynomials in $R(n,r)$, $1 \le i,j,t \le m \le r$ which skew-symmetrically depends on the indices i,j,t. Suppose $z_i z_j z_t g_{ijt} = 0$ for all values of indices. Then there exist polynomials $g_1, \dots, g_m$ such that $z_i z_j g_{ij} = z_i z_j (g_i - g_j)$, $1 \le i,j \le m$.*

Proof. If already $g_{ijt} = 0$ then it is sufficient to set $g_i = g_{i1}$, $1 \le i \le m$. It is easy to find g_i also in the case where $z_i g_{ijt} = 0$. Indeed, we note that the following is always true:

$$g_{ijt} = g_{lij} + g_{ljt} + g_{lti}, \tag{7.26}$$

hence

$$z_l g_{ijt} = z_l g_{lij} + z_l g_{ljt} + z_l g_{lti} = 0.$$

This shows that the elements g_{ijt} belong to the monolith of $R(n,r)$. It is clear that adding to g_i an element in the monolith of $R(n,r)$ does not affect the equation required in the statement of Lemma 7.53.

A more general case is where we know only that $z_i z_j g_{ijt} = 0$. Since $z_j z_t g_{ijt} = z_t z_i g_{ijt} = 0$, for $i < j < t$ we may write

$$g_{ijt} = h_{jt}^i + h_{ti}^j + h_{ij}^t, \tag{7.27}$$

where

$$h_{jt}^i = z_j^{n-1} z_t^{n-1} e_{jt}^i, h_{ti}^j = z_t^{n-1} z_i^{n-1} e_{ti}^j, h_{ij}^t = z_i^{n-1} z_j^{n-1} e_{ij}^t, \tag{7.28}$$

with $e_{jt}^i, e_{ti}^j, e_{ij}^t \in R(n,r)$. We extend the definition of elements h_{ij}^t to all choices of indices to make them skew-symmetric. From equation (7.27), we get $z_j h_{jt}^i = z_t h_{jt}^i = 0$. Using equation (7.26), we write $z_i z_l g_{ijt} = z_i z_l g_{ljt}$. Hence we have a relation of the form

$$z_i z_l h_{jt}^i = z_i z_l h_{jt}^l. \tag{7.29}$$

We fix a pair of indices $j_0 < t_0$ and consider a collection of elements of the form $h_{j_0 t_0}^i$, $i \neq j_0, t_0$. Let $\bar{B}$ be a subalgebra in B generated by the elements z_i, $i \neq j_0, t_0$. Also let $\bar{R}$ denote the $\bar{B}$-submodule generated by $z_{j_0}^{n-1} z_{t_0}^{n-1}$. Since $z_{j_0} h_{j_0 t_0}^i = z_{t_0} h_{j_0 t_0}^i = 0$, we see that the elements of the form $h_{j_0 t_0}^i$, $i \neq j_0, t_0$ are in $\bar{R}$. Applying equation (7.29) and Lemma (7.21) gives us an element $h_{j_0 t_0}$ such that for all $i \neq j_0, t_0$ we have $z_i h_{j_0 t_0}^i = z_i h_{j_0 t_0}$. Since $h_{j_0 t_0} \in \bar{R}$, we also have $z_{j_0} h_{j_0 t_0} = z_{t_0} h_{j_0 t_0} = 0$. Since the choice $i_{0},_0$ is arbitrary, considering the skew symmetry in the indices i, j, t, $1 \leq i, j, t \leq m$, we arrive at the relations of the form

$$z_i h_{jt}^i = z_i h_{jt}, \quad z_j h_{jt} = z_t h_{jt} = 0. \tag{7.30}$$

We set $\bar{g}_{jt} = g_{jt} - h_{jt}$. For this newly obtained skew-symmetric collection, we have $z_j z_t \bar{g}_{jt} = z_t z_t g_{jt}$. Now

$$\bar{g}_{ijt} = g_{ijt} - (h_{ij} + h_{jt} + h_{ti}) = (h_{jt}^i - h_{jt}) + (h_{ti}^j - h_{ti}) + (h_{ij}^t - h_{ij}).$$

Using equation (7.30) and similar relations for h_{jt}^i, we get $z_i \bar{g}_{ijt} = 0$, which brings us back to the already considered case.

Now we consider the general situation where all we know is $z_i z_j z_t g_{ijt} = 0$. Then for $i < j < t$ (hence, as previously, for all $1 \leq i, j, t \leq m$) we can select the elements p_{jt}^i such that

$$g_{ijt} = p_{jt}^i + p_{ti}^j + p_{ij}^t, \quad p_{jt}^i = z_i^{n-1} q_{jt}^i, \, p_{ti}^j = z_j^{n-1} q_{ti}^j, \, p_{ij}^t = z_t^{n-1} q_{ij}^t. \tag{7.31}$$

It is easy to observe that $z_i p_{jt}^i = 0$. Fixing $i = i_0$, we consider a subalgebra $\widetilde{B}$ with a basis $\{z_i \mid i \neq i_0\}$ and a $\widetilde{B}$-submodule R generated by $z_{i_0}^{n-1}$. We see that $p_{it}^{i_0} \in \widetilde{R}$. Applying induction over r with obvious basis gives elements $p_j^{i_0} \in \widetilde{R}$ $(j \neq i_0)$ such that

$$z_i z_t p_{jt}^{i_0} = z_j z_t (p_j^{i_0} - p_{t_0}^i). \tag{7.32}$$

Since these elements are in $\widetilde{R}$, we also have $z_{i_0} p_j^{i_0} = 0$. Putting $p_i^i = 0$, we arrive at a family of elements p_j^i with the following property:

$$z_j z_t p_{jt}^i = z_j z_t (p_j^i - p_t^i).$$

Now we set $\tilde{g}_{ij} = g_{ij} - (p_j^i - p_i^j)$. Then

$$\begin{aligned} z_j z_t \tilde{g}_{ijt} &= z_j z_t (g_{ijt} - (p_j^i - p_i^j + p_t^j - p_j^t + p_i^t - p_t^i)) \\ &= z_j z_t (p_{jt}^i - (p_j^i - p_t^i) + p_{ti}^j - (p_t^j - p_i^j) + p_{ij}^t - (p_i^t - p_j^t)) = 0. \end{aligned}$$

Since $z_i z_j \tilde{g}_{ij} = z_i z_j g_{ij}$, we are under the conditions of the preceding case. □

Lemma 7.53 enables us to complete our argument on the splitting of algebras in $\mathcal{G}(n,r,s) \cap \mathbf{M}_n$. Indeed, using the notation introduced before Lemma 7.53 and Lemma 7.53 itself, we can find polynomials $g_i \in R$, $i = 1,\dots,r$, such that $z_i z_j g_{ij} = z_i z_j (g_i - g_j)$. Then $f_{ij} = z_j(z_i g_i) - z_i(z_j g_j)$. Hence f turns out to be a coboundary of the 1-cocycle mapping each z_i to the polynomial $z_i g_i$. This proves that all algebras in $\mathcal{G}(n,r,s) \cap \mathbf{M}_n$ split and provides full proofs for Theorems 7.48 and 7.50.

7.10 Exercises

7.10.1. Let $\mathbf{V} = \mathbf{A}^2$ be the variety of all metabelian Lie algebras over a field k, char $k \neq 2$, $\mathbf{W} = \mathbf{V}^{(2)}$. Prove that the basis for identities of $\mathbf{W}$ consists of a single identity $(x^2y)xy \equiv 0$.

7.10.2. Find a basis for the identities of a Lie algebra $M = \langle x, y \mid xy = y \rangle$ over a field k, $|k| = q$.

7.10.3. Let G be a critical algebra which can be generated by n elements but has no generating set of $n-1$ elements. Let H be an m-generator algebra with $\mathrm{var}(G) = \mathrm{var}(H)$. Prove that $n \leq m$.

7.10.4. Let G, H be two finite simple algebras such that $\mathrm{var}(G) = \mathrm{var}(H)$. Prove that $G \cong H$.

7.10.5. Let G, H be two critical nilpotent algebras such that $\mathrm{var}(G) = \mathrm{var}(H)$. We denote by $\mathbf{U}, \mathbf{V}$ the varieties generated by all proper factors of G, H, respectively. Prove that $\mathbf{U} = \mathbf{V}$.

7.10.6. Prove that a finite algebra cannot belong to the variety generated by its quotient algebra over its socle.

7.10.7. Let G, H be two critical algebras with nonabelian monoliths, $\mathbf{U}, \mathbf{V}$ the varieties generated by proper factors of G, H, respectively. If $\mathrm{var}(G) = \mathrm{var}(H)$ then $\mathbf{U} = \mathbf{V}$.

7.10.8. Let k_2 denote the Lie algebra of all matrices of order 2 over a field k such that $|k| = 2^n$. Prove that k_2 is a critical algebra. Does this claim hold valid in the case of finite fields of characteristic different from 2?

7.10.9. Let L be a critical algebra whose monolith is nonabelian. If $L \in \mathrm{var}(G)$, where G is a finite algebra, then L is isomorphic to a factor of G.

7.10.10. For what n can the subvariety of $\mathbf{S}_l$ determined by the standard identity of degree n over the Galois field $\mathbb{Z}_p$ be generated by a finite algebra?

7.10.11. Furnish with a complete proof the fact that the class $\mathbf{S}_l(f, c)$ defined in Section 7.8.6 is a Cross variety.

7.10.12. Suppose that $v(x_1, \ldots, x_n) \equiv 0$ is not satisfied by $\mathbf{M}$ over a finite field of characteristic $p > 0$. Then, for any natural l it defines within $\mathbf{S}_l$ a variety of the form $\mathbf{V} = \mathrm{var}(G)$ where G is a finite algebra. Find an explicit formula for the bound of the nilpotent steps of nilpotent factors of G in terms of p, n and l.

7.10.13. Give an explicit formula for an upper bound of the orders of chief factors in algebras from $\mathbf{S}_l(f, c)$.

7.10.14. Give an explicit formula for an upper bound of orders of critical algebras in $\mathbf{C}(f, m, c)$.

7.10.15. (see Exercise 7.10.11). Describe an algorithm for constructing an algebra G such that $\mathrm{var}(G)$ is equal to the variety determined within $\mathbf{S}_l$ by an identity $v(x_1, \ldots, x_n) \equiv 0$.

7.11 Comments

The main source for this chapter are the papers by Bahturin and Olshanskii [51, 52]. Just-non-Cross varieties in the case of groups have been studied in [250, 299]. Section 7.9 is based on Ritter [321]. In the paper of Volkov and Gein [366] there is a generalization of the main result of this chapter to the case of some finite extensions of nilpotent Lie rings. However, it remains unknown whether the identities of an arbitrary nilpotent-by-finite Lie ring have finite basis. In the papers of Sheina, there is a wealth of results on metabelian locally finite varieties of A-algebras, that is, in which all nilpotent algebras are abelian (see Sheina [332]). In particular, it has been shown there that the lattice of such varieties is not distributive (in contrast to similar, even more general, setting in the varieties of groups). Drensky [106] has found a basis for the identities and the list of critical algebras in the variety generated by the Lie algebra of all matrices of order 2 over a finite field of characteristic 2. Sheina has found bases for the identities in certain metabelian finite Lie algebras (Sheina [331]). These are the only cases where we know the bases for identities of finite Lie algebras. It would be interesting to determine the bases for the identities of some known finite simple Lie algebras since these completely determine the algebras themselves. An interesting question is to find necessary and sufficient conditions on a finite Lie algebra G under which we have $\mathrm{var}(G) = \mathrm{qvar}(G)$. In the case of varieties of groups this has been done by Olshanskii [251]. As a comment to Section 7.7, we remark that the variety $\mathbf{M}$ was introduced by Artamonov [11] in connection with his study of chain varieties of linear algebras. See also his papers [13, 15]. To conclude, we list some papers on locally finite varieties in some other classes of algebras: Kruse [187]; L'vov [207–209]; Malcev and Parfenov [220]; Medvedev [223, 224]. For a brief survey of the problem: when is a finite algebra finitely based, see MacDonald, Sheila Oates [212].

8 Applications to group theory

8.1 Constructing a group of a Lie algebra

8.1.1 Projective limit

In this subsection the word algebra does not mean a linear algebra over a ring but merely a nonempty set with a fixed collection of operations Ω. Now let I be a well-ordered set and let $\{T_\alpha, \varphi_{\alpha\beta} \mid \alpha, \beta \in I\}$ be the set of algebras with the same collection of operations and epimorphisms $\varphi_{\alpha\beta} : T_\beta \to T_\alpha$, where $\alpha \le \beta$, and such that $\varphi_{\alpha\beta}\varphi_{\beta\gamma} = \varphi_{\alpha\gamma}$, $\varphi_{\alpha\alpha} = 1_T$. We consider the Cartesian product

$$C = \prod_{\alpha \in I} T_\alpha.$$

The set of all functions $f \in C$ such that $f(\alpha) = \varphi_{\alpha\beta}(f(\beta))$ for all α, β with $\alpha \le \beta$ is called the *projective limit* of the above family $\{T_\alpha, \varphi_{\alpha\beta}\}$ and is denoted by $\lim_{\longleftarrow} T_\alpha$. Clearly, the componentwise operations on $\lim_{\longleftarrow} T_\alpha$ turn the direct limit to an algebra with the same operations Ω.

As an example of a projective limit, we consider a group G with a descending series of normal subgroups of the form $\{N_n\}_{n\ge 1}$. Set $G_n = G/N_n$, $n = 1, 2, \ldots$ For all $m \le n$, we define $\varphi_{mn} : G_n \to G_m$ as a natural homomorphism, that is, $\varphi_{mn}(gN_n) = gN_m$. Then $\widehat{G} = \lim_{\longleftarrow} G_n$ is also a group. With each $g \in G$ we associate a function $\bar{g} \in \widehat{G}$ such that $\bar{g}(n) = gN_n$. The map $g \mapsto \bar{g}$ is then a homomorphism from G into G whose kernel is the subgroup

$$N = \bigcap_{n=1}^{\infty} N_n.$$

8.1.2 Completion in one class of topological algebras

In this subsection the reader is supposed to have a certain familiarity with basic concepts of topological algebras, e. g., see Chapter 20 of van der Waerden [367]. Now let R be a linear algebra with a descending chain of ideals

$$R \supseteq R^{(0)} \supseteq R^{(1)} \supseteq \cdots \supseteq R^{(m)} \supseteq \cdots \supseteq \bigcap_{m=0}^{\infty} R^{(m)} = \{0\}.$$

A topology on R can be defined as follows. An element $r \in R$ is called the *limit* of a sequence $\{r_n\}$ if for any $m = 1, 2, \ldots$ there exists a natural number N such that for all $n \ge N$ we have $r_n - r \in R^{(m)}$. A subset $S \subseteq R$ is called *closed* if S contains any limit of a sequence of its elements having limit in R. It is clear that the ideals $R^{(m)}$, $m = 1, 2, \ldots$, themselves are both closed and open subsets of R. A sequence $\{r_n\}$ is

https://doi.org/10.1515/9783110566659-008

called *fundamental* if for any $m = 1, 2, \dots$ there exists N such that for all $n_1, n_2 \geq N$ we always have $r_{n_1} - r_{n_2} \in R^{(m)}$. If any fundamental sequence has limit then we say that R is a *complete* topological algebra. A subset S is called *dense* in R if its closure is equal to R, that is, any element in R is a limit of a sequence of elements in S. If S is a dense subalgebra of a complete topological algebra R then any continuous homomorphism from S into a complete topological algebra R' extends to a continuous homomorphism from R into R'.

We note that if the topology in R is defined by a descending chain $\{R^{(m)}\}$ as above and in R_1 by a descending chain $\{R_1^{(m)}\}$ then a homomorphism φ is continuous if and only if for any $n = 0, 1, 2, \dots$ there exists n' such that $\varphi(R^{(n')}) \subseteq R_1^{(n)}$. If a topological algebra S is a dense subset in a complete topological algebra R then we say that R is a *completion* of S and write $R = \widehat{S}$. The completion is unique up to a continuous isomorphism. The completion of an algebra can be built using the notion of the projective limit (see Section 8.1.1). We set $R_{(m)} = R/R^{(m)}$, $m = 0, 1, 2, \dots$ For any $m < n$, let φ_{mn} denote the natural map from $R_{(n)}$ into $R_{(m)}$ mapping each element of the form $r + R^{(n)}$ into an element of the form $r + R^{(m)}$. We set $\widehat{R} = \lim_{\longleftarrow} R_{(m)}$. Then we consider a descending chain of ideals $\widehat{R}^{(m)}$ such that $f \in R^{(m)}$ if $f(1) = f(2) = \cdots = f(m) = 0$. Now $\widehat{R}$ is a complete topological algebra with topology defined by the above chain $\widehat{R}^{(m)}$.

Indeed, if $f_1, f_2, \dots, f_n, \dots$ is a fundamental sequence in R then its limit $\tilde{f}$ is a function which can be defined as follows. Let $\{c(m)\}$ be a sequence of natural numbers such that for all $n_1, n_2 \geq c(m)$ we have $f_{n_1} - f_{n_2} \in \widehat{R}^{(m)}$. Then we put $\tilde{f}(m) = f_{c(m)}(m)$. If $m < n$ then $\varphi_{mn}(f_{c(n)}(n)) = f_{c(n)}(m) = f_{c(m)}(m)$; this shows that, indeed, $\tilde{f} \in \widehat{R}$. Now let us show that $\{f_n\} \to \tilde{f}$. For any $m = 0, 1, 2, \dots$, we choose $N = c(m)$. If $n > N$ and $l \leq m$ then $n > c(l)$, hence $f_n(l) - \tilde{f}(l) = f_n(l) - f_{c(l)}(l) = 0$. Finally, R is continuously isomorphic to a subalgebra $\bar{R}$ of R whose elements are all functions $\bar{r}$ such that $\bar{r}(n) = r + R^{(n)}$, $n = 0, 1, 2, \dots$ It is clear that if $f \in \widehat{R}$ and $r_m + R^{(m)} = f(m)$ then $\{\bar{r}_m\} \to f$. Thus $\widehat{R}$ is a completion of R. In a complete topological algebra, one may, in a usual manner, define the convergence and divergence of infinite series. We remark that a series $\sum_{n=1}^{\infty} x_n$ is convergent if and only if $\{x_n\} \to 0$. If $\widehat{R}$ is a completion for R then any element $a \in \widehat{R}$ can be represented in the form $a = \sum_{n=1}^{\infty} x_n$ where $x_1, x_2, \dots$ are some elements in R. We also remark that if R has another chain

$$R = \bar{R}^{(0)} \supseteq \bar{R}^{(1)} \supseteq \cdots \supseteq \bar{R}^{(m)} \supseteq \cdots \supseteq \bigcap_{m=0}^{\infty} \bar{R}^{(m)} = \{0\}$$

such that $\bar{R}^{(m)} \supseteq R^{(m)}$, $m = 0, 1, 2, \dots$, and $\widetilde{R}$ is a completion with respect to the topology given by this series then the identity map from R to R extends to a continuous epimorphism ψ from $\widehat{R}$ onto $\widetilde{R}$. This can be defined as follows. If $f \in \widehat{R}$ then $\psi(f) = f'$ where $f'(s) = r_s + \bar{R}^{(s)}$ as soon as $f(s) = r_s + R^{(s)}$. We note that if, conversely, each member of the former chain contains a member of the latter one then ψ is injective. Indeed, suppose f is a nonzero element in R and let s be such that $f(s) = r_s + R^{(s)}$ is nonzero. Let t be such that $R^{(s)} \supseteq \bar{R}^{(t)}$. If $f(t) = r_t + R^{(t)}$ then $f'(t) = r_t + \bar{R}^{(t)}$. Assuming that

$f'(t) = 0$, we will get $r_t \in \bar{R}^{(t)} \subseteq R^{(s)}$. But $r_s + R^{(s)} = r_t + R^{(s)}$. Since $R^{(s)} \supseteq R^{(t)}$, we arrive at $r_s + R^{(s)} = R^{(s)}$, contradicting our assumption.

Finally, if R is an algebra in a certain variety then R belongs to the same variety since it is a subalgebra of a Cartesian product of quotient algebras of R.

8.1.3 Completion of graded algebras

Let $R = \sum_{n=0}^{\infty} R_n$ be a graded algebra. Then we set $R^{(m)} = \sum_{n>m} R_n$. We have $R/R^{(m)} \cong \sum_{l=1}^{m} R_l$. Let φ_{mn} denote the natural projection of $\sum_{l=1}^{n} R_l$ onto $\sum_{l=1}^{m} R_l$. Then $\lim_{\longleftarrow} R/R^{(m)}$ consists of all functions f such that $f(m) = r_1 + \cdots + r_m, f(m+1) = r_1 + \cdots + r_m + r_{m+1}$, where $r_i \in R_i$. It is clear that this function is equal to the sum of an infinite series of the form $\sum_{m=0}^{\infty} r_m$, where $r_m \in R_m$. This is the form which will be used for representing elements of the completion $\widehat{R}$ of R. Infinite series of this kind can be multiplied using distributivity, since any component of the product depends on finitely many elements of the factors. If S is a homogeneous subalgebra in R then, obviously, $\widehat{R} \supset \widehat{S}$. If R and S are two graded algebras then their tensor product (Section 6.1.4) is also a graded algebra $T = R \otimes_k S$ whose homogeneous components T_n have the form

$$T = \sum_{m=0}^{n} R_m \otimes_k S_{n-m}.$$

As above, we can speak about $\widehat{T}$, the completion of T with respect to this grading. Finally, if $\varphi : R \to S$ is a homomorphism such that $\varphi(R^{(n)}) \subseteq S^{(n)}$ for all $n = 0, 1, 2, \ldots$ then φ extends to a homogeneous continuous homomorphism $\hat{\varphi} : \widehat{R} \to \widehat{S}$, viz., if

$$a = \sum_{n=0}^{\infty} a_n \quad \text{then } \varphi(a) = \sum_{n=0}^{\infty} \varphi(a_n).$$

8.1.4 Topology on free associative algebras and free Lie algebras. Friedrichs' criterion

The argument as just above applies to the free associative algebra $A(X)$ and to its Lie subalgebra $L(X)$ which is a free Lie algebra. Thus the topology of $L(X)$ is defined by its lower central series. As in the preceding subsection, we arrive at completions $\widehat{A} = A(X)$ and $\widehat{H} = L(X)$. It is customary to call $A(X)$ the *Magnus algebra*. In the case where $X = \{t\}$, $A = k[t]$, the polynomial ring in one variable, and $\widehat{A} = k[[t]]$, the formal power series ring in one variable. Now we want to establish a criterion for an element of $a \in \widehat{A}$ to be in $\widehat{L}$. For this we consider a homogeneous homomorphism $\Delta : A \to A \otimes A$, extending the (diagonal) map $x \mapsto x \otimes 1 + 1 \otimes x$, $x \in X$. Then, as is noted above, Δ extends to a continuous homomorphism $\Delta : \widehat{A} \to \widehat{A \otimes A}$.

Theorem 8.1. *An element $a \in \widehat{A}$ is in $\widehat{L}$ if and only if $\widehat{\Delta}(a) = a \otimes 1 + 1 \otimes a$.*

Proof. We first consider the case where $a \in A$. Then $\widehat{\Delta}(a) = \Delta(a)$. With no loss of generality, we may restrict ourselves to the case of homogeneous elements. The sufficiency is almost obvious. If $a \mapsto a \otimes 1 + 1 \otimes a$ and $b \mapsto b \otimes 1 + 1 \otimes b$ then

$$[a, b] \mapsto [a \otimes 1 + 1 \otimes a, b \otimes 1 + 1 \otimes b] = [a, b] \otimes 1 + 1 \otimes [a, b]$$

since $1 \otimes u$ and $v \otimes 1$ commute for all $u, v \in A$.

Suppose, conversely, $\Delta(a) = a \otimes 1 + 1 \otimes a$, $a \in A$. We choose in L a well-ordered basis W whose elements are monomials (Section 2.3.1). It should be recalled that a basis in A is formed by all products of the form $b = w_1 w_2 \cdots w_s$ where $w_1 \geq w_2 \geq \cdots \geq w_s$ are the elements of the chosen basis of L. Then $\Delta(b) = (w_1 \otimes 1 + 1 \otimes w_1)(w_2 \otimes 1 + 1 \otimes w_2) \cdots (w_s \otimes 1 + 1 \otimes w_s)$ by the first part of the proof. It is obvious that

$$\Delta(b) = \sum_{r+t=s} w_{i_1} \cdots w_{i_t} \otimes w_{j_1} \cdots w_{j_r},$$

and $i_1 \geq \cdots \geq i_t$, $j_1 \geq \cdots \geq j_r$, that is, we see that $\Delta(b)$ can be written as a sum of basic elements of $A \otimes A$ (we use the canonical basis of tensor products). Now let $b' = w_1' w_2' \cdots w_s'$, $w_1' \geq w_2' \geq \cdots \geq w_s'$. Now suppose $\Delta(b)$ and $\Delta(b')$ have like terms in their representations through basic elements of $A \otimes A$. Then $w_{i_1} \cdots w_{i_t} \otimes w_{j_1} \cdots w_{j_r} = w'_{m_1} \cdots w'_{m_{t'}} \otimes w'_{t_1} \cdots w'_{t_{r'}}$. It is obvious that $t = t'$, $r = r'$ and thus $s = s'$. Consequently, the sets $\{w_1, \ldots, w_s\}$ and $\{w_1', \ldots, w_s'\}$ coincide; now since the sets are ordered, we must have $b = w_1 \cdots w_s = w_1' \cdots w_s' = b'$. Thus if a basic expression for a includes a monomial b with $s > 1$ then $\Delta(a)$ in its basic expression includes monomials different from $w_1 w_2 \cdots w_s \otimes 1$ and $1 \otimes w_1 w_2 \cdots w_s$; hence $\Delta(a) \neq a \otimes 1 + 1 \otimes a$. Thus $s = 1$, that is, $a \in L$.

Now we come back to $\widehat{\Delta}$. If $a = \sum_{n=0}^{\infty} a_n \in \widehat{L}$, that is, if each $a_n \in L_n$, $n = 0, 1, 2, \ldots$ then

$$\widehat{\Delta}(a) = \sum_{n=0}^{\infty} \Delta(a_n) = \sum_{n=0}^{\infty} (a \otimes 1 + 1 \otimes a) = \left(\sum_{n=0}^{\infty} a_n \right) \otimes 1 + 1 \otimes \left(\sum_{n=0}^{\infty} a_n \right)$$
$$= a \otimes 1 + 1 \otimes a.$$

If, conversely, $\widehat{\Delta}(a) = a \otimes 1 + 1 \otimes a$ then the above argument shows $a_n \in L_n$ for all $n = 0, 1, 2, \ldots$, hence $a \in \widehat{L}$. □

8.1.5 Magnus' group

Let $k[[t]]$ be the ring of formal power series, where k is a field of characteristic zero, and e^t, $\ln(1 + t)$ its elements of the form

$$e^t = \sum_{n=0}^{\infty} \frac{t^n}{n!}, \quad \ln(1 + t) = \sum_{n=1}^{\infty} (-1)^{n-1} \frac{t^n}{n}.$$

Now let B be a complete topological algebra as in Section 8.1.2 and $\varphi : k[[t]] \to B$ a continuous homomorphism. Then

$$\varphi(e^t) = \varphi\left(\sum_{n=0}^{\infty} \frac{t^n}{n!}\right) = \sum_{n=0} \frac{(\varphi(t))^n}{n!}$$

is a well-defined element of B and it is denoted by $e^{\varphi(t)}$. Similarly, we define $\ln(1+\varphi(t))$. Using a remark at the end of Section 8.1.3, we see that the maps $t \mapsto \ln(1+t)$ and $t \mapsto e^t - 1$ extend to continuous homomorphisms of $k[[t]]$ into itself. An immediate verification shows that $e^{\ln(1+t)} = 1 + t$ and $\ln e^t = t$.

Similarly, if $a \in A(X)$ is an element with zero constant term then the map $t \mapsto a$ extends to a continuous homomorphism of $k[[t]]$ into itself. Let M denote the ideal of $A(X)$ whose elements are as above. Also let $G = G(X) = 1+M$ be the set of all series with constant term 1. Then G is a group under multiplication since $(1-a)^{-1} = 1+a+a^2+\cdots$. By previous remarks, the maps $a \mapsto e^a$, $1+a \mapsto \ln(1+a)$ are bijections of M onto G and of G onto M, inverses of each other. We may also consider $e^{\widehat{L}}$, which is well-defined since $\widehat{L} \subseteq M$. Now we formulate a theorem which is a key point of our arguments.

Theorem 8.2. *$e^{\widehat{L}}$ is a subgroup of G.*

Proof. Since $e^a e^b = e^c$ for all $a, b \in \widehat{L}$ and some $c \in M$ it is sufficient to show that $c \in \widehat{L}$. We remark first that if a and b commute then, obviously, $e^a e^b = e^{a+b}$. In this case $c = a + b \in \widehat{L}$ and all is proved. In the general case, we write $c = \ln(e^a e^b)$ and then we apply to both sides of this equation the continuous homomorphism $\widehat{\Delta}$ from Section 8.1.4. It is easy to verify the following equations (some minor details are left to the reader):

$$\begin{aligned}\widehat{\Delta}(c) &= \widehat{\Delta}(\ln(e^a e^b)) = \ln(e^{\widehat{\Delta}(a)} e^{\widehat{\Delta}(b)}) = \ln(e^{a\otimes 1+1\otimes a} e^{b\otimes 1+1\otimes b}) \\ &= \ln e^{a\otimes 1} e^{1\otimes a} e^{b\otimes 1} e^{1\otimes b} = \ln((e^a e^b \otimes 1)(1\otimes e^a e^b)) \\ &= \ln(e^a e^b \otimes 1) + \ln(1\otimes e^a e^b) = (\ln(e^a e^b)) \otimes 1 + 1 \otimes (\ln(e^a e^b)) \\ &= c \otimes 1 + 1 \otimes c.\end{aligned}$$

By Theorem 8.1, we have $c \in \widehat{L}$. □

8.1.6 Group associated with complete topological Lie algebra

We fix, for the sequel, $X = \{x, y\}$, $\widehat{A} = \widehat{A}(X)$, and $\widehat{L} = \widehat{L}(X)$. Let $x \circ y$ denote the unique element of $\widehat{L}$ such that $e^x e^y = e^{x\circ y}$. If B is a Lie algebra with topology given by a descending chain of ideals B_n such that the quotient algebras B/B_n are nilpotent then we can introduce a new operation on B by setting $b \circ c = \widehat{\varphi}(x \circ y)$ where $\widehat{\varphi}$ is a continuous extension of a homomorphism φ from L into B such that $x \mapsto b$ and $y \mapsto c$. The existence of such an extension will be shown below.

Proposition 8.3. *$\langle B, \circ\rangle$ is a group.*

Proof. We remark first that the homomorphism $\varphi : L \to B$ as above is continuous. Indeed, since $\mathrm{alg}(b,c) + B_n/B_n$ is nilpotent of step, say, $c(n)$, we have $\varphi(L_{c(n)+1}) = \varphi(L^{c(n)+1}) \subseteq B$ for all $n = 0,1,2,\dots$ Hence φ is continuous, i. e., it admits an extension to a continuous homomorphism $\hat{\varphi}$ from $\widehat{L}$ into B. It is sufficient to verify the axioms of the group in the case where $B = \widehat{L}(x,y,z) \subseteq \widehat{A}(x,y,z)$. Since $x \circ 0 = 0 \circ x = x$ and $x \circ (-x) = (-x) \circ x = 0$, it remains to prove $x \circ (y \circ z) = (x \circ y) \circ z$. Now we write $(x \circ y) \circ z = \ln e^{(x\circ y)\circ z} = \ln e^{x\circ y} e^{z} = \ln (e^x e^y) e^z = \ln e^x (e^y e^z) = \ln e^x e^{y\circ z} = \ln e^{x\circ (y \circ z)} = x \circ (y \circ z)$, since the exponents are the elements of an associative algebra $\widehat{A}$. □

We will call $\langle B, \circ\rangle$ the *group of the topological Lie algebra B* and it will be denoted B°.

Remark.

(1) Let C be a closed subalgebra in B. Then C is a subgroup of B°; moreover, this subgroup is identically isomorphic with C°.

(2) Let $\varphi : B \to D$ be a continuous epimorphism of complete algebras B and D. Then φ preserves multiplication $\circ$ and hence is a homomorphism of B° into D°. This will be denoted by φ°.

(3) Any closed ideal C can be considered as a kernel of a continuous homomorphism $\varphi : B \to B/C$. According to (2), its kernel $\operatorname{Ker}\varphi = \operatorname{Ker}\varphi^\circ$ is a normal subgroup in B°. Applying the homomorphism theorem once more, we get $B^\circ/C^\circ \cong (B/C)^\circ$.

8.1.7 Campbell–Hausdorff formula

It is often useful to have an explicit form for $x \circ y$ as an element in $\widehat{A}(x,y)$ that is a series in x, y. This form can be found using a criterion for an element $a \in A(X)$ to be in $L(X)$. We recall (see Section 2.5.5) that if π is a homogeneous map from $A(X)$ into $L(X)$ such that

$$\pi(1) = 0, \quad \pi(x_1 \cdots x_n x_{n+1}) = [x_1, \dots, x_n, x_{n+1}]$$

then $a \in A_n$, $n \ge 1$, is an element of L_n if and only if $\pi(a) = na$.

Now the promised explicit form results by a computation of the form

$$\begin{aligned} x \circ y = \ln(e^x e^y) &= \ln\Big(1 + \sum_{n+m\ge 1} \frac{x^n y^m}{n!m!}\Big) = \sum_{s>0} (-1)^{s-1} \frac{1}{s} \Big(\sum_{n+m\ge 1} \frac{x^n}{n!}\frac{y^m}{m!}\Big)^s \\ &= \sum_{s>0} (-1)^{s-1} \frac{1}{s} \sum_{n_i+m_i\ge 1} \frac{x^{n_1} y^{m_1} x^{n_2} y^{n_2} \cdots x^{n_s} y^{n_s}}{n_1! m_1! n_2! m_2! \cdots n_s! m_s!}. \end{aligned}$$

Let $(x \circ y)_r$ denote the homogeneous component of degree r. Then applying the above criterion, we see that $r(x \circ y)_r$ is a linear combination of monomials of the form

$$(\mathrm{ad}(x))^{n_1} (\mathrm{ad}(y))^{m_1} \cdots (\mathrm{ad}(x))^{n_s} (\mathrm{ad}(y))^{m_s - 1}(y),$$

provided that $m_s > 0$, and of the form

$$(\mathrm{ad}(x))^{n_1}(\mathrm{ad}(y))^{m_1}\cdots(\mathrm{ad}(x))^{n_s-1}(x)$$

otherwise. Using this we obtain the Campbell–Hausdorff formula

$$x\circ y=\sum_r\frac{1}{r}\sum_{t=1}^{m}\frac{(-1)^{s-1}}{s}\Bigg(\sum_{\substack{n_1+\cdots+n_s=p,\\ m_1+\cdots+m_s=q,\\ p+q=r,n_i+m_i\ge 1}}\frac{(\mathrm{ad}(x))^{n_1}\cdots(\mathrm{ad}(y))^{m_s-1}(y)}{n_1!\cdots m_s!}$$

$$+\sum_{\substack{n_1+\cdots+n_s=p,\\ m_1+\cdots+m_{s-1}=q,\\ p+q=r,n_i+m_i\ge 1}}\frac{(\mathrm{ad}(x))^{n_1}\cdots(\mathrm{ad}(y))^{m_{s-1}}(\mathrm{ad}(x))^{n_s-1}(x)}{n_1!\cdots m_{s-1}!n_s!}\Bigg).$$

In particular,

$$(x\circ y)_0=0,\quad (x\circ y)_1=x+y,\quad (x\circ y)_2=\frac{1}{2}[x,y],$$

i. e.,

$$x\circ y=x+y+\frac{1}{2}[x,y]+\frac{1}{12}[x,x,y]+\frac{1}{12}[y,y,x]+\cdots.$$

8.2 Magnus' theory of free groups

8.2.1 Free groups

Let X and X^{-1} denote two nonempty sets and $x\mapsto x^{-1}$ a bijection of X onto X^{-1}. We consider the semigroup $W = W(X\cup X^{-1})$ of all words in the alphabet $X\cup X^{-1}$, with respect to juxtaposition. Let 1 stand for the empty word. We say that two words u and v are equivalent, $u\sim v$, if u can be obtained from v by a finite (possibly zero) number of insertions and deletions of subwords of the form xx^{-1} and $x^{-1}x$ for some $x\in X$. It is easy to see that this relation is an equivalence. The equivalence class of an element u is designated by $[u]$. We say that u is *reduced* if it does not contain subwords of the above form.

Lemma 8.4. *Any class* $[u]$ *contains precisely one reduced word.*

Proof. Let $\rho(u)$ denote the unique reduced word resulting from u by consecutive deletions of subwords of the form $x^{\varepsilon}x^{-\varepsilon}$ (that is, xx^{-1} if $\varepsilon=1$ and $x^{-1}x$ if $\varepsilon=-1$) from the right to the left. Then, clearly, $\rho(u)\sim u$ and $\rho(u)=u$ if u is reduced. Now $\rho(uv)=\rho(u\rho(v))$. This gives $\rho(ux^{\varepsilon}x^{-\varepsilon}v)=\rho(uv)$, $\varepsilon=\pm1$. Arguing by induction over the length of u, we finally get $\rho(uv)=\rho(u)\rho(v)$. Now suppose that u and v are two reduced words such that $u\sim v$. Then, by definition, there exists a sequence

$$u=u_1,u_2,\dots,u_m=v$$

such that u_{i+1} results from u_i by one deletion or one insertion of a subword of the form $x^\varepsilon x^{-\varepsilon}$, $\varepsilon = \pm 1$, $x \in X$. By our formulas, then $\rho(u_{i+1}) = \rho(u_i)$. From this, easily, $\rho(u) = \rho(v)$. Since u and v are irreducible, we finally get $u = v$. □

Now we introduce multiplication on the set of classes of words $F(X)$ by setting $[u][v] = [uv]$. This definition is correct since from $u \sim u'$, $v \sim v'$ it follows that $\rho(u) = \rho(u')$, $\rho(v) = \rho(v')$ and, using $\rho(uv) = \rho(\rho(u)\rho(v))$, we get $\rho(uv) = \rho(u'v')$. Hence $uv \sim u'v'$. The associativity of $F(X)$ follows from that of $W(X \cup X^{-1})$; moreover, [1] is the unity of $F(X)$ and $[u^{-1}]$ is an inverse of u where u^{-1} is the image of u under the antihomomorphism of W induced by the map $x \mapsto x^{-1}$, $y^{-1} \mapsto y$, where $x, y \in X$.

Proposition 8.5. *$F(X)$ is generated by all classes of the form $[x]$, $x \in X$. Moreover, given a group G and a map $\varphi : X \to G$, there exists a unique homomorphism $\psi : F(X) \to G$ such that $\psi([x]) = (\varphi(x))$.*

Proof. Only the second claim needs verification. Let $\chi : X \cup X^{-1} \to G$ be given by $\chi(x^\varepsilon) = (\varphi(x))^\varepsilon$ where $\varepsilon = \pm 1$. Then there exists a unique homomorphism of semigroups $\bar{\chi}$ from W into G extending χ. The kernel of this homomorphism $\operatorname{Ker}\bar{\chi}$ as an equivalence relation obviously contains $\sim$. Hence $\bar{\chi}$ induces a unique homomorphism $\psi : F = W/\sim \to G$. It is clear that ψ is already a homomorphism of groups with the above condition. □

We will be omitting square brackets in the expression of elements of F. Thus we obtain the free group $F(X)$ with the free generating set X. By our convention, each element of F can be uniquely written in the form

$$x_1^{\varepsilon_1} x_2^{\varepsilon_2} \cdots x_n^{\varepsilon_n},\ x_i \in X,\ \varepsilon_i = \pm 1 \text{ and } \varepsilon_i \neq -\varepsilon_{i+1}, \quad \text{if } x_i = x_{i+1},\ i = 1, \dots, n-1.$$

Corollary 8.6. *Any group is isomorphic to a quotient group of a free group.*

In this section we treat only that part of the theory of free groups that relates to Lie rings. Other examples and theorems can be found in many known books on group theory (e. g., Kurosh [196]; Magnus et al. [215]; Lyndon and Schupp [210]; Kargapolov and Merzlyakov [160]).

We will prove first that Magnus' group $G = G(X) = 1 + M \subseteq \widehat{A}(X)$ is abundant with free subgroups.

Theorem 8.7. *Let k be an arbitrary commutative ring with 1. We fix, for each $x \in X$, an element $y(x) \in \widehat{A}(X)$ without components of degree 0 and 1. Then the set of the form*

$$\{a_x = 1 + x + y(x) \mid x \in X\}$$

freely generates a subgroup of $G(X)$.

Proof. Passing to k/M, M being a maximal ideal of k, we may restrict ourselves to the case where k is a field. Suppose first that $y(x) = 0$ for all $x \in X$. We consider a homomorphism from $F(X)$ into G that extends the mapping $x \mapsto a_x$, $x \in X$. Denoting by $\bar{b}$ the image of an element $b \in F$, it is enough to verify $\bar{b} \neq 1$ as soon as $b \neq 1$. To do this, we represent b in the form $b = x_1^{m_1} x_2^{m_2} \cdots x_n^{m_n}$, where $m_i \neq 0$, $x_i \neq x_{i+1}$, $i = 1, 2, \ldots, n-1$. Then

$$\bar{b} = (1 + x_1)^{m_1}(1 + x_2)^{m_2} \cdots (1 + x_n)^{m_n}.$$

According to binomial decomposition, we have

$$(1 + x_i)^{m_i} = 1 + u_i x + v_i x_i^2 + \cdots.$$

There exists $s_i \neq 0$ such that the coefficient w_i of $x_i^{s_i}$ is nonzero. Thus $\bar{b}$ contains $x_1^{s_1} x_2^{s_2} \cdots x_n^{s_n}$ with the coefficient $w_1 w_2 \cdots w_n \neq 0$. Hence $\bar{b}$ is indeed nontrivial and, for the particular case, the proof is complete. Now let H denote the free subgroup of G generated by the elements $1 + x$, $x \in X$. We consider the continuous homomorphism $\varphi : \widehat{A}(X) \to \widehat{A}(X)$ extending the map $x \mapsto x + y(x)$, $x \in X$. It is easy to see that φ is injective. Thus $\varphi : H \to \mathrm{gr}(\{a_x \mid x \in X\})$ is an isomorphism of groups. □

Corollary 8.8. *We consider the group $\widehat{L}^\circ$ of the complete topological algebra $\widehat{L}$, where $L = L(X)$ is a free Lie algebra. Let $C = C(X)$ be a subgroup generated by X. Then C is a free group with free generating set X.*

Proof. By definition, $\widehat{L}^\circ$ is isomorphic to $e^{\widehat{L}}$. Under this isomorphism, C is isomorphic to the subgroup $T = \mathrm{gr}\{e^x \mid x \in X\}$ in $e^{\widehat{L}}$. By Theorem 8.7, T is a free group freely generated by the above set. □

8.2.2 Commutator calculus

Let G be a group and let a, b be its elements. We denote by (a, b) the commutator of a, b, that is, the expression of the form $(a, b) = a^{-1}b^{-1}ab$. Setting $a^{\pm b} = b^{-1}a^{\pm 1}b$, we may write $(a, b) = a^{-1}a^b$. Using this notation, we have the following Hall's identities:

$$(ab, cd) = (a, d)^b (a, c)^{db} (b, d)(b, c)^d, \tag{8.1}$$

$$((a, b), c^a)((c, a), b^c)((b, c), a^b) = 1. \tag{8.2}$$

To prove them, we note that if $y \in G$ then the map $x \mapsto x^y$ is an automorphism of G. Then

$$\begin{aligned}(ab, cd) &= (ab)^{-1}(ab)^{cd} = b^{-1}a^{-1}a^d bb^{-1}a^{-d}a^{cd}b^{cd} \\ &= (a, d)^b (a, c)^{db} b^{-1} b^d b^{-d} b^{cd} = (a, d)^b (a, c)^{db} (b, d)(b, c)^d,\end{aligned}$$

proving equation (8.1). Equation (8.2) is verified by computing directly

$$\begin{aligned}((a, b), c^a) &= (a, b)^{-1} c^{-a} (a, b) c^a = ((b, a)) a^{-1} c^{-1} a (a, b) a^{-1} c a \\ &= b^{-1} a^{-1} b c^{-1} b^{-1} a b a^{-1} c a.\end{aligned}$$

Permuting the letters cyclically, we get

$$((c,a),b^c) = a^{-1}c^{-1}ab^{-1}a^{-1}cac^{-1}bc,$$
$$((b,c),a^b) = c^{-1}b^{-1}ca^{-1}c^{-1}bcb^{-1}ab.$$

Now the product of these three expressions after cancelations reduces to 1, proving the identity.

Now suppose we are given two subgroups A, B of a group G. Their *mutual commutator subgroup* (A,B) is the subgroup generated by all possible commutators of the form (a,b), where $a \in A$, $b \in B$. Since $(a,b)^{-1} = (b,a)$, we have $(A,B) = (B,A)$. If A and B are normal in G then (A,B) is also a normal subgroup in G.

Lemma 8.9. *Let A, B, C be normal subgroups of a group G. Then*

$$((A,B),C) \subseteq ((B,C),A)((C,A),B). \tag{8.3}$$

Proof. It is sufficient to show that any element on the left-hand side of the form $u = (u_1,u_2)$, where $u_1 \in (A,B)$, $u_2 \in C$, is on the right-hand side of (8.3). To do this, we represent u in the form $u_1 = v_1v_2\cdots v_t$, where $v_i = (x_i,y_i)$, and x_i, y_i are in A, B (or in B, A). Using equation (8.1), we may write

$$(v_1v_2\cdots v_t, u_2) = (v_1,u_2)^{v_2\cdots v_t}(v_2,u_2)^{v_3\cdots v_t}\cdots(v_t,u_2).$$

But

$$(v_i,u_2) = ((x_i,y_i),(u_2^{x_i^{-1}})^{x_i}) = ((y_i,u_2^{x_i^{-1}}),x_i^{y_i})^{-1},$$
$$((u_2^{x_i^{-1}},x_i),y^{u_2^{x_i^{-1}}})^{-1} \in ((B,C),A)((C,A),B).$$

To complete the proof, it remains to note that, given two normal subgroups X, Y of a group G, we always have $XY = YX$. □

8.2.3 Central filtrations

Similar to the definitions in the Lie algebra case, one may introduce the notions of the lower central, upper central, and the derived series of a group (cf. Section 1.7).

Thus we put $G = \gamma_1(G) = \delta_0(G)$ and then $\gamma_s(G) = ((G,\gamma_{s-1}(G)))$, $s > 1$; $\delta_r(G) = ((\delta_{r-1}(G),\delta_{r-1}(G)))$, $r > 0$. We also set $\zeta_0(G) = \{1\}$ and define $\zeta_t(G)$ as a subgroup in G containing $\zeta_{t-1}(G)$ and such that $\zeta_t(G)/\zeta_{t-1}(G)$ is the center of $G/\zeta_{t-1}(G)$. The series thus defined are respectively called *lower central*, *derived*, and *upper central series* of G. The former two are descending and the latter is ascending. Besides, the consecutive factors of each of the three series are abelian.

The lower central series of a group is a particular case of the notion of the central filtration of a group. Namely, a descending series $f = \{G_n\}_{n\geq 1}$ of normal subgroups of a group G is called *central filtration* in G if for all $s, r = 1, 2, \ldots$ we have $(G_s, G_r) \subseteq G_{r+s}$. Using Lemma 8.9, we find

$$\begin{aligned}(\gamma_s(G), \gamma_r(G)) &= ((G, \gamma_{s-1}(G)), \gamma_r(G)) \\ &\subseteq ((\gamma_{s-1}(G), \gamma_r(G)), G)((\gamma_r(G), G), \gamma_{s-1}(G)).\end{aligned}$$

Applying induction over s with trivial basis $s = 1$, we derive from here that the lower central series $\{\gamma_s(G)\}_{n\geq 1}$ is a central filtration of G.

Now we consider an arbitrary central filtration $f = \{G_n\}_{n\geq 1}$ of a group G. Then, obviously, $L_n^f(G) = G_n/G_{n+1}$ is an abelian group. We introduce an additive abelian group of the form

$$L^f(G) = \bigoplus_{n\geq 1} L_n^f(G).$$

We will turn this group into a Lie ring by defining the commutator of two homogeneous elements, as follows. If $\xi = xG_{s+1}$, $\eta = yG_{r+1}$ then we put $[\xi, \eta] = (x, y)G_{s+r+1} \in L_{s+r}^f$. The definition does not depend on the choice of representatives in ξ and η. For if $x' = xz$, $y' = yw$, $z \in G_{s+1}$, and $w \in G_{r+1}$ then, according to equation (8.1),

$$(x', y') = (x, w)^z (x, y)^{wz} (z, w)(z, y)^w \equiv (x, y) \pmod{G_{s+r+1}}$$

by the definition of the central filtration. Now the Jacobi identity and the distributivity can be verified for the homogeneous components only. In fact, these relations are immediate from equations (8.1) and (8.2). For instance, arguing as in the proof of Lemma 8.9, we can easily prove that $n[\xi, \eta] = [n\xi, \eta] = [\xi, n\eta]$. Indeed,

$$n[\xi, \eta] = (x, y)^n G_{s+r+1} = (x^n, y)G_{s+r+1} = (x, y^n)G_{r+s+1}.$$

In a particular case of the lower central filtration, we obtain a Lie ring which is called the *Lie ring of a group G*, denoted by $L(G)$.

The main aim of this section is to use the Lie ring $L(F)$ of a free group F for the description of the structure of the lower central series of F.

8.2.4 Lower central series of a free group

Let $F = F(X)$ be a free group with free generating set X, and consider $A = A(X)$, $L = L(X)$, $\widehat{A}$, $\widehat{L}$, M, the algebras introduced in the preceding sections. For each $x \in X$, we fix an element $\gamma(x)$ of $\widehat{A}(X)$ without components of the zeroth and first degree. We identify F with a subgroup $1 + M \subseteq A\widehat{A}$ by means of the isomorphism extending the map $x \mapsto 1 + x + \gamma(x)$, $x \in X$. Now let M^n denote the set of the elements of A of the form

$a = \sum_{s\geq n} a_s$, $a_s \in A_s$. We will show that $\gamma_n(F) \subseteq 1 + M^n$. Indeed, $\gamma_1(F) = F \subseteq 1 + M$. Now if, further, $c = (g, h)$ where $g \in \gamma_{n-1}(F)$, $h \in F$, then by the induction hypothesis $g = 1 + g_{n-1} + g_n + \cdots$, $h = 1 + h_1 + h_2 + \cdots$. An immediate calculation shows

$$\begin{aligned} c = g^{-1}h^{-1}gh &= (1 - g_{n-1} + g_n^* + \cdots)(1 - h_1 + h_2^* + \cdots) \\ &\quad \cdot (1 + g_{n-1} + g_n + \cdots)(1 + h_1 + h_2 + \cdots) \\ &= (1 + g_{n-1}h_1 - h_1 g_{n-1} + c_{n+1} + \cdots), \quad c_{n+1} \in A_{n+1}. \end{aligned} \tag{8.4}$$

Now let F_n denote the intersection of F and $1 + M^n$.

Proposition 8.10. *$\{F_n\}_{n\geq 1}$ is a central filtration of F such that $F_n \supseteq \gamma_n(F)$. The intersection $\bigcap_{n=1}^{\infty} \gamma_n(F)$ of the terms of the lower central series in a free group is trivial.*

Proof. Write $g = 1 + g_s + g_{s+1} + \cdots \in F_s$ and $h = 1 + h_r + h_{r+1} + \cdots \in F_r$. An argument similar to that used in equation (8.4) shows the validity of the following:

$$c = (g, h) = 1 + (g_s h_r - h_r g_s) + c_{s+r+1} + \cdots, \quad c_{s+r+1} \in A_{s+r+1}. \tag{8.5}$$

Hence $c \in F_{s+r}$. The inclusion $F_n \supseteq \gamma_n(F)$ has been proved in equation (8.4). Now the principal claim of Proposition 8.10 becomes trivial since $\bigcap_{n=1}^{\infty} \gamma_n(F) \subseteq \bigcap_{n=1}^{\infty}(1 + M^n) = \{1\}$. □

The main result of this section is the following.

Theorem 8.11. *Let $L = L(X)$ be a free Lie ring, $F = F(X)$ a free group and $L(F)$ the Lie ring of F. Let ξ_x be an element in $L(F)$ of the form $\xi_x = x\gamma_2(F)$, $x \in X$. Then the map $x \mapsto \xi_x$, extends to an isomorphism of graded Lie rings.*

Proof. By the definition of a free Lie ring, such a homomorphism φ exists. We must prove that φ is bijective. Using Proposition 8.10, we first consider the Lie ring $L^f(F)$ by the filtration $f = \{F_n\}_{n\geq 1}$ introduced in that proposition. Since $F_n \supseteq \gamma_n(F) \supseteq F_{n+1} \supseteq \gamma_{n+1}(F)$, there exists a natural homomorphism of abelian groups $L_n(F) \to L_n^f(F)$ that naturally extends to a homomorphism ψ of graded Lie rings from $L(F)$ into $L^f(F)$. We will also need another map, $\chi : L^f(F) \to A$, which can be described as follows. Let $g = 1 + g_s + g_{s+1} + \cdots \in F_s$ and $\mu = gF_{s+1}$ (we recall that F has been identified with a subgroup in $1 + M$). So we set $\chi(\mu) = g_s$. It is easy to see that this definition does not depend on the choice of representatives. Besides, by equation (8.5), we see that χ is a homomorphism of Lie rings. By our construction, $\tau = \chi\psi\varphi : L \to A$ is a homomorphism of Lie rings such that for any $x \in X$ we have

$$\tau(x) = \chi\psi\varphi(x) = \chi\psi(\xi_x) = \chi((1 + x + \gamma(x))F_2) = x.$$

Applying Section 2.3, the corollary of Theorem 2.10, we see that τ is an injective map. Hence φ is injective, too. Moreover, by the definition of the Lie ring of a group, it is easy to see that $L(F)$ is generated by $\{\xi_x \mid x \in X\}$. This shows that φ is bijective. Since $\varphi(L_n) \subseteq L_n(F)$, we see that φ is a homomorphism of graded Lie rings. □

Corollary 8.12. *(i) For any $n = 1, 2, \ldots$, $\gamma_n(F)/\gamma_{n+1}(F)$ is a free abelian group which is canonically isomorphic to $L_n(X)$.*

(ii) $\gamma_n(F) = F_n$, $n = 1, 2, \ldots$

Proof. The first claim is immediate from Theorem 8.11 and Section 2.3, Theorem 2.10. Let us prove the second one. Clearly, $\gamma_1(F) = F_1 = F$. Now suppose $\gamma_{n-1}(F) = F_{n-1}$. We note that $F_n \supseteq \gamma_n(F)$ and that the map $L_n(F) \to L_n^f(F)$ is injective. But, by the definition, its kernel is equal to $F_n/\gamma_n(F)$. Therefore, certainly, $F_n = \gamma_n(F)$. □

8.3 Isomorphism of categories of nilpotent D-groups and of nilpotent rational Lie algebras

8.3.1 Nilpotent groups

A group G is called *c-step nilpotent* if for some $c \geq 1$ we have $\gamma_c(G) \neq \{1\}$ and $\gamma_{c+1}(G) = \{1\}$. Adapting the arguments of Section 1.7, we easily see that a group G is c-step nilpotent if and only if c is the least natural number such that for any $g_1, g_2, \ldots, g_{c+1}$ in G and any arrangement of brackets we have

$$(g_1, g_2, \ldots, g_{c+1}) = 1,$$

or, equivalently,

$$(x_1, x_2, \ldots, x_{c+1}) = (x_1, (x_2, (\cdots (x_c, x_{c+1}) \cdots))) \equiv 1$$

is an identity in G. As in Section 1.7, this gives that any subgroup, any quotient group, and Cartesian product of any family of c-step nilpotent groups are themselves nilpotent of nilpotent step at most c. Saying it differently, the class of all groups with nilpotent step at most c is a variety of groups, denoted by $\mathbf{N}_c$.

We will need, for the sequel, a fairly detailed study of finitely generated torsion-free nilpotent groups. We start with a proposition concerning arbitrary finitely generated nilpotent groups.

Proposition 8.13. *Let G be a nilpotent group generated by a finite set of elements of the form $g_1, g_2, \ldots, g_n$. Then for any $s \geq 1$ the subgroups $\gamma_s(G)$ can be generated by a finite set of all nontrivial commutators of the form*

$$((g_{i_1}, g_{i_2}, \ldots, g_{i_r})), \quad 1 \leq i_1, \ldots, i_r \leq m,\ i = 1, 2, \ldots, r;\ r \geq s. \tag{8.6}$$

Moreover, any subgroup of G is finitely generated.

Proof. Indeed, it is sufficient to show that any commutator of the form $w = (u, v) \in \gamma_{n+1}(G)$, where $u \in \gamma_n(G)$, $v \in G$, can be expressed through commutators (8.6) and the set of their inverses. So, let $u = u_1 u_2 \cdots u_t$ where u_i have the form (8.6) with $r \geq n$ and

$v = v_1 v_2 \cdots v_q$, where v_i are the generators g_j or their inverses. Using formulas of the commutator calculus in Section 8.2.2, we can represent w as a product of commutators of the form (8.6) and of their conjugates by elements in G. Now we note that $a^b = a(a, b)$ and

$$a^{b_1 b_2} = (a^{b_1})^{b_2} = a^{b_2}(a, b_1)^{b_2} = a(a, b_2)(a, b_1)((a, b_1), b_2).$$

These generators enable us to represent w in the desired from except possibly having inverses of generators. To handle them, one may use the relation

$$(a^{-1}, b) = (b, a)((b, a), a^{-1}),$$

which, in a nilpotent group, enables us to eliminate the inverses, proving the first claim of our proposition.

To prove the second one, we will apply induction over c with trivial basis $c = 0, 1$. Now we assume $c > 1$ and pick a subgroup A in G. Then $A\gamma_c(G)/\gamma_c(G)$ is a finitely generated subgroup of $G/\gamma_c(G)$. Since the latter group is $(c-1)$-step nilpotent, we can find a finite set $\{a_1, a_2, \ldots, a_q\} \subseteq A$ such that $\{a_1\gamma_c(G), \ldots, a_1\gamma_c(G)\}$ generates $A\gamma_c(G)/\gamma_c(G)$. Also let $a_{q+1}, \ldots, a_t$ be generators of $A \cap \gamma_c(G)$, the subgroup of a finitely generated abelian group $\gamma_c(G)$ (this is finitely generated by the first part of our proof). Then, if we are given an element $w \in A$, we have $w\gamma_c(G) = w(a_1\gamma_c(G), \ldots, a_q\gamma_c(G)) = w(a_1, \ldots, a_q)\gamma_c(G)$. Thus $w = w(a_1, \ldots, a_q)w'$ where $w' \in \gamma_c(G)$. Obviously, $w' \in \gamma_c(G) \cap A$. Hence $w' = w'(a_{q+1}, \ldots, a_t)$, proving that A is generated by the finite set $\{a_1, \ldots, a_t\}$. □

Corollary 8.14. *Any ascending chain of subgroups of a finitely generated nilpotent group stabilizes at some finite step. That is to say, if $H_1 \subseteq H_2 \subseteq \cdots \subseteq H_q \subseteq H_{q+1} \subseteq \cdots$ is such a chain then, for some q, we must have $H_q = H_{q+1} = \cdots$.*

8.3.2 Monolithic nilpotent groups

A group G is called *monolithic* if the intersection of all its nontrivial normal subgroups is itself nontrivial. The least nontrivial normal subgroup in this case is called the *monolith* of G. Before we prove a theorem describing finitely generated nilpotent monolithic groups, we formulate some auxiliary results. Their proof is quite similar to some results on nilpotent Lie algebras which can be found in Section 1.7.

(1) In any nilpotent group the lower and the upper central series have the same length equal to the nilpotent step of G.

(2) Let H be a subgroup of a nilpotent group G and $N_G(H)$ its normalizer, that is, the set of all $x \in G$ such that $x^{-1}Hx \subseteq H$. Then $H \neq G$ implies $N_G(H) \neq H$.

(3) Let H be a subgroup of a group G such that $H\gamma_2(G) = G$. Then $H = G$.

Besides, we will need another assertion.

Lemma 8.15. *Any finitely generated nilpotent group with finite center is itself finite.*

Proof. We use induction over the nilpotent step c of G as in the statement. If $c = 1$ then $G = \zeta_1(G)$. Since $|\zeta_1(G)| < \infty$, all is proved. Otherwise we consider $G_1 = G/\zeta_1(G)$ and prove that the center $\zeta_2(G)/\zeta_1(G)$ of G_1 is finite. Then it would be possible to apply the induction hypothesis. But $\zeta_2(G)/\zeta_1(G)$ is a finitely generated abelian group, so it suffices to show that all elements of this group have finite order. To this end, we set $n = |\zeta_1(G)|$ and pick an arbitrary element x in $\zeta_2(G)$. We are going to prove $x^n \in \zeta_1(G)$. Indeed, using the commutator calculus, we may write

$$((x^n, g)) = ((x, g))^{x^{n-1}}((x, g))^{x^{n-2}} \cdots ((x, g)) = ((x, g))^n = 1.$$

Here we have used the centrality of $((x, g))$ and the fact that the order of an element in the center divides n. □

Theorem 8.16. *Any finitely generated monolithic nilpotent group is a finite p-group for some prime number p.*

Proof. We will show first that a finite monolithic nilpotent group is a p-group for a suitable prime p. Indeed, if this is not the case then there exist two distinct primes p, q such that the Sylow subgroups S_p and S_q are nontrivial, hence distinct from G. By a Sylow theorem (Kurosh [196, Chapter 13]) if, say, N_p is a normalizer of S_p in G then $N_G(N_p) = N_p$. Using (2) above, we see that $N_p = G$, that is, S_p is a normal subgroup of G. Obviously, the same is true for S_q. However, this is a contradiction since, on the one hand, $S_p \cap S_q = \{1\}$ and, on the other, $S_p \cap S_q \supseteq M \neq \{1\}$ since G is monolithic. Continuing the proof in the general case, we proceed by induction over c. Our claim is trivial if $c = 0$. Thus we may assume $\zeta_1(G) \neq \{1\}$ and let M be the monolith of G. Since any subgroup of the center is normal in G, either $\zeta_1(G)$ is cyclic of order p^t or it is infinite cyclic. Since $\xi_1(G)$ must have a minimal subgroup, the latter is impossible. By Lemma 8.15, G is a finite group. Applying the first part of the proof, we see that G is indeed a finite p-group. □

A useful remark is the following (implied by Proposition 8.13 and Lemma 8.15).

Corollary 8.17. *Any finitely generated periodic nilpotent group is finite.*

8.3.3 Extraction of roots in nilpotent groups

We will define another class of groups called D-groups. A group D is called a *D-group* if for any nonzero integer n and any $g \in G$ the equation $x^n = g$ is uniquely solvable in G. The importance of D-groups for us is clear since the groups of complete topological Lie algebras over fields of characteristic zero are D-groups. Indeed, the explicit form of the Campbell–Hausdorff formula shows that $x^n = nx$, hence $(1/n)g$ is the unique solution of the equation $x^n = g$, $n \neq 0$, in such a group.

One of the principal results of this section is to show that any c-step nilpotent torsion-free subgroup is isomorphic to a subgroup of a c-step nilpotent D-group. To formulate this result in a more precise form, we need a concept of an isolated subgroup in a group. Namely, we say that a group H is *isolated* in a group G if for any $g \in G$ such that $g^n \in H$, n being a nonzero integer, we always have $g \in H$. The least isolated subgroup of G containing H is called the *isolator* of H in G, and it is denoted by $I_G(H)$. We also denote by $\sqrt{H}$ the set of elements $g \in G$ such that $g^n \in H$, h as above. It is obvious that if $\sqrt{H}$ is a subgroup in G then $I_G(H) = \sqrt{H}$. We are going to show that this is always the case for nilpotent groups.

Lemma 8.18. *Let H be a subgroup of a nilpotent group G. Then $I_G(H) = \sqrt{H}$.*

Proof. As noted just above, it is sufficient to prove that $\sqrt{H}$ is a subgroup. We will proceed by induction over the nilpotent step c of G. The basis of induction $c = 1$ is trivial. Hence we may assume $c > 1$. Choose x, y so that $x^n, y^m \in H$. We set $A = \mathrm{gp}(\{x, y\})$ and consider the lower central series of A, $\{\gamma_i(A)\}_{1\leq i\leq c}$. The induction hypothesis can be applied to $\overline{A \cap H} = (A\cap H)\gamma_c(A)/\gamma_c(A)$ as a subgroup of $A/\gamma_c(A)$. Then for any $\bar{a} = a\gamma_c(A)$ there exists an integer $s \neq 0$ such that $\bar{a}^s \in \overline{A \cap H}$. In other words, $a^s \in (A \cap H)\gamma_c(A)$. Let us prove that $\gamma_c(A) \subseteq \sqrt{A \cap H}$. First suppose that some z in $\gamma_c(A)$ has the form (a, b) where $a \in \gamma_{c-1}(A)$, $b \in A$. Then $a^s \in (A \cap H)\gamma_c(A)$, $b^r \in (A \cap H)\gamma_c(A)$, that is, $a^s = h_1z_1$, $b^r = h_2z_2$. Using formula (8.1) and the centrality of all commutators emerging in the process of its application, we see that

$$(a, b)^{sr} = (a^s, b^r) = (h_1z_1, h_2z_2) = (h_1, h_2) \in A \cap H.$$

Hence $(a, b) \in \sqrt{A \cap H}$. Any element in $\gamma_c(A)$ is the product of mutually commuting commutators of the above form so that our claim about $\gamma_c(A)$ follows.

Now we recall that for any $a \in A$ we have found a nonzero integer s such that $a^s = hz$, $h \in A \cap H$, $z \in \gamma_c(A)$. But now we know that for some $r \neq 0$ we have $z^r \in A \cap H$. Since h and z commute, we finally get $a^{sr} = h^r z^r \in A \cap H$. □

Some important properties of the extraction of roots in torsion-free nilpotent groups are given in what follows.

Proposition 8.19. *Let G be a torsion-free nilpotent group.*
(1) *If $x^n = y^n$ for some $x, y \in G$, n being a nonzero integer, then $x = y$.*
(2) *Any term of the upper central series $\zeta_s(G)$ of G is an isolated subgroup, $s = 0, 1, 2, \ldots$*

Proof. (1) We apply induction over the nilpotent step c of G with trivial basis $c = 1$. By the induction hypothesis applied to $G/\zeta_1(G)$, we have $x\zeta_1 = y\zeta_1$, that is, $x = yz$ for some $z \in \zeta_1(G)$. But then $y^n = x^n = y^n z^n$, that is, $z^n = 1$. Since G is torsion-free, we have $z = 1$, hence $x = y$.

(2) We may restrict ourselves to the case $s = 1$, because otherwise we could apply induction over c. So, let $g^n \in \zeta_1(G)$ for some $g \in G$, n being a nonzero integer. Then for

any $a \in G$, we may write $g^n = a^{-1}g^n a = (a^{-1}ga)^n$ hence, by (1), $g = a^{-1}ga$. Therefore $g \in \zeta_1(G)$. □

A consequence for arbitrary groups is that for any n, $\sqrt{\gamma_n(G)} = I_G(\gamma_n(G))$.

An interesting way of obtaining D-groups is the contents of the following.

Proposition 8.20. *Let π be an infinite set of prime numbers and, for each $p \in \pi$, G_p a nontrivial finite p-group whose nilpotent step does not exceed a fixed natural number $c \geq 1$. We consider the* Cartesian *product $T = \prod_{p\in\pi} G_p$ of all G_p, $p \in \pi$, and $S = \prod_{p\in\pi} G_p$, its subgroup equal to the* direct *product of the same set of groups (alternatively, the subgroup of all functions with finite support in T). Then T/S is a D-group which is nilpotent of nilpotent step at most c.*

Proof. We remark first that, given a finite group G and an equation of the form $x^n = a$, $a \in G$, such that n is coprime to the order of G, this equation is uniquely solvable in G. Another remark is that a group G is a D-group if and only if the unique solubility takes place for the equations of the form $x^q = a$, $a \in G$, q being an arbitrary prime number. Using these remarks, it is sufficient to prove that in T/S the unique solvability takes place for the equations of the form $x^p = a$, $p \in \pi$, $a \in T/S$. Let $a = fS$, where f is an arbitrary function in T. Multiplying both sides by an element in S, we may think that $f(p) = 1$. By the first remark in this proof, there exists an element $b \in T$ such that $b^p = f$, hence bS is a solution of $x^p = a$ in T/S. To prove the uniqueness, we note that, by our hypothesis, T is nilpotent of nilpotent step at most c and that $S = \sqrt{\{1\}}$ is an isolated subgroup in T. Hence T/S is torsion-free nilpotent, and Proposition 8.19 applies. □

In anticipation of Theorem 8.22, we require one more result.

Proposition 8.21. *Let G be a finitely generated torsion-free nilpotent group, p a prime number, G^{p^s} a subgroup of G generated by all elements of the form g^{p^s}, $g \in G$, $s = 0, 1, 2, \ldots$ Then $\bigcap_{s=0}^{\infty} G^{p^s} = \{1\}$.*

Proof. We will show that for any $1 \neq x \in G$ there exists $s > 0$ such that $x \notin G^{p^s}$. We apply induction over the nilpotent step of G. By Proposition 8.19, the induction hypothesis applies to $G/\zeta_1(G)$. Hence if $x \notin \zeta_1(G)$ then $x\zeta_1(G) \notin (G/\zeta_1(G))^{p^s} = G^{p^s}\zeta_1(G)/\zeta_1(G)$ for some $s \geq 1$. Clearly, then $x \notin G^{p^s}$ and all is finished. So, we may assume that $x \in \zeta_1(G)$ and $x \neq 1$. Let $\{K_i\}$ denote a nonempty ascending chain of normal subgroups $K_i \lhd G$, $i = 1, 2, \ldots$, such that $x \notin K_i$ and $x^p \in K_i$ (such a chain exists since $\text{gp}(x^p)$, the cyclic subgroup generated by x^p has the above property). By the corollary of Proposition 8.13, any such chain possesses a maximal element K. By the maximality of K, the quotient group G/K is monolithic. Indeed, any normal subgroup of this group contains $\{1\} \neq M = \text{gp}(x)K/K$. By Theorem 8.16, G/K is a finite q-group for some prime q. On the other hand, G/K contains an element xK of order p. Hence G/K, in fact, is a finite p-group of order, say, p^l, $l \geq 1$. It is clear then $K \supseteq G^{p^l}$, thus $x \notin G^{p^l}$. □

In the terminology of residual properties (cf. Section 6.6.3) we have proved that *for any prime p any finitely generated nilpotent torsion-free group* is residually a finite p-group.

Theorem 8.22. *Any torsion-free nilpotent group G of nilpotent step c can be embedded in a nilpotent D-group G^* of the same nilpotent step and such that any homomorphism from G into a D-group Q uniquely extends to a homomorphism from G^* to Q.*

Proof. Suppose first that G is finitely generated. In this case G is countable. Then there exists a bijection $x \mapsto p_x$ of G onto an infinite set π of prime numbers. We construct the Cartesian product $T = \prod_{p\in\pi} G_p$ and direct product $S = \prod_{p\in\pi} G_p$ as in Proposition 8.20 in such a way that for a $p = p_x$ the group G_p is precisely G/G^{p^s} where s is an integer such that $x \notin G^{p^s}$ (see Proposition 8.21). Using the corollary of Theorem 8.16 we see that, for any $p \in \pi$, G_p is a finite p-group. Now, by Proposition 8.20, $P = T/S$ is a D-group. We define $i : G \to P$ by setting $i(y) = f_y S$ where $f_y(p)$ is the image of y under the natural homomorphism of G onto G_p. By definition i is a homomorphism. Suppose $i(y) = 1$ for some $y \in G$. Then $f_y \in S$ so that $f_y(p) \neq 1$ only for a finite subset of π. Hence there exists a nonzero integer n such that $f_{y^n}(p) = 1$. By our choice of subgroups G_p, this gives $y^n = 1$, hence $y = 1$. Thus i is an injective homomorphism. Using i we may identify G with its image in P. By Proposition 8.20, the nilpotent step of P is at most c.

Now we set $G^* = I_P(G)$, the isolator of G in P. Then G^* is a D-group whose nilpotent step is at most c. Thus, by Lemma 8.18, $G^* = \sqrt{G}$. Let $g^{1/n}$ denote the unique $x \in P$ such that $x^n = g$. Then any elements in G^* has the form $g^{1/n}$, $g \in G$, $n \neq 0$. Now let Q be a D-group and $\varphi : G \to Q$ a homomorphism from G to Q. We define $\varphi^* : G^* \to Q$ by setting $\varphi^*(g^{1/n}) = (\varphi(g))^{1/n}$. The definition is correct. Indeed, if $g^{1/n} = h^{1/m}$ then $(\varphi^*(g^{1/n}))^{mn} = (\varphi(g^{1/n}))^{mn} = \varphi(g)^m = \varphi(h)^n = (\varphi^*(h^{1/m}))^{mn}$. Since Q is a D-group, we must have $\varphi^*(g^{1/n}) = \varphi^*(h^{1/m})$. Similarly, one can check that φ^* is a homomorphism, completing the proof in the case of finite generation.

Now let G be an arbitrary torsion-free nilpotent group. We consider the set of all symbols of the form $g^{1/n}$, $g \in G$, n being a natural number. We say that $g^{1/n}$ is equivalent to $h^{1/m}$, $g^{1/n} \sim h^{1/m}$, if $g^m = h^n$. Let G^* denote the set of equivalence classes of the above symbols. If g_1, g_2 are two elements of G, $G_1 = \mathrm{gp}(\{g_1, g_2\})$, G_1^* a completion of G_1 then the product of elements in G^* will be introduced if we put $g_1^{1/m} g_2^{1/n} = g_3^{1/l}$ as soon as this equation holds valid in G_1^*. That the product is well defined can be verified as follows. Suppose $g^{1/m} = \bar{g}^{1/m'}$, $g^{1/n} = \bar{g}^{1/n'}$ and let $\bar{g}^{1/m'}\bar{g}^{1/n'} = \bar{g}^{1/l'}$. We may consider $G_2 = \mathrm{gp}(\{g_1, g_2, \bar{g}_1, \bar{g}_2\}) \subseteq G$ and $G_2^* = \sqrt{G_2}$. Then $(g_1^{1/m})^{mm'} = g_1^{m'} = \bar{g}_1^m = (\bar{g}_1^{1/m'})^{mm'}$. Hence $g_1^{1/m} = \bar{g}_1^{1/m'}$ in G_2^*. Similarly, $g_2^{1/n} = \bar{g}_2^{1/n'}$ in G^*. But then also $g_3^{1/l'} = \bar{g}_3^{1/l}$ in G^*. Thus $g_3^l = \bar{g}_3^{l'}$, so that $g_3^{1/l} \sim \bar{g}_3^{1/l'}$. All the remaining verifications of the same nature are left to the reader. □

The unique up to isomorphism D-group G^* of this theorem is called the *Malcev completion* of G.

8.3.4 The group of a nilpotent Lie algebra

Let S be a nilpotent Lie algebra over a field k of characteristic zero. Coming back to our argument in Section 8.1.2, we consider in S the topology determined by the series $S \supset \{0\}$. Then S becomes a discrete topological algebra. As before, let S° denote the group S (Section 8.1.6). Using remarks in Sections 8.1.6 and the discreteness of the topology in S, we can see that if R is an ideal in S then R° is a normal subgroup in S° and $S^\circ/R^\circ \cong (S/R)^\circ$. Moreover, $\gamma_n(S^\circ) = (S^n)^\circ$, $n = 1, 2, \ldots$ Indeed, if c is the least number with $S^{c+1} = \{0\}$ then the direct computation using the Campbell–Hausdorff formula shows

$$(a_1, a_2, \ldots, a_n) = [a_1, a_2, \ldots, a_n] + t, \quad t \in S^{n+1}, n = 1, 2, \ldots, \tag{8.7}$$

the arrangement of brackets on both sides being the same. From this equation we immediately derive that $(S^c)^\circ = \gamma_c(S^\circ)$. Using induction over the nilpotent step of S, we get

$$\begin{aligned}\gamma_r(S^\circ)/\gamma_c(S^\circ) &= \gamma_r(S^\circ/\gamma_c(S^\circ)) = \gamma_r((S/S^c)^\circ) = (S^r)^\circ/(S^c)^\circ \\ &= (S^r)^\circ/\gamma_c(S^\circ)\end{aligned}$$

so that $\gamma_r(S^\circ) = (S^r)^\circ$ for any $r = 1, 2, \ldots, c$.

Proposition 8.23. *Any isolated subgroup T in S° has the form R° where R is a subalgebra of S; moreover, if R° is a normal subgroup of S° then R is an ideal of S.*

Proof. Let T be the subgroup in the statement. We denote by R the subalgebra generated by T. By the Campbell–Hausdorff formula, $T + R^2$ is a subalgebra of S. Then $R = T + R^2$ and

$$R^\circ/\gamma_2(R^\circ) = R^\circ/(R^2)^\circ \cong (R/R^2)^\circ.$$

Thus, modulo $\gamma_2(R^\circ)$ the subgroup R° is generated by T. By Burnside's lemma (see (3) in Section 8.3.2), we derive $R^\circ = T$. Now suppose R° is normal in S° and choose $x \in S$, $y \in R$. Let $P = \operatorname{alg}(\{x, y\})$. Then $R^\circ \cap P^\circ$ is a normal subgroup of P° with abelian quotient group $R^\circ/R^\circ \cap P^\circ$. This gives $R^\circ \cap P^\circ \supseteq \gamma_2(P^\circ) = (P^2)^\circ$ and $[x, y] \in R \cap P \subseteq R$, that is, $R \triangleleft S$. □

8.3.5 Free nilpotent D-group

We will need, for the sequel, two more concepts. First, given a set X and a natural number c, the quotient group $F/\gamma_{c+1}(F)$, $F = F(X)$ being a free group with free generating set X, is called the *free c-step nilpotent group* with free generating set X. This is well justified since (cf. Section 8.2.1) any map from X into a group of the nilpotent step at most c uniquely extends to a homomorphism from G into that group. From Magnus' theory of free groups (see the corollary of Theorem 8.11), we derive that G is a torsion-free group, hence it possesses a Malcev completion G^*. The latter group is called the *free c-step nilpotent D-group* with free generating set X. Clearly (see Theorem 8.22), G^*, in the class of D-groups of nilpotent step at most c, enjoys the same universal property as G in the class of all nilpotent groups of nilpotent step at most c. Now let S be a free nilpotent Lie algebra which is c-step nilpotent and whose free generating set is the same X as above, over the field $\mathbb{Q}$ of rational numbers.

Lemma 8.24. *$S^\circ \cong G^*$, the isomorphism extending the identity map of X.*

Proof. We will employ the notation from Section 8.2. By the corollary of Theorem 8.7, the subgroup $C = \mathrm{gp}(e^x \mid x \in X)$ of $1 + M \subseteq \hat{A}$ generated by the set of all exponentials of the form e^x, $x \in X$, is canonically isomorphic to $F = F(X)$ and $\gamma_n(C) = C \cap (1 + M^n)$ for all $n \geq 1$. Let D be a subgroup generated by X with respect to the circle multiplication $\circ$. Then $D \cong F(X)$. Taking logarithms of both sides of the equation for $\gamma_n(C)$, we get $\gamma_n(D) = D \cap (\hat{L}^n)^\circ$. Therefore

$$D/\gamma_n(D) = (D \circ (\hat{L}^n)^\circ)/(\hat{L}^n)^\circ \subseteq \hat{L}^\circ/(\hat{L}^n)^\circ \cong (\hat{L}/\hat{L}^n)^\circ = S^\circ.$$

This shows that the subgroup T of S generated by X is a free nilpotent group isomorphic to G. It remains to show that the isolator of this subgroup coincides with S. As is noted in Section 8.3.4, $\gamma_2(S^\circ) = (S^2)^\circ$. Hence, modulo $\gamma_2(S^\circ)$, S° is generated by the elements of the form $(m_1/m_2)x = x^{m/m_2}$, $x \in X$, $m_i \in Z$, that is, $S^\circ = \sqrt{T}\gamma_2(S)$. But we always have $\sqrt{T} \subseteq I_{S^\circ}(T)$, hence $S^\circ = I_{S^\circ}(T)\gamma_2(S^\circ)$. Applying Burnside's lemma once more, we get $S^\circ = I_{S^\circ}(T)$. By the uniqueness of the completion, we finally get $S^\circ \cong G^*$. □

Now we are in a position to prove the principal result of this section.

Theorem 8.25. *Every nilpotent D-group is a group of a unique up to isomorphism nilpotent Lie algebra over the field of rational numbers.*

Proof. Let B be an arbitrary nilpotent D-group. It can be represented as a quotient group of a free nilpotent D-group. By Lemma 8.24, there exists a free nilpotent Lie algebra S over the field of rational numbers such that $B \cong S^\circ/R^\circ \cong (S/R)^\circ$, R being an ideal of S. This shows that the map $T \mapsto T^\circ$ associating with each nilpotent Lie algebra over the field of rational numbers its D-group is surjective. Assume now that U and T are two nilpotent Lie algebras over Q such that an $f : T^\circ \to U^\circ$ is a homomorphism of groups. We represent T° in the form $T^\circ \cong S^\circ/R^\circ$ and let $\varphi : S \to T$ denote the natural epimorphism of S onto T as Lie algebras. Using remarks in Section 8.1.6, the map $\varepsilon^\circ : S^\circ \to T^\circ$ is a natural epimorphism of groups. We set $u_x = (f\varepsilon^\circ)(x)$, $x \in X$. Let π be a unique homomorphism $\pi : S \to U$ extending the map $x \mapsto u_x$, $x \in X$. Then π° is a unique homomorphism of groups $S^\circ \to U^\circ$ extending the same map. Hence $\pi^\circ = f\varepsilon^\circ$. This enables us to define correctly a homomorphism of Lie algebras $\varphi : T \to U$ by putting $\varphi(t) = \pi(\varepsilon^{-1}(t))$. It is clear that $\varphi^\circ = f$, φ being the unique map such that

$\varphi^\circ = f$. Now it follows that, given two nilpotent Lie algebras T, U with $T^\circ \cong U^\circ$, we have $T \cong U$. □

Corollary 8.26.
(1) *Any torsion-free nilpotent group G is isomorphic to a subgroup of a group of the form S°, S being a nilpotent Lie algebra over $\mathbb{Q}$. Moreover, $\sqrt{G} = S^\circ$.*
(2) $I_G(\gamma_n(G)) = \gamma_n(G^*) \cap G$.

Proof. The first claim is immediate from Theorems 8.22 and 8.25. Now $S = \mathbb{Q}G$, hence $\gamma_n(G^*) = S^n = \mathbb{Q}\gamma_n(G)$, $n = 1, 2, \ldots$ □

Corollary 8.27. *Categories $\mathcal{N}_L$ and $\mathcal{N}_D$ of respectively nilpotent Lie algebras over $\mathbb{Q}$ and nilpotent D-groups are isomorphic.*

Proof. Two categories are isomorphic if there exists mutually inverse functors from one category into the other (see Section 3.4.1). Now, by Theorem 8.25, there exists a one-to-one correspondence $T \mapsto T^\circ$ of their objects and $\varphi \mapsto \varphi^\circ$ of their morphisms such that $(1_T)^\circ = 1_{T^\circ}$ and $(\varphi\psi)^\circ = \varphi^\circ\psi^\circ$. □

8.3.6 Lie-type groups

The last result of this section relates to groups which are *residually torsion-free nilpotent*. By this we mean a group G in which for any $1 \neq g \in G$ there exists a normal subgroup N_g such that $g \notin N_g$ and G/N_g is a torsion-free nilpotent group. According to Magnus' theory in Section 8.2, this class of groups includes free groups. We will see below that (and this is one of the main aims of the next section) any free solvable group is residually torsion-free nilpotent. Now the condition of being residually torsion-free nilpotent is equivalent to the following one. Let $I_n(G)$ denote the isolator of the nth term of the lower central series for G, $I_n(G) = I_G(\gamma_n(G))$, $n = 1, 2, \ldots$ Then the equivalent condition is $\bigcap_{n=1}^{\infty} I_n(G) = \{1\}$. We leave the proof to the reader. We will also call the groups of this class *Lie-type groups*. The motivation of such terminology is clear from what follows.

Proposition 8.28. *Any Lie-type group is isomorphic to a subgroup of a group of the form T° where T is a Lie algebra over the field $\mathbb{Q}$ of rational numbers.*

Proof. Let $\{G^{(n)} = G/I_n(G), \varphi_{nm} : G^{(n)} \to G^{(m)} \mid n \geq m \geq 1\}$ be a projective system of groups and their homomorphism constructed using G in the same way as is done in Section 8.1.1. Let $T^{(n)}$ denote a Lie algebra such that $(T^{(n)})^\circ = (G^{(n)})^*$, $n = 1, 2, \ldots$ We denote by ψ_{nm} the epimorphism $\psi_{nm} : T^{(n)} \to T^{(m)}$ such that $\psi^\circ_{nm} = \psi^*_{nm}$. Now set $\widetilde{T} = \lim_{\leftarrow} T^{(n)}$ and turn $\widetilde{T}$ into a topological algebra in the same way as in Section 8.1.2. Then

$$G \subseteq \widetilde{G} = \varprojlim G^{(n)} \subseteq \varprojlim (T^{(n)})^\circ = \widetilde{T}^\circ. \qquad \square$$

8.4 Varieties of groups of Lie type

8.4.1 Free group of a variety

All definitions and simple results for varieties of Lie algebras in Section 4.1 can be easily adapted to the group case (this is not surprising since their natural setting is much wider, namely algebraic systems). Thus a *variety of groups* is a class $\mathbf{V}$ of all groups satisfying a fixed system of identical relations $\{v(x_1, x_2, \dots, x_n) = 1 \mid v \in V \subseteq F(X)\}$ or, simply, identities. A nonempty class $\mathbf{V}$ of groups is a variety if and only if $\mathbf{V}$ is closed under taking homomorphic images, subgroups and Cartesian products (cf. Section 4.1, Theorem 4.3). The verbal subgroup $\mathbf{V}(G)$ of a group G corresponding to a set V of words $v \in F$ is a subgroup generated by all elements of the form $v(g_1, g_2, \dots, g_n)$, $g_1, g_2, \dots, g_n \in G$. The full set of left-hand sides of identities holding in a variety $\mathbf{V}$ forms a verbal subgroup of the free group F_∞ of countable rank. A subgroup V of a free group F is verbal if and only if it is stable under all endomorphisms of this group. If we are given a variety $\mathbf{V}$ and its determining subset $V \subseteq F_\infty$ and a set $X \neq \varnothing$ then $F(X, \mathbf{V}) = F(X)/V(F(X))$ is called a *free group of* $\mathbf{V}$ with free generating set X. This is the class of groups which will be our main concern in the present section.

A variety of groups is called a *variety of Lie type* if its free groups of arbitrary rank are Lie type groups. By Theorem 8.11, the variety $\mathbf{O}$ of all groups and the varieties of the form $\mathbf{N}_c$, $c \geq 1$, of all nilpotent groups of nilpotent step at most c are varieties of Lie type. As we have learned in Section 8.3.6, there is a close connection between groups of Lie type and Lie algebras over rational numbers. A similar connection will be established between varieties of groups of Lie type and varieties of Lie algebras over the field of rational numbers. We will need some auxiliary results concerning isolated lower central series. We define the *rank of finitely generated abelian group* $\mathrm{rk}(A)$ of a finitely generated abelian group A as the number of infinite cycles in any decomposition of A as the direct product of cyclic groups.

Proposition 8.29. *Let G be an arbitrary group. We denote by I_n, $n \geq 1$, the subgroup of the form $I_n = I_G(\gamma_n(G))$. Then the following are true:*
(1) *if G is a $\mathbf{V}$-free group then all I_n are verbal subgroups;*
(2) *the set $\{I_n\}_{n\geq 1}$ is a central filtration in G;*
(3) *if G is finitely generated then $\mathrm{rk}(I_n/I_{n+1}) = \mathrm{rk}(\gamma_n(G)/\gamma_{n+1}(G))$.*

Proof. As we have seen in Proposition 8.19, $I_s = \sqrt{\gamma_s(G)}$. Since, obviously $\sqrt{\gamma_s(F)}$, F being free, is stable under all endomorphisms of F and since any endomorphism of a free group of a variety is induced by an endomorphism of a suitable F, we can see that I_s is indeed verbal. Now we are going to show $(I_r, I_s) \subseteq I_{r+s}$. Indeed, choose $x \in I_r$, $y \in I_s$. Abbreviating $\gamma_s(G)$ to γ_s, we may write $x^n \in \gamma_r$, $y^m \in \gamma_s$ for some $m, n \neq 0$, hence $(x^n, y^m) \in \gamma_{r+s} \subseteq I_{r+s}$. Let $\tilde{x}, \tilde{y}$ denote the images of x, y in the torsion-free nilpotent group $\widetilde{G} = G/I_{r+s}$. Then we have $\tilde{x}^n = \tilde{y}^{-m}\tilde{x}^n\tilde{y}^m = (\tilde{y}^{-m}\tilde{x}\tilde{y}^m)^n$. From Proposition 8.19

it follows that $y = xyx^{-1}$ or, in other words, $(x, y) = 1$. Finally, we get $(x, y) \in I_{r+s}$ and the proof of (2) is complete. To prove (3), we recall that if A is a free abelian group of some finite rank and B its subgroup of finite index then $\mathrm{rk}(A) = \mathrm{rk}(B)$. Now the group $\gamma_n \cap I_{n+1}/\gamma_{n+1}$ is finite and $\gamma_n/\gamma_n \cap I_{n+1} \cong \gamma_n I_{n+1}/I_{n+1} \subseteq I_n/I_{n+1}$. Hence $\gamma_n/\gamma_n \cap I_{n+1}$ is free abelian and

$$\gamma_n/\gamma_{n+1} \cong \gamma_n \cap I_{n+1}/\gamma_{n+1} \cong \gamma_n I_{n+1}/I_{n+1}.$$

So it remains to show that $\mathrm{rk}(\gamma_n I_{n+1}/I_{n+1}) = \mathrm{rk}(I_n I_{n+1})$. But this is immediate from the above remark since $I_n/\gamma_n I_{n+1}$ is periodic, hence finite. □

Theorem 8.30. *Let X be a countable set, $R = L(X, \mathbf{V})$ a free Lie algebra in a variety $\mathbf{V}$ of Lie algebras over the field $\mathbb{Q}$ of rational numbers. Suppose we are given a descending central series $R = R^{(1)} \supseteq R^{(2)} \supseteq \cdots$ of ideals of R such that for any $t = 0, 1, 2, \ldots$ there exists l such that $R^t \supseteq R^{(l)}$. Then there exists a variety $\mathbf{V}^\circ$ of groups of Lie type such that, for any series as above, the subgroup $G = \mathrm{gp}(X) \subseteq \tilde{R}$, where $\tilde{R}$ is the completion of R in the topology given by the series, is $\mathbf{V}$-free with free generating set X.*

Proof. By a remark at the end of Section 8.1.2, the completion $\hat{R}$ of R with respect to the topology given by the lower central series is isomorphically embedded in the completion $\tilde{R}$ of R with respect to the topology given by a series as in the statement, the isomorphism being induced by the identity map of R. Since $X \subset \hat{R}$ then the Campbell–Hausdorff formula shows that $G = \mathrm{gp}(X) \subseteq \hat{R}$. This enables us to restrict ourselves to the case of the completion with respect to the lower central series.

To prove that G is a free group of a certain variety of groups of Lie type it is sufficient to show that:

(1) $\bigcap_{n=1}^{\infty} I_n(G) = \{1\}$;

(2) any relation among the free generators in X is identically true in G.

But $\gamma_n(G) \subseteq \hat{R}^n$. Since $\hat{R}^n$ is closed under the multiplication by rational numbers, we have $I_n(G) \subseteq \hat{R}^n$. According to Section 4.2.4, we have $\bigcap_{n=1}^{\infty} R^n = \{0\}$, proving (1). Now $\hat{R}/\hat{R}^n \cong R/R^n$ is a free algebra of a certain variety of nilpotent Lie algebras, hence, by Corollary 8.27 we have $\hat{R}^\circ/(\hat{R}^n)^\circ = (\hat{R}/\hat{R}^n)^\circ$. So this is a free group of a variety of nilpotent D-groups. Hence if $v(x_1, x_2, \ldots, x_n) = 1$ is a relation among the elements $x_1, x_2, \ldots, x_n$ then it identically holds in any of the groups $(\hat{R}^\circ/\hat{R}^n)^\circ$. But $\bigcap_{n=1}^{\infty}(\hat{R}^n)^\circ = \{1\}$, that is, $\hat{R}^\circ \in \mathrm{var}(\{\hat{R}^\circ/(R^n)^\circ \mid n = 1, 2, \ldots\})$. Consequently, $v(x_1, x_2, \ldots, x_n) \equiv 1$ is an identity in $\hat{R}^\circ$ proving (2). So, $G = \mathrm{gp}(X)$ is a free group of a variety of groups which will be denoted by $\mathbf{V}^\circ$. □

The above theorem establishes a corresponding $g : \mathbf{V} \to \mathbf{V}^\circ$ between varieties of Lie algebras over the field of rational numbers and the varieties of groups of Lie type. We want to show that this correspondence is surjective.

Theorem 8.31. *Suppose we are given a variety* $\mathbf{V}$ *of groups of Lie type. Then there exists a variety* $\mathbf{U}$ *of Lie algebras over the field* $\mathbb{Q}$ *of rational numbers such that, for any nonempty set* X*, the free group* $F(X, \mathbf{V})$ *of* $\mathbf{V}$ *is isomorphic to a subgroup of a group of the form* $(\widehat{L}(X, \mathbf{U}))^\circ$ *generated by* X*, the completion being taken with respect to the lower central series of* $L(X, \mathbf{U})$.

Proof. It is a standard trick to restrict to the case where X is countable. Let G be a group of the form $F(X, \mathbf{V})$. We apply a construction from Section 8.3.6. Our objective is to prove that a subalgebra L of the algebra $\widetilde{T}$ constructed in that subsection, generated by X, is $\mathbf{U}$-free for a certain variety of Lie algebras over $\mathbb{Q}$. It is enough to verify that any map of X into $\widetilde{T}$ extends to a homomorphism of L into $\widetilde{T}$, that is, if for some v, $v(x_1, x_2, \dots, x_m) = 0$ then $v(z_1, z_2, \dots, z_m) \equiv 0$ holds in $\widetilde{T}$ identically. We recall that any element x in X considered as a subset of the projective limit $\widetilde{T}$ has the form of a function $\bar{x}$, $\bar{x}(n) \in T^{(n)}$, $n = 1, 2, \dots$ Therefore, if $v(\bar{x}_1, \bar{x}_2, \dots, \bar{x}_m) = 0$ then $v(x_1(n), x_2(n), \dots, x_m(n)) = 0$ in each $T^{(n)}$. By Proposition 8.29, each group $T^{(n)}$ is a free group in a variety of nilpotent groups freely generated by $X(n)$. Therefore $v(z_1, z_2, \dots, z_m) \equiv 0$ is an identity in $T^{(n)}$. By the construction of the projective limit $\widetilde{T} \in \operatorname{var}(\{T^{(n)} \mid n = 1, 2, \dots\})$, hence $v(z_1, z_2, \dots, z_m) \equiv 0$ is an identity in $\widetilde{T}$ and we have proved that L is a $\mathbf{U}$-free algebra for some variety $\mathbf{U}$ of Lie algebras over $\mathbb{Q}$. Now since the groups comprising our projective family were constructed as quotient groups over the terms of the isolated lower central series, we see that the topology in $\widehat{L}$, the closure of L in $\widetilde{T}$, is determined by the lower central series. Hence, by the preceding theorem, G is a free group in $\mathbf{U}^\circ$. □

The free Lie algebra L of $\mathbf{U}$ in the just proved theorem can also be constructed in an alternative way. By Proposition 8.29, in an arbitrary group G, the series $\{I_n(G)\}_{n\geq 1}$ is a central filtration. We can form a graded Lie ring $L^f(G)$, using this filtration. Since the homogeneous components of $L^f(G)$ are free abelian groups, this ring is a subring in the Lie algebra $\mathscr{L}(G) = \mathbb{Q} \otimes_{\mathbb{Z}} L^f(G)$, which is a graded algebra whose homogeneous components are the completions of the groups of the form I_n/I_{n+1}, $n = 1, 2, \dots$

Theorem 8.32. *Let* $F = F(X, \mathbf{V})$ *be the free group in a variety* $\mathbf{V}$ *groups of Lie type freely generated by* X*. Then* $M = \mathscr{L}(F)$ *is a free algebra of a variety* $\mathbf{U}$ *of Lie algebras over* $\mathbb{Q}$ *freely generated by the image of* X*. Having denoted* $\mathbf{U} = \mathscr{L}(\mathbf{V})$*, we may write* $(\mathscr{L}(\mathbf{V}))^\circ = \mathbf{V}$.

Proof. As in Theorem 8.31, we consider F embedded in the projective limit $\widetilde{T}$. Let $L = \operatorname{alg}(X)$ in $\widetilde{T}$. Let us show first that $L^n = L \cap \widetilde{T}_n$, $n = 1, 2, \dots$ The inclusion $L^n \subseteq L \cap \widetilde{T}_n$ being obvious, we consider a Lie element $w(x_1, x_2, \dots, x_s) \in \widetilde{T}_n$. We expand w in the form $w = w_1 + w_2 + \cdots + w_m$, where w_i is the homogeneous component of a degree i. By our construction and using Corollary 8.27, we can see that any algebra $T^{(n)}$ is varietally free generated by $X(n)$. Hence all $w_i \equiv 0$, $1 \leq i \leq n-1$, are identical relations holding in all $T^{(n)}$. Since $\widetilde{T}$ is residually $\{T^{(n)}, n = 1, 2, \dots\}$, we can see that $w_1 + w_2 + \cdots + w_{n-1} \equiv 0$ is an identity in $\widetilde{T}$, hence in L. Now $w \in L^n$, and then $L^n \supseteq L \cap T_n$.

Now since both L and M are graded and by the "likeness" of the multiplication in these algebras a homomorphism from L into M can be given by

$$w(x_1, \ldots, x_s) \mapsto \tilde{w}(x_1, \ldots, x_s)I_{n+1}$$

where $w(x_1, x_2, \ldots, x_s)$ is a linear combination with integral coefficients of basic monomials of degree n in X and $\tilde{w}(x_1, x_2, \ldots, x_s)$ is the corresponding product of commutators in F. By the above equation, $L^n = L \cap \widetilde{T}_n$, this definition is correct and the mapping is injective. Its surjectivity follows since both L and M are generated by (the image of) X. □

It is obvious from the proof that $L(F)$ is also free in some variety of Lie rings. This variety is denoted by $L(\mathbf{V})$.

8.4.2 Product varieties of groups

A variety $\mathbf{W}$ of groups is called the *product of varieties* $\mathbf{U}$ *and* $\mathbf{V}$ (denoted by $\mathbf{UV}$) if $\mathbf{W}$ is the class of all groups G such that G possesses a normal subgroups $H \in \mathbf{U}$ with $G/H \in \mathbf{V}$. Suppose $\mathbf{U}$ is determined by a set of words U and $\mathbf{V}$ by a set of words V. Let also F_∞ denote the free group of countable rank. Then $\mathbf{W} = \mathbf{UV}$ is determined by the set of words $U(V(F_\infty))$. For, $G = F_\infty/U(V(F_\infty)) \in \mathbf{UV}$ (in contrast to the case of Lie algebras any verbal subgroup of a normal subgroup is always normal in the whole group). Hence $W(F) \subseteq U(V(F))$. Similarly, one can prove the converse. Moreover, given a set X, we always have $F(X, \mathbf{UV}) = F(X)/U(V(F(X)))$.

Our nearest goal is to express a free group of a product of varieties in terms of free groups (of the same rank) in the factors of this product.

8.4.3 Wreath products of groups

We start with a construction which is important in the theory of groups. Let A and B be two groups. We consider the Cartesian product $P = \prod_{b\in B} A_b$, $A_b \cong A$ for all $b \in B$, and then turn P into a B-set by putting $f^{b_1}(b) = f(b_1 b)$, $f \in P$, $b, b_1 \in B$. The split extension S of B with kernel P relative to the above action of B is called the *Cartesian wreath product* of A and B, denoted by $S = A \operatorname{Wr} B$. Thus, any element of S has the form bf, $b \in B, f \in P$, and $bfb_1f_1 = bb_1f^{b_1}f_1$.

Proposition 8.33. *Let G be an extension of A by B, that is, $A \lhd G$, $G/A = B$. Then there exists a monomorphism $\mu : G \to S = A \operatorname{Wr} B$ such that $\mu(A) \subseteq P$ and, if $\bar{\mu} : G/A \to S/A$, we have $\bar{\mu} = 1_B$.*

Proof. Let $s : B \to G$ be an imbedding such that b^s is a representative of $b \in B$ and $1^s = 1$ (this is sometimes called a *transversal*). We define $\mu : G \to S$ by setting $\mu(g) = \bar{g}f_g$

where $\bar{g} = gA$ and $f_g(b) = ((\bar{g}b)^s)^{-1}gb^s$. It is clear that $f_g(b) \in A$ for all g and b. Now we will verify that μ is a homomorphism. Indeed, $\mu(g_1g_2) = \bar{g}_1\bar{g}_2f_{g_1g_2}$ where $f_{g_1g_2}(b) = ((\overline{g_1g_2}b)^s)^{-1}g_1g_2b^s$. On the other hand,

$$\mu(g_1)\mu(g_2) = \bar{g}_1f_{g_1}\bar{g}_2f_{g_2} = \bar{g}_1\bar{g}_2f_{g_1}^{\bar{g}_2}f_{g_2},$$

so it remains to compare the "P-components" of the sides in the equation sought:

$$\begin{aligned}(f_{g_1}^{\bar{g}_2}f_{g_2})(b) &= f_{g_1}^{\bar{g}_2}(b)f_{g_2}(b) = f_{g_1}(\bar{g}_2b)f_{g_2}(b)\\ &= ((\bar{g}_1\bar{g}_2b)^s)^{-1}\bar{g}_1(\bar{g}_2b)^s((\bar{g}_2b)^s)^{-1}\bar{g}_2b^s = (\overline{g_1g_2}b)^s\overline{g_1g_2}b^s = f_{g_1g_2}(b).\end{aligned}$$

Now from $\mu(g) = 1$ it follows that $\bar{g} = 1$, that is, $g \in A$. But then $f_g(1) = (\bar{g}1)^sg1^s = g$, hence $g = 1$. □

A similar construction which is close to that considered in Section 4.4 in the case of Lie algebras is the *verbal wreath product* (**V**-*wreath product*) of groups. So, let **V** be a variety of groups, A a group in **V**, and B an arbitrary group. A group $W = A \operatorname{wr}_{\mathbf{V}} B$ is called a **V**-wreath product of A and B if the following conditions hold:

(1) $W = \operatorname{gp}(A, B)$;
(2) let P be a normal subgroup in W generated by A, then $P \in \mathbf{V}$;
(3) if $\varphi : A \to G$, $\psi : B \to G$ be homomorphisms of A and B into G such that:
   (a) $G = \operatorname{gp}(\varphi(A), \psi(B))$,
   (b) the normal subgroup T generated by $\varphi(A)$ is in **V**,

then there exists a homomorphism $\chi : W \to G$ such that $\chi|_A = \varphi$ and $\chi|_B = \psi$.

The existence of the verbal wreath product similar to the Lie algebra case can be proved using free products (the definition can be given as in Section 2.6). So let Q be a normal subgroup of the free product $\Phi = A * B$ generated by A, and $V(Q)$ be the verbal subgroup of Q corresponding to **V**. Then $V(Q)$ is a normal subgroup in Φ and $\Phi/V(Q)$ is a split extension of B by $Q/V(Q)$ (cf. Section 4.4.1). Now all the defining properties of the wreath product are immediate and $\Phi/V(Q)$ is the desired wreath product.

One of important results in the theory of varieties of groups is as follows.

Theorem 8.34. *Consider $G = F(X, \mathbf{UV})$, $A = F(Y, \mathbf{U})$, $B = F(Z, \mathbf{V})$ where X, Y, Z have the same cardinality, the bijections being designated as $x \mapsto y \mapsto z$. Form the* **U***-wreath product $W = A \operatorname{wr}_{\mathbf{U}} B$. Then the map $x \mapsto xy$ extends to a monomorphism of G into W.*

Proof. By construction, $W \in \mathbf{UV}$; hence the map $x \mapsto yz$ extends to a homomorphism α from G into W. We also consider $S = \mathbf{V}(G) \operatorname{Wr} B$. By the above proposition, there exists a monomorphism $\mu : G \to S$ such that $\mu(V(G)) \subseteq P$. We set $\mu(x) = b_xf_x$. Then the maps $y \mapsto f_x$ and $z \mapsto b_x$ extend to homomorphisms $\varphi : A \to V(G)$ and $\psi : B \to B$. Since P is a direct product of groups isomorphic to $V(G)$, it is a **U**-group. Now by the definition of the verbal wreath product, there exists a homomorphism $\chi : W \to S$ extending φ and ψ. The diagram

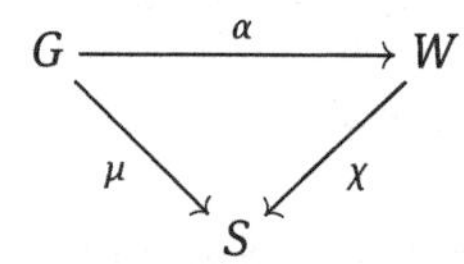

being commutative, we immediately see that α is a monomorphism. □

Now we remark that using the universal property of the verbal wreath product $R = A \operatorname{wr}_{\mathbf{U}} B$ it is easy to see that this latter is a split extension of a $\mathbf{U}$-free group P freely generated by the set $\{x^b \mid x \in X, b \in B\}$ by B, the action of B in P being given by $b_1^{-1} x^b b_1 = x^{bb_1}$, $b, b_1 \in B$. Before we formulate the next theorem, we recall that, given a $\mathbf{V}$-wreath product $W = L \operatorname{wr}_{\mathbf{U}} B$ of a free algebra in a variety $\mathbf{V}$ of Lie algebras and a nilpotent Lie algebra B, by Section 6.6, Proposition 6.48, we have $\bigcap_{n=1}^{\infty} W^n = \{0\}$ so that we may speak about the completion of W with respect to the central series formed by these ideals. The closure $\bar{K}$ of the ideal $K = \operatorname{id}_W(L)$ in this topology is its completion with respect to the central series formed by the ideals $K_n = K \cap W^n$. Let Y be an arbitrary finite subset of the canonical generating set for K (see Section 6.6, Proposition 6.48) and R a subalgebra generated by Y. Then the central series of the ideals $R^{(m)} = K_m \cap R$ has the property that for any $t = 0, 1, \ldots$ there exists an l such that $R^t \supseteq R^{(l)}$. Then, by Theorem 8.30, the subgroup A generated by Y in $\bar{K}$ is a free algebra in a variety of groups $\mathbf{V}^\circ$. Since Y is an arbitrary finite subset, the same can be said about any subgroup generated by canonical generators of $\bar{K}$.

Theorem 8.35. *Let $L = L(X, \mathbf{V})$ be a $\mathbf{V}$-free Lie algebra over $\mathbb{Q}$ and B a nilpotent Lie algebra, $W = L \operatorname{wr}_{\mathbf{U}} B$ their $\mathbf{V}$-wreath product, $\overline{W}$ its completion by the lower central series. Then the subgroup T generated by X and by B° in $\overline{W}^\circ$ is isomorphic to $A \operatorname{wr}_{\mathbf{U}} B^\circ$ where A is a $\mathbf{U}$-free group generated by X, $\mathbf{U} = \mathbf{V}^\circ$.*

Proof. The proof will be complete if we manage to show that the elements $x^b = b^{-1}xb$, $x \in X$, $b \in B$ form a free generating set of a subgroup they generate in the completion of the wreath product W. From the Campbell–Hausdorff formula (Section 8.1.7), we can derive the following equation (see Exercise 8.7.1):

$$x^{y^{-1}} = yxy^{-1} = \sum_{n=0}^{\infty} \frac{(\operatorname{ad} y)^n}{n!}(x) = e^{\operatorname{ad} y}(x), \tag{8.8}$$

that is, the inner automorphism of $\widehat{L}^\circ$ is an inner automorphism of $\widehat{L}$. According to Section 4.4.1, K/K^2 is a free B-module, hence a free $U(B)$-module. Clearly, then its completion $\overline{K/K^2}$ by the series $K^2 + K_n/K_n$, $n = 1, 2, \ldots$, is a free $\overline{U(B)}$-module, $\overline{U(B)}$ being the completion of $U(B)$ by the series of ideals U_s, $s = 1, 2, \ldots$ (see Section 6.6, Proposition 6.48). To prove the linear independence of elements x^b, $b \in B$, it is sufficient to show that they are linearly independent when x is fixed. This can be done by induction on the dimension of B. Let z be a nonzero element in the center of B. Suppose there exists nonzero elements $\lambda_1, \lambda_2, \dots, \lambda_m \in k$ such that

$$\lambda_1 e^{b_1} + \lambda_2 e^{b_2} + \dots + \lambda_m e^{b_m} = 0.$$

Let B_1 denote the quotient algebra B/kz. Since the canonical map $U(B) \to U(B_1)$ is continuous, we may write

$$\lambda_1 e^{\bar{b}_1} + \lambda_2 e^{\bar{b}_2} + \dots + \lambda_m e^{\bar{b}_m} = 0, \tag{8.9}$$

where $\bar{b}_i = b_i + kz$, $i = 1, 2, \dots, m$. By the induction hypothesis, we may then assume that all $\bar{b}_i$ are the same and still some of λ_i, $i = 1, 2, \dots, m$, are nonzero in equation (8.9). In this case we may write $b_i = u + \mu_i z$. Since z is central, this can be presented as

$$e^u(\lambda_1 e^{\mu_1 z} + \lambda_2 e^{\mu_2 z} + \dots + \lambda_m e^{\mu_m z}) = 0,$$

hence

$$\lambda_1 e^{\mu_1 z} + \lambda_2 e^{\mu_2 z} + \dots + \lambda_m e^{\mu_m z} = 0. \tag{8.10}$$

Taking derivatives of the both sides in equation (8.10) sufficiently many times, we conclude that equation (8.10) is possible only if some of μ_i coincide.

Now we may notice that x^b depending on X to the first degree, the linear combination of such element is in K^2 if and only if it equals zero. Hence for any $x_1, x_2, \dots, x_n \in X$ and $b_1, b_2, \dots, b_n \in B$ one can choose an integer $s \geq 1$ and the elements $u_1, u_2, \dots, u_n$ in K such that $x_i^{b_i} - u_i \in K_s$ and u_i are linearly independent modulo K^2. (The elements u_i are sufficiently long beginnings of the series (8.8) constructed for each $x_i^{b_i}$, $i = 1, 2, \dots, n$.) By Section 4.2, Theorem 4.14, the elements $u_1, u_2, \dots, u_n$ are free generators of a subalgebra they generate. Now let $v(x_1^{b_1}, x_2^{b_2}, \dots, x_n^{b_n}) = 0$. Then also $v(u_1, u_2, \dots, u_n) = 0$. Indeed, having written $x_i^{b_i} = u_i + u_i'$, we find (see equation (8.8)) that u_i' expressed through free generators of K which are different from those used for expressing the elements u_j. But in a free algebra of a variety, if an element is zero then the same is true of its homogeneous components (the base field is of characteristic 0). Hence we derive $v(u_1, u_2, \dots, u_n) = 0$. Thus $v \equiv 0$ is an identity in $\mathbf{V}$. Therefore the elements $x_1^{b_1}, x_2^{b_2}, \dots, x_n^{b_n}$ are free generators in a $\mathbf{V}$-free subalgebra. Hence they generate a $\mathbf{V}$-free subgroup. □

The main consequence of this theorem is as follows.

Theorem 8.36. *Let* $\mathbf{U}$ *and* $\mathbf{V}$ *be two varieties of groups of Lie type. Then their product* $\mathbf{UV}$ *is a variety of groups of Lie type.*

Proof. By Theorem 8.34, it is sufficient to verify that $T = A \operatorname{wr}_{\mathbf{U}} B$ is a group of Lie type, A and B being free groups in **U** and **V**. If **V** is a nilpotent variety then all follows using Theorem 8.35 and a simple fact that $I_n(\overline{W}^{\circ}) \subseteq \overline{W}^n$. Now a reduction to the nilpotent case can be carried out as follows. To prove that T is a group of Lie type, it is sufficient, for each $a \neq 1$ in T, to find a normal subgroup Q_a such that T/Q_a is a torsion-free nilpotent group and $aQ_a \neq 1$. If $a \notin P$ then $aP \in I_s(B)$ for some $s > 1$ and the desired subgroup is $PI_s(B)$. Suppose now $a \in P$. Then $a = w(x_1^{b_1}, x_2^{b_2}, \ldots, x_n^{b_n})$, where $x_i^{b_i}$ is a free generator of P. Now let s be such that all elements of the form b_i and $b_i b_j^{-1}$ are not in $I_s(B)$, $1 \le i \neq j \le m$, and let B_1 denote the quotient group $B/I_s(B)$. We consider also $T_1 = A \operatorname{wr}_{\mathbf{U}} B_1$. Then, denoting $bI_s(B)$ by $\bar{b}$, we see that the map $b \mapsto \bar{b}$, $x^b \mapsto x^{\bar{b}}$ extends to a homomorphism of T onto T_1 such that $a \mapsto w(x_1^{\bar{b}_1}, x_2^{\bar{b}_2}, \ldots, x_m^{\bar{b}_m}) \neq 1$ since $x_i^{\bar{b}_i}$, for different i, are different generators of a varietally free group P_1. This gives the desired reduction of the nilpotent case. Now the proof is complete also in the general case. □

In particular, we may say that, given a free solvable group of solvable step l, we have $\bigcap_{n=1}^{\infty} \gamma_n(G) = \{1\}$. Another group of interest is a free polynilpotent group R, that is, the free group of a variety $\mathbf{N}_{c_1}\mathbf{N}_{c_2} \cdots \mathbf{N}_{c_t}$. According to our theorem, we may write $\bigcap_{n=1}^{\infty} \gamma_n(R) = \{1\}$.

8.4.4 Magnus groups

A group G is called a *Magnus group* if:
(i) $\bigcap_{n=1}^{\infty} \gamma_n(G) = \{1\}$;
(ii) $\gamma_n(G)/\gamma_{n+1}(G)$ is a torsion-free group, $n = 1, 2, \ldots$

Adapting this to Lie rings, we arrive at the concept of a *Magnus Lie ring*. A variety of groups (respectively, Lie rings) is called a *Magnus variety* if all its free groups are Magnus (respectively, if all its free Lie rings are Magnus).

Proposition 8.37. *Let* **U** *and* **V** *be two varieties of Lie rings. Then their product is a Magnus variety provided* **U** *and* **V** *are multihomogeneous Magnus varieties.*

Proof. Arguing as in Section 4.3.1, with the help of Section 4.2, Theorem 4.12, we see that the verbal ideal of a free Lie ring defining **UV** is precisely $\mathbf{U}(\mathbf{V}(L_\infty))$. We will prove the Magnus property for a free Lie algebra of the product variety in the case of finite rank, which is not a loss of generality. So let $L = L(X)$ be a free Lie algebra of finite rank. Since **V** is determined by multihomogeneous identities $\mathbf{V}(L)$ is a homogeneous subring in $L(X)$ such that $L/\mathbf{V}(L)$ is a torsion-free abelian group. By Section 2.4, Theorem 2.20, then $\mathbf{V}(L)$ is a free Lie ring. Hence $\mathbf{V}(L)/\mathbf{U}(\mathbf{V}(L))$ is a Magnus ring. Thus, as abelian groups, $L/\mathbf{U}(\mathbf{V}(L))$ is an extension of a free abelian group $L/\mathbf{V}(L)$ with kernel

$\mathbf{V}(L)/\mathbf{U}(\mathbf{V}(L))$ which is torsion-free. We see that $L/\mathbf{U}(\mathbf{V}(L))$ is torsion-free. Now $\mathbf{UV}$ is determined by multihomogeneous identities, hence $T = L/\mathbf{U}(\mathbf{V}(L))$ is a graded ring with respect to grading by degrees. Thus we may write $\bigcap_{n=1}^{\infty} T^n = \{0\}$ and $T = \bigoplus_{n=1}^{\infty} T^n/T^{n+1}$. Now each T^n/T^{n+1} is torsion-free. □

Lemma 8.38. *Let* $\mathbf{U}$ *and* $\mathbf{V}$ *be two varieties of groups of Lie type,* $\mathscr{L}(\mathbf{U})$ *and* $\mathscr{L}(\mathbf{V})$ *their associated varieties of Lie algebras over the field* $\mathbb{Q}$ *of rational numbers (Theorem 8.32). Then* $\mathscr{L}(\mathbf{UV}) = \mathscr{L}(\mathbf{U})\mathscr{L}(\mathbf{V})$.

Proof. Any variety of Lie algebras over a field of characteristic zero is generated by its nilpotent algebras. Thus let $P \in \mathscr{L}(\mathbf{UV})$ where P is a nilpotent algebra. According to Theorem 8.32, we have $P^\circ \in \mathbf{UV}$. Then $P^\circ \rhd Q^\circ \in \mathbf{U}$ and $P^\circ/Q^\circ \in \mathbf{V}$. Applying Theorem 8.32 once more, we see that $Q \in \mathscr{L}(\mathbf{U})$ and $P/Q \in \mathscr{L}(\mathbf{V})$, hence $P \in \mathscr{L}(\mathbf{U})\mathscr{L}(\mathbf{V})$. The converse inclusion is obvious. □

Theorem 8.39. *Let* $\mathbf{U}$ *and* $\mathbf{V}$ *be Magnus varieties of groups. Then their product* $\mathbf{UV}$ *is a Magnus variety.*

Proof. Let $L(\mathbf{U})$ and $L(\mathbf{V})$ denote Magnus varieties of Lie rings determined by $\mathbf{U}$ and $\mathbf{V}$. Since by the very definition the free algebras of these varieties are graded and torsion-free, these varieties are determined by multihomogeneous identities. By Proposition 8.37, the product variety $L(\mathbf{U})L(\mathbf{V})$ is a Magnus variety.

Now let $F = F(X)$ be a free ring of this variety and $G = G(X)$ be a free group of the group variety $\mathbf{UV}$ of the same rank. To prove our theorem, it is sufficient to establish that $L(G) \cong F$ under the canonical map induced by the identity map of X. Let us first prove that $L(G) \in L(\mathbf{U})L(\mathbf{V})$. Indeed, for each $n > 1$, $\gamma_n(G)/\gamma_{n+1}(G)$ is an extension of $\gamma_{n+1}(G)(\mathbf{V}(G) \cap \gamma_n(G))/\gamma_{n+1}(G)$ by $\gamma_n(G)/\gamma_{n+1}(G)(\mathbf{V}(G) \cap \gamma_n(G))$. Now

$$\begin{aligned}\gamma_n(G)/\gamma_{n+1}(G)(\mathbf{V}(G) \cap \gamma_n(G)) &= \gamma_n(G)/\gamma_n(G) \cap (\gamma_{n+1}(G)\mathbf{V}(G))\\ &\cong \gamma_n(G)\mathbf{V}(G)/\gamma_{n+1}(G)\mathbf{V}(G) \cong \gamma_n(G/\mathbf{V}(G))/\gamma_{n+1}(G/\mathbf{V}(G)).\end{aligned}$$

Hence the latter of the above-mentioned factors is isomorphic to

$$\gamma_n(G/\mathbf{V}(G))/\gamma_{n+1}(G/\mathbf{V}(G))$$

and the former to

$$\mathbf{V}(G) \cap \gamma_n(G)/\mathbf{V}(G) \cap \gamma_{n+1}(G).$$

Thus $L(G)$ is an extension of $L(G/\mathbf{V}(G)) \in L(\mathbf{V})$ with the kernel equal to the ideal T which is a Lie ring of $\mathbf{V}(G)$ constructed (see Section 8.2.3) by the series

$$\mathbf{V}(G) \supset \mathbf{V}(G) \cap \gamma_2(G) \supset \mathbf{V}(G) \cap \gamma_3(G) \supset \cdots. \tag{8.11}$$

But $L(\mathbf{V}(G)) \in L(\mathbf{U})$. Now using the fact that $\mathbf{V}(G) \cap \gamma_n(G) \supseteq \gamma_n(\mathbf{V}(G))$ for each n we see that any multihomogeneous identity valid in $L(\mathbf{V}(G))$ is also valid in a Lie ring T

constructed by series (8.11). Since $L(\mathbf{U})$ is determined by its multihomogeneous identities, we finally get $T \in L(\mathbf{U})$. Thus $L(G) \in L(\mathbf{U})L(\mathbf{V})$. As a result, we may state that the identity map of X extends to an epimorphism $\varphi : F \to L(G)$. We want to show that φ is an isomorphism.

To start with, we remark that $F_{\mathbb{Q}} = \mathbb{Q} \otimes_{\mathbb{Z}} F$ is a free algebra of $\mathscr{L}(\mathbf{U})\mathscr{L}(\mathbf{V})$ (additively F is the direct sum of two free abelian groups isomorphic to $L(\mathbf{U})$ and $L(\mathbf{V})$). By Lemma 8.38, $\mathscr{L}(\mathbf{U})\mathscr{L}(\mathbf{V}) = \mathscr{L}(\mathbf{UV})$ so that $F_{\mathbb{Q}}$ is also a free algebra of $\mathscr{L}(\mathbf{UV})$. Hence the nth homogeneous component of $F_{\mathbb{Q}}$ is equal to the completion of $I_n(G)/I_{n+1}(G)$, $n = 1, 2, \ldots$ By Proposition 8.29, the rank of F^n/F^{n+1} is equal to the rank of $\gamma_n(G)/\gamma_{n+1}(G)$. But, on the other hand, φ maps F^n/F^{n+1} is mapped onto $\gamma_n(G)/\gamma_{n+1}(G)$ and $\bigcap_{n=1}^{\infty} F^n = \{0\}$. It follows now that φ is indeed an isomorphism. □

In particular, free solvable groups of arbitrary ranks are Magnus groups. The same is true for free polynilpotent groups.

8.5 Varieties and Razmyslov's α-function

8.5.1 Construction and relations of one matrix pair

Let $\mathscr{P}$ be a commutative algebra over the field $\mathbb{C}$ of complex numbers with 1, generated by a system of elements $\{a_r, b_r, c_r \mid r = 1, 2, \ldots\}$ modulo the relations of the form

$$\begin{aligned} a_r b_r = a_r c_r = b_r c_r = 0, \\ a_r^2 = b_r^2 = c_r^2. \end{aligned} \tag{8.12}$$

It is obvious that, by these relations, $a_r^3 = b_r^3 = c_r^3 = 0$. Let X be an infinite set of matrices in $\mathfrak{sl}(2, \mathscr{P}) \subseteq \mathscr{P}_2$ of the form

$$x_r = \begin{pmatrix} a_r i & b_r + c_r i \\ -b_r + c_r i & -a_r i \end{pmatrix}, \quad r = 1, 2, \ldots$$

Let (R, G) denote the pair whose components are respectively the associative and Lie algebras generated by X in $\mathscr{P}_2$ with respect to usual multiplication and commutator of matrices. It is clear that (R, G) belongs to the variety of pairs generated by $(\mathbb{C}_2, \mathfrak{sl}(2, \mathbb{C}))$ (see Section 5.6.1). Thus the following weak identities hold valid in (R, G):

(i) $[x \circ y, z] = 0$;

(ii) $[x, y, z] = 2(x \circ y)z - 2(x \circ z)y$;

(iii) $x(y \circ [z, u]) = (x \circ y)[z, u] - (x \circ z)[y, u] + (x \circ u)[y, z]$.

An immediate consequence of (i) is the following:

(iv) $[x, y] \circ z = x \circ [y, z]$.

The following are also true in our pair, provided $v \in G$:

(v) $x_r v x_r = -\frac{1}{3} v x_r^2$, $(x_r \circ v) x_r = \frac{2}{3} v x_r^2$, $[x_r, x_r, v] = \frac{8}{3} v x_r^2$.

Indeed, any matrix $v \in G$ has the form $v = p_1 e_2 + p_2 e_2 + p_3 e_3$ where

$$e_1 = \begin{pmatrix} i & 0 \\ 0 & -i \end{pmatrix}, \quad e_2 = \begin{pmatrix} 0 & 1 \\ -1 & 0 \end{pmatrix}, \quad e_3 = \begin{pmatrix} 0 & i \\ i & 0 \end{pmatrix}.$$

Using equation (8.12), we may write

$$\begin{aligned} x_r e_1 x_r &= (a_r e_1 + b_r e_2 + c_r e_3) e_1 (a_r e_1 + b_r e_2 + c_r e_3) \\ &= a_r^2 e_1^3 + b_r^2 e_2 e_1 e_2 + c_r^2 e_3 e_1 e_3 = -a_r^2 e_1. \end{aligned}$$

Similarly,

$$x_r e_2 x_r = a_r^2 e_2, \quad x_r e_3 x_r = a_r^2 e_3.$$

On the other hand, $x_r^2 = -3a_r^2 I$ where I is the identity matrix, proving the first of the relations in (v). The remaining two are its simple consequences. From (v) it is a simple matter to derive the following:
(vi) $[x_r, x_s] a x_r = x_r a [x_s, x_s]$, $a \in R$, $r, s = 1, 2, \ldots$

Indeed, any element in R is a linear combination of elements of the form $a = v_1 \cdots v_t$, where $v_1, \ldots, v_t \in G$. We will use induction on t with its basis $t = 0$ following from (i). If $t = 1$ then, using (v), we get

$$\begin{aligned} [x_r, x_s] v x_r - x_r v [x_s, x_r] &= x_r [x_s, v] x_r - x_s x_r v x_r + x_r v x_r x_s \\ &= -\frac{1}{3} x_r^2 ([x_2, v] - x_s v + v x_s) = 0. \end{aligned}$$

If $t > 1$ then we apply (i), the induction hypothesis and an obvious equation $2v_1 v_2 = v_1 \circ v_2 + [v_1, v_2]$:

$$\begin{aligned} 2[x_r, x_s] v_1 v_2 v_3 \cdots v_t x_r &= [x_r, x_s](v_1 \circ v_2) v_3 \cdots v_t x_r + [x_r, x_s][v_1, v_2] v_3 \cdots v_t x \\ &= 2 x_r v_1 \cdots v_t [x_s, x_r]. \end{aligned}$$

Now we introduce an important concept of the 2-*word*. This is a monomial of a free associative algebra in X depending on each of its letters to the second degree. A multi-homogeneous polynomial is called a 2-*element* if it is a linear combination of 2-words. We will need another relation:
(vii) if $a x_r b$ is a 2-element then $[a, x_r] b = a [x_r, b]$, $a, b \in A$.

This is equivalent to the equation of the form $[ab, x_r] = 0$. For any $s = 1, 2, \ldots, s \neq r$, either ab does not contain x_s or ab has the form $ab = c_s x_s d_s x_s e_s$. In the latter case we write

$$[ab, x_r] = \sum_s (c_s [x_s, x_r] d_s x_s e_s + c_s x_s d_s [x_s, x_r] e_s).$$

By (vi), each summand on the right side is trivial, proving (vii).

8.5.2 Definition and some properties of Razmyslov's α-function

We can write elements of R in a canonical form. We remark that by the defining relations (8.12) of $\mathscr{P}$ the degree of a nonzero polynomial in X is at most 2 with respect to any variable.

Proposition 8.40. *Let $a \in R$ be a multihomogeneous element in R of degree 2 in $x_{r_1}, \ldots, x_{r_l}$ and 1 in $x_{s_1}, \ldots, x_{s_m}$. Then a can be uniquely written as follows:*

$$a = \pi(a)\frac{x_{r_1}^2}{3} \cdots \frac{x_{r_l}^2}{3}, \tag{8.13}$$

where $\pi(a)$ is a multilinear polynomial depending on $x_{s_1}, \ldots, x_{s_m}$.

Proof. The existence of such a presentation will be proved for a monomial a', which is obviously sufficient. Thus, applying (i) and (v) in Section 8.5.1, we will have

$$\begin{aligned} x_r x_s a' x_r &= -x_s x_r a' x_r + (x_r \circ x_s) a' x_r = -x_s x_r a' x_r + a'(x_r \circ x_s) x_r \\ &= -x_s x_r a' x_r + 2a' x_s \frac{x_r^2}{3}. \end{aligned}$$

Such a transformation shortens the distance between two neighboring occurrences of x_r. Thus the possibility of presenting a' in the form (8.13) does exist. We further remark that $(x_{r_1}^2/3) \cdots (x_{r_l}^2/3)$ is a scalar matrix on the diagonal of which there is an element of the form $(-1)^l a_{r_1}^2 \cdots a_{r_l}^2$. On the other hand, $\pi(a)$ is a matrix whose entries do not depend on the variables $a_{r_1}, b_{r_1}, c_{r_1}, \ldots, a_{r_l}, b_{r_l}, c_{r_l}$. Then the defining relations of $\mathscr{P}$ enable us to conclude that if $a = 0$ then $\pi(a) = 0$, too. □

The restriction of π to the space of 2-elements is a map to $\mathbb{C}$ and it is denoted by α. Thus, any 2-element a can be uniquely represented in the form

$$a = \alpha(a)\frac{x_{r_1}^2}{3} \cdots \frac{x_{r_l}^2}{3}.$$

Since $\alpha(a)$ does not depend on the way of reducing a to the canonical form, we may derive the following properties of α:

(i) If f is a multilinear polynomial and g is a polynomial such that fg is a 2-element then $\alpha(fg) = \alpha(f\pi(g))$.

(ii) If σ is a permutation on the set $\{1, 2, \ldots, l\}$ and v a 2-element in $x_1, \ldots, x_l$ then $\alpha(v(x_1, \ldots, x_l)) = \alpha(v(x_{\sigma(1)}, \ldots, x_{\sigma(l)}))$.

8.5.3 Bilinear function associated with α-function

We define subsets C and $\widetilde{C}$ in R as the least subsets such that:

(a) $X \subseteq C, \widetilde{C}$;
(b) if $v_1, v_2 \in C$ then $\frac{1}{2}[v_1, v_2] \in C, \widetilde{C}$;
(c) if $v_1 \in \widetilde{C}$, $v_2, v_3 \in C$ then $\frac{1}{2}v_1(v_2 \circ v_3) \in \widetilde{C}$.

Let $S, \widetilde{S}$ denote the additive subgroups in R generated by $C, \widetilde{C}$, respectively.

Lemma 8.41. *As an abelian group S is generated by the elements of the form*

$$x_n \frac{1}{2}(x_{r_1} \circ x_{s_1}) \cdots \frac{1}{2}(x_{r_l} \circ x_{s_l}), \tag{8.14}$$

$$\frac{1}{2}[x_n, x_m] \frac{1}{2}(x_{r_1} \circ x_{s_1}) \cdots \frac{1}{2}(x_{r_l} \circ x_{s_l}). \tag{8.15}$$

For any multihomogeneous $w \in \widetilde{S}$, we have $\pi(w) \in \widetilde{S}$.

Proof. We will apply an argument similar to that used in Section 5.6.1. Thus, we denote by M the subgroup generated by elements (8.14) and (8.15). Let $w \in C$. Applying induction over the degree in X, we see that $w = \frac{1}{2}[v_1, v_2]$, $v_1, v_2 \in M$. Since $x_r \circ x_s$ is in the center of R, it is sufficient to verify that the following elements are in M:

$$\frac{1}{2}\left[x_r, \frac{1}{2}[x_s, x_t]\right], \quad \frac{1}{2}\left[\frac{1}{2}[x_r, x_s], \frac{1}{2}[x_t, x_u]\right].$$

This is a simple matter if we use the weak identity (ii) in Section 8.5.1. Now we pass to $w = v_1\frac{1}{2}(v_2 \circ v_3) \in \widetilde{C}$, $v_1, v_2, v_3 \in M$. By the same reason as previously, it is sufficient to consider the case where v_1, v_2, and v_3 are commutators of degrees 1 or 2 in C. Here we will have to apply relations (ii), (iii), and (iv) from Section 8.5.1 and, in the case where v_2 and v_3 have degree 2, to apply the following consequence of relations (iv) and (ii):

$$[x_r, x_s] \circ [x_t, x_u] = x_r \circ [x_s, x_t, x_u] = 2(x_r \circ x_u)(x_s \circ x_t) - 2(x_r \circ x_t)(x_s \circ x_u).$$

For the proof of the second part, it is sufficient to restrict to the case where w has one of the forms (8.14) or (8.15). We use induction on the number l of round brackets in the expression for w. If $l = 0$ then $\pi(w) = w$. If $l > 0$ then all is clear in the case where w is multilinear since then also $\pi(w) = w$. If x_r occurs in w twice then there is a choice of three options:
(1) $n = m = r$. Then $w = \pi(w) = 0$.
(2) $w = w'\frac{1}{2}(x_r \circ x_r)$. Then $\pi(w) = \pi(3w'\frac{x_r^2}{3}) = 3\pi(w') \in S$ by induction.
(3) $w = w'(x_r)\frac{1}{2}(x_r \circ x_s)$, $r \neq s$. Applying (v) and the centrality of $x_r \circ x_s$, we get $w = w'(x_s)\frac{x_r^2}{3}$ and $\pi(w) = \pi(w'(x_s)) \in \widetilde{S}$ by induction. □

Now we consider a free Lie ring $L = L(y_1, y_2, \ldots)$. Let L_λ be the space of multilinear polynomials of multidegree λ. We define M_λ as a free abelian group with free generators of the form

$$y_m(y_{r_1}y_{s_1}) \ldots (y_{r_t}y_{s_t}),$$

where $\operatorname{supp}\lambda = \{y_m, y_{r_1}, y_{s_1}, \dots, y_{r_t}, y_{s_t}\}$ in the case where $|\lambda|$ is odd, and of the form

$$y_m y_n (y_{r_1} y_{s_1}) \dots (y_{r_t} y_{s_t}),$$

where $\operatorname{supp}\lambda = \{y_m, y_n, y_{r_1}, y_{s_1}, \dots, y_{r_t}, y_{s_t}\}$ if $|\lambda|$ is even. We note that the formal symbols used above as free generators do not mean products or commutators.

We consider a map $\varphi : y_r \mapsto x_r$, $r = 1, 2, \dots$ This map extends to a homomorphism $\bar{\varphi}$ of L onto S (which is considered as a Lie ring with respect to the operations $\frac{1}{2}[x, y]$), as well as to a homomorphism $\tilde{\varphi}$ of abelian group M_λ into $\widetilde{S}$, if one puts

$$\tilde{\varphi}(y_m(y_{r_1}y_{s_1})\dots(y_{r_t}y_{s_t})) = x_m \frac{1}{2}(x_{r_1} \circ x_{s_1}) \cdots \frac{1}{2}(x_{r_t} \circ x_{s_t}),$$

$$\tilde{\varphi}([y_m y_n](y_{r_1}y_{s_1})\dots(y_{r_t}y_{s_t})) = \frac{1}{2}[x_m, x_n] \frac{1}{2}(x_{r_1} \circ x_{s_1}) \cdots \frac{1}{2}(x_{r_t} \circ x_{s_t}).$$

It should be noted that, by Lemma 8.41, $\tilde{\varphi}$ is surjective. Now we introduce a bilinear map $\langle \cdot, \cdot \rangle : L_\lambda \times M_\lambda \to \mathbb{C}$ given by

$$\langle f, g \rangle = \frac{1}{6}\alpha(\bar{\varphi}(f)\tilde{\varphi}(g)).$$

Proposition 8.42. *If $|\lambda| > 1$ then the above bilinear map takes only integral values.*

Proof. We use induction on $|\lambda|$. If $|\lambda| = 2$ then we have $f = [y_r, y_s]$, $g = [y_r y_s]$. Using (ii), (v), and (vii) from Section 8.5.1, we may write

$$\begin{aligned} \frac{1}{6}\alpha\left(\frac{1}{2}[x_r, x_s]\frac{1}{2}[x_r, x_s]\right) &= \frac{1}{6}\alpha\left(\frac{1}{4}x_r[x_s, x_r, x_s]\right) \\ &= \frac{1}{24}\alpha(2x_r(x_s \circ x_r)x_s - 2x_r(x_s \circ x_s)x_r) \\ &= \frac{1}{24}\alpha\left(2\frac{2}{3}x_r^2x_s^2 - 4x_r^2x_s^2\right) = -1. \end{aligned}$$

Now suppose $|\lambda| > 2$. By additivity, it is sufficient to restrict to the case where f is a right-normed commutator and g is a generator in M. Then $f = [y_r, f_1]$ for some y_r such that $\lambda(y_r) = 1$. Applying properties of the function α and (vii) from Section 8.5.1, we get

$$\frac{1}{6}\alpha(\bar{\varphi}(f)\tilde{\varphi}(g)) = -\frac{1}{6}\alpha\left(\bar{\varphi}(f_1)\frac{1}{2}[x_r, \tilde{\varphi}(y)]\right) = -\frac{1}{6}\alpha\left(\bar{\varphi}(f_1)\pi\left(\frac{1}{2}[x_r, \tilde{\varphi}(g)]\right)\right).$$

By Lemma 8.41, $\pi(\frac{1}{2}[x_r, \tilde{\varphi}(g)]) \in \widetilde{S}$. Since $\tilde{\varphi}$ is surjective, we see that for some $g_1 \in M_\lambda$ we have $\langle f, g \rangle = \langle f_1, g_1 \rangle$ where f_1, g_1 depend on a smaller number of variables. By induction, $\langle f_1, g_1 \rangle$ is an integer. □

We remark that, in fact, a stronger result has been obtained.

Corollary 8.43. *If $f \in S$ is linear in $x_{r_1}, \dots, x_{r_l}$ and is quadratic in some other set of variables and $g \in S$ is linear in $x_{r_1}, \dots, x_{r_l}$ and does not depend on other variables then $\frac{1}{6}\alpha(fg)$ is an integer.*

8.5.4 Variety associated with an α-function

We fix further a prime number p and a free Lie algebra $\mathscr{L}$ over $\mathbb{Z}_p$. We then have $\mathscr{L} \cong L/pL$. If $f_1 - f_2 = pf_3$ then for any $g \in M_\lambda$ we will have $\langle f_1, g\rangle - \langle f_2, g\rangle = p\langle f_3, g\rangle$. Hence it is correct to define $\langle f, g\rangle, f \in \mathscr{L}_\lambda$, as a residue modulo $\mathbb{Z}_p$. Let Ann M denote the set of polynomials f in $\mathscr{L}_\lambda$ such that $\langle f, g\rangle = 0$ for any $g \in M_\lambda$.

Lemma 8.44. *Let $|\lambda| > 1$, f a multilinear polynomial in $\mathscr{L}_\lambda$ and $f(\mathscr{L})$ a verbal ideal in $\mathscr{L}$ corresponding to the identity $f \equiv 0$. If $f \in \operatorname{Ann} M_\lambda$ then, for any μ such that $\mathscr{L}_\mu$ is multilinear, we have*

$$f(\mathscr{L}) \cap \mathscr{L}_\mu \subseteq \operatorname{Ann} M_\mu.$$

Proof. It is enough to consider the two cases: (1) $|\mu| = |\lambda|$; (2) $|\mu| = |\lambda| + 1$. In the case (1), one may use property (ii) of α (see Section 8.5.2). In the case (2), it is sufficient to consider a sequence of the form $f' \equiv 0$ where $f' = [f, y_{l+1}]$ or $f' = f([y_1, y_{l+1}], y_2, \dots, y_l)$. The first of these options reduces to the second one using the Jacobi identity (cf. Section 4.2.3). In the remaining case we have

$$\begin{aligned}
&\langle f([y_1, y_{l+1}], y_2, \dots, y_1), g(y_1, y_2, \dots, y_{l+1})\rangle \\
&= \frac{1}{6}\alpha\Big(\bar{\varphi}(f)\Big(\frac{1}{2}[x_1, x_{l+1}], x_2, \dots, x_l\Big)\tilde{\varphi}(g)(x_1, x_2, \dots, x_{l+1})\Big) \\
&= \frac{1}{6}\alpha\Big(\bar{\varphi}(f)\tilde{\varphi}(g)\Big(\frac{1}{2}[x_i, x_{l+1}], x_2, \dots, x_{l+1}\Big)\Big) \\
&= \Big\langle f, \tilde{\varphi}^{-1}\pi(\tilde{\varphi}(g))\Big(\frac{1}{2}[x_{+1}, x_1], x_2, \dots, x_{l+1}\Big)\Big\rangle = 0.
\end{aligned}$$

We use here (vi) from Section 8.5.1, property (i) of α (Section 8.5.2), and the surjectivity of $\tilde{\varphi}$. □

Now we may define a variety $\mathbf{R}_p$ of Lie algebras over a field $\mathbb{Z}_p$ by setting $\mathbf{R}_p(\mathscr{L}) \cap \mathscr{L}_\lambda = \operatorname{Ann} M_\lambda$ for any λ such that $|\lambda| > 1$ and $\mathscr{L}_\lambda$ multilinear. This definition is correct by the just proved Lemma 8.44 and it completely defines $\mathbf{R}_p$.

Theorem 8.45. *Lie algebra variety $\mathbf{R}_p$ is unsolvable.*

Proof. Let $\delta_n(x_1, \dots, x_{2^n})$ denote the left-hand side of the identity determining the solvability of step precisely n. By induction over n, we are going to show that $\delta_n \notin \mathbf{R}_p(\mathscr{L})$. If $n = 1$ then, by Proposition 8.42, we have

$$\langle [y_1, y_2], [y_1 y_2]\rangle = \frac{1}{6}\alpha\Big(\frac{1}{2}[x_1, x_2]\frac{1}{2}[x_1, x_2]\Big) = -1.$$

If $n > 1$ then, applying (v) and (ii) from Section 8.5.1, we will get

$$\begin{aligned}\frac{1}{2}\left[\frac{1}{2}[x_r, x_s], \frac{1}{2}[x_t, x_u]\right]\frac{1}{2}(x_s \circ x_t) &= \frac{x_t^2}{3}\frac{1}{2}\left[\frac{1}{2}[x_r, x_s], \frac{1}{2}[x_s, x_u]\right] \\ &= -\frac{x_t^2}{3}\frac{1}{4}([x_r, x_s] \circ x_u)x_s \\ &= \frac{x_t^2}{3}\frac{1}{4}(x_s \circ [x_r, x_u])x_s \\ &= \frac{x_t^2}{3} \circ \frac{x_s^2}{3}\frac{1}{2}[x_r, x_u]. \end{aligned} \tag{8.16}$$

Now suppose $\operatorname{supp}\lambda = \{x_1, x_4, \ldots, x_{2^n-3}, x_{2^n-2}\}$. We represent δ_n in the form

$$\delta_n = \delta_{n-2}([[y_1, y_2], [y_3, y_4]], \ldots, [[y_{2^n-3}, y_{2^n-2}], [y_{2^n-1}, y_{2^n}]]).$$

By induction, there exists a $g_1 \in M_\lambda$ such that

$$\langle \delta_{n-2}([y_1, y_4], \ldots, [y_{2^n-3}, y_{2^n}]), g_1\rangle \neq 0.$$

We get $g = g_1(y_2y_3)\ldots(y_{2^n-2}y_{2^n-1})$. Applying (8.16), we get

$$\begin{aligned}\langle \delta_n, g\rangle &= \frac{1}{6}\alpha\Big(\pi\Big(\bar{\varphi}(\delta_n)\frac{1}{2}(x_2 \circ x_3)\cdots\frac{1}{2}(x_{2^n-2} \circ x_{2^n-1})\Big)\bar{\varphi}(g_1)\Big) \\ &= \langle f_1, g_1\rangle \neq 0. \end{aligned}$$

□

8.5.5 Some identities in $\mathbf{R}_2$

We are going to introduce one class of identities in $\mathbf{R}_2$ closely connected with Engel's identity. Let $\Delta_{m,l}(z_1, \ldots, z_l)$ denote a polynomial of the form

$$\sum_{a_1,\ldots,a_m} a_1 \cdots a_m \tag{8.17}$$

where each nonempty word a_i has the form $z_{i_1}z_{i_2}\cdots z_{i_t}$ ($i_1 > i_2 > \cdots > i_t$), the word $a_1 \cdots a_m$ being multilinear of degree l. An important property of $\Delta_{m,l}$ is the following:

$$\begin{aligned}&\Delta_{m,l}(z_1, \ldots, z_i, z_{i+1}, \ldots, z_l) - \Delta_{m,l}(z_1, \ldots, z_{i+1}, z_i, \ldots, z_l) \\ &= \Delta_{m,l-1}(z_1, \ldots, [z_i, z_{i+1}], \ldots, z_l). \end{aligned} \tag{8.18}$$

Indeed, we may split equation (8.17) into two parts: f_1 is the sum of all those products in which z_i and z_{i+1} are in different factors $a_1, \ldots, a_m$ and f_2 is the sum of all remaining products. Now f_1 is symmetric with respect to z_i, z_{i+1}; so the summand from f_1 will be canceled on the left-hand side of equation (8.18). Now if a_s depends on both z_i, z_{i+1}

and a'_s results from a_s by a transposition of these letters then the set of products on the left-hand side of equation (8.18) which remain uncanceled splits into the pairs of the form

$$a_1 \cdots \underbrace{z_{i_1} \cdots z_i z_{i+1} \cdots z_{i_t}}_{a_s} \cdots a_m - a_1 \cdots \underbrace{z_{i_1} \cdots z_{i+1} z_i \cdots z_{i_t}}_{a'_s} \cdots a_m$$
$$= a_1 \cdots \underbrace{z_{i_1} \cdots [z_i, z_{i+1}] \cdots z_{i_t}}_{a''_s} \cdots a_m,$$

where a''_s is an ordered word in new variables $z_1, \dots, [z_i, z_{i+1}], \dots, z_l$, the set of variables being ordered as follows:

$$z_1 < \cdots < z_{i-1} < [z_i, z_{i+1}] < z_{i+2} < \cdots < z_l.$$

Another property of $\Delta_{m,l}$ is the following:

$$\Delta_{m,l}(t, t, z_3, \dots, z_l) = \Delta_{m,l-1}(t^2, z_3, \dots, z_l) + 2h_m(t, t, z_3, \dots, z_l), \tag{8.19}$$

where $h_m(z_1, \dots, z_l)$ is an integral multilinear polynomial and t an element of the free generating set. The proof is similar to that just above. Now we set

$$q_{m,l}(y_1, y_2, \dots, y_{l+1}) = \Delta_{m,l}(\operatorname{ad} y_1, \dots, \operatorname{ad} y_l)(y_{l+1}).$$

Lemma 8.46. *For all $m > 2$, $l \geq m$, $q_{m,l}(y_1, y_2, \dots, y_{l+1}) \equiv 0$ is an identity of $\mathbf{R}_2$.*

Proof. We employ induction on l. If $l = m$ then we want to verify that $q_{m,m}(y_1, y_2, \dots, y_{m+1}) \equiv 0$, $m > 2$. It is immediate from (8.17) that any such identity is a consequence of $q_{3,3}(y_1, y_2, y_3, y_4) \equiv 0$ (all these identities are linearizations of Engel's identities of the same degree). Writing $q_{3,3} \equiv 0$ in the form

$$[[x_1, x_2], [x_3, x_4]] + [[x_1, x_3], [x_2, x_4]] + [[x_2, x_3], [x_1, x_4]] \equiv 0$$

shows that, over a field of characteristic 2, it is symmetric over all four variables. Thus, to prove the validity of this identity in $\mathbf{R}_2$, it is sufficient to show that $\langle q_{3,3}, [y_1 y_4](y_2 y_3)\rangle$ is an even number. Applying relations from Section 8.5.1, we will write

$$\begin{aligned}
&\langle q_{3,3}, [y_1 y_4](y_2 y_3)\rangle \\
&= \frac{1}{6}\alpha\left(2\left\{\frac{1}{8}[x_1, x_2, x_2, x_4] + [x_2, x_1, x_2, x_4] + [x_2, x_2, x_1, x_4]\right\}\frac{1}{2}[x_1, x_4]\right) \\
&= \frac{1}{6}\alpha\left(2[x_1, x_4]^2 + \frac{1}{8}(2(x_2 \circ x_1)[x_2, x_4] - 2(x_2 \circ [x_2, x_4])x_1)[x_1, x_4]\right) \\
&= \frac{1}{6}\alpha\left(2[x_1, x_4]^2 + \frac{1}{2}[x_1, x_4]^2\right) = -10.
\end{aligned}$$

To proceed by induction on l, we will use relation (8.18). It follows from this relation that for any permutation σ on $\{1, 2, \dots, l\}$ the identity

$$q_{m,l}(y_1, \dots, y_l, y_{l+1}) - q_{m,l}(y_{\sigma(1)}, \dots, y_{\sigma(l)}, y_{l+1}) \equiv 0$$

is a consequence of the system of identities of the form $q_{m,t}(y_1, y_2, \dots, y_t) = 0$, $t \le l$. By property (ii) of α (Section 8.5.2), the left-hand side of this identity is in $\operatorname{Ann} M_\lambda$, where $\operatorname{supp}\lambda = \{y_1, \dots, y_{l+1}\}$. Since in our case $l > 2$, the above argument shows that it is enough to show that the following number is even:

$$\langle q_{m,l}(y_1, y_2, \dots, y_{l+1}), g_1(y_1 \circ y_2)\rangle,$$

where g_1 does not depend on y_1, y_2. Applying (v) from Section 8.5.1 and equation (8.19), we get

$$\begin{aligned}\langle q_{m,l}, g_1(y_1 \circ y_2)\rangle &= \frac{1}{6}\alpha\Big(q_{m,l}\Big(\frac{1}{2}x_1, \frac{1}{2}x_1, \dots, \frac{1}{2}x_l, x_{l+1}\Big)\tilde{\varphi}(g_1)\Big)\\ &= \frac{1}{6}\alpha\Big(\Delta_m\Big(\frac{1}{4}(\operatorname{ad} x_1)^2, \dots, \frac{1}{2}(\operatorname{ad} x_l)\Big)(x_{l+1})\tilde{\varphi}(g_1)\Big)\\ &\quad + 2\cdot\frac{1}{6}\alpha\Big(h_m\Big(\frac{1}{2}\operatorname{ad} x_1, \frac{1}{2}\operatorname{ad} x_1, \dots, \frac{1}{2}\operatorname{ad} x_l\Big)(x_{l+1})\tilde{\varphi}(g_1)\Big).\end{aligned}$$

Since any value of $\frac{1}{6}(\tilde{\varphi}(u)\tilde{\varphi}(v))$ is integral (see the corollary of Proposition 8.42) the second summand above is even. Now each right-normed commutator of the first summand has the form $[\frac{1}{2}x_1, \frac{1}{2}x_1, v] = \frac{2}{3}x_1^2 v$ (see (v) from Section 8.5.1). Then for some multilinear polynomial d we have

$$\langle q_{m,l}(y_1, \dots, y_{l+1}), g_1(y_1 \circ y_2)\rangle = 2\langle d, g_1\rangle.$$

By Section 8.5.3, $\langle d, g_1\rangle$ is integral. □

Now we are in a position to pass to the principal result of this section. To make it clear, we remark that, due to a classical result by Sanov [323] (the reader may attempt to prove it using some hints in Section 8.7), any finitely generated group of exponent 4, that is, satisfying $x^4 \equiv 1$, is nilpotent, hence finite.

Theorem 8.47. *There exists an unsolvable group of exponent 4.*

Proof. We consider a (varietally) free Lie algebra $L = L(Y, \mathbf{R}_2)$ over $\mathbb{Z}_2$ with the free generating set $Y = \{y_1, y_2, \dots\}$. Since $\mathbf{R}_2$ is determined by multilinear identities, $L = L_1 \oplus L_2 \oplus \cdots \oplus L_n \oplus \cdots$ where L_n is a homogeneous component of some degree n in Y. We consider an associative subalgebra in the adjoint algebra $\operatorname{Ad} L$ generated by all $z_i = \operatorname{ad} y_i$, $i = 1, 2, \dots$ This will be denoted by F. We have $F = F_1 \oplus F_2 \oplus \cdots \oplus F_n \oplus \cdots$ where F_n is the homogeneous component of degree n with respect to $z_1, z_2, \dots$ Let I be an ideal of F generated by all elements of the form $z_i a z_i$, $i = 1, 2, \dots$, $a \in F$. The intersection of this ideal with the space of multilinear (in z_i) elements of F is equal to zero. We consider the quotient algebra $\bar{F} = F/I$ and adjoin a unity 1 to it. The elements $g_i = 1 + z_i$, $i = 1, 2, \dots$ belong to the group $\bar{F}^*$ of invertible elements of $\bar{F}$. Let R denote the group generated by these elements. If $c(g_1, g_2, \dots, g_t)$ is a multilinear group commutator in $g_1, g_2, \dots, g_t$, $\bar{c}(z_1, z_2, \dots, z_t)$ its corresponding Lie algebra commutator, then we have

$$c(g_1, g_2, \dots, g_t) = 1 + \bar{c}(z_1, z_2, \dots, z_t).$$

Thus, if R is solvable then, for some l, we must have $\delta_l(z_1, \dots, z_{2^l}) = 0$, that is, $\operatorname{ad}\delta_l(y_1, \dots, y_{2^l}) = 0$, in particular, then we would have $[\delta_l(y_1, \dots, y_{2^l}), y_{2^l+1}] = 0$ (we recall that I contains no multilinear elements). In this case L would be solvable which is not the case by Theorem 8.45. Thus R is an unsolvable group. Now it remains to show that $x^4 \equiv 1$ holds in R identically. We will compute $g^4 - 1$ for an arbitrary $g \in R$. The base field being $\mathbb{Z}_2$, we always have $g^4 - 1 = (g - 1)^4$ and $(1 + z_i)^{-1} = 1 + z_i$. Hence

$$g^4 - 1 = ((1 + z_{i_1}) \cdots (1 + z_{i_l}) - 1)^4$$

for some $i_1, \dots, i_l$. Using the relations of the form $z_i a z_i = 0$, we will see that the right-hand side is the sum of products of the form

$$a_1 a_2 a_3 a_4 \quad \text{where } a_i = z_{i_{s_1}} \cdots z_{i_{s_t}} \quad (s_1 < \cdots < s_t),\ i = 1, 2, 3, 4,$$

with $a_1 a_2 a_3 a_4$ being multilinear. Fixing the multidegree of such products, we arrive at elements of the form $\Delta_4(z_{i_{p_1}}, \dots, z_{i_{p_m}})$ which are trivial in $\bar{F}$ by Lemma 8.46. Thus $g^4 = 1$ and R is an unsolvable group of exponent 4. □

8.5.6 An application to Engel Lie algebras

One of the very important consequences of the above construction is a counterexample due to Razmyslov, as follows. Let $\mathbf{E}_{p-2,p}$ denote the variety of Lie algebras over a field of characteristic $p > 0$ satisfying an identity of the form $(\operatorname{ad} x)^{p-2} \equiv 0$. As has been remarked, according to a theorem of Kostrikin [177], $\mathbf{E}_{p-2,p}$ is a locally nilpotent variety.

Theorem 8.48. *If $p \geq 5$ then $\mathbf{E}_{p-2,p}$ is a nonnilpotent variety.*

Proof. We restrict ourselves here to a very particular case $p = 5$, the base field being equal to $\mathbb{Z}_5$. In this case it is sufficient to prove that the Engel identity holds in $\mathbf{R}_5$, since by Section 8.5.4 this is an unsolvable variety. Repeating the argument at the beginning of the proof of Lemma 8.46, we note that in the case of $\mathbf{Z}_5$ the identity $q_{3,3}(y_1, y_2, y_3, y_4) \equiv 0$ is symmetric only in y_1, y_2, y_3. Thus, to the verification made in Lemma 8.46, we must add the verification of $\langle q_{3,3}, [y_1 y_2](y_3 y_4)\rangle$ being divisible by

5. As a matter of fact, computing with the use of relations in Section 8.5.1 gives us the following:

$$\begin{aligned}\langle q_{3,3}, [y_1y_2](y_3y_4)\rangle &= \frac{1}{6}\alpha\Big(\frac{1}{8}\{[x_1,x_4,x_2,x_4] + [x_2,x_4,x_1,x_4]\\ &\quad + [x_4,x_1,x_2,x_4] + [x_4,x_2,x_1,x_4]\}\cdot\frac{1}{2}[x_1,x_2]\Big)\\ &= \frac{1}{6}\alpha\Big(\Big\{-\frac{1}{2}[x_1,x_2] - \frac{1}{2}[x_2,x_1] + \frac{1}{8}[[x_4,x_1],[x_2,x_4]]\\ &\quad + \frac{1}{8}[[x_4,x_2],[x_1,x_4]]\Big\}\cdot\frac{1}{2}[x_1,x_2]\Big) = 0.\end{aligned}$$

Thus in either case we have a number divisible by 5. This shows that $\mathbf{R}_5$ is a subvariety in $\mathbf{E}_{p-2,p}$. Hence $\mathbf{E}_{p-2,p}$ is an unsolvable Engel variety of Lie algebras. □

8.6 Lattices of varieties of nilpotent groups

8.6.1 Lattices

In Sections 4.7 and 4.8 we have considered some topics related to lattices. The elements of the lattice theory which are quite sufficient for our purposes can be found in some books, e. g., Birkhoff [74] and Skornyakov [344, 345]. For convenience, we recall that the *lattice* is a nonempty set with two associative and commutative operations $\wedge$ and $\vee$ (intersection and union) satisfying the following laws:

$$x \wedge x = x = x \vee x \quad \text{and} \quad (x \wedge y) \vee y = y = (x \vee y) \wedge y.$$

Any lattice becomes a partially ordered set if one puts $x \le y$ as soon as $x \wedge y = x$ (equivalently, $x \vee y = y$). Conversely, any partially ordered set in which any two-element subset has supremum and infimum becomes a lattice if one puts

$$x \vee y = \sup\{x, y\}, \quad x \wedge y = \inf\{x, y\}.$$

A lattice is called *distributive* if it satisfies either of the two identities

$$(x \vee y) \wedge z = (x \wedge z) \vee (y \wedge z), \quad (x \wedge y) \vee z = (x \vee z) \wedge (y \vee z). \tag{8.20}$$

An example of a distributive lattice is the lattice of all subsets of an arbitrary set with respect to usual operations of intersection and union of subsets. A lattice L is called *modular* (Dedekind) if for any $x, y, z \in L$ such that $x \le z$ we have $(x \vee y) \wedge z = x \vee (y \wedge z)$. The class of modular lattices includes such lattices as the lattice of submodules of a given module, or the lattice of ideals of a given ring. Since the class of modular lattices can be distinguished in the class of all lattices by an identity (for instance, replace z by $z \vee x$ in equation (8.20)) there exist free modular lattices. We will draw (see Birkhoff

[74, Section III.6]) the diagram of the free modular lattice with free generators (it is convenient for the sequel to denote these generators by U, V, N). If $U_1, U_2 \in L$ and $U_1 \leq U_2$ then we say that (U_1, U_2) is an *interval* of L. An interval (U_1, U_2) is said to be *perspective* to (V_1, V_2) if $U_1 = U_2 \wedge V_1$, $U_2 \vee V_1 = V_2$. The least equivalence relation on the set of intervals with respect to which all perspective intervals are equivalent is called *projectivity*.

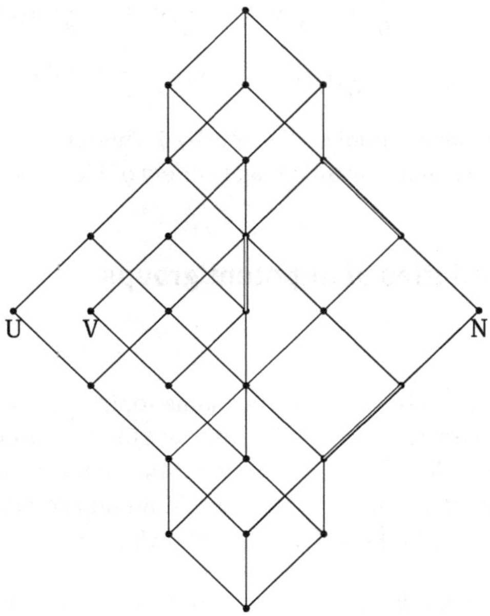

8.6.2 Verbal subgroups and lattices

Let c be a natural number and consider F, a free nilpotent group of nilpotent step c and of rank c, with free generating set X. Any nilpotent variety of groups with nilpotent step at most c can be determined within $\mathbf{N}_c$ by a set of identical relations depending on at most c variables (cf. Section 5.1.2). Hence there exists an antiisomorphism of the lattice $\mathbf{N}_c$ of verbal subgroups in F and the lattice $\mathscr{L}(\mathbf{N}_c)$ of all subvarieties of $\mathbf{N}_c$. We will be interested in the set $\mathscr{L}_0(\mathbf{N}_c)$ of those varieties $\mathbf{V}$ whose free groups of arbitrary rank are torsion-free. It is easy to verify (using Birkhoff's theorem) that this is equivalent to $\mathbf{V}$ being generated by its torsion-free groups. Hence it is obvious that $\mathbf{U}, \mathbf{V} \in \mathscr{L}_0(\mathbf{N}_c)$ implies also $\mathbf{U} \cap \mathbf{V} \in \mathscr{L}_0(\mathbf{N}_c)$. We remark that if V is a verbal subgroup of F determining $\mathbf{V} \in \mathscr{L}_l(\mathbf{N}_c)$ then F/V is torsion-free, that is, $I_F(V) = V$. This enables us to define $\mathbf{U} \vee \mathbf{V}$ as a variety whose normal subgroup is $I_F(UV)$. The lattice of such verbal subgroups of F with operations $U \wedge V = U \cap V$ and $U \vee V = I_F(UV)$ is denoted by $\mathscr{N}_c^0$. This lattice will be our main concern in the last section of this book. It is easy to verify that given three isolated verbal subgroups U, V, W, we always have $I_F(UV) \cap W = I_F(U(V \cap W))$ provided that $U \subset W$. Hence $\mathscr{N}_c^0$ is a modular lattice. Now, according to Theorem 8.11, all groups of the form $F/\gamma_t(F)$, $t = 1, 2, \dots, c$ are torsion-free, i. e., $\gamma_t(F) \in \mathscr{N}_\gamma^0$. We set $N = \gamma_c(F)$.

Proposition 8.49. *Suppose $U, V \in \mathscr{N}_c^0$. Then the sublattice $\mathscr{D}$ generated in $\mathscr{N}_c^0$ by U, V, N is distributive.*

Proof. Suppose not. By a remark in Section 8.6.1, then none of the equations in (8.20) is satisfied if we set $x = U$, $y = V$, $z = N$. Consequently, $U_1 = (U \wedge N) \vee (V \wedge N)$ is a proper subgroup in $U_2 = (U \vee V) \wedge N$ and $V_1 = (U \wedge N) \vee N$ is a proper subgroup in $V_2 = (U \vee N) \wedge (V \vee N)$.

We represent $\mathscr{D}$ as a homomorphic image of the free modular lattice drawn on the diagram under a homomorphism extending the identity map of U, V and N. Then the inverse images of intervals (U_1, U_2) and (V_1, V_2) are shown by double edges on the boundary of the diagram on the right. It is clear that they are projective to the same interval containing the double edge in the middle of the diagram. Thus (U_1, U_2) and (V_1, V_2) are projective also in the original lattice $\mathscr{D}$.

We want to show that in this case U_2/U_1 contains a subgroup of finite index U_2'/U_1 which is isomorphic to a subgroup of finite index in V_2/V_1 which we denote by V_2'/V_1. Indeed, it is sufficient to restrict to the case where the intervals are perspective. Then $V_2 = I_F(U_2V_1)$ and $V_1 = U_1 \cap V_2$. The map $uU_1 \mapsto uV_1$ is an embedding. Since all the subgroups we consider are verbal, this map permutes with all endomorphisms of F; that is, it is an $\operatorname{End} F$-endomorphism.

Now V_2 is a finitely generated nilpotent group and $I_F(U_2V_1) = \sqrt{U_2V_1}$. Hence V_2/U_2V_1 is a finitely generated periodic nilpotent group. By Section 8.3.1, V_2/U_2V_1 is a finite group. One may set $U_2' = U_2$, $V_2' = U_2V_1$. Now we notice that $V_1 \supseteq N$ and $U_2' \subseteq N$.

According to what we have just shown, there exists an $\operatorname{End} F$-isomorphism of a factor of F/N and of N. In particular, V_2'/V_1 is abelian torsion-free. Replacing V_2'/V_1 with factors of the form $V_2' \cap V_1\gamma_{t-1}(F)/V_2' \cap V_1 \cdot \gamma_t(F)$, $t = 1, 2, \dots, c$, we will find that there exists a number t (maximal for which such a factor is nontrivial) and an $\operatorname{End} F$-isomorphism φ from a torsion-free factor of the group $\gamma_{t-1}(F)/\gamma_t(F)$, $1 \le t \le c$, to a torsion-free factor of $\gamma_c(F)$.

Now we consider an endomorphism α of F transforming each generator of F into its square. If w belongs to a factor of $\gamma_{t-1}(F)/\gamma_t(F)$ then, by Hall's formulas (Section 8.2.1), we may write $\alpha(w) = w^{2t-1}$. Since $\varphi(\alpha(w)) = \alpha(\varphi(w))$, we derive $(\varphi(w))^{2^{t-1}} = (\varphi(w))^{2^c}$. Since $t \le c$ and we deal with torsion-free groups, we conclude $2^{t-1} = 2^c$, a contradiction. □

We introduce a sublattice $\mathfrak{l}_t$ in $\mathscr{N}_c^0$ whose elements are those normal subgroups V of F for which we have $\gamma_t(F) \supseteq V \supseteq \gamma_{t+1}(F)$.

Corollary 8.50. *$\mathscr{N}_c^0$ is equal to a subdirect product of lattices $\mathfrak{l}_1, \dots, \mathfrak{l}_c$.*

Proof. We will use the induction on c with trivial basis. Obviously, it is enough to verify that $\mathcal{N}_c^0$ is the subdirect product of $\mathcal{N}_{c-1}^0$ and $\mathfrak{l}_c$. We remark that in a distributive lattice if x, y and z are such that $x \vee z = y \vee z$ and $x \wedge z = y \wedge z$ then $x = y$. Now we may define $\varphi : \mathcal{N}_c^0 \to \mathcal{N}_{c-1}^0 \times \mathfrak{l}_c$ by setting $\varphi(U) = (U \vee N, U \wedge N)$. It is a simple matter to verify that φ preserves lattice operations. By the above property of distributive lattices, φ is injective. Finally, the compositions of φ and projections onto $\mathcal{N}_{c-1}^0$ and $\mathfrak{l}_c$ are surjective. All these properties show that, indeed, $\mathcal{N}_c^0$ is isomorphic to a subdirect product of $\mathcal{N}_{c-1}^0$ and $\mathfrak{l}_c$. □

8.6.3 Description of $\mathcal{N}_c^0$

Let $D_n(q)$ denote the set of all Young diagrams of order n with not more than q rows. We consider the group $G = \mathrm{GL}(\mathbb{Q}^q)$ of all nondegenerate linear operators of a q-dimensional vector space over the field $\mathbb{Q}$ of rational numbers. Let $L_n(\mathbb{Q}^q)$ denote the nth homogeneous component of the free Lie algebra generated by $\mathbb{Q}^q$ in the tensor algebra $T(\mathbb{Q}^q)$ with respect to commutator (see Section 3.4.1). For any $d \in D_n(q)$, we denote by $m(d)$ the multiplicity of the canonical simple G-module $\mathbb{Q}_d^q$ corresponding to d in the decomposition of the semisimple G-module $L_n(\mathbb{Q}^q)$ (see Section 3.3.5). If l is a natural number then $\mathcal{S}(l)$ denotes the lattice which is isomorphic to the lattice of all subspaces in an l-dimensional vector space over $\mathbb{Q}$. Finally, let $\mathcal{L}_n$, $n = 1, 2, \ldots, c$, denote the direct product of lattices of the form $\mathcal{S}(m(d))$, $d \in D_n(c)$. Now we are in a position to prove the main result of this section (we remark that the numbers $m(d)$ can be computed using the formulas of Section 3.4, Proposition 3.22).

Theorem 8.51. *For any $c \geq 1$, the lattice $\mathcal{N}_c^0$ of all nilpotent varieties of groups of nilpotent step at most c whose free groups are torsion-free is isomorphic to a subdirect product of lattices $\mathcal{L}_1, \mathcal{L}_2, \ldots, \mathcal{L}_c$.*

Proof. Let F^* denote the Malcev completion of F (Theorem 8.22). By Section 8.3.5, F^* can be canonically considered as a free Lie algebra over $\mathbb{Q}$ of the same rank as F, i. e., of rank c. Now verbal subgroups of F^* are precisely the verbal ideals of Lie algebra F^*. Those verbal subgroups which are between $\gamma_n(F^*)$ and $\gamma_{n+1}(F^*)$ are the verbal ideals which are between $(F^*)^n$ and $(F^*)^{n+1}$. By multihomogeneity (see Section 4.2), such an ideal has the form $V = V_n + (F^*)^{n+1}$ where V_n is the space of polynomials of degree n in V. The subspace V of such a form is a verbal ideal if and only if V_n is stable under all endomorphisms of F^* induced by linear operators of $\mathbb{Q}X$, X being a free generating set for the group F and for the Lie algebra F^*. Since $G = \mathrm{GL}(\mathbb{Q}X)$ is irreducible in $\mathbb{Q}X$, we may restrict ourselves to nondegenerate operators.

Thus we arrive at a bijection of the lattice of verbal ideals between $(F^*)^n$ and $(F^*)^{n+1}$ and the lattice of G-submodules in $L_n(\mathbb{Q}^c)$. Furthermore, if H is a verbal subgroup of F which is between $\gamma_n(F)$ and $\gamma_{n+1}(F)$ and such that F/H is torsion-free then

$H^* = I_{F^*}(H)$ is a subgroup of F^* between $\gamma_n(F^*)$ and $\gamma_{n+1}(F^*)$. We mention that if $H_1^* = H_2^*$ for such subgroups then $H_1 = H_2$. This is obvious from $H^* = \mathbb{Q}H$.

Now let, conversely, T be a verbal subgroup of F^* form the interval $\gamma_n(F^*) \supset \gamma_{n+1}(F^*)$. Then the subgroup $H = T \cap F$ is isolated in F. If $\varphi \in \operatorname{End} F$ then φ uniquely extends to an endomorphism $\varphi^* \in \operatorname{End} F^*$ such that $\varphi^*|_F = \varphi$. Since $\varphi^*(T) \subseteq T$ and $\varphi^*(F) \subseteq \varphi(F) \subseteq F$, we have $\varphi(H) = \varphi^*(T \cap F) \subseteq T \cap F = H$, that is, H is a verbal subgroup. Hence $H \mapsto H^* \mapsto H_n^*$ is an isomorphism of $\mathfrak{l}_n$ and the lattice of G-submodules in $L_n(\mathbb{Q}^c)$, $1 \le n < c$.

According to Section 3.3.4, $I_n(\mathbb{Q}^c) \subseteq T_n(\mathbb{Q}^c)$ decomposes as the direct sum of simple G-modules of the form $\mathbb{Q}_d^c$ corresponding to Young diagrams $d \in D_n(c)$ with multiplicities $m(d)$. From Section 3.3, Theorem 3.16, it is clear that $\operatorname{End}_{\mathbb{Q}G} \mathbb{Q}^c = \mathbb{Q}$. Let M_d be the sum of all G-submodules isomorphic to $\mathbb{Q}_d^c$, d being fixed. Then M_d is a semisimple R-module where R is the linear span of the image of G in $\operatorname{End} \mathbb{Q}_d^c$ and $l = l_R(M_d) = m(d)$. We have $R \cong \mathbb{Q}_s$, $s = \dim_{\mathbb{Q}} \mathbb{Q}_d^c$. Now R is a Frobenius ring and M_d is a regular R-module. Let $C = \operatorname{End}_R(M_d)$. We apply the double centralizer property, Section 3.3, Lemma 3.11 and Theorem 3.13, changing R and C places. Then we will see that the lattice of R-submodules in M_d is isomorphic to the lattice of left ideals in $C = \mathbb{Q}_l$. But $\mathbb{Q}_l \cong \operatorname{End}_{\mathbb{Q}} \mathbb{Q}^l$ and this lattice is isomorphic to the lattice of subspaces in $\mathbb{Q}^l$. Thus, for d fixed, the lattice of submodules in M_d is isomorphic to $\mathscr{S}(m(d))$. By semisimplicity, we see $\mathfrak{l}_n \cong \mathscr{L}_n$. Applying the corollary of Proposition 8.49, we see that our proof is complete. □

It should be remarked that the description obtained in Theorem 8.51 is incomplete since being a subdirect product of certain lattices does not uniquely determine our lattice. Still some properties may be derived showing that the description is meaningful.

Corollary 8.52. *$\mathscr{N}_c^0$ is distributive if and only if $c \le 5$.*

Proof. As was noted in Section 4.8, for any $d \in D_n(c)$, $c \le 5$, we always have $m(d) \le 1$ and for some d_0, $d_0 \in D_6(6)$, we have $m(d_0) > 1$. Since the lattice of subspaces in a two-dimensional space is nondistributive, our corollary follows. □

As a matter of fact (see Kovács [182]), it is known that $\mathscr{N}_5^0$ consists of 39 elements. Another remark is that a result similar to Section 8.6.3 can be proved concerning the lattices of varieties of nilpotent p-groups of nilpotent step at most c, $c < p$. These results are due to Klyačko [170] and can also be found in Kovács [182].

8.7 Exercises

8.7.1. Let $\widehat{A}(x,y)$ be the completion of free associative algebra $A(x,y)$. Show that

$$e^x y e^{-x} = e^{\operatorname{ad} x}(y).$$

Hint: Write $\operatorname{ad} x = l_x - r_x$.

8.7.2. Give an example of a group G in which the isolator $I_n(G)$ of a term of lower central series with number n is not a verbal subgroup.

8.7.3. Compute the ranks of factors of isolated lower central series in the case of the free metabelian group of rank n.

8.7.4. Let G be a 2-step nilpotent group, x, y its elements. Prove that

$$(xy)^n = x^n y^n (x, y)^{n(n-1)/2}.$$

Derive a similar formula for 3-step nilpotent groups.

8.7.5. Let G, G^* be a torsion-free nilpotent group and its Malcev completion, $\mathscr{L}(G)$ and $\mathscr{L}(G^*)$ Lie algebras of these groups over the field of rational numbers. Then $\mathscr{L}(G) \cong \mathscr{L}(G^*)$.

8.7.6. Let G be a group of exponent 4, D its finite subgroup, c an element such that $c \notin D$, $c^2 \in D$. Show that for any $R \in D$ we always have $cRc = R^{-1}cR^*cR^*$, where $R^* \in D$.

8.7.7. (see 8.7.6) Show that, using relations from the preceding exercise, one can reduce a subword of the form

$$cS_1cR_1cS_2cR_2c$$

to the form

$$cS_1S_2R_1^{-1}R_2^{-1}cXcYcZ.$$

8.7.8. Prove that if $t > d$ then, reducing the word of the form

$$cS_1cR_1c \dots cS_{t-1}cR_tc$$

(see preceding exercises) where $d = |D|$, it is possible to obtain in it a subword of one of the forms

$$cS_iS_{i+1} \dots S_pR_p^{-1} \dots R_{i+1}^{-1}R_i^{-1}c, \quad cR_iR_{i+1} \dots R_pS_p^{-1} \dots S_{i+1}^{-1}S_i^{-1}c$$

and this subword is equal to c^2 (here S_i, R_j are in D).

8.7.9. Derive from Exercise 8.7.8 that the subgroup $B = \mathrm{gp}(c, D)$, c, D as in Exercise 8.7.6 is finite.

8.7.10. Using induction on the number of generators prove that any finitely generated group of exponent 4 is finite, hence solvable (Sanov's theorem). Hint: Use Exercises 8.7.6–8.7.9.

8.7.11. Apply Razmyslov's α-function to construct a nonnilpotent variety with identity $(\mathrm{ad}\, x)^{p-2} \equiv 0$ over a field of characteristic p, $p \geq 7$. For instance, consider the case $p = 7$.

8.7.12. Prove that if $c \to \infty$ then the basic rank of the group variety $\mathbf{A}^3 \cap \mathbf{N}_c$ tends to infinity.

8.8 Comments

Results in Sections 8.1 and 8.2 are classical. The correspondence between nilpotent D-groups and nilpotent Lie algebras over $\mathbb{Q}$ is due to Malcev [216, 217]. See also Lazard's fundamental paper [203]. For more general results about groups see classical M. Hall's [137], for nilpotent groups see [66]. Section 8.4 is based on a paper of Shmelkin [342]. See also Shmelkin's [343]. The correspondence between varieties of groups has been studied by Andreev [7, 8], Eĭdel'kind [113, 114], and Bahturin [29]. Theorem 8.32 is due to Andreev. In [7, 8] there are examples showing that the correspondence $\mathbf{V} \mapsto \mathbf{V}^{\circ}$ is not injective. Proposition 8.33 is due to Kaloujnine and Krasner [157]. Section 8.5 is based on Razmyslov [298, 300, 305]. It is worth noting that in Razmyslov [305] there are examples of nonsolvable groups of exponent p^2, p being an arbitrary prime number. See also [18] for an example of a nonsolvable group of exponent 5. The importance of Lie algebras for the problems of Burnside type was first exhibited by Kostrikin and this relates to an important topic of Lie algebras with Engel identity, see [175, 176, 179, 180]. More results on this topic are given in Chapter 13. Results in Section 8.6 are due to Klyačko [170], see also [171]. Our present text is close to Kovács [182]. For more results about lattices of subvarieties see [15, 365].

9 Codimension growth

The notion of codimension growth was defined in Section 4.8.6. An early survey of the work in the area was written in 1990 by S. P. Mishchenko, see [235]. A survey of much more recent achievements, with some proofs, is presented in A. Giambruno and M. Zaicev's monograph [123, Chapter 12]. In the majority of papers about the codimension growth, the base field $\mathbb{F}$ has characteristic zero. The case of positive characteristic is also interesting, as shown, for example, by of V. M. Petrogradsky's paper [267].

Recall that if $\mathbb{F}$ is a field of characteristic zero and M a Lie algebra over $\mathbb{F}$ then the *codimension sequence* $\{c_n(M)\}$ is defined as follows. Let $L(X)$ be a free Lie algebra over $\mathbb{F}$, generated by a countable set $X = \{x_1, x_2, \ldots\}$, and $\mathrm{Id}(M)$ the verbal ideal of $L(X)$, consisting of all identities of M. Let P_n be the space of all multilinear Lie polynomials in the variables $x_1, \ldots, x_n$ inside $L(X)$. Since $\operatorname{char}\mathbb{F} = 0$, the ideal $\mathrm{Id}(M)$ is uniquely determined by its multilinear components, that is, $\mathrm{Id}(M) \cap P_n$, $n = 1, 2, \ldots$. We denote by $c_n(M)$ the nth *codimension of identities* of M if

$$c_n(M) = \dim(P_n / P_n \cap \mathrm{Id}(M)).$$

In a similar manner, if $\mathbf{V}$ is a variety of Lie algebras over $\mathbb{F}$ then the nth codimension of $\mathbf{V}$ is defined by

$$c_n(\mathbf{V}) = \dim(P_n / P_n \cap V),$$

where V is the space of all identities of $\mathbf{V}$.

The asymptotic behavior of the sequence $\{c_n(M)\}, n = 1, 2, \ldots$, is an important numerical characteristic of M. In 1980 paper [67] I. Benediktovich and A. Zalesskii gave a characterization of the varieties with polynomial growth of codimensions in the language of Young diagrams. In particular, they showed that such varieties have finite bases of identities. In [234], Mishchenko proved the following important result.

Theorem 9.1 (S. P. Mishchenko). *The codimensions $c_n(\mathbf{V})$ have polynomial growth if and only if*

$$\mathbf{N}_2\mathbf{A} \not\subset \mathbf{V} \subset \mathbf{N}_c\mathbf{A},$$

for some c.

Corollary 9.2 ([67, 229]). *Any variety of polynomial growth has finite basis for its identities.*

Proof. This is true because, by Krasil'nikov's theorem (Theorem 5.32), the variety $\mathbf{N}_c\mathbf{A}$ is Specht, that is, every subvariety of $\mathbf{N}_c\mathbf{A}$ has the finite basis property. □

An important consequence of this and other Mishchenko's results is the following.

Corollary 9.3. *If the growth of the codimension sequence of a variety is subexponential then actually this growth is polynomial.*

https://doi.org/10.1515/9783110566659-009

9.1 Exponential growth

Many important classes of Lie algebras M have exponentially bounded growth of the sequence $c_n(M)$. A very general result, holding in many classes of algebras (Lie, Jordan, etc.), is the following.

Theorem 9.4 (Bahturin–Drensky). *If B is an n-dimensional Lie algebra then $c_n^L(B) \leq (\dim B)^{n+1}$.*

Proof. Let $\mathcal{B} = \{b_1, b_2, \ldots, b_m\}$ be a basis of a Lie algebra B. Consider an arbitrary multilinear Lie polynomial in n variables $x_1, \ldots, x_n$. It has the form

$$w(x_1, \ldots, x_n) = \sum_{\sigma \in \mathrm{Sym}(n-1)} \xi_\sigma [x_{\sigma(1)}, x_{\sigma(2)}, \ldots, x_{\sigma(n-1)}, x_n],$$

where the commutators on the right side are left-normed.

The equation $w(x_1, \ldots, x_n) = 0$ is an identity in B if and only if $w(b_{i_1} \ldots, b_{i_n}) = 0$ for any n-tuple $\beta = (b_{i_1} \ldots, b_{i_n}) \in \mathcal{B}^n$.

Now

$$w(b_{i_1}, \ldots, b_{i_n}) = \sum_{\sigma \in \mathrm{Sym}(n-1)} \xi_\sigma [b_{i_{\sigma(1)}}, b_{i_{\sigma(2)}}, \ldots, b_{i_{\sigma(n-1)}}, b_{i_n}].$$

If

$$[b_{i_{\sigma(1)}}, b_{i_{\sigma(2)}}, \ldots, b_{i_{\sigma(n-1)}}, b_{i_n}] = \sum_{i=1}^{m} \lambda^i_{\sigma,\beta} b_i$$

then

$$w(b_{i_1}, \ldots, b_{i_n}) = \sum_{\sigma \in \mathrm{Sym}(n-1)} \xi_\sigma \left(\sum_{i=1}^{n} \lambda^i_{\sigma,\beta} b_i \right) = \sum_{i=1}^{n} \left(\sum_{\sigma \in \mathrm{Sym}(n-1)} \xi_\sigma \lambda^i_{\sigma,\beta} \right) b_i.$$

So $w(x_0, \ldots, x_n) = 0$ is an identity in B if and only if the set of its coefficients $\{\xi_\sigma \mid \sigma \in \mathrm{Sym}(n)\}$ satisfies all linear equations

$$\sum_{\sigma \in \mathrm{Sym}(n-1)} \xi_\sigma \lambda^i_{\sigma,\beta} = 0,$$

for all tuples $\beta \in \mathcal{B}^m$ and all $i = 1, \ldots, m$. The rank of this system of m^{n+1} equations equals $c_n^L(B)$. Thus $c_n^L(B) \leq m^{n+1}$. □

A similar idea can be used to prove the following result by S. P. Mishchenko [233].

Theorem 9.5. *The codimension growth $c_n(W_k)$ of the kth Cartan simple Lie algebra W_k is exponential.*

Proof. Let us choose a basis of the derivation algebra W_k of the polynomial algebra $\mathbb{F}[x_1, \dots, x_k]$ consisting of the differential operators

$$\mathcal{D}_i(s_1, \dots, s_k) = x_1^{s_1} \cdots x_k^{s_k} \frac{\partial}{\partial x_i} \quad \text{where } s_i \text{ is an integer for any } i = 0, 1, \dots, k. \tag{9.1}$$

Let n be a natural number and $\sigma \in \operatorname{Sym}(n)$. The result of evaluation of a left normed monomial $x_{\sigma(1)} \cdots x_{\sigma(k)}$ on the elements

$$\mathcal{D}_{i_1}(s_{11}, \dots, s_{1k}), \dots, \mathcal{D}_{i_n}(s_{n1}, \dots, s_{nk})$$

is a linear combination of the elements of the form (9.1), where each s_i is a polynomial in s_{pq} of degree t_i and $s_1 + \cdots + s_k = n - 1$. It is well-known that the number of such monomials of degree n equals

$$\binom{(n+1)k + n - 1}{(n+1)k - 1} < 2^{(n+1)k+n-1}.$$

Thus $c_n(W_k) \le (4^k)^n$, for all $n > 1$, and the proof is complete. □

Other varieties of exponential growth include affine Kac–Moody Lie algebras [376], special Lie algebras, their generalizing API Lie algebras (see Section 14.4.2 below), and others.

In the case where the growth of $c_n(L)$ is at most exponential, the sequence $\sqrt[n]{c_n(L)}$ is bounded and one can write its upper and lower limits,

$$\overline{\operatorname{EXP}}(L) = \varlimsup_{n\to\infty} \sqrt[n]{c_n(L)}, \quad \underline{\operatorname{EXP}}(L) = \varliminf_{n\to\infty} \sqrt[n]{c_n(L)},$$

which are called the *upper* and *lower exponents*, respectively. If they coincide, there exists the limit of the sequence

$$\operatorname{EXP}(L) = \overline{\operatorname{EXP}}(L) = \underline{\operatorname{EXP}}(L),$$

called the *exponent of the growth of identities L* or simply the *exponent of L*.

For an associative PI-algebra A, its nth codimension, upper, lower, and ordinary exponents are defined in a similar matter. A. Regev proved that $\{c_n(A)\}$ is always exponentially bounded [314, 315]. In [47] the authors study relations between the growth of associative, proper associative and Lie identities of an associative PI-algebra.

Several decades ago, S. Amitsur conjectured that, given a PI-algebra A, its exponent $\operatorname{EXP}(A)$ always exists and is an integer. This conjecture was proven to be true in the papers of A. Giambruno and M. Zaicev (see this and many other interesting results in their monograph [123]).

In the case of Lie algebras, there are many examples where $\operatorname{EXP}(L)$ exists and equals an integer. For instance, if L is an algebra with nilpotent commutator subalgebra, $(L^2)^m = 0$, then $\operatorname{EXP}(L)$ is an integer bounded by m [241].

At the same time, for arbitrary Lie algebras, even with an exponentially bounded growth of codimensions, the answer to Amitsur's question is negative. It was shown by S. P. Mishchenko and M. V. Zaicev (see the proof of their result later in Section 9.32). In [380], it is proven that there exists an infinite-dimensional Lie algebra L, for which

$$3.1 < \underline{\mathrm{EXP}}(L) \leq \overline{\mathrm{EXP}}(L) < 3.9.$$

In [243] the authors mention that actually for this algebra $\mathrm{EXP}(L)$ exists and is a number close to 3.61. In fact, the integrality of the exponent does not hold for simple Lie algebras of Cartan type, as shown by an example of S. S. Mishchenko [244]: $13.1 < \mathrm{EXP}(W_2) < 13.5$. In [243] the following was conjectured:

$$\mathrm{EXP}(W_k) = k(k+1)\left(1 + \frac{1}{k}\right)^k.$$

In the case of finite-dimensional Lie algebras, Amitsur's problem was solved by M. Zaicev in 2001 [379].

Theorem 9.6. *Let L be a finite-dimensional Lie algebra over a field of characteristic zero. Then $\mathrm{EXP}(L)$ exists and is an integer.*

It should be mentioned that the techniques suggested in the proof allow one to explicitly compute $\mathrm{EXP}(L)$, if one knows the structure of L. The proof of Zaicev's theorem will be given later is this chapter, see Theorem 9.31.

The integrality of the exponent of the codimension growth cannot be extended from the finite-dimensional Lie algebras to special Lie algebras. In [125] the authors produce an example of a special Lie algebra, which turns to be the Grassmann envelope of a simple finite-dimensional dimensional Lie superalgebra of type $P(1)$. Recall that simple Lie superalgebras $P(n)$, $n \geq 1$, form an infinite series of "strange" superalgebras in Victor Kac's classification [155]. They consists of the matrices of the form

$$\begin{pmatrix} A & B \\ C & -{}^tA \end{pmatrix} \quad \text{where } A, B, C \in M_n(\mathbb{F}), \operatorname{tr} A = 0,\ {}^tB = B,\ {}^tC = -C.$$

If $L = L_{\bar{0}} \oplus L_{\bar{1}}$ is a Lie superalgebra (see Section 6.9) and $G = G_{\bar{0}} \oplus G_{\bar{1}}$ is an infinite-dimensional Grassmann algebra, then a Lie algebra

$$G(L) = (L_{\bar{0}} \otimes G_{\bar{0}}) \oplus (L_{\bar{1}} \otimes G_{\bar{1}}),$$

with the natural commutator $[x \otimes \xi, y \otimes \eta] = [x, y] \otimes \xi\eta$, $x, y \in L$, $\xi, \eta \in G$, is called the *Grassmann envelope* of L. We call L *special* if it is a subalgebra of an associative superalgebra which is a PI-algebra. Clearly, if L is special, in particular finite-dimensional, then its Grassman envelope $G(L)$ is a special Lie algebra (SPI) (see Section 6.3).

In an earlier paper [124] the authors proved that the exponent of the codimension growth of $P(1)$ is not an integer, actually,

$$6.26 < c_n(P(1)) < 6.47.$$

This contrasts with Theorem 9.6. The same inequalities hold also for the exponent of the codimension growth of $G(F(1))$

Grassman envelopes of Lie superalgebras play an important role in the study of special Lie algebras. One of the reasons is the following result of A. Ya. Vais [352].

Theorem 9.7. *Let $\mathbb{F}$ be a field of characteristic zero and $\mathbf{V}$ a special variety of Lie algebras over $\mathbb{F}$. Then $\mathbf{V}$ can be generated by the Grassman envelope $G(L)$ of a finitely generated Lie superalgebra $L = L_{\bar{0}} \oplus L_{\bar{1}}$.*

OPEN PROBLEM. Is it true that for any variety of Lie algebras $\mathbf{V}$ with exponentially bounded growth we always have

$$\underline{\mathrm{EXP}}(\mathbf{V}) = \overline{\mathrm{EXP}}(\mathbf{V})?$$

9.2 Overexponential growth

In distinction with the case of associative PI-algebras, the sequence of codimensions of a Lie algebra, satisfying a nontrivial identity, has much more involved behavior. As we mentioned earlier in Section 4.8.6, I. B. Volichenko in [364] showed that a Lie L algebra can have overexponential growth of codimensions already if L satisfies the identity

$$[[x_1, x_2, x_3], [x_1, x_2, x_3]] = 0,$$

that is, belongs to the product variety $\mathbf{AN}_2$. Note that in Section 4.8.6 we called varieties with *overexponential* growth *superexponential.*

Still, there is an upper bound for the codimension growth of a Lie algebra satisfying a nontrivial identity. This is given in the following theorem by A. Grishkov [132].

Theorem 9.8 (A. N. Grishkov). *For any positive number a, the growth of a nontrivial variety of Lie algebras is at most $\frac{n!}{a^n}$. In other words, for any $a > 0$ there is a natural number N such that for any $n \geq N$ one has $c_n(V) < \frac{n!}{a^n}$.*

In Section 11.2.1.5, Lemma 11.18, we prove a result about G-graded codimensions, which is a straightforward generalization of Theorem 9.8.

Razmyslov [313] has associated with any nontrivial variety $\mathbf{V}$ of (Lie) algebras a power series

$$C(\mathbf{V}, z) = \sum_{n=1}^{\infty} \frac{c_n(\mathbf{V})}{n!} z^n, \tag{9.2}$$

which he called the *complexity function* of $\mathbf{V}$.

The following is a theorem of Razmyslov [313] that implies the above result of Grishkov:

Theorem 9.9 (Yu. P. Razmyslov). *For any nontrivial variety* $\mathbf{V}$*, the series* $C(\mathbf{V}, z)$ *defines an entire function of complex variable.*

The reader can find the proof of this important result in [313, Chapter 1, § 2].

Petrogradsky found a better general bound for the complexity function; this allowed him to prove a sharper upper bound for the codimension growth. For the definition of multiple exponent and logarithm functions see (10.1).

Theorem 9.10 ([264, 266, 275]). *Let* L *be a Lie algebra satisfying a nontrivial identity of degree* m*. Then*

1. *The following coefficientwise bound for the series holds:*

$$C(\mathbf{V}, z) \prec z \underbrace{\exp(z \exp(\cdots (z \exp(z \exp}_{m-2 \text{ times } \exp}(z))) \cdots)).$$

2. *There exists an infinitesimal* $o(1)$ *such that*

$$c_n(L) \leq \frac{n!}{(\ln^{(m-3)} n)^n} (1 + o(1))^n, \quad n \to \infty, \ m \geq 3.$$

These upper bounds are good because free solvable Lie algebras have similar growth, as we are going to show below.

9.2.1 Explicit formulae for the complexity functions

The results about the explicit form of the complexity function for certain varieties of Lie algebras have been obtained by Petrogradsky, using his exponential Schreier's formula, which will be discussed in detail in Section 10.1.4.

Theorem 9.11 ([269]). *Suppose that* $\mathbf{MV}$ *is the product of varieties of Lie algebras, where* $\mathbf{M}$ *is multihomogeneous. Then*

$$C(\mathbf{MV}, z) = C(\mathbf{V}, z) + C(\mathbf{M}, 1 + (z - 1) \exp(C(\mathbf{V}, z))).$$

This result says that $C(\mathbf{MV}, z)$ behaves "almost" as a composition of three functions $C(\mathbf{M}) \circ \exp \circ C(\mathbf{V})$. Recall that the polynilpotent variety is defined by $\mathbf{V} = \mathbf{N}_{s_q} \cdots \mathbf{N}_{s_1}$ and it depends on the sequence of natural numbers $s_q, \ldots, s_1$.

Theorem 9.12 ([271, 274]). *Consider the polynilpotent variety* $\mathbf{V} = \mathbf{N}_{s_q} \cdots \mathbf{N}_{s_1}$, $q \geq 1$. *Define the following functions:*

$$\beta_1(z) = \sum_{m=1}^{s_1} \frac{z^m}{m};$$

$$\beta_i(z) = \beta_{i-1}(z) + \sum_{m=1}^{s_i} \frac{(1 + (z - 1) \exp(\beta_{i-1}(z)))^m}{m}; \quad 2 \leq i \leq q.$$

Then $C(\mathbf{V}, z) = \beta_q(z)$.

We give some particular cases of this formula.

Corollary 9.13. *Let $d \geq 1$. Then*

$$C(\mathbf{AN}_d, z) = z + \frac{z^2}{2} + \cdots + \frac{z^d}{d} + 1 + (z-1)\exp\left(z + \frac{z^2}{2} + \cdots + \frac{z^d}{d}\right).$$

Corollary 9.14 ([266]). *Let $q \geq 1$. Then*

$$C(\mathbf{A}^q, z) = \beta_q(z), \quad \text{where}$$
$$\beta_1(z) = z, \quad \beta_{i+1}(z) = \beta_i(z) + 1 + (z-1)\exp(\beta_i(z)), \quad i = 1, 2, \ldots, q-1.$$

Corollary 9.15 ([266]). *Let $c \geq 1$. Then*

$$C(\mathbf{N}_c\mathbf{A}, z) = z + \sum_{m=1}^{c} \frac{1}{m}\big(1 + (z-1)\exp(z)\big)^m.$$

9.2.2 Application to codimension functions

Complexity functions are useful for the computation of the codimension growth. For example, using the latter formula, one has the following sharp asymptotics, as compared with the simple polynomiality of the growth of any algebra with nilpotent commutator subalgebra.

Corollary 9.16 ([266]). $c_n(\mathbf{N}_c\mathbf{A}) \approx c^{n-c-1}n^c$.

Let $f(z)$ be an entire function and denote $\mathrm{M}_f(r) = \max_{|z|=r} |f(z)|$. For the complexity functions, one has $\mathrm{M}_f(r) = f(r)$, $r \in \mathbb{R}^+$, since all the coefficients of $f(z)$ are nonnegative.

By the results above, $C(\mathbf{N}_{s_q} \cdots \mathbf{N}_{s_1}, z)$ is essentially the $(q-1)$th iteration of the exponent applied to $\beta_1(z) = \sum_{m=1}^{s_1} \frac{z^m}{m}$, something like $\exp^{(q-1)}(z^{s_1}/s_1)$. By induction, one derives the following result.

Theorem 9.17 ([266]). *Let $\mathbf{V} = \mathbf{N}_{s_q} \cdots \mathbf{N}_{s_1}$, $q \geq 2$, be the polynilpotent variety and $f(z) = C(\mathbf{V}, z)$ its complexity function. Then*

$$\lim_{r\to\infty} \frac{\ln^{(q-1)} \mathrm{M}_f(r)}{r^{s_1}} = \frac{s_2}{s_1}.$$

This theorem is an important step in the proof of the following asymptotics.

Theorem 9.18 ([266]). *Let $\mathbf{V} = \mathbf{N}_{s_q} \cdots \mathbf{N}_{s_1}$, $q \geq 2$, be a polynilpotent variety. Then there exists an infinitesimal $o(1)$ such that*

$$c_n(\mathbf{V}) = \begin{cases} (n!)^{\frac{s_1-1}{s_1}}(s_2 + o(1))^{n/s_1}; & q = 2; \\ \frac{n!}{(\ln^{(q-2)} n)^{n/s_1}}\left(\frac{s_2+o(1)}{s_1}\right)^{n/s_1}; & q = 3, 4, \ldots \end{cases}$$

In the following cases, one has somewhat more precise asymptotics, but they look rather complicated.

Theorem 9.19 ([271]). *Let $d \geq 1$. Then*

$$c_n(\mathbf{AN}_d) \approx \mu(n!)^{1-1/d} \exp\left(\sum_{k=1}^{d-1} \lambda_k n^{1-k/d}\right) n^{\frac{3-d}{2d}}, \quad \text{where}$$

$$\lambda_k = \frac{(\frac{k}{d}+1)\cdots(\frac{k}{d}+k-1)}{k!(d-k)}, \quad k = 1,\dots,d-1;$$

$$\mu = \exp\left(-\frac{1}{d}\sum_{k=2}^{d}\frac{1}{k}\right)(2\pi)^{\frac{1-d}{2d}} d^{-1/2}.$$

It is well known that $c_n(\mathbf{A}^2) = n-1 \approx n$, which coincides with the above asymptotics. The particular cases are as follows:

$$c_n(\mathbf{AN}_2) \approx \sqrt{n!}\,\frac{\exp(\sqrt{n})\sqrt[4]{n}}{\sqrt[4]{8\pi e}};$$

$$c_n(\mathbf{AN}_3) \approx (n!)^{2/3}\,\frac{\exp(\frac{1}{2}n^{2/3}+\frac{5}{6}n^{1/3})}{\sqrt{3}\sqrt[3]{2\pi}e^{5/18}}.$$

Theorem 9.20 ([271]). *Let $c_n = c_n(\mathbf{A}^3)$. Then*

$$c_n = \frac{n!}{(\ln n)^n}\exp\left(\frac{n}{\ln n}\left(2\ln\ln n + 1 + \frac{2(\ln\ln n)^2 - 2\ln\ln n - 1}{\ln n} + o\left(\frac{1}{\ln n}\right)\right)\right).$$

9.3 Colength

The study of the growth of codimensions is done through the representation theory of the symmetric group S_n. Recall Section 4.8 that defines an S_n-action on P_n, where S_n is the symmetric group on n symbols, by setting

$$\sigma \circ f(x_1,\dots,x_n) = f(x_{\sigma(1)},\dots,x_{\sigma(n)}),$$

where $\sigma \in S_n$, $f \in P_n$. Under this action, $P_n \cap \mathrm{Id}(A)$ is an S_n-submodule of P_n and we consider the induced action on the quotient space $P_n(A) = P_n/(P_n \cap \mathrm{Id}(A))$.

Let $\chi_n(A) = \chi(P_n(A))$ be the S_n-character of $P_n(A)$. The character $\chi_n(A)$ is called the nth cocharacter of A. It can be decomposed as the sum of irreducible characters

$$\chi_n(A) = \sum_{\lambda\vdash n} m_\lambda \chi_\lambda,$$

where λ is a partition of n, χ_λ is the associated irreducible S_n-character and the integer m_λ is the corresponding multiplicity. If $d_\lambda = \deg\chi_\lambda$ then

$$c_n(A) = \sum_{\lambda\vdash n} m_\lambda d_\lambda.$$

Now the sequence of colengths is defined as follows. First of all, the nth colength $\ell_n(A)$ of A is defined as the sum of the multiplicities m_λ, as above. Hence

$$\ell_n(A) = \sum_{\lambda \vdash n} m_\lambda.$$

We can also speak about the sequence of colengths for any variety of nonassociative algebras, $\{\ell_n(\mathbf{V})\}$. An easy fact is the following:

$$\ell_n(\mathbf{V}) \geq \frac{c_n(\mathbf{V})}{\sqrt{n!}}.$$

Now it follows easily from Petrogradsky's formula for codimensions (9.2.2) that the growth of the sequence of colengths for $\mathbf{N}_b\mathbf{N}_a$ is overexponential as soon as $a \geq 3$.

The following theorem is proven in [121].

Theorem 9.21 (A. Giambruno–S. P. Mishchenko–M. V. Zaicev). *If A is finite-dimensional algebra over a field of characteristic zero and* $\dim A = d$, *then* $\ell_n(A) \leq d(n+1)^{d^2+d}$.

Note that it is not required in this theorem for A to be a Lie algebra. Finally, in the same paper, the authors find the following:

$$\ell_n(\mathbf{AN}_2) \sim \exp\left(\sqrt{\frac{2}{3}n\pi}\right).$$

Thus the *growth of the sequence of colengths for a variety of Lie algebras can be intermediate between polynomial and exponential.*

9.4 Proof of Zaicev's theorem

In this section we give the proof of M. V. Zaicev's theorem (Theorem 9.6) in the case of a simple Lie algebra L over a field $\mathbb{F}$ of characteristic zero. The proof in many parts uses both Lie and associative polynomials. Therefore, it will be convenient to make a distinction between these two situations by adding superscript ^{L} in the case we deal with Lie algebras. For instance, we denote by P_n the space of multilinear associative polynomials of degree n in the variables $x_1, \dots, x_n$, and by P_n^L the space of multilinear Lie polynomials of degree n in the variables $x_1, \dots, x_n$.

The upper bound for the codimension growth is easy and it holds for all general algebras in which any element of a subalgebra generated by a set X can be written as a linear combination of left-normed product (commutators) of the elements in X. See Theorem 9.4 above.

Here we study the upper and lower exponents of the codimension growth of a Lie algebra B in the case where B is a simple finite-dimensional Lie algebra over a field $\mathbb{F}$, $\operatorname{char}(\mathbb{F}) = 0$. The main result is the following.

Theorem 9.22. *Let* $\mathbb{F}$ *be algebraically closed,* $\mathrm{char}(\mathbb{F}) = 0$. *If a Lie algebra* B *is simple then*

$$\underline{\mathrm{EXP}}(B) = \overline{\mathrm{EXP}}(B) = \mathrm{EXP}(B) = \dim B.$$

To prove that $\mathrm{EXP}(B) = \alpha$, it is enough to find constants $a_1, a_2 > 0$, g_1, and g_2 such that

$$a_2 n^{g_2} \alpha^n \le c_n(B) \le a_1 n^{g_1} \alpha^n$$

for all n. The upper bound was established above (Theorem 9.4). The lower bound follows from the study of the cocharacters $\chi_n^L(B)$.

An important tool for the construction of the lower bounds is a theorem of Razmyslov [313, Theorem 12.1], which implies the existence in $\chi_n^L(B)$ of a certain rectangular diagram of arbitrary length such that the corresponding array of Lie polynomials are not identities of B.

9.4.1 Various codimensions

Here we look more closely at Lie codimensions and cocharacters. Recall that $c_n(B) = \deg \chi_n(B)$, where $\chi_n(B)$ is the S_n-character of the left $\mathbb{F}S_n$-module $P_n^L/(\mathrm{Id}^L(B) \cap P_n^L)$. We write

$$\chi_n(B) = \sum_{\lambda \vdash n} m_\lambda(B)\chi_\lambda.$$

Here λ is a partition of n, χ_λ is the corresponding S_n-irreducible character, whose multiplicity in $\chi_n^L(B)$ equals $m_\lambda^L(B)$.

We know that

$$P_{n+1}^L = P_{n+1}^L(x_1, \dots, x_n, x_0) = \mathrm{Span}_{\mathbb{F}}\{[x_{\sigma(1)}, \dots, x_{\sigma(n)}, x_0] \mid \sigma \in S_n\}.$$

Both P_n and P_{n+1}^L are left $\mathbb{F}S_n$ modules in an obvious way, and the map

$$\varphi : x_{\sigma(1)} \cdots x_{\sigma(n)} \to [x_{\sigma(1)}, \dots, x_{\sigma(n)}, x_0]$$

is an $\mathbb{F}S_n$-module isomorphism $P_n \to P_{n+1}^L$. In particular, $\dim P_{n+1}^L = n!$. Given $f \in P_n$, denote

$$\varphi(f) = f^L = f^L(x_1, \dots, x_n, x_0) \in P_{n+1}^L.$$

Recall the ad map for Lie algebras, $\mathrm{ad} : B \to \mathrm{ad}\, B \subset \mathrm{End}_{\mathbb{F}}(B)$ (see Section 1.3.1. We say that the associative polynomial $f(x_1, \dots, x_n) \in \mathbb{F}\langle X \rangle$ is an identity of $\mathrm{ad}\, B$ if $f(\mathrm{ad}\, b_1, \dots, \mathrm{ad}\, b_n) = 0$ for all $b_1, \dots, b_n \in B$, that is, f is an identity of the pair $(\mathrm{ad}\, B, \mathrm{End}_{\mathbb{F}}(B))$ [313].

An easy lemma is the following.

Lemma 9.23. *Let B be a Lie algebra and let $f(x_1,\dots,x_n) \in P_n$. Then f is an identity of* $\operatorname{ad} B$ *if and only if $f^L(x_1,\dots,x_n,x_0)$ is an identity of L.*

Recall the decomposition $\mathbb{F}S_n = \bigoplus_{\lambda\vdash n} I_\lambda$, where each I_λ is a minimal two-sided ideal, corresponding to partition λ.

Lemma 9.24. *Let B be a Lie algebra. Let $\lambda \vdash n$ with the corresponding two sided ideal $I_\lambda \subset \mathbb{F}S_n$. Let $f(x_1,\dots,x_n) \in I_\lambda$ be such that $f^L(x_1,\dots,x_n,x_0)$ is a nonidentity of B (or equivalently, f is a nonidentity of* $\operatorname{ad} B$*). Then $c^L_{n+1}(B) \ge d_\lambda$ (where $\dim I_\lambda = d_\lambda^2$).*

Proof. The $\mathbb{F}S_n$-left module isomorphism $P_n \to P^L_{n+1}$ restricts to an isomorphism $(\mathbb{F}S_n)\cdot f \to (\mathbb{F}S_n)\cdot f^L$. Since f is a nonidentity, $(\mathbb{F}S_n)\cdot f$ contains a minimal left ideal $0 \ne J_\lambda \subset I_\lambda$, all of whose nonzero elements are nonidentities of $\operatorname{ad} B$. Hence, by Lemma 9.23, $J^L_\lambda \cap \operatorname{Id}^L(B) = 0$, where $J^L_\lambda = \varphi(J_\lambda)$. But $\dim J^L_\lambda = \dim J_\lambda = d_\lambda$, hence $c^L_{n+1}(B) \ge d_\lambda$. □

We shall need the following extension of the previous result:

Lemma 9.25. *Let B be a Lie algebra, $\lambda \vdash n$, $f(x_1,\dots,x_n) \in I_\lambda$, and $0 \le q \in \mathbb{Z}$, such that $f(x_1,\dots,x_n)x_{n+1}\cdots x_{n+q}$ is a nonidentity of* $\operatorname{ad} B$*. Then $c^L_{n+q+1}(B) \ge d_\lambda$.*

Proof. Let $f(x) \in J_\lambda \subset I_\lambda$, J_λ as in the proof of Lemma 9.24. Then

$$f(x_1,\dots,x_n)x_{n+1}\cdots x_{n+q} \in \mathbb{F}S_{n+q}J_\lambda.$$

By the branching rule for $\mathbb{F}S_n$, $\mathbb{F}S_{n+q}J_\lambda$ is a sum of irreducible left ideals $J_\mu \subset \mathbb{F}S_{n+q}$, for some partitions μ such that $\lambda \subset \mu \vdash (n+q)$. By irreducibility, it follows that there exist such μ and a left ideal J_μ all of whose nonzero elements are nonidentities of $\operatorname{ad} B$. As in Lemma 9.24, this implies that $c^L_{n+q+1}(B) \ge d_\mu$. Finally, again by the branching rule, $d_\mu \ge d_\lambda$. □

9.4.2 A theorem of Razmyslov and multialternating polynomials

We begin with an important theorem of Razmyslov. This theorem is essential for establishing the lower bounds for the codimensions.

Theorem 9.26 ([313, Theorem 12.1, p. 73]). *Let B be a semisimple Lie algebra over an algebraically closed field $\mathbb{F}$ of characteristic zero, and let $\dim B = m$. Let B be a Lie subalgebra in an associative algebra U such that B generates U as an associative algebra. Assume that U is simple, with a nonzero center.*

Then there exist a natural number k and a multilinear polynomial

$$f = f(x^1_1,\dots,x^1_m,\dots,x^k_1,\dots,x^k_m),$$

alternating in each group of variables $\{x^i_1,\dots,x^i_m\}$, $1 \le i \le k$, such that
1. *f does not vanish on B, and*
2. *for all $\overline{x}^i_j \in B$, $f(\overline{x}^1_1,\dots,\overline{x}^1_m,\dots,\overline{x}^k_1,\dots,\overline{x}^k_m)$ belongs to the center of U.*

We call the above f "multialternating" and introduce the following definition.

Definition 9.27. Let $mk \le n$. Denote by $Q_{m,k,n}$ a subspace in P_n spanned by all polynomials that are alternating in k disjoint subsets of variables $\{x_1^i, \dots, x_m^i\} \subset \{x_1 \dots, x_n\}$, $1 \le i \le k$. Also set $Q_{m,k} = \bigcup_n Q_{m,k,n}$.

Next we derive some properties of such polynomials.

Let $n = mk$ and $f \in Q_{m,k,mk}$. Clearly, such a polynomial is a linear combination of "monomial" multialternating polynomials $p_M(x)$. Here

$$M = M(x) = M(x_1^1, \dots, x_m^1, \dots, x_1^k, \dots, x_m^k)$$

is a multilinear monomial, and

$$p_M(x) = \sum_{\sigma^{(i)} \in S_m} \operatorname{sgn} \sigma^{(1)} \cdots \operatorname{sgn} \sigma^{(k)} \sigma^{(1)} \cdots \sigma^{(k)} M(x),$$

where $\sigma^{(i)}$ permutes the variables $\{x_1^i, \dots, x_m^i\}$. Now, the right action of a permutation on a monomial is a position rearrangement of the variables. Hence there exists $\rho \in S_{mk}$ such that

$$M(x) = (x_1^1 \cdots x_m^1 \cdot x_1^2 \cdots x_m^2 \cdots x_1^k \cdots x_m^k)\rho.$$

It follows that

$$p_M(x) = (s_m(x_1^1, \dots, x_m^1) \cdots s_m(x_1^k, \dots, x_m^k))\rho,$$

where $s_m(x_1^1, \dots, x_m^1)$ is the standard polynomial. By taking linear combinations, we obtain the following.

Lemma 9.28. *Let $f(x) \in Q_{m,k,mk}$, as above. Then there exists $a \in \mathbb{F}S_{km}$ such that*

$$f(x) = (s_m(x_1^1, \dots, x_m^1) \cdots s_m(x_1^k, \dots, x_m^k))a.$$

Remark. Note that $s_m(x_1, \dots, x_m) \in I_{(1^m)}$. By Young's (or Pieri's) rule, in $\mathbb{F}S_{km}$, one has

$$s_m(x_1^1, \dots, x_m^1) \cdots s_m(x_1^k, \dots, x_m^k) \in \bigoplus_{\lambda \vdash km,\, \lambda_1 \le k} I_\lambda,$$

where I_λ is the two-sided ideal in $\mathbb{F}S_{km}$, corresponding to the partition λ.

As a result, we have proved the following.

Proposition 9.29. *In $\mathbb{F}S_{km}$, one has*

$$Q_{m,k,mk} \subset \bigoplus_{\lambda \vdash km,\, \lambda_1 \le k} I_\lambda.$$

Note that if $\lambda \vdash km$ and $\lambda_1 \le k$ then either $\lambda = (k^m)$, the $k\times m$-rectangle, or $\lambda'_1 \ge m+1$, where λ' is the conjugate partition of λ. Thus the following is true.

Corollary 9.30. *Let A be an associative algebra and let $B \subset A$ be a subspace with $\dim B = m$. Let $p(x) \in Q_{m,k,mk}$ and write $p(x) = g(x) + r(x)$, where $g(x) \in I_{(k^m)} = I_{(k\times m)}$ and $r(x) \in \bigoplus_{\lambda\vdash km, \lambda'_1>m} I_\lambda$. Then $p(b) = g(b)$ for all $(b) = (b_1^1, \dots, b_m^k)$, $b_j^i \in B$.*

Proof. Recall the structure of irreducible $\mathbb{F}S_n$-modules. Given a partition $\lambda = (\lambda_1, \dots, \lambda_t) \vdash n$ of n, one can construct a Young diagram D_λ containing n boxes with λ_1 boxes in the 1st row, λ_2 boxes in the 2nd row, and so on. Filling up the boxes of D_λ with the integers $1, \dots, n$, we get a Young tableau T_λ. To a Young tableau T_λ there corresponds an essential idempotent e_{T_λ} of $\mathbb{F}S_n$, $e_{T_\lambda} = R^+_{T_\lambda} C^-_{T_\lambda}$ where

$$R^+_{T_\lambda} = \sum_{\sigma\in R_{T_\lambda}} \sigma, \quad C^-_{T_\lambda} = \sum_{\tau\in C_{T_\lambda}} (\operatorname{sign}\tau)\tau,$$

and $R_{T_\lambda}, C_{T_\lambda}$ are the subgroups of S_n stabilizing the rows and the columns of T_λ, respectively. It is known that $\mathbb{F}S_n e_{T_\lambda}$ is a minimal left ideal of $\mathbb{F}S_n$ and any irreducible left S_n-module is isomorphic to one $\mathbb{F}S_n e_{T_\lambda}$. Moreover, if M is an $\mathbb{F}S_n$-module and

$$\chi(M) = \sum_{\lambda\vdash n} q_\lambda \chi_\lambda$$

is its decomposition into irreducible components then $q_\lambda \ne 0$ for some $\lambda \vdash n$ (i. e., one of the irreducible summands of M is isomorphic to $\mathbb{F}S_n e_{T_\lambda}$) if and only if $e_{T_\lambda} M \ne 0$.

Now let $\lambda \vdash n = md$ with $\lambda'_1 > m$ and let $f(x) \in J_\lambda$, minimal left $\mathbb{F}S_n$ ideal. Then $f_0(x) = C^-_{T_\lambda} e_{T_\lambda} f(x) \ne 0$ since $e^2_{T_\lambda} = \gamma e_{T_\lambda}$ where γ is a nonzero scalar. Moreover, f_0 is PI-equivalent to f. Note that $f_0(x)$ is alternating on at least $m+1$ variables. Hence $f_0(b) = f(b) = 0$, and so $r(b) = 0$, for any such $(b) = (b_1^1, \dots, b_m^k)$, $b_j^i \in B$. □

Remark. Let $\lambda \vdash n \ge km$ be such that $\lambda_m \ge k$ (i. e. λ contains the $k \times m$ rectangle). By (now) standard arguments it follows that $I_\lambda \subset Q_{m,k,n}$. (See, for example, [316].)

9.4.3 Conclusion

Theorem 9.31. *Let $\mathbb{F} = \overline{\mathbb{F}}$ (i. e., $\mathbb{F}$ is algebraically closed) and let B be a simple Lie algebra, $\dim B = m$. Then there exist constants $C_1, C_2 > 0$, r_1 and r_2 such that*

$$C_2 \cdot n^{r_2} \cdot m^n \le c_n^L(B) \le C_1 \cdot n^{r_1} \cdot m^n$$

for all n. In particular, $\lim_{n\to\infty} \sqrt[n]{c_n^L(B)} = m = \dim B$.

To prove the lower bound, we apply Theorem 9.26. Note that by the simplicity ad B generates the associative (enveloping) algebra $\operatorname{End}(B) = U$, and U has a nontrivial center. By Theorem 9.26 (with ad B replacing B), there exists a multilinear polynomial $f = f(x_1^1, \dots, x_m^1, \dots, x_1^k, \dots, x_m^k)$, alternating in each group of variables $\{x_1^i, \dots, x_m^i\}$, $1 \le i \le k$, which is a central nonidentity of ad B. Thus there exist $b_j^i \in B$ and $a \in B$ such that

$$f(\operatorname{ad} b_1^1, \dots, \operatorname{ad} b_m^1, \dots, \operatorname{ad} b_m^k, \dots, \operatorname{ad} b_m^k)(a) = \alpha a \neq 0, \quad \alpha \in \mathbb{F}.$$

Given n, write $n - 1 = tkm + q$, $0 \le q < km$. Also write

$$f(x_1^1, \dots, x_m^1, \dots, x_1^k, \dots, x_m^k) = f(x_1, \dots, x_{km}).$$

Let

$$g(x_1, \dots, x_{n-1}) = g_{tkm}(x) \cdot x_{tkm+1} \cdots x_{n-1},$$

where

$$g_{tkm}(x) = f(x_1, \dots, x_{km}) f(x_{km+1}, \dots, x_{2km}) \cdots f(x_{(t-1)km+1}, \dots, x_{tkm}).$$

It easily follows that $g(x_1, \dots, x_{n-1})$ is not an identity of ad B. Clearly, $g_{tkm}(x)$ is a polynomial alternating in tk disjoint sets of m variables each. As in Corollary 9.30, we can write $g_{tkm} = g_R + g^*$, where $g_R \in I_{((tk)^m)}$ and $g_{tkm}(x) = g_R(x)$ on ad B. Thus g_R is a nonidentity on ad B. Therefore, by Lemma 9.24, $c_n^L(B) \ge d_R = d_{((tk)^m)}$. Note that, as $n \to \infty$, $n \sim tkm = |R|$, where R is the partition $R = ((tk)^m)$. The proof of Theorem 9.31 now follows, since by [317, F.1.1], as $n \to \infty$, we have $d_R \sim u \cdot n^v \cdot m^n$ for some (explicit) constants u and v.

9.5 Mishchenko–Zaicev's nonintegrality example

As we mentioned earlier, see Corollary 9.3, in the case of varieties of Lie algebras, the intermediate growth [236] of codimension series is impossible, namely there is no variety $\mathbf{V}$ such that the sequence $\{c_n(\mathbf{V})\}, n = 1, 2, \dots$ is asymptotically less than $\{a^n\}, n = 1, 2, \dots$ for any $a > 1$, and asymptotically greater than $\{n^k\}$ for any $k > 0$. So, if $\overline{\mathrm{EXP}}(\mathbf{V}) = 1$, then $\mathbf{V}$ has polynomial growth. In [241], both this property and the integrality of the exponents are proved for the varieties of Lie algebras with nilpotent commutator subalgebra and in [122] the same result was obtained for any variety generated by any finite dimensional Lie algebra. The exponents of the following varieties are also integral: all varieties of almost polynomial growth (see [240]), the varieties generated by the Lie algebra of vector fields on the straight line, and some other.

In the case of associative algebras, it was known that the upper and lower exponents of any nontrivial variety coincide and have integral value (see [318, 69, 123]) for many examples in the associative case. Finally, the integrality of the exponent was established by A. Giambruno and M. V. Zaicev, and the proof of this remarkable result is given in [123, Theorem 6.5.2].

In this section we produce an example of Lie algebras' variety with fractional exponent.

The question of coincidence of the upper and lower exponents of this example has been solved in the positive in [358]; the common value is between 3.610718613 and 3.610718615.

Let $\mathbf{A}^2$ be the variety of all metabelian Lie algebras. We denote by $M = L_3(\mathbf{A}^2)$ the relatively free algebra of this variety with the set of free generators $X = \{x_1, x_2, x_3\}$.

Consider a linear map d acting on the vector space $\langle x_1, x_2, x_3\rangle$ by the rule

$$d(x_1) = x_2, \quad d(x_2) = x_3, \quad d(x_3) = x_1.$$

This map can be extended to a derivation of M, which we also denote by d.

Let us consider a one-dimensional subalgebra $\langle d\rangle$ generated by d in the algebra of all derivations of M. Finally, let us consider the semidirect product $L = M \rtimes \langle d\rangle$ of algebras M and $\langle d\rangle$. Note that M is a metabelian ideal of codimension 1 in L. As a result, L belongs to the variety $\mathbf{A}^3$ of solvable Lie algebras of solvability step at most 3.

Now we can state the main result of this section.

Theorem 9.32. *For the variety $\mathbf{V} = \mathrm{var}(L)$ of Lie algebras over a field $\mathbb{F}$ of zero characteristic, the following strict inequalities hold:*

$$3 < \underline{\mathrm{EXP}}(\mathbf{V}) \le \overline{\mathrm{EXP}}(\mathbf{V}) < 4.$$

In other words, neither the upper nor the lower exponent of $\mathbf{V}$ is an integer. It then follows that even if they coincide, *the exponent of $\mathbf{V}$ is not an integer.*

Proof. The main tool in the proof is the representation theory of the symmetric groups and its applications to the varieties of Lie algebras, as given in Chapters 3 and 4 above.

To start with, we decompose the space $P_n(\mathbf{V})$, with n a natural number, as the sum of irreducible S_n-modules corresponding to certain Young diagrams:

$$P_n(\mathbf{V}) = \bigoplus_{c=1}^{l_n} W_c, \tag{9.3}$$

where $l_n = l_n(\mathbf{V})$ is a number of summands and W_c is an irreducible module over S_n corresponding to the Young diagram d_c, associated with the partition $\lambda = (\lambda_1, \dots, \lambda_k)$ of n. □

Lemma 9.33. *Any component W_c of the sum (9.3) corresponds to the diagram which has less than two boxes out of the first four rows. In addition, the rows of λ satisfy the following inequality:*

$$\lambda_1 - \lambda_3 \ge 2(\lambda_4 - 3).$$

Proof. Consider an n-partition λ such that corresponding Young diagram has more than 2 boxes outside of the first 4 rows.

Now look at the irreducible S_n-module corresponding to λ. This module can be generated by some multilinear Lie polynomial f. Let $\lambda_1', \dots, \lambda_{l(\lambda)}'$ be the heights of the columns of the corresponding Young diagram. According to [239], f is equal to a linear combination of several summands, and every summand contains $l(\lambda)$ skew-symmetric sets with λ_i' variables in the ith set.

So it is sufficient to prove that any multilinear polynomial containing either 6 skew-symmetric variables or two sets with 5 skew-symmetric variables is equal to zero identically. Note that the algebra L contains an abelian ideal M^2 of codimension 4. Hence, in both cases we either substitute one and the same element from a basis of L instead of two skew-symmetric variables or replace two variables by the elements from the ideal M^2. All these values are zero, hence, the first statement of our lemma is proven.

Now consider a partition $\lambda = (\lambda_1, \lambda_2, \lambda_3, \lambda_4, \dots)$ of n with $\lambda_1 + \lambda_2 + \lambda_3 + \lambda_4 \geq n - 2$. As before, let $\lambda_1', \dots, \lambda_{l(\lambda)}'$ be the heights of columns of the corresponding Young diagram, where $l(\lambda) = \lambda_1$.

Suppose that $\lambda_1 - \lambda_3 < 2(\lambda_4 - 3)$. We will show that this diagram determines an identity on L. It is sufficient to prove that any multilinear polynomial f depending on $l = \lambda_1$ skew-symmetric sets of variables of lengths $\lambda_1', \dots, \lambda_l'$ respectively takes only zero values on L.

Let $d, z_1, z_2, \dots$ be a basis of the algebra L. We will substitute some of these elements for the variables in f. First note that we can only substitute d once into any skew-symmetric set of variables, otherwise we get zero. Let us replace all the variables in f, where we substituted d, by the new variable b. All other variables will be marked as $y_1, \dots, y_k$. We may always rewrite the polynomial f as a linear combination of the products

$$(y_{s_1} d^{\alpha_1})(y_{s_2} d^{\alpha_2}) \cdots (y_{s_k} d^{\alpha_k}), \tag{9.4}$$

with $\alpha_1, \dots, \alpha_k \geq 0$. Note that $\alpha_1 + \dots + \alpha_k \leq l = \lambda_1$. The elements $y_i b^{\alpha}$ may be considered as new variables. Element f is not multilinear in these variables but it may be written as a sum $f = f_1 + \dots + f_m$ where each component is a multilinear polynomial on some subset of the set of these new variables.

Let f be skew-symmetric on y_i and y_j. Then those elements from $f_1, \dots, f_m$ that depend on $y_i b, y_j b$ are skew-symmetric in these new variables. For example, if $f = c y_i b b y_j - c y_j b b c y_i$ then only $2cb(y_i b)(y_j b) - 2cb(y_j b)(y_i b)$ depends on $y_i b, y_j b$. Similarly, the skew symmetry in $y_i b^2$ and $y_j b^2$ will be preserved.

Now we will prove that any component of $f_i, i = 1, \dots, m$ takes only zero values. For that purpose, we write, say, f_1, as a linear combination of the elements (9.4), fix indices s_1, s_2, and show that the partial sum f_1^1 of f_1 with only these fixed indexes equals zero. Indeed, let f be alternating in y_i, y_j, y_k and $i, j, k \neq s_1, s_2$. Suppose that f_1^1 depends on $y_ib^{\alpha_i}, y_jb^{\alpha_j}, y_kb^{\alpha_k}$, and $\alpha_i + \alpha_j + \alpha_k \le 2$. Then f_1^1 contains at least one of the skew-symmetric pairs of the new variables, namely, $y_ib^{\alpha_i}, y_jb^{\alpha_j}$, $y_ib^{\alpha_i}y_kb^{\alpha_k}$ or $y_jb^{\alpha_j}y_kb^{\alpha_k}$. By our agreement, the variables y_i, y_j, y_k will be evaluated in the metabelian ideal M. Since $i, j, k \ne s_1, s_2$, the value of f_1^1 is zero.

Similarly, if f is skew-symmetric in y_1, y_2 then $\alpha_1 + \alpha_2 \ge 1$. Recall that the original polynomial f depends on λ_4 skew-symmetric sets of length 4 or 5 and on $\lambda_3 - \lambda_4$ skew-symmetric sets of length 3. The variables y_{s_1} and y_{s_2} appear in one or two of these sets, so we have at least $\lambda_4 - 2$ skew-symmetric sets of length 4 and $\lambda_3 - \lambda_4$ sets of length 3 without y_{s_1}, y_{s_2}. Only one variable from each skew-symmetric set can be replaced by d and hence the element f_1^1, as a polynomial in the new variables y_ib^j, contains at least than $\lambda_4 - 2$ triple and $\lambda_3 - \lambda_4$ double skew-symmetric sets. And, as it was shown above, f_1^1 can take nonzero values only if the following condition holds:

$$\alpha_1 + \cdots + \alpha_k \ge 3(\lambda_4 - 2) + (\lambda_3 - \lambda_4) = 2(\lambda_4 - 3) + \lambda_3.$$

But the condition $\alpha_1 + \cdots + \alpha_k \le \lambda_1$ implies the inequality $\lambda_1 - \lambda_3 \ge 2(\lambda_4 - 3)$. Therefore, if $\lambda_1 - \lambda_3 < 2(\lambda_4 - 3)$ then all of f_1^1, f_1 and f take zero values. □

Lemma 9.34. *The following inequalities hold for the variety* $\mathbf{V}$*:*

$$3 < \underline{\mathrm{EXP}}(\mathbf{V}) \le \overline{\mathrm{EXP}}(\mathbf{V}).$$

Proof. Consider a Lie polynomial

$$\sum_{p_1 \in S_4} \cdots \sum_{p_k \in S_4} (-1)^{p_1} \cdots (-1)^{p_k} x_{01}x_{02}x_{1p_1(1)}(x_{1p_1(2)}x_{1p_1(3)})(x_{1p_1(4)}x_{15}x_{16})$$
$$\cdots x_{kp_k(1)}(x_{kp_k(2)}x_{kp_k(3)})(x_{kp_k(4)}x_{k5}x_{k6}).$$

We denote this element by f_n, where $n = \deg f_n = 6k+2$. One can see from the structure of f_n that it contains k skew-symmetric sets, with 4 variables in each. It follows from the restrictions on the diagram shape set in Lemma 9.33 that this element generates a module which, being decomposed as a sum of irreducible submodules, includes only such nonzero components that correspond to the Young diagrams involving a $4\times(k-2)$ rectangle.

Let us prove that at least one of these irreducible submodules is not zero in $P_n(\mathbf{V})$. For this, we will make the following substitution:

$$x_{01} = x_1, \quad x_{02} = x_2, \quad x_{s1} = x_1, \quad x_{s2} = x_2, \quad x_{s3} = x_3,$$

where $1 \le s \le k$, and all other variables will be replaced by d.

Taking into account the properties of L, we derive that the result of this substitution will be equal to

$$x_1x_2[\bar{x}_1(\bar{x}_2\bar{d})(\bar{x}_3d^2)]^k,$$

where the bar denotes alternation.

First, we consider the summands where d appears in the position of x_1. All these summands are zero since in this case the value of the parenthesis $(\bar{x} *)$ belongs to the metabelian ideal M. We will also get zero if the element d appears in position of x_3. According to the definition of the action of the derivation d, the result is nonzero. For example, some summands will contain the element of the basis $x_1x_2x_3^{3k}$ of M, with coefficient 1.

Because we have substituted $3k$ variables in place of the d, the diagram corresponding to a nonzero module has the first row of length at least $3k$. Besides, this diagram, as it was mentioned above, must contain a $4 \times (k-2)$ rectangle. Thus, our diagram includes a subdiagram $(3m, m, m, m)$, where $m = k - 2$.

The last step is to estimate the dimension of the corresponding irreducible module. Using Hook formula (3.4), we get that the dimension (maybe modulo some factor which is a rational function on n) is not less than

$$\frac{(6m)!}{(m!)^3(3m)!}.$$

In turn, the last number is not less than $(\frac{6}{\sqrt{3}})^n\varphi$, where φ is a rational function of n. As a result, the inequality $3 < \underline{\mathrm{EXP}}(\mathbf{V}) \leq \overline{\mathrm{EXP}}(\mathbf{V})$ holds, and the proof of Lemma 9.34 is complete. □

Lemma 9.35. *The dimension of any irreducible term of the sum* (9.3) *is restricted by the number* $(3.9)^n$ *for sufficiently great* n.

Proof. According to Lemma 9.33, the components of the sum (9.3) correspond to the partitions λ such that $\lambda_1 + \lambda_2 + \lambda_3 + \lambda_4 \geq n - 2$ and $\lambda_1 - \lambda_3 \geq 2\lambda_4 - 6$. Now consider two partitions λ, λ' of n, n', respectively, where $n' = n + k > n$, corresponding to the Young diagrams d, d' and irreducible $S_n, S_{n'}$-modules M, M'. From the Hook formula (3.4), it follows that if d' contains d then

$$(n+k)^{-k} \dim M' \leq \dim M \leq \dim M'.$$

Hence, we can restrict the dimension of any summand in (9.3) by

$$a_n = n^6 \frac{n_0!}{n_1!n_2!n_3!n_4!},$$

where n_1, n_2, n_3, n_4 are lengths of the diagram rows, n_0 is their sum. Besides, the following inequalities take place: $n_1 - n_3 \geq 2n_4$, $n - 2 \leq n_0 \leq n + 6$.

Let $n_4 < n_3 - 1$. It is easy to see that a_n will increase if we replace n_3 by $n_3 - 1$ and n_4 by $n_4 + 1$. Similarly, one can change n_1 and n_2 if $n_2 < n_1 - 1$. So, we can replace $n_1, \dots, n_4$ in a_n by $\bar{n}_1, \dots, \bar{n}_4$ such that
1. $\bar{n}_1 - \bar{n}_3 \geq \bar{n}_4$,
2. $0 \leq \bar{n}_1 - \bar{n}_2 \leq 1$,
3. $0 \leq \bar{n}_3 - \bar{n}_4 \leq 1$.

Now if $\bar{n}_2 < \bar{n}_1$ or $\bar{n}_4 < \bar{n}_3$, we can increase $\bar{n}_2$ or $\bar{n}_4$ by larger values, as before. It means that a_n can be bounded by the following expression:

$$b_n = n^2 \frac{\bar{n}!}{(m!)^2((2m)!)^2},$$

where $\bar{n} = 6m$, $n - 6 \le \bar{n} \le n + 6$.

Using Stirling's formula, we obtain that the exponential part of b_n has the form $(\frac{6}{4^{1/3}})^n$. Since $\frac{6}{4^{1/3}} < 3.9$, the proof is complete. □

Lemma 9.36. *The number of summands l_n in (9.3) is restricted by a polynomial function in n of degree at most 33.*

Proof. By Lemma 9.33, it is sufficient to consider an arbitrary Young diagram, including at most 6 rows and to prove that the number of corresponding isomorphic terms in the sum (9.3) is restricted by a polynomial in n.

First, we estimate the number of partitions of n into at most 6 components. This number does not exceed the number of different monomials of degree n in 6 variables, which is equal to C^5_{n+5}. This is a polynomial of degree 5 in n.

So, let $\lambda = (\lambda_1, \lambda_2, \lambda_3, \lambda_4, \lambda_5, \lambda_6)$ be one of these partitions. Suppose that the multilinear elements $f_1, f_2, \dots, f_s$ generate different isomorphic irreducible submodules that are direct summands in (9.3) and correspond to this λ. Without lost of generality, one may assume that each element f_r, $r = 1, \dots, s$, is symmetric in 6 groups of variables with λ_j, $j = 1, \dots, 6$, variables in the jth group. Applying, if necessary, some permutation $\sigma_r \in S_n$, to each element f_r, one gets an equivalent identity $\sigma_r f_r$ such that all elements $\sigma_1 f_1, \dots, \sigma_s f_s$ are symmetric in the same sets of variables. Hence, we may assume that the variables $x_1, \dots, x_n$ are separated into 6 disjoint sets and each f_r is symmetric in the variables of any set.

We will prove that if s is sufficiently large (this phrase will acquire exact mathematical sense at the end of the proof) then the elements $f_1, f_2, \dots, f_s$ will be linearly dependent, which contradicts to the fact that they are from different direct summands of (9.3).

Let us consider the following elements of the adjoint associative algebra $\operatorname{Ad} L$:

$$a_j = (\operatorname{ad} x_1)^{p_1} (\operatorname{ad} x_2)^{p_2} (\operatorname{ad} x_3)^{p_3} (\operatorname{ad} x_1 d)^{q_1} (\operatorname{ad} x_2 d)^{q_2} (\operatorname{ad} x_3 d)^{q_3} \cdot (\operatorname{ad} x_1 d^2)^{r_1} (\operatorname{ad} x_2 d^2)^{r_2} (\operatorname{ad} x_3 d^2)^{r_3},$$

with

$$p_1 + p_2 + p_3 + 2q_1 + 2q_2 + 2q_3 + 3r_1 + 3r_2 + 3r_3 \le n - 1.$$

It is easy to verify that the number m of such elements is restricted by a polynomial of degree 8 on n.

When evaluating $f_1, \dots, f_s$ in L, we need to consider only 2 types of substitutions. In the first, all variables are changed to the elements x_1, x_2, x_3, d of L. In the second type, some element from M^2 is placed instead of one of the variables. Any substitution of more than 2 elements from M^2 leads to the zero result because M^2 is abelian ideal of L.

The substitution of the first type is completely determined by the number of basis elements x_1, x_2, x_3, d, being placed instead of the variables from each of the symmetric sets. Thus, there exists no more than $Q = \prod_{j=1}^{6} C^3_{\lambda_j+3}$ different substitutions of the first type and no more than nQ different substitutions of the second type. The last number is obtained from the fact that there exists n choices for replacing n variables by an element t from M^2. For our purposes, it is sufficient to note that the total number of different substitutions is restricted by a polynomial of degree 19 in n.

Let us consider the result of the first substitution in greater detail. This is a linear combination of at most $81m$ elements of the form:

$$(x_{j_1} d^{\alpha_1})(x_{j_2} d^{\alpha_2}) a_j,$$

where a_j is an element of the adjoint associative algebra, defined earlier, $1 \le j_1, j_2 \le 3$; $0 \le \alpha_1, \alpha_2 \le 2$; $j = 1, \dots, m$. We will order these elements and denote them as b_j, $j = 1, \dots, 81m$.

Let $t \in M^2$. The result obtained after the second type of substitution and corresponding to a given t is a linear span of at most than m elements of the form ta_j where a_j is an element of adjoint algebra determined above and t is an element from M^2 placed instead of one of the variables.

We will compose a tableaux of scalars that consists of s rows. The rth row corresponding to the element f_r is filled in the following way. Let us enumerate all possible substitutions of the first and second types. The element f_r can be written as a linear combination of $b_1, \dots, b_{81m}$ with coefficients $\beta_1^j, \dots, \beta_{81m}^j$ under the jth substitution of the first type, and as a linear combination of the elements $ta_1, \dots, ta_m$ with coefficients $\mu_1^j, \dots, \mu_m^j$ under the jth substitution of the second type. Let us write coefficients $\beta_1^1, \dots, \beta_{81m}^1$ for the first substitution in the rth row of element, then coefficients $\beta_1^2, \dots, \beta_{81m}^2$ for the next substitution, and so on. Then, we will write $\mu_1^1, \dots, \mu_m^1, \mu_1^2, \dots, \mu_m^2, \dots$ in our row.

The length of any row of the tableaux is restricted by a polynomial in n whose the degree at most 27, as follows: degree 19 due to the number of different substitutions plus 8 from the restriction on the number of coefficients corresponding to one particular substitution.

Therefore, if the number s exceeds the length of the rows thus obtained then the rows will be linearly dependent. Now suppose that s is sufficiently large, i. e., $s \ge n^{28}$. Then there exists a nontrivial linear relation between the rows of our tableaux for some n. We will denote the coefficients of this relation by $\alpha_1, \dots, \alpha_s$.

We consider a multilinear element of the free Lie algebra of the form

$$f = \sum_{j=1}^{s} \alpha_j f_j.$$

It follows from the choice of the coefficients α_j that f will be zero for all values of the variables in L, that is, $f \equiv 0$ is an identity of L.

This means that the number s of different isomorphic $\mathbb{F}S_n$-submodules in (9.3) corresponding to same Young diagram is bounded by a polynomial in n of degree 28.

Thus, we have proved that the number of terms in sum (9.3) is restricted by a certain polynomial of degree at most 33 (5 at the expense of different diagrams and 28 at the expense of the restriction of the number of isomorphic terms for a fixed diagram). □

The proof of the theorem directly follows from these four lemmas.

9.6 Exponents of Lie algebras

In this section we give the results of calculations of the exponents of the codimension growth for some important Lie algebras. First of all, by Zaicev's theorem (Theorem 9.22),

- *For any simple finite-dimensional Lie algebra over a field of characteristic* 0, *we have* $\mathrm{EXP}(L) = \dim L$.

An important example of infinite-dimensional Lie algebras is the Volichenko algebra

$$V = \begin{pmatrix} G & G \\ 0 & 0 \end{pmatrix} \quad \text{where } G \text{ is an infinite-dimensional Grassman algebra,}$$

with respect to the matrix commutator operation. Since V is closed under the matrix multiplication which satisfies $[x_1, x_2, x_3][x_4, x_5, x_6] \equiv 0$, V is an SPI-algebra.

One of its properties is that $\mathbf{V}_1 = \operatorname{var} V$ does not satisfy any standard identity, while every proper subvariety does. Also, $\mathbf{V}_1$ has *almost polynomial growth*, in the sense that its growth is exponential but any of its subvarieties has polynomial growth:

- $\mathrm{EXP}(\mathbf{V}_1) = 2$.

Two more algebras arise when one studies varieties of almost polynomial growth (see Section 6.3.4) First, let $G = \langle x, y, z \mid xy = z, xz = yz = 0 \rangle$, $M = \mathbb{F}[t]$. We give $M = \mathbb{F}[t]$ a G-module structure by setting

$$x \circ f(t) = f'(t),\ y \circ f(t) = tf(t),\ z \circ f(t) = f(t).$$

If $\operatorname{char} \mathbb{F} = 0$, then M is an absolutely irreducible G-module and $B = G \rtimes M$ is a Lie algebra of interest to us.

Now let $G = \langle x, y \mid xy = y \rangle$. Then again we choose $M = \mathbb{F}[t]$, but in this case we set $x \circ f(t) = tf(t)$, $y \circ f(t) = f(t-1)$. Consider $C = G \rtimes M$. The exponents of B and C are as follows:

- $\mathrm{EXP}(B) = 3$,
- $\mathrm{EXP}(C) = 2$.

We denote $\mathbf{V}_0 = \mathrm{var}(\mathfrak{sl}_2(\mathbb{F}))$, $\mathbf{V}_1 = \mathbf{N}_2\mathbf{A}$, $\mathbf{V}_2 = \mathrm{var}(V)$, $\mathbf{V}_3 = \mathrm{var}(B)$, $\mathbf{V}_4 = \mathrm{var}(C)$.

Theorem 9.37 (S. P. Mishchenko [235]). *The varieties $\mathbf{V}_0, \mathbf{V}_1, \mathbf{V}_2, \mathbf{V}_3, \mathbf{V}_4$ have exponential, almost polynomial growth. If V is a solvable variety of almost polynomial growth then, for some i, $1 \le i \le 4$, $\mathbf{V} = \mathbf{V}_i$.*

As we mentioned earlier, one more large class of Lie algebras whose codimension growth is exponential is the class of Lie algebras with nilpotent commutator subalgebra:
- $\mathrm{EXP}(\mathbf{N}_c\mathbf{A}) = c$.

We also mentioned that $\mathbf{AN}_2$ has overexponential growth. Actually, its growth is almost exponential, as shown in the paper by Mishchenko–Regev–Zaicev [242].

Theorem 9.38. *Let $\mathbf{V}$ be a proper subvariety of $\mathbf{AN}_2$ Then $\mathrm{EXP}(\mathbf{V})$ exists an is an integer. If $L_m = L_m(\mathbf{AN}_2)$ is the free algebra of rank m in $\mathbf{AN}_2$ then $\mathrm{EXP}(L_m) = m + 1$ for even m, and $\mathrm{EXP}(L_m) = m$ for m odd.*

10 Subalgebra growth in free Lie algebras

In this chapter we apply the approach of generating functions and Hilbert series to the study of free Lie algebras.

10.1 Schreier's formula, with applications

Suppose that G is a free group of rank n and $H \subset G$ a subgroup of finite index. It is well-known that then H is also a free group and the rank m of H is determined by O. Schreier's formula [325]

$$m - 1 = (n - 1) \cdot |G : H|.$$

By Sirshov–Witt theorem (see Chapter 2), any subalgebra of a free Lie algebra over a field is also free. But a straightforward analogue of Schreier's formula for free Lie algebras is not possible since it is easy to see that any subalgebra of finite codimension has an infinite number of generators.

Still, an appropriate Schreier's formula for free Lie algebras does exist, although in terms of the formal power series. There exists also a version for the exponential generating functions.

We start by surveying these formulas and their applications. Note that, for example, these formulas allow one to write down explicit formulas for the generating functions for free solvable (or more generally, polynilpotent) Lie algebras. These explicit formulas for generating functions can applied to study various asymptotics.

10.1.1 Schreier's formula for free Lie algebras

As usual, by $\mathbb{F}$ we denote the ground field of coefficients. Let X be an at most countable set, with a weight function $\mathrm{wt} : X \to \mathbb{N}$ such that

$$X = \bigcup_{i=1}^{\infty} X_i, \quad X_i = \{x \in X \mid \mathrm{wt}\, x = i\}; \quad |X_i| < \infty, \ i \in \mathbb{N}.$$

We call such sets *finitely graded*. If A is an algebra generated by X then we define the *weight* of a monomial $a \in A$ in X, in a natural way. Suppose that Y is a set of monomials on X, then we define the Hilbert series of Y as

$$\mathcal{H}_X(Y) = \mathcal{H}_X(Y, t) = \sum_{i=1}^{\infty} |Y_i| t^i; \quad Y_i = \{y \in Y \mid \mathrm{wt}\, y = i\}, \ i \in \mathbb{N}.$$

If $V \subset A$ is a subspace then $\mathcal{H}_X(V)$ is defined using a homogeneous basis of $\operatorname{gr} V$, where $\operatorname{gr} V$ is the associated graded space (see more details in Section 10.1.4).

https://doi.org/10.1515/9783110566659-010

Next, we consider an operator $\mathcal{E}$ on the power series $\varphi(t) = \sum_{n=1}^{\infty} b_n t^n$, $b_n \in \{0,1,2,\dots\}$ defined by

$$\mathcal{E} : \varphi(t) = \sum_{n=1}^{\infty} b_n t^n \to \mathcal{E}(\varphi(t)) := \sum_{n=0}^{\infty} a_n t^n = \prod_{n=1}^{\infty} \frac{1}{(1-t^n)^{b_n}}.$$

Let L be a Lie algebra generated by X and $U(L)$ its universal enveloping algebra. Suppose that

$$\mathcal{H}_X(L,t) = \sum_{n=1}^{\infty} b_n t^n, \quad \mathcal{H}_X(U(L),t) = \sum_{n=0}^{\infty} a_n t^n.$$

It is well known that $\mathcal{H}_X(U(L)) = \mathcal{E}(\mathcal{H}_X(L))$. Now we can state Victor Petrogradsky's analogue of Schreier's formula.

Theorem 10.1 ([272]). *Let L be a free Lie algebra generated by a finitely graded set X. Suppose that H is a subalgebra and Y is a set of free generators for H. Then*

$$\mathcal{H}(Y,t) - 1 = (\mathcal{H}(X,t) - 1) \cdot \mathcal{E}(\mathcal{H}(L/H), t).$$

The proof of this theorem will be given later in Section 10.1.4.

Denote by $\mu(*)$ the Möbius function.

Theorem 10.2 ([272]). *Suppose that $L = F(\mathbf{N}_{s_q} \cdots \mathbf{N}_{s_1}, k)$, where $q \geq 1$. We set $\beta_0(z) = 0$, $\alpha_0(z) = kz$, and recursively define functions*

$$\beta_i(z) = \beta_{i-1}(z) + \sum_{m=1}^{s_i} \frac{1}{m} \sum_{a|m} \mu\left(\frac{m}{a}\right) (\alpha_{i-1}(z^{m/a}))^a, \quad 1 \leq i \leq q;$$
$$\alpha_i(z) = 1 + (kz - 1) \cdot \mathcal{E}(\beta_i(z)), \quad 1 \leq i \leq q.$$

Then $\mathcal{H}(L,z) = \beta_q(z)$.

Consider a particular case.

Corollary 10.3 ([272]). *Let $L = F(\mathbf{AN}_d, k)$. Then*

$$\mathcal{H}(L,z) = \psi_k(1)z + \cdots + \psi_k(d)z^d + 1 + \frac{kz-1}{(1-z)^{\psi_k(1)} \cdots (1-z^d)^{\psi_k(d)}},$$

where $\psi_k(m) = \frac{1}{m} \sum_{a|m} \mu(m/a) k^a$.

10.1.2 Growth of finitely generated algebras

Now we discuss the applications of the above result to the growth of the free polynilpotent Lie algebra of finite rank $L = F(\mathbf{N}_{s_q} \cdots \mathbf{N}_{s_1}, k)$. Suppose that L is freely generated by $X = \{x_1, \dots, x_k\}$, all letters being of weight one. Then we have a natural grading $L = \bigoplus_{n=1}^{\infty} L_n$. We define the *growth function* with respect to X as $\gamma_L(X, n) = \sum_{s=1}^{n} \dim_{\mathbb{F}} L_s$.

Define by iteration the functions

$$\begin{aligned} \ln^{(0)} x = x, &\quad \ln^{(s+1)} x = \ln(\ln^{(s)} x), \quad s = 0, 1, 2, \dots; \\ \exp^{(0)} x = x, &\quad \exp^{(s+1)} x = \exp(\exp^{(s)} x), \quad s = 0, 1, 2, \dots \end{aligned} \tag{10.1}$$

We use the standard notation: $f(x) \approx g(x)$, $x \to \infty$ iff $\lim_{x\to\infty} f(x)/g(x) = 1$; $f(x) = o(g(x))$ iff beginning with some number $f(x) = \alpha(x)g(x)$, $\lim_{x\to\infty} \alpha(x) = 0$. Denote by $\zeta(x)$ the zeta function.

Theorem 10.4 ([270]). *Let* $\mathbf{V} = \mathbf{N}_{s_q} \cdots \mathbf{N}_{s_1}$, $q \geq 2$, *be the polynilpotent variety of Lie algebras. Suppose that* $L = F(\mathbf{V}, k)$ *is its free algebra of rank* k, $k \geq 2$, *generated by* $X = \{x_1, \dots, x_k\}$. *Then*

$$\gamma_L(X, n) = \begin{cases} \frac{A+o(1)}{N!} n^N, & q = 2, \\ \exp((C+o(1))\, n^{N/(N+1)}), & q = 3, \\ \exp((B^{1/N}+o(1)) \frac{n}{(\ln^{(q-3)} n)^{1/N}}), & q \geq 4, \end{cases}$$

where the constants are as follows:

$$N = s_2 \dim_K F(\mathbf{N}_{s_1}, k), \quad A = \frac{1}{s_2} \left(\frac{k-1}{\prod_{q=2}^{s_1} q^{\psi_k(q)}} \right)^{s_2},$$

$$B = s_3 A \zeta(N+1), \quad C = (1 + 1/N)(BN)^{\frac{1}{1+N}}.$$

The key step in the proof of this theorem is the study of the behavior of $\mathcal{H}(L, t)$ as $t \to 1{-}0$.

Theorem 10.5 ([270]). *Let* $\mathbf{V} = \mathbf{N}_{s_q} \cdots \mathbf{N}_{s_1}$, $q \geq 2$ *be the polynilpotent variety. Suppose that* $L = F(\mathbf{V}, k)$ *is its free algebra of rank* k. *Then*

$$\begin{aligned} \lim_{t\to 1-0} (1-t)^N \mathcal{H}(L, t) = A, &\quad q = 2, \\ \lim_{t\to 1-0} (1-t)^N \ln^{(q-2)}(\mathcal{H}(L, t)) = s_3\, \zeta(N+1)\, A, &\quad q \geq 3; \end{aligned}$$

where N, A *are the same as in the previous theorem.*

Let us comment on Petrogradsky's approach to proving this result.

The explicit formula of the generating function above yields that $\mathcal{H}(L, t)$ is essentially the $(q-1)$th iteration of the application of the operator $\mathcal{E}$ to $\beta_1(z)$:

$$\beta_1(z) = \psi_k(1)z + \psi_k(2)z^2 + \cdots + \psi_k(s_1)z^{s_1}, \quad \psi_k(m) = \frac{1}{m} \sum_{a|m} \mu(m/a) k^a;$$

$$\mathcal{E}(\beta_1(z)) = \frac{1}{(1-z)^{\psi_k(1)} \cdots (1-z^{s_1})^{\psi_k(s_1)}} \approx \frac{\mu}{(1-t)^M}, \ t \to 1{-}0;$$

$$M = \psi_k(1) + \cdots + \psi_k(s_1) = \dim_K F(\mathbf{N}_{s_1}, k), \quad \mu = \prod_{q=2}^{s_1} q^{-\psi_k(q)}.$$

This observation shows that $\mathcal{E}$ is "almost" an exponent, so $\mathcal{H}(L,t)$ behaves like

$$\exp^{(q-2)}\left(\frac{\lambda + o(1)}{(1-t)^N}\right), \quad t \to 1\text{–}0.$$

Another step in the proof of the main theorem is to use the results connecting the growth of a function analytic in the unit circle with the asymptotics of its coefficients. For the details, we refer the reader to [270].

There exists a version of Schreier's formula for free Lie superalgebras as well [272, 274], and the above asymptotics is extended to the case of the free solvable (polynilpotent) Lie *super*algebras of finite rank [167].

Since the scale for the superexponential codimension growth of varieties of Lie algebras is rather complete (see below), Petrogradsky has conjectured that the scale for the intermediate growth of finitely generated Lie PI-algebras is complete as well.

Conjecture ([265]). *Suppose that L is a finitely generated Lie PI-algebra. Then there exist natural numbers q, N_0 such that*

$$\gamma_L(n) \le \exp\left(\frac{n}{\ln^{(q)} n}\right), \quad n \ge N_0.$$

Such a bound was established for almost solvable Lie algebras [168, 169].

10.1.3 Exponential Schreier's formula

We recall (see (9.2)) the *exponential generating functions*, also called the *complexity functions*. A version of this function is as follows.

Suppose that we are given a set A of monomials Y in the alphabet $X = \{x_i \mid i \in \mathbb{N}\}$. For any set of distinct elements $\widetilde{X} = \{x_{i_1}, \ldots, x_{i_n}\} \subset X$, by $P_n(A, \widetilde{X})$ we denote the set of all multilinear monomials of degree n on $\widetilde{X}$ belonging to A. Suppose that the number of elements $c_n(A, \widetilde{X})$ does not depend on the choice of $\widetilde{X}$, but depends only on n. In this case we write $c_n(A) = c_n(A, \widetilde{X})$ and say that A is X-*uniform*. We define the *complexity function with respect to X* as

$$\mathcal{C}_X(A, z) = \sum_{n=1}^{\infty} \frac{c_n(A)}{n!} z^n, \quad z \in \mathbb{C},$$

where the sum is taken from $n = 0$, $c_0 = 1$, in the case of associative algebras or groupoids with unity. Remark that A need not necessarily consist of multilinear elements. These definitions are naturally extended to algebras (and spaces) with respect to generating sets.

Consider some examples. Suppose that X is a countable set, $A = A(X)$ is the free associative algebra, and $L = L(X)$ is the free Lie algebra. One easily has

$$\mathcal{C}_X(A,z) = \sum_{n=0}^{\infty} z^n = \frac{1}{1-z},$$
$$\mathcal{C}_X(L,z) = \sum_{n=1}^{\infty} \frac{z^n}{n} = -\ln(1-z),$$
$$\mathcal{C}(\mathbf{N}_s,z) = \sum_{n=1}^{s} \frac{z^n}{n}.$$

An exponential version of Schreier's formula for the free Lie algebras now reads as follows.

Theorem 10.6 ([271]). *Let L be a free Lie algebra with a countable generating set X. Suppose that H is an X-uniform subalgebra. Then H has a uniform set of free generators, and for any such set Y one has*

$$\mathcal{C}_X(Y,z) - 1 = (z-1)\cdot \exp(\mathcal{C}_X(L/H,z)).$$

10.1.4 Petrogradsky's proof of Schreier's formula

In what follows, we consider all power series as the elements of the ring of the formal power series $\mathbb{F}[[t]]$, where as usual $\mathbb{F}$ is the ground field of coefficients. Let X be at most countable set with a weight function $\mathrm{wt} : X \to \mathbb{N}$ such that

$$X = \bigcup_{i=1}^{\infty} X_i, \quad X_i = \{x \in X \mid \mathrm{wt}\, x = i\}; \quad |X_i| < \infty,\ i \in \mathbb{N}.$$

We call such a set *finitely graded*. For any monomial $y = x_{i_1} \cdots x_{i_n}$, $x_j \in X$, we set $\mathrm{wt}\, y = \mathrm{wt}\, x_{i_1} + \cdots + \mathrm{wt}\, x_{i_n}$. If Y is a set of monomials in X then we set $Y_i = \{y \in Y \mid \mathrm{wt}\, y = i\}$, $i \in \mathbb{N}$. Also we define the Hilbert–Poincaré series of Y as

$$\mathcal{H}_X(Y,t) = \sum_{i=1}^{\infty} |Y_i| t^i.$$

Let $A = A(X)$ be an algebra generated by a finitely graded set X and assume that A inherits the grading from X. Let $V \subset A$ be a homogeneous subspace, that is $V = \bigoplus_{i=1}^{\infty} V_i$ with $V_i = \{v \in V \mid \mathrm{wt}\, v = i\}$. As usual, we define the Hilbert–Poincaré series of V by setting

$$\mathcal{H}_X(V,t) = \sum_{i=1}^{\infty} (\dim V_i)\, t^i.$$

We sometimes omit t, X and write $\mathcal{H}(Y)$, $\mathcal{H}(V)$.

Now let $Y \subset A$ be a not necessarily homogeneous subset. For $v \in A$, we consider its decomposition as the sum of homogeneous components $v = v_1 + v_2 + \cdots + v_n$, $\operatorname{wt} v_i = i$, $i = 1, \ldots, n$; $v_n \neq 0$; in this case we write $\deg v = n$, $\operatorname{gr} v = v_n$. We define

$$\mathcal{H}_X(Y,t) = \sum_{n=1}^{\infty} |Y_n|\, t^n; \quad \text{where } Y_n = \{y \in Y \mid \deg y = n\},\ n = 1, 2, \ldots$$

Given a subspace $V \subset A$, one has a filtration $0 = V^0 \subset V^1 \subset V^2 \subset \cdots$ with $V^i = \{v \in V \mid \deg v \leq i\}$, $i = 0, 1, 2, \ldots$ Let

$$\operatorname{gr} V = \bigoplus_{n=1}^{\infty} \operatorname{gr}_n V, \quad \operatorname{gr}_n V = \{\operatorname{gr} v \mid v \in V,\ \deg v = n\} \cup \{0\} \cong V^n / V^{n-1},$$

denote the associated graded space. In the nonhomogeneous case we define the Hilbert–Poincaré series as

$$\mathcal{H}_X(V,t) = \mathcal{H}(\operatorname{gr} V, t) = \sum_{n=1}^{\infty} \dim_{\mathbb{F}}(\operatorname{gr}_n V) t^n;$$
$$\mathcal{H}_X(A/V,t) = \mathcal{H}(A/\operatorname{gr} V, t) = \sum_{n=1}^{\infty} \dim_{\mathbb{F}}(A_n/\operatorname{gr}_n V)\, t^n. \tag{10.2}$$

We introduce the operator $\mathcal{E}$ on the power series $\varphi(t) = \sum_{n=1}^{\infty} b_n t^n$, $b_n \in \{0, 1, 2, \ldots\}$, as

$$\mathcal{E} : \varphi(t) = \sum_{n=1}^{\infty} b_n t^n \longrightarrow \mathcal{E}(\varphi(t)) = \sum_{n=0}^{\infty} a_n t^n = \prod_{n=1}^{\infty} \frac{1}{(1-t^n)^{b_n}}. \tag{10.3}$$

Let L be a Lie algebra generated by X and $U(L)$ its universal enveloping algebra. Suppose that

$$\mathcal{H}_X(L,t) = \sum_{n=1}^{\infty} b_n t^n, \quad \mathcal{H}_X(U(L),t) = \sum_{n=0}^{\infty} a_n t^n.$$

It is well known that $\mathcal{H}_X(U(L)) = \mathcal{E}(\mathcal{H}_X(L))$ [351]. The relation (10.3) is important for many combinatorial problems (see [10]), and plays a key role in the study of the *growth* of Lie algebras [277, 273]. Our first lemma is easy.

Lemma 10.7.
1. $\mathcal{E}(0) = 1$;
2. $\mathcal{E}(\varphi_1(t) + \varphi_2(t)) = \mathcal{E}(\varphi_1(t)) \cdot \mathcal{E}(\varphi_2(t))$.

Now we can formulate our main statement on the subalgebras of free Lie algebras.

Theorem 10.8. *Let L be a free Lie algebra freely generated by a finitely graded set X. Suppose that H is a subalgebra. Then there exists a set of free generators Y for H such that*

$$\mathcal{H}(Y) - 1 = (\mathcal{H}(X) - 1) \cdot \mathcal{E}(\mathcal{H}(L/H)).$$

A special case is the following.

Lemma 10.9. *Let L be a free Lie algebra freely generated by a finitely graded set X. Suppose that $H \subset L$ is a homogeneous subalgebra with $L = H \oplus \langle Z \rangle_{\mathbb{F}}$, $Z \subset X$, $|Z| < \infty$. Suppose that Y is a free generating set for H. Then*

$$\mathcal{H}(Y) - 1 = (\mathcal{H}(X) - 1) \cdot \mathcal{E}(\mathcal{H}(L/H)).$$

Proof. We prove this statement by induction on $|Z|$. Let $Z = \{z\}$, $\operatorname{wt} z = a$. Then $\mathcal{H}(L/H) = t^a$, $\mathcal{E}(\mathcal{H}(L/H)) = 1/(1-t^a)$. In this case H is freely generated by the following set (see Section 2.4.2):

$$Y = \{[x, z^m] \mid x \in X \backslash \{z\},\ m = 0, 1, 2, \dots\}, \tag{10.4}$$

where $[x, z^m] = [\dots[x, \underbrace{z], \dots, z}_{m \text{ times}}]$. Then

$$\begin{aligned}\mathcal{H}(Y) &= \mathcal{H}(X \backslash \{z\}) \cdot (1 + t^a + t^{2a} + \dots + t^{ma} + \dots)\\ &= (\mathcal{H}(X) - t^a)\frac{1}{1-t^a} = (\mathcal{H}(X) - 1 + (1 - t^a))\frac{1}{1-t^a}\\ &= 1 + (\mathcal{H}(X) - 1)\frac{1}{1-t^a} = 1 + (\mathcal{H}(X) - 1) \cdot \mathcal{E}(\mathcal{H}(L/H)).\end{aligned}$$

Now suppose that $|Z| > 1$, take $z \in Z$ and $Z' = Z \backslash \{z\}$. Then $H' = H \oplus \langle Z' \rangle_{\mathbb{F}}$ is a subalgebra. It is free by Shirshov–Witt's theorem (Theorems 2.18 and 2.20). Suppose that H' is freely generated by Y'. We apply the inductive hypothesis to the inclusions $H \subset H'$, $H' \subset L$:

$$\begin{aligned}\mathcal{H}(Y') - 1 &= (\mathcal{H}(X) - 1) \cdot \mathcal{E}(\mathcal{H}(L/H')),\\ \mathcal{H}(Y) - 1 &= (\mathcal{H}(Y') - 1) \cdot \mathcal{E}(\mathcal{H}(H'/H)).\end{aligned}$$

Finally, we apply the multiplicative property of $\mathcal{E}$ to get

$$\begin{aligned}\mathcal{H}(Y) - 1 &= (\mathcal{H}(X) - 1) \cdot \mathcal{E}(\mathcal{H}(L/H')) \cdot \mathcal{E}(\mathcal{H}(H'/H))\\ &= (\mathcal{H}(X) - 1) \cdot \mathcal{E}(\mathcal{H}(L/H') + \mathcal{H}(H'/H))\\ &= (\mathcal{H}(X) - 1) \cdot \mathcal{E}(\mathcal{H}(L/H)),\end{aligned}$$

and the inductive step is proved. □

Now we can prove Theorem 10.8.

Proof. First, we assume that H is homogeneous. We remark that $\mathcal{H}(Y) = \mathcal{H}(H/[H, H])$, hence our series does not depend on the choice of a homogeneous Y. We prove our claim for a specifically constructed free generating set Y. In our construction, we follow the proof of Theorem 2.17.

By assumption, $L = \bigoplus_{i=1}^{\infty} L_i$, $H = \bigoplus_{i=1}^{\infty} H_i$ where $H_i = H \cap L_i$, $i \in \mathbb{N}$. Let us consider subalgebras

$$H^{(0)} = L; \quad H^{(s)} = H_1 \oplus \cdots \oplus H_s \oplus L_{s+1} \oplus L_{s+2} \oplus \cdots, \quad s = 1, 2, \ldots$$

Then $H = \bigcap_{s=0}^{\infty} H^{(s)}$. Suppose that for any $H^{(s)}$ we have constructed a free homogeneous generating set $Y^{(s)}$ and the following condition is satisfied: $Y_i^{(s)} = Y_i^{(s-1)}$, $i = 1, \ldots, s-1$. Then $Y = \bigcup_{i=1}^{\infty} Y_i^{(i)}$ is a free generating set for H. Indeed, let $\mathcal{R}(Y)$ be a basis built on Y. If some monomials of $\mathcal{R}(Y)$ are linearly dependent then this is the linear dependence among the basis elements of some $Y^{(s)}$, a contradiction to the fact that $Y^{(s)}$ freely generates $H^{(s)}$. Let us consider a set $\widetilde{Y}$ which is in one-to-one correspondence with Y and construct the free Lie algebra $L(\widetilde{Y})$ whose basis consists of the elements $\mathcal{R}(\widetilde{Y})$. We have a natural homomorphism $L(\widetilde{Y}) \to L$, under which the basis of $L(\widetilde{Y})$ is mapped onto the basis of H. Therefore, $H \cong L(\widetilde{Y})$.

Now let us consider the inclusion $H^{(s-1)} \supset H^{(s)}$; we temporarily assume that $L = H^{(s-1)}$, $H = H^{(s)}$. In this case we have $H_i = L_i$, $i \neq s$, $L_s = H_s \oplus \overline{L}_s$. Our goal now is to construct a free generating set Y for H such that $Y_1 = X_1, \ldots, Y_{s-1} = X_{s-1}$. Let G be a homogeneous subalgebra in H generated by $X_1 \cup \cdots \cup X_{s-1}$. Since X is the free generating set for L, we have $L_s = \langle X_s \rangle_{\mathbb{F}} \oplus G_s$. Also $G_s \subset H_s$ and, by the modular law, $H_s = (\langle X_s \rangle_{\mathbb{F}} \cap H_s) \oplus G_s$. In the vector space $\langle X_s \rangle_{\mathbb{F}}$, we take another basis $\overline{X}_s = P \cup Z$ such that $\langle \overline{X}_s \rangle_{\mathbb{F}} \cap H_s = \langle P \rangle_{\mathbb{F}}$. If we set $\overline{X}_i = X_i$, $i \neq s$ then $\overline{X} = \bigcup_{i=1}^{\infty} \overline{X}_i$ is a free generating set for L. Remark also that $L = H \oplus \langle Z \rangle_{\mathbb{F}}$ and $Z \subset \overline{X}$. So, we are in position to apply Lemma 10.9. By our construction, $\overline{X}$ is the free homogeneous generating set for L with $\mathcal{H}(\overline{X}) = \mathcal{H}(X)$. We apply Lemma 10.9 and obtain a free generating set Y for H with

$$\mathcal{H}(Y) - 1 = (\mathcal{H}(\overline{X}) - 1) \cdot \mathcal{E}(\mathcal{H}(L/H)) = (\mathcal{H}(X) - 1) \cdot \mathcal{E}(\mathcal{H}(L/H)). \tag{10.5}$$

Since $\operatorname{wt} z_i = s$, $z_i \in Z$, we see that the iteration of the process (10.4) in the proof of Lemma 10.9 gives us $Y_i = \overline{X}_i = X_i$, $i = 1, \ldots, s-1$.

Now we return to our chain of subalgebras $H^{(s)}$ with the free generating sets $Y^{(s)}$, $s = 0, 1, 2, \ldots$ We rewrite (10.5) as

$$\mathcal{H}(Y^{(s)}) = 1 + (\mathcal{H}(Y^{(s-1)}) - 1) \cdot \mathcal{E}(\mathcal{H}(H^{(s-1)}/H^{(s)})). \tag{10.6}$$

Let us prove by induction that

$$\mathcal{H}(Y^{(i)}) = 1 + (\mathcal{H}(X) - 1) \cdot \mathcal{E}(\mathcal{H}(L/H^{(i)})), \quad i = 0, 1, 2, \ldots \tag{10.7}$$

For $i = 0$, this relation is trivial. Suppose that it holds true for $i = s - 1$, that is, $\mathcal{H}(Y^{(s-1)}) = 1 + (\mathcal{H}(X) - 1) \cdot \mathcal{E}(\mathcal{H}(L/H^{(s-1)}))$. Using the inductive assumption, (10.6), and Lemma 10.7,

$$\begin{aligned}
\mathcal{H}(Y^{(s)}) &= 1 + (\mathcal{H}(Y^{(s-1)}) - 1) \cdot \mathcal{E}(\mathcal{H}(H^{(s-1)}/H^{(s)})) \\
&= 1 + (\mathcal{H}(X) - 1) \cdot \mathcal{E}(\mathcal{H}(L/H^{(s-1)})) \cdot \mathcal{E}(\mathcal{H}(H^{(s-1)}/H^{(s)})) \\
&= 1 + (\mathcal{H}(X) - 1) \cdot \mathcal{E}(\mathcal{H}(L/H^{(s-1)}) + \mathcal{H}(H^{(s-1)}/H^{(s)})) \\
&= 1 + (\mathcal{H}(X) - 1) \cdot \mathcal{E}(\mathcal{H}(L/H^{(s)})),
\end{aligned}$$

the inductive step is proved.

By construction,

$$\mathcal{H}(Y) = \mathcal{H}(Y^{(i)}) \mod (t^{i+1}), \quad i \in \mathbb{N};$$
$$\mathcal{H}(L/H) = \mathcal{H}(L/H^{(i)}) \mod (t^{i+1}), \quad i \in \mathbb{N};$$
$$\mathcal{E}(\mathcal{H}(L/H)) = \mathcal{E}(\mathcal{H}(L/H^{(i)})) \mod (t^{i+1}), \quad i \in \mathbb{N}.$$

Therefore, we can take the limit $i \to \infty$ in (10.7) and conclude that

$$\mathcal{H}(Y) - 1 = (\mathcal{H}(X) - 1) \cdot \mathcal{E}(\mathcal{H}(L/H)).$$

Now we suppose that $H \subset L$ is an arbitrary subalgebra. We consider the associated graded subspace

$$\operatorname{gr} H = \bigoplus_{n=1}^{\infty} \operatorname{gr}_n H, \quad \operatorname{gr}_n H = \{\operatorname{gr} h \mid h \in H, \deg h = n\} \cup \{0\},\ n = 1, 2, \dots \tag{10.8}$$

We also observe that $\operatorname{gr} H$ is a homogeneous subalgebra. Then the above arguments yield a homogeneous free generating set $\overline{Y}$ with $\mathcal{H}(\overline{Y}) - 1 = (\mathcal{H}(X) - 1) \cdot \mathcal{E}(\mathcal{H}(L/\operatorname{gr} H))$.

By (10.8), there exists $Y \subset H$ with $\overline{Y} = \operatorname{gr} Y$ and we have the bijection $Y \to \overline{Y}$. Let $\mathcal{R}(Y) = \bigcup_{n=1}^{\infty} \mathcal{R}_n(Y)$ be a formally constructed basis family on Y with $\mathcal{R}_n(Y) = \{a \in \mathcal{R}(Y) \mid \deg a = n\}$, $\mathcal{R}_n(Y) \subset \mathcal{R}_{n+1}(Y)$, $n = 1, 2, \dots$ This family is in a bijective correspondence with $\mathcal{R}(\overline{Y}) = \bigcup_{n=1}^{\infty} \mathcal{R}_n(\overline{Y})$, $\mathcal{R}_n(\overline{Y}) = \{b \in \mathcal{R}(\overline{Y}) \mid \operatorname{wt} b = n\}$. Denote $H^n = \{h \in H \mid \deg h \le n\}$, $n = 1, 2, \dots$ Then under the natural isomorphism of vector spaces $H^n/H^{n-1} \to \operatorname{gr}_n H$ the set $\mathcal{R}_n(Y)$ is mapped onto $\mathcal{R}_n(\overline{Y})$. This proves that $\mathcal{R}(Y)$ is linearly independent, thus Y is the free generating set for H; it is obvious that $\mathcal{H}(Y) = \mathcal{H}(\overline{Y})$. Also, by definition (10.2), $\mathcal{H}(L/H) = \mathcal{H}(L/\operatorname{gr} H)$. Finally, we have

$$\mathcal{H}(Y) - 1 = \mathcal{H}(\overline{Y}) - 1 = (\mathcal{H}(X) - 1) \cdot \mathcal{E}(\mathcal{H}(L/\operatorname{gr} H)) = (\mathcal{H}(X) - 1) \cdot \mathcal{E}(\mathcal{H}(L/H)). \qquad \square$$

Recall that a function $f(n) : \mathbb{N} \to \mathbb{C}$ is called *quasipolynomial* if $f(n) = \sum_{j=0}^{s} \lambda_j(n)\, n^j$, where $\lambda_j(n) : \mathbb{N} \to \mathbb{C}$ are periodic functions, $j = 0, 1, \dots, s$ [346]. If $\lambda_s(n) \not\equiv 0$, then s is referred to as its degree.

Corollary 10.10. *Let L be a free Lie algebra freely generated by $X = \{x_1, \dots, x_k\}$, $\operatorname{wt} x_i = 1$, $i = 1, \dots, k$. Suppose that H is a homogeneous subalgebra with $\dim L/H = s$ and Y is a free homogeneous generating set for H with $\mathcal{H}(Y, t) = \sum_{n=1}^{\infty} d_n t^n$. Then*

1. *$\mathcal{H}(Y, t)$ is a rational function and*

$$\mathcal{H}(Y, t) = 1 + (kt - 1) \cdot \prod_{n=1}^{\infty} \frac{1}{(1 - t^n)^{c_n}}, \tag{10.9}$$

where $\sum_{n=1}^{\infty} c_n = s$ and only finitely many integers c_n, $n \in \mathbb{N}$ are nonzero;

2. *d_n is a quasipolynomial in n of degree s − 1,*

$$d_n = \mu\, n^{s-1} + \sum_{j=0}^{s-2} \lambda_j(n)\, n^j, \quad \textit{where } \mu = \frac{k-1}{(s-1)! \prod_{j=1}^{\infty} j^{c_j}};$$

3. *Y is infinite.*

Proof. One has $\mathcal{H}(X,t) = kt$. Suppose that $\mathcal{H}(L/H,t) = \sum_{n=1}^{\infty} c_n t^n$. Then $\sum_{n=1}^{\infty} c_n = \dim L/H = s$ and only finitely many c_n are nonzero. Also, by (10.3),

$$\mathcal{E}(\mathcal{H}(L/H,t)) = 1/\prod_{n=1}^{\infty} (1-t^n)^{c_n}.$$

Remark that 1 is a root of the denominator (10.9) with multiplicity s and also $\lim_{t\to 1}(1-t)^s \mathrm{H}(Y,t) = \mu(s-1)!$. Claim 2 follows from the well known facts on the coefficients of rational functions [346]. Finally, it is clear that claim 3 follows from claim 2. □

If X, or $\dim L/H$, is infinite, then the "size" of Y can be described with the help of q-dimensions [276–278]. They were suggested by Victor Petrogradsky to describe the growth of the universal enveloping algebra, provided that the growth of a Lie algebra is known.

Remark. We could also consider a more general weight function $\mathrm{wt} : X \to \mathbb{N}^m$ with

$$X = \bigcup_{i_1=1,\dots,i_m=1}^{\infty} X_{i_1\dots i_m}, \quad X_{i_1\dots i_m} = \{x \in X \mid \mathrm{wt}\, x = (i_1,\dots,i_m)\};$$
$$|X_{i_1\dots i_m}| < \infty, \ i_j \in \mathbb{N}.$$

For any monomial $y = x_{i_1} \cdots x_{i_n}$, $x_j \in X$, we set $\mathrm{wt}\, y = \mathrm{wt}\, x_{i_1} + \cdots + \mathrm{wt}\, x_{i_n}$. For a set of monomials Y (or subspace), we write

$$\mathcal{H}_X(Y,t_1,\dots,t_m) = \sum_{i_1=1,\dots,i_m=1}^{\infty} |Y_{i_1\dots i_m}| t_1^{i_1} \cdots t_m^{i_m};$$
$$Y_{i_1\dots i_m} = \{y \in Y \mid \mathrm{wt}\, y = (i_1,\dots,i_m)\}.$$

In this case, $\mathcal{E}$ looks as

$$\mathcal{E}\left(\sum_{i_1=1,\dots,i_m=1}^{\infty} b_{i_1\dots i_m} t_1^{i_1} \cdots t_m^{i_m} \right) = \prod_{i_1=1,\dots,i_m=1}^{\infty} \frac{1}{(1-t_1^{i_1}\cdots t_m^{i_m})^{b_{i_1\dots i_m}}}.$$

All claims of this section remain valid for superalgebras and p-algebras, see [272].

10.2 Growth and cogrowth functions in free Lie algebras

According to the classical Shirshov–Witt theorem (Theorems 2.18 and 2.20), every subalgebra of a free Lie algebra is free. There are a number of ways to draw distinction between different subalgebras. In this section we are interested in the (relative) growth and cogrowth of subalgebras, following the papers of Yu. A. Bahturin and A. Yu. Olshanskii [56, 55].

Any free basis (i. e., free generating set) X of a free Lie algebra $L = L(X)$ defines an increasing filtration

$$L^{(1)} \subset L^{(2)} \subset \cdots \subset L^{(n)} \subset \cdots,$$

where $\dim L^{(i)} < \infty$, if X is finite. The (relative) growth function of a subspace $H \subset L$ is an integer-valued function of an integral argument whose nth value equals $\dim(H \cap L^{(n)})$. The nth value of the cogrowth function equals $\dim L^{(n)}/(H \cap L^{(n)})$.

Let us recall that if $|X| = m \geq 2$ then the growth of the whole of $L = L(X)$ is exponential. This easily follows from the classical Witt's formula (Theorem 3.2) that the dimension of the nth homogeneous component L_n of L is

$$\dim L_n = \frac{1}{n} \sum_{d|n} \mu(d) m^{\frac{n}{d}} \sim \frac{1}{n} m^n. \tag{10.10}$$

Here μ is the Möbius function and the equivalence means that the ratio of the left- and right-hand sides tends to 1 as $n \to \infty$. The exponentiality holds for the growth and cogrowth of the most of subalgebras encountered in this section but a more precise comparison shows that the degree of exponentiality can be rather different.

In this section, we call a function $f(n)$ of the natural argument with nonnegative values *exponential* if the limit of the function $\sqrt[n]{f(n)}$ exists and is greater than 1. We call $\lim_{n\to\infty} \sqrt[n]{f(n)}$ the (exponential) *base* of $f(n)$.

Let us say that a (growth) function $f_1(n)$ is *exponentially negligible* compared to another (growth) function $f_2(n)$ if the ratio $\frac{f_2(n)}{f_1(n)}$ grows exponentially. We say that the growth or the cogrowth of a subalgebra H of a free Lie algebra $L = L(X)$, with a fixed free basis X, is exponentially negligible if the respective relative growth (cogrowth) function for H is exponentially negligible compared with the growth function for L. The exponential base of the relative growth function for H in L with respect to a fixed free basis X is denoted by β_H.

In the study of the growth and cogrowth of subalgebras of free Lie algebras, the best explored is the case of ideals. The cogrowth of an ideal I of a free Lie algebra L is actually the growth of the factor-algebra L/I. This is one of the main notions in the theory of infinite-dimensional Lie algebras. Many classes of algebras are defined in terms of their growth. For a general monograph on the topic, see [186]. In the theory of simple Lie algebras, confirming a conjecture by V. Kac [154] (see also his monograph [156]), O. Mathieu classified simple Lie algebras of polynomial growth [222]. Earlier in this chapter, we have seen the results of Petrogradsky on the growth of solvable Lie algebras and algebras satisfying polynomial identities. For a survey of this area, see [268].

Note that in the case of ideals, the cogrowth of a nonzero ideal I of a free Lie algebra $L = L(X)$ of rank ≥ 2 is exponentially negligible compared with the growth of L. This easily follows from the results of [54, Section 2.2] where the respective claim has been established for free associative algebras. Indeed, consider the free associative algebra $A = A(X)$ as the universal enveloping algebra of L and the two-sided ideal J of A generated by I. Then $L/I \subset A/J$ by Poincaré–Birkhof–Witt theorem and the growth of A/J by [54, Proposition 8] is negligible. Since by Witt's formula (10.10) the exponent of the growth of L is the same as that of A, it follows that the growth of L/I is negligible, as well.

Much less explored is the case of subalgebras which are not ideals. In this section we start with general finitely generated subalgebras.

Theorem 10.11. *The (relative) growth function $g_H(n)$ of any proper finitely generated subalgebra H of a free Lie algebra L of finite rank m is exponentially negligible.*

In the case of subalgebras which are *subideals*, the situation is entirely opposite.

Let us recall that a subalgebra M of a Lie algebra L is named an *ℓ-subideal*, $\ell = 1, 2, \ldots$, if there is a decreasing sequence

$$L \rhd L_1 \rhd \cdots \rhd L_{\ell-1} \rhd M,$$

where each term is an ideal in the preceding term.

The general theory of infinite-dimensional Lie algebras with a sizeable pprtion of material devoted to subideals is exposed in the book [5]; however, our present topic, the *growth and cogrowth of subideals*, is not covered there.

Theorem 10.12. *The cogrowth of any nonzero subideal of a free Lie algebra of rank $m \geq 2$ is exponentially negligible.*

The result of Theorem 10.12 contrasts with the situation in the group theory where the cogrowth of a nontrivial subnormal subgroup of the free group F_r, $r \geq 2$, can be equivalent to the growth of F_r.

Compared to the proof of the same result for the ideals, as given earlier, the proof of this result requires much more sophistication and even the development of a new piece of technique. This technique is presented in Section 10.2.4. One of the immediate applications of the technique is the following result, which provides essential details to our claim of exponential negligibility of finitely generated subalgebras in Theorem 10.11.

Theorem 10.13. *Let L be a free Lie algebra of rank m, with a fixed basis X. Then the following are true:*

(i) *The relative growth function $g_H(n)$ of a proper nonabelian subalgebra H of L, with respect to X, is exponential.*
(ii) *If H is finitely generated nonabelian subalgebra of L then the exponential base β_H of g_H is an algebraic integer. If H is proper then $\beta_H < m$.*
(iii) *For any real $m_0 \in [1, m]$, there is a subalgebra H of L such that $\beta_H = m_0$. One can choose H as a retract of an ideal of codimension 1 in L.*
(iv) *For any real $m_0 \in [1, m]$, there is a 2-subideal S of L such that the exponential base $\beta_{L/S}$ of the relative cogrowth function $g_{L/S}$ equals m_0. One can choose S as an ideal in an ideal M of codimension 1 in L, with free factor algebra M/S.*

We do not know if for any $m_0 \in [1, m]$ there is an ideal S with $\beta_{L/S} = m_0$.

To further emphasize the potential of the combinatorial approach to the questions of the growth/cogrowth, in Section 10.2.5.1, as an example, we calculate the cogrowth function for the minimal 2-subideal containing one of the free generators of a free Lie algebra of rank 2.

Theorem 10.11 gives a quantitative evaluation of the "size" of a finitely generated subalgebra H compared to the "size" of the whole of the free Lie algebra L. An example showing how the "quantitative" Theorems 10.11 and 10.12 can be applied to obtain purely "qualitative" consequences is as follows.

Corollary 10.14. *No proper finitely generated subalgebra K of a free Lie algebra L can contain a nonzero subideal H of L.*

Proof. We may assume that L has a finite rank $m \geq 2$. By Theorem 10.11, K is negligible. If $H \subset K$ then also H is negligible. However, this contradicts Theorem 10.12. □

Apart from the sharp difference in the character of growth/cogrowth, there are other properties where finitely generated subalgebras and subideals differ drastically. Let us say that, given a subalgebra H of a Lie algebra G, the *idealizer* of H in G is the unique maximal subalgebra K containing H as an ideal. Clearly, any proper subideal is different from its idealizer in any Lie algebra. At the same time, an easy consequence of our Theorem 10.11 is the following, probably known fact: *any nonzero finitely generated subalgebra of a nonabelian free Lie algebra is self-idealizing, that is, equal to its idealizer.* Using this property, one could derive Corollary 10.14 using a more traditional argument, like Theorem 10.15 below. This result constitutes a natural counterpart to M. Hall's theorem about finitely generated subgroups in the free groups [135].

Theorem 10.15. *Let H be a finitely generated subalgebra of a free Lie algebra L of finite rank. Then a free basis B_0 of H can be complemented to produce a free basis B of a subalgebra C of finite codimension in L. In other words, if K is the subalgebra generated by $B \setminus B_0$ then $C \cong H * K$, the free product of H and K.*

One can always choose $B \setminus B_0$, hence K, homogeneous. If H is itself homogeneous, one can choose B so that C contains the tth term of the lower central series L^t, for some t.

For the details about the free products of Lie algebras, see Section 2.6.2.
From this theorem we easily obtain the following consequence.

Corollary 10.16. *Every proper finitely generated subalgebra H in a free Lie algebra L of finite rank is a free factor in strictly greater subalgebra of L; in addition, if homogeneous, H is a free factor of a* subideal *of L.*

A comparison with group theory reveals the following differences. First, in Hall's theorem [135], saying that a finitely generated subgroup H of a free group F_r of rank r is a free factor in a greater subgroup of F_r, one must stipulate that H is of *infinite index* in F_r. Second, in the case of Lie algebras, if H is proper nonzero, the free factor K appearing in Theorem 10.15 can never be chosen finitely generated; whereas in the case of groups, Hall's theorem asserts that the complementary free factor K is *always finitely generated.*

When dealing with the cogrowth of subideals, the following situation comes to mind quite naturally. If a subideal S contained a nonzero ideal I then the cogrowth of S would be exponentially negligible, without any further argument. The result that follows shows that this such situation is highly unlikely. In the statement of the result, we will use the following notation.

We denote the ideal of an algebra L generated by a set S, $\mathrm{id}_L\, S$, is the minimal ideal of L that contains S. We also say that $\mathrm{id}_L\, S$ is an *ideal closure* of S in L. If ℓ is a natural number then an ℓ-subideal of L generated by S is the subalgebra $\mathrm{id}_L^\ell\, S$ of L which is defined by induction as follows. We set $\mathrm{id}_L^1\, S = \mathrm{id}_L\, S$ and if $\ell > 1$ then define $\mathrm{id}_L^\ell\, S$ as the ideal of $\mathrm{id}_L^{\ell-1}\, S$ generated by S. An easy inductive argument shows that $\mathrm{id}_L^\ell\, S$ is contained in every ℓ-subideal of L containing S. We also say that $\mathrm{id}_L^\ell\, S$ is an *ℓ-subideal closure* of S in L.

Theorem 10.17. *Let J be a proper ideal of a free Lie algebra L and S a finite subset of J. Then the ideal closure I of S in J does not contain nonzero ideals of L.*

In the case of subideal closures we have the following.

Corollary 10.18. *Let L be a free Lie algebra, J a proper ideal of L, and S a finite subset of J. Then for no $\ell \geq 1$ the ℓ-subideal closure $\mathrm{id}_J^\ell\, S$ contains a nonzero ℓ-subideal of L.*

Corollary 10.19. *Let L be a free Lie algebra and S a finite subset of L whose ideal closure in L is different from L. Then for any $\ell \geq 1$, the $(\ell + 1)$th subideal closure of S in L does not contain any nonzero ℓ-subideal of L.*

The restriction $\mathrm{id}_L(S) \neq L$ in the statement of the above result is necessary because, otherwise, for every $\ell = 1, 2, \ldots$, the ℓ-subideal closure of S equals L and the claim is not valid.

We conclude this section with few remarks about the differences of the situation with the growth and cogrowth in the case of algebras as opposed to the case of groups. We will also give some hints as for the methods we have used or developed.

Two observations make the situation in the case of algebras so different from the case of groups (and semigroups!). In the case of algebras, we measure the growth using the *degree* of the elements while in the case of groups we deal with the *length*. So our first observation is that the degree of the sum of two elements never exceeds the degree of the summands, while the length of the product does not need to obey this rule. As for the second, the reader probably already noticed that in the proof of Corollary 10.14 we have used a simple general purely "algebraic" property: *The sum of the (relative) growth and cogrowth functions of a subalgebra always equals to the growth function of the whole algebra*. Again, no analogue of this claim holds valid in the case of groups. As a result, dealing with the growth and cogrowth functions is by necessity more delicate in the case of groups.

Still, to prove the results about the growth of subalgebras in the free Lie algebras here, it was necessary to suggest at least two pieces of technique we think new.

One of them, suggested in Section 10.2.5, deals with the following issue. Although we define the growth functions for subalgebras in the free Lie algebra $L = L(X)$ using a finite fixed free basis X, in our proofs we need to switch from this fixed basis to other bases, which could be by necessity infinite. To preserve growth functions, we need to develop an approach to *the growth with respect to infinite graded bases* of free Lie algebras. This is done in Section 10.2.5 where certain analytic finiteness conditions are suggested that allow us to estimate the cogrowth when we use induction to pass to infinitely generated ideals and subideals and at the same time have the degrees of elements unchanged. The second piece of technique is produced in Lemma 10.39 where we will examine what can be naturally called *shifting derivations* in associative and Lie algebras.

10.2.1 Preliminaries

In this section we give necessary definitions concerning the relative growth of subspaces in algebras. We also recall some results about the connection between associative and Lie algebras, (linear) bases of free Lie algebras and their subalgebras.

10.2.1.1 Relative growth of subspaces in finitely filtered spaces

Let V be a vector space over a field F with the filtration

$$\alpha: \quad \{0\} \subset V^{(0)} \subset V^{(1)} \subset \cdots \subset V^{(n)} \subset \cdots, \qquad V = \bigcup_{n=0}^{\infty} V^{(n)}, \tag{10.11}$$

We say that $\alpha = \{V^{(n)}\}$ is a *finite ascending filtration*, if all subspaces $V^{(n)}$ are finite-dimensional. The *growth function* of α, g_α, is then given by $g_\alpha(n) = \dim V^{(n)}$. The *graded growth function* of α, d_α, is given by $d_\alpha(n) = \dim V^{(n)}/V^{(n-1)}$ (assuming $V^{(-1)} = \{0\}$).

The filtration α also defines a degree function $\deg_\alpha$ on the set of nonzero elements of L given by $\deg_\alpha(a) = n$ provided that $a \in V^{(n)} \setminus V^{(n-1)}$.

If W is a subspace then the *subfiltration* $\alpha \cap W$ is a filtration formed by the subspaces $W^{(n)} = W \cap V^{(n)}$. The factor-space V/W acquires a *factor-filtration* α/W given by the subspaces $(V/W)^{(n)} = (V^{(n)} + W)/W \cong V^{(n)}/(W \cap V^{(n)})$.

In each of these cases, the filtrations of the respective spaces are finite. Once α and g_α are fixed, we can define the *growth function relative to* α, $g_{\alpha\cap W}(n) = \dim W^{(n)}$, and *cogrowth function relative to* α, $g_{\alpha/W}(n) = \dim(V/W)^{(n)}$. If the filtrations are fixed then we denote the respective growth functions as g_V, g_W, and $g_{V/W}$.

In each of these cases, we have also *relative graded functions* d_V, d_W, and $d_{V/W}$. Say, $d_{V/W}(n) = g_{V/W}(n) - g_{V/W}(n-1)$.

It often so happens that the space V is $\mathbb{Z}$-graded, $V = V_0 \oplus V_1 \oplus V_2 \oplus \cdots \oplus V_n \oplus \cdots$, and for each term of the filtration α one has $V^{(n)} = V_0 \oplus V_1 \oplus V_2 \oplus \cdots \oplus V_n$, for all $n = 1, 2, \ldots$ The nonzero elements of V_n are called *homogeneous of degree n* and V_n the *homogeneous subspace of the grading of degree n*. If $v \in V$ then v can be uniquely written as the sum of some elements $v_n \in V_n$. We call v_n a homogeneous components of v of degree n. If n is the greatest such that $v_n \neq 0$ then we call v_n the *leading part* of u and write $v_n = \mathrm{Lp}(u)$.

A subspace $U \subset V$ is called *graded* if U contains all homogeneous components of its elements. Equivalently, $U = (U \cap V_0) \oplus (U \cap V_1) \oplus (U \cap V_2) \oplus \cdots \oplus (U \cap V_n) \oplus \cdots$. If W is not graded then the linear span of the leading parts of nonzero elements of W is a graded subspace of V, which is denoted by $\mathrm{gr}\, W$. For each natural number n, the homogeneous component $(\mathrm{gr}\, W)_n$ of this space is the linear span of the leading parts of all elements of degree n in W.,

An easy but useful technical result is the following.

Lemma 10.20. *For the relative growth functions, one has*
(a) $g_W = g_{\mathrm{gr}\, W}$;
(b) $g_{V/W} = g_{V/\mathrm{gr}\, W}$.

Proof. (a) It follows from the definition of $(\mathrm{gr}\, W)_n$ that the mapping $w + W^{(n-1)} \mapsto \mathrm{Lp}(w)$ for $w \in W^{(n)} \setminus W^{(n-1)}$ and $w \mapsto 0$ for $w \in W^{(n-1)}$ is a well-defined linear isomorphism between $W^{(n)}/W^{(n-1)}$ and $(\mathrm{gr}\, W)_n$. Hence

$$g_W(n) = \dim W^{(n)} = \sum_{i=0}^{n} \dim W^{(i)}/W^{(i-1)} = \sum_{i=0}^{n} \dim\, (\mathrm{gr}\, W)_i = g_{\mathrm{gr}\, W}(n).$$

(b) Follows from (a) and the equality $\dim V^{(n)}/W^{(n)} = \dim V^{(n)} - g_W(n)$. □

10.2.1.2 Linear and free bases of subalgebras in free Lie algebras

As usual, given an alphabet X, we denote by $W(X)$ the monoid of all words in X, including the empty word 1. In this section we will need to count the words in the subsets of $W(X)$, where X is not necessarily finite. If we do not impose any conditions, the number of words of the same length can be infinite. To cope with this, we consider the alphabets which are *finitely graded*, that is, $X = \bigsqcup_{i=1}^{\infty} X_n$ with each set X_n finite. Now in addition to the *lengths* of words, we can speak about the *degrees* of words. We define the degree of a word w in $W(X)$ by induction on its length if we set $\deg(1) = 0$, $\deg(x) = n$ for $x \in X_n$ and if $w = x_1 \cdots x_k$ then $\deg(w) = \deg(x_1) + \cdots + \deg(x_k)$. Let $W(X)_n$ be the set of all words of degree n. Now each number $d_X(n) = \operatorname{Card} W(X)_n$ is finite. We have $W(X) = \bigsqcup_{n=0}^{\infty} W(X)_n$ where $W(X)_n$ is the set of all words of degree n in X.

In its turn, the free associative algebra $A(X)$, which is a vector space with basis $W(X)$, acquires a $\mathbb{Z}$-grading $A(X) = \bigoplus_{n=0}^{\infty} A(X)_n$, where $A(X)_n$ is the linear span of all words in $W(X)_n$. Now in addition to the grading, $A(X)$ have a natural *degree filtration*

$$A(X)^{(0)} \subset A(X)^{(1)} \subset \cdots \subset A(X)^{(n)} \subset \cdots,$$

where $A(X)^{(n)} = A(X)_0 \oplus A(X)_1 \oplus \cdots \oplus A(X)_n$, for each $n = 0, 1, 2, \ldots$

As mentioned earlier, every element $f \in A(X)$ acquires a degree if we set $\deg(f) = n$ in the case where $f \in A(X)^{(n)}$ but $f \notin A(X)^{(n-1)}$. In this case $f = \sum_{i=0}^{n} f_i$, where $f_i \in A(X)_i$, $f_n \neq 0$; $f_n = \operatorname{Lp}(f)$ is the *leading part* of f.

In what follows we are going to study the growth and cogrowth of certain subalgebras H of a free Lie algebra $L = L(X)$. We view L as a graded subspace of $A = A(X)$ which is generated by X with respect to the bracket operation $[a, b] = ab - ba$. The cardinality $\operatorname{Card} X$ of X is often called the *rank* of any of the system appearing in this text: $W(X)$, $A(X)$, or $L(X)$. Being a graded subspace of A, L becomes $\mathbb{Z}$-graded, $L = \bigoplus_{n=1}^{\infty} L_n$, where $L_n = L \cap A_n$. Similarly, we have a degree filtration of L as follows:

$$L^{(1)} \subset L^{(2)} \subset \cdots \subset L^{(n)} \subset \cdots, \tag{10.12}$$

where $L^{(n)} = L \cap A^{(n)}$ or $L^{(n)} = L_1 \oplus \cdots \oplus L_n$, for each $n = 1, 2, \ldots$ Every element $u \in L$ has degree which is the same whether we use the degree filtration of A or L. When we apply the notions of subsection 10.2.1.1 to the free Lie algebra, we view (10.12) as the "base" filtration α of the formula (10.11) and then we know the degree, the leading part of an element of L, and, given a subalgebra $H \subset L$, we also know the associated graded subalgebra $\operatorname{gr} H$. This is, indeed, a subalgebra because the commutator of two leading parts is ether zero or the leading part of the commutator of two elements of H.

Using Lemma 10.20, one can see that the growth and cogrowth of H and $\operatorname{gr} H$ are the same.

Before we go on with the proof, we make a quick notational remark. As in several other parts of this book, when we deal with the interaction of Lie and associative algebras, we use the "commutator" notation in place of general "nonassociative" notation. Having in mind that the theory of free Lie algebras developed in Section 2.4 used "nonassociative" notation, we choose here to recall some of the notation and facts from Chapter 2 using the "commutator" notation.

In the next lemmas we consider the elements of the free Lie algebra $L(X)$ called the *commutators in* X of certain length. The commutators of length 1 are just the elements of X. If the commutators of all lengths less than $n > 1$ have been defined then the commutators of length n are all the elements $[c, d]$ where c is a commutator of length k and d a commutator of length $n - k$.

Since we always view $L(X)$ as a subalgebra in the free associative algebra $A(X)$, recalling $[c, d] = cd - dc$, one can write each commutator as a linear combination of associative words, each of which has the same degree. Thus each commutator acquires a uniquely defined degree and $\deg([c, d]) = \deg(c) + \deg(d)$.

If we forget all the brackets and commas in a commutator c then the result is an associative word w, we write $c = [w]$, called the *associative support of* c. We can also say that any commutator can be obtained by replacing brackets (and the inherent commas) on an associative word. Replacing (arranging) brackets in a certain way on certain associative words may produce a linear basis of the free Lie algebra $L(X)$. Let us recall a process due to A. I. Shirshov from Chapter 2.

Suppose we have a linear ordering $\leq$ on the set X of free generators of the free Lie algebra $L(X)$. We extend this ordering to the lexicographical ordering of the free monoid $W(X)$. Under this ordering, given two words u and v, then we say that $u \geq v$ if u is a prefix of v. If not, we say that $u < v$ if $u = wxu'$, $v = wyv'$ and $x < y$ where $x, y \in X$ and $w, u, v, u', v' \in W(X)$.

If c is a commutator then the *leading word* $\bar{c}$ of c is defined as the greatest word nontrivially entering the expansion of c as a linear combination of the elements of $W(X)$. Let us say that w is a *Lyndon–Shirshov word*, or simply *LS-word*, if whenever $w = uv$, with u and v nonempty, we have $w > vu$. (Notice that in this case we must also have $w > v$, that is, every LS-word is greater than its proper suffix.)

Let us call a commutator c an *LS-commutator* if $c = [w]$ where w is an LS-word and the following two conditions hold:

(a) If $c = [c_1, c_2]$ then each c_i is an LS-commutator with LS-support w_i, $i = 1, 2$, and $w_1 > w_2$;

(b) If c is as in (a) and $c_1 = [c_1', c_1'']$ where w_1'' is the associative support of c_1'' then $w_1'' \leq w_2$.

In the remainder of this section, for an easier reference, we state several results of A. I. Shirshov, which are now conveniently available in the collection [79]. They can also be found in our Chapter 2 and in the book of L. A. Bokut' and G. P. Kukin [78].

Lemma 10.21. *In each LS-word w, one can replace brackets in a unique way so that the resulting commutator $c = [w]$ is an LS-commutator (such that $\bar{c} = w$). The set of all LS-commutators is a linear basis of $L(X)$.*

Lemma 10.22. *Any associative word $v \neq 1$, which is not necessarily an LS-word, can be uniquely written as the product of LS-words $v = u_1 u_2 \cdots u_s$ so that $1 \neq u_1 \leq u_2 \leq \cdots \leq u_s$.*

To list a consequence of this fact, of importance to us, we introduce a convenient notation $[c_1, c_2, \ldots, c_k]$, called the *left-normed commutator* of $c_1, c_2, \ldots, c_k$. This is defined by induction, starting with just c_1, if $k = 1$. If $k > 1$ then one sets $[c_1, c_2, \ldots, c_{k-1}, c_k] = [[c_1, c_2, \ldots, c_{k-1}], c_k]$.

Corollary 10.23. *If c is an LS-commutator, $w = \bar{c} = yv$ for some $y \in X$, and $w \in W(X)$ then y is the maximal letter among all letters involved in the expression of w and $c = [y, [u_1], [u_2], \ldots [u_s]]$, $u_1 \le u_2 \le \cdots \le u_s$, where each $[u_i]$ is the (unique) LS-commutator with the associative support u_i, $i = 1, 2, \ldots, s$.*

A standard technique in the theory of free Lie algebras is the *elimination* of one or more elements of a free basis (sometimes called Lazard elimination [84, 203]).

Lemma 10.24. *Let $L(X)$ be a free Lie algebra with free basis $X = Y \bigsqcup \{z\}$, where $Y \neq \emptyset$. Let $J = \mathrm{id}_L\, Y$ be the ideal of L generated by Y. Let B be the set of all left normed commutators $[y, z, \ldots, z]$, where $y \in Y$. Then J is a free Lie algebra with free basis B.*

10.2.2 Relative growth of finitely generated subalgebras

As before, $L = L(X)$ stands for a free Lie algebra over a field F with a free basis X. This section is devoted to the proof of Theorem 10.11.

Let us call a subset S of a free Lie algebra L *irreducible* if no leading part of any of its elements belongs to the subalgebra generated by the leading parts of the remaining elements of S. In his proof of the theorem on the freeness of subalgebras of free Lie algebras in Section 2.4, Shirshov shows that any subalgebra M can be generated by an irreducible set S. Then he shows that any irreducible set is *independent*, that is, may serve as a free basis for a subalgebra it generates. We can summarize and slightly complement this as follows.

Lemma 10.25. *Let S be a subset of a free Lie algebra L, S' the set of leading parts of the elements in S. If S is irreducible then S' is independent. If S' is independent then S is also independent. Any subalgebra of a free Lie algebra can be generated by an irreducible set.*

Proof. We only need to explain the second claim. However, it is true in any algebra which is free in a variety of algebras and endowed with a grading with respect to a free basis that any nontrivial relation between the elements of a subset S entails a nontrivial relation between the elements of S'. □

Lemma 10.26. *Let M be a subalgebra of a free Lie algebra L and $M' = \mathrm{gr}\, M$ the associated graded subalgebra. Let S be an irreducible free basis of M and $S' = \{\mathrm{Lp}(s) \mid s \in S\}$ the set of leading parts of the elements in S. Then the map $\varphi : S \to S'$ given by $\varphi(s) = \mathrm{Lp}(s)$ extends to a degree preserving isomorphism $\overline{\varphi} : M \to M'$.*

Proof. According to Lemma 10.25, S' is an independent set, hence a free basis of a subalgebra N. As a result, $\overline{\varphi}$ is an isomorphism from M to N. We only need to show that $N = M'$. Since S' and M' are homogeneous, it is sufficient to establish $\mathrm{Lp}(\overline{\varphi}(u)) = \mathrm{Lp}(u)$, for any $u \in M \setminus \{0\}$. To do this, let us express u as a Lie polynomial $u = f(s_1, \dots, s_m)$, in $s_1, \dots, s_m \in S$, and write $f = f_1 + f_2$, where f_1 is a linear combination of monomials of the highest degree if we count each variable s_i with the same degree as $\deg s_i$ in L. Then

$$\begin{aligned}\overline{\varphi}(u) &= \mathrm{Lp}(\overline{\varphi}(f_1(s_1,\dots,s_m)) + \overline{\varphi}(f_2(s_1,\dots,s_m)))\\ &= \mathrm{Lp}(f_1(s'_1,\dots,s'_m) + f_2(s'_1,\dots,s'_m)) = f_1(s'_1,\dots,s'_m) = \mathrm{Lp}(u),\end{aligned}$$

as needed. □

Now we proceed to the proof of Theorem 10.11.

Proof. According to Lemmas 10.20 and 10.26, given a finitely generated subalgebra M of L, the growth and cogrowth of a finitely generated homogeneous subalgebra $M' = \operatorname{gr} M$ are the same as the growth and cogrowth of M. If M is proper then the same is true for M'. Therefore, in proving our theorem, we may restrict ourselves to the case where M is a nonzero graded subalgebra. This entails $m \geq 2$.

Let $\{z_1, \dots, z_k\}$ be a free homogeneous basis of M. Since M is proper in L, the elements of degree 1 among $\{z_1, \dots, z_k\}$ form a basis Z of $M \cap L_1$, which is proper in L_1. Let us complement Z to a linear basis Y of L_1. Then we will obtain a new free basis Y of L, which properly includes Z. As a result, the main degree filtration (10.12) does not change and so the relative growth functions of subalgebras do not change. The elements of degree 1 of this set are a proper part Z of a free basis Y. Hence there is an element of the new free basis, say y, which is not an element of Z.

As a result, without loss of generality, we may assume from the very beginning that the elements of degree 1 in the free homogeneous basis $\{z_1, \dots, z_k\}$ of M are a part of the fixed free basis $\{x_1, \dots, x_m\}$ of L and x_1 is not in M. Let us consider L as a subalgebra in the free associative algebra $A = A(x_1, \dots, x_m)$. Clearly, $z_1, \dots, z_k$ are the elements of the subalgebra B of A which is generated by finitely many monomials $u_1, \dots, u_t$ (all the monomials used to write the generators $z_1, \dots, z_k$). By our assumption, we know that none of these monomials equals x_1. Moreover, we can actually assume that none of $u_1, \dots, u_t$ is a power of x_1. Indeed, no such monomials can appear while writing the elements of the free Lie algebra $L(x_1, \dots, x_m)$ in the free associative algebra $\mathcal{A}(x_1, \dots, x_m)$. It remains to show that the exponent of the growth of B is less than m.

Let us choose an integer d which is at least the double maximum of the degrees of all $u_1, \dots, u_t$ with respect to $x_1, \dots, x_m$. Since any of these monomials contains as a factor a letter different from x_1, no product can contain a subword x^d. It is known (see, e. g., [54, Lemma 8]) that for every nonempty word w in $x_1, \dots, x_m$ there exist $C, \varepsilon > 0$ such that the number of words of length n which do not have w as a subword is bounded by $C(m - \varepsilon)^n$. □

Remark. There are some consequences of Theorem 10.11 that are well-known (see, e. g., [37, Chapter 3]). For instance, because the growth of a subspace of finite codimension is equivalent to the growth of the whole space, we conclude that *in a free Lie algebra the proper nonzero subalgebras of finite codimension cannot be finitely generated.* Another consequence, already mentioned in the introduction, is the following: *in a free Lie algebra any nonzero finitely generated subalgebra is self-idealizing.*

Remark. Theorem 10.11 fails in the case of associative algebras. A simple example of a proper finitely generated subalgebra whose growth is not exponentially negligible is the subalgebra B generated by all monomials of degree 2 and 3 in the free associative algebra A of rank $m \geq 1$. In this case $\dim A/B = m + 1$, following because any number $n > 1$ can be written as $n = 2k + 3l$, where k and l are nonnegative integers.

However, even if we assume A/B infinite-dimensional, we still can have examples of not exponentially negligible finitely generated subalgebras. For instance, one can proceed as follows. Let α be the standard filtration on A of rank $m \geq 1$ and B a subalgebra of A generated by all monomials of degree 2. We obviously have $\dim A/B = \infty$, but at the same time the value of the growth function for the filtration $\beta = \alpha \cap B$ is $g_\beta(n) = \frac{m^{n+2}-1}{m^2-1}$ if n is even. For α we have $g_\alpha(n) = \frac{m^{n+1}-1}{m-1}$. Clearly, B is *not* exponentially negligible.

10.2.3 Distortion of finitely generated subalgebras

If $\beta = \{V_n\}$ and $\beta' = \{V'_n\}$ are two filtration on the same vector space V, as in (10.11), we say that β is *majorized* by β' if there is an integer $t > 0$ such that $B_n \subset B'_{tn}$, for all $n \geq 0$. We then write $\beta \preceq \beta'$. If $\beta \preceq \beta'$ and $\beta' \preceq \beta$ then we say that β and β' are *equivalent* and write $\beta \sim \beta'$.

If B is an (associative, Lie, etc.) algebra generated by a finite set S then B has a natural filtration α, called the *degree filtration* $\{B(n)\}$, where $B(1) = \langle S\rangle$ and $B^{(n)}$ is the linear span of the monomials of degree at most n. If S' is another finite generating set and α' is another degree filtration with respect to S' then it is easy to show that $\alpha \sim \alpha'$.

Now suppose B is a finitely generated subalgebra in a finitely generated algebra A. Then given a degree filtration α on A, one can compare the *relative degree filtration* $\beta' = \alpha \cap B$, for B in A and the degree filtration β on B. If these two are equivalent then one says that B is embedded in A without distortion. If this is not the case, one defines the distortion function by

$$\operatorname{dist}_{\beta'}^{\beta}(n) = \min\{m \mid B'_n \subset B_m\}.$$

No distortion occurs if both $\operatorname{dist}_{\beta'}^{\beta}(n)$ and $\operatorname{dist}_{\beta}^{\beta'}(n)$ are bounded from the above by linear functions.

It is easy to show that a subgroup ring $\mathbb{F}H$ is embedded without distortion in a group ring $\mathbb{F}G$ if and only if H is embedded without distortion as a subgroup of the group G. A classical example is a Baumslag–Solitar group G with presentation $G = \langle a, b \mid aba^{-1} = b^2\rangle$. The distortion of H in G is exponential: the element b^{2^n} has length 2^n in H and $2n + 1$ in G. Hence the distortion of $\mathbb{F}H$ in $\mathbb{F}G$ is exponential.

In the case of the free associative of free group algebras the following can be found in [55].

Theorem 10.27. *Let A be a free associative or a free group algebra of rank at least 2 over a field $\mathbb{F}$. Then one can choose a finitely generated distorted subalgebra B in A in such a way that none of the distortion functions for B in A is bounded from above by any recursive function.*

This makes the following result from [55] quite surprising.

Theorem 10.28. *The embedding of a finitely generated subalgebra in a free Lie algebra $L = L(X)$ of finite rank has no distortion.*

Proof. We know that B is a free Lie algebra and we can choose a free generating set Y in B so that the set $\overline{Y} = \{\mathrm{Lp}(y) \mid y \in Y\}$ is the free generating set of the subalgebra it generates in L.

Now let us assume that B is a finitely generated subalgebra of $L = L(X)$ and that $Y = \{f_1, \ldots, f_n\}$ is the set of free generators of B. Let the degrees of $\mathrm{Lp}(f_1) = f_1', \ldots, \mathrm{Lp}(f_n) = f_n'$ be $d_1, \ldots, d_n$, respectively.

We consider another free Lie algebra $L(Z) \subset A(Z)$, where $Z = \{z_1, \ldots, z_n\}$ and $A(Z)$ is the free associative algebra with free basis Z. In addition to the ordinary degree filtration defined by Z, with degree function $\deg_Z$, we can endow $L(Z)$ with another filtration, with degree function dg, defined on the words in Z by induction on the length as follows. We set $\mathrm{dg}\, z_1 = d_1, \ldots, \mathrm{dg}\, z_k = d_k$, for the letters of the alphabet Z. If $w = uv$ is a word of degree > 1 then by induction $\mathrm{dg}\, u$ and $\mathrm{dg}\, v$ are already defined and we set $\mathrm{dg}\, uv = \mathrm{dg}\, u + \mathrm{dg}\, v$. Given an element $h(z_1, \ldots, z_n) \in A(Z)$ which is a linear combination of words $u_1, \ldots, u_s$ with nonzero coefficients, we set $\mathrm{dg}\, h(z_1, \ldots, z_n) = \max\{\mathrm{dg}\, u_1, \ldots, \mathrm{dg}\, u_s\}$. Finally, we denote by $\mathrm{Lp}'(h)$ the "leading part" of $h(z_1, \ldots, z_n)$ with respect to the new degree dg, that is, the homogeneous component of the above decomposition of degree $\mathrm{dg}\, h(z_1, \ldots, z_n)$.

Now, for any nonzero Lie polynomial $h(z_1, \ldots, z_n)$, we consider the equality $g(x_1, \ldots, x_m) = h(f_1, \ldots, , f_n)$. Then the leading part of $g(x_1, \ldots, x_m)$ will be $\mathrm{Lp}'(h)(f_1', \ldots, f_n')$. This polynomial in X is nonzero since $\{f_1', \ldots, f_n'\}$ is a free generating set.

It follows that $\deg_X g = \mathrm{dg}\, h \geq \deg_Z h$. Let α be the degree filtration of $L(X)$ defined by X and β the degree filtration of B defined by $Y = \{f_1, \ldots, f_n\}$. We just showed that $\mathrm{dist}^{\beta}_{\alpha \cap B}(n) \leq n$. Thus B is an undistorted subalgebra in $L(X)$. □

10.2.4 Some properties of words in infinite alphabets

Infinite alphabets naturally appear when one applies a standard technique of elimination (see Lemma 10.24). Although it is not our goal to formally generalize certain results to the case of free Lie algebras of infinite rank, the logic of the proofs makes it necessary to consider such algebras, as an auxiliary tool.

In what follows, we will be considering only the alphabets with at *least two letters*. Let $k_n = |X_n|$ be the number of letters of degree n. In our analysis, we will be imposing on X some conditions, as follows:

Condition G. *Either $k_2 = k_3 = \cdots = 0$ or, for every i, if $k_i > 0$ then $k_{i+1} > 0$.*

To formulate another condition, we first introduce a function of one real variable ζ (z is a positive parameter) as

$$F(\zeta) = F_z^X(\zeta) = \sum_{i=1}^{\infty} \frac{k_i}{(z-\zeta)^i}. \tag{10.13}$$

Its domain is the set of real values of ζ for which the series on the right-hand side converges.

Recall (Section 10.2.1.2) that $d_X(n)$ stands for the number of words of degree n in the free monoid $W(X)$. A simple fact is the following.

Lemma 10.29. *If $F(0) \leq 1$ then $d_X(n) \leq z^n$, for all natural values of n.*

Proof. We use induction on $n = 0, 1, \ldots$ We have $d_X(0) = 1$ and, for $n \geq 1$, the number of words of degree n with the last letter from X_i equals to $d_X(n-i)k_i$. Applying induction and our hypothesis, we obtain

$$d_X(n) = \sum_{i=1}^{n} k_i d_X(n-i) \leq \sum_{i=1}^{n} k_i z^{n-i} \leq z^n \sum_{i=1}^{\infty} \frac{k_i}{z^i} = z^n F(0) \leq z^n. \qquad \square$$

Lemma 10.30. *Let the series $\sum_{i=1}^{\infty} \frac{k_i}{z^i}$ converge for some real $z > 1$ to a number $\alpha > 1$. Denote by δ the greatest common divisor of all integers in the set $I = \{i \mid k_i \neq 0\}$. Then there is a positive constant c such that, for every sufficiently large n divisible by δ, we have $d_X(n) > cz^n$.*

Proof. Since the series converges to $\alpha > 1$, there is an N such that $\sum_{i=1}^{N} \frac{k_i}{z^i} > 1$. It is well known that there is an integer M such that every $n > M$ divisible by δ is a linear combination of the numbers from I with nonnegative integral coefficients. Without loss of generality, let us assume that both M and N are divisible by δ. It follows that there exists a word of degree n over the graded alphabet X, and so there is a small real number $c > 0$ such that $d_X(n) > cz^n$ for every $n = \delta t$ belonging to the segment $[M, M+N]$, where t is a positive integer.

The assertion of the lemma will be proved by induction on t, where $n = \delta t$, with base $t = M/\delta$ guaranteed above by the choice of c. Moreover, for the inductive step, we may assume that $n > M + N$ and so $t \geq (M + N)/\delta$. Now we have

$$d_X(n) = \sum_{i=1}^{n} k_i d_X(n-i) \geq \sum_{i=1}^{N} k_i d_X(n-i). \tag{10.14}$$

The right-hand side of (10.14) can be rewritten as $\sum_{i=1}^{N/d} k_{\delta i} d_X(\delta t - \delta i)$ because, obviously, $d_X(j) = 0$ if j is not a multiple of δ. Note that $t > t - i \geq t - N/\delta > M/\delta$ for $1 \leq i \leq N/\delta$, and so, by the inductive hypothesis, $d_X(\delta t - \delta i) > cz^{\delta t - \delta i} = c\frac{z^n}{z^{\delta i}}$. Taking (10.14) into account, we obtain $d_X(n) > cz^n \sum_{i=1}^{N/\delta} \frac{k_{\delta i}}{z^{\delta i}}$. Again, the sum on the right-hand side is equal to $\sum_{i=1}^{N} \frac{k_i}{z^i}$, that is, greater than 1 by the choice of N. Hence $d_X(n) > cz^n$, as required. □

In the next result, we denote by $g_A(n)$ the growth function of $A(X)$ with respect to the degree filtration defined by the graded set X.

Lemma 10.31. *The function $g_A(n)$ is superexponential if the series $F(0)$ diverges for every positive z. Otherwise, provided that $|X| > 1$, the growth of $g_A(n)$ is exponential. The base of the exponent of the growth can be determined as z_0 given by*

$$\lim_{n\to\infty} \sqrt[n]{g_X(n)} = z_0 = \inf\left\{ z > 1 \,\middle|\, \sum_{i=1}^{\infty} \frac{k_i}{z^i} \leq 1 \right\}. \tag{10.15}$$

Moreover, $d_X(n) \leq z^n$ for any $n \geq 0$.

Proof. By Lemma 10.29, $d_X(n) \leq z^n$ for any $z > z_0$, whence $g_X(n) \leq c_1 z^n$ for a positive constant c_1. By Lemma 10.30, $d_X(n) > c(z')^n$ for any sufficiently large n divisible by δ if $z' < z_0$. Therefore $g_X(n) > c_2(z')^n$ for a positive c_2 and every large enough n. To finish the proof of formula (10.15), we should extract the nth roots from the inequalities obtained for $g_X(n)$ and pass to the limit, using that z and z' can be chosen arbitrary close to z_0. The second statement is proved in Lemma 10.29. □

To further proceed we need to impose one more condition on the set X.

Condition $\mathbf{W_z}$ ($z > 1$ is a real number). *$F(\zeta)$ is defined in a neighborhood of 0 and $F(0) \leq 1$.*

Our main lemma in this section deals with the following sets. Given a nonempty word $u \in W(X)$, let $N(u)$ be the set of all words in $W(X)$ that do not include u as a subword. We also set $N(u)_n = N(u) \cap W(X)_n$ and define $f_u(n) = |N(u)_n|$, that is, $f_u(n)$ is the number of all words of degree n without subword u.

Lemma 10.32. *Let X satisfy Conditions G and W_z. Then for every nonempty word $u \in W(X)$ there exist positive constants C and ε such that*

$$f_u(n) < C(z - \varepsilon)^n.$$

Proof. With u fixed, let us write $f(n) = f_u(n)$. If u is a subword of u' then, clearly, $N_n(u) \subset N_n(u')$. Since X has at least two letters, it is easy to include u as a subword in a word u' such that no proper prefix of u' is a suffix of u'. Therefore we may assume from the very beginning that the word u itself enjoys this property.

It follows from Condition W_z that $F(\zeta)$ is continuous in a neighborhood of 0. Let us choose a small positive $\varepsilon \in (0,1)$ such that $F(\varepsilon) < 1 + \frac{1}{z^\ell}$, where $\ell = \deg(u)$. Clearly, one can choose C so that $f(n) < C(z-\varepsilon)^n$ for every $n = 1, \ldots, \ell - 1$. To show that the same inequality also holds for all $n \geq \ell$, we first apply the argument of Lemma 10.29. Namely, we use

$$N(u)_n \subset \bigsqcup_{i=1}^{n} N(u)_{n-i}X_i \tag{10.16}$$

to produce an upper bound

$$f(n) \leq \sum_{i=1}^{n} f(n-i)k_i.$$

Let us note that the union on the right-hand side of (10.16) contains some words with subword u. For example, such are the words in $N(u)_{n-\ell}u$. Indeed, if $u = u'x$ with $x \in X_s$, and $v \in N_{n-\ell}(u)$ then $vu = (vu')x \in N_{n-s}(u)X_s$. This follows because, considering v has no subword u, if vu' has such a subword then $v = v'p$, $u = pq$, and $u' = qr$, for some words v', p, q, r, with q nonempty. Then a nonempty word q would be a suffix and a prefix of u at the same time, which contradicts our assumption about u. Since the number of the words in $N(u)_{n-\ell}u$ is $f(n-\ell)$, a stronger inequality holds:

$$f(n) \leq -f(n-\ell) + \sum_{i=1}^{n} k_i f(n-i). \tag{10.17}$$

Now, according to our Condition G, we either have (G1) $k_2 = k_3 = \cdots = 0$ or (G2) for every i, if $k_i > 0$ then $k_{i+1} > 0$. Let us handle these cases separately.

In the case (G1), all letters are contained in X_1 and $f(n-1) \leq f(n-\ell)z^{\ell-1}$ by Lemma 10.29 since in this case every word of degree $n-1$ is a product of a word of degree $n-\ell$ and a word of degree $\ell-1$. Therefore we obtain from (10.17) that

$$f(n) \leq -f(n-1)z^{1-\ell} + \sum_{i=1}^{n} k_i f(n-i) \leq \sum_{i=1}^{n} k_i' f(n-i), \tag{10.18}$$

where $k_1' = k_1 - z^{1-l} \geq 0$ and $k_j' = k_j$ for $j \neq 1$. Note that in this case

$$\frac{k_1'}{z-\varepsilon} = \frac{k_1}{z-\varepsilon} - \frac{1}{z^{\ell-1}(z-\varepsilon)} \leq \frac{k_1}{z-\varepsilon} - \frac{1}{z^\ell}. \tag{10.19}$$

In the case (G2), since the last letter x of u is in X_s, we have $k_s > 0$. Also $\ell = \deg(u) = \deg(u'x) \geq \deg(x) = s$. As a result, we have $k_\ell > 0$ and so can write

$$f(n) \leq -f(n-\ell) + \sum_{i=1}^{n} k_i f(n-i) \leq \sum_{i=1}^{n} k_i' f(n-i), \tag{10.20}$$

where $k_\ell' = k_\ell - 1 \geq 0$ and $k_j' = k_j$ for $j \neq \ell$. In this case we have

$$\frac{k_\ell'}{(z-\varepsilon)^\ell} \leq \frac{k_\ell}{(z-\varepsilon)^\ell} - \frac{1}{z^\ell}. \tag{10.21}$$

Now we can make conclusions which are the same both for (G1) and (G2). Since the coefficients k_i' are nonnegative, we conclude from the inductive hypothesis and equations (10.18) and (10.20) that

$$f(n) \leq \sum_{i=1}^{n} k_i' f(n-i) \leq \sum_{i=1}^{n} C k_i' (z-\varepsilon)^{n-i} = C(z-\varepsilon)^n \sum_{i=1}^{n} \frac{k_i'}{(z-\varepsilon)^i}. \tag{10.22}$$

Hence, both in cases (G1) and (G2), it follows from equation (10.22), the definition of the coefficients k_i', and equations (10.19) and (10.21) that

$$\begin{aligned} f(n) &\leq C(z-\varepsilon)^n \left(-\frac{1}{z^\ell} + \sum_{i=1}^{n} \frac{k_i}{(z-\varepsilon)^i} \right) \\ &\leq C(z-\varepsilon)^n \left(-\frac{1}{z^\ell} + F(\varepsilon) \right) \\ &< C(z-\varepsilon)^n \end{aligned}$$

because $F(\varepsilon) < 1 + \frac{1}{z^\ell}$. The proof is now complete. □

10.2.5 Cogrowth of subideals

This section contains the proof of Theorem 10.12. We will be using the notions and results about free associative and Lie algebras given in Section 10.2.1.2.

Lemma 10.33. *Let $L = L(X)$ be a free Lie algebra with free linearly ordered basis X. Let w be an LS-word, $x \in X$ the maximal letter involved in w, and $c = [w]$ the (unique) LS-commutator with associative support w. If x^ℓ is a subword of w for some $\ell > 0$ then c is an element of the ℓ-subideal closure $\mathrm{id}_L^\ell x$ of L generated by x.*

Proof. By the definition of the lexicographical order, it follows that $w = x^\ell w'$. According to Corollary 10.23, $c = [x, [u_1], [u_2], \ldots, [u_s]]$, where $u_1 \leq u_2 \leq \cdots \leq u_s$. If $\ell = 1$ then c is an element of the ideal generated by x. So we can proceed by induction on ℓ with basis $\ell = 1$. Assume $\ell > 1$. Since $[u_1]$ and c are LS-commutators, u_1 cannot be a power of x, hence u_1 has a proper prefix $x^{\ell-1}$. By the property of the lexicographical order, $x^{\ell-1}$ is also a prefix in each $u_2, \ldots, u_s$. Then by induction $[u_1], \ldots, [u_s]$ are all in the $(\ell - 1)$-subideal generated by x, hence, c is an element of the ℓ-subideal $\mathrm{id}_L^\ell x$ generated by x. □

Lemma 10.34. *Let $L = L(X)$ be a free Lie algebra with the graded free basis X satisfying Conditions G and W_m. Let c be a nonzero linear combination of several free generators of the same degree in L. Then, for any $\ell \geq 1$, there are $C, \varepsilon > 0$ such that the cogrowth function of the ℓ-subideal H generated by c in L does not exceed $C(m - \varepsilon)^n$.*

Proof. We will use the fact that given a linear order on X, a linear basis of L_n is given by LS-commutators of degree n. Hence the linear basis of L mod H can be chosen as a subset of the set of LS-commutators.

Let us first consider the case where $c = y$, an element of X. We can order X so that y is the greatest element with respect to this order. According to Lemma 10.33, then any LS-commutator whose associative support contains a subword y^ℓ is an element of H. According to Lemma 10.32, there are positive C, ε such that the number of associative words without a subword y^ℓ is bounded from above by $C(m - \varepsilon)^n$, for any n. Thus $d_{L/H}(n) < C(m - \varepsilon)^n$. Since $g_{L/H}(n) = d_{L/H}(1) + \cdots + d_{L/H}(n)$, we immediately observe that there is $C' > 0$ such that $g_{L/H}(n) \leq C'(m - \varepsilon)^n$, as needed.

Now assume c is not necessarily one letter and c involves a generator y with a nonzero coefficient. We have $\deg(c) = \deg(y)$. According to [37, Lemma 2.4.1], the set $X' = (X \setminus \{y\}) \cup \{c\}$ is another set of free generators of L. Notice that the degrees of elements of L with respect to X' remain to be the same as with respect to X. By the previous paragraph, $d_{L/H}(n) < C(m - \varepsilon)^n$, and the proof is complete. □

Lemma 10.35. *Let $L = L(X)$ be a free Lie algebra with the graded free basis X satisfying Conditions G and W_m. Let $w \in L$ be any nonzero homogeneous element. Then for any $\ell \geq 1$, there are $C, \varepsilon > 0$ such that the cogrowth function of the ℓ-subideal H generated by w in L does not exceed $C(m - \varepsilon)^n$.*

Proof. Let us consider a linear order on X such that $x < y$ if $\deg(x) < \deg(y)$. We choose the minimal generator $x \in X$. The minimal ideal L' of L containing all other generators has codimension 1 in L. By Lemma 10.24, the Lie algebra L' has a free basis X' consisting of all commutators z of the form $[y, x, \ldots, x]$, where $y \in X \setminus \{x\}$. The free basis X' of L' is a graded subset of L and therefore the degree of every element of L' with respect to X' is equal to its degree with respect to X.

Note that also X' satisfies Condition W_m. Indeed, assume $\deg(x) = s$. Removing x from X, we decrease the value $F(0)$ by $\frac{1}{m^s}$. Then adding all the commutators $[y, x, \ldots, x]$ to $X \setminus \{x\}$, we multiply the contribution of every such $y \in X \setminus \{x\}$ by $(1 + \frac{1}{m^s} + \frac{1}{m^{2s}} + \cdots)$. Hence the $F_m^{X'}(0)$ does not exceed

$$\left(1 - \frac{1}{m^s}\right)\left(1 + \frac{1}{m^s} + \frac{1}{m^{2s}} + \cdots\right) = 1.$$

Also the function $F_m^{X'}(\zeta)$ for X' remains defined in a neighborhood of 0. Indeed, we obtain this new function $F_m^{X'}(\zeta)$ by first subtracting $\frac{1}{(m-\zeta)^s}$ from the old function $F_m^X(\zeta)$. Then we multiply the difference obtained by the function

$$\left(1+\frac{1}{(m-\zeta)^s}+\frac{1}{(m-\zeta)^{2s}}+\cdots\right)=\left(1-\frac{1}{(m-\zeta)^s}\right)^{-1},$$

which is defined in the neighborhood of zero. Thus all components of Condition W_m remain in place.

Because we remove x of the smallest degree in X, also Conditions G holds for X'.

We will repeat the above construction applying it to L' and X', and so on. After t steps, we obtain the Lie algebra $L^{(t)}$ and its free basis $X^{(t)}$.

Now let us show that $\bigcap_{t=1}^{\infty} L^{(t)} = \{0\}$. Indeed, by our construction, the number of generators of degree 1 in X' is strictly less (if any) than in X (at the same time it is possible that we get more generators of degree 2). After several steps of application of our construction, there are no more generators of degree 1. Then, in the same manner, we get rid of all generators of degree 2, and so on. As a result, if u is an element of degree k and on the tth step we got rid of all generators of degree less than or equal to k, we have $u \notin L^{(t)}$.

So let us assume that $w \in L^{(t)} \setminus L^{(t+1)}$. We want to show that w is a linear combination of some elements of $X^{(t)}$. Suppose to the contrary that the expression for w in $X^{(t)}$ nontrivially includes some commutators of length n where $n > 1$. Because w is homogeneous, it cannot include commutators of length 1 equal to the generators of the minimal degree, and so $w \in L^{(t+1)}$ by the definition of the transition $L^{(t)} \to L^{(t+1)}$. This contradiction proves that a finite sequence of transitions described above will bring us to the case where w is a linear combination of the free generators of some $L^{(t)}$. This case has been dealt with in Lemma 10.34.

We have that the ℓ-subideal H' generated by w in $L^{(t)}$ has codimension growth in $L^{(t)}$ with respect to $X^{(t)}$ at most $C(m-\varepsilon)^n$. The same is true with respect to X, since the degrees of the elements in $L^{(t)}$ with respect to these bases are equal. Since the codimension of $L^{(t)}$ in L is t, the codimension growth of H' with respect to X in the entire L is bounded by $t + C(m-\varepsilon)^n$, where t does not depend on n. Since $H' \subset H$, we have the same upper bound for the cogrowth of H, and altering C we complete the induction. □

We are now in position to complete the proof of one of the main results of this chapter, Theorem 10.12.

Proof. Every nonzero subideal S contains, for some $\ell \geq 1$, an ℓ-subideal generated by one nonzero element w. If we prove our claim for such subideals S, we will be finished. Since we already have this proven in Lemma 10.35 for the case where w is homogeneous (in that case, with respect to the standard grading associated with the finite free basis of L), using part (b) of Lemma 10.20, it is sufficient to show the following.

Let w be a nonzero element of subalgebra H and u its leading part. Then for $\ell \geq 1$ the ℓ-subideal K_ℓ generated by u in $K = \operatorname{gr} H$ is contained in $\operatorname{gr} H_\ell$, where H_ℓ is an ℓ-subideal generated by w in H.

Indeed, by definition of K_1, any nonzero homogeneous element $v \in K_1$ is a linear combination of nonzero commutators $[u, u_1, \dots, u_s]$, where all u_i are homogeneous elements in K. In this case each u_i is a leading part of an element $w_i \in H$. Therefore, v itself is a leading part of a respective linear combination of commutators $[w, w_1, \dots, w_s] \in H_1$, so that $K_1 \subset \operatorname{gr} H_1$.

If $\ell > 1$ then we have $K_{\ell-1} \leq \operatorname{gr} H_{\ell-1}$, by induction. Since K_ℓ is an ideal in $K_{\ell-1}$ generated by the element u, it is contained in the ideal M generated by u in $\operatorname{gr} H_{\ell-1}$. But H_ℓ is an ideal generated by w in $H_{\ell-1}$. Hence, if we apply what we have proved for the ideals (that is, when $\ell = 1$), we obtain $M \subset \operatorname{gr} H_\ell$. As a result, $K_l \subset \operatorname{gr} H_\ell$, as needed. □

10.2.5.1 An example

In this section, based on Lemmas 10.33 and 10.36, we find the cogrowth function for the 2-subideal $\operatorname{id}_L^2 x$ generated by x in the free Lie algebra $L = L(x, y)$. We will need the following converse to Lemma 10.33 in our particular case.

Lemma 10.36. *Let $L = L(x, y)$. Then the set T of LS-commutators, different from x, whose associative supports do not have x^2 as a subword, are linearly independent modulo $H = \operatorname{id}_L^2 x$.*

Proof. Any LS-commutator without x^2 in the support (excluding x and y) is an element of the subalgebra M of L with free basis $\{[x,y], [x,y,y], \dots\}$. At the same time, M is a free factor in the ideal $N = \langle M, x\rangle = \operatorname{id}_L^1 x$. Both claims follow by Lemma 10.24. Therefore, M trivially intersects the ideal generated by x in N which is exactly $H = \operatorname{id}_L^2 x$. It then follows that all the set T of commutators in question, excluding y, being linearly independent in L and M, remains linearly independent modulo $\operatorname{id}_L^2 x$. Restoring y to the set does not change the situation because $y \notin \langle M, x\rangle$, while all the rest (including $\operatorname{id}_L^2 x$) is a subset of $\langle M, x\rangle$. □

We will use a linear basis of L composed of LS-commutators. It follows from Lemmas 10.33 and 10.36 that we need to find the growth of the set of associative LS-words different from x that do not have x^2 as a subword.

We first count the number of words of length n each being the product of the subwords y and yx. Let a_n denote the number of such words with suffix x and b_n with suffix y. Then $a_n = a_{n-2} + b_{n-2}$ (these are obtained by attaching yx to the words of length $n-2$) and $b_n = a_{n-1} + b_{n-1}$. Then also $b_{n-1} = a_{n-2} + b_{n-2} = a_n$. It follows that b_n is the nth term of the Fibonacci sequence $1, 1, 2, 3, 5, \dots$, while a_n is the $(n-1)$th term of the same sequence (we assume the zeroth term being 0).

The total number of these words of length n is now $a_n + b_n = b_{n-1} + b_n$, hence the $(n+1)$th term of the Fibonacci sequence. Actually, we have counted the number of words w of degree n, starting with y, such that no cyclic shift of w contains x^2, as a subword. We will say that w does not contain a *cyclic subword* x^2.

The set of words we have just counted is the set of all words of length n without subword x^2, with the exclusion of those words that have prefix x. It is obvious that the number of words of length n without cyclic subword x^2 with prefix x is equal to the number of words without cyclic subword x^2 with suffix x. Each of these latter can already be written via y and yx and has suffix x. Earlier, we denoted the number of such words as a_n. As a result, we have the total number c_n of the words of length n without cyclic subwords x^2 equal to $(a_n + b_n) + a_n$ (with the exception of x). Thus, c_n is the sum of the $(n+1)$th and $(n-1)$th terms of the Fibonacci sequence.

Now we want to delete from the above number those subwords that are powers of shorter words. Notice that every proper power is a proper prime power. Note that if v is a word without cyclic subwords x^2 then each power of v enjoys the same property. Now the computation of the number of words which are prime powers is routine. If $p|n$, where p is prime, then $c_{n/p}$ of our words will be pth proper powers. These must be removed, for each prime divisor p of n.

If $d = p_1p_2$ is the product of two different primes, then the dth powers have appeared in the above process twice, so while considering such powers we have to correct this error, etc. All this leads to a formula for the number d_n of words of length n without cyclic subwords x^2, where the Möbius function plays the role of the "corrector", $d_n = \sum_{d|n} \mu(d)c_{n/d}$. Although we considered only those divisors d which split as the product of pairwise different prime divisors, we have included in the preceding formula all divisors $d \mid n$ simply because $\mu(d) = 0$ as soon as d is divisible by a square of a prime number.

The set of all words we just counted splits into n-element subsets consisting of cyclic shifts of one of its elements. The greatest word in this subset, in the sense of the LEX-grading such that $x > y$, is an LS-word (see Section 10.2.1.2). As a result, the number of all LS-words of length n in the alphabet x, y, without cyclic subwords x^2 is now given by the formula $(1/n)\sum_{d|n} \mu(d)c_{n/d}$ (as earlier, x is not counted here). Since no LS-word of length ≥ 2 can have suffix x, this formula also gives the total number of LS-words of length n, without subword x^2.

Proposition 10.37. *Let $L = L(X)$ be a free Lie algebra of rank 2, with free basis X, x a letter in X, and S a 2-subideal of L generated by x. Let* Fib(n) *denote the nth Fibonacci number. Then the graded function $d_{L/S}$ of the relative cogrowth of S in L is given by*

$$d_{L/S}(n) = \frac{1}{n}\sum_{d|n} \mu(d)\left(\mathrm{Fib}\left(\frac{n}{d} - 1\right) + \mathrm{Fib}\left(\frac{n}{d} + 1\right)\right). \tag{10.23}$$

We can see from (10.23) that the growth of the functions $d_{L/S}$ and $g_{L/S}$ is exponential, with the base of exponent $\lim \sqrt[n]{g_{L/S}(n)}$ being the same number as for the Fibonacci sequence, that is, $\frac{1+\sqrt{5}}{2}$. Recall (see (10.10)) that for the whole of L the base of exponent is 2.

Table 10.1: Cogrowth of the 2-subideal generated by x in $L(x,y)$.

n	1	2	3	4	5	6	7	8	9	10
$d_{L/S}(n)$	1	1	1	1	2	2	4	5	8	11
$g_{L/S}(n)$	1	2	3	4	6	8	12	17	25	36
n	**11**	**12**	**13**	**14**	**15**	**16**	**17**	**18**	**19**	**20**
$d_{L/S}(n)$	18	25	40	58	90	135	210	316	492	750
$g_{L/S}(n)$	54	79	119	177	267	402	612	928	1420	2170

The first values of the functions $d_{L/S}$ and $g_{L/S}$ are given in Table 10.1.

10.2.6 Exponential base of relative growth

The approach suggested in Section 10.2.4 allows us to prove our claims about the exponential base of the growth of (not necessarily finitely generated) subalgebras in free Lie algebras, as stated in Theorem 10.13. To be able to apply the results of Section 10.2.4, we need an auxiliary lemma.

Lemma 10.38. *Let A and L be a free associative algebra and a free Lie algebra, with the same graded free basis Y. Suppose $\lim_{n\to\infty}\sqrt[n]{g_A(n)}$ exists and >1. Then $\beta_L = \lim_{n\to\infty}\sqrt[n]{g_L(n)}$ exists and is equal to $\lim_{n\to\infty}\sqrt[n]{g_A(n)}$.*

Proof. If $d|n$ then the product of any d words of degree n/d is a word of degree n, which implies $d_A(n/d)^d \le d_A(n)$. If, additionally, $d \ge 2$ then $d_A(n/d) \le \sqrt{d_A(n)} \le \sqrt{g_A(n)}$. But $d_A(n/d)$ is the number of words of degree n which are d-powers. After summation over all divisors $d \ge 2$, we will find that the number of words of degree n does not exceed $n\sqrt{g_A(n)}$.

Now the existence of $\lim_{n\to\infty}\sqrt[n]{g_A(n)} > 1$ means that $\sqrt{g_A(m)} < g_A(n)^{2/3}$, for all sufficiently great n and all $m \le n$. Therefore, after summation over $m \le n$, we will have for large enough n that the number of proper powers among the words of degree at most n is bounded from above by $n^2 g_A(n)^{2/3}$, which is $o(g_A(n))$, because $g_A(n)$ is exponential. It follows that the number of words of degree $\le n$, which are not proper powers is $g_A(n)(1-o(1))$.

Since among the cyclic shifts of any such word of degree $m \le n$ precisely one is an LS-word, the number of these latter takes the form of $g_A(n)(1-o(1))\gamma(n)$, where $\gamma(n) \in [1/n, 1]$. By Lemma 10.21, the number of LS-words of length at most n is $g_L(n)$. Taking the nth roots of these numbers, we arrive at the following:

$$\lim_{n\to\infty}\sqrt[n]{g_L(n)} = \lim_{n\to\infty}\sqrt[n]{g_A(n)},$$

as claimed. □

Now we can proceed to the proof of Theorem 10.13.

Proof. Recall that we are dealing with the relative growth a subalgebra H of a free Lie algebra L of rank m with a nongraded free basis X. Using Lemmas 10.20 and 10.26 allows us to always assume that H is a homogeneous subalgebra generated by an irreducible free basis Y, in which the number of elements of degree i equals k_i, for $i = 1, 2, \dots$ The relative growth of H in L is the same as the absolute growth of a free Lie algebra $L(Y)$, where Y is viewed as an abstract graded set. The degree of each $y \in Y$ equals the degree of y with respect to X in L. Let us embed $L(Y)$ in a free associative algebra $A(Y)$. Lemma 10.38 allows us to translate our claims about H to the claims about $A(Y)$.

We start with claim (i), where H is not necessarily finitely generated but is nonabelian. Let us use Lemma 10.31 (notice that in that lemma the graded set of generators is denoted by $X!$). Since $\dim L_i \le m^i$, it follows that $k_i \le m^i$, for each $i = 1, 2, \dots$ It follows that for any $z > m$ the series $F(0)$, as a function of z, converges, hence z_0 in formula (10.15) exists and provides us with the exponential base for the growth of $A(Y)$, hence for the growth of $L(Y)$ (Lemma 10.38), hence for the relative growth of H in L. As noted in Lemma 10.31, if $|Y| > 1$, the growth is exponential. Thus we have proved claim (i).

In the proof of claim (ii), the function $F(\zeta) = F_z^Y(\zeta)$ takes the form of

$$F(\zeta) = \frac{k_1}{z-\zeta} + \frac{k_2}{(z-\zeta)^2} + \dots + \frac{k_d}{(z-\zeta)^d}, \tag{10.24}$$

where d is the maximal degree of the elements in Y, and each k_i is the number of elements of degree i, $i = 1, 2, \dots, d$. As in the proof of claim (i), we use Lemma 10.31. To be able to apply formula (10.15), we need to determine zeros of $F(0) - 1$, as a function of z. One can write $F(0)$, as a function of the real argument $z > 0$, in the form $\frac{f(z)}{z^d}$, where $f(z)$ is a polynomial of degree less than d, with nonnegative coefficients and positive free term k_d. Using the Descartes rule of signs, the equation $z^d - f(z) = 0$ has a unique positive root z_0. Since z_0 is a root of a monic polynomial with integral coefficients, our claim about the algebraic integrality of the exponential base of H follows. The exponential base is less than m, which follows from the exponential negligibility of Theorem 10.11.

We prove (iii) for the case $m = 2$, the general case being quite analogous. Choose $m_0 \in (1, 2)$. Let M be the subalgebra of codimension 1, with free basis $u_1 = x$, $u_2 = [x, y], \dots$, given in Lemma 10.24. For this subalgebra, $\sum_{i=1}^{\infty} \frac{1}{2^i} = 1$. Suppose we could discard few summands in the series so that

$$\sum_{i=1}^{\infty} \frac{k_i}{m_0^i} = 1, \tag{10.25}$$

where $k_i = 0$ if u_i is discarded and $k_i = 1$ otherwise. Let B be the subalgebra of M generated by those u_i for which $k_i = 1$. Then, according to our previous argument, m_0 will be the exponential base for the B.

To find the sequence $k_1, k_2, \dots$ so that (10.25) holds, let us assume by induction that we have chosen $k_1, \dots, k_j$ so that $1 - \sum_{i=1}^{j} \frac{k_i}{m_0^i} \in [0, \frac{1}{m_0^j})$. Let us set $a_j = 1 - \sum_{i=1}^{j} \frac{k_i}{m_0^i}$. If

$a_j < \frac{1}{m_0^{j+1}}$, then we set $k_{j+1} = 0$, otherwise, $k_{j+1} = 1$. Since $m_0 < 2$, in either case, a_{j+1} is in the desired interval. Since $m_0 > 1$, it follows that $a_j \to 0$. Hence $\sum_{i=1}^{\infty} \frac{k_i}{m_0^i} = 1$, proving claim (iii).

Claim (iv) is an easy consequence of the construction in the proof of claim (iii). Indeed, the subalgebra H in that proof is generated by a subset of the set of the canonical free basis of M. Let S be the ideal of M generated by the complement of that set. We have a vector space decomposition $L = H \oplus S \oplus \langle y \rangle$, where all subspaces are homogeneous. In this case, for any $n \geq 1$, $L^{(n)} = (L^{(n)} \cap H) \oplus (L^{(n)} \cap S) \oplus \langle y \rangle$. Hence,

$$g_{L/S}(n) = \dim L^{(n)}/(L^{(n)} \cap S) = \dim(L^{(n)} \cap H) + 1 = g_H(n) + 1.$$

It follows that the exponential base of $g_{L/S}$ is he same as that for g_H. Finally, $M/S \cong H$, a free Lie algebra. □

We conclude this section by exhibiting a positive algebraic integer which is not the exponential base of any finitely generated subalgebra. For instance, if H is generated in $L(x, y)$ by x, $[x, y]$ then by claim (ii) in Theorem 10.13, we have to solve $1 + z = z^2$. As a result, we obtain $z_0 = \frac{1+\sqrt{5}}{2}$. As a side remark, not every positive algebraic integer $< m$ is the exponential base for the growth of a finitely generated subalgebra in a free Lie algebra of rank m. For instance, $\lambda = (5 - \sqrt{5})/2$, which is a positive root of $z^2 - 5z + 5$, cannot serve as such a base. Indeed, if λ is the solution to the equation with integral coefficients $f(z) = z^d$ then another root $\mu = (5 + \sqrt{5})/2 > 0$ of $z^2 - 5z + 5$ is also a solution. At the same time, by the argument in the proof of claim (ii), the positive solution z_0 of $f(z) = z^d$ must be unique!

10.2.7 More about subalgebras and subideals

In this section we prove additional results about subalgebras and subideals in free Lie algebras mentioned in Section 10.2.

10.2.7.1 Free complements

The main contents of this section is the proof of Theorem 10.15.

Proof. We first consider the case where H is a homogeneous subalgebra in L with a homogeneous set B_0 of free generators. Since B_0 is finite, there is a t equal to the maximum of degrees of all elements in B_0. For the homogeneous component L_t of L, we have $L_t = (H \cap L_t) \oplus M_t$, for some (homogeneous) complementary subspace M_t. Let us denote by B_1 a linear basis of M_t and consider the subalgebra H_1 generated by H and B_1. We want to prove that $B(1) = B_0 \sqcup B_1$ is a free basis for H_1.

For this, it is necessary and sufficient (see [37, Theorem 4.2.11]) to show that $B(1)$ generates H_1 (this is clear) and that the elements of $B(1)$ are linearly independent modulo the commutator subalgebra $[H_1, H_1]$. Let us denote by L^m the mth term of the lower central series for L. We have $L^m = L_m \oplus L_{m+1} \oplus \cdots$. Since $[M_t, H] \subset L^{t+1}$, we have that $[H_1, H_1] \subset [H, H] \oplus L^{t+1}$. So a nontrivial linear dependence of the elements of $B(1)$ modulo $[H_1, H_1]$ would imply a nontrivial linear dependence of the elements of $B(1)$ modulo $[H, H] + L^{t+1}$. The subspaces $[H, H]$ and L^{t+1} are homogeneous. It follows then that we can consider the linear dependence only for homogeneous elements of $B(1)$ of the same degree d.

If $d < t$ then neither L^{t+1} nor B_1 has homogeneous elements of degree d. Thus we have a nontrivial linear dependence of some elements of B_0 modulo $[H, H]$. This is not possible because B_0 is a free basis of H. The case $d > t$ is vacuous because neither B_0 nor B_1 has elements of such degree. Now if $d = t$ then neither B_0 nor L^{t+1} has elements of degree t. As a result, we have a nontrivial linear dependence between the elements of B_1 modulo $[H, H] \cap L_t \subset H \cap L_t$. This is not possible by the choice of B_1 as a basis of a direct complement to $H \cap L_t$ in L_t. Thus we have proved that $B(1)$ is indeed a free basis for H_1.

Since H_1 is homogeneous and the degrees of elements in $B(1)$ are less than $t + 1$, we can repeat the previous construction by replacing t with $t + 1$. So we write $L_{t+1} = (H \cap L_{t+1}) \oplus M_{t+1}$, choose a linear basis B_2 in M_{t+1}, set $B(2) = B_0 \sqcup B_1 \sqcup B_2$, introduce the subalgebra H_2 generated by $B(2)$, prove that $B(2)$ is a free basis of H_2, etc. The union $C = \bigcup_{i=1}^{\infty} H_i$ is a free Lie subalgebra with basis $B = \bigsqcup_{i=0}^{\infty} B_i$. Clearly, H is a free factor in C.

By construction, any homogeneous polynomial of degree $\geq t$ is an element of C. It follows that the codimension of C in L is at most the dimension of the tth term $L^{(t)}$ of the degree filtration of L. Since L is finitely generated, we have $\dim L^{(t)} < \infty$, proving that C is of finite codimension in L.

Now suppose that H is not necessarily homogeneous. Recalling Section 10.2.2, we choose a finite irreducible set S of free generators for H and consider the set S' of leading parts of the elements of this set. Let H' be the subalgebra generated by S'. By Lemma 10.25, S' is independent. Let us set $B_0 = S'$ and proceed as just above to produce a homogeneous subalgebra C' of finite codimension in L with the set B' of free generators such that $B' = B_0 \bigsqcup \overline{B_0}$, where $\overline{B_0} = \bigsqcup_{i=1}^{\infty} B_i$. Now let us consider $B = S \bigsqcup \overline{B_0}$. Then the set of the leading parts of elements in B is B'. Since B' is independent, by Lemma 10.25, the same is true for B.

Hence B is the free basis of a subalgebra C. Invoking Lemma 10.26, we have that $C' = \operatorname{gr} C$. If K is a subalgebra generated by $\overline{B_0}$, we have $C = H * K$. Notice that by construction the subalgebra K is homogeneous.

It remains to note that $\dim L/C < \infty$, which follows by part (b) of Lemma 10.20. Now the proof is complete. □

It is worth noting that in a particular case where the subalgebra H of a free Lie algebra L is generated by its homogeneous component of degree t, it is true that H is a free factor in the tth term L^t of the lower central series of L. Probably, this result was already known to A. I. Shirshov.

10.2.7.2 Triviality of ideals in subideals

We start this section with a fairly general result about the derivations of free algebras, which we think could be of interest in its own.

Lemma 10.39. *Let A (resp., L) be the free associative (resp., Lie) algebra with free generators $x_1, x_2, \dots$ over an arbitrary field and D its derivation given on the generators by the rule $Dx_i = x_{i+1}$ $(i = 1, 2, \dots)$. Suppose a is a nonzero element of A (of L). Then for every $k \geq 1$, there exists $n \geq 0$ such that the element $D^n(a)$ does not belong to the ideal I_k (to the Lie ideal J_k) of A (resp., of L) generated by the set $\{x_1, \dots, x_k\}$.*

Proof. Let us begin with the associative case. We may assume that a is a homogeneous element of degree $c \geq 1$ and a is an element of a subalgebra A_l generated by $x_1, \dots, x_l$ for some $l \geq k$. We can write

$$a = \sum_{1 \leq i_1 \leq \ell, \dots, 1 \leq i_c \leq \ell} \alpha_{i_1,\dots,i_c} x_{i_1} \cdots x_{i_c}$$

for some scalars $\alpha_{i_1,\dots,i_c}$. We say that a c-tuple $(i_1, \dots, i_c)$ is an element of the *support* $\operatorname{Supp} a$ if $\alpha_{i_1,\dots,i_c} \neq 0$.

Let us define a sequence of numbers $K_1, K_2, \dots, K_c$ by setting

$$K_d = (1/2)(2k+2)^{2^{c-d}}, \quad \text{for any of } d = 1, 2, \dots, c.$$

Then $K_c = k + 1$ and $K_d = 2K_{d+1}^2$, for any $1 \leq d < c$. Our lemma is an easy consequence of the following.

Claim. *For any d, $1 \leq d \leq c$, there is $n \geq 1$ such that the support of $D^n(a)$ has a c-tuple $(i_1, \dots, i_c)$ with $\min\{i_1, \dots, i_d\} \geq K_d$.*

The estimate for the value of n can be recovered from the proof that follows. Once proven, our lemma follows because $K_c > k$.

For the proof, let us use induction with respect to d.

In the case $d = 1$, our task is easy. Indeed, let $(i_1, \dots, i_c) \in \operatorname{Supp} a$ with the greatest possible i_1. Then the only c-tuple $(i_1 + 1, i_2, \dots, i_c)$ comes to $\operatorname{Supp} D(a)$ from

$$D(\alpha_{i_1,i_2,\dots,i_c} x_{i_1} x_{i_2} \cdots x_{i_c}) = \alpha_{i_1,i_2,\dots,i_c}(x_{i_1+1} \cdots x_{i_c} + x_{i_1} x_{i_2+1} \cdots x_{i_c} + \cdots),$$

and $i_1 + 1$ is the greatest first index in the members of $\operatorname{Supp} D(a)$. Then we repeat the argument and have $(i_1 + K_1, i_2, \dots, i_c) \in \operatorname{Supp} D^n(a)$ if $n = K_1$.

Now assume that the statement is true for some d $(1 \leq d < c)$. In this case, there is n such that the support of $b = D^n(a)$ includes a c-tuple $(i_1, \dots, i_c)$ with $\min\{i_1, \dots, i_d\} \geq K_d$. Let us split $\operatorname{Supp} b$ as the disjoint union

$$\operatorname{Supp} b = \Big(\bigsqcup_{j \geq K_{d+1}} S_j\Big) \bigsqcup S,$$

where $(i_1,\dots,i_c) \in S_j$ if $\min\{i_1,\dots,i_d\} = j$, $j \geq K_{d+1}$. The c-tuples in S satisfy $\min\{i_1,\dots,i_d\} < K_{d+1}$. The maximal j with nonempty S_j satisfies the inequality $j \geq K_d$ because of the way we have obtained b.

To proceed, we introduce a function $f(j)$ ($j \geq K_{d+1}$) whose jth value equals to the maximal value of the $(d+1)$th coordinate of the tuples in the subset S_j; we set, by definition, $f(j) = 1$ if the subset S_j is empty.

If $f(j) \geq K_{d+1}$ for some $j \geq K_{d+1}$, then there is a c-tuple $(i_1,\dots,i_c) \in S_j$ such that $\min(i_1,\dots,i_{d+1}) \geq K_{d+1}$, which is sufficient to complete the induction step. Therefore, we further assume that $f(j) < K_{d+1}$ for every $j \geq K_{d+1}$, and so $f(j)$ takes less than K_{d+1} different values.

It follows that the length of any decreasing sequence $j(1) > j(2) > \cdots$ such that the values $f(j(1)) < f(j(2)) < \cdots$ increase is less than K_{d+1}. Recall that the maximal value of j with nonempty S_j is at least K_d. Let us choose a particular decreasing sequence as follows. There exists $j(1) \geq K_d$ such that $f(j(1)) \geq f(q)$ for every $q \geq j(1)$. If for all $j < j(1)$ we have $f(j) \leq f(j(1))$ then we are done. Otherwise we keep working and define $j(2)$ as the maximal index such that $j(2) < j(1)$ while $f(j(2)) > f(j(1))$. Proceeding in the same way, we eventually arrive at a finite sequence $j(1) > j(2) > \cdots > j(s)$ with $f(j(1)) < f(j(2)) < \cdots < f(j(s))$, which cannot be extended any further. Here we additionally have $s < K_{d+1}$ and $f(j(t)) \geq f(q)$, for arbitrary $t \leq s$ and $q \geq j(t)$.

Notice that it is not possible that every number in the sequence $j(1) - j(2),\dots,j(s-1) - j(s), j(s)$ is strictly less than $2K_{d+1}$. Indeed, if this were the case, we would be able to write

$$K_d \leq j(1) = (j(1) - j(2)) + \cdots + (j(s-1) - j(s)) + j(s) < 2K_{d+1}K_{d+1}.$$

Therefore, either $j(t) - j(t+1) > 2K_{d+1}$ for some $t < s$ or $j(s) > 2K_{d+1}$. In both cases, we can conclude that there is a number m such that the following are true:

$$m \geq 2K_{d+1} \quad \text{and} \quad f(m) \geq f(q) \quad \text{for any } q \geq m - K_{d+1}.$$

Now we take $(i_1,\dots,i_d, f(m),\dots,i_c) \in S_m$, consider the respective monomial $u = x_{i_1}\cdots x_{i_d} x_{f(m)} \cdots x_{i_c}$, entering b with nonzero coefficient, and compute the elements $D^{K_{d+1}}(u)$ and $D^{K_{d+1}}(b)$. The support of $D^{K_{d+1}}(u)$ includes the c-tuple

$$\mathbf{i} = (i_1,\dots,i_d, f(m) + K_{d+1}, i_{d+2},\dots,i_c).$$

This follows exactly as in the case $d = 1$ above. The first $d+1$ indexes of this tuple are not less than K_{d+1}, by the definition of S_m. Therefore to complete the induction, it suffices to prove that for any monomial v occurring in b with nonzero coefficient, the tuple $\mathbf{i}$ does not appear in the support of $D^{K_{d+1}}(v)$. Let $\operatorname{Supp} v = \{(i'_1,\dots,i'_d,\dots,i'_c)\}$.

Case 1. $\operatorname{Supp} v \in S_q$, where $q \geq m - K_{d+1}$. By the choice of m, $f(m) \geq f(q)$. Our argument, as just above, shows that $D^{K_{d+1}}(v)$ has at most one monomial with multiindex of the form

$$\mathbf{i}' = (i'_1,\dots,i'_d, f(m) + K_{d+1},\dots,i'_c).$$

Since $\mathbf{i}' \neq \mathbf{i}$, we have $\mathbf{i} \notin \operatorname{Supp} D^{K_{d+1}}(v)$.

Case 2. $\operatorname{Supp} v \in S_q$, where $q < m - K_{d+1}$. According to the definitions of S_q and S_m, we have $\min(i'_1, \dots, i'_d) = q < m - K_{d+1}$. Each of the K_{d+1} derivations can increase the minimum of the first d indices in the members of the support at most by 1. Hence $\mathbf{i} \notin \operatorname{Supp} D^{K_{d+1}}(v)$.

Case 3. $\operatorname{Supp} v \in S$. Then by the definitions of D and S, one of the first d indices of any term of $D^{K_{d+1}}(v)$ is less than $2K_{d+1}$. Hence, by the definition of m and S_m, $\mathbf{i} \notin \operatorname{Supp} D^{K_{d+1}}(v)$.

Thus the proof is complete in the case of associative algebras. The case of Lie algebras reduces to the associative case because the free generators of A also freely generate L with respect to the bracket operation, $J_k \subset I_k$, and the derivation D of L extends to a derivation of A (see Section 2.5.5). □

Theorem 10.40. *Let J be a proper ideal of a free Lie algebra L and S a finite subset of J. Let $I = \operatorname{id}_J S$ be the ideal closure of S in J. Then for every $z \in L \backslash J$ and for every nonzero $a \in I$ there is n such that $[a, z, \dots, z]$ (n times) is not an element of I.*

Proof. One may assume that $L = \langle J, z \rangle$ where J is an ideal of codimension 1 in L. Then an element $z' = z - v$, where $v \in J$ can be included in a free basis of L. If for any number n of occurrences of z' we had $b_n = [a, z', \dots, z'] \in I$ then we would also have that all $a_n = [a, z, \dots, z] \in I$. To prove this, we first note that if we replace z by $z' + v$ then we will have the sum of left-normed commutators $w = [a, u_1, u_2, \dots, u_n]$, where each u_i is either z' or v. Using the Jacobi identity $[v, s, t] = [v, t, s] + [v, [s, t]]$, we can exchange z' and v in w if $z = t$ follows $v = s$ in w. The resulting commutators would still be of the form of w but now u_i would be either z' or some elements of J. If there is still some z' to the right of some u_i, we repeat the process. When all z' are to the left of all u_i we have the elements of the form $[a, z', \dots, z', u_1, \dots, u_k]$. The initial portion $[a, z', \dots, z']$ is in I, by our assumption. Since I is an ideal of J, the whole commutator is in I, as well.

Hence we may assume that z is an element of the free basis of L. Then the inner derivation $\operatorname{ad} z$ shifts the free generators of J (see Lemma 10.24), so we are tempted to apply Lemma 10.39. However, a direct application is not possible because there can be many different "orbits" for the action of $\operatorname{ad} z$ on the free basis of J. Hence we need an easy modification of Lemma 10.39, as follows.

Lemma 10.41. *Let A (resp., L) be the free associative (resp., Lie) algebra with the set X of free generators split as a disjoint union of subsets $X^{(j)} = \{x_1^{(j)}, x_2^{(j)}, \dots\}$ over arbitrary field and δ its derivation given on the generators by the rule $\delta x_i^{(j)} = x_{i+1}^{(j)}$ ($i = 1, 2, \dots$). Suppose a is a nonzero element of A (of L). Then for every finite subset S of X, there exists $n \geq 0$ such that the element $\delta^n(a)$ does not belong to the ideal I_S (to the Lie ideal J_S) of A (resp., of L) generated by S.*

Proof. This is a direct consequence of Lemma 10.39. Indeed, considering, say, the associative case, let us assume that, for any n, we have $\delta^n(a) \in I_S$. Let us consider a homomorphism of associative algebras φ from $A(X)$ to the free associative algebra $A(x_1, x_2, \ldots)$ of Lemma 10.39, defined in the following way. Suppose an expression for a in $A(X)$ includes only at most m first letters from each of subalphabets $X^{(j)}$, $j = 1, 2, \ldots$ Then we set $\varphi(x_i^{(1)}) = x_i$, $\varphi(x_i^{(2)}) = x_{i+m}$, $\varphi(x_i^{(3)}) = x_{i+2m}$, etc., where $i = 1, 2, \ldots$ Clearly, under this homomorphism, $\varphi(a) \neq 0$ because we simply applied a bijective change of the sets of free variable involved in a. Now both algebras $A(X)$ and $A(x_1, x_2, \ldots)$ are equipped with derivations, δ in the former case and D, as in Lemma 10.39, in the latter, that is, $D(x_i) = x_{i+1}$, $i = 1, 2, \ldots$ Clearly, φ is a *differential homomorphism*, that is, $\varphi \circ \delta = D \circ \varphi$.

Since we assumed that $\delta^n(a) \in I_S$, for all n, applying φ would mean that $D^n(\varphi(a)) \in I_{\varphi(S)}$. This comes as a contradiction to Lemma 10.39. □

Theorem 10.40 follows from Lemma 10.41. □

Clearly, Theorem 10.17 from the introduction is the direct consequence of the just proven Theorem 10.40.

Now we came closer to the proof of Corollary 10.18 from the introduction.

Proof. Let $J(i)$ be the ith subideal closure of a finite set S of elements of $J = J(0)$, as in Corollary 10.18, and $L \supset M_1 \supset \cdots$ an arbitrary subideal series in L with nonzero terms. We need to show that M_ℓ is not contained in $J(\ell)$.

Let us first prove that setting $N_i = J(i-1) \cap M_i$, $i = 1, 2, \ldots$, the series $L \supset N_1 \supset N_2 \supset \cdots$ is also a subideal series with nonzero terms. First of all, if $P \cap Q = \{0\}$ for $\{0\} \neq P, Q \lhd L$, $a \in P$, $b \in Q$ then $[a, b] = 0$. Now the subalgebra M generated by a, b is free; since a, b commute, M must be abelian, hence $\dim M = 1$. So a, b are scalar multiples of each other, a contradiction to $P \cap Q = \{0\}$. Therefore, the intersection N_1 of two nonzero ideals M_1 and $J = J(0)$ of a free Lie algebra L is again nonzero. Now since N_1 and $J(1)$ are two nonzero ideals of $J(0)$, their intersection $N_1 \cap J(1)$ is nonzero. Similarly, $M_2 \cap N_1$ is nonzero. Hence $(M_2 \cap N_1) \cap (N_1 \cap J(1))$ is nonzero, being the intersection of two nonzero ideals of N_1. As a result, $N_2 = M_2 \cap J(1) \neq \{0\}$. Continuing in the same way, we find that all $N_i = M_i \cap J(i-1)$ are nonzero.

As a result, when we resume proving Corollary 10.18, we may assume that $M_\ell \subset J(\ell - 1)$, for all ℓ. So we need to prove that $M_\ell \not\subset J(\ell)$, for any $\ell = 1, 2, \ldots$ Assume the contrary, that is, $M_\ell \subset J(\ell)$, for some $\ell = 1, 2, \ldots$ If $\ell = 1$, we can apply Theorem 10.17 because J is a proper ideal in L while $J(1)$ is an ideal generated in $J = J(0)$ by a finite subset of elements. It follows that $J(1)$ does not contain an ideal M_1 of the whole algebra L.

Thus, if there is ℓ such that $M(\ell) \subset J(\ell)$ then $\ell > 1$. In the case this happens, let ℓ be minimal with this property. Consider the ideal closure R of $M(\ell)$ in $J(\ell - 1)$. Then R is contained in $J(\ell)$, which is the ideal closure of the finite set $S \subset J(\ell - 1)$. By the minimality of the choice of ℓ, there is $z \in M_{l-1}$, which is not an element of $J(l - 1)$. As established earlier, $z \in M(l - 1) \subset J(l - 2)$. Since $[z, M(\ell)] \subset M(\ell)$ and $[z, J(\ell - 1)] \subset J(\ell - 1)$, we have that $[z, R] \subset R$. This easily follows using the Jacobi identity because every element of R is a linear combination of the commutators of the form $[y, x_1, \ldots, x_m]$, where $y \in M(\ell)$ and $x_1, \ldots, x_m \in J(\ell - 1)$. As a result, R is a nonzero ideal in the subalgebra P generated by $J(\ell - 1)$ and z. This contradicts Theorem 10.17 applied to the algebra P and its proper ideal $J(\ell - 1)$. Hence, the proof of Corollary 10.18 is complete. □

Remark. In the case of associative algebras, we have a very simple example when there is an ideal I generated by finitely many elements in a proper ideal J of a free associative algebra A, and still containing a nonzero ideal of A. For this, one can take A with free basis $\{x\}$, J the principal ideal of A generated by x^2, $S = \{x^2, x^3\}$, and I the ideal of J generated by S. The reader would easily check that the nonzero ideal of A contained in I is I itself!

11 Identities of graded Lie algebras

11.1 Graded Lie algebras and identities

Let L be a Lie algebra over a field $\mathbb{F}$ and G a (multiplicative) group. By definition (see Section 1.10.1), L is G-graded if

$$L = \bigoplus_{g \in G} L_g \tag{11.1}$$

so that for any $g, h \in G$ one has $[L_g, L_h] \subset L_{gh}$. The subset $\operatorname{Supp} L = \{g \in G \mid L_g \neq 0\}$ is called the *support* of the grading. If L is a simple algebra, a well-known results by Bahturin–Zaicev or Draper–Martin (see [115, Proposition 1.12]) that $gh = hg$ for any $g, h \in \operatorname{Supp} L$. So if $\overline{G}$ is a subgroup generated by $\operatorname{Supp} L$, this is an abelian group and $L = \bigoplus_{g \in \overline{G}} L_g$ is the same grading as above, but by an abelian group. So, usually we consider gradings of Lie algebras by abelian groups. A homomorphism $\varphi : L = \bigoplus_{g \in G} L_g \to M = \bigoplus_{g \in G} M_g$ is called a *G-homomorphism* if $\varphi(L_g) \subset M_g$ for all $g \in G$. If φ is an isomorphism then we say that L and M are *isomorphic as G-graded algebras* or simply *G-isomorphic*.

For the wealth of material about the gradings on algebras, in particular, simple Lie algebras, we refer the reader to the monograph of Alberto Elduque and Mikhail Kochetov [115]. At this time, the classification theory of gradings on many types of finite-dimensional simple Lie algebras over algebraically closed fields is complete, often even in the case of the fields of prime characteristic. Note that one can compare two gradings not only by isomorphism but also by equivalence. Algebras L and M as above are called *G-equivalent* if there exists an isomorphism $\varphi : L \to M$ such that for any $g \in G$ there is $\alpha(g) \in G$ such that $\varphi(L_g) = M_{\alpha(g)}$. In many cases the map $\alpha : G \to G$ turns out to be an automorphism G and then we say that L and M are *weakly isomorphic*. So the classification of gradings on simple Lie algebras can be up to isomorphism or up to equivalence. For more details, see [115].

11.1.1 Graded identities

The notion of a G-graded identity appears if we consider the *free G-graded Lie algebra* $L_G(X)$, where X is a G-graded set, that is, $X = \bigsqcup_{g \in G} X_g$ (the disjoint union of subsets X_g). This is just the ordinary free Lie algebra generated by X, turned to a G-graded algebra in an obvious way: $\deg x = g$ if $x \in X_g$ and by induction, if $w = [w_1, w_2]$ then $\deg w = (\deg w_1)(\deg w_2)$ (or $\deg w = \deg w_1 + \deg w_2$, if G is written additively). Similarly, one can define a free G-graded associative algebra $A_G(X)$. Clearly, $L_G(X)$ is a graded subspace (even a Lie subalgebra of $A_G(X)$) under the commutator structure of this latter algebra. Clearly, any Hall or Shirshov bases of $L_G(X)$ are graded bases of $L_G(X)$. Now given a nonzero element $f(x_1, \dots, x_n) \in L_G(X)$, with $\deg x_i = g_i$, $i = 1, \dots, n$, and a G-graded Lie algebra M, we say that $f(x_1, \dots, x_n) \equiv 0$ is a *G-identity* (*G*-

https://doi.org/10.1515/9783110566659-011

graded identity) in M if for any homogeneous elements $a_1 \in M_{g_1}, \dots, a_n \in M_{g_n}$, we have $f(a_1,\dots,a_n) \equiv 0$.

Let $\theta : G \to K$ be a homomorphism of groups and A a G-graded algebra. Denote by Γ the G-grading $A = \bigoplus_{g\in G} A_g$ on A. The corresponding graded identities will also be denoted by $\mathrm{Id}_G(\Gamma)$. We may obtain a K-grading on A, called *quotient grading induced by* θ and denoted by ${}^{\theta}\Gamma$, as ${}^{\theta}\Gamma : A = \bigoplus_{k\in K} A_k$, where the homogeneous components are $A_k = \bigoplus_{g\in\theta^{-1}(k)} A_g$. The following result gives a useful "folklore" criterion for the coincidence of the identities.

Proposition 11.1. *Let A and B be two algebras with G-gradings Γ_A and Γ_B, respectively, and $\theta : G \to K$ a homomorphism of groups. If $\mathrm{Id}_G(\Gamma_A) = \mathrm{Id}_G(\Gamma_B)$ then $\mathrm{Id}_K({}^{\theta}\Gamma_A) = \mathrm{Id}_K({}^{\theta}\Gamma_B)$.*

Proof. Let $f(x_{k_1 1},\dots,x_{k_n n})$ be a K-graded identity for A. To prove that the polynomial f is a K-graded identity for B, let $b_1,\dots,b_n$ be homogeneous elements of B such that $b_i \in B_{k_i}$, $i = 1,\dots,n$. We claim that $f(b_1,\dots,b_n) = 0$. To prove this, decompose b_i into its homogeneous components in the G-grading

$$b_i = b_{i1} + b_{i2} + \dots + b_{il_i},$$

for some integer l_i, and denote by g_{ij} the homogeneous degree of b_{ij}. Note that, since b_i has homogeneous degree k_i in the K-grading induced by θ, we have $\theta(g_{ij}) = k_i$. Let $S = \{(i,j) \mid 1 \le i \le n, 1 \le j \le l_i\}$, $m = |S|$ and $\nu : S \to \{1,\dots,m\}$ a bijection. Construct the polynomial

$$f' = f\left(\sum_{j=1}^{l_1} x_{g_{1j}\nu((1,j))}, \dots, \sum_{j=1}^{l_n} x_{g_{nj}\nu((n,j))}\right),$$

obtained from f by replacing $x_{k_i 1}$ by the sum $\sum_{j=1}^{l_i} x_{g_{ij}\nu((i,j))}$. Since $\theta(g_{ij}) = k_i$ and f is a K graded identity for A, we conclude that f' is a G-graded identity for A. Hence from the equality $\mathrm{Id}_G(A) = \mathrm{Id}_G(B)$ it follows that f' is a G-graded identity for B. Thus the result of the substitution $x_{g_{ij}\nu((i,j))} \mapsto b_{ij}$ is zero. Note that by the construction of f' the result of this substitution is $f(b_1,\dots,b_n)$. Hence $f(b_1,\dots,b_n) = 0$ and f is a K-graded identity for B. We proved the inclusion $Id_K(A) \subseteq Id_K(B)$. The reverse inclusion can be proved analogously. □

Note that the determination of the graded identities can be easier than that of the ordinary ones. For example, obtaining the ordinary identities of the associative matrix algebras $M_n(\mathbb{F})$ is a difficult task except when $n = 1, 2$. At the same time, the bases for the identities of any $M_n(\mathbb{F})$ endowed with either the standard division grading by the group $\mathbb{Z}_n \times \mathbb{Z}_n$ or fine elementary grading are well-known (see [39, Theorems 3.4, 4.5]). Moreover, the degree of identities forming these bases in the case of $M_3(\mathbb{F})$ are 1 and 2, while for the ordinary identities, the least degree by Amitsur–Levitzky's theorem equals to 6, see [123, Theorem 1.7.7]. There are a number of papers where people find bases of graded identities for a matrix algebra, say an algebra of upper triangular or block-triangular matrices, see, e. g., [97, 100]

11.1.1.1 Graded identities of $\mathfrak{sl}_2(\mathbb{F})$

Although there is a classification of gradings on virtually all simple Lie algebras with nondegenerate Killing form over algebraically closed fields, graded identities, in contrast with the associative situation (see [39]), are not well explored. A few bright spots, which, incidentally, include nonsimple Lie algebras and associative-Lie pairs of algebras, are the following.

Up to equivalence, there are exactly three nontrivial gradings on $\mathfrak{g} = \mathfrak{sl}_2(\mathbb{F}) = \langle e, h, f\rangle$, where

$$e = \begin{pmatrix} 0 & 1 \\ 0 & 0 \end{pmatrix}, \quad h = \begin{pmatrix} 1 & 0 \\ 0 & -1 \end{pmatrix}, \quad f = \begin{pmatrix} 0 & 0 \\ 1 & 0 \end{pmatrix},$$

namely:

1. ($\mathbb{Z}_2$-grading) $\mathfrak{g}_0 = \langle h\rangle$, $\mathfrak{g}_1 = \langle e, f\rangle$,
2. ($\mathbb{Z}_2 \oplus \mathbb{Z}_2$-grading) $\mathfrak{g}_{(0,0)} = \{0\}$, $\mathfrak{g}_{(1,0)} = \langle h\rangle$, $\mathfrak{g}_{(0,1)} = \langle e+f\rangle$, $\mathfrak{g}_{(1,1)} = \langle e-f\rangle$,
3. ($\mathbb{Z}$-grading) $\mathfrak{g}_{-1} = \langle f\rangle$, $\mathfrak{g}_0 = \langle h\rangle$, $\mathfrak{g}_1 = \langle e\rangle$.

The description of the graded identities given below is due to Plamen Koshlukov [173]. The main result deals with the $\mathbb{Z}_2$-grading, and easy consequences are for $\mathbb{Z}$- and $\mathbb{Z}_2 \oplus \mathbb{Z}_2$-grading. The highly nontrivial part is the following.

Theorem 11.2. *Let $\mathbb{F}$ be a field of characteristic 0, or an infinite field of characteristic $p \neq 2$. The $\mathbb{Z}_2$-graded identities for the Lie algebra $\mathfrak{g} = \mathfrak{sl}_2(\mathbb{F})$ follow from the identity $[x_1^0, x_2^0] \equiv 0$.*

Now we give two immediate corollaries of Theorem 11.2.

Theorem 11.3. *Let $\mathbb{F}$ be an infinite field of characteristic $\neq 2$. The $\mathbb{Z}$-graded identities for $\mathfrak{g} = \mathfrak{sl}_2(\mathbb{F})$ follow from $[x_1^0, x_2^0] \equiv 0$ and $x^i \equiv 0$ when $|i| \geq 2$.*

Theorem 11.4. *Let $\mathbb{F}$ be an infinite field of characteristic $\neq 2$. The $\mathbb{Z}_2 \times \mathbb{Z}_2$-graded identities of $\mathfrak{g} = \mathfrak{sl}_2(\mathbb{F})$ follow from $x^{(0,0)} \equiv 0$.*

11.1.1.2 $\mathbb{Z}$-graded identities of W_1

The definition and the standard basis for the algebra W_n are given in Section 1.2.3 and the standard $\mathbb{Z}$-grading on W_n is described in Section 1.10.2. The $\mathbb{Z}$-graded identities are the elements of a free $\mathbb{Z}$-graded Lie algebra generated by the graded set $X = \bigcup_{j\in\mathbb{Z}} X^j$ where $X^j = \{x_i^j \mid i = 1, 2, \ldots\}$ is the jth graded component of X. To state the central result about the $\mathbb{Z}$-graded identities of W_1, we introduce the graded polynomials

$$f_{ijk}(x_1^i, x_2^j, x_3^k) = (k-i)(j-k-i)[x_1^i, x_2^j, x_3^k] - (j-i)(k-j-i)[x_1^i, x_3^k, x_2^j],$$

the commutator being right normed. The following theorem is due to José Freitas, Plamen Koshlukov, and Alexei Krasil'nikov [117].

Theorem 11.5. *Let $\mathbb{F}$ be a field of characteristic 0. The graded identities*

$$[x_1^i, x_2^i] \equiv 0 \quad \textit{for all } i \geq -1, \tag{11.2}$$

$$(k-i)(j-k-i)[x_1^i, x_2^j, x_3^k] - (j-i)(k-j-i)[x_1^i, x_3^k, x_2^j] \equiv 0, \tag{11.3}$$

$$x^i \equiv 0, \quad \textit{where } i \leq -2, \tag{11.4}$$

form a basis for $\mathbb{Z}$-graded identities of the Lie algebra W_1.

The set of all identities (11.2), (11.4), and those in (11.3) with $i > j > k \geq -1$, $i \neq j + k$, $j \neq i + k$ form an independent basis for $\mathbb{Z}$-graded identities of W_1. In particular, the set of Z-graded identities of W_1 is infinitely based.

There is no difficulty with the finite basis problem for G-graded identities if G is infinite, like we have seen in Theorem 11.5. Since we do not know examples of infinitely based varieties of Lie algebras in the case of zero characteristic field, it is of interest to ask:

Can one find infinitely based varieties of G-graded Lie algebras in the case where G is a finite abelian group?

11.2 Graded algebras where the identity component satisfies an identical relation

The identity component L_e of the grading (11.1) is a Lie subalgebra and one of the natural questions is the following:

Suppose that $L = \bigoplus_{g \in G} L_g$ is graded by a finite group G so that L_e satisfies a nontrivial identity. Is the same true for the whole of L?

The restriction on the finiteness of the number of homogeneous components is necessary: a free Lie algebra $L(X)$ is naturally graded by $G = \mathbb{Z}$, with $L(X)_0 = \{0\}$.

A positive answer to the above question, asked by A. Zalessky, was given in 1996 in [60].

Theorem 11.6 (Y. Bahturin, M. Zaicev). *Let L be a Lie algebra over an arbitrary field $\mathbb{F}$ and G a finite group. If the component L_e is a Lie algebra with a nontrivial identity then so is L.*

Actually, the theorem deals with so called "Lie-type algebras", which includes associative algebras, Lie algebras, and Lie superalgebras, as particular cases (see Chapter 12 below).

One of the consequences of the above theorem deals with fixed points of automorphisms in group theory. It was proved in [60] for solvable groups but later V. Linchenko [206] provided a proof without this restriction.

Theorem 11.7 (V. Linchenko). *Let G be a finite subgroup in the automorphism group of a Lie algebra L over a field $\mathbb{F}$ where* char $\mathbb{F}$ *is not a divisor of $|G|$ and let the subalgebra of G-invariants L^G satisfy a nontrivial identity. Then also L satisfies a nontrivial identity*

Having in mind Theorem 11.7, one asks:

Is there a connection between the identities satisfied by L_e and L?

For example, a famous G. Higman's problem [143] asks:

What is the nilpotency class of a Lie algebra graded by a finite group so that the neutral component of the grading is trivial?

In connection with the study of groups admitting a fixed-points-free automorphism of prime-power order, Higman proved the following.

Theorem 11.8 (G. Higman). *To each prime p corresponds an integer $k(p)$ such that if a Lie ring L has an automorphism of order p which leaves fixed no element except zero then L is nilpotent of class at most $k(p)$.*

It is easy to see that $k(2) = 1$ and $k(3) = 2$; Higman proves that $k(5) = 6$, and that, for any odd p, $k(p) > \frac{p^2-1}{4}$. However, it was not possible to find any general upper bound for $k(p)$, and Higman writes *"it appears to be quite difficult even to find its order of magnitude."*

One of the consequences of Theorem 11.8 is the following

Theorem 11.9 (G. Higman). *If a finite solvable group has an automorphism of prime order which leaves no element fixed except the identity, it is nilpotent of class at most $k(p)$, where $k(p)$ is as in the previous theorem.*

11.2.1 Proof of Bahturin–Zaicev's Theorem 11.6

11.2.1.1 Introduction

It was shown in [71] that if we have a nontrivial identity satisfied in the identity component A_1 of an associative algebra A, graded by the elements of a finite group G, over an arbitrary field, then a nontrivial identity is satisfied also in $A = \bigoplus_{g\in G} A_g$. In the paper [41] this result has been extended in the sense that the sum $A = \sum_{g\in G} A_g$ need not be direct and the degree of identity in A depends only on the order of group G and the degree of identity in A_1.

In the present section we make use of the fact that in [41], in contrast with [71], the proof is not using any structure theory but is purely of combinatorial nature. Thus it seemed natural to extend the main idea of [41] into the case of Lie algebras and then even to a much wider class of algebras including as subclasses all associative algebras, Lie algebras, and color Lie superalgebras. Unfortunately, in our case the conclusion of the main Theorem 11.12 does not contain an evaluation of the degree of the identity satisfied by A in terms of the identity satisfied by A_1 and the order of the group G. The most “classic” is the form of our first theorem about Lie algebras:

Theorem 11.10. *Let $L = \sum_{g\in G} L_g$ be a Lie algebra over an arbitrary field $\mathbb{F}$ and G a finite group. If the component L_1 is a Lie algebra with a nontrivial identity then such is also L.*

The question about the validity of such theorem even in the case of $|G| = 2$ was asked by A. E. Zalesskii in an oral conversation with the first author in the beginning of the 1980s. This result can be applied to Lie algebras with an action of a finite automorphism group G. (The most correct formulation is in the case of Lie algebras with actions of finite-dimensional Hopf algebras). We have here a result on such algebras as well.

Theorem 11.11. *Let L be a Lie algebra over a field $\mathbb{F}$, G a finite solvable subgroup in the automorphism group* Aut G, *and suppose* char $\mathbb{F}$ *does not divide $|G|$. If the invariant subalgebra L^G is a* PI*-algebra then also L is a* PI*-algebra.*

In the case of associative algebras, this result belongs to V. K. Kharchenko [166] and S. Montgomery [245] (no need to assume G solvable).

11.2.1.2 Statement of results

We introduce the notation and the notions we will need here. Let G be a semigroup with 1. An algebra A over a field $\mathbb{F}$ is called *G-graded in this section* if A is the sum of the subspaces $A_g, g \in G$, and $A_gA_h \subset A_{gh}$. Generally, there is no need to assume that the sum is direct but, to avoid ambiguity, we will be assuming the sum is direct (although the reduction from the general sums to the case of direct sums is very easy). We say that the G-grading is *finite* if the support Supp A of the grading is finite. In other words, there is a finite subset H in G such that $A_g = \{0\}$ for all $g \notin H$. (Without loss of generality, one may assume $1 \in H$.) In other words,

$$A = \sum_{g\in H} A_g, \quad |H| = s < \infty.$$

Further, we say that $A = \sum A_g$ is a *Lie-type* algebra if for any $g, h, k \in G$ there exist constants $\alpha, \beta \in \mathbb{F}$ such that

$$a(bc) = \alpha(ab)c + \beta(ac)b, \tag{11.5}$$

for all $a \in A_g$, $b \in A_h$, $c \in A_k$, where $\alpha \neq 0$. If $\alpha = 1$, $\beta = 0$ for all $g, h, k \in G$ then A is an ordinary associative algebra with a G-grading, and if $\alpha = 1$, $\beta = -1$ then A is a Lie algebra over $\mathbb{F}$. The relations of the type (11.5) are satisfied also by Lie superalgebras with a $\mathbf{Z}_2$-grading, as well as more general color Lie superalgebras, see below. The so-called *quantum* Lie algebras (see [205]) also fit into the pattern of these relations. Now the main result of this section can be formulated as follows.

Theorem 11.12. *Let*

$$L = \sum_{g \in G} L_g$$

be a Lie-type algebra over an arbitrary field $\mathbb{F}$ *with a finite G-grading where G is a semigroup with identity and cancelation. Suppose that in the identity homogeneous component* L_1 *we have a nontrivial identity of the form*

$$x_0 x_1 \cdots x_{d-1} \equiv \sum_{\sigma \in \mathrm{Sym}(d-1), \sigma \neq e} \alpha_\sigma x_0 x_{\sigma(1)} \cdots x_{\sigma(d-1)}, \tag{11.6}$$

in which $x_0 x_1 \cdots x_{d-1}$ *and all* $x_0 x_{\sigma(1)} \cdots x_{\sigma(d-1)}$ *are left-normed monomials and* $\alpha_\sigma \in \mathbb{F}$. *Then also in L a nontrivial identity of the same form* (11.6) *is satisfied.*

From this theorem, we easily derive the result of [41] about the associative algebras graded by a finite group. But more important here are the applications to Lie algebras and their generalizations.

We recall that if Q is an abelian group then the function $\varepsilon : Q \times Q \to \mathbb{F}^*$ is called an alternating bicharacter if the following two identities hold:

$$\varepsilon(x+y,z) = \varepsilon(x,z)\varepsilon(y,z), \quad \varepsilon(x,x) = 1.$$

An algebra with a Q-grading

$$L = \bigoplus_{q \in Q} L_q$$

is called *a color Lie superalgebra* if for all homogeneous elements $x \in L_p, y \in L_q, z \in L_t$ the following equations hold:

$$xy = -\varepsilon(p,q)yx,$$
$$x(yz) = (xy)z + \varepsilon(p,q)y(xz).$$

More detailed information about color Lie superalgebras can be found in [46] and below in Section 14.4.1.2. If we denote by 0 the neutral element of Q then L_0 is a usual Lie algebra. Let G be a semigroup. Then by a G-grading of L we understand a decomposition

$$L = \sum_{g \in G} L^{(g)},$$

in which $L^{(g)}L^{(h)} \subset L^{(gh)}$, and all the subspaces $L^{(g)}$ are homogeneous under the Q-grading, that is,

$$L^{(g)} = \bigoplus_{q \in Q} (L^{(g)} \cap L_q).$$

In particular, G may be Q and $L^{(g)} = L_g$. The following two results are immediate from Theorem 10.40; it suffices to pass to the grading by the direct product $Q \times G$.

Theorem 11.13. *Let $L = \bigoplus_{q\in Q} L_q$ be a color Lie superalgebra over a field $\mathbb{F}$ with a finite grading by a semigroup G with identity and cancelation and such that Q is a finite group and the Lie algebra $L_0^{(1)}$ satisfies a nontrivial Lie identity. Then also L satisfies a nontrivial nongraded identity.*

A standard example of color Lie superalgebras is an ordinary Lie superalgebra where $Q = \mathbf{Z}_2$, $\varepsilon(0,0) = \varepsilon(0,1) = 1$, $\varepsilon(1,1) = -1$ (see Section 1.10.2). In this case our theorem takes the following form.

Theorem 11.14. *We consider a Lie superalgebra*

$$L = L_0 \oplus L_1 = \sum_{g\in G} (L_0^{(g)} \oplus L_1^{(g)}),$$

with a finite grading by a semigroup G with identity and with cancelation. If $L_0^{(1)}$ satisfies a nontrivial Lie algebra identity then also L is a Lie superalgebra with a nontrivial identity.

Remark. The explicit form of the relation (11.6) tells us that the nongraded identity $f \equiv 0$, which should be satisfied in L according to Theorems 11.13 and 11.14, is nontrivial in the sense that the nonassociative polynomial f is not identically zero even in the free Lie algebra.

Remark. Under the restrictions of Theorems 11.13 and 11.14, the grading $L = \sum L^{(g)}$ satisfying $L^{(g)} = \bigoplus_{q\in Q}(L^{(g)} \cap L_q)$ can be replaced by the condition

$$L_q = \sum_{g\in G} (L_q)^{(g)}$$

for all $q \in Q$.

Remark. If in the statement of Theorem 11.13 we take Q as the grading semigroup then we arrive at a corollary as follows. Let Q be a finite abelian group and

$$L = \bigoplus_{q\in Q} L_q$$

be a color Lie superalgebra over a field $\mathbb{F}$. If L_0 satisfies a nontrivial Lie identity then L has a nontrivial nongraded identity.

We remark that if $Q = \mathbf{Z}_2$ and $L = L_0 \oplus L_1$ is an ordinary Lie superalgebra then L satisfies the identity

$$J(x_1x_2 + x_2x_1, y_1y_2 + y_2y_1, z_1z_2 + z_2z_1) \equiv 0,$$

where $J(x,y,z) = (xy)z + (yz)x + (zx)y$, even if L_0 is a free Lie algebra. For an arbitrary finite group Q with the form ε, the question about the existence of a nontrivial identity in a free color Lie superalgebra remains open.

Finally, the simplest example of a nonassociative algebra satisfying (11.5) is a Lie algebra. Thus Theorem 11.10 is an immediate consequence of Theorem 11.12. We will show in what follows that Theorem 11.11 also easily follows from Theorem 11.12.

11.2.1.3 Main techniques

For the proof of Theorem 11.7, we will need some constructions.

For each g in G, we consider a countable set

$$Z_g = \{z_1^{(g)}, \dots, z_m^{(g)}, \dots\}$$

and denote by Z the union of all $Z_g, g \in G$. Then the *free nonassociative algebra* $\mathbb{F}\langle Z\rangle$ over $\mathbb{F}$, generated by the set Z, is naturally endowed by a G-grading. Let us fix a finite subset H in G and set

$$x_i = \sum_{g\in H} z_i^{(g)}, \quad i = 1, 2, \dots$$

The subalgebra $\mathbb{F}\langle X\rangle$ generated by $X = \{x_1, x_2, \dots\}$ in $\mathbb{F}\langle Z\rangle$ is not G-graded but it is also a free nonassociative $\mathbb{F}$-algebra.

We denote by V_n the linear span of all products of the form $[x_{\sigma(1)} \cdots x_{\sigma(n)}]$ in $\mathbb{F}\langle X\rangle$ with all possible arrangements of brackets, where σ runs through the whole of the symmetric group $\mathrm{Sym}(n)$. Similarly, for any selection $g = (g_1, \dots, g_n)$ in which all $g_i \in H$, $i = 1, \dots, n$, by V_n^g we shall denote the subspace in $\mathbb{F}\langle Z\rangle$, spanned by all products

$$[z_{\sigma(1)}^{(g_{\sigma(1)})} \cdots z_{\sigma(n)}^{(g_{\sigma(n)})}].$$

Finally, set

$$V_n^G = \bigoplus_{g\in H} V_n^g.$$

By $T(L)$ we will denote the ideal of identities of algebra L in $\mathbb{F}\langle X\rangle$, and by $T^G(L)$ the ideal of graded identities of L in $\mathbb{F}\langle Z\rangle$. We recall the codimension sequence

$$c_n(L) = \dim \frac{V_n}{V_n \cap T(L)},$$

and define the *G-graded codimension sequence*

$$c_n^G(L) = \dim \frac{V_n^G}{V_n^G \cap T^G(L)},$$

characterizing the number of ordinary and graded (non)identities of L.

Lemma 11.15. $c_n(L) \le c_n^G(L)$.

Proof. Let $H = \{h_1, \dots, h_s\}$ and $x_i = z_i^{(h_1)} + \cdots + z_i^{(h_s)}$, $i = 1, 2, \dots$ We obtain an embedding $\mathbb{F}\langle X\rangle \subset \mathbb{F}\langle Z\rangle$. Clearly, $V_n \subset V_n^G$. Notice that $T(L) \subset T^G(L)$. To see this, we consider arbitrary graded homomorphism $\varphi : \mathbb{F}\langle Z\rangle \to L = \sum_{h\in H} L_h$. The image of the element $f(x_1, \dots, x_n)$ in $T(L)$ is equal to

$$\begin{aligned} &f(\varphi(z_1^{(h_1)} + \cdots + z_1^{(h_s)}), \dots, \varphi(z_n^{(h_1)} + \cdots + z_n^{(h_s)})) \\ &= f(\varphi(z_1^{(h_1)}) + \cdots + \varphi(z_1^{(h_s)}), \dots, \varphi(z_n^{(h_1)}) + \cdots + \varphi(z_n^{(h_s)})) = 0. \end{aligned}$$

The latter equation is true since this is the image under a homomorphism of $\mathbb{F}\langle X\rangle$ in L mapping x_i into $\varphi(z_i^{(h_1)}) + \cdots + \varphi(z_i^{(h_s)})$. Moreover, we have $V_n \cap T(L) = V_n \cap T^G(L)$. Indeed, if $f(z_i^{(h_j)}) \in V_n \cap T^G(L)$ then this element has the form $\tilde{f}(z_1^{(h_1)} + \cdots + z_1^{(h_s)}, \ldots, z_n^{(h_1)} + \cdots + z_n^{(h_s)})$, where $\tilde{f}(x_1, \ldots, x_n) \in V_n$, and it vanishes under all substitutions of $z_i^{(h_j)}$ by homogeneous elements from $L = \sum_h L_h$. If

$$a_i = \sum_{j=1}^{s} a_i^{(h_j)}, \quad i = 1, \ldots, n,$$

is an arbitrary element in L then $\tilde{f}(a_1, \ldots, a_n) = f(a_i^{(h_j)}) = 0$, that is, $f = \tilde{f}$, where $\tilde{f} \in T(L)$.

Furthermore,

$$c_n(L) = \dim(V_n/V_n \cap T(L)),\ c_n^G(L) = \dim(V_n^G/V_n^G \cap T^G(L)).$$

Let us write

$$\begin{aligned} V_n/V_n \cap T(L) &= V_n/V_n \cap T^G(L) = V_n/V_n^G \cap V_n \cap T^G(L) \\ &\cong V_n + V_n^G \cap T^G(L)/V_n^G \cap T^G(L) \subset V_n^G/V_n^G \cap T^G(L). \end{aligned}$$

Hence, $c_n(L) \le c_n^G(L)$, and the proof of the lemma is complete. □

To compute the products of elements in a graded algebra, we will need the following.

Lemma 11.16. *Let G be a multiplicative semigroup with identity 1 and cancelation. Assume that H is a finite subset consisting of s elements. Let, further, $u = abc \ldots d$ be an arbitrary word of length $N = ms$ in the alphabet G, such that $a \in H$, $ab \in H$, $abc \in H$, $\ldots, abc \ldots d \in H$. Then there are at least $m - 1$ subsequent subwords of u whose value is* 1.

Proof. The total number of initial subwords is equal to ms, and each value is equal to one of s elements of the subset H. Let g be that value in H which is taken at least m times. If $g = 1$ then we are done. If 1 is not taken at least m times then some value g is taken at least $m + 1$ times. This means that we have $u = gg_1g_2 \ldots g_mg'$ and $gg_1 = g$, $gg_1g_2 = g, \ldots, gg_1 \ldots g_m = g$. It follows from the restrictions imposed on G that $g_1 = g_2 = \cdots = g_m = 1$. The proof of Lemma 11.16 is thus complete. □

11.2.1.4 Main lemma

The sequence $(a_1, \ldots, a_n)$ composed of pairwise distinct integers is called *m-decomposable* if there exist indices $1 \le i_1 < i_2 < \cdots < i_m \le n$ such that $a_{i_k} > a_j$ for all $j = i_k + 1, \ldots, i_{k+1}$ where $k = 1, \ldots, m - 1$. From m-decomposability it follows that $a_{i_1} > a_{i_2} > \cdots > a_{i_m}$. If the sequence $(a_1, \ldots, a_n)$ has no m-decompositions then we call it *m-indecomposable*.

Lemma 11.17. *Let $L = \sum L_g$ be a Lie-type algebra with a finite G-grading and s the number of nonzero components L_g. We consider all homogeneous in the G-grading elements $y_0, y_1, \ldots, y_n$ and denote by P the linear span of all products of the form $y_0(y_{i_1} \cdots y_{i_n})$ with all possible arrangements of brackets in the right factor $(y_{i_1} \cdots y_{i_n})$ such that $(i_1, \ldots, i_n)$ is a permutation of $1, \ldots, n$. Suppose that the identity component L_1 satisfies a multilinear identity of the form (11.6). Then any element in P is a linear combination of left-normed products of the form $y_0 y_{j_1} \cdots y_{j_n}$ for which $\{j_1, \ldots, j_n\} = \{1, \ldots, n\}$ and the sequence $(j_1, \ldots, j_n)$ is m-indecomposable for any $m \geq sd$.*

Proof. It follows from (11.5) that P is a linear span of left-normed products of elements $x_0, x_1, \ldots, x_n$, starting with x_0. Everywhere in the sequel we will omit brackets in the expression of left-normed products of the elements of algebra L.

Remark that we easily derive from (11.6) the existence of a graded identity of the form

$$z_0 z_1 \cdots z_d \equiv \sum_{\sigma \in \mathrm{Sym}(d), \sigma \neq e} \beta_\sigma z_0 z_{\sigma(1)} \cdots z_{\sigma(d)}, \tag{11.7}$$

in which $z_i = z_i^{(1)}$, $\beta_\sigma \in \mathbb{F}$, and z_0 is an arbitrary element. Just multiply z_0 by the left- and right-hand sides of (11.6), in which x_i are replaced by z_{i+1}, and reduce all products to the left-normed expressions using (11.5). In doing so, the product $z_0 z_1 \cdots z_d$ appears only once with the coefficient $\alpha^{d-1} \neq 0$, while the other summands form the right-hand side in (11.7).

Now we introduce on the integer tuples a lexicographical ordering by comparing them from the left to the right. Let $u = y_0 y_{j_1} \cdots y_{j_n}$ be a left-normed product. If $j_1 = 1, j_2 = 2, \ldots, j_n = n$ then $(j_1, \ldots, j_n)$ is m-indecomposable. Arguing by induction, we may assume that all products $y_0 y_{k_1} \cdots y_{k_n}$ with $(k_1, \ldots, k_n)$ strictly less than $(j_1, \ldots, j_n)$ are linear combinations of left-normed products $y_0 y_{i_1} \cdots y_{i_n}$ with m-indecomposable tuples $(i_1, \ldots, i_n)$.

We assume that the tuple $(j_1, \ldots, j_n)$ is m-decomposable and the indices $t_1, \ldots, t_m$ determine the m-decomposition, that is,

$$j_{t_1} > j_{t_1+1}, \ldots, j_{t_2}; \ldots; j_{t_{m-1}} > j_{t_{m-1}+1}, \ldots, j_{t_m},$$

and let us show that u can also be expressed through m-indecomposable left-normed products. We denote by v_i, $i = 1, \ldots, m$, a left-normed product of the elements y_k, with k running through all values from j_{t_i} to $j_{t_{i+1}} - 1$, and by v_0 the product of the first y_k with $k < j_{t_1}$. We denote by B the operator of subsequent multiplication by the remaining elements y_k where $k = j_{t_m} + 1, \ldots, j_n$, if $j_{t_m} < j_n$. With the use of (11.5), the element

$$u' = v_0 v_1 \cdots v_m B$$

can be written as a linear combination of u with a nonzero coefficient and the products $y_0 y_{q_1} \cdots y_{q_n}$, with $(q_1, \ldots, q_n) < (j_1, \ldots, j_n)$. Since all y_i were assumed homogeneous, we have $v_0 \in L_{g_0}, v_1 \in L_{g_1}, \ldots, v_m \in L_{g_m}$ for some $g_0, g_1, \ldots, g_m \in G$. Owing to the choice of m, by Lemma 11.16, there exist $0 \le l_0 < l_1 < \cdots < l_d \le m$ such that $g_{l_{i-1}+1} \cdots g_{l_i} = 1$ are in G for all $i = 1, \ldots, d$. We introduce some notation for new left-normed products:

$$w_0 = v_0 \cdots v_{l_0}, \; w_i = v_{l_{i-1}+1} \cdots v_{l_i}, \; i = 1, \ldots, d.$$

Then $w_0, \ldots, w_d \in L_1$ and the element

$$u'' = w_0 w_1 \cdots w_d C,$$

in which C stands for the operator of right multiplications by $v_{l_d+1}, \ldots, v_m$, if $l_d < m$, and then the action of the operator B, differs from u by a scalar factor plus a linear combination of some $y_0 y_{q_1} \cdots y_{q_n}$ with $(q_1, \ldots, q_n) < (j_1, \ldots, j_n)$. Applying (11.7), we arrive at

$$u'' = \sum_{\sigma \in \mathrm{Sym}(d), \sigma \neq e} \beta_\sigma w_0 w_{\sigma(1)} \cdots w_{\sigma(d)} C. \tag{11.8}$$

Now let us express $w_0 w_{\sigma(1)} \cdots w_{\sigma(d)} C$ on the right-hand side of (11.8) through left-normed monomials $y_0 y_{q_1} \cdots y_{q_n}$, by expanding the brackets with the help of (11.5) first on w_i, and then on v_i. Since $\sigma \neq e$ in $\mathrm{Sym}(d)$, we are going to obtain only the summands with $(q_1, \ldots, q_n) < (j_1, \ldots, j_n)$. As a result, u'' and then also u' and u have the required form, and the proof of Lemma 11.17 is complete. □

11.2.1.5 Proof of Theorem 11.12

We start with a lemma about codimensions.

Lemma 11.18. *If the subalgebra L_1 satisfies an identity of the form* (11.6) *then for any $b > 0$ and sufficiently large n the following inequality is satisfied:*

$$c_n^G(L) < \frac{n!}{b^n}.$$

Proof. Let $g = (g_1, \ldots, g_n)$ be a tuple in which $g_1, \ldots, g_n \in H$. Then $V_n^G = \bigoplus_{g \in H^n} V_n^g$ and

$$V_n^G / V_n^G \cap T^G(L) = \sum_{g \in H^n} V_n^g / V_n^g \cap T^G(L). \tag{11.9}$$

The number of summands on the right-hand side of (11.9) is equal to s^n. For the proof of the lemma, it is sufficient to show that the dimension of each summand is less than $\frac{n!}{(bs)^n}$. For convenience, we replace n by $n + 1$. We fix a tuple $g = (g_0, g_1, \ldots, g_n)$ and set $z_i = z_i^{(g_i)}$. Modulo $T^G(L)$, the space $V_{n+1}^G = V_{n+1}(z_0, z_1, \ldots, z_n)$ is equal to the sum $Q_0 + Q_1 + \cdots + Q_n$, in which Q_i is a linear span of the left-normed products $z_0, z_1, \ldots, z_n$, starting with z_i. It is sufficient to show that the dimension of each summand is asymptotically less than $\frac{n!}{(2bs)^n}$. We consider, for example, $Q = Q_0$, that is, the linear span of left-normed products $z_0 z_{\sigma(1)} \cdots z_{\sigma(n)}$. By Lemma 11.17, the tuple $(\sigma(1), \dots, \sigma(n))$ can be assumed m-indecomposable, where $m = sd$.

We denote by $a_m(n)$ the number of m-indecomposable tuples composed of the numbers $1, 2, \dots, n$. Then

$$\overline{\lim_{n\to\infty}} \sqrt[n]{\frac{a_m(n)}{n!}} = 0 \tag{11.10}$$

(see, e. g., [313, Chapter 1, Section 2.1]). From (11.10), it follows that for all sufficiently large n the following inequality is satisfied:

$$\sqrt[n]{\frac{a_m(n)}{n!}} < \frac{1}{2bs},$$

from which it follows that for $n \geq 2bs$ we have

$$\dim Q \leq a_m(n) < \frac{n!}{(2bs)^n} < \frac{(n+1)!}{(2bs)^{n+1}},$$

and the proof of Lemma 11.18 is complete. □

Proof of Theorem 11.12. From Lemmas 11.15 and 11.18, it follows that the left-normed products $x_0 x_{\sigma(1)} \cdots x_{\sigma(n)}$, $\sigma \in \mathrm{Sym}(n)$, are linearly dependent modulo the ideal $T(L)$ of identities of algebra L in the free nonassociative algebra $\mathbb{F}\langle X\rangle$. This means that also L satisfies a nontrivial identity of the same type as (11.6), and the proof of Theorem 11.12 is complete. □

Theorems 11.10, 11.13, and 11.14 are now immediate corollaries of Theorem 11.12. Let us show how the last unproved theorem can be derived.

11.2.1.6 Automorphisms of graded algebras

This short section is devoted to the derivation of Theorem 11.11.

Proof. Recall that by

$$L^G = \{x \in L \mid g(x) = x \quad \text{for any } g \in G\}$$

we denote the set of fixed points of the action of a group G by Lie automorphisms of L. To start with, we assume that G is a cyclic group of prime order p and g is a generator of G. Then g^p is an identity mapping of L and $p \neq \operatorname{char} \mathbb{F}$. Any identity of Lie algebras has a nontrivial multilinear consequence, see, e. g., [37, Theorem 4.2.1]. Besides, any multilinear Lie polynomial f in $x_0, x_1, \dots, x_n$ can be written in the form

$$f = \sum_{\sigma\in\mathrm{Sym}(n)} \alpha_\sigma x_0 x_{\sigma(1)} \cdots x_{\sigma(n)}.$$

In other words, L^G satisfies an identity of the form (11.6).

We consider first the case where $\mathbb{F}$ is an algebraically closed field. Since $p \neq \operatorname{char} \mathbb{F}$, the polynomial $x^p - 1$ has p pairwise different roots $1, \lambda, \lambda^2, \dots, \lambda^{p-1}$ in $\mathbb{F}$ and the automorphism g is diagonalizable on L, that is,

$$L = L_0 + L_1 + \cdots + L_{p-1}, \tag{11.11}$$

where $L_j = \{a \in L \mid g(a) = \lambda^j a\}$. It is easy to see that for all $a \in L_i$, $b \in L_j$, $g(ab) = \lambda^{i+j} ab$, that is, the decomposition (11.11) determines on L a grading by a cyclic group of prime order p, with $L_0 = L^G$. Hence, for the case $|G| = p$ for an algebraically closed field $\mathbb{F}$, our theorem follows from Theorem 11.13.

If $\mathbb{F}$ is not algebraically closed then it is sufficient to pass to the algebra $\overline{L} = L \otimes_{\mathbb{F}} \overline{\mathbb{F}}$ over the algebraic closure $\overline{\mathbb{F}}$ of $\mathbb{F}$. The multilinear identity of algebra L^G holds valid for $\overline{L}^G$, and in the latter we then have a nontrivial multilinear identity $f \equiv 0$ with the coefficients in $\overline{\mathbb{F}}$. Decomposing the coefficients of f through an arbitrary basis $\overline{\mathbb{F}}$ over $\mathbb{F}$ gives us a nontrivial identity of the $\mathbb{F}$-algebra $\overline{L}$ and its subalgebra L.

Let now G be any finite solvable group of automorphisms of L, and assume its order n is not divisible by char $\mathbb{F}$. We may assume that n is not a prime number. This means that in G there is a nontrivial normal subgroup H. We consider the subalgebra L^H of its fixed points. Since $H \triangleleft G$, we have $g(L^H) \subset L^H$ for any $g \in G$. Accordingly, $\operatorname{Aut}(L^H)$ has a finite solvable subgroup G/H, whose order is strictly less than $|G|$ and not divisible by char $\mathbb{F}$. Arguing by induction, we may assume that L^H is a Lie algebra with nontrivial identity. It remains to apply induction once again to L and the subgroup H of $\operatorname{Aut}(L)$, and then the proof of Theorem 11.11 is complete. □

Note that using techniques of Hopf algebras V. Linchenko removed the solvability restriction in the latter theorem. We discuss this in Chapter 12.

11.3 Ω-algebras

In this section we deal with the so-called Ω-algebras, where Ω is a set, called *signature*. One has $\Omega = \bigcup_{m=0}^{\infty} \Omega_m$. An algebra A is called an *Ω-algebra*, if A is a vector space such that every $\omega \in \Omega_m$ defines an m-ary operation on A, that is, an m-linear map $\omega : \underbrace{A \times \cdots \times A}_{m \text{ times}} \to A$. In a natural way, one can define the standard notions of subalgebras, homomorphisms of algebras, ideals, and so on.

Given a nonempty set X, one can define the *free Ω-algebra* $F = F_\Omega(X)$ as follows. First, we build the set $W = W_\Omega(X)$ of Ω-monomials in X as the union of subsets W_n, $n = 0, 1, 2, \dots$ given by $W_0 = \Omega_0 \cup X$ and for $n > 0$,

$$W_n = \bigcup_{m=1}^{\infty} \bigcup_{\omega \in \Omega_m} \bigcup_{i_1 + \cdots + i_m + 1 \leq n} \omega(W_{i_1}, \dots, W_{i_m}).$$

From this definition, it follows that for any $\omega \in \Omega_m$ and any $a_1, \dots, a_m \in W$ the expression $\omega(a_1, \dots, a_m)$ is a well-defined element of W. The elements of W_n are called *monomials of degree n*.

Then we consider the linear span $F = F_\Omega(X)$ of $W = W_\Omega(X)$. If $F_n = \mathrm{Span}\{W_n\}$ then $F = \bigoplus_{n=0}^{\infty} F_n$. The elements of F_n are called *homogeneous polynomials of degree n*. In a usual way, one defines the degree of an arbitrary nonzero polynomial. By linearity, every $\omega \in \Omega_m$ defines an m-linear operation on F. Also, it follows that for any Ω-algebra A any map $\varphi : X \to A$ uniquely extends to a homomorphism $\bar{\varphi} : F_\Omega(X) \to A$. The Ω-algebra $F = F_\Omega(X)$ is called the *free Ω-algebra with the basis* (i. e., set of free generators) X.

An equation of the form $f(x_1, \dots, x_n) \equiv 0$, where $0 \neq f(x_1, \dots, x_n) \in F_\Omega(X)$ is called a *(polynomial) identity* in an Ω-algebra A if under any map $\varphi : X \to A$ one has $\bar{\varphi}(f(x_1, \dots, x_n)) = 0$. In other way, $f(a_1, \dots, a_n) = 0$, for any $a_1, \dots, a_n \in A$. An ideal I of an Ω-algebra A closed under all endomorphisms of A is called a *T-ideal*. If $A = F_\Omega(X)$ then I is a T-ideal if with every $f(x_1, x_2, \dots, x_n) \in I$ also $f(a_1, a_2, \dots, a_n) \in I$, for any $a_1, \dots, a_n \in F_\Omega(X)$. Given a T-ideal I, the algebra $F_\Omega(X)/I$ is called a *relatively free* algebra. The set of all (left-hand sides) of the identical relations in an algebra A, depending on the variables in X, is a T-ideal $T(A)$ of $F_\Omega(X)$. We denote the relatively free algebra $F_\Omega(X)/T(A)$ by $F_\Omega^A(X)$. If A and B satisfy the same identities then $F_\Omega^A(X) = F_\Omega^B(X)$. Finally, we denoted by $\mathrm{var}_\Omega A$ the variety of Ω-algebras generated by A.

For now on, for each set Ω, we assume that $\bigcup_{m=2}^{\infty} \Omega_m \neq \emptyset$. In his book [313, Corollary 1 of Theorem 5.3] Yu. P. Razmyslov proves the following remarkable result.

Theorem 11.19. *Two simple finite-dimensional Ω-algebras over an algebraically closed field, satisfying the same polynomial identities, are isomorphic. In other words, in the variety* $\mathrm{var}_\Omega A$ *generated by a simple algebra A there are no other simple Ω-algebras.*

It is clear that this theorem generalizes previous Razmyslov's results about simple ungraded algebras. Moreover, Corollary 2 of Theorem 5.3 of [313] states the same result for prime algebras.

In several papers people treated the isomorphism of finite-dimensional simple algebras, satisfying the same identities, without using the approach of Ω-algebras. For example, see [112, 333].

In this section, following Bahturin–Yasumura's [61, 62] we will show that any two algebras graded by the same group can be regarded as ungraded Ω-algebras of the same signature Ω, and their graded identities are identities of Ω-algebras. This will allow us to prove an analogue of Razmyslov's theorem in the case of finite-dimensional graded-simple (and graded-prime) algebras.

11.3.1 Graded algebras as Ω-algebras

Let $\mathbb{F}$ be an arbitrary field, G any semigroup, and A an algebra over $\mathbb{F}$, in the usual sense, that is with one binary operation. It is not necessary that A is associative or finite-dimensional. Consider a G-grading on A,

$$A = \bigoplus_{g \in G} A_g.$$

Then, for any $g \in G$, we have the natural projection $\pi_g : A \to A$ given by

$$\pi_g\Big(\sum_{h \in G} a_h\Big) = a_g, \quad \text{where } a_h \in A_h, \text{ for any } h \in G. \tag{11.12}$$

Consider the set $\Omega = \Omega_1 \cup \Omega_2$, where $\Omega_1 = \{\pi_g \mid g \in G\}$ and $\Omega_2 = \{\mu\}$. We then view A as an Ω-algebra, whose underlying vector space is A itself, μ a binary operation $\mu(x, y) = xy$, the original product on A, and each π_g is a unary operation given by (11.12). The following two propositions are immediate.

Proposition 11.20. *Let $I \subset A$. Then I is a G-graded ideal of A if and only if I is an ideal of A, as an Ω-algebra.*

In particular, A is simple as a G-graded algebra if and only if A is simple as an Ω-algebra.

Proposition 11.21. *Let A and B be two G-graded algebras and let $\varphi : A \to B$ be any map. Then φ is a homomorphism of G-graded algebra if and only if φ is a homomorphism of Ω-algebras.*

In particular, A is isomorphic to B as a G-graded algebra if and only if A is isomorphic to B as an Ω-algebra.

Since we have only one binary operation in Ω, there is no confusion when writing ab instead of $\mu(a, b)$ in all Ω-algebras. It is then easy to prove that the following identities of Ω-algebras hold in A ($x, y \in X$):

(i) If $g, h, k \in G$, then

$$\pi_g(\pi_h(x)\pi_k(y)) = \begin{cases} \pi_h(x)\pi_k(y), & \text{if } g = hk, \\ 0, & \text{if } g \neq hk. \end{cases}$$

(ii) Let $g, h \in G$, then

$$\pi_g(\pi_h(x)) = \begin{cases} \pi_g(x), & \text{if } g = h, \\ 0, & \text{if } g \neq h. \end{cases}$$

From now on, we will be assuming that the G-grading on A is finite, that is, $\operatorname{Supp} A = \{g \in G \mid A_g \neq \{0\}\}$ is finite. This always holds if A is finite-dimensional or if G is finite. In this case, if $S = \operatorname{Supp} A$, then the following are identities in A:

$$\text{(iii)} \quad x = \sum_{g \in S} \pi_g(x), \quad \pi_h(x) = 0, \text{ for } h \notin S.$$

Let $M = \bigcup_{k=1}^{\infty} M_k$ be a subset of the set of monomials in $W = W_\Omega(X)$ defined by induction on the degree, as follows. If $k = 1$ then $M_1 = \{\pi_g(x) \mid g \in G,\ x \in X\}$. If $k > 1$ then $w = w_1 w_2$, where $w_1 \in M_\ell$, $w_2 \in M_{k-\ell}$, with $\ell \geq 1$.

Now let T be the T-ideal in $F_\Omega(X)$ defined by the identities (i), (ii), and (iii).

Lemma 11.22. *$F_\Omega(X)$ is the vector space span of M and T.*

Proof. In other words, using transformations (i), (ii), and (iii), one can reduce any monomial $u \in W = W_\Omega(X)$ to a linear combination of monomials w in M. Actually, we also prove that $\pi_g(w) = w$ for an appropriate $g \in G$, for any $w \in M$. Let us call such monomials homogeneous.

Let us use induction by $\deg u$. A monomial of degree 0 is an $x \in X$. Using (iii), we can write $x = \sum_{g \in S} \pi_g(x)$, proving the claim. The monomials of degree 1 are $\pi_g x$, for some $g \in G$ and $x \in X$, and xy, for $x, y \in X$. The former are in M_1 by definition. The latter, using (iii), can be written as linear combinations of monomials $w = \pi_g(x)\pi_h(y)$. Again by definition, these are in M. In addition, by (i), $\pi_{gh}(w) = w$. Now assume $\deg u = k > 1$. Then either $u = \pi_g(w)$, where $\deg w = k-1$, or $u = w_1 w_2$, where $\deg w_i < k$ for $i = 1, 2$. By induction, w, w_1, w_2 are linear combinations of monomials in M, that is, either $\pi_h(x)$, $h \in G$, $x \in X$, or $v_1 v_2$, where $\pi_{g_1}(v_1) = v_1$, $\pi_{g_2}(v_2) = v_2$ are homogeneous monomials in M, $g_1, g_2 \in G$. Thus if $u = w_1 w_2$ then u is a linear combination of monomials in M. Now $\pi_g(\pi_h(x)) = \delta_{gh}\pi_g(x)$ by (i),

$$\pi_g(v_1 v_2) = \pi_g(\pi_{g_1}(v_1)\pi_{g_1}(v_1)) = \delta_{g,g_1 g_2}\pi_{g_1}(v_1)\pi_{g_1}(v_1) = \delta_{g,g_1 g_2} v_1 v_2.$$

This proves that $u = \pi_g(w)$ is a linear combination of monomials in M, as well. Also, the latter calculation proves our claim that $\pi_{g_1 g_2}(v_1 v_2) = v_1 v_2$. □

Now we denote by $\mathbb{F}\langle X^G \rangle$ the free nonassociative G-graded algebra, with free generators $x^{(g)}$, where $x \in X$, $g \in G$. Consider the homomorphism $\bar{\psi} : \mathbb{F}\langle X^G \rangle \to F_\Omega^A(X)$ extending the map $\psi(x^{(g)}) = \pi_g(x)$. By the previous lemma, $\bar{\psi}$ is surjective.

Lemma 11.23. *Let $f \in \mathbb{F}\langle X^G \rangle$. Then $f \equiv 0$ is a graded identity of A if and only if $\bar{\psi}(f) \equiv 0$ is a polynomial identity of A as an Ω-algebra.*

Proof. Assume $f \equiv 0$ is not a G-graded identity of A. Then there exists an evaluation

$$e : x_i^{(g)} \mapsto a_i^{(g)} \in A_g$$

such that $e(f) \neq 0$. Define the evaluation $e' : x_i \mapsto \sum_{g \in G} a_i^{(g)}$ (since the grading is finite, the sum is well defined). Let $w = w(x_1^{(g_1)}, \dots, x_n^{(g_n)}) \in \mathbb{F}\langle X^G \rangle$ be a monomial. Then $\bar{\psi}(w) = w(\pi_{g_1}(x_1), \dots, \pi_{g_n}(x_n))$. An easy induction with respect to $\deg w$ proves that $e'(\bar{\psi}(w)) = e(w)$, hence we obtain $e'(\bar{\psi}(f)) = e(f) \neq 0$, proving that $\bar{\psi}(f)$ is not a polynomial identity of A, as an Ω-algebra.

Conversely, if $\bar{\psi}(f) \equiv 0$ is not a polynomial identity of A, as an Ω-algebra, then there exists an evaluation $e_2' : x_i \mapsto \sum b_i^{(g)}$ such that $e_2'(\bar{\psi}(f)) \neq 0$. So we can define the evaluation $e_2 : x_i^{(g)} \mapsto b_i^{(g)}$, and the same argument shows that $e_2(f) = e_2'(\bar{\psi}(f)) \neq 0$, concluding that $f \equiv 0$ is not a G-graded polynomial identity of A. □

An immediate consequence is the following

Corollary 11.24. *Let A and B be two algebras endowed with finite G-gradings. Then A and B satisfy the same graded polynomial identities if and only if A and B satisfy the same polynomial identities as Ω-algebras.*

Now, using Proposition 11.20 and Corollary 11.24, applying Razmyslov's Theorem 11.19 and concluding with Proposition 11.21, we obtain our main result.

Theorem 11.25. *Let A and B be finite-dimensional G-graded algebras which are graded simple over an algebraically closed field $\mathbb{F}$, where G is any semigroup. Then A and B are isomorphic as G-graded algebras if and only if they satisfy the same G-graded polynomial identities.*

A particular case was settled in a paper [72]: under the restriction that G is an abelian group, the authors proved that any two finite-dimensional nonassociative G-graded simple algebras having the same G-graded identities must be isomorphic as G-graded algebras. Also, the case of graded simple associative algebras was settled in [174, 4].

In particular, if we have two different (i. e., nonisomorphic) G-gradings of a finite-dimensional simple Lie algebra then their graded identities are different. Thus it could be interesting to find identities that separate nonisomorphic gradings.

11.3.2 Graded Ω-algebras

Let A be an Ω-algebra and let G be any semigroup. Consider a vector space G-grading on A, that is, we fix a vector space decomposition $A = \bigoplus_{g\in G} A_g$. For any $m \in \mathbb{N}$, one obtains naturally a G-grading on

$$\otimes^m A := \underbrace{A \otimes \cdots \otimes A}_{m \text{ times}}$$

defining the homogeneous component of degree g by

$$(\otimes^m A)_g = \sum_{g_1 g_2 \cdots g_m = g} A_{g_1} \otimes \cdots \otimes A_{g_m},$$

see [115, Chapter 1, p. 11]. We say that A is a G-graded Ω-algebra if every m-ary operation $\omega : \otimes^m A \to A$ is G-homogeneous, that is, $\omega((\otimes^m A)_g) \subset A_g$. This notion generalizes the common notion of a semigroup grading on an algebra.

We can turn any G-graded Ω-algebra A into an Ω_G-algebra with signature $\Omega_G = \Omega \cup \{\pi_g \mid g \in G\}$, as we did earlier in this chapter. This gives a full faithful functor from the category of G-graded Ω-algebras to the category of Ω_G-algebras.

Similarly to the previous case, we obtain

Lemma 11.26. *Let A and B be two G-graded Ω-algebras.*
1. *A is simple as an Ω_G-algebra if and only if A is simple as a G-graded Ω-algebra.*
2. *A is isomorphic to B as an Ω_G-algebra if and only if A is isomorphic to B as a G-graded Ω-algebra.*

Denote by F_{Ω_G} the free Ω_G-algebra with a set X of free generators. The following are polynomial identities of a given G-graded Ω-algebra:
(i) $\pi_g(\pi_h(x)) = 0$, if $g \neq h$,
(ii) $\pi_g(\pi_g(x)) = \pi_g(x)$,
(iii) $\omega(\pi_{g_1}x_1,\dots,\pi_{g_n}x_n) = \pi_g(\omega(\pi_{g_1}x_1,\dots,\pi_{g_n}x_n))$, where $g = g_1\cdots g_n$, $\omega \in \Omega$.

Moreover, if the G-grading is finite, then the following is also an identity:
(iv) $x = \sum_{g\in G}\pi_g x$ and $\pi_h(x) = 0$, if $h \notin \operatorname{Supp} A$.

Let T be the T-ideal generated by the identities (i)–(iv). Let $Y = \{\pi_g(x) \mid x \in X, g \in G\}$. Consider the set of monomials $M = W_\Omega(Y)$ which consists of all monomials generated by Y using the operations of Ω.

Lemma 11.27. *The sets M and T span F_{Ω_G}.*

Proof. Let $w \in F_{\Omega_G}$ be a monomial. We work modulo T. We prove by induction on the degree of w that
(a) w is a linear combination of monomials in M,
(b) if $w \in M$ then there exists $g \in G$ such that $\pi_g(w) = w$.

If $\deg w = 0$ then we use identities (iv) and (ii) to conclude (a) and (b). So assume $\deg w > 0$. Then we can write $w = \omega(w_1,\dots,w_m)$, where $m \geq 1$, and $\deg w_i < \deg w$ for all i; or $w = \pi_g(u)$, for some $g \in G$ and $\deg u = \deg w - 1$. For the first case, by induction step, every w_i is a linear combination of monomials in M. Hence, w is a linear combination of monomials in M. In addition, suppose $w_i \in M$ and $\pi_{g_i}w_i = w_i$, for some $g_i \in G$, for each i. Then identity (iii) implies that w satisfies (b).

For the last case $w = \pi_g(u)$, we can apply induction step on u. Thus u is a linear combination of monomials in M, say $u = \sum u_i$. The induction hypothesis says that $\pi_{h_i}u_i = u_i$, for some $h_i \in G$. By identities (i)–(ii), one has $\pi_g\pi_{h_i}u_i = \delta_{gh_i}u_i$. Hence, w satisfies (a). If it happens that $w \in M$ then $u \in M$, and the last computation implies that w satisfies (b), as well. □

Let $X^G = \{x^{(g)} \mid x \in X, g \in G\}$. With each $\omega_0 \in \Omega_0$, we associate an arbitrary degree, $\deg_G \omega_0 \in G$. Also, we write $\deg_G x^{(g)} = g$. Then F_Ω^{gr}, the free Ω-algebra with the basis X^G, acquires a G-grading as follows. We already defined the homogeneous degree for the elements of degree 0. Now, given $\omega(w_1,\dots,w_n)$, where $\omega \in \Omega$, $n \geq 1$, we set $\deg_G \omega(w_1,\dots,w_n) = \deg_G w_1 \cdots \deg_G w_n$. This is a well-defined G-grading on F_Ω.

There is no doubt that we should not exclude 0-ary operations in the free Ω-algebra since polynomial identities with 1 or without 1 play an essential role in the theory. However, it is interesting to mention one example. In the context of associative algebra A with a unit (where 1 is a 0-ary operation), it is not possible to find a graded homomorphism $F_\Omega \to A$, unless we impose $\deg_G 1 = 1 \in G$ in Ω_0.

We can then consider the G-graded evaluations and speak about the G-graded polynomial identities of G-graded Ω-algebras. Let $F_\Omega^{\mathrm{gr}}(A)$ be a relatively free G-graded Ω-algebra. Then the same argument as in the previous section can be used to obtain

Lemma 11.28. *The map $\psi_G : F_\Omega^{\mathrm{gr}}(A) \to F_{\Omega_G}(A)$, given by $\psi_G(x_i^{(g)}) = \pi_g(x_i)$, is a bijective homomorphism of Ω-algebras. Moreover, $\psi_G(\mathrm{Id}_\Omega^{\mathrm{gr}}(A)) = \mathrm{Id}_{\Omega_G}(A)$.*

Hence A and B satisfy the same G-graded polynomial identities as Ω-algebras if and only if they satisfy the same polynomial identities as Ω_G-algebras. As a consequence, we can apply Razmyslov's theorem in the setting of G-graded Ω-algebras.

Corollary 11.29. *Let A and B be two finite-dimensional G-graded Ω-algebras over an algebraically closed field, which are simple as G-graded Ω-algebras, where G is any semigroup. Then A is isomorphic to B, as a G-graded Ω-algebra, if and only if both satisfy the same G-graded polynomial identities as G-graded Ω-algebras.*

Applying [313, Corollary 2 of Theorem 5.3] instead of [313, Corollary 1 of Theorem 5.3], we obtain the same statement of the previous lemma but for prime algebras.

Corollary 11.30. *Let A and B be two finite-dimensional G-graded Ω-algebras over an algebraically closed field, which are* prime *as G-graded Ω-algebras, where G is any semigroup. Then A is isomorphic to B, as a G-graded Ω-algebra, if and only if both satisfy the same G-graded polynomial identities as G-graded Ω-algebras.*

11.4 Further applications of Razmyslov's theorem

Recently, many researches were done concerning polynomial identities, codimension growth and its asymptotics, exponent and related properties of algebras with additional structure. To cite a few examples, see [3, 45, 99, 98, 120, 119, 128, 129, 147, 148]. It turns out that all these algebras can be described in terms of gradings on some Ω-algebras, as we will illustrate bellow.

11.4.1 Algebras with involution

Let $(A, *)$ be an algebra with involution. Let $\Omega_* = \{\mu, *\}$, where μ defines the original algebra product on A and $*$ defines the involution on A. Thus an algebra with involution is an Ω_*-algebra. Let $F^*(X)$ be the relatively free algebra satisfying the identities $x^{**} = x$ and $(xy)^* = y^*x^*$.

In the classical theory, the free algebra with involution is defined as the free algebra $\mathbb{F}\langle X\rangle$, together with the application of a symbol $*$. It turns out that an equivalent description is considering the free algebra $\mathbb{F}\langle Y^+, Y^-\rangle$, with free generators $Y^+ = \{y_i^+\}$ and $Y^- = \{y_i^-\}$, where we assume the variables in Y^+ and Y^- symmetric and skew-symmetric with respect to $*$, respectively. The classical notion of free algebra with involution coincides with $F^*(X)$. In particular, the polynomial identities with involution of an algebra with involution coincide with the polynomial identities as Ω_*-algebra. Hence, Razmyslov's Theorem 11.19 (or Corollary 11.29) translates as follows.

Theorem 11.31. *Two finite-dimensional algebras with involution over an algebraically closed field, which are involution-simple, satisfying the same polynomial identities with involution are isomorphic as algebras with involution.*

11.4.2 Superalgebras with involution and superinvolution

Superalgebras are examples of $\mathbb{Z}_2$-graded algebras. Given a superalgebra $(A, *)$ with involution (or superinvolution), we can consider $\Omega_{*,2}$, containing a binary product, the $\mathbb{Z}_2$-grading structure and $*$. A superalgebra with involution (or superinvolution) can be viewed as an $\Omega_{*,2}$-algebra. Using the same considerations of the previous example, one concludes that the graded polynomial identities with involution of A coincide with the graded polynomial identities as Ω_*-algebra. Hence, Corollary 11.29 translates as follows.

Theorem 11.32. *Two finite-dimensional superalgebras with involution (or superinvolution) over an algebraically closed field, which are simple as superalgebras with involution (or superinvolution), satisfying the same graded polynomial identities with involution (or superinvolution) are isomorphic as graded algebras with involution (or superinvolution).*

11.4.3 Color Lie superalgebras

The definition of color Lie superalgebras and color Lie p-superalgebras is given in Section 14.4.1.2. We can view color Lie superalgebras as nonassociative graded algebras, hence color Lie superalgebras is a particular example of Ω-algebras.

Another related example, given that $\operatorname{char}\mathbb{F} = p > 0$, is that of a color Lie p-superalgebra. Note that $x \mapsto x^{[p]}$ is not always linear. Hence, we cannot always see the "raising to pth power" as a unary operation. So, it is not obvious how we can describe a color Lie p-superalgebra as an Ω-algebra. However, in the context of simple algebras, the second identity completely defines $\operatorname{ad} x^{[p]}$. Since ad is an isomorphism for a simple finite-dimensional algebra, we conclude that the pth power map is completely defined by the product of the algebra. In this way, an isomorphism of color Lie

p-superalgebras preserving the product will preserve the pth power as well. Hence, Theorem 11.25 (or Corollary 11.29) reads as follows.

Theorem 11.33. *Two finite-dimensional color Lie (p-)superalgebras over an algebraically closed field, which are simple as color Lie superalgebras, satisfying the same graded polynomial identities are isomorphic as color Lie (p-)superalgebras.*

11.4.4 Trace identities

Another important example to consider is trace identities. We consider the signature $\Omega_{\mathrm{tr}} = \{\mathrm{tr}, \mu\}$, where μ is a binary operation and tr is a unary operation. For a matrix algebra M_n, the signature Ω_{tr} consists of the following operations: μ is the usual matrix multiplication and $\mathrm{tr}(A) = a \cdot I$, where a is the usual trace of A and I is the $n \times n$ identity matrix.

The classical free algebra with trace is defined as the relatively free algebra with signature Ω_{tr} satisfying the following polynomial identities (see [302]):
- $\mathrm{tr}(x)y = y\,\mathrm{tr}(x)$,
- $\mathrm{tr}(xy) = \mathrm{tr}(yx)$,
- $\mathrm{tr}(x\,\mathrm{tr}(y)) = \mathrm{tr}(x)\,\mathrm{tr}(y)$.

So, Razmyslov's Theorem 11.19 (or Corollary 11.29) translates into:

Theorem 11.34. *Two finite-dimensional algebras over an algebraically closed field, which are simple as algebra with trace, satisfying the same trace polynomial identities are isomorphic as algebras with trace.*

Traces of generic matrices are related to the invariants of matrix algebras. Also, it is known that trace polynomial identities of matrix algebras are consequences of the Cayley–Hamilton identity [302, 295]. It is worth mentioning that *ordinary* polynomial identities for the matrix algebra M_n are known only when $n \leq 2$, for infinite fields.

Let $A_1 = M_{n_1} \oplus \cdots \oplus M_{n_r}$ and $A_2 = M_{n'_1} \oplus \cdots \oplus M_{n'_s}$, and assume $n_1 \geq n_2 \geq \cdots \geq n_r$, $n'_1 \geq \cdots \geq n'_s$, $r > 1$ and $s > 1$. It is clear that $A_1 \cong A_2$ if and only if $r = s$ and $n_i = n'_i$ for all i. Moreover, assume $n_1 = n'_1$. In this case, A_1 and A_2 satisfy the same polynomial identities, namely, the polynomial identities of the matrix algebra M_{n_1}. Both are not simple as ordinary algebras, but they are tr-simple if we define the trace as the induced trace from M_n, where $n = n_1 + \cdots + n_r$. Our results say that we can find a trace identity satisfied by one algebra, but not by the other.

12 Identities of algebras with action of Hopf algebras

12.1 Hopf algebras

Definition 12.1. Let H be an associative algebra with 1 over a field $\mathbb{F}$. We denote by $m : H \otimes H \to H$ the *product* map given by $m(a \otimes b) = ab$, where $a, b \in H$. We also denote by $u : \mathbb{F} \to H$ the unit map given by $u(\alpha) = \alpha \cdot 1$. If there is no confusion, we write $\alpha \cdot 1 = \alpha$. Suppose H is endowed with two algebra homomorphisms $\Delta : H \to H \otimes H$, $\varepsilon : H \to \mathbb{F}$, called *coproduct* and *counit*, and one antiautomorphism $S : H \to H$, called *antipode*, satisfying the following conditions:
1. (coassociativity) $(\Delta \otimes \mathrm{id}_H) \circ \Delta = (\mathrm{id}_H \otimes \Delta) \circ \Delta$;
2. (counit) $(\varepsilon \otimes \mathrm{id}_H) \circ \Delta = \mathrm{id}_H = (\mathrm{id}_H \otimes \varepsilon) \circ \Delta$;
3. (antipode) $m \circ (S \otimes \mathrm{id}_H) \circ \Delta = u \circ \varepsilon = m \circ (\mathrm{id}_H \otimes S) \circ \Delta$.

Among the most popular examples of Hopf algebras are the group algebra $\mathbb{F}G$ of a group G, the universal enveloping algebra $U(L)$ for a Lie algebra L, and so on (see [246]). In $\mathbb{F}G$, one has $\Delta(g) = g \otimes g$, $S(g) = g^{-1}$, and $\varepsilon(g) = 1$, for all $g \in G$. In $U(L)$, one has $\Delta(x) = x \otimes 1 + 1 \otimes x$, $S(x) = -x$, and $\varepsilon(x) = 0$, for all $x \in L$. These two examples are *cocommutative* Hopf algebras in the sense that for any $h \in H$ the tensor $\Delta(h)$ is symmetric.

The value $\Delta(h)$ of the coproduct Δ on an element $h \in H$ is a tensor of degree 2:

$$\Delta(h) = h_1^1 \otimes h_2^1 + h_1^2 \otimes h_2^2 + \cdots + h_1^s \otimes h_2^s \quad \text{for some } h_i^j \in H. \tag{12.1}$$

People often use the so-called Sweedler notation to simplify the above expression for the coproduct of an element $h \in H$ and write

$$\Delta(h) = \sum_h h_{(1)} \otimes h_{(2)} \quad \text{or} \quad \Delta(h) = \sum_h h_1 \otimes h_2, \quad \text{or even} \quad \Delta(h) = h_1 \otimes h_2.$$

By $h_{(i)}$ they mean a sequence of elements $h_i^1, \ldots, h_i^s$ and by $h_{(1)} \otimes h_{(2)}$ the expression (12.1). Using coassociativity of the coproduct in H, one can correctly define $\Delta_1 = \Delta$, $\Delta_{m+1} = (\Delta_m \otimes \mathrm{id}_H) \circ \Delta = (\mathrm{id}_H \otimes \Delta_m) \circ \Delta$, for $m \geq 1$, and write

$$\Delta_{n-1}(h) = \sum_h h_{(1)} \otimes h_{(2)} \otimes \cdots \otimes h_{(n)}.$$

Then the remaining two conditions in Definition 12.1 can be written as follows:
2. $\sum_h \varepsilon(h_{(1)})h_{(2)} = h = \sum h_{(1)}\varepsilon(h_{(2)})$,
3. $\sum_h S(h_{(1)})h_{(2)} = \varepsilon(h) \cdot 1 = \sum_h h_{(1)}S(h_{(2)})$.

One can simply write $\varepsilon(h_{(1)})h_{(2)} = h = h_{(1)}\varepsilon(h_{(2)})$ and $S(h_{(1)})h_{(2)} = \varepsilon(h) \cdot 1 = h_{(1)}S(h_{(2)})$.

https://doi.org/10.1515/9783110566659-012

12.2 *H*-algebras

Definition 12.2. The following are the definitions of an H-Lie algebra and invariant subalgebra:

1. A Lie algebra L is called an *H-algebra* if L is a left H-module and
$$h * [a,b] = \sum_h [h_1 a, h_2 b].$$
2. For any H-algebra A, the set
$$A^H = \{a \in A \mid \forall h \in H \Rightarrow ha = \varepsilon(h)a\}$$
is called the *invariant subalgebra* of A, denoted by A^H.

For example, $\mathbb{F}G$ acts on L in such a way that the elements of G act as automorphisms thanks to $\Delta g = g \otimes g$. Also, in $H = U(\mathfrak{g})$, one has $\Delta x = x \otimes 1 + 1 \otimes x$, for $x \in \mathfrak{g}$. As a result in a $U(\mathfrak{g})$-algebra L, the elements of $\mathfrak{g}$ act as derivations. If $H = (\mathbb{F}G)^*$ is the dual to the group algebra of a group G then the basis of H is formed by the projections $p_g, g \in G$ such that
$$\Delta p_g = \sum_{hk=g} p_h \otimes p_k.$$

One has $p_g[a,b] = \sum_{hk=g}[p_h a, p_k b]$. If we set $L_g = p_g L$ then $L = \bigoplus_{g\in G} L_g$ is a grading of L by the group G.

12.2.1 Free *H*-algebras

To define the identities of H-algebras, we need to define the free H-algebra with the free set X of generators. For this, we consider the algebra $T(H) \otimes \mathbb{F}\langle X\rangle$, where $T(H)$ is the tensor algebra of a vector space H and $\mathbb{F}\langle X\rangle$ a free nonassociative algebra with free generating set X. First, we make $T(H)$ an H-algebra as follows, and define
$$h(a_1 \otimes a_2 \otimes \cdots \otimes a_n) = \sum_h (h_{(1)}a_1) \otimes (h_{(2)}a_2) \otimes \cdots \otimes (h_{(n)}a_n). \tag{12.2}$$

Now we choose inside $T(H) \otimes \mathbb{F}\langle X\rangle$ a subalgebra $\mathbb{F}_H(X)$ spanned by $T^n(H) \otimes (\mathbb{F}\langle X\rangle)_n$, $n = 1, 2, \ldots$, where $(\mathbb{F}\langle X\rangle)_n$ is spanned by the monomials of degree n in $\mathbb{F}\langle X\rangle$. This will be an absolutely free H-algebra with free generators X. Any map $\varphi : X \to A$, with A an H-algebra, uniquely extends to a homomorphism $\overline{\varphi} : \mathbb{F}_H(X) \to A$ if one sets
$$\overline{\varphi}((h_1 \otimes \cdots \otimes h_n)x_1 \cdots x_n) = (h_1\varphi(x_1)) \cdots (h_n\varphi(x_n)).$$

Finally, a free H-Lie algebra $L_H(X)$ is the factor-algebra of $\mathbb{F}_H(X)$ by an H-invariant ideal generated by all $\{u(vw) + v(wu) + w(uv) \mid u, v, w \in \mathbb{F}_H(X)\}$ and $\{u^2 \mid u \in \mathbb{F}_H(X)\}$.

If H is cocommutative and char $\mathbb{F} \neq 2$ then $L_H(X)$ is an ordinary free Lie algebra with free generators $h_\beta \otimes x$, where h_β runs through a basis of H and x runs through X. Otherwise, it does not need to be a Lie algebra. So when we use this approach to deal with the identities of graded algebra, we must assume that the grading group G is abelian, which is quite natural in the case of Lie algebras.

In the case of G-graded identities, G a group or even a semigroup, people normally choose an alphabet $X = \bigcup_{g\in G} X^g$, where $X^g = \{x_1^g, x_2^g, \dots\}$, consider a free algebra $L(X)$. If $w(x_{i_1}^{g_1}, \dots, x_{i_n}^{g_n})$ is an element in $L(X)$ then $w(x_{i_1}^{g_1}, \dots, x_{i_n}^{g_n}) = 0$ is an identity in a G-graded algebra A if $w(a_1, \dots, a_n) = 0$ whenever $a_1 \in A_{g_1}, \dots, a_n \in A_{g_n}$. In terms of H-identities, one can rewrite $w(x_{i_1}^{g_1}, \dots, x_{i_n}^{g_n})$ as follows. The grading is given by the action of the dual Hopf algebra $H = (\mathbb{F}\, G)^*$. A natural basis of H consists of the projections $p_g, g \in G$. The graded monomial $x_{i_1}^{g_1} \cdots x_{i_n}^{g_n}$ in $L(X)$ can now be associated with the element $(p_{g_1} \otimes \cdots \otimes p_{g_n})(x_{i_1} \cdots x_{i_n})$ in $L_H(X)$. This enables one to rewrite graded identities in terms of H-identities, and vice versa.

12.2.2 *H*-algebras, whose invariant subalgebras are PI-algebras

The subalgebra $T(H)^H$ of invariants of the action of H on $T(H)$ via the left regular action (12.2) turns out to be important in the study of identities. In [42] the authors introduced a technical condition on H which, as they proved, was equivalent to the following *Condition (*)*:

$$\dim\ T(H)/\mathrm{Ideal}(T(H)^H) < \infty. \tag{12.3}$$

Suppose that A is an H-algebra. We denote by $I(A)$ the subalgebra of H-invariants in A. So $T(H)^H = I(H)$. Then the following is true.

Theorem 12.3 (Y. Bahturin, V. Linchenko). *Let H be a finite-dimensional Hopf algebra over a field $\mathbb{F}$. The following conditions are equivalent:*

1. *For any associative algebra A over $\mathbb{F}$ with an action of H, it follows from $I(A)$ being a PI-algebra that also A is a PI-algebra.*
2. *There exists a function $f : \mathbb{N} \to \mathbb{N}$ such that for any associative algebra A with an action of H if $I(A)$ satisfies a nontrivial identity of degree t then A satisfies a nontrivial identity of degree $f(t)$.*
3. *There exists a function $g(t)$ such that for any natural t and any H-algebra A with $I(A)^t = \{0\}$ one has $A^{g(t)} = \{0\}$.*
4. *There exists a number N such that any H-algebra A, where $I(A)$ has zero multiplication, satisfies $A^N = \{0\}$.*
5. *Condition (12.3) is satisfied.*

Each of the above conditions implies that H is semisimple.

It is noted in [42] that (12.3) holds for the dual group algebras $(\mathbb{F}G)^*$, so that the result applies to algebras graded by a finite group. It also holds for semisimple group algebras and for the crossed product of algebras, satisfying (12.3). In particular, all semisimple Hopf algebras of dimension p^n, with p a prime number, over a field of characteristic zero satisfy (12.3). Thus, from the above results, one can recover well-known results on identities of graded associative algebras [41] and algebras with action of a group by automorphisms [165].

Condition (12.3) is crucial also in the case of Lie H-algebras. In the same way as in the case of Theorem 11.6, the result applies not only to Lie algebras with action of a Hopf algebras, but to a wider class of the so-called Lie-type algebras, which incidentally includes associative algebras, Lie supearalgebras, and so on.

Definition 12.4. A variety of algebras $\mathbf{M}$ given by multilinear identities is called a *Lie-type variety* if the following conditions are satisfied:

1. There exist $\lambda, \mu \in \mathbb{F}$, $\lambda \neq 0$, such that in any algebra $M \in \mathbf{M}$ one has

$$a(bc) = \lambda(ab)c + \mu(ac)b, \quad \text{where } \lambda, \mu \in \mathbb{F},\ \lambda \neq 0.$$

2. The codimension growth $c_{\mathbf{M}}(n)$ of $\mathbf{M}$ is faster than a function of the form $\frac{n!}{b^n}$, $b \in \mathbb{N}$.

An important theorem by V. Linchenko (see the proof below in Section 12.3) says the following.

Theorem 12.5 (V. Linchenko). *For a finite-dimensional Hopf algebra H, the following conditions are equivalent:*

1. $\dim T(H)/\mathrm{Ideal}(I(H)) \leq \infty$.
2. *For any H-algebra L in a Lie-type variety $\mathbf{M}$, if L^H satisfies a nontrivial identity then so does L.*

As a corollary from Linchenko's theorem, one gets Bahturin–Zaicev's theorem [60]. These theorems have been actively applied in group theory while studying, for example, groups with automorphisms such that the subgroups of fixed points satisfy some finiteness conditions. As one of the latest references, see [1].

12.3 Proof of Linchenko's theorem

In this section we give the proof of V. Linchenko's theorem (Theorem 12.5).

From the point of view of applications mentioned just above, the main concern is the proof of Theorem 11.7 stating that any Lie algebra L satisfies a nontrivial identity provided that the subalgebra of fixed elements L^G under the action of a finite group of automorphisms G satisfies a nontrivial identity and the characteristic of the base field is not a divisor of the order of G.

This result generalizes Theorem 11.11, where the group of automorphisms had to be solvable. We obtain this result as a consequence of an analogous result on associative algebras, the proof of which can be found in [166] or [245]. Recalling Theorem 12.3 (see [42]), we get an assertion about the structure of $T(\mathbb{F}[G])$, the tensor algebra of the vector space $\mathbb{F}[G]$, and apply it to Lie algebras.

The action of a finite group of automorphisms on an algebra is a particular case of the action of a finite-dimensional Hopf algebra on an algebra. In this latter case the question of the existence of an identity on the whole of algebra provided the subalgebra of invariants satisfies an identity can be reduced to a condition on the Hopf algebra only. The case of associative algebras was considered in [42]. For a larger class of algebras, which includes associative algebras and Lie algebras, the condition on the Hopf algebra remains the same.

12.3.1 Definitions

Example 12.6. Let G be a finite subgroup of the group of automorphisms of an algebra A then the structure of an $\mathbb{F}[G]$-algebra on A is defined by

$$g \circ a = g(a),$$

for any $g \in G$, $a \in A$. In this case, $\mathbb{F}[G]$ is a Hopf algebra with $\Delta(g) = g \otimes g$ and $\varepsilon(g) = 1$ for any $g \in G$, and

$$A^{\mathbb{F}[G]} = A^G = \{a \in A \mid \forall g \in G \Rightarrow g(a) = a\},$$

i. e., $A^{\mathbb{F}[G]} = A^G$ is the subalgebra of fixed points for the action of G.

Example 12.7. Let $A = \bigoplus_{g\in G} A_g$ be an algebra graded by a finite group G (A may be nonassociative), i. e., with the condition

$$A_g A_h \subset A_{gh},$$

for all $g,\ h \in G$. Then A is an H-algebra for $H = (\mathbb{F}[G])^*$, the dual Hopf algebra of $\mathbb{F}[G]$. We recall that, in H, there is the basis of $\{p_g \mid p_g = \delta_{g,h},\ g, h \in G\}$ and

$$\Delta(p_g) = \sum_{hk=g} p_h \otimes p_k, \quad \varepsilon(p_g) = 0 \quad \text{for all } g \in G,\ g \neq e \text{ and } \varepsilon(p_e) = 1.$$

In this case, p_g acts as the projection of A onto A_g, and the subalgebra of invariants is A_e.

In Section 11.2.1.2 we have introduced the class of Lie-type algebras. We slightly change the definition so that it suits our purposes and still call them of Lie type.

Definition 12.8. Let $\mathcal{C}$ denote the class of varieties of algebras such that for any $\mathbf{M} \in \mathcal{C}$ the following conditions are satisfied:

1) Variety $\mathbf{M}$ is of Lie type.
2) Any system of identities in the varieties from $\mathcal{C}$ is equivalent to a system of multilinear identities.
3) The growth of the codimension sequence $c_n(\mathbf{M})$ of $\mathbf{M} \in \mathcal{C}$ is faster than a function of the form $n!/b^n$, $b \in \mathbf{N}$.

The two most popular varieties in $\mathcal{C}$ are: $\mathbf{M}_1$, the variety of all associative algebras, which belongs to $\mathcal{C}$ because $c_n(\mathbf{M}_1) = n!$, and $\mathbf{M}_2$, the variety of Lie algebras because $c_n(\mathbf{M}_2) = (n-1)!$.

By analogy with the case of associative algebras, where instead of the free associative algebra $\mathbb{F}\langle X\rangle$ with the set of free generators X one may consider the tensor algebra $T(V)$ of the vector space V with $\dim V = \operatorname{card} X$, we will consider free algebras of varieties freely generated by vector spaces. We use $\mathbb{F}(V, \mathbf{M})$ to denote these algebras.

Let X contain the subset $\{x_i \mid i \in \mathbf{N}\}$. We define the following subspaces in $D = \mathbb{F}(H \otimes \langle X\rangle, \mathbf{M})$ for a Lie-type variety $\mathbf{M}$:

$$\begin{aligned} V_{n,H} &= \langle\{[(h_1 \otimes x_{\sigma(1)}) \cdots (h_n \otimes x_{\sigma(n)})] \mid \sigma \in S_n, h_1, \dots, h_n \in H\}\rangle, \\ T(A) &= \{y \in D \mid \varphi(y) = 0\ \forall \varphi : D \to A, \\ &\qquad \varphi \text{ is an algebra homomorphism with } \varphi(h \otimes x) = h\varphi(e \otimes x)\}, \\ V_n &= \langle\{[(e \otimes x_{\sigma(1)}) \cdots (e \otimes x_{\sigma(n)})] \mid \sigma \in S_n\}\rangle. \end{aligned}$$

Notice that $V_n \cap T(A)$ obviously forms the subspace of all multilinear identities of degree n of algebra A. Therefore in the case when $A \in \mathbf{M} \in \mathcal{C}$, proving that A satisfies a nontrivial in $\mathbf{M}$ identity is equivalent to showing that for any natural number n there exists a natural number N such that for any $n \geq N$ one has $\dim V_n / V_n \cap T(A) < n!/b$.

An important tool in the proof in this section is Condition (*) below, which is equivalent to Condition (12.3) introduced earlier in the introductory Section 12.

Definition 12.9. The following condition will be referred to as

$$\begin{aligned} &\text{CONDITION } (*): \\ &(\forall t \in \mathbf{N})(\exists N(t)) : (\forall n \geq N(t)) \Rightarrow \\ &H^{\otimes n} = \sum_{\substack{k_0+\cdots+k_{t+1}=n,\\ k_1,\dots,k_t \geq 1}} H^{\otimes k_0} \otimes \operatorname{Inv}(H^{\otimes k_1}) \otimes \cdots \otimes \operatorname{Inv}(H^{\otimes k_t}) \otimes H^{\otimes k_{t+1}}. \end{aligned}$$

Recall that the H-algebra structure on $T(H)$ is defined as follows:

$$\begin{aligned} &\forall h, h_1, \dots, h_k \in H : \\ &\quad h(h_1 \otimes \cdots \otimes h_k) = \sum h_{(1)} h_1 \otimes \cdots \otimes h_{(k)} h_k \end{aligned}$$

and $\mathrm{Inv}(H^{\otimes k}) = H^{\otimes k} \cap T^H(H)$.

In [42] it was shown that Condition (*) is equivalent to its particular case when $t = 1$.

Recall the following definition (compare with that in Section 11.2.1.4 above).

Definition 12.10. A permutation $\sigma \in S_n$ is said to be m-decomposable if there exist $1 \le i_1 < \cdots < i_m \le n$ such that

$$\sigma(i_j) > \sigma(i_j + 1), \ldots, \sigma(i_{j+1}),$$

$j = 1, \ldots, m-1$, otherwise σ is called m-indecomposable.

12.3.2 Main theorem

We are now ready to prove Theorem 12.5. It will be convenient to use the following more explicit form.

Theorem 12.11. *For a finite-dimensional Hopf algebra H, the following conditions are equivalent:*
1) *Algebra H satisfies Condition (*).*
2) *For an H-algebra A such that $A \in \mathbf{M} \in \mathcal{C}$, if A^H satisfies a nontrivial identity in $\mathbf{M}$ then so does A.*

Proof. 2) $\Rightarrow$ 1). Note that $\mathbf{M}_1 \in \mathcal{C}$ and in [42] it was shown that in this case H satisfies Condition (*).

Let us show 1) $\Rightarrow$ 2).

Lemma 12.12. *For any $n \in \mathbf{N}$, one has*

$$\dim V_n/V_n \cap T(A) \le \dim V_{n,H}/V_{n,H} \cap T(A).$$

Proof. Observe that

$$\begin{aligned} V_n/V_n \cap T(A) &\cong V_n + T(A)/T(A) \\ &\subset V_{n,H} + T(A)/T(A) \cong V_{n,H}/V_{n,H} \cap T(A). \end{aligned}$$ □

Now let $\mathbf{M} \in \mathcal{C}$, $A \in \mathbf{M}$, and let A^H satisfy a nontrivial identity in $\mathbf{M}$. By Lemma 12.12, it suffices to show that for any $b \in \mathbf{N}$ there exists $N \in \mathbf{N}$ such that for any $n \ge N$ one has

$$\dim V_{n,H}/V_{n,H} \cap T(A) < n!/b^n.$$

Now observe that if

$$\sum_s a_1^s \otimes \cdots \otimes a_k^s \in \mathrm{Inv}(H^{\otimes k})$$

then for any algebra homomorphism φ, $\varphi : D \to A$, with $\varphi(h \otimes x) = h\varphi(e \otimes x)$, one has

$$\varphi\Big(\sum_s ((a_1^s \otimes x_{i_1}) \cdots (a_k^s \otimes x_{i_k}))\Big) \in A^H.$$

Here $(\cdots)$ denotes a fixed arrangement of brackets.

In order to see this, we consider

$$\psi : H^k \to A,$$

where H^k is the kth Cartesian power of H,

$$\psi(h_1, \ldots, h_k) = ((h_1 a_1) \cdots (h_k a_k)).$$

Here $\{a_i\}$ is an arbitrary set of elements in A and $(\cdots)$ is the fixed arrangement of brackets. Then ψ is a k-linear map so that there exists

$$\overline{\psi} : H^{\otimes k} \to A, \quad \overline{\psi}(h_1 \otimes \cdots \otimes h_k) = \psi(h_1, \ldots, h_k).$$

If $\varphi(e \otimes x_{i_j}) = a_j$, $h \in H$, then

$$\begin{aligned}
h\varphi\Big(\sum_s ((a_1^s \otimes x_{i_1}) \cdots (a_k^s \otimes x_{i_k}))\Big) &= h \sum_s ((a_1^s a_1) \ldots (a_k^s a_k)) \\
&= \sum \sum_s ((h_{(1)} a_1^s a_1) \cdots (h_{(k)} a_k^s a_k)) \\
&= \overline{\psi}\Big(\sum \sum_s h_{(1)} a_1^s \otimes \cdots \otimes h_{(k)} a_k^s\Big) \\
&= \varepsilon(h)\overline{\psi}\Big(\sum_s a_1^s \otimes \cdots \otimes a_k^s\Big) \\
&= \varepsilon(h)\varphi\Big(\sum_s (a_1^s \otimes x_{i_1}) \cdots (a_k^s \otimes x_{i_k})\Big).
\end{aligned}$$

Lemma 12.13. *Under the hypotheses of Theorem* 12.11, $V_{n,H}$ *is generated by the set*

$$\{[(h_1 \otimes x_{\sigma(1)}) \cdots (h_n \otimes x_{\sigma(n)})] \mid \sigma \text{ is } m\text{-indecomposable}, \ h_1, \ldots, h_n \in H\}$$

modulo $T(A)$, *where* $n \geq m \geq N(t)$, t *is the degree of a multilinear identity in* A^H, *and* $N(t)$ *is the number from Condition* (*).

Proof. We write explicitly the identity satisfied by A^H, namely

$$\sum_{\rho \in S_t} \alpha_\rho [x_{\rho(1)} \cdots x_{\rho(t)}] \equiv 0. \tag{12.4}$$

By the remark before Lemma 12.13, we have

$$\sum_{\rho \in S_t} \alpha_\rho (w[u_{\rho(1)} \cdots u_{\rho(t)}]) \in T(A),$$

for any $w \in D$ and u_k of the form

$$u_k = \sum_{S_k} [(a_{j_k}^{S_k} \otimes x_{\sigma(j_k)}) \cdots (a_{j_{k+1}-1}^{S_k} \otimes x_{\sigma(j_{k+1}-1)})],$$

where $\sum_{S_k} a_{j_k}^{S_k} \otimes \cdots \otimes a_{j_{k+1}-1}^{S_k} \in T(H)^H$.

Applying (11.5), we get an element in $T(A)$ of the form

$$\sum_{\pi \in S_t} \beta_\pi [w\, u_{\pi(1)} \cdots u_{\pi(t)}], \tag{12.5}$$

where $\beta_{\text{id}} \neq 0$ (here we have used the requirement of $\lambda \neq 0$ in the definition of Lie-type algebras).

Now take any w from $V_{j_1-1,H}$ of the form $[(a_1 \otimes x_{\sigma(1)}) \cdots (a_{j_1-1} \otimes x_{\sigma(j_1-1)})]$ and $\sigma(j_k) > \sigma(j_k+1), \ldots, \sigma(j_{k+1})$, $k = \overline{1, t-1}$, where $1 \leq j_1 < \cdots < j_m \leq n$, $\sigma \in S_n$. Again, with the help of (11.5), we can transform (12.5) into an element of $T(A)$ of the form

$$\begin{aligned} &\sum_{S_1,\ldots,S_t} \overline{\alpha}[(a_1 \otimes x_{\sigma(1)}) \cdots (a_{j_1-1} \otimes x_{\sigma(j_1-1)})(a_{j_1}^{S_1} \otimes x_{\sigma(j_1)}) \\ &\cdots (a_{j_2-1}^{S_1} \otimes x_{\sigma(j_2-1)}) \cdots (a_{j_t}^{S_t} \otimes x_{\sigma(j_t)}) \cdots (a_{j_{t+1}}^{S_t} \otimes x_{\sigma(j_{t+1})})] + z, \end{aligned} \tag{12.6}$$

where z is the sum of some elements of lexicographically lesser order.

By definition of a Lie type variety, $V_{n,H}$ is spanned by left-normed monomials. So we take an arbitrary left-normed monomial $[(h_1 \otimes x_{\sigma(1)}) \cdots (h_n \otimes x_{\sigma(n)})]$ in $V_{n,H}$, with σ being m-decomposable. That is, there are $1 \leq i_1 < \cdots < i_m \leq n$ such that $\sigma(i_j) > \sigma(i_j+1), \ldots, \sigma(i_{j+1})$, $j = \overline{1, m-1}$.

From Condition (*) we obtain that $e^{\otimes m}$ can be presented as

$$e^{\otimes m} = \sum_r w^r \otimes v_1^r \otimes \cdots \otimes v_t^r \otimes u^r. \tag{12.7}$$

Here

$$v_i^r \in \text{Inv}(H^{\otimes m_i^r}),\ m_i^r \geq 1,\ w^r \in H^{\otimes m_0^r},\ u^r \in H^{\otimes m_{t+1}^r},\ \sum_{i=0}^{t+1} m_i^r = m.$$

In [42], the following inclusion was proved:

$$(\Delta \otimes \text{id}^{\otimes k})(\text{Inv}(H^{\otimes k+1})) \subset \text{Inv}(H^{\otimes k+2}).$$

Consider now the map

$$\psi : H^{\otimes m} \to H^{\otimes n},$$
$$\psi(v) = e^{\otimes i_1 - 1} \otimes ((\Delta_{i_2-i_1-1} \otimes \cdots \otimes \Delta_{n-i_m})(v)).$$

Here $\Delta_0 = \mathrm{id}$. Applied to (12.7), it gives

$$e^{\otimes n} = \sum_r \overline{w}^r \otimes \overline{v_1}^1 \otimes \cdots \otimes \overline{v_t}^r \otimes \overline{u}^r,$$

where the invariants $\overline{v_j}^r$ now start at the selected points $\{i_j\}$. The map $\varphi : H^n \to V_{n,H}$ given by

$$\varphi(g_1, \dots, g_n) = [(g_1 \otimes x_{\sigma(1)}) \cdots (g_n \dots x_{\sigma(n)})]$$

is n-linear, hence there is

$$\begin{aligned} \overline{\varphi} &: H^{\otimes n} \to V_{n,H}, \\ \overline{\varphi}(g_1 \otimes \cdots \otimes g_n) &= \varphi(g_1, \dots, g_n), \end{aligned}$$

where $\overline{\varphi}$ is now linear. Therefore we can write

$$\begin{aligned} &[(h_1 \otimes x_{\sigma(1)}) \cdots (h_n \otimes x_{\sigma(n)})] \\ &\quad = \overline{\varphi}(h_1 \otimes \cdots \otimes h_n) \\ &\quad = \overline{\varphi}(e^{\otimes n} \circ (h_1 \otimes \cdots \otimes h_n)) \\ &\quad \text{(here } \circ \text{ is the multiplication in } H^{\otimes n}) \\ &\quad = \overline{\varphi}\Big(\Big(\sum_r \overline{w}^r \otimes \overline{v_1}^r \otimes \cdots \otimes \overline{v_t}^r \otimes \overline{u}^r\Big) \circ (h_1 \otimes \cdots \otimes h_n)\Big) \\ &\quad (\mathrm{Inv}(H^{\otimes k}) \text{ is a right ideal in } H^{\otimes k}) \\ &\quad = \overline{\varphi}\Big(\Big(\sum_q \overline{\overline{w}}^q \otimes \overline{\overline{v_1}}^q \otimes \cdots \otimes \overline{\overline{v_t}}^q \otimes \overline{\overline{u}}^q\Big)\Big) \\ &\quad = \sum_q \sum_{s_1^q, \dots, s_t^q} [(a_1^q \otimes x_{\sigma(1)}) \cdots (a_{j_1^q - 1}^q \otimes x_{\sigma(j_1^q - 1)}) \\ &\qquad (a_{j_1^q}^{s_1^q} \otimes x_{\sigma(j_1^q)}) \cdots (a_{j_2^q - 1}^{s_1^q} \otimes x_{\sigma(j_2^q - 1)}) \cdots (a_{j_t^q}^{s_t^q} \otimes x_{\sigma(j_t^q)}) \\ &\qquad \cdots (a_{j_{t+1}^q - 1}^{s_t^q} \otimes x_{\sigma(j_{t+1}^q - 1)})(a_{j_{t+1}^q}^q \otimes x_{\sigma(j_{t+1}^q)}) \cdots (a_n^q \otimes x_{\sigma(n)})]. \end{aligned} \tag{12.8}$$

Here

$$\overline{\overline{v_p^q}} = (a_{j_p^q}^{s_p^q} \otimes x_{\sigma(j_p^q)}) \cdots (a_{j_{p+1}^q - 1}^{s_p^q} \otimes x_{\sigma(j_{p+1}^q - 1)}),$$

i. e., the coefficients $a_{j_p^q}^{s_p^q}$ are as in (12.6). We apply (12.6) to (12.8) and get that $[(h_1 \otimes x_{\sigma(n)}) \cdots (h_n \otimes x_{\sigma(n)})]$ is expressed through some elements labeled by permutations lesser than σ. The induction on the lexicographical ordering yields the desired result. □

The completion of the proof is very similar to the argument in Section 11.2.1.5. Recall that if $a_m(n)$ is the number of m-indecomposable permutations in S_n then

$$\lim_{n\to\infty} \sqrt[n]{a_m^n/n!} = 0.$$

If we set $\dim H = d$, then, for any $b \in \mathbf{N}$ and sufficiently large n, the following holds:

$$\sqrt[n]{a_m(n)/n!} < 1/b^d,$$

so that

$$\dim V_{n,H}/V_{n,H} \cap T(A) \leq a_m(n)d^n < n!/b^n. \qquad \square$$

Now we state the main corollaries of Theorem 12.11.

Theorem 12.14 (See Theorem 11.11 for G solvable). *Let G be a finite subgroup in the automorphism group of a Lie algebra L over a field $\mathbb{F}$ where* $\operatorname{char} \mathbb{F}$ *is not a divisor of $|G|$ and let L^G satisfy a nontrivial identity. Then L satisfies a nontrivial identity.*

Proof. In [42] it was shown that an analogous theorem on associative algebras proved in [166] and [245] makes it necessary for H to satisfy Condition (*). □

The following corollary was proved earlier (Theorem 11.10).

Theorem 12.15. *Let $L = \bigoplus_{g\in G} L_g$ be a Lie algebra over a field $\mathbb{F}$, graded by a finite group G. If L_e satisfies a nontrivial identity then so does L.*

Proof. In [42], with the help of [41], it was shown that $(\mathbb{F}[G])^*$ satisfies Condition (*). □

One may produce analogous (well-known) corollaries for associative algebras.

12.4 Codimension growth of H-algebras

Alexei Gordienko [128] has proved a number of results on the codimension growth of identical relations in finite-dimensional Lie algebras. The most general of them are the following. In their statement there is the notion of an *H-nice Lie algebra L*, L being a finite-dimensional Lie algebra over an algebraically closed field $\mathbb{F}$ of characteristic zero. One says that L is H-nice if either L is semisimple or the following conditions hold:

(1) The nilpotent radical N and the solvable radical R of L are H-invariant;
(2) (Levi decomposition) There exists an H-invariant maximal semisimple subalgebra $B \subset L$ such that $L = B \oplus R$ (direct sum of H-modules);

(3) (Wedderburn–Mal'cev decompositions) For any H-submodule $W \subset L$ and associative H-module subalgebra $A_1 \subset \mathrm{End}_{\mathbb{F}}(W)$, the Jacobson radical $J(A_1)$ is H-invariant and there exists an H-invariant maximal semisimple associative subalgebra $\widetilde{A}_1 \subset A_1$ such that $A_1 = \widetilde{A}_1 \oplus J(A_1)$ (direct sum of H-submodules);

(4) For any H-invariant Lie subalgebra $L_0 \subset \mathfrak{gl}(L)$ such that L_0 is an H-module algebra and L is a completely reducible L_0-module disregarding H-action, L is a completely reducible (H, L_0)-module.

The following is Gordienko's main result on the codimension growth of H-Lie algebras,

Theorem 12.16. *Let L be a nonnilpotent H-nice Lie algebra over an algebraically closed field $\mathbb{F}$ of characteristic 0. Then there exist constants $C_1, C_2 > 0$, $r_1, r_2 \in \mathbb{R}$, $d \in \mathbb{N}$ such that $C_1 n^{r_1} d^n \le c_n^H(L) \le C_1 n^{r_2} d^n$ for all $n \in \mathbb{N}$.*

Theorem 12.17. *Let $L = L_1 \oplus \cdots \oplus L_s$ (direct sum of H-invariant ideals) be an H-module Lie algebra over an algebraically closed field $\mathbb{F}$ of characteristic 0 where H is a Hopf algebra. Suppose L_i are H-nice algebras. Then* $\mathrm{PI}\exp^H(L) = \max_{1\le i\le s} \mathrm{PI}\exp^H(L_i)$.

12.5 Isomorphism of simple H-algebras

In the study of identities of H-algebras, such as graded algebras, algebras with fixed group of automorphisms or algebras with derivation, and superalgebras, one can use a corollary of the results of Razmyslov in Section 14.3.

We always assume that the signature of Ω-algebras contains at least one operation of arity at least 2. Recall that by Razmyslov's Theorem 11.19, two simple finite-dimensional Ω-algebras over an algebraically closed field, satisfying the same polynomial identities, are isomorphic.

An H-algebra A can be turned to an Ω-algebra in the following way. Let μ be the original operation of A and suppose that, for each $h \in H$, ρ_h is a unary operation given by $\rho_h(a) = h * a$. Set $\Omega_H = \{\mu\} \cup \{\rho_h \mid h \in H\}$. Now consider the relatively free Ω_H-algebra given by the identities:

1. $\rho_h(x) + \rho_g(x) = \rho_{h+g}(x)$;
2. $\rho_{\alpha g}(x) = \alpha \rho_g(x)$;
3. $\rho_1(x) = x$;
4. $\rho_g(\rho_h(x)) = \rho_{gh}(x)$;
5. $\rho_h(\mu(x,y)) = \sum_h \mu(\rho_{h_{(1)}}(x), \rho_{h_{(2)}}(y))$.

Much in the same way as we did just above for the graded identities, one can turn H-identities of A into its identities as Ω_H-algebra. Thus we obtain the following.

Theorem 12.18 (Y. Bahturin, F. Yasumura, [61]). *Two finite-dimensional H-algebras over an algebraically closed field, which are simple as H-algebras and satisfy the same H-identities, are isomorphic as H-algebras.*

There are many applications of Theorem 12.18.

12.5.1 Action by automorphisms and antiautomorphisms

Let G be a subgroup of the group of automorphisms and antiautomorphisms of the algebra A. It is well-known that the group algebra KG is a Hopf algebra. Then the action of G on A is a particular case of Hopf action by the group algebra KG.

12.5.2 Action by derivations

Let D be a Lie subalgebra of the algebra of derivations of A. Then the universal enveloping algebra $U(D)$ is a Hopf algebra. Thus the action of D on A can be viewed as a Hopf action of $U(D)$. As a classsical example, see R. Block's paper [75].

12.5.3 Generalized action

Let $\mathcal{H} = (H, \Delta^{(1)}, \Delta^{(2)})$ be a triple where H is a unital associative algebra and $\Delta^{(1)}, \Delta^{(2)}$ are two linear maps, called coproducts $\Delta^{(1)}, \Delta^{(2)} : H \to H \otimes H$. Using Sweedler's notation, we can write $\Delta^i(h) = h^{(i)}_{(1)} \otimes h^{(i)}_{(2)}$, meaning that $\Delta^i(h)$ are arbitrary tensors of degree 2. In distinction with Hopf algebras, we impose no restrictions on the coproducts.

An algebra A is called an *$\mathcal{H}$-algebra* if A is a left H-module via $(h, a) \to h * a$, for any $h \in H$ and $a \in A$, and for any $a, b \in A$ one has

$$h * (ab) = (h^{(1)}_{(1)} * a)(h^{(1)}_{(2)} * b) + (h^{(2)}_{(1)} * b)(h^{(2)}_{(2)} * a).$$

Such algebras have appeared in [68, 128]. In a natural way, one can define the notions of the homomorphisms of $\mathcal{H}$-algebras, simple $\mathcal{H}$-algebras, and so on. The construction of a free $\mathcal{H}$-algebra also does not create any problems (see algebras with H-action above), and so one can speak about $\mathcal{H}$-identities. As earlier, one can define the set $\Omega_{\mathcal{H}}$ consisting of one binary operation μ and the set of unary operations ρ_h, for each $h \in H$. As earlier, if A is an $\mathcal{H}$-algebra then $\rho_h(a) = h * a$. The variety of $\Omega_{\mathcal{H}}$-algebras is distinguished by the family of identical relations, one for each $h \in H$:

$$\rho_h(\mu(a, b)) = \mu(\rho_{h^{(1)}_{(1)}}(a), \rho_{h^{(1)}_{(2)}}(b)) + \mu(\rho_{h^{(2)}_{(1)}}(b), \rho_{h^{(2)}_{(2)}}(a)).$$

13 Engel Lie algebras

An element x of a Lie algebra L is called ad-nilpotent, or Engel, or simply nil, if there is a natural number n such that $(\operatorname{ad} x)^n = 0$. The least $n = n(x)$ with the above property is called the nil-index of x. A Lie algebra in which all elements are nil is called nil or Engel Lie algebra. By Engel's theorem (Theorem 1.14) in Section 1.7.3, any finite-dimensional Engel's algebra is nilpotent. The direct sum $L = \bigoplus_{n=2}^{\infty} \mathfrak{nt}(n, \mathbb{F})$, where $\mathfrak{nt}(n, \mathbb{F})$ is a Lie algebra of upper nil-triangular matrices of order n, is an example of Engel but not nilpotent Lie algebra. Of course, neither L is finitely generated nor the nil-indexes of the elements of L are totally bounded. So, by analogy with Burnside problem in group theory and in view of some needs for the solution of this problem, people began getting interested in the following problems. We denote by $\mathbf{E}_l$ the variety of Lie algebras in which $(\operatorname{ad} x)^l(y) \equiv 0$ is identically true. We will often write simply $(\operatorname{ad} x)^l \equiv 0$.

GLOBAL ENGEL PROBLEM: Is it true that a Lie algebra with identity $(\operatorname{ad} x)^n \equiv 0$ is nilpotent? In other words, does there exit $c = c(n)$ such that $\mathbf{E}_l \subset \mathbf{N}_c$?

LOCAL ENGEL PROBLEM: Is it true that a finitely generated Engel Lie algebra is nilpotent?

RESTRICTED ENGEL PROBLEM: Is it true that for any natural n, r there is a natural $g(n, r)$ such that any r-generator algebra in $\mathbf{E}_n$ is nilpotent of class $g(n, r)$?

In Chapters 1 and 8 the following results on these problems have been provided.

P. M. Cohn's example in Section 1.8.3 showed that over a field of characteristic $p > 0$ there are nonnilpotent algebras satisfying $(\operatorname{ad} x)^{p+1} = 0$. Yu. P. Razmyslov provided a much sharper result saying that over a field of characteristic $p \geq 5$ there are nonnilpotent algebras satisfying $(\operatorname{ad} x)^{p-2} = 0$ (see Section 8.5). In the case of solvable Lie algebras, a very important theorem (Theorem 1.25) is due to P. J. Higgins: any solvable Lie algebra of solvable step ℓ over a field of characteristic 0 or $p \geq n$, satisfying an identical relation $(\operatorname{ad} x)^n \equiv 0$, is nilpotent of step at most c, where $c = \frac{n^\ell - 1}{n-1}$.

We will discuss further developments on the global Engel problem in Section 13.2 below. In Section 13.2.2 we will discuss the restricted Engel problem, especially in connection to the celebrated restricted Burnside problem in group theory.

13.1 Local Engel problem

The answer to the local Engel problem is well-known to be in the negative. The first famous E. S. Golod's example dates back to 1963 (see [127]). Golod produced examples of infinite finitely-generated periodic groups, by first constructing infinite-dimensional finitely-generated associative nil-algebras. As a byproduct, he solved in the negative the local Engel problem for Lie algebras:

Over any field there are infinite-dimensional finitely-generated Engel algebras.

https://doi.org/10.1515/9783110566659-013

There are also a number of examples of finitely-generated nil *restricted* infinite-dimensional (hence nonnilpotent) Lie algebras (p-Lie algebras). In each of these algebras L over a field $\mathbb{F}$ of characteristic $p > 0$, we have a finite number of generators $a_1, \dots, a_d$ of a p-algebra, and for every $x \in L$, there is a natural $n(x)$ such that $(\operatorname{ad} x)^{p^{n(x)}} = 0$. Note that an easy argument (see, for example, the proof of [54, Corollary 3] and also the last paragraph of the proof of Theorem 13.5 below) shows that a Lie algebra $\widetilde{L}$ generated by $a_1, \dots, a_d$, as an ordinary Lie algebra, is infinite-dimensional. Of course, for each $x \in \widetilde{L}$ we also have $(\operatorname{ad} x)^{p^{n(x)}} = 0$.

In 2006 Petrogradsky [279] has constructed examples of self-iterating Lie algebras similar to the Grigorchuk group. In the case of characteristic 2, his example gives a two-generator restricted Lie algebra of polynomial growth that has a nil-p-mapping. It has a low Gelfand–Kirillov dimension. Similar examples exist now in any characteristic $p > 0$ (see [288] and the reference therein). These examples are based on the self-similarity idea due to R. I. Grigorchuk and N. Gupta–S. Sidki in group theory; see, for example, [38, Chapter II]. Even closer connected to Grigorchuk's group is the Lie algebra of Grigorchuk's group in characteristic 2. Bartholdi has shown that this Lie algebra is finitely generated, nil, and infinite-dimensional [65].

Another collection of examples was given in a 2007 paper by Bahturin–Olshanskii [53]. This time the techniques were similar to those people use while constructing infinite Burnside groups. One needs to add sufficiently many relations to a free restricted Lie algebra to make it nil, but not too many so it remains infinite-dimensional. Note that if we deal with the restricted Lie algebras over a field of prime characteristic p (see the definition in [37]) then nil elements are those for which $x^{[p^n]} = 0$, for some n, which makes them closer to the associative nil-algebras. To state this result, we call an algebra "large" if it admits a surjective homomorphism onto a nonabelian free Lie algebra.

Theorem 13.1. *Let $\mathbb{F}$ be a perfect, at most countable field of characteristic $p > 0$. Then for any finitely-generated restricted Lie algebra G with an ideal P of finite codimension that can be mapped homomorphically onto a nonabelian free restricted Lie algebra there exists an infinite-dimensional homomorphic image $\widetilde{G}$ in which the image $\widetilde{P}$ of P is a* nil *restricted Lie algebra, and $\widetilde{G}/\widetilde{P} \cong G/P$. One can choose $\widetilde{G}$ residually finite-dimensional or the direct limit of large restricted Lie algebras.*

Corollary 13.2. *Let $\mathbb{F}$ be a perfect, at most countable field of characteristic $p > 0$. Then for any restricted Lie algebra G, with a presentation where the number of generators exceeds the number of relations at least by two, there exists an infinite-dimensional homomorphic image $\widetilde{G}$ which is a* nil *restricted Lie algebra. One can choose $\widetilde{G}$ residually finite-dimensional nilpotent or the direct limit of large restricted Lie algebras.*

Corollary 13.3. *Let $\mathbb{F}$ be a perfect, at most countable field of characteristic $p > 0$. Then there exist infinite-dimensional finitely generated* nil *restricted Lie algebras. One can* *choose such algebras residually finite-dimensional or the direct limits of large restricted Lie algebras.*

13.1.1 Olshanskii's proof of Golod's example

Let $\mathbb{F}$ be a field. Let $A = A(x, y)$ be the free associative algebra with 1 on the free generating set $\{x, y\}$ over $\mathbb{F}$. As usual, we write $A = \bigoplus_{n=0}^{\infty} A_n$, where A_n is the nth homogeneous component of A, $A_0 = \mathbb{F} \cdot 1$. We know that $\dim A_n = 2^n$. Recall that a subspace $V \subset A$ is called *homogeneous* if $V = \bigoplus_{n=1}^{\infty}(V \cap A_n)$. In this section, given a finite set S, we denote by $\#S$ the number of elements in S.

Lemma 13.4. *Let I be a homogeneous ideal of A generated by the set $R = R_9 \cup R_{10} \cup \cdots \cup R_n \cup \cdots$ of homogeneous elements such that $R_n \subset A_n$, for all $n = 9, 10, \ldots$ Suppose that for all $k \geq 9$, we have $(\#R_k)(2/3)^k < (3/4)^k$. Then $I_n \neq A_n$, for all $n \geq 0$.*

Proof. Every component I_n, $n \geq 9$, is a linear span of homogeneous products urv, where $r \in R_j$, $9 \leq j \leq n$, and u, v are (possibly empty) monomials. There are two cases:
(a) $\deg v > 0$, so that $urv \in I_{n-1}A_1$;
(b) v is an empty word so that there is $9 \leq j \leq n$ such that $urv = ur \in A_{n-j}R_j$.

Taking both cases into account, one obtains

$$I_n \subset I_{n-1}A_1 + \sum_{j=9}^{n} A_{n-j}R_j. \tag{13.1}$$

For any natural $n \geq 1$, we choose a subspace $C_n \subset A_n$ such that $A_n = I_n \oplus C_n$. Then

$$A_{n-j}R_j = I_{n-j}R_j \oplus C_{n-j}R_j. \tag{13.2}$$

Using (13.2) and an obvious inclusion $I_{n-j}R_j \subset I_{n-1}A_1$ allows us to rewrite (13.1) as follows:

$$I_n \subset I_{n-1}A_1 + \sum_{j=9}^{n} C_{n-j}R_j. \tag{13.3}$$

We set $c_n = \dim C_n$, $d_n = \dim I_n$. We need to prove $c_n > 0$ for all $n \geq 1$. Note that $c_n = \dim A_n = 2^n$ for $1 \leq n \leq 8$.

Let us use induction on n to prove $c_n \geq \frac{3}{2}c_{n-1}$ for all $n = 1, 2, 3, \ldots$ The claim is obviously true for all $n = 1, 2, , \ldots, 8$. Indeed, by (13.3), we have $d_n \leq 2d_{n-1} + \sum_{j=9}^{n} c_{n-j}(\#R_j)$. Since $d_n = \dim A_n - c_n = 2^n - c_n$, we can write

$$2^n - c_n \leq 2(2^{n-1} - c_{n-1}) + \sum_{j=9}^{n} c_{n-j}(\#R_j), \tag{13.4}$$

or

$$c_n \geq 2c_{n-1} - \sum_{j=9}^{n} c_{n-j}(\#R_j). \tag{13.5}$$

By induction hypothesis, $c_{n-9} \leq (\frac{2}{3})^8 c_{n-1}$, $c_{n-10} \leq (\frac{2}{3})^9 c_{n-1}, \ldots$ One can write $c_{n-j} \leq (\frac{2}{3})^{j-1} c_{n-1}$. By our hypotheses, $(\#R_k)(2/3)^k < (3/4)^k$. Hence

$$c_{n-j}(\#R_j) \leq (\#R_j)\left(\frac{2}{3}\right)^{j-1} \leq \frac{3}{2}\left(\frac{3}{4}\right)^j.$$

So (13.5) implies

$$\begin{aligned} c_n &\geq 2c_{n-1} - \sum_{j=9}^{n} c_{n-j}(\#R_j) \\ &\geq 2c_{n-1} - \frac{3}{2}c_{n-1}\left(\left(\frac{3}{4}\right)^8 + \left(\frac{3}{4}\right)^9 + \cdots\right) \\ &\geq c_{n-1}\left(2 - \left(\frac{3}{2}\right)\frac{(\frac{3}{4})^9}{1-\frac{3}{4}}\right) \\ &= c_{n-1}\left(2 - 6\left(\frac{3^9}{4^9}\right)\right) > c_{n-1}(2 - 0.45) > \frac{3}{2}c_{n-1}. \end{aligned}$$

□

Theorem 13.5. *Let $\mathbb{F}$ be a field. Then there exists a 2-generator nonnilpotent Engel Lie algebra over $\mathbb{F}$.*

Proof. Let $\{w_1, w_2, \ldots\}$ be the list of all monomials of A. We need to form the set of homogeneous relations R satisfying the conditions of Lemma 13.4 and such that for any $m = 1, 2, \ldots$ there is n_i such that for any $\lambda_1, \ldots, \lambda_m$ we have $(\lambda_1 w_1 + \cdots \lambda_m w_m)^{n_m} = 0$. This can be done by induction on m, as follows. We set $R^{(1)} = \{w_1^9\}$. Suppose we have created disjoint sets of relations of growing degrees $R^{(1)}, \ldots, R^{(m-1)}$ and the numbers $n_1 < n_2 < \cdots < n_{m-1}$ such that the conditions for all $R_k \subset R^{(1)} \cup \cdots \cup R^{(m-1)}$ are satisfied and also $(\lambda_1 w_1 + \cdots + \lambda_m w_i)^{n_i} = 0$, for all $i = 1, \ldots, m-1$. We now write

$$(\lambda_1 w_1 + \cdots + \lambda_m w_m)^{n_m} = \sum_{\substack{k_1, \ldots, k_m, \\ k_1 + \cdots + k_m = n_m}} \lambda_1^{k_1} \cdots \lambda_m^{k_m} W_{k_1, \ldots, k_m}.$$

Here, for every choice of $k_1, \ldots, k_m$, the polynomial $W_{k_1,\ldots,k_m}$ is homogeneous of degree $k_1 \deg w_1 + \cdots + k_m \deg w_m$. Let us choose n_m greater than the degree of any elements in $R_k \subset R^{(1)} \cup \cdots \cup R^{(m-1)}$. Additionally, n_m must be such that for the number $K_m(n_m)$ of monomials $\lambda_1^{k_1} \cdots \lambda_m^{k_m}$ of total degree n_m we have $K_m(n_m)(2/3)^{n_m} \leq (3/4)^{n_m}$. This is possible since the growth of any polynomial ring in finitely many variables is polynomial. Now we form $R^{(m)}$ as the set of all $W_{k_1,\ldots,k_m}$. Clearly, for any $R_k \subset R^{(m)}$, we still have $(\#R_k)(2/3)^{n_m} \leq (3/4)^{n_m}$. Note that $n_m \leq k$. Multiplying both sides of the previous inequality by the sides of the latter, we obtain the desired $(\#R_k)(2/3)^k \le (3/4)^k$. Now we take $R = \bigcup_{m=1}^{\infty} R^{(m)}$. Then R satisfies all the conditions of Lemma 13.4, and so the algebra $B = A/I$ is an infinite-dimensional associative algebra. By construction, for any $b \in B$ there is n such that $b^n = 0$. It should be now standard that $(\operatorname{ad} b)^{2n_i-1} = 0$.

Let L be a Lie subalgebra of B generated by $\bar{x}$ and $\bar{y}$, with respect to the commutator. From the above, L is an Engel algebra. If L is finite-dimensional then L spanned by $b_1, \ldots, b_m$, for some m. Suppose n_i is such that $b_i^{n_i} = 0$ for $i = 1, \ldots, m$. We can now use Proposition 2.7 to conclude that B is spanned as a vector space by $b_1^{k_1} \cdots b_m^{k_m}$, where each $k_i < n_i$, $1 = 1, \ldots, m$. Then B is finite-dimensional, which is not the case. Thus L finitely generated, Engel and infinite-dimensional. Of course, L is nonnilpotent because finitely generated nilpotent algebras are finite-dimensional. □

13.1.2 Petrogradsky's Fibonacci restricted Lie algebra

Note that the growth of algebras in Golod's examples, as well as in the examples provided in Corollary 13.3, is exponential. Much "lighter" examples have been suggested by Victor Petrogradsky. In this section we study his *restricted Fibonacci Lie algebra* **L**, which is a natural analogue of the famous Grigorchuk's infinite finitely generated periodic 2-group [131].

The exposition provided by Victor Petrogradsky is based on [279, 285, 286]. We start with describing this algebra in terms of its basic relations, find an explicit basis, show that it is $\mathbb{Z}^2$-graded by multidegree in the generators, not PI, of slow *polynomial* growth, and with a nil p-mapping.

13.1.2.1 Pivot elements and relations

Temporarily, let $\mathbb{F}$ be an arbitrary field. We consider the polynomial ring $R = \mathbb{F}[t_i | i \ge 0]$. Let $\partial_i = \frac{\partial}{\partial t_i} \in \operatorname{Der} R$, $i \ge 0$, be the respective partial derivations. We define two derivations:

$$\begin{aligned} v_1 &= \partial_1 + t_0(\partial_2 + t_1(\partial_3 + t_2(\partial_4 + t_3(\partial_5 + t_4(\partial_6 + \cdots))))), \\ v_2 &= \qquad\quad \partial_2 + t_1(\partial_3 + t_2(\partial_4 + t_3(\partial_5 + t_4(\partial_6 + \cdots)))). \end{aligned}$$

The action on R and arbitrary commutators of such operators are well-defined; these operators are so called *special derivations*, see [297, 285, 284]. Denote by $\tau : R \to R$ the *shift endomorphism* $\tau(t_i) = t_{i+1}$ for $i \ge 0$. Let also $\tau(\partial_i) = \partial_{i+1}$, $i \ge 1$. We write these derivations recursively:

$$\begin{aligned} v_1 &= \partial_1 + t_0 \tau(v_1), \\ v_2 &= \tau(v_1). \end{aligned}$$

Similarly, we define

$$\begin{aligned} v_i = \tau^{i-1}(v_1) &= \partial_i + t_{i-1}(\partial_{i+1} + t_i(\partial_{i+2} + t_{i+1}(\partial_{i+3} + \cdots))) \\ &= \partial_i + t_{i-1}v_{i+1}, \quad i \geq 1. \end{aligned} \tag{13.6}$$

We refer to $\{v_i \mid i \geq 1\}$ as *pivot elements*. The commutators of the neighbors look like the Fibonacci relation.

Lemma 13.6. $[v_i, v_{i+1}] = v_{i+2}$ *for* $i = 1, 2, \ldots$

Proof. Using (13.6) yields

$$[v_i, v_{i+1}] = [\partial_i + t_{i-1}v_{i+1}, v_{i+1}] = [\partial_i, v_{i+1}] = [\partial_i, \partial_{i+1} + t_i v_{i+2}] = v_{i+2}.$$ □

In general, the commutators of arbitrary pivot elements are as follows.

Lemma 13.7. $[v_i, v_j] = t_{i-1}t_i \cdots t_{j-3}v_{j+1}$ *for* $1 \leq i < j$.

Proof. Using (13.6), we get

$$\begin{aligned} v_i &= \partial_i + t_{i-1}\partial_{i+1} + t_{i-1}t_i\partial_{i+2} + \cdots + t_{i-1}\cdots t_{j-3}\partial_{j-1} + t_{i-1}\cdots t_{j-2}v_j, \\ [v_i, v_j] &= [\partial_i + t_{i-1}\partial_{i+1} + \cdots + t_{i-1}\cdots t_{j-3}\partial_{j-1} + t_{i-1}\cdots t_{j-2}v_j,\ v_j] \\ &= [\partial_i + t_{i-1}\partial_{i+1} + \cdots + t_{i-1}\cdots t_{j-3}\partial_{j-1},\ \partial_j + t_{j-1}v_{j+1}] \\ &= t_{i-1}\cdots t_{j-3}\partial_{j-1}(t_{j-1})v_{j+1} = t_{i-1}\cdots t_{j-3}v_{j+1}. \end{aligned}$$ □

Lemma 13.8. *For all* $n \geq 1, j \geq 0$, *we have the action*

$$v_n(t_j) = \begin{cases} t_{n-1}t_n\cdots t_{j-2}, & n < j, \\ 1, & n = j, \\ 0, & n > j. \end{cases}$$

Proof. A direct check. □

Let $\mathcal{L} = \mathrm{Lie}(v_1, v_2)$ be the Lie subalgebra of $\operatorname{Der} R$ generated by v_1, v_2.

Lemma 13.9. *Define Lie algebras* $L_i := \mathrm{Lie}(v_i, v_{i+1})$ *generated by* v_i, v_{i+1} *for* $i \geq 1$. *Then*

1. $L_i \cong \mathcal{L}, i \geq 1$.
2. *We get an infinite chain of isomorphic subalgebras, inclusions being proper:*

$$\mathcal{L} = L_1 \supset L_2 \supset \cdots \supset L_n \supset \cdots.$$

3. $\mathcal{L}$ *is infinite-dimensional.*

Proof. From $\tau^{i-1}(v_1) = v_i$, $\tau^{i-1}(v_2) = v_{i+1}$, and the commutator relations of Lemma 13.7, we get $L_i = \tau^{i-1}(\mathcal{L})$. Also, $v_{i-1} \notin L_i$. □

13.1.2.2 Weights and $\mathbb{Z}^2$-grading

The field $\mathbb{F}$ is still arbitrary. Construct a *weight function* defined on all variables and respective derivations in a correlated way: $\mathrm{wt}(\partial_i) = -\mathrm{wt}(t_i) = a_i$, $i \ge 0$, the latter being arbitrary complex numbers. We extend these values to the finite products (of any type) of ∂_i and t_j. Observe that this weight is additive due to the correlation.

Namely, let a, b be the monomials as above, then $\mathrm{wt}(a \cdot b) = \mathrm{wt}([a, b]) = \mathrm{wt}(a) + \mathrm{wt}(b)$. We say that a not necessarily finite linear combination of monomials is *homogeneous* provided that all monomials have the same weight. The additivity extends to Lie or associative products of homogeneous elements, as well.

We want all summands in (13.6) to have the same weight, so we assume that

$$a_i = \mathrm{wt}\, v_i = \mathrm{wt}\, \partial_i = \mathrm{wt}\, t_{i-1} + \mathrm{wt}\, v_{i+1} = -a_{i-1} + a_{i+1}.$$

We get the Fibonacci recurrence relation $a_{i+1} = a_i + a_{i-1}$. Its two basic solutions yield the *weight* and *superweight* functions:

$$\begin{aligned} \mathrm{wt}\, v_n &= \mathrm{wt}\, \partial_n = -\mathrm{wt}\, t_n = \lambda^n, \quad \lambda = \frac{1+\sqrt{5}}{2}, \\ \mathrm{swt}\, v_n &= \mathrm{swt}\, \partial_n = -\mathrm{swt}\, t_n = \bar{\lambda}^{n-2}, \quad \bar{\lambda} = \frac{1-\sqrt{5}}{2}, \end{aligned} \qquad n \in \mathbb{N}.$$

Define also the *vector weight function* for the homogeneous elements:

$$\mathrm{Wt}(v) = (\mathrm{wt}(v), \mathrm{swt}(v)) \in \mathbb{R}^2.$$

Lemma 13.10. *The functions* $\mathrm{wt}(*)$, $\mathrm{swt}(*)$, $\mathrm{Wt}(*)$ *are additive for the products of homogeneous elements of* $\mathcal{L}$.

Lemma 13.11. *Let* $\mathcal{L} = \mathrm{Lie}(v_1, v_2) \subset \mathrm{Der}\, R$, *the field being arbitrary. We have a* $\mathbb{Z}^2$*-grading* $\mathcal{L} = \bigoplus_{a,b \ge 0} \mathcal{L}_{a,b}$, *where* $\mathcal{L}_{a,b}$ *is spanned by the monomials in the generators with a factors* v_1 *and b factors* v_2.

Proof. By construction, $\mathrm{Wt}(v_1) = (\lambda, \bar{\lambda})$, $\mathrm{Wt}(v_2) = (\lambda^2, \bar{\lambda}^2)$, they are linearly independent vectors of $\mathbb{R}^2$. Let $w \in \mathcal{L}_{a,b}$, then by additivity, $\mathrm{Wt}(w) = a\,\mathrm{Wt}(v_1) + b\,\mathrm{Wt}(v_2)$. Consider a pair of integers $(a_1, b_1) \ne (a, b)$ and $w_1 \in \mathcal{L}_{a_1,b_1}$, we have $\mathrm{Wt}(w_1) \ne \mathrm{Wt}(w_2)$, thus w, w_1 are expressed via different homogeneous basis elements of $\mathcal{L}$. □

13.1.2.3 Fibonacci restricted Lie algebra ($p = 2$), a basis, not PI

From now we assume that $p = \mathrm{char}\, \mathbb{F} = 2$ and $R = \mathbb{F}[t_i | i \ge 0]/(t_i^p | i \ge 0)$, the truncated polynomial ring, ∂_i its partial derivations, and $R_1 \subset R$ the ideal of polynomials without constant term. Let $\{v_i \mid i \ge 0\}$ be the pivot elements as above, and $\mathbf{L} = \mathrm{Lie}_p(v_1, v_2)$ the restricted Lie algebra generated by $v_1, v_2 \in \mathrm{Der}\, R$ with respect to the pth power in $\mathrm{End}\, R$. We call $\mathbf{L}$ the *Fibonacci restricted Lie algebra*. We draw the reader's attention to the fact that we now deal with the derivations of the *truncated* polynomial ring R.

Lemma 13.12. $v_i^2 = t_{i-1}v_{i+2}$ *for all* $i \geq 1$.

Proof. In the case $p = 2$, any associative algebra satisfies the identity:

$$(a+b)^2 = a^2 + b^2 + [a,b]. \tag{13.7}$$

Thus,

$$\begin{aligned} v_i^2 &= (\partial_i + t_{i-1}v_{i+1})^2 = \partial_i^2 + t_{i-1}^2 v_{i+i}^2 + [\partial_i, t_{i-1}v_{i+1}] \\ &= t_{i-1}[\partial_i, v_{i+1}] = t_{i-1}[\partial_i, \partial_{i+1} + t_i v_{i+2}] = t_{i-1}v_{i+2}. \end{aligned}$$

□

Define *standard monomials* of *length* n as the elements of the set

$$W_n := \{t_0^{\alpha_0} \cdots t_{n-4}^{\alpha_{n-4}} v_n \mid \alpha_i \in \{0,1\}\}, \quad n \geq 1.$$

We call $t_0^{\alpha_0} \cdots t_{n-4}^{\alpha_{n-4}}$ the *tail* and, for brevity, denote the tail by $t_0^* \cdots t_{n-4}^*$, where $*$ stands for all possible powers $\alpha_i \in \{0,1\}$. By r_{n-4} we denote a tail with an additional scalar factor, including zero.

Lemma 13.13. *Let* $p = 2$ *and* $\mathcal{L} = \mathrm{Lie}(v_1, v_2) \subset \operatorname{Der} R$, *the Lie algebra generated by* v_1, v_2 *(i. e., we use the Lie bracket only). Then* $W := \bigcup_{n=1}^{\infty} W_n$ *is a basis of* $\mathcal{L}$.

Proof. Let us check by induction on n that W_n belong to $\mathcal{L}$. As the base of induction, we have $W_i = \{v_i\}$ for $i = 1,2,3$. Let $n \geq 3$ and assume that $W_i \subset \mathcal{L}$ for $i \leq n$. Using the commutation relations (* denotes all possible powers below),

$$\begin{aligned} [v_{n-1}, W_n] &= \{t_0^* \cdots t_{n-4}^* [v_{n-1}, v_n]\} = \{t_0^* \cdots t_{n-4}^* v_{n+1}\} \subset \mathcal{L}, \\ [v_{n-2}, W_n] &= \{t_0^* \cdots t_{n-4}^* [v_{n-2}, v_n]\} = \{t_0^* \cdots t_{n-4}^* t_{n-3} v_{n+1}\} \subset \mathcal{L}. \end{aligned}$$

These formulas provide the induction step.

Using Lemmas 13.7 and 13.8, the product of two standard monomials of lengths $n < m$ is expressed via the standard monomials:

$$\begin{aligned} [r_{n-4}v_n, r'_{m-4}v_m] &= r_{n-4}v_n(r'_{m-4})v_m + r_{n-4}r'_{m-4}[v_n, v_m] \\ &= r_{n-4}v_n(r'_{m-4})v_m + r_{n-4}r'_{m-4}t_{n-1}\cdots t_{m-3}v_{m+1} \\ &= r''_{m-4}v_m + r''_{m-3}v_{m+1}. \end{aligned} \tag{13.8}$$

□

Corollary 13.14. *We have disjoint unions*

$$W_{n+1} = [v_{n-1}, W_n] \cup [v_{n-2}, W_n], \quad n \geq 3.$$

Proof. Follows from the proof above. □

Lemma 13.15. *Let* $p = 2$ *and* $\mathbf{L} = \mathrm{Lie}_p(v_1, v_2) \subset \operatorname{Der} R$, *the restricted Lie algebra generated by* v_1, v_2. *Then* $\widetilde{W} := (\bigcup_{n=1}^{\infty} W_n) \bigcup \{t_{n-3}v_n \mid n \geq 3\}$ *is a basis of* $\mathbf{L}$.

Proof. By the properties of restricted Lie algebras, in order to obtain the restricted hull $\mathbf{L}$ of a Lie algebra $\mathcal{L}$, it is sufficient to add the pth powers of elements of W. These squares are nontrivial only in case of the pivot elements, the latter given by Lemma 13.12. □

Theorem 13.16. *Let $p = 2$ and $\mathcal{L} = \mathrm{Lie}_p(v_1, v_2) \subset \mathrm{Der}\,R$, the Fibonacci Lie algebra. Then $\mathcal{L}$ does not satisfy any Lie identity.*

Proof. Suppose that $\mathcal{L}$ satisfies a nontrivial multilinear polynomial identity, which can be written as (commutators are left-normed):

$$\sum_{\pi\in S_n} \alpha_\pi [X_{\pi(n)}, \dots, X_{\pi(1)}, X_0] \equiv 0, \quad \alpha_\pi \in \mathbb{F},\ \alpha_e = 1.$$

Substitute $X_0 := v_1, X_i := v_{2i}$ for $i = 1, \dots, n$. Since $p = 2$, using Lemma 13.6, the term for the identity permutation is

$$[v_{2n}, \dots, v_6, v_4, v_2, v_1] = [v_{2n}, \dots, v_6, v_4, v_3] = [v_{2n}, \dots, v_6, v_5] = \dots = v_{2n+1}.$$

Suppose that $\pi(1) = k > 1$. By Lemma 13.7, $[X_{\pi(1)}, X_0] = [v_{2k}, v_1] = t_0 \cdots t_{2k-3} v_{2k+1}$. Observe that t_0 cannot disappear after subsequent products with the remaining $\{v_{2n}, \dots, \widehat{v_{2k}}, \dots v_2\}$. Hence, the products for all such π have a factor t_0.

If $\pi(1) = 1$ then $[X_{\pi(1)}, X_0] = [v_2, v_1] = v_3$, so we get the monomials of the form

$$\alpha_\pi [v_{2\pi(n)}, \dots, v_{2\pi(3)}, v_{2\pi(2)}, v_3].$$

Similarly, letting $\pi(2) = k > 2$, such products contain $[v_{2k}, v_3] = t_2 \cdots t_{2k-3} v_{2k+1}$ and, moreover, t_2 cannot disappear after subsequent products with the remaining $\{v_{2n}, \dots, \widehat{v_{2k}}, v_4\}$, etc. We get a contradiction because the term for the identity permutation cannot cancel with any other. □

13.1.2.4 Growth of Fibonacci restricted Lie algebra

The following result shows that the weight function "separates and stratifies" the sets W_n.

Lemma 13.17. $\lambda^{n-1} < \mathrm{wt}(W_n) \le \lambda^n$ *for $n \ge 1$.*

Proof. Since $\mathrm{wt}(t_i) < 0$, the upper bound is trivial. Since $\mathrm{wt}(t_i^*) \ge \mathrm{wt}(t_i) = -\lambda^i$ and $\lambda^2 - \lambda = 1$, we get

$$\begin{aligned}\mathrm{wt}(t_0^* \cdots t_{n-4}^* v_n) &\ge \mathrm{wt}(v_n) - 1 - \lambda - \dots - \lambda^{n-4} = \lambda^n - \frac{\lambda^{n-3} - 1}{\lambda - 1} \\ &= \lambda^n - \frac{\lambda^{n-2} - \lambda}{\lambda^2 - \lambda} = \lambda^n - \lambda^{n-2} + \lambda \\ &= \lambda^{n-2}(\lambda^2 - 1) + \lambda = \lambda^{n-1} + \lambda > \lambda^{n-1}.\end{aligned}$$

□

We show that the Fibonacci (restricted) algebra has a slow polynomial growth.

Theorem 13.18. *Let* char $\mathbb{F} = 2$, $\mathcal{L} = \mathrm{Lie}(v_1, v_2)$, $\mathbf{L} = \mathrm{Lie}_p(v_1, v_2)$. *Denote* $\lambda = (1 + \sqrt{5})/2$. *Then the following are true:*

$$\mathrm{GKdim}\,\mathbf{L} = \underline{\mathrm{GKdim}}\,\mathbf{L} = \mathrm{GKdim}\,\mathcal{L} = \underline{\mathrm{GKdim}}\,\mathcal{L} = \log_\lambda 2 \approx 1.44.$$

Proof. Consider $x \geq 0$. In order to specify the growth, we use a *weight growth function*

$$\tilde{\gamma}_{\mathcal{L}}(x) := \dim\langle w \mid w \in W,\ \mathrm{wt}(w) \leq x\rangle_{\mathbb{F}}, \quad x \geq 0.$$

Let $w \in W$ be of length n of weight at most x. By Lemma 13.17, $x \geq \mathrm{wt}(w) > \lambda^{n-1}$. Hence, $n < n_0 := 2 + [\log_\lambda x]$. Remark that $|W_n| = 2^{n-3}$ for all $n \geq 3$. We evaluate $\tilde{\gamma}_{\mathcal{L}}(x)$ by the number of all standard monomials of length at most n_0 as

$$\begin{aligned}\tilde{\gamma}_{\mathcal{L}}(x) &\leq |W_0 \cup \cdots \cup W_{n_0}| = 1 + 1 + 1 + 2 + \cdots + 2^{n_0-3} \\ &= 1 + 2^{n_0-2} \leq 1 + 2^{\log_\lambda x} = 1 + x^{\log_\lambda 2}.\end{aligned}$$

Using the same x, n_0 and Lemma 13.17, $\mathrm{wt}(W_{n_0-2}) \leq \lambda^{n_0-2} = \lambda^{[\log_\lambda x]} \leq x$. We get a lower bound

$$\begin{aligned}\tilde{\gamma}_{\mathcal{L}}(x) &\geq |W_0 \cup \cdots \cup W_{n_0-2}| = 1 + 1 + 1 + 2 + \cdots + 2^{n_0-5} \\ &> 2^{n_0-4} > 2^{\log_\lambda x - 3} = \frac{1}{8}x^{\log_\lambda 2}.\end{aligned}$$

These bounds yield the required Gelfand–Kirillov dimension of $\mathcal{L}$.

In case of the restricted Lie algebra $\mathbf{L}$, consider a square $u = t_{n-3}v_n = v_{n-2}^2$ of weight at most x. Then $x \geq \mathrm{wt}(u) = 2\,\mathrm{wt}(v_{n-2}) = 2\lambda^{n-2}$. So, $n \leq n_1(x) := 3 + [\log_\lambda(x/2)]$. We obtain bounds $\tilde{\gamma}_{\mathcal{L}}(x) \leq \tilde{\gamma}_{\mathbf{L}}(x) \leq \tilde{\gamma}_{\mathcal{L}}(x) + n_1(x)$, yielding the same Gelfand–Kirillov dimension. □

13.1.2.5 Nillity of the Fibonacci restricted Lie algebra

The following fact is an analogue of the periodicity of the Grigorchuk and Gupta–Sidki groups.

Theorem 13.19. *Let* char $\mathbb{F} = 2$, *and consider the restricted Lie algebra* $\mathbf{L} = \mathrm{Lie}_p(v_1, v_2)$. *Then* $\mathbf{L}$ *has a nil p-mapping.*

Proof. By Lemma 13.15, $\mathbf{L}$ is spanned by the standard monomials and squares of pivot elements. So, we write any $a \in \mathbf{L}$ roughly as

$$a = \sum_{n=1}^{s-1} r_{n-3}v_n + h_{s-3}v_s, \quad r_i \in R,\ h_{s-3} \in R_1, \tag{13.9}$$

where we assume that the senior tail has no scalar terms (i. e., $h_{s-3} \in R_1$, it may be zero). Compute a^2 using (13.7). Mutual commutators of terms (13.9) are computed similar to (13.8), using Lemmas 13.7 and 13.8:

$$\begin{aligned}
[r_{n-3}v_n, r_{m-3}v_m] &= r_{n-3}v_n(r_{m-3})v_m + r_{n-3}r_{m-3}[v_n, v_m] \\
&= r'_{m-3}v_m + r''_{m-3}t_{n-1}\cdots t_{m-3}v_{m+1} \\
&= r'_{m-3}v_m + r'''_{m-3}v_{m+1}, \quad n < m < s, \\
[r_{n-3}v_n, h_{s-3}v_s] &= r_{n-3}v_n(h_{s-3})v_s + r_{n-3}h_{s-3}[v_n, v_s] \\
&= r'_{s-3}v_s + h'_{s-3}t_{n-1}\cdots t_{s-3}v_{s+1} \\
&= r'_{s-3}v_s + h''_{s-3}v_{s+1}, \quad n < s,
\end{aligned}$$

where $r^*_* \in R$, $h^*_* \in R_1$. The squares of the terms (13.9) are

$$\begin{aligned}
(r_{n-3}v_n)^2 &= r^2_{n-3}t_{n-1}v_{n+2} = h'_{n-1}v_{n+2}, \quad h'_{n-1} \in R_1, \quad n < s, \\
(h_{s-3}v_s)^2 &= h^2_{s-3}v^2_s = 0,
\end{aligned}$$

because $h_{s-3} \in R_1$. In case $n = s - 1$, the first line above yields $h'_{s-2}v_{s+1}$. Combining all these cases, we get a presentation similar to (13.9)

$$a^2 = \sum_{n=1}^{s} \bar{r}_{n-3}v_n + \bar{h}_{s-2}v_{s+1}, \quad \bar{r}_i \in R,\ \bar{h}_{s-2} \in R_1.$$

Repeating the process, we get

$$a^{2^m} = \sum_{n=1}^{s+m-1} \tilde{r}_{n-3}v_n + \tilde{h}_{s+m-3}v_{s+m}, \quad \tilde{r}_i \in R,\ \tilde{h}_{s+m-3} \in R_1,\ m \geq 1.$$

Using Lemma 13.17, we evaluate the weight $\mathrm{wt}(a^{2^m}) \leq \mathrm{wt}(v_{s+m}) = \lambda^{s+m}$, the inequality denoting the bound on the weights of all homogeneous summands. The weights of two generators of the algebra are $\mathrm{wt}(v_1) = 1$ and $\mathrm{wt}(v_2) = \lambda > 1$. By additivity of the weight function, $\mathrm{wt}(a) \geq 1$ and $\mathrm{wt}(a^{2^m}) \geq 2^m$. Hence, $2^m \leq \lambda^{m+s}$, where $\lambda \approx 1.618$, which is false for sufficiently large m. This contradiction proves that $a^{2^m} = 0$ for sufficiently large m (depending on a). □

13.1.2.6 More facts about Fibonacci algebras

Theorem 13.20 ([285, 286]). *Let $p = \operatorname{char} \mathbb{F}$ be one of 2, 3, and $R = \mathbb{F}[t_i | i \geq 0]/(t_i^p | i \geq 0)$, the truncated polynomial ring. Consider the Lie algebra $\mathcal{L} = \mathrm{Lie}(v_1, v_2)$, the restricted Lie algebra $\mathbf{L} = \mathrm{Lie}_p(v_1, v_2)$, and the associative hull $\mathbf{A} = \mathrm{Int}(v_1, v_2) \subset \operatorname{End} R$ generated by v_1, v_2. Denote $\lambda = (1 + \sqrt{5})/2$. Then*
1. $\operatorname{GKdim} \mathbf{L} = \underline{\operatorname{GKdim}}\, \mathbf{L} = \operatorname{GKdim} \mathcal{L} = \underline{\operatorname{GKdim}}\, \mathcal{L} = \log_\lambda p$.
2. $\operatorname{GKdim} \mathbf{A} = \underline{\operatorname{GKdim}}\, \mathbf{A} = 2 \log_\lambda p$.

3. **L** *has a nil-p-mapping.*
4. *Let us put the homogeneous components of the above algebras on the plane, using the multidegree in the generators as the coordinates. Then the respective points of* $\mathbb{Z}^2$ *for* **L** *and* **A** *lie in the strips. The border lines for these stirps are determined by the points of fixed superweight values.*
5. *The points for the restricted enveloping algebra* $\mathbf{u} = u(\mathbf{L})$ *lie in a region bounded by a parabola-like curve.*
6. *Using the geometric observations of the previous items,* **L**, **A**, *and the augmentation ideal of the restricted enveloping algebra* $\mathbf{u} = u_0(\mathbf{L})$ *are the direct sums of two locally nilpotent subalgebras*

$$\mathbf{L} = \mathbf{L}_+ \oplus \mathbf{L}_-, \quad \mathbf{A} = \mathbf{A}_+ \oplus \mathbf{A}_-, \quad \mathbf{u} = \mathbf{u}_+ \oplus \mathbf{u}_-.$$

The lower central series, the derived series, and the generating functions (recursive relations and formulas) for $\mathcal{L}$ and **L** in case $p = 2$ are described in [286].

Remark that the correct analogue of the Grigorchuk group [131] and Gupta–Sidki group [134] in the world of associative algebras is not yet known.

Conjecture. *Let* $p = 2$. *Then* (1) *the associative hull* $\mathbf{A}(v_1, v_2)$ *of the Fibonacci Lie algebra is not nil;* (2) *the restricted Poisson algebra that arises from the natural filtration of* **A** *is not nil.*

It is known that the Grigorchuk group is infinitely presented [211]. The following conjecture was somewhat discussed in [286].

Conjecture. *Let* $p = 2$. *The Fibonacci Lie algebra* $\mathcal{L}$ *and the restricted Fibonacci Lie algebra* **L** *are infinitely presented.*

In the case of $p = 0$, little is known about $\mathcal{L}$. There is evidence that the growth of $\mathcal{L}$ is greater than polynomial and the arguments above show that also $\mathcal{L}$ is $\mathbb{Z}^2$-graded by multidegree in the generators. It would be interesting to determine the growth of the Fibonacci Lie algebra $\mathcal{L} = \mathrm{Lie}(v_1, v_2) \subset \operatorname{Der} R$, where $R = \mathbb{F}[t_i \mid i \geq 0]$.

13.1.2.7 Similar constructions of restricted Lie algebras and Lie superalgebras

In the case of an arbitrary prime characteristic, Shestakov and Zelmanov constructed an example of a 2-generator restricted Lie algebra with a nil p-mapping [334]. An example of a p-generator nil restricted Lie algebra L, characteristic p being arbitrary, was studied in [288]. These infinite-dimensional restricted Lie algebras and their restricted enveloping algebras, too, have different decompositions as the direct sums of two locally nilpotent subalgebras [288] (similar to the decompositions of the theorem above). Unlike the Fibonacci restricted Lie algebra, in these latter examples, the elements of a Lie algebra are linear combinations of monomials; working with such linear combinations is sometimes an essential technical difficulty, see, e. g., [334, 288, 283].

A systematic approach to the construction of (restricted) Lie (super)algebras having *good monomial bases* was developed in [280] where the author introduced the second Lie superalgebra. Lie superalgebras over the fields of arbitrary characteristic constructed in [280, 96] can be viewed as natural analogues of the Grigorchuk and Gupta–Sidki groups in the class of Lie superalgebras. This idea was extended to the restricted Lie algebras, and now we have examples of nil restricted Lie algebras for any prime characteristic, with clear monomial bases and slow polynomial growth, including quasilinear growth [281, 282]. Next, along with a specially constructed just infinite fractal Lie superalgebra over an arbitrary field, fractal Poisson and Jordan superalgebras have been constructed, supplying analogues of the Grigorchuk and Gupta–Sidki groups in the respective classes of algebras [287].

13.2 Global Engel problem

In 1987, E. I. Zelmanov proved the following major result [381, 382].

Theorem 13.21 (E. I. Zelmanov). *An n-Engel Lie algebra over a field of characteristic 0 is nilpotent.*

For the proof of this result and some other results in this and the following sections, we refer the reader to the monograph of A. I. Kostrikin [181, Chapter 6, § 4].

Corollary 13.22. *There is a function of natural argument $f(n)$ such that n-Engel algebra over a field of characteristic $p > f(n)$ is nilpotent.*

Zelmanov's result implies that there is a function $f(n)$ such that $\mathfrak{E}_n \subset \mathfrak{N}_{f(n)}$ (take as $f(n)$ the nilpotent class of the free algebra of countable rank in $\mathfrak{E}_n$), but it does not provide such a number. In his proof, Zelmanov uses transfinite induction, which some specialists find inconvenient. In a paper [350], G. Traustason provides the reader with a constructive proof of Zelmanov's theorem. He also gives the bounds for the nilpotent class of n-Engel algebras over a field of characteristic zero, as follows.

Theorem 13.23 (Gunnar Traustason). *If the characteristic of the ground field is zero or big, say, $p > \ell(n) = 1024 \cdot 10^{46} \cdot n^{40} \cdot 3^{[(n-5)/2]}$, then n-Engel Lie algebras are nilpotent of class $T(n, \ell(n))$ where $T(n, s)$ is given recursively by $T(n, 1) = n$, $T(n, s + 1) = n^{T(n,s)}$, for $s > 1$.*

It is more or less obvious that the numbers $\ell(n)$ and $T(n, \ell(n))$ in Traustason's theorem are too big. For instance, Zelmanov's guess was that $\ell(n)$ could be replaced by just $2n$.

13.2.1 Mishchenko's exponentially growing Engel Lie algebras

One more mathematician was close to the solution of Engel's problem. In [231, 233], S. P. Mishchenko considered Engel Lie algebras in the varieties of exponential growth (see Section 9.1). His result is as follows.

Theorem 13.24. *Let* $\mathbf{V}$ *be a variety of exponential growth over* $\mathbb{F}$, *satisfying an Engel identity*

$$(\operatorname{ad} x)^l \equiv 0. \tag{13.10}$$

Then $\mathbf{V}$ *is nilpotent.*

This result was included as an addendum to [36] but not to the first edition of this book, so we found it appropriate to include it in the current text.

Recall that $\mathbf{E}_l$ denotes the variety of Lie algebras, satisfying (13.10).

In the proof of Theorem 13.24, we will follow the material about the representation theory of the symmetric group from Section 3.2 of Chapter 3.

We start with the following general observation:

Remark. Given a Young diagram d of order m, let b_d denote the dimension of any irreducible $\mathrm{Sym}(m)$-module corresponding to d. Then for any natural p, there exists a Young diagram d of order m such that $b_d > p^m$.

Indeed, according to Section 3.2.1(ii),

$$m! = \sum_d b_d^2.$$

At the same time, the number of different Young diagrams of order m is at most 2^m. If $b_{d_0} = \max\{b_d \mid |d| = m\}$ then $b_{d_0}^2 > \frac{m!}{2^m}$. It is clear that, for m sufficiently large, we have $\frac{m!}{2^m} > p^{2m}$. For such m, we will have $b_{d_0}^2 > p^m$, as claimed.

Lemma 13.25. *Let the growth of* $c_n(\mathbf{V})$ *be at most exponential and (13.10) hold in* $\mathbf{V}$. *Then there is a natural number m and a Young tableau* τd *of degree m with the associated idempotent of the form* $e_{\tau d} = \sum_{\sigma\in\mathrm{Sym}(n)} \alpha_\sigma \sigma$ *such that for any words* $A_1, \dots, A_m$ *in adjoint derivations of free variables, the following are identities of* $\mathbf{V}$:

$$\sum_{\sigma\in\mathrm{Sym}(n)} \alpha_\sigma x_{\sigma(1)} A_1 x_{\sigma(2)} A_2 \cdots x_{\sigma(m)} A_m \equiv 0. \tag{13.11}$$

Proof. By our hypotheses, there is $q > 1$ such that $c_n(V) < q^n$, for all $n = 1, 2, \dots$ By the above remark, there is a diagram d_0 of some degree m such that $b_{d_0} > (ql)^m$. If d is a Young diagram containing d_0 as a subdiagram then $b_d > b_{d_0} > q^{lm} \geq q^n$. In this case, for any $\tau \in \mathrm{Sym}(n)$ we will have that $f_{\tau d} \equiv 0$ is an identity in $\mathbf{V}$.

Moreover, it follows from Richardson–Littlewood rule (see Section 3.2.5) that if $w = x_1 \cdots x_n$ is an arbitrary nonassociative monomial and $\mathrm{Sym}(m)$ is the group of permutation of the symbols $\{i_1, i_2, \ldots, i_m\}$ then $e_{\tau d_0} w \equiv 0$ is an identity in V. The rule in question needs to be applied to an $\mathrm{Sym}(n)$-module induced from the irreducible module T_{d_0} for the subgroup $\mathrm{Sym}(m)$ mentioned above. The identical relation thus obtained looks like what we need in the statement of our lemma, however, at this time, the degrees of the associative monomials A_i are restricted by $l-1$. To lift this restriction, we can use the Engel identity $(\mathrm{ad}\, x)^l \equiv 0$. Assume, for instance, that $A_1 = a_1 a_2 \cdots a_l$. Using the full linearization of $(\mathrm{ad}\, x)^l \equiv 0$, we get that

$$a_1 a_2 \cdots a_l = -\sum_{\sigma \neq e} a_{\sigma(1)} a_{\sigma(2)} \cdots a_{\sigma(l)}.$$

We also have

$$a_1 a_2 \cdots a_l = a_2 a_1 \cdots a_l + (a_1 a_2) \cdots a_l.$$

We now can see that

$$a_1 a_2 \cdots a_l b = \Big(\frac{1}{l!} \sum_{\sigma \in \mathrm{Sym}(l)} a_{\sigma(1)} \cdots a_{\sigma(l)} \Big) b + \sum_i A_1^i b = \sum_i A_1^i b$$

by Engel identity, where for each i, we have $\deg A_1^i < l$. Now one can use induction on the length of the monomial A_i, and the proof is complete. □

The direct consequence of the above lemma is the fact that if T_d is a simple submodule in $L_n(\mathbf{V})$ corresponding to a Young diagram d then d does not contain our fixed d_0 as a subdiagram. Using a rougher estimate, we can say that in our situation there is natural number m such that no diagram d corresponding to the module T_d that occurs in $L_n(V)$ contains an $m \times m$ square.

As a result, to complete the proof of our theorem, it is sufficient to find a natural number N such that for any Young diagram d of order $n > N$ not containing an $m \times m$ square, any identity $f_{\tau d} \equiv 0$ holds in $\mathbf{V}$. In other words, we need to prove that any multilinear identity of sufficiently great degree, in which all variables are split into a limited number of symmetric and skew-symmetric collections, is an identity in $\mathbf{V}$.

13.2.1.1 New variables

Let us consider an alphabet of the form $Y \cup \overline{Y}$, with $Y \cap \overline{Y} = \varnothing$. We associate with every nonassociative left-normed monomial w in this new alphabet an element $\pi(w)$ of the free Lie algebra $L(X)$ of countable ran,k by the following rule. If r is the number of occurrences of $y \in Y$ in w, say $w = A_1 y A_2 y \cdots A_r y A_{r+1}$, then we replace w by

$$w' = \sum_{\sigma \in \mathrm{Sym}(r)} A_1 x_{\sigma(1)} A_2 x_{\sigma(2)} \cdots A_r x_{\sigma(r)} A_{r+1}.$$

If t is the number of occurrences of $\overline{y} \in \overline{Y}$ in w, say $w = B_1\overline{y}B_2\overline{y}\cdots B_t\overline{y}B_{t+1}$, then we replace w by

$$w'' = \sum_{\sigma\in\mathrm{Sym}(t)} \mathrm{sgn}(\sigma)B_1x_{\sigma(1)}B_2x_{\sigma(2)}\cdots B_tx_{\sigma(t)}B_{t+1}.$$

We apply these two operations to all the letters in w, every time using new letters of X. The resulting element of $L(X)$ is our $\pi(w)$.

Conversely, suppose we have a multilinear identity $f \equiv 0, f \in L(X)$, in which the set of all letters is a disjoint union of subsets of symmetric and skew-symmetric variables. Then we replace all letters from each symmetric subset by a unique $y \in Y$. Likewise, we replace all letters from each skew-symmetric subset by a unique $\overline{y} \in Y$. If w is a monomial thus obtained then $\pi(w) \equiv 0$ is equivalent to $f \equiv 0$.

Since every monomial can be written as a linear combination of left-normed monomials, the described procedure can be applied to any multihomogeneous polynomials in the variables $Y \cup \overline{Y}$. Now we say that $w \equiv 0$, where $w \in L(Y \cup \overline{Y})$, holds in A Lie algebra G if G satisfies $\pi(w) \equiv 0$. Using this terminology enables one to write certain consequences of the Engel identity in a compact form.

Lemma 13.26. *Let w be a nonassociative monomial in the alphabet $Y \cup \overline{Y}$, $y \in Y \cup \overline{Y}$. Then $w^{2l}y \equiv 0$ is an identity in $\mathbf{E}_l$.*

Proof. Let $u = w^{2l}y$. We first consider the case where w is a letter. If $w \in Y$ then $\pi(u) \equiv 0$ takes the form of

$$\sum_{\sigma\in\mathrm{Sym}(2l)} x_{\sigma(1)}\cdots x_{\sigma(2l)}x_0 \equiv 0,$$

which is a linearization of the Engel identity, hence holds in $\mathbf{V}$.

However, if $w \in \overline{Y}$ then

$$\pi(u) = \sum_{\sigma\in\mathrm{Sym}(2l)} \mathrm{sgn}(\sigma)x_{\sigma(1)}\cdots x_{\sigma(2l)}x_0,$$

the left-hand side of a standard identity. As we did earlier in Chapter 5, this can be rewritten as

$$\pi(u) = \sum_{\sigma\in\mathrm{Sym}(2l)\sigma(2l-1)>\sigma(2l)} \mathrm{sgn}(\sigma)(x_{\sigma(1)}x_{\sigma(2)})\cdots(x_{\sigma(2l-1)}x_{\sigma(2l)})x_0. \tag{13.12}$$

If we switch to the generators of the commutator subalgebra $z_{ij} = x_ix_j$, where $i > j$, then we can rewrite (13.12) as follows:

$$\pi(u) = \sum \mathrm{sgn}(\sigma)z_{\sigma(1)\sigma(2)}\cdots z_{\sigma(2l-1)\sigma(2l)}x_0. \tag{13.13}$$

Since $\mathrm{sgn}(\sigma)$ does not change if we exchange z_{ij}, each multihomogeneous component of (13.13) will look like

$$\pi(u) = \sum_{\sigma\in\mathrm{Sym}(l)} z_{\tau(i_1j_1)}\cdots z_{\tau(i_lj_l)}x_0, \tag{13.14}$$

which vanishes on the algebras from $\mathbf{E}_l$.

Now we need to switch to an arbitrary w, where the letters from Y enter r times while those from $\overline{Y}$ enter s times, that is, $w = w(y_1, \dots, y_r, \overline{y}_1, \dots, \overline{y}_s)$. In this case, $\pi(u)$ has the form of a linear combination of the products of the form $w(X_1)w(X_2)\cdots w(X_{2l})x_0$ where the total set of the variables used in $\pi(u)$ equals the disjoint union $X_1 \sqcup \cdots \sqcup X_{2i} \sqcup \{x_0\}$. Let us fix such a partition. Now if s is even then $\pi(u)$ is a linear combination of the elements

$$\sum_{\tau \in \mathrm{Sym}(2l)} w(X_{\tau(1)}) \cdots w(X_{\tau(2l)})x_0 \equiv 0.$$

If s is odd then we get

$$\sum_{\tau \in \mathrm{Sym}(2l)} \mathrm{sgn}(\tau) w(X_{\tau(1)}) \cdots w(X_{\tau(2l)})x_0 \equiv 0.$$

In both cases it remains to apply the arguments of the particular case considered earlier. □

13.2.1.2 Minimal nonidentity

Let us totally order the alphabet $Y \cup \overline{Y}$ and use it to lexicographically order the words in this alphabet. The same order will be used on the right-normed nonassociative monomials. As mentioned earlier, we need to prove that there is a natural number N such that any monomial of degree $n > N$ in the alphabet $Y \cup \overline{Y}$, depending on a bounded number of variables vanishes in $\mathbf{V}$. Assuming the opposite, for each sufficiently large N there is a monomial M in the alphabet $Y \cup \overline{Y}$ such that $M \equiv 0$ is not an identity in $\mathbf{V}$. Let us assume M is the least possible with respect to the lexicographical order.

To prove Theorem 13.24, we plan to rearrange the parentheses on M in such a way that it remains to be a nonidentity in $\mathbf{V}$. The new parentheses would select in M a set of m factors such that if we permute them in a nontrivial way, we get lexicographically lesser monomials. Then one would be able to apply the identical relations (13.11). Remember that we have a number m at our disposal such that the number of the symmetric and skew-symmetric subsets is bounded by $2m$. Since the length of M is unbounded, one may assume that M contains m arbitrarily long pairwise disjoint subwords $B_1, \dots, B_m$ such that the following hold:

(i) The maximal letter x_0 occurring in B_1 occurs unboundedly many times;
(ii) The last letter of B_i equals x_0;
(iii) The first letter of B_i is different from x_0;
(iv) The first letter following B_i equals x_0.

Obviously, these restrictions can be satisfied considering that the lengths are unbounded and because, by Lemma 13.26, none of the letter can appear in M more than $2l - 1$ times in a row.

13.2.1.3 Variables of the second level

Each word B_i can be partitioned as the product of the subwords $x_1x_2\cdots x_tx_0^s$ where $x_0 > x_1,\ldots,x_t$. Let us set the parentheses on each of the subwords in the following way:

$$x_1(x_2(\cdots(x_{t-1}(((x_tx_0)x_0)\cdots x_0))\cdots)).$$

Let us call the monomials thus obtained the *variables of the second level*, the *variables of the first level* being the elements of $Y\cup\overline{Y}$. On the variables of the second level, there is a total order resulting from the order of their underlying associative monomials.

Let us transition from the identity $M\equiv 0$ to the identity $\widetilde{M}\equiv 0$ by setting parenthesis on the variables of the second level in the described way. Then, considering the resulting monomial as an element of $L(Y\cup\overline{Y})$, transition to $\pi(\widetilde{M})$. For instance, if $y_2 > y_1$ then from $y_1y_2y_1y_2y_2\equiv 0$ we transition to $(y_1y_2)((y_1y_2)y_2)\equiv 0$, that is, to

$$y_1y_2y_1y_2y_2 - y_2y_1y_1y_2y_2 + y_2y_1y_2y_1y_2 - y_1y_2y_2y_1y_2 + y_1y_2y_2y_2y_1 + \cdots \equiv 0.$$

It is easy to see that, with respect to the lexicographical ordering, the leading term of the resulting linear combination is the original word M. Hence the new identity is still a nonidentity in $\mathbf{V}$.

Lemma 13.27. *The resulting number of the variables of the second level is at most* $m-1$.

Proof. Let us assume that the entries in the words $B_1,\ldots,B_m$ are numbered from left to right. Assume that B_1 has m variables of the second level: $u_1 < u_2 < \cdots < u_m$. We select in each B_i one of the entries of u_i. Drop the parentheses in all entries of u_i except those in the selected ones. Then we will get a monomial M_1 in $L(Y\cup\overline{Y})$. As earlier, the identity $M_1\equiv 0$ does not hold in $\mathbf{V}$. If $M_1 = U_1u_1U_2u_2\cdots U_mu_mU_{m+1}$ then by (13.11) M_1 is a linear combination of the monomials of the form $\sigma(M_1) = U_1u_{\sigma(1)}U_2u_{\sigma(2)}\cdots U_mu_{\sigma(m)}U_{m+1}$, where $\sigma\neq e$. Let M_s be the rightmost subword in $\sigma(M_1)$, in which u_s is replaced by u_t, $s > t$. Then $\overline{\sigma(M_1)}$ is lexicographically greater than $\overline{M}_1 = \overline{M}$ (here the bars mean that we take the underlying associative words or left normed nonassociative monomials). This claim is obvious when u_s and u_t have the same length but if u_s is a right subword in u_t then we have to invoke restriction (iv) above. Then $\sigma(M_1)\equiv 0$ holds in V, hence $M_1\equiv 0$ holds in V, which is a contradiction. □

13.2.1.4 Completion of the proof of Theorem 13.24

Let us view B_1 as written in terms of the variables of the second level. According to our hypotheses and Lemma 13.27, the length of the word obtained with respect to the new variables is arbitrarily high, none of the variables is in the row longer than $2l$, and the number of the variables is restricted by m. This enables us to pass to the variables of the third level. These are obtained by setting brackets on the words written through the variables of the second level in the same way as while introducing the variables of the second level. The variables will be obtained by setting brackets on the words $u_1u_2\cdots u_tu_0^s$

If we repeat the process of introducing the variables of the new levels, then we will eventually arrive at the variables of level $2m$. Let us fix in M, from left to right, the entries of m equal subwords $T_1,\ldots,T_m$, which become the same variable of level $2m$. Let us write T_i in terms of the variables of level $2i-1$. Let us denote the rightmost variable (of level $2i-1$) as v_i. Let us drop the brackets in all the variables except $v_1,\ldots,v_m$ and denote by M' the monomial arising in this way. We can apply to $v_1,\ldots,v_m$ the argument from the proof of Lemma 13.27. The desired contradiction will be obtained if we show that $\sigma(M')\equiv 0$ is an identity in $\mathbf{V}$, for any nonidentity permutation $\sigma\in\mathrm{Sym}(m)$.

So let i be the rightmost symbol such that $\sigma(i)>i$ and $\sigma(j)=j$, for all $j>i$. Then, in the word $\sigma(M')$, a variable of level $2i-1$ will be replaced by a variable of a higher level $2k-1$. It is clear that the number of the maximal variables of the old level standing in a row decreases, so that the element obtained is lexicographically lesser than M'. As a result, $\sigma(M')\equiv 0$ is an identity in $\mathbf{V}$. But then, according to (13.11), also $M'\equiv 0$ is an identity in V, which is a contradiction. Now the proof of Theorem 13.24 is complete. □

An upper bound for the nilpotent step of Lie algebras satisfying $(\operatorname{ad}x)^n\equiv 0$ was determined in the paper of A. Ya. Vais [352], as a consequence of his Theorem 9.7 in Chapter 9. He introduces a function

$$\varphi(n,k,d)=(2n+d)d^{k^{d-1}+1}+1,\quad \text{where } n,k,d \text{ are natural numbers.}$$

Vais calls a (special) Lie algebra *r-normal* if it is isomorphic to a Lie subalgebra of an associative algebra, satisfying a nontrivial identity of degree r. Then he proves the following

Theorem 13.28 (A. Ya. Vais). *Let L be a Lie algebra over a field $\mathbb{F}$ of characteristic 0, satisfying the Engel identity* $(\operatorname{ad}x)^n\equiv 0$. *Then $L^{\varphi(n,k,d)}=\{0\}$ where the values of the parameters are as follows:*
1. *If L is an r-normal algebra then $k=8(r-1)^8$, $d=(r+2)^2$;*
2. *If L has exponential codimension growth $c_n(L)<b^n$ (b a natural number) then $k=8(d-1)^8$, $d=(3(4b)^{n-1}+2)^2$.*

13.2.2 Local nilpotence. Restricted Burnside problem

In 1994, E. I. Zelmanov was awarded the Fields medal for his groundbreaking solution of the restricted Burnside problem, which reads as follows:

For what values of r and n, is there an upper bound on the orders of finite r-generator groups of exponent n?

The answer turns out to be that such an upper bound exists for all r and n. From the classification of finite simple groups and the work of P. Hall and G. Higman, it follows that it is sufficient to consider n in the case where n is a power of a prime. In 1959, Kostrikin [177] proved that there is an upper bound if n is a prime. In 1989, Zelmanov [383] showed that an upper bound exists if n is a power of an odd prime. In 1991, Zelmanov [384] showed that an upper bound exists also if n is a power of 2. Thus he solved the restricted Burnside problem in its full generality.

The theorems of Kostrikin and Zelmanov are, in fact, theorems about Lie algebras. The reason is that the following are equivalent:

1. There is a largest finite r-generator group of exponent p^m;
2. The associated Lie ring of $B(r, p^m)$ is nilpotent.

Here $B(r, n)$ is the (relatively) free r-generator group of exponent n. Now the associated Lie ring of $B(r, p)$ (for the definition of a Lie ring of a group, or associated Lie ring, see Section 8.2.3) satisfies $(p-1)$ Engel identity and has characteristic p, and so we can think of it as an Engel Lie algebra over the field $\mathbb{Z}/p\mathbb{Z}$. So it was sufficient for A. I. Kostrikin to prove the following.

Theorem 13.29 (A. I. Kostrikin). *Let L be a finitely generated Engel $(p-1)$-Engel Lie algebra over a field of characteristic p. Then L is nilpotent.*

In a later paper of Braun [85] one can find the strengthening of the above result to the case where L is a $p+1$-Engel algebra.

The original proof of Kostrikin's theorem was not constructive. A constructivization of the proof is given in the paper of S. I. Adian and A. A. Razborov [2].

A more general result, that the associated Lie ring of $B(r, p^m)$ is nilpotent, follows from

Theorem 13.30 (E. I. Zelmanov). *Let L be a finitely generated Lie algebra. Suppose that there exist positive integers s and t such that*

$$\sum_{\sigma \in \mathrm{Sym}(s)} (\operatorname{ad} x_{\sigma(1)})(\operatorname{ad} x_{\sigma(2)}) \cdots (\operatorname{ad} x_{\sigma(s)})(x) = 0,$$

for all $x, x_1, x_2, \ldots, x_s \in L$, and

$$(\operatorname{ad} u)^t = 0,$$

for all Lie monomials $u \in L$, in terms of the generators of L. Then L is nilpotent.

A somewhat more general result reported at Kyoto ICM in 1991 and widely used in group theory is the following (see [385]). We call a subset $S \subset L$ a *Lie set* if, for arbitrary elements $a, b \in S$, we have $[a, b] \in S$. For a subset $X \subset L$, the Lie set generated by X is the smallest Lie set $S\langle X\rangle$ containing X. It consists of X and of all iterated commutators in elements from X.

Theorem 13.31 (E. I. Zelmanov). *Let L be a Lie algebra satisfying a polynomial identity and generated by elements $a_1, \ldots, a_m$. If an arbitrary element $s \in S\langle a_1, \ldots, a_m\rangle$ is* ad*-nilpotent then the Lie algebra L is nilpotent.*

One more related theorem deals with Lie algebras of finite width. These are graded algebras for which there is a natural d by a group Γ (usually abelian) such that $\dim L_\gamma \leq d$, for all $\gamma \in \Gamma$. It is well known that Grigorchuk's infinite finitely generated 2-group has finite width.

Theorem 13.32 (Martinez and Zelmanov [221]). *Let $L = \bigoplus_{\gamma\in\Gamma} L_\gamma$ be a Lie algebra over a field $\mathbb{F}$ of characteristic zero graded by an abelian group Γ. Suppose that:*
1. *There exists $d > 0$ such that $\dim_{\mathbb{F}} L_\gamma \leq d$ for all $\gamma \in \Gamma$;*
2. *Every homogeneous element $a \in L_\gamma, \gamma \in \Gamma$, is* ad*-nilpotent.*

Then the Lie algebra L is locally nilpotent.

14 Miscellaneous topics

In this chapter we present some other branches of the theory of identical relations in Lie algebras.

14.1 Finite basis problem

We start by mentioning one of the core problems of the theory, which remains open, at least in an important case of Lie algebras over the fields of characteristic zero.

FINITE BASIS PROBLEM

Is it true that any variety of Lie algebras over a field of characteristic zero can be given by a finite number of identical relations?

Equivalently,

Is it true that any verbal ideal of a free Lie algebra of countable rank over a field of characteristic zero can be generated, as a verbal ideal, by a finite number of elements?

Many results in the positive direction are presented in Chapter 5 of this book. Let us mention a consequence of Krasil'nikov's theorem in Section 5.5.2 which says that $\mathbf{N}_c\mathbf{A}$ is a Specht variety over any field of characteristic 0:

Corollary 14.1. *Identities of a finite-dimensional solvable Lie algebra over a field of characteristic zero are finitely based.*

A more recent Krasi'lnikov's result [185] deals with $\mathbf{N}_c\mathbf{A}$ over the field of characteristic $p > 0$.

Theorem 14.2. *Over a field $\mathbb{F}$ of arbitrary characteristic $p > 0$, every Lie algebra with $(p-1)$-nilpotent commutator subalgebra has a finite basis of identities.*

Examples of Vaughan–Lee and Drensky, see Theorem 5.32, do not allow us to expect that the statement of Corollary 14.1 is true in the case of the field $\mathbb{F}$ of positive characteristic. Of course, if $\mathbb{F}$ is finite, the identities of a finite-dimensional algebra have finite basis by Theorem 7.25. Without restrictions on the field, the best we have seems to be the following.

Corollary 14.3. *Let G be a Lie algebra represented by triangular $n \times n$ matrices over a field $\mathbb{F}$ of characteristic $p > 0$, where $n \leq p$. Then G has a finite basis of identities.*

It would be interesting to know what other restrictions on a finite-dimensional Lie algebra G over a field $\mathbb{F}$ of characteristic $p > 0$ allow one to claim that the identities of G have finite basis.

In Section 5.8 we mentioned a result of V. V. Stovba [348], whose more precise form says that a subvariety $\mathbf{V}$ in $\mathbf{N}_c\mathbf{N}_d$ over a field $\mathbb{F}$ such that $|\mathbb{F}| \geq c$ is a finitely based *if and only if* $\mathbf{V}$ has finite axiomatic rank. It follows from Corollary 5.31 that if $\mathbb{F}$ has

https://doi.org/10.1515/9783110566659-014

characteristic 0 then any subvariety of $\mathbf{N}_c\mathbf{A}$ has finite axiomatic rank. Hence Theorem 5.32 follows.

Another consequence is that Vaughan-Lee-Drensky's infinitely based variety $\mathbf{V} = \operatorname{var} L$ from Theorem 5.32 is an example of a variety of infinite axiomatic and finite basic rank. The first claim follows because $\mathbf{V} \subset \mathbf{N}_p\mathbf{A}$, the second because $\dim L < \infty$.

One of the most general results concerning the finite basis problem is published in a paper [146] of A. Iltyakov. We define identities of representations of Lie algebras in Section 14.3.

Theorem 14.4 (A. Iltyakov). *Identities of finite-dimensional representations of any Lie algebra over a field of characteristic zero are finitely based.*

Corollary 14.5. *Identities of a finite-dimensional Lie algebra over a field of characteristic zero are finitely based.*

We conclude by mentioning that the finite basis property for associative algebras over a field of characteristic zero has been established by Alexander Kemer [163]. He also showed that this is true for finitely generated associative algebras over infinite fields of positive characteristic. In the general case where the field is of positive characteristic, there are counterexamples by Belov, Grishin and Shchigolev (see [133] with a somewhat simpler counterexample over the field of characteristic 2).

14.2 Otto Kegel's problem

If R is an associative ring, which is the sum of two nilpotent subrings, $R = N + K$, then O. Kegel in [161] proved that R is itself nilpotent. The following question is known as

KEGEL'S PROBLEM

Is it true that a Lie algebra $L = A + B$ over a field $\mathbb{F}$, which is the sum of two nilpotent subalgebras A and B, is solvable?

An example of a two-dimensional nonabelian algebra says that we should not expect that L must be nilpotent. A. N. Petravchuk [259] gave an example of a semidirect product Lie algebra $L = D \rtimes S$ over a field of characteristic 2, which is the sum of two nilpotent subalgebras. Here $S \cong \mathfrak{sl}(2, \mathbb{F})$ while D is an abelian 2-dimensional subalgebra in $\operatorname{Der} S$. So, in Kegel's problem, we must require $\operatorname{char} \mathbb{F} \neq 2$. In the case of finite-dimensional Lie algebras over $\mathbb{F}$ with $\operatorname{char} \mathbb{F} \neq 2$, V. Panyukov [253] proved that the answer to Kegel's problem is "yes." Under the restriction $p > 5$, this result was obtained by P. Zusmanovich [387], using different techniques. The case of $L = A + B$, where B is any nilpotent and A is finite-dimensional abelian, was settled in the positive by Cornelius Pillen [290]. Later, A. N. Petravchuk [260] proved that a Lie algebra $L = A + B$ over any associative commutative ring in which 2 is invertible, A abelian, and B nilpotent, must be solvable. For some further results on the topic, see [261] and [262].

In [43] the authors used an old idea of N. Itô's theorem from group theory to prove the following result

Theorem 14.6. *Let $L = \bigoplus_{g\in G} L_g$ be a color Lie superalgebra (see Section 14.4.1.2) over a commutative ring $\mathbb{F}$ with unity and $L = A + B$ where A and B are homogeneous abelian subalgebras. Then L is metabelian, that is, satisfies the identity*

$$[[x,y],[z,t]] \equiv 0.$$

The commutators used above are the color commutators, corresponding to an alternating bicharacter $\beta : G \times G \to \mathbb{F}^*$. If the grading is trivial, we have the direct analogue of Itô's theorem: a Lie algebra L, which is the sum of two abelian subalgebras, is metabelian. A. I. Kostrikin attributes this result to Ya. Krylyuk (unpublished).

Proof. This is a direct calculation, following Itô's idea from [149]. It is sufficient to consider the case of the commutator of the form $[[a_1,b_1],[a_2,b_2]]$, where $a_1,a_2 \in A$ and $b_1,b_2 \in B$. We set

$$[a_1,b_2] = a_3 + b_3 \quad \text{and} \quad [b_1,a_2] = a_4 + b_4 \quad \text{with } a_3,a_4 \in A \text{ and } b_3,b_4 \in B.$$

Let also $a_i \in A_{g_i}$, and $b_i \in B_{h_i}$, $i = 1,2$, where $g_i,h_j \in G$. Now we use these conventions, and the defining identities in Section 14.4.1.2 of a color Lie superalgebra to obtain the following:

$$\begin{aligned}
[[a_1,b_1],[a_2,b_2]] &= [a_1,b_1,a_2,b_2] - \beta(g_1,h_1)[b_1,a_1,a_2,b_2]\\
&= [a_1,[b_1,a_2],b_2] - \beta(g_1,h_2)[a_1,a_1,b_1,b_2]\\
&\quad - \beta(g_1,h_1)[b_1,[a_1,a_2],b_2] - \beta(g_1,h_1)\beta(g_1,g_2)[b_1,a_2,a_1,b_2]\\
&= [a_1,a_4+b_4,b_2] - \beta(g_1,h_1)\beta(g_1,g_2)[b_1,a_2,a_3+b_3]\\
&= [a_1,a_4,b_2] + [a_1,b_4,b_2] - \beta(g_1,h_1)\beta(g_1,g_2)[b_1,a_2,a_3]\\
&\quad - \beta(g_1,h_1)\beta(g_1,g_2)[b_1,a_2,b_3]\\
&= [[a_1,a_4],b_2] + \beta(g_1,g_2h_1)[a_4,a_1,b_2]\\
&\quad - \beta(g_1,h_1)\beta(g_1,g_2)[[b_1,a_2],b_3]\\
&\quad - \beta(g_1,h_1)\beta(g_1,g_2)\beta(h_1,g_2)[a_2,b_1,b_3]\\
&= \beta(g_1,g_2h_1)[a_4,a_3+b_3] - \beta(g_1,h_1)\beta(g_1,g_2)[a_4+b_4,b_3]\\
&= \beta(g_1,g_2)\beta(g_1,h_1)[a_4,b_3] - \beta(g_1,h_1)\beta(g_1,g_2)[a_4,b_3] = 0.
\end{aligned}$$ □

An even further generalization has been obtained by Y. Bahturin, D. Fischman, and S. Montgomery in [40] for Lie algebras in the category of comodules over a cotriangular cocommutative Hopf algebra H, which are the sums of two H-abelian subalgebras. The conclusion is the same: they must satisfy the metabelian identity $[[x,y]_H,[z,t]_H]_H \equiv 0$. Of course, here $[x,y]_H$ is a much more complex operation, compared with the ordinary Lie bracket (see (14.6) below).

14.3 Identities of simple Lie algebras

Many of the results in this section belong to Yu. P. Razmyslov. We do not give them with proofs because the reader can easily find them in Razmyslov's monograph [313] "Identities of Algebras and their Representations", whose English translation appeared as volume 138 of AMS Translations of Mathematical Monographs in 1994.

14.3.1 The isomorphism problem

In 1983 paper [197] (see also [313, Chapter 1]), the following important result was proven. We used this in Section 11.3 mostly for graded algebras. We recall this result.

Theorem 14.7. *Over an algebraically closed field, any simple finite-dimensional Lie algebra is completely determined, up to isomorphism, by its identical relations.*

Actually, this result was a consequence of previous deep results by Yu. P. Razmyslov [313] dealing with Ω-algebras and their representations. Associative and Lie algebras are natural examples of Ω-algebras, with just one, binary, operation. For basic facts about Ω-algebras in this book, see Section 11.3 above.

Let us fix in $F_\Omega(X)$ a subset Ω' of multilinear polynomials. Elements of Ω' have natural arity and can be viewed as operations on any subspace $\mathfrak{G}$ of an Ω-algebra B, which is closed under these operations. Razmyslov calls $(B, \mathfrak{G})$ an (Ω, Ω')-pair. Again, natural examples are associative Lie pairs, in particular, the pairs which arise when we consider a representation $\rho : \mathfrak{G} \to \operatorname{End} V$ of a Lie algebra $\mathfrak{G}$ by linear transformations of a vector space V and consider an associative algebra B generated by $\rho(\mathfrak{G})$.

The elements of $F_\Omega(X)$ are called Ω-polynomials, where an Ω-polynomial $w(x_1, \dots, x_n)$ is called an identity of an (Ω, Ω')-pair $(B, \mathfrak{G})$ if $w(g_1, \dots, g_n) = 0$, for any $g_1, \dots, g_n \in \mathfrak{G}$. Identities of a Lie algebra $\mathfrak{G}$ can be viewed as identities of the pair $(B, \mathfrak{G})$ where B is an associative subalgebra of $\operatorname{End} L$ generated by all $\operatorname{ad} g$, $g \in \mathfrak{G}$.

In [308] Razmyslov proves the following

Theorem 14.8. *Let $(B_1, \mathfrak{G}_1)$ and $(B_2, \mathfrak{G}_2)$ be two (Ω, Ω')-algebras over an infinite field $\mathbb{F}$ such that B_1 and B_2 are centrally prime and $\mathfrak{G}_1$ and $\mathfrak{G}_2$ are finite-dimensional over $\mathbb{F}$. The identities of $(B_1, \mathfrak{G}_1)$ and $(B_2, \mathfrak{G}_2)$ are the same if and only if there exists an $\mathbb{F}$-linear automorphism σ of an algebraically closed extension $\mathbb{F}_1$ of infinite transcendental degree over $\mathbb{F}$ such that the pairs $(B_1, \mathfrak{G}_1)$ and $(B_2, \mathfrak{G}_2)$ are $\mathbb{F}$-isomorphic and this isomorphism is σ-semilinear.*

The above Theorem 14.7 is a corollary of Theorem 14.8. Another corollary is

Theorem 14.9. *Let $\mathbb{F}$ be an algebraically closed field. Let $\rho_i : \mathfrak{G}_i \to \operatorname{End} V_i$, $i = 1, 2$, are two faithful irreducible not necessarily finite-dimensional representations of finite-dimensional Lie algebras $\mathfrak{G}_1$ and $\mathfrak{G}_2$, respectively. Let us denote by B_i the associative* *subalgebra in* End V_i *generated by* $\rho(\mathfrak{G}_i)$, $i = 1, 2$. *If* $(B_1, \mathfrak{G}_1)$ *and* $(B_2, \mathfrak{G}_2)$ *have the same identities then they are isomorphic. If, moreover,* V_1 *and* V_2 *are finite-dimensional then there exist a Lie isomorphism* $\varphi : \mathfrak{G}_1 \to \mathfrak{G}_2$ *and a linear isomorphism* $\psi : V_1 \to V_2$ *such that* $\psi(gv) = \varphi(g)\psi(v)$, *for all* $g \in \mathfrak{G}_1$ *and all* $v \in V_1$.

Note that Theorem 14.7 is just one possible corollary of Theorem 14.8. It also follows that

Any two (nonassociative) finite-dimensional simple algebras over an algebraically closed field have the same identities if and only if they are isomorphic.

Easy examples, say $\mathfrak{g}_1 = \mathfrak{su}_2(\mathbb{R})$ and $\mathfrak{g}_2 = \mathfrak{so}_3(\mathbb{R})$, show that the requirement of $\mathbb{F}$ being algebraically closed is essential. Another example by Razmyslov [312] shows that the following conjecture is false:

Given two finite-dimensional simple Lie algebras $\mathfrak{g}_1$ *and* $\mathfrak{g}_2$ *over an algebraically closed field* $\mathbb{F}$, *does it follow from* $\operatorname{var}\mathfrak{g}_1 \subset \operatorname{var}\mathfrak{g}_2$ *that* $\mathfrak{g}_1$ *is isomorphic to a subalgebra in* $\mathfrak{g}_2$?

Razmyslov proves the following

Theorem 14.10. *Let* $\operatorname{char}\mathbb{F} = p > 5$, $\mathfrak{g}_1$ *be a simple Cartan* $\mathbb{F}$*-algebra of Hamiltonian type* H_n *while* $\mathfrak{g}_1$ *a simple* $\mathbb{F}$*-algebra of classical type* A_{p^n-1}. *Then*

$$\operatorname{var}\mathfrak{g}_1 \subset \operatorname{var}\mathfrak{g}_1, \quad \dim_{\mathbb{F}}\mathfrak{g}_1 = \dim_{\mathbb{F}}\mathfrak{g}_2$$

and, at the same time, $\mathfrak{g}_1$ *and* $\mathfrak{g}_2$ *are nonisomorphic algebras.*

Corollary 14.11. $\operatorname{var}\mathfrak{g}_1 \subset \operatorname{var}\mathfrak{g}_2$, *but* $\mathfrak{g}_1$ *is not a subalgebra in* $\mathfrak{g}_2$.

It would be interesting to see what happens if $\operatorname{char}\mathbb{F} = 0$.

14.3.2 Identities of Cartan-type Lie algebras

This section mostly consists of the results of Yu. P. Razmyslov. Again, for the proofs we refer the reader to [313, Chapter 6]. In [311], Razmyslov states the following conjecture, the answer to which is unknown even now.

RAZMYSLOV'S CONJECTURE

Any simple Lie algebra satisfying a nontrivial Lie identity is an algebra of Cartan type.

In the case of the fields of characteristic zero, Cartan-type Lie algebras have been known in geometry for more than 100 years and belong to one of the types W_n, S_n, H_n, and K_n. Their analogues in the case of positive characteristic appear later in Kostrikin–Shafarevich conjecture for the classification of simple Lie algebras. However, the latter algebras are finite-dimensional. One of the basic properties of Cartan Lie algebras of rank n is the presence of a proper subalgebra of codimension n. Razmyslov chooses this property for the definition of simple Cartan-type algebras. As a result, any finite-dimensional simple algebras are of Cartan type, which is at odds with the terminology accepted in the classification theory of modular simple Lie algebras.

Definition 14.12. A simple Lie algebra $\mathfrak{g}$ over an arbitrary field $\mathbb{F}$ is called a *Cartan-type Lie algebra* if there is an extension $\overline{\mathbb{F}}$ of the centroid C of $\mathfrak{g}$ such that the extended $\overline{\mathbb{F}}$-Lie algebra $\overline{\mathfrak{g}} = \overline{\mathbb{F}} \otimes_C \mathfrak{g}$ has a proper $\overline{\mathbb{F}}$-subalgebra of finite codimension.

It follows from the classical Cartan's result that, if $\mathbb{F}$ is a field of characteristic zero, any simple Lie algebra $\mathfrak{g}$ such that $\overline{\mathfrak{g}} = \overline{\mathbb{F}} \otimes \mathfrak{g}$ has a proper $\overline{\mathbb{F}}$-subalgebra of finite codimension n can be embedded in a Lie algebra $\widetilde{W}_n(\overline{\mathbb{F}})$ of all derivations of the power series $\overline{\mathbb{F}}[[t_1, \dots, t_n]]$; in $\widetilde{W}_n(\overline{\mathbb{F}})$ there is a unique Lie subalgebra of codimension n.

Thus the algebras $\widetilde{W}_n(\overline{\mathbb{F}})$ are unique "universal" simple algebras containing arbitrary simple Cartan-type Lie algebras. Similar universal simple algebras can be defined in the case of the fields of positive characteristic.

The following Razmyslov's result is among the strongest towards the solution of the above conjecture [311].

Theorem 14.13. *Let* $\mathbf{V}$ *be an arbitrary variety of Lie algebras over a field* $\mathbb{F}$ *of characteristic zero. Assume the growth of* $\mathbf{V}$ *is at most exponential. Then any simple* $\mathbb{F}$*-algebra in* $\mathbf{V}$ *is a Cartan-type algebra.*

As we mentioned in Section 9.1 earlier, S. P. Mishchenko proved that

$$\operatorname{var} W_n(\mathbb{F}) \quad \text{and hence } \operatorname{var} \widetilde{W}_n(\overline{\mathbb{F}})$$

have exponential growth.

If in Theorem 14.13 we take $\mathbf{V} = \operatorname{var} W_n(\mathbb{F})$ then Razmyslov proves that any simple algebra $\mathfrak{g}$ in $\mathbf{V}$ such that $\dim \mathfrak{g} > n^2 + 2n$ is isomorphic to a subalgebra of $\widetilde{W}_n(\overline{\mathbb{F}})$.

An earlier result by Razmyslov, in the case of $n = 1$, gives an additional interesting piece of information:

Theorem 14.14. *Let* $\mathfrak{g}$ *be a simple Lie algebra over a field* $\mathbb{F}$, $\operatorname{char} \mathbb{F} \neq 2$. *Let* C *be the centroid of* $\mathfrak{g}$*. Then the following properties of* $\mathfrak{g}$ *are equivalent:*

1. *There is an extension* $\mathbb{F}_1$ *of* C *such that the simple* $\mathbb{F}_1$*-algebra* $\mathfrak{g}_1 = \mathbb{F}_1 \otimes_V \mathfrak{g}$ *has an* $\mathbb{F}_1$*-subalgebra of codimension 1;*
2. $\mathfrak{g} \in \operatorname{var} W_1(\mathbb{F})$*;*
3. $\mathfrak{g}$ *satisfies a standard Lie identity of degree 5,*

$$\sum_{\sigma \in S_4} (\operatorname{sgn} \sigma)[x_{\sigma(1)}, x_{\sigma(2)} x_{\sigma(3)}, x_{\sigma(4)}, x_5] = 0. \tag{14.1}$$

This brings us to another open question:

Open problem

Is it true that the standard Lie identity of degree 5 forms a basis for the identities of $W_1(\mathbb{F})$?

To answer this question in the positive, we would need to prove that, for any algebra $\mathfrak{g}$ over $\mathbb{F}$, if $\mathfrak{g}$ satisfies (14.1) then $\mathfrak{g} \in \operatorname{var} W_1(\mathbb{F})$. In [310] Razmyslov proved this in the case where $\mathfrak{g}$ is simple. This is now generalized in a recent paper [291].

Theorem 14.15. *Let L be a semiprime Lie algebra over a field $\mathbb{F}$ of zero characteristic satisfying the standard Lie identity of degree 5. Then, L satisfies all identities satisfied by the algebra W_1.*

Note that L is a *prime Lie algebra* if $[A,B] \neq \{0\}$ for nonzero ideals A, B, and a *semiprime Lie algebra* if $[A,A] \neq \{0\}$ for any nonzero ideal A.

M. V. Zaicev [378] noted that the version of the above problem for $m \geq 2$ has a negative answer, namely, the identities of W_m do not follow from a standard identity.

We conclude with one more result aimed at the solution of Razmyslov's conjecture. To state it, we recall that an algebra L is called *locally finite-dimensional* if every finite subset of L generates a finite-dimensional subalgebra.

The following result is proven in [25, 58].

Theorem 14.16 (Y. Bahturin–H. Strade). *Let L be a simple locally finite-dimensional Lie algebra over a field $\mathbb{F}$, satisfying a nontrivial identical relation. Assume* $\operatorname{char} \mathbb{F} = 0$ *or* $\operatorname{char} \mathbb{F} = p > 7$. *Then L is an algebra of Cartan type.*

The restriction $p > 7$ is related to the state of the classification theory of simple modular Lie algebras in 1994. Since then there was essential progress by Premet–Strade [294] in the case of $\operatorname{char} \mathbb{F} = 5, 7$, which raises the question about the validity of the above theorem in the case of these smaller characteristics.

14.3.3 Capelli identities

For some time, people believed that infinite-dimensional Cartan-type Lie algebras do not satisfy nontrivial identities. But in 1974 I. Sumenkov proved that W_1 satisfies a standard identity of degree 5. Later Razmyslov found that standard identities of appropriate degrees are satisfied in all W_n, $n = 1, 2, \ldots$ Standard identities are a particular case of the so-called *Capelli identities* (see Section 5.5.1) which can be alternatively defined as follows.

Definition. Let $w(x_1, \ldots, x_m; x_{m+1}, \ldots, x_n) \in L(x_1, \ldots, x_n)$ be a Lie polynomial which is multilinear and alternating with respect to the variables $x_1, \ldots, x_m$, that is,

$$\begin{aligned} &w(x_1, \ldots, x_i, \ldots, x_j, \ldots, x_m; x_{m+1}, \ldots, x_n) \\ &\quad = -w(x_1, \ldots, x_j, \ldots, x_i, \ldots, x_m; x_{m+1}, \ldots, x_n) \end{aligned} \tag{14.2}$$

for any $1 \leq i < j \leq m$.

Then we say that $w(x_1, \dots, x_m; x_{m+1}, \dots, x_n) = 0$ is a *Capelli identity* of order m.

Any finite-dimensional (Lie) algebra L satisfies all Capelli identities of order n, $n > \dim L$. In [307] Yu. P. Razmyslov proves several results for algebras and pairs of algebras far more general than (the representations of) Lie algebras. These results show that, in a certain sense, Lie algebras satisfying all Capelli identities of order $m + 1$ can be viewed as m-dimensional over an extended domain of "scalars." One of the corollaries of the main theorems, closest to Lie algebras, says the following. Recall (see Section 6.3.3) that, given a Lie algebra L over a field $\mathbb{F}$, we denote by $\operatorname{Ad} L$ the associative subalgebra of $\operatorname{End}_{\mathbb{F}} L$ generated by all $\operatorname{ad} x$ where $x \in L$. Sometimes one calls $\operatorname{Ad} L$ the *adjoint algebra* of L.

Theorem 14.17 (Yu. P. Razmyslov). *Suppose that a Lie algebra L satisfies all Capelli identities of order $m + 1$. Then there is an ideal J in L satisfying $J^{m-2} = \{0\}$, with the following property. The adjoint algebra $D = \operatorname{Ad}(L/J)$ is a PI-algebra, moreover, D satisfies all identities of a matrix algebra of certain order.*

Capelli identities enable one to define the *rank* $r(A, V)$ *of a subspace V of an algebra A* as follows. Suppose we have a Capelli polynomial

$$w(x_1, \dots, x_m; x_{m+1}, \dots, x_n),$$

as before. A natural number k is called the rank of V in A if k is the smallest number m such that $w(v_1, \dots, v_m; a_{m+1}, \dots, a_n) = 0$ as soon as $v_1, \dots, v_m \in V$ and $a_{m+1}, \dots, a_n \in A$. The following rank theorem appears in the proof of the fact that simple algebras are determined by their identities (see Section 14.3.1). Given a semiprime algebra (could be an Ω-algebra!) A, one can view A as a module over the associative algebra $D(A)$ of multiplications of A ($\operatorname{Ad} A$ if A is a Lie algebra), take the injective hull P of this module, consider $E = \operatorname{End}_{D(A)} P$ and define the central closure $Q(A) = E * A$. If $p : E \to \operatorname{End}_{D(A)} Q(A)$ is the natural representation then $C(A) = \operatorname{Im} p = E / \operatorname{Ker} p$ is called the *Martindale centroid* of A. We have $Q(A) = C(A) * A$. For A simple, this is just the ordinary centroid of A. If A is prime then $C(A)$ is a field.

Theorem 14.18 (Yu. P. Razmyslov [313]). *Let V be an $\mathbb{F}$-vector subspace of a prime algebra A. If* $\operatorname{rank}(A, V) < \infty$, *then*

$$\dim_{C(A)} C(A)V = \operatorname{rank}(A, V) - 1.$$

In [307] the author asks if, for any m, the set of all Capelli identities of order m is finitely based. He mentions that in the case of weak identities of a pair (A, L), where A is associative and L is Lie, the answer is "yes." No proof substantiated this claim. We conclude this section by an extended version of a theorem of Stovba [347] (see Section 5.30).

Theorem 14.19 (V. V. Stovba). *Let $\mathbb{F}$ be a field of characteristic zero. The following systems of identical relations in Lie and associative algebras are finitely based:*
1. *The set of Capelli identities of any given order m.*
2. *The set of symmetric identities of any given order m.*
3. *Any set of identities in finitely many variables such that the degrees of all variables, but one, are bounded.*

14.4 Special Lie algebras and generalizations

Recall that a Lie algebra L is called special, or SPI, if it is isomorphic to a Lie subalgebra of an associative PI-algebra. We treated these algebras in a greater detail in Chapter 6. Later, an excellent survey of this area was written by M. V. Zaicev [377]. A number of results about SPI-algebras are also contained in Giambruno–Zaicev's monograph [123, Sections 12.3, 12.4].

One important open question mentioned in the comments to Chapter 6 is no longer open. We speak about V. N. Latyshev's questions: *Is a homomorphic image of an SPI-algebra itself SPI?* This is equivalent to the following: *Is it true that a central extension of an SPI-algebra is an SPI algebra?* For a counterexample, the following Lie algebra was suggested in [44].

Let $\mathfrak{g}$ be a finite-dimensional semisimple Lie algebra and consider $\mathbb{F}[t,t^{-1}]$, the Laurent polynomial ring in one variable t and coefficients from a field $\mathbb{F}$ of zero characteristic. Let us consider a well-known affine Kac–Moody Lie algebra

$$\widehat{\mathfrak{g}} = \mathfrak{g} \otimes \mathbb{F}[t,t^{-1}] \oplus \mathbb{F}z \oplus \mathbb{F}d,$$

where $\mathbb{F}z$ is a 1-dimensional center of $\widehat{\mathfrak{g}}$. The commutator in $\widehat{\mathfrak{g}}$ is given by

$$\begin{aligned}[][a \otimes t^m, b \otimes t^n] &= [a,b] \otimes t^{m+n} + (a,b)m\delta_{m,-n}z,\\ [d, a \otimes t^m] &= m(a \otimes t^m),\end{aligned}$$

where $a, b \in \mathfrak{g}$, $m, n \in \mathbb{Z}$, $[a,b]$ is computed in $\mathfrak{g}$ and (a,b) is the value of the Killing form of $\mathfrak{g}$ on a, b. The subalgebra $\widetilde{\mathfrak{g}} = \mathfrak{g} \otimes \mathbb{F}[t,t^{-1}] \oplus \mathbb{F}z$ is a central extension of the *loop Lie algebra* $\mathfrak{g} \otimes \mathbb{F}[t,t^{-1}]$. Since $\mathfrak{g}$ is finite-dimensional, there is a matrix algebra $M_n(\mathbb{F})$ such that $\mathfrak{g} \subset M_n(\mathbb{F})$. As a result, $\mathfrak{g} \otimes \mathbb{F}[t,t^{-1}] \subset M_n(\mathbb{F}[t,t^{-1}])$ is an SPI-algebra.

Theorem 14.20 (Yuly Billig [73]). *A Lie algebra $\widetilde{\mathfrak{g}}$ is a central extension of an SPI-algebra but not an SPI-algebra.*

Corollary 14.21. *There exist special varieties of Lie algebras (Section 6.5.1), that is, generated by a special algebra, which contain a non-SPI-algebra. If* $\mathbf{S} = \mathrm{var}(\mathfrak{sl}_2(\mathbb{F}))$ *then* $\mathbf{V} = [\mathbf{S}, \mathbf{E}]$ *is one of such varieties.*

An essential step in the proof of Theorem 14.20 is the description of the so-called PI-envelopes of finite-dimensional semisimple Lie algebras. If L is an SPI-algebra contained in a PI-algebra A, we say that A is a *PI-envelope* of L if L generates A as an associative algebra. In [35] we find the following result.

Theorem 14.22. *Let L be a finite-dimensional reductive Lie algebra over a field $\mathbb{F}$ of characteristic zero. Let m be the dimension of the center of L. Assume that A is a PI-envelope of L. Then there are natural numbers $n_1, \dots, n_t$ and rings $R_1, \dots, R_t$, which are factor-rings of the polynomial ring in m variables over the division algebras $\Delta_1, \dots \Delta_t$ of finite dimension over $\mathbb{F}$, such that $A \cong M_{n_1}(R_1) \oplus \cdots \oplus M_{n_t}(R_t)$.*

Here $M_n(R)$ is the matrix ring of order n over the ring R.

There is a much more general version of this result for the color Lie superalgebras [48], which in particular says that a PI-envelope of a finite-dimensional Lie superalgebra $\mathfrak{g} = \mathfrak{g}_- \oplus \mathfrak{g}_+$ is finite-dimensional if and only if $\mathfrak{g} = \mathfrak{g}_+$ is a *perfect color Lie algebra* ($\mathfrak{g} = [\mathfrak{g}, \mathfrak{g}]$).

Yuly Billig in [73] reverses Theorem 14.22 and proves the following

Theorem 14.23. *An associative algebra $A = M_{n_1}(\mathbb{F}) \oplus \cdots \oplus M_{n_t}(\mathbb{F})$ is a PI-envelope of a semisimple finite-dimensional Lie algebra L if and only if L has pairwise inequivalent irreducible finite-dimensional representations $\rho_i : L \to \operatorname{End}(V_i)$, $i = 1, \dots, t$, such that $\dim V_i = n_i$ and $\bigcap_{i=1}^{t} \operatorname{Ker} \rho_i = \{0\}$.*

This is an appropriate place to mention one open problem from [102] and a partial solution from the above survey.

OPEN PROBLEM

Is it true that any variety of Lie algebras is locally solvable if it does not contain $\mathfrak{sl}_2$?

Theorem 14.24 (A. Ya. Vais). *Let $\mathbf{V}$ be a special variety of Lie algebras over a field $\mathbb{F}$ with $\operatorname{char} \mathbb{F} = 0$. If $\mathbf{V}$ has no finite-dimensional semisimple Lie algebras then $\mathbf{V}$ is solvable.*

In the case of arbitrary infinite field, the “yes” answer follows from the following general result from Zaicev’s survey [377].

Theorem 14.25. *Let $\mathbb{F}$ be an infinite field and L a finitely generated SPI-algebra over $\mathbb{F}$. Then L has the largest solvable ideal $S(L)$ such that $L/S(L)$ is a subcartesian product of finite-dimensional semisimple algebras of bounded dimension. If $\operatorname{char} L = 0$ then $[L, S(L)]$ is nilpotent.*

One can also say that L is representable by matrices of finite order over a commutative $\mathbb{F}$-algebra.

14.4.1 Identities in the enveloping algebras

In Sections 6.7 and 6.9, we gave some results on the identities in the universal enveloping algebras of Lie algebras over any field and Lie superalgebras over a fields of characteristic zero.

Here we give more recent results on the identities in the enveloping algebras of the restricted Lie algebras, color Lie superalgebras, and color Lie p-superalgebras. The proofs of most of these results can be found in the monograph [46]. For a most recent contribution, see [59].

14.4.1.1 Restricted enveloping algebra

Restricted Lie algebras were defined in Section 1.11. It was noted that any associative algebra over a field $\mathbb{F}$ of characteristic $p > 0$ can be turned into a Lie p-algebra, denoted by $[A]_p$ if, in addition to the commutator $[a, b] = ab - ba$, we consider the p-operation on A given by $x^{[p]} = x^p$.

If L is a restricted Lie algebra over a field $\mathbb{F}$ of characteristic $p > 0$ (we also say a Lie p-algebra) then its *restricted enveloping algebra* $u(L)$ is defined as an associative algebra A over $\mathbb{F}$ enjoying an expected universal property for the Lie p-maps, that is, Lie homomorphisms φ from one Lie-p-algebra L to another M such that for any $x \in L$ one has $\varphi(x^{[p]}) = (\varphi(x))^{[p]}$. So, by definition, $u(L)$ is an associative algebra over $\mathbb{F}$ such that there is a canonical Lie p-map $\iota : L \to [A]_p$ such that for any associative algebra B and any Lie p-map $\alpha : L \to [B]_p$ there is a unique homomorphism $f : A \to B$ such that $f \circ \iota = \alpha$. Note that if $f : A \to B$ is a homomorphism of associative algebras then it is also a Lie p-map of Lie p-algebras $[A]_p$ to $[B]_p$. Clearly, two universal enveloping algebras must be isomorphic.

If $U(L)$ is an ordinary universal enveloping algebra then it follows from the identities of p-algebras that $x^p - x^{[p]}$ is in the center of $U(L)$, for any $x \in L$, identified with its image in $U(L)$. So one easily sees that the algebra $U(L)/P$, where P is the two-sided ideal generated by all $x^p - x^{[p]}$, $x \in L$, and $\iota = \varepsilon \circ \mathrm{id}$, ε being a natural epimorphism from $U(L)$ to $U(L)/P$, satisfies the universal property, hence $u(L) = U(L)/P$. From this argument it follows easily that if E is a totally ordered basis of L then 1 and the set of ordered monomials

$$e_1^{k_1} e_2^{k_2} \cdots e_m^{k_m}, \quad \text{where } e_1 < e_2 < \cdots < e_m,\ m \geq 1, \text{ are in } E,\ 0 < k_1, \ldots, k_m < p, \tag{14.3}$$

form a basis of $u(L)$.

As in the case of ordinary Lie algebras, the restricted enveloping algebras are important in the study of representations of Lie p-algebras. A representation of a Lie p-algebra is a Lie p-map $\rho : L \to [\mathrm{End}(V)]_p$. So any representation of L is a representation of $u(L)$, and vice versa. If L is finite-dimensional, it is seen from (14.3) that $u(L)$ is finite-dimensional. So,

All irreducible representations of a finite-dimensional Lie p-algebra are finite-dimensional.

The following theorem is independently proved by V. M. Petrogradsky [258] and D. S. Passman [263].

Theorem 14.26. *A restricted enveloping algebra $u(L)$ of a restricted Lie algebra L is PI if and only if the following conditions hold:*
1. *L has restricted ideals $N \subset M \subset L$ such that* $\dim L/M$, $\dim N \le \infty$*;*
2. *$[M, M] \subset N$, $[N, N] = 0$;*
3. *There is natural n such that $x^{[p^n]} = 0$, for each $x \in N$.*

As in Section 6.9 earlier, we could derive the consequences for the p-algebras with bounded degrees of irreducible representations. However, in [46, Chapter 5] this result, as well as the proof of Theorem 14.26 in [46, Chapter 4], has been given in the context of color Lie superalgebras, which we discuss in the next section.

14.4.1.2 Color Lie superalgebras

We now elaborate on the definition *color Lie superalgebras* treated in Section 11.4.3. To start with, we mention that this subject was treated in Bahturin–Mikhalev–Petrogradsky–Zaicev's book [46] and Mikhalev–Zolotykh's book [227]. This is also an approprite place to mention A. A. Mikhalev's papers on free Lie superalgebras [225, 226].

Let T be an abelian group and let $\beta : T \times T \to \mathbb{F}^\times$ be a skew-symmetric bicharacter, sometimes called "color." A T-graded algebra $L = \bigoplus_{t \in T} L_t$ with a commutator product $(x, y) \to [x, y]$ is called a *β-Lie superalgebra*, or a *color Lie superalgebra*, if the following hold for any $x \in L_t$, $y \in L_u$, $z \in L_v$:

$$[x, y] = \beta(t, u)[y, x] \quad \text{(color anticommutativity)}, \tag{14.4}$$

$$[[x, y], z] = [x, [y, z]] - \beta(t, u)[y, [x, z]] \quad \text{(color Jacobi identity)}. \tag{14.5}$$

A T-graded associative algebra $A = \bigoplus_{t \in T} A_t$ becomes a color Lie superalgebra if one sets

$$[x, y]_\beta = xy - \beta(t, u)yx, \quad \text{for any homogeneous } x \in A_t,\ y \in A_u. \tag{14.6}$$

Given a color (or β-) Lie superalgebra L, the (universal) enveloping algebra $U(L)$ for L is defined as follows. First of all, $U(L)$ is a T-graded associative algebra, generated by L, which is a β-subsuperalgebra of $U(L)$ (that is, closed under β-commutator (14.6)). Second, given a T-graded associative algebra A and any homomorphism of β-superalgebras $\varphi : L \to A$, there exists a (unique) homomorphism of associative algebras $f : U(L) \to A$ whose restriction to L is φ.

Given an alternating bicharacter $\beta : T \times T \to \mathbb{F}^\times$, for any $t \in T$, we must have either $\beta(t,t) = 1$, and then t is called *even*, or $\beta(t,t) = -1$, and then t is called *odd*. The set of all even elements in T is a subgroup T_+, the set of odd elements is a coset of T_+, which we denote by T_-. Respectively, if $V = \bigoplus_{t\in T} V_t$ is a T-graded space then we write $V_+ = \bigoplus_{t\in T_+} V_t$ and $V_- = \bigoplus_{t\in T_-} V_t$. If L is a β-superalgebra with $L = L_+$, then we call L a β-Lie algebra.

Finally, if $\mathbb{F}$ is a field of characteristic $p > 0$ then we define a color Lie p-superalgebra as a color Lie superalgebra such that for any homogeneous $x \in L_+$ there is a homogeneous element $x^{[p]} \in L_+$ such that the operation $x \mapsto x^{[p]}$ satisfies the usual properties from Section 1.11.

The structure group T is a finite abelian group.

The proof of Theorems 14.27 and 14.28 can be found in [46, Chapter 4].

Theorem 14.27. *Let $\mathbb{F}$ be a field of characteristic zero and L is a β-Lie algebra. Then $U(L)$ is PI if and only if L is abelian. If L is a general β-superalgebra, $L = L_+ \oplus L_-$, then $U(L)$ is PI if and only if there exists a homogeneous L_+-submodule $M \subset L_-$ such that*
1. $\dim L_-/M$, $\dim[L_+,M] < \infty$,
2. $[L_+,L_+] = [M,M] = 0$.

Theorem 14.28. *Let $\mathbb{F}$ be a field of characteristic $p > 2$ and L is a restricted β-Lie algebra. Then $u(L)$ is PI if and only if there exist homogeneous ideals $Q \subset R \subset L$ such that*
1. $\dim L/R$, $\dim Q < \infty$,
2. $R^2 \subset Q$, $Q^2 = 0$,
3. *Q has a nilpotent p-map, that is, for each homogeneous $x \in L_+$ we have $x^{[p^n]} = 0$, for some natural number n.*

In the case, where $\mathbb{F}$ is a perfect field, if $u(L)$ satisfies a nontrivial identity of degree d, we have some estimates of the dimensions in Theorem 14.28:

$$\dim L/R < 4^{4|T|d^4}, \ \dim Q < 2^{4|T|d^4}.$$

14.4.1.3 Dimensions of irreducible representations

Here we extend the results of Section 6.9 to superalgebras. Using the fact that an associative algebra satisfying a nontrivial identity cannot have irreducible modules of high degrees, we use Theorems 14.27 and 14.28 to get the following.

Theorem 14.29. *Let T be a finite abelian group and L a color Lie superalgebra over an algebraically closed field of characteristic 0. Then the dimensions of all irreducible L-representations are bounded by some finite number if and only if*
1. $\dim L_-/M < \infty$,
2. $[L_+,L_+] = [M,M] = 0$,

3. $\dim_{\mathbb{F}} L_+ < |\mathbb{F}|$.

Theorem 14.30. *Let T be a finite abelian group and L a color Lie superalgebra over an algebraically closed field of characteristic $p > 0$. Then the dimensions of all irreducible L-representations are bounded by some finite number if and only if there exists homogeneous ideals $R \subset L$ such that*
1. $\dim L/R < \infty$,
2. $R^2 \subset R_-$,
3. *For all homogeneous $x \in L_+$, the inner derivations $\operatorname{ad} x|_{L_+}$ are algebraic of bounded degree,*
4. $\dim_{\mathbb{F}} L_+ < |\mathbb{F}|$.

It is interesting that, in distinction with the respective theorem from Section 6.9, the conditions necessary for the L-dimensions to be bounded are less restrictive than for the existence of polynomial identity in $U(L)$.

14.4.2 API-algebras

A meaningful generalization of special Lie algebras was suggested by S. P. Mishchenko in [237]. First, we introduce the so-called *diluted identities of degree n*. Let X be a set of free generators of a free Lie algebra of countable rank, $x_1, \ldots, x_n \in X$, $\{y_{ij} \in X \setminus \{x_1, \ldots, x_n\}$, with $i, j = 1, 2, \ldots\}$. Let $\overline{w} = (w_1, w_2, \ldots, w_{n-1})$ be an n-tuple of associative words $w_i = y_{i1} \cdots y_{is_i}$. Then the diluted identity of degree n is a left normed expression of the form

$$\sum_{\sigma \in \operatorname{Sym}(n)} \alpha_\sigma(\overline{w}) x_{\sigma(1)} y_{11} \cdots y_{1s_1} x_{\sigma(2)} y_{21} \cdots y_{2s_2} \cdots y_{n-1,1} \cdots y_{n-1,s_{n-1}} x_{\sigma(n)} \equiv 0. \tag{14.7}$$

Here $\{\alpha_\sigma(\overline{w})\}$ is the set of coefficients depending on the choice of $\overline{w}$. A particular case is the Capelli identity of order n.

A Lie algebra L satisfying (14.7) for any choice of $\overline{w}$ is called an *API-algebra*. One can show that any special Lie algebras is API.

Proposition 14.31. *A Lie algebra L with nilpotent ideal I of finite codimension is API.*

Proof. Actually, one can easily see that if $\dim L/I = n$ and $I^{c+1} = 0$ then L satisfies the Capelli identities of order $n + c + 1$. □

For instance, if M is a finite-dimensional algebra and V is an M-module, viewed as an abelian Lie algebra, then the semidirect product $L = M \rtimes V$ is an API-algebra. Any nonabelian finite-dimensional Lie algebra M has an irreducible module V of infinite dimension. All these algebras are API-algebras. As shown in Section 6.3.4, none of these algebras is special. So the following is a genuine extension of a simple fact that the codimension sequence of a special Lie algebras has exponential growth.

Theorem 14.32. *Let A be an API Lie algebra. Then the sequence of codimensions $\{c_n(A)\}$ is exponentially bounded and the sequence $\{\ell_n(A)\}$ of colengths is polynomially bounded.*

Another characterization of API-algebras is this. We define an infinite hook $H(d,l)$ by setting

$$H(d,l) = \bigcup_{n\geq 1}\{\lambda = (m_1, m_2, \ldots) \vdash n \mid m_{d+1} \leq l\}. \tag{14.8}$$

We write $\chi_\lambda \subset H(d,l)$ if $\lambda \subset H(d,l)$. By the fundamental Amitsur–Regev's hook theorem (see [123, Theorem 4.5.1]), for any PI-algebra A there are natural numbers d,l with the following property. If $\{\chi_n(A)\}$ is the cocharacter sequence of A and

$$\chi_n(A) = \sum_{\lambda\vdash n} m_\lambda \chi_\lambda$$

then $m_\lambda \neq 0$ only if $\lambda \subset H(d,l)$.

Next we say that a Lie algebra L is AT, or of associative type, if there exist d,l such that $\chi_n(L) \subset H(d,l)$ for all $n \geq 1$. It is quite clear that any SPI-Lie algebra is of associative type. The following is known.

Theorem 14.33 ([238]). *The class of API algebras coincides with the class AT of associative type Lie algebras.*

One more result showing the similarity of API-algebras with the associative algebras deals with the so-called *heights* in Lie algebras. For the definition in the case of associative algebras see Chapter 6.

Definition 14.34. One says that a Lie algebra L has a *finite height at most h* with respect to a set of its elements $a_1, \ldots, a_m$ if every element of L can be written as a linear combination of left normed product of the form

$$a_{i_1}^{k_1} \cdots a_{i_s}^{k_s} a_{i_{s+1}},$$

where $s+1 \leq h$ and $1 \leq i_1, \ldots, i_{s+1} \leq m$.

Theorem 14.35 ([237]). *Any finitely generated API-algebra has finite height with respect to a finite set of its elements.*

14.5 Automorphisms and derivations

Two survey papers of V. S. Drensky [109, 110] are devoted to the study of automorphisms of relatively free algebras, which includes Lie algebras. Here we give some results mostly on the automorphisms of relatively free Lie algebras (see the definitions in Chapter 4).

If $L_n(\mathbf{V})$ is such an algebra freely generated by $X = \{x_1, \dots, x_n\}$ then any automorphism of $L_n(\mathbf{V})$ is completely determined by its restriction to X. Conversely, any map $\varphi : X \to L_n(\mathbf{V})$ extends to an endomorphism $\bar{\varphi} : L_n(\mathbf{V}) \to L_n(\mathbf{V})$. So we "only" need to determine for what φ the endomorphism $\bar{\varphi}$ is injective and surjective. If $L = L_n(X)$ is an (absolutely) free Lie algebra generated by X then $L_n(\mathbf{V}) \cong L(X)/\mathbf{V}(L(X))$, where $\mathbf{V}(L(X))$ the ideal of identities of $\mathbf{V}$ in $L(X)$.

We know that $\mathbf{V}(L(X))$ is stable under all endomorphisms of L. Therefore, any endomorphism $\theta : L \to L$ induces an endomorphism $\widetilde{\theta}$ of $L_n(\mathbf{V})$. If $\theta \in \operatorname{Aut} L$ then, clearly, $\widetilde{\theta} \in \operatorname{Aut}(L_n(\mathbf{V}))$. Such automorphisms are called *tame*, while all the rest are called *wild*.

14.5.1 Automorphisms of the algebras L/R^k

A basis for any research on tame and wild automorphisms of Lie algebras is formed by the paper [92] of P. M. Cohn. Let $L = L_n(X)$ be the free Lie algebra of a finite rank $n > 2$ over a field $\mathbb{F}$ of characteristic 0. Then the following is true.

Theorem 14.36 (Cohn, P. M.). *The generators of the group* $\operatorname{Aut} L$ *of automorphisms of L are given explicitly as:*
(i) *The automorphisms which are induced by the action of the general linear group* $\mathrm{GL}_n(\mathbb{F})$ *on the linear subspace of L spanned by* $X = \{x_1, \dots, x_n\}$, *the free generators of L;*
(ii) *The automorphism of the form* $x_1 \mapsto x_1 + f(x_2, \dots, x_n)$; $x_k \mapsto x_k$, $k \geq 2$, *where the Lie polynomial* $f(x_2, \dots, x_n)$ *does not depend on* x_1.

Alternative proofs of this theorem were given in the works of other people; see, for example, [189, 319].

Given an arbitrary characteristic ideal R of L (i. e., an ideal invariant under all automorphisms of L), we say again that an automorphism φ of L/R is tame if there is an automorphism ψ of L such that $\psi(R) = R$ and $\varphi = \widetilde{\psi}$, that is, $(x + R) = \psi(x) + R$, for any $x \in L$.

It follows that in the case $n = 2$ the tame automorphisms of any Lie algebra L/R could only be those induced by nondegenerate linear transformations of the linear span of $\{x_1, x_2\}$. A general theorem on wild automorphisms of 2-generator algebras is due to Drensky [108]. Recently Drensky and Gupta [111] investigated the structure of the automorphism groups of some relatively free nilpotent Lie algebras over a field of characteristic 0. One of the consequences of their thorough calculations is that any such noncommutative Lie algebra has wild automorphisms.

Our main result in this section is the following result by Bahturin–Shpilrain from [57].

Theorem 14.37. *Let R be a nonzero ideal of the free Lie algebra L over a field* $\mathbb{F}$ *of characteristic 0, and let* $c \geq 1$ *be such that* $R \subset L^c$, *but* $R \not\subset L^{c+1}$. *Let* $k \geq 2$ *be an integer.*

Suppose there is an element $u \in R$, $u \notin L^{kc-2}$, which does not depend on x_1. Then there exists a wild automorphism of Lie algebra L/R^k.

The main tool in the proof of Theorem 14.37 is the *free differential calculus* by R. H. Fox (see [116]).

14.5.1.1 Endomorphisms and Jacobians

Let $L = L_n$ be the free Lie algebra over a field $\mathbb{F}$ of characteristic 0, with the set $X = \{x_i\}$, $1 \leq i \leq n$ of free generators, and $U(L)$ its universal enveloping algebra, which is the free associative algebra over $\mathbb{F}$ with the same set X of free generators. By Δ we denote the augmentation ideal of $U(L)$, i. e., the kernel of the augmentation map $\varepsilon : U(L) \to \mathbb{F}$ (defined by $\varepsilon(x_i) = 0$, $1 \leq i \leq n$). If $R \neq L$ is an ideal of L then by Δ_R we denote the kernel of the natural homomorphism $\varepsilon_R : U(L) \to U(L/R)$. If T is an ideal of L then by T^m we denote the mth term of the lower central series of T; for J an (associative) ideal of $U(L)$, by J^m we denote the "usual" mth associative power of J, without ambiguity.

In [116], R. H. Fox gave a detailed account of the differential calculus in the free group ring. Since any free associative algebra is naturally embedded in a free group algebra (over the same field), most of the technical results remain valid for free associative algebras.

To be more specific, we introduce here free derivations as the mappings $d_i : U(L) \to U(L)$, $1 \leq i \leq n$, satisfying the following conditions for all $\alpha, \beta \in \mathbb{F}$, $u, v \in U(L)$:

(1) $d_i(x_j) = \delta_{ij}$ (Kronecker delta);
(2) $d_i(\alpha u + \beta v) = \alpha d_i(u) + \beta d_i(v)$;
(3) $d_i(uv) = d_i(u)\varepsilon(v) + u d_i(v)$.

Instead of $d_i(u)$, we sometimes write $\frac{\partial u}{\partial x_i}$. It is easy to prove (see [116]) that these derivations have another nature as well. The ideal Δ is a free left $U(L)$-module with free basis $\{x_i\}$, $1 \leq i \leq n$, and the mappings d_i are the projections to the corresponding free cyclic direct summands. Thus any element $u \in \Delta$ can be uniquely written in the form $u = \sum_{1 \leq i \leq n} d_i(u) x_i$.

It follows from the definitions that $d_i(1) = 0$, $1 \leq i \leq n$. Straightforward calculation shows that

$$d_i([u,v]) = (\varepsilon(v) - v)d_i(u) - (\varepsilon(u) - u)d_i(u), \quad \text{for any } u, v \in U(L).$$

We need some technical lemmas. The first lemma is an immediate consequence of the definitions.

Lemma 14.38. *Let J be an arbitrary ideal of $U(L)$ and let $u \in \Delta$. Then $u \in J\Delta$ if and only if $d_i(u) \in J$ for each i, $1 \leq i \leq n$.*

We also need the "chain rule" for the Fox derivations (see [116]):

Lemma 14.39. *Let φ be an endomorphism of L (it can be naturally extended to $U(L)$, defined by $\varphi(x_k) = y_k$, $1 \le k \le n$), and let $v(y_1, \dots, y_n) = \varphi(v(x_1, \dots, x_n))$, for some $v \in \Delta$. Then*

$$d_j(v(y_1, \dots, y_n)) = \sum_{l \le k \le n} \frac{\partial v}{\partial y_k} \cdot \frac{\partial y_k}{\partial x_j},$$

where by $\frac{\partial v}{\partial y_k}$ we mean $\varphi(d_k(v))$.

For an endomorphism φ of L, let $J_\varphi = ||d_i(\varphi(x_j))||_{1 \le i,j \le n}$ be the *Jacobian matrix of* φ. Lemma 14.39 implies, in particular, that for $\varphi \in \operatorname{Aut} L$ this matrix is left invertible and, therefore, belongs to $\mathrm{GL}_n(U(L))$ (see [17]). The proof of the following claim can be found in Chapter 6, see Theorem 6.20.

Lemma 14.40. *Let $u \in L$. Then $u \in \Delta^c$ if and only if $u \in L^c$. Thus $L^c = L \cap \Delta^c$.*

Now we are going to elaborate on a necessary condition for an endomorphism of L to be an automorphism. We will use a generalization of the determinant in the sense of Dieudonné. We refer to [17] for an exposition of the original construction of Dieudonné.

Let us just briefly summarize what we need here.

Let S be an associative ring with 1 and with the following property: every invertible square matrix with the entries from S (i. e., a matrix from $\mathrm{GL}_k(S)$ for some $k \ge 1$) has at least one invertible element in every row and in every column. For instance, every local ring (i. e., a ring in which all noninvertible elements form an ideal) has this property. Over such a ring, any invertible matrix M can be written in a special canonical form $M = TD(\mu)$, where T is a product of transvections, and $D(\mu)$ is the matrix that is obtained from the identity matrix by replacing the "1" in the right lower corner with $\mu \in S, \mu \neq 0$. Take now the group S^* of all invertible elements of the ring S, and the quotient group $\overline{S}$ of S^* by its commutator subgroup. Then the image $\bar{\mu}$ of μ in $\overline{S}$ is defined uniquely and is called the *determinant* of the matrix M in the sense of Dieudonné. It has some usual properties of the determinant; in particular, the determinant of the product of two matrices from $\mathrm{GL}_k(S)$ is equal to the product of their determinants.

An application of this construction to the study of the automorphisms of L is as follows. Given a matrix $M \in \mathrm{GL}_k(U(L))$, we take a ring $S_m = U(L)/\Delta^m$, $m \ge 2$, and come up with the determinant $\delta_m(M)$, in accordance with the construction described above, using the fact that in the ring S_m every element with nonzero augmentation is invertible. More specifically, given an element $u \in U(L)$ of the form $u = \alpha - w$, $0 \neq \alpha \in \mathbb{F}$, $\varepsilon(w) \neq 0$, one has modulo Δ^m:

$$u^{-1} = \alpha^{-1} + \alpha^{-2} w + \dots + \alpha^{-m} w^{m-1}.$$

Hence the commutator subgroup of the group S_m^* is generated as a group (actually as a semigroup) modulo Δ^m by the elements of the form

$$(1 - v)(l - w)(l + v + \dots + v^{m-1})(l + w + \dots + w^{m-1}), \tag{14.9}$$

with $\varepsilon(v) = \varepsilon(w) = 0$. Denote by P_m the (multiplicative) subsemigroup of $U(L)$ generated by all elements of the above form.

Now we make the following crucial observation. Every automorphism of L is a product of elementary automorphisms described in the introduction. Hence given the Jacobian matrix J_φ of some $\varphi \in \operatorname{Aut} L$, we have (by Lemma 14.39) $\delta_m(J_\varphi)$ equal to the product of the determinants of elementary Jacobian matrices each of which is readily seen to be equal to the image $\bar{\alpha}$ of some $\alpha \in \mathbb{F}$, $\alpha \neq 0$. This observation yields the following necessary condition for φ to be an automorphism. By $\det_m(J_\varphi)$ we mean an arbitrary preimage of $\delta_m(J_\varphi)$ in $U(L)$.

Proposition 14.41. *Let φ be an automorphism of L and J_φ the Jacobian matrix of φ. Then for an arbitrary $m \geq 2$, one has $\det_m(J_\varphi) = \alpha p_m + w_m$, for some $\alpha \in \mathbb{F}$, $\alpha \neq 0$, $p_m \in P_m$, $w_m \in \Delta^m$.*

14.5.1.2 Wild automorphisms

We begin this section with the proof of Theorem 14.37. Let us agree to denote by $\overline{u}$ the image of an element $u \in L$ in a quotient-algebra of L when it is clear what quotient-algebra we are talking about.

Proof. For an element u provided by the conditions of the theorem, we construct the automorphism φ of Lie algebra L/R^k as follows:

$$\varphi(\overline{x}_1) = \overline{x}_1 + [\overline{u}, \overline{x}_1, \overline{x}_1] \quad \text{and} \quad \varphi(\overline{x}_i) = \overline{x}_i, \quad \text{for } 2 \leq i \leq n.$$

This φ is readily seen to be an automorphism of Lie algebra L/R^k since it has the inverse φ^{-1} defined by $\varphi^{-1}(\overline{x}_1) = \overline{x}_1 - [\overline{u}, \overline{x}_1, \overline{x}_1]$, and $\varphi^{-1}(\overline{x}_i) = \overline{x}_i$ for $2 \leq i \leq n$. Hence φ is an automorphism of any Lie algebra L/R^k, $k \geq 2$, as well. Were φ a tame automorphism, there would exist $v_i \in R^k$, $1 \leq i \leq n$, such that in the free Lie algebra L, the mapping $\psi : x_1 \to x_1 + [u, x_1, x_1] + v_1$, $x_i \to x_i + v_i$ could be extended to an automorphism of L. In that case, for every $m \geq 1$ we would have the Dieudonné determinant of the matrix $J_\psi = \|d_j(y_i)\|_{1 \leq i,j \leq n}$ with $y_i = \psi(x_i)$, satisfying the necessary condition of Proposition 14.41. Evaluate $d_1(y_1)$ first:

$$d_1(y_1) = 1 - 2x_1u + ux_1 + d_1(v_1),$$
$$d_j(y_1) = x_1^2 d_j(u) + d_j(v_1), \quad j \neq 1,$$

and when $i \neq 1$ we have $d_j(y_i) = \delta_{ij} + d_j(v_i)$. Since $v_i \in R^k$ and $R \subset L^c$, we have $v_i \in \Delta^{kc}$, so $d_j(v_i) \in \Delta^{kc-1}$, $1 \leq i, j \leq n$. Under the conditions of the theorem, $u \notin L^{kc-2}$, so $ux_1 \notin \Delta^{kc-1}$. Let $m < kc$ be the maximal integer such that $u \in \Delta^{m-2}$. Then, evaluating the Dieudonné determinant of J_ψ modulo Δ^m, we would have by Proposition 14.41:

$$\det_m(J_\psi) = 1 - 2x_1u + ux_1 = \alpha p_m + w_m \quad \text{for some } \alpha \in \mathbb{F},\ \alpha \neq 0;$$
$$p_m \in P_m;\ w_m \in \Delta^m.$$

This implies (by applying the augmentation map) $\alpha = 1$, so $1 - 2x_1u + ux_1 = p_m + w_m$ or

$$1 - 2x_1u + ux_1 \in P_m. \tag{14.10}$$

But this is not the case; indeed, every element of P_m has a nonconstant homogeneous component of the smallest weight equal to the sum of some elements of the form $(vw - wv)$ for $v, w \in U(L)$. This is readily seen to be true for the generating elements (14.9), and hence for every element of P_m, as well. The element $t = -2x_1u + ux_1$, however, has the homogeneous component of degree $(m-1)$ equal to $t_{m-1} = -2x_1u_{m-2} + u_{m-2}x_1$, where u_{m-2} is the homogeneous component of weight $(m-2)$ of the element u. Since u is an element of L, the element $-x_1u_{m-2} + u_{m-2}x_1$ is also in L (and it is homogeneous!), so it is equal to the sum of elements of the form $(vw - wv)$. Hence, were the element t_{m-1} equal to this kind of sum, the element $-x_1u_{m-2}$ would have the same property. But this is not the case since u and hence u_{m-2} does not depend on x_1, so the element $-x_1u_{m-2}$ has only monomials with the single entry of x_1 on the first place. Thus the inclusion (14.10) yields a contradiction, and this completes the proof of the theorem □

The case of the free metabelian Lie algebra, which is not covered by this theorem, will be treated in Section 14.5.2, based on the paper of Y. Bahturin and S. Nabiev [49], one of the results of which is that for the free nonabelian metabelian Lie algebra $M_m = L_m/(L_m^2)^2 = L_m(\mathbf{A}^2)$, $m \geq 2$, over any *integral domain* (!) $\mathbb{F}$, every inner automorphism of the form $\exp(\operatorname{ad} z)$, $z \in M_m^2$, is wild. This, together with Theorem 14.37, yields a consequence about the automorphism groups of any free polynilpotent Lie algebras, that is, relatively free algebras in the varieties $\mathbf{N}_{c_t} \cdots \mathbf{N}_{c_2}\mathbf{N}_{c_1}$ (see Section 4.5).

Corollary 14.42. *Any nonabelian Lie algebra over a field of characteristic 0 which is free in some variety of all polynilpotent Lie algebras with a given set of nilpotency steps has wild automorphisms.*

It should be mentioned that the result of this corollary has been independently obtained by Papistas [254] in a different context and by using different methods. We emphasize one more corollary.

Corollary 14.43 ([111]). *Let $L = L_n$ be a free algebra of rank $n \geq 2$ in an arbitrary nilpotent nonabelian variety of Lie algebras over a field of characteristic 0. Then L has wild automorphisms.*

A topological approach to the automorphism group $\operatorname{Aut} M_m$, where $M_m = L_m(\mathbf{A}^2)$, the free metabelian group of rank m, has been taken in Bryant - Drensky's paper [87]. The group $\operatorname{Aut} M_m$ becomes a *metric space* if we define the distance between the automorphisms φ and ψ as $\exp(-n)$ where n is the maximal number such that $\varphi(x) + M_m^{n+1} = \psi(x) + M_m^{n+1}$, for all $x \in M_m$.

They proved the following highly nontrivial result.

Theorem 14.44 (R. M. Bryant and V. S. Drensky). *Let T be the group of tame automorphisms of the free metabelian Lie algebra $M_m = L_m(\mathbf{A}^2)$ of finite rank $m > 2$.*

1. *For $m > 4$, T is dense in $A = \mathrm{Aut}(L_m(\mathbf{A}^2))$.*
2. *For $m = 2$ and $m = 3$, T is not dense in A, and so $L_m(\mathbf{A}^2)$ possesses nontame automorphisms.*

There are some results where the automorphism group of a relatively free Lie algebra can be completely described, for example, the following two are from [111].

Let us define an endomorphism δ of the free Lie algebra $L_m = L(x_1, \dots, x_m)$ by setting

$$\delta(x_1) = x_1 + [x_1, x_2],\ \delta(x_i) = x_i, \quad \text{if } i = 2, \dots, m.$$

If R is a characteristic ideal of L_m then the endomorphism $\widetilde{\delta}$ of L_m/R induced by δ will be still denoted by δ.

Theorem 14.45 (V. S. Drensky and C. K. Gupta). *The following are true:*
1. $\mathrm{Aut}\, L_m(\mathbf{N}_c \cap \mathbf{A}^2)$ *is generated by* $\mathrm{GL}_m(\mathbb{F})$ *and* δ;
2. *Let* $m \geq c \geq 2$. *Then* $\mathrm{Aut}\, L_m(\mathbf{N}_c)$ *is generated by* $\mathrm{GL}_m(\mathbb{F})$ *and* δ.

14.5.2 Shmelkin's wreath products and automorphisms

The case $m = 2$ in Theorem 14.44 is due to A. L. Shmelkin [342], to whom we are indebted for a very useful notion of the wreath products in Lie algebras (see Section 4.4.1). It works, in particular, for Lie algebras of the form L/R^2, where R is an ideal in a free Lie algebra L. Recall Theorem 4.26.

If R be an ideal of an absolutely free Lie algebra $L = L(x_1, \dots, x_n)$ over a commutative ring $\mathbb{F}$, $B = L/R$, and let F be a left free B-module (same as $U(B)$-module) with basis $a_1, \dots, a_n$, and trivial commutator. The semidirect product $W = B \rtimes F = B \oplus F$ is an (abelian) wreath product of B and an abelian Lie algebra with basis $a_1, \dots, a_n$. By Shmelkin's theorem, if B is $\mathbb{F}$-free then the map $\varphi(x_i) = \bar{x}_i + a_i$, $1 \leq i \leq n$, induces an isomorphic embedding $\tilde{\varphi} : L/M^2 \to W$. Here $\overline{x}_i = x_i + R$.

To determine the automorphism group $\mathrm{Aut}\, L/R^2$, we need a subgroup $\overline{\mathrm{Aut}\, W}$ of $\mathrm{Aut}\, W$ consisting of all elements leaving F and B invariant. In $\overline{\mathrm{Aut}\, W}$, we select a normal subgroup $\mathrm{I\,Aut}\, W$ whose elements are identical on B and a subgroup P whose elements fix each $a_i \in F$, $1 \leq i \leq n$. It is obvious from the definition of the abelian wreath product (see Section 4.4.1) that any mapping of $\{a_i\}$ into F and any automorphism of B into itself extend to an endomorphism of W and this is an automorphism of W (lying in $\overline{\mathrm{Aut}\, W}$) if and only if the image of $\{a_i\}$ is a free generating set of F as a B-, hence $U(B)$-module. Thus we derive that $\mathrm{I\,Aut}\, W$ is isomorphic to $\mathrm{GL}(n, U(B))$, the group of all invertible $n \times n$ matrices with entries from (a noncommutative) ring $U(B)$. It is obvious that the second of the subgroups mentioned above is isomorphic to $\mathrm{Aut}\, B$. So we have the following

Proposition 14.46. *Using the above notation,*

$$\overline{\operatorname{Aut} W} = P \rtimes \operatorname{I}\operatorname{Aut} W \quad \text{and if } \operatorname{rank} F = n \text{ then } \overline{\operatorname{Aut} W} \cong \operatorname{Aut} B \rtimes \operatorname{GL}_n(U(B)).$$

14.5.2.1 Automorphisms of L/R^2

We are going to study the group $\overline{\operatorname{Aut} L/R^2}$ of those automorphisms of a Lie algebra L/R^2 which leave R/R^2 invariant. In many applications, this is always the case. The following proposition plays a key role.

Proposition 14.47. *There exists an embedding* $\nu : \overline{\operatorname{Aut} L/R^2} \to \overline{\operatorname{Aut} W}$ *such that if* $\alpha \in \overline{\operatorname{Aut} L/R^2}$ *then* $\nu(\alpha) = \tilde{\alpha}$ *satisfies* $\tilde{\alpha}\mu = \mu\alpha$ *where* μ *is the canonical embedding* $\mu : L/R^2 \to W$.

Proof. First, we observe that if $w \in L$ and $\overline{w}, \overline{\overline{w}}$ are canonical images of w in L/R^2, L/R, respectively, then $\mu(\overline{w}) = \overline{\overline{w}} + f$ where $f \in F$. Now given $\alpha : L/R^2 \to L/R^2$ with $\alpha(R/R^2) \subset R/R^2$, we define $\tilde{\alpha}$ by first writing

$$\mu\alpha(\overline{x}_i) = \overline{\overline{b}}_i + p_i \in B \oplus F, \tag{14.11}$$

with $B = L/R$, and then setting

$$\tilde{\alpha}(\overline{\overline{x}}_i) = \overline{\overline{b}}_i, \quad \tilde{\alpha}(a_i) = p_i. \tag{14.12}$$

This automatically makes $\tilde{\alpha}\mu = \mu\alpha$ satisfied. It is obvious that $\tilde{\alpha} : L/R \to L/R$ is an isomorphism since it is induced by α which leaves R/R^2 invariant. By definition of the wreath product and using $\tilde{\alpha}\mu = \mu\alpha$, $\tilde{\alpha}$ extends to a uniquely defined map of W. Also, if α^{-1} is the inverse of α then

$$\mu\alpha\alpha^{-1} = \tilde{\alpha}\mu\alpha^{-1} = \tilde{\alpha}\widetilde{\alpha^{-1}}\mu$$

and $\tilde{\alpha}\widetilde{\alpha^{-1}} = \widetilde{\operatorname{id}_{L/R^2}}$. Since $\operatorname{id}_W = \widetilde{\operatorname{id}_{L/R^2}}$, we have $\tilde{\alpha}\widetilde{\alpha^{-1}} = \operatorname{id}_W$, so that $\tilde{\alpha}$ is an automorphism of W, and the proof is complete. □

We will be particularly interested in the normal subgroup $\operatorname{I}\operatorname{Aut} L/R^2$ of $\operatorname{Aut} L/R^2$ whose elements are the automorphisms of L/R^2 which are identity modulo R/R^2. This is quite understandable since L/R is an arbitrary Lie algebra. The explicit matrix form for the elements of $\operatorname{I}\operatorname{Aut} L/R^2$ will be found in the case where L/R is a free $\mathbb{F}$-module finitely related as a Lie algebra over $\mathbb{F}$. So let $L = L(x_1, \ldots, x_n)$ and let R/R^2 be generated by $\overline{r}_1, \ldots, \overline{r}_m$ as an ideal, that is, as a B-module, where $B = L/R$. We denote the module action of $u \in U(B)$ on $\overline{r} \in R/R^2$ by $u\overline{r}$. If $u \in B$, i. e., $u = \overline{\overline{w}}$, $w \in L$, then $u\overline{r} = [\overline{u}, \overline{r}]$. Consider $\alpha \in \operatorname{I}\operatorname{Aut} L/R^2$. Then

$$\alpha(\overline{x}_i) = \overline{x}_i + \sum_{j=1}^{m} q_{ij}\overline{r}_j, \quad i = 1, \ldots, n, \; q_{ij} \in U(B). \tag{14.13}$$

Apply μ to both sides of (14.13). Then

$$\mu\alpha(\overline{x}_i) = \overline{\overline{x_i}} + a_i + \sum_{j=1}^{m} \mu(q_{ij}\overline{r}_j) = \overline{\overline{x_i}} + a_i + \sum_{j=1}^{m} q_{ij}\mu(\overline{r}_j).$$

Since F is an abelian ideal, all summands in $\mu(q_{ij})$ except q_{ij} contain factors in F and thus their action on $\mu(\overline{r}_j)$ is trivial. Now we have

$$\mu(\overline{r}_j) = \sum_{k=1}^{n} a_{jk}a_k, \quad j = 1, \ldots, m. \tag{14.14}$$

We denote by $A_n = (a_{jk})$ an $(m \times n)$-matrix with entries in $U(L/R)$. Turning back to (14.12) and recalling Proposition 14.46, we find that the matrix of $\tilde{\alpha}$ restricted to F has the form

$$I_n + QA_n, \quad Q = (q_{ij}),\ q_{ij} \in U(L/R);\ 1 \le i \le n,\ 1 \le j \le m. \tag{14.15}$$

This matrix is two-sided invertible. It is clear, conversely, that given Q as in (14.15) with $I_nm + QA_n$ invertible, we have that (14.13) determines an element in $\mathrm{I\,Aut}\,L/R^2$. This becomes obvious since

$$\tilde{\alpha} : \overline{x}_i \mapsto \overline{x}_i, \quad \tilde{\alpha} : a_i \mapsto a_i + \sum_j \sum_k q_{ij}q_{jk}a_k$$

defines an automorphism of W and under this automorphism

$$\mu(\overline{x}_i) \mapsto \mu\left(\overline{x}_i + \sum_{j=1}^{m} q_{ij}\overline{r}_j\right) \in \mu(L/R^2),$$

meaning that $\alpha : L/R^2 \to L/R^2$ is an automorphism. Clearly, $I_n + QA_n$ is determined by α and determines it uniquely. Thus we have proved the following.

Theorem 14.48. *If $L = L(x_1, \ldots, x_n)$ is a free Lie algebra over a ring $\mathbb{F}$, R an ideal of L such that L/R is a free $\mathbb{F}$-module, L/R is a finitely presented algebra given by the defining relations $r_1 = 0, \ldots, r_m = 0$ then $\mathrm{I\,Aut}\,L/R^2$ is isomorphic to the group G of all invertible matrices of the form $I_n + QA_n$ where A_n is a fixed matrix of order $m \times n$ defined in (14.14) and Q is an arbitrary $n \times m$ matrix, both with coefficients in $U(L/R)$, such that $I_n + QA_n$ is invertible in the matrix ring $U(L/R)_n$.*

It follows from the above construction that the jth row of A_n is formed by the coefficients of $\mu(r_j)$ in its unique expression through $a_1, \ldots, a_n$.

14.5.2.2 Automorphisms of free metabelian algebras

From Section 2.4.2 we know that the commutator subalgebra M_n^2 of the free metabelian Lie algebra of rank n, $M_n = L_n/(L_n^2)^2$, is generated by the elements $r_{ij} = [x_i, x_j]$ where

$1 \le i < j \le n$. For convenience, we order the set of pairs (i,j) by setting $(i,j) < (k,l)$ if either $j < l$ or $j = l$ and then $i < k$. Write the generators ordered by indices as $r_1, r_2, \dots, r_m$ where $m = n(n-1)/2$. The expression of each $\mu(r_{ij})$ in terms of $a_l, \dots, a_n$ then takes the form

$$\mu(r_{ij}) = [x_i + a_i, x_j + a_j] = x_i a_j - x_j a_i,$$

where for brevity we write x_i in place of $\overline{\overline{x}}_i$. Now the expression for A_n can be given inductively if we set

$$B_n = \begin{pmatrix} x_1 & x_2 & \dots & x_n \end{pmatrix}^{\top}, \quad A_2 = \begin{pmatrix} -x_2 & x_1 \end{pmatrix}, \quad A_n = \begin{pmatrix} A_{n-1} & 0 \\ -x_n \mathrm{I} & B_{n-1} \end{pmatrix}, \tag{14.16}$$

which can be verified immediately. It follows from Theorem 14.48 that the automorphism group I Aut M_n is isomorphic to the group of all invertible matrices of the form $I + QA_n$ where A_n is defined by (14.16) whereas Q is an $n \times m$ matrix with entries from the polynomial ring $\mathbb{F}[x_l, \dots, x_n]$. In the particular case of $n = 2$ with $Q = \binom{q_1}{q_2}$, $A_2 = \begin{pmatrix} -x_2 & x_1 \end{pmatrix}$, we have

$$I_2 + QA_2 = \begin{pmatrix} 1 - q_1 x_2 & q_1 x_1 \\ -q_1 x_1 & 1 + q_2 x_1 \end{pmatrix}.$$

The latter matrix is invertible if and only if its determinant is invertible in $\mathbb{F}[x_l, x_2]$. Now

$$\det(I_2 + QA_2) = 1 - q_l x_2 + q_2 x_l.$$

Since $q_l x_2 + q_2 x_l$ has constant term 0, we have $q_l x_2 = q_2 x_l$. If $\mathbb{F}$ is factorial, it is easy to derive from this the existence of q with $q_l = q x_1$, $q_2 = q x_2$, so that our matrix takes the form

$$I_2 + q \begin{pmatrix} -x_1 x_2 & x_1^2 \\ -x_2^2 & x_2 x_1 \end{pmatrix}.$$

But there is an automorphism whose associated matrix has the above form. It is equal to $e^{\operatorname{ad} q[x_2,x_1]}$. Indeed,

$$e^{\operatorname{ad} q[x_2,x_1]}(x_1) = x_1 + \big[q[x_2,x_1], x_1\big] = x_1 + q x_1 [x_1, x_2],$$
$$e^{\operatorname{ad} q[x_2,x_1]}(x_2) = x_2 + q x_2 [x_1, x_2].$$

Then $Q = \binom{qx_1}{qx_2}$ and hence our automorphism is the same as the initial one. Thus we have the second part of the result to follow, which is due to A. L. Shmelkin and dates back to 1973. As for the first part, it is immediate since any mapping of free generators of a relatively free Lie algebra into itself extends to a homomorphism.

Theorem 14.49. *For any $n > 1$, we have*

$$\operatorname{Aut} M_n \cong \mathrm{GL}(n, \mathbb{F}) \rtimes \mathrm{I}\operatorname{Aut} M_n. \tag{14.17}$$

For $n = 2$, it follows that

$$\operatorname{Aut} M_2 \cong \mathrm{GL}(2, \mathbb{F}) \rtimes e^{\operatorname{ad} M_2^2}. \tag{14.18}$$

Here $\mathbb{F}$ is an arbitrary commutative and associative ring of coefficients with unity. In formula (14.18), $\mathbb{F}$ is assumed to be factorial.

Since $e^{\operatorname{ad} M_2^2} \cong (\mathbb{F}[x_1, x_2], +)$, we can also write

$$\operatorname{Aut} M_2 \cong \mathrm{GL}(2, \mathbb{F}) \rtimes \mathbb{F}[x_1, x_2]. \tag{14.19}$$

The structure of $\mathrm{I}\operatorname{Aut} M_n$, $n > 2$, is far from being as clear as in (14.19). But the subgroup $e^{\operatorname{ad} M_n^2}$ still exists and it can be described in the following way.

Proposition 14.50. *Let $\mathbb{F}$ be a ring without zero divisors. A matrix $P = I_n + QA_n$ is a matrix of an inner automorphism induced by an element of M_2^2 (i. e., an element in $e^{\operatorname{ad} M_n^2}$) if and only if it belongs to the right stabilizer of A_n, that is, $A_n P = A_n$.*

Proof. This is a direct calculation, we leave it to the reader. □

14.5.2.3 Wild and tame automorphisms of free metabelian algebras

In the case $n = 2$, we know that $\operatorname{Aut} L_2 = \mathrm{GL}_2(\mathbb{F})$ and then, of course, all nonidentity elements of the group $\mathrm{I}\operatorname{Aut} L_2/R^2$ are nontame. For instance, any inner automorphism $e^{\operatorname{ad} z}$, $0 \neq z \in M_2^2$, of the free metabelian algebra M_2 of rank 2 is nontame. The existence of wild automorphisms of M_n, $n \geq 3$, requires much more work.

We start our discussion with considering a Cohn's automorphism (ii) in Theorem 14.36, for which f has zero linear part. We write therefore the matrix in $\mathrm{GL}(n, \mathbb{F}[x_1, \dots, x_n])$ corresponding to (ii) in the form

$$I_n + efA_n, \quad \text{where } e = (1 \quad 0 \quad \dots \quad 0)^\top, f = (0 \quad \dots \quad f_{23} \quad \dots \quad f_{n-1,n}). \tag{14.20}$$

It is easy to observe that the determinant of this matrix equals 1. (The first column of A_n does contain nonzero elements $x_l, \dots, x_n$ but exactly at the places where $(0 \quad \dots \quad f_{23} \quad \dots \quad f_{n-1,n})$ has $f_{12}, \dots, f_{1n}$, the latter being zero by the definition of the tame automorphism.)

Now each element of $\mathrm{I}\operatorname{Aut} M_n$, which is tame, is equal to the product of the elements which are the conjugates of various (14.20) by all possible β induced by the linear transformations of $\langle x_1, \dots, x_n \rangle$. The latter automorphisms, in fact, generate a subgroup of $\operatorname{Aut} M_n$ isomorphic to $\mathrm{GL}(n, \mathbb{F})$. A characteristic property of (14.20) is that this matrix has determinant 1 and it is representable in the form $I + \Phi\Psi$ where Φ, replaces e as a column of height n, and Ψ replaces fA_n as a row of length n. We want to show that each tame element has the form

$$(I_n + \Phi_1\Psi_1)(I_n + \Phi_2\Psi_2)\cdots(I_n + \Phi_m\Psi_m). \tag{14.21}$$

To do this, we have to compute $\beta\alpha\beta^{-1}$, where $\beta \in \mathrm{GL}(n, \mathbb{F})$, $\alpha \in \mathrm{I\,Aut}\, M_n$. If B is a matrix of β acting on $\langle \overline{x}_1, \ldots, \overline{x}_n\rangle$, hence on $\mathbb{F}[\overline{\overline{x}}_1, \ldots, \overline{\overline{x}}_n]$, hence on the matrices over this latter algebra, and $I + QA_n$ is the matrix of α acting on the module generated by $a_1, \ldots, a_n$ then $\beta\alpha\beta^{-1}$ is $IE + Q'A_n$ where Q' can be computed using the procedure described in Section 14.5.2.1. If $d_{(l,p),(s,t)} = b_{ls}b_{pt} - b_{lt}b_{ps}$ are the entries of a square $n(n-1)/2$-matrix which we denote by D then quite tedious calculations show that

$$Q' = B^{-1}\beta(Q)D,$$

and the matrix associated with $\beta\alpha\beta^{-1}$ takes the form

$$I + B^{-1}\beta(Q)DA_n.$$

If α is tame and of the form (ii) then $Q = ef$. In this case we arrive at

$$I + \Phi\Psi, \quad \text{where } \Phi = B^{-1}\beta(e) = B^{-1}e \text{ and } \Psi = \beta(f)DA_n. \tag{14.22}$$

As for the determinant of this matrix, then we have

$$\widetilde{\beta\alpha\beta^{-1}}\mu = \mu\beta\alpha\beta^{-1} = \widetilde{\beta}\mu\alpha\beta^{-1} = \widetilde{\beta}\widetilde{\alpha}\mu\widetilde{\beta}^{-1} = \widetilde{\beta}\widetilde{\alpha}\widetilde{\beta}^{-1}\mu,$$

whence $\widetilde{\beta\alpha\beta^{-1}} = \widetilde{\beta}\widetilde{\alpha}\widetilde{\beta}^{-1}$, so that the determinant does not change.

We start with a simple result.

Proposition 14.51. *Let v be a nilpotent element in a commutative and associative ring $\mathbb{F}$ with 1. Then the mapping*

$$x_1 \mapsto x_1 + v[x_1, x_2], \quad x_2 \mapsto x_2, \ldots, x_n \mapsto x_n$$

induces an automorphism in $\mathrm{I\,Aut}\, M_n$ *which is not tame. Its matrix takes the form*

$$I + e\,(v \quad 0 \quad \ldots \quad 0)\,A_n = \begin{pmatrix} 1 + v\overline{\overline{x}}_2 & -\overline{\overline{x}}_1 & 0 & \ldots & 0 \\ 0 & 1 & 0 & \ldots & 0 \\ \vdots & \vdots & \ddots & \ldots & \vdots \\ 0 & 0 & 0 & \ldots & 1 \end{pmatrix}.$$

This matrix has determinant $1 + v\overline{\overline{x}}_2$ which is an invertible element different from 1. That means that we indeed have an automorphism, a non-tame one.

In the case of an integral domain $\mathbb{F}$, the following general result holds.

Theorem 14.52 (Y. Bahturin and S. Nabiev). *Let $\mathbb{F}$ be an integral domain, M_n a free metabelian Lie algebra over $\mathbb{F}$ of rank n. Then any inner automorphism $e^{\operatorname{ad} z}$, $z \in M_n^2$, is not tame.*

Proof. Notice that for the inverse of (14.22) one has $(I + \Phi\Psi)^{-1} = I - \Phi\Psi$ which follows directly from the definition of the tame automorphism (ii). Also, $\Psi\Phi = 0$ since

$$I = (I + \Phi\Psi)(I - \Phi\Psi) = I - \Phi\Psi\Phi\Psi = I - \lambda\Phi\Psi,$$

where $\lambda \in \mathbb{F}[x_1, \dots, x_n]$. From this, our claim is immediate even if $\Phi\Psi = 0$. Now if we assume that $e^{\operatorname{ad} z}$ is tame for some $z \in M_n^2$ then, by Proposition 14.50, we would have an equation of the form

$$A_n(I + \Phi_1\Psi_1)(I + \Phi_2\Psi_2) \cdots (I + \Phi_m\Psi_m) = A_n. \tag{14.23}$$

By the above remarks, we can rewrite (14.21) in the form

$$I + \sum_{i=1}^{k} \Phi_k\Psi_k + \sum_{1 \le i_1 < \cdots < i_s \le m} \lambda_{i_1i_2}\lambda_{i_2i_3} \cdots \lambda_{i_{s-1}i_s}\Phi_{i_1}\Psi_{i_s}, \tag{14.24}$$

with $\Phi_k\Psi_l = \lambda_{kl}$. If we denote $\beta(f)D$ in (14.22) by $\overline{\Psi}$ then $\Psi_k = \overline{\Psi}_k A_n$, for each $k = 1, \dots, m$. We rewrite (14.23) with the help of (14.24) in the form

$$A_n\left(\sum_{k=1}^{m} \Phi_k\Psi_k + \sum_{\substack{1 \le i_1 < \cdots < i_s \le m \\ s \ge 2}} \lambda_{i_1i_2}\lambda_{i_2i_3} \cdots \lambda_{i_{s-1}i_s}\Phi_{i_1}\Psi_{i_s}\right) = 0. \tag{14.25}$$

Our next aim is to show that all λ_{kl} under these conditions are zero. To do this, we multiply (14.25) on the left by $\overline{\Psi}_l$. Then it is easy to see that

$$\sum_{k=1}^{m} \lambda_{1k}\Phi_k + \sum_{\substack{1 \le i_1 < \cdots < i_s \le m, \\ s \ge 2}} \lambda_{i_1i_2}\lambda_{i_2i_3} \cdots \lambda_{i_{s-1}i_s}\Psi_{i_s} = 0,$$

or, by changing notation,

$$\sum_{2 \le i_2 < \cdots < i_s \le m} \lambda_{1i_2}\lambda_{i_2i_3} \cdots \lambda_{i_{s-1}i_s}\Psi_{i_s} = 0. \tag{14.26}$$

Now we rewrite our matrix in the form

$$\begin{aligned}&(I + \Phi_1\Psi_1)(I + \Phi_2\Psi_2)(I - \Phi_1\Psi_1)(I + \Phi_1\Psi_1)(I + \Phi_3\Psi_3) \cdots (I + \Phi_m\Psi_m)\\ &\quad = (I + (\Phi_2 + \lambda_{12}\Phi_1)(\Psi_2 - \lambda_{21}\Psi_1))(I + \Phi_1\Psi_1)(I + \Phi_3\Psi_3) \cdots (I + \Phi_m\Psi_m).\end{aligned}$$

Then the relation (14.26) for the latter product satisfying the same condition as (14.21) takes the form

$$\sum_{3\le i_2<\cdots<i_s\le m} \lambda_{2i_2}\lambda_{i_2i_3}\cdots\lambda_{i_{s-1}i_s}\Psi_{i_s} + \sum_{3\le i_3<\cdots<i_s\le m} \lambda_{21}\lambda_{1i_3}\lambda_{i_3i_4}\cdots\lambda_{i_{s-1}i_s}\Psi_{i_s} - \sum_{3\le i_2<\cdots<i_s\le m} \lambda_{21}\lambda_{1i_2}\lambda_{i_2i_3}\cdots\lambda_{i_{s-1}i_s}\Psi_{i_s} = 0,$$

whence

$$\sum_{3\le i_2<\cdots<i_s\le m} \lambda_{2i_2}\lambda_{i_2i_3}\cdots\lambda_{i_{s-1}i_s}\Psi_{i_s} = 0. \tag{14.27}$$

Similarly, rewriting (14.21) in the form

$$(I+\Phi_1\Psi_1)(I+(\Phi_3+\lambda_{23}\Phi_2)(\Psi_3-\lambda_{32}\Psi_2)) \cdot (I+\Phi_2\Psi_2)(I+\Phi_4\Psi_4)\cdots(I+\Phi_m\Psi_m)$$

and working as just above, we can write

$$\sum_{4\le i_2<\cdots<i_s\le m} \lambda_{3i_2}\lambda_{i_2i_3}\cdots\lambda_{i_{s-1}i_s}\Psi_{i_s} = 0. \tag{14.28}$$

Continuing in the same way, we finally get

$$\sum_{1\le i_1<\cdots<i_s\le m} \lambda_{i_1i_2}\lambda_{i_2i_3}\cdots\lambda_{i_{s-1}i_s}\Phi_{i_1}\Psi_{i_s} = \sum_{k=1}^{m-1}\Phi_k\Big(\sum_{k\le i_2<\cdots<i_s\le m}\lambda_{ki_2}\lambda_{i_2i_3}\cdots\lambda_{i_{s-1}i_s}\Psi_{i_s}\Big) = 0.$$

Then (14.21) is equal to $I+\sum_k \Phi_k\Psi_k$. This shows that, in fact, the factors in (14.21) commute and

$$A_n\Big(\sum_k \Phi_k\Psi_k\Big) = 0,$$

so that (see Proposition 14.50)

$$\sum_k \Phi_k\Psi_k = B_n\Phi. \tag{14.29}$$

It follows from (14.26)–(14.28) that we have

$$\lambda_{m-1,m}\Psi_m = 0. \tag{14.30}$$

It follows from (14.30) that $\lambda_{m-1,m} = 0$, i. e., $\Psi_{m-1}\Phi_m = 0$. Since the factors in (14.30) commute, we conclude that $\Psi_k\Phi_l = 0$ for all k, l. Now we have $\Psi_1\Phi_1 = \Psi_1\Phi_2 = \cdots = \Psi_1\Phi_m = 0$. If the dimension of $\langle\Phi_1, \dots, \Phi_m\rangle$ is n (recall that each Φ_i is an n-tuple) then it is immediate that $\Psi_1 = 0$, which is not the case. Thus we have no more than $n-1$ of $\Phi_1, \dots, \Phi_m$ linearly independent. Suppose $\Phi_{i_1}, \dots, \Phi_{i_p}$ form a basis. Then it follows from (14.29) that there exist polynomials $f, f_l, \dots, f_p, f \neq 0$, such that

$$B_n f + \Phi_{i_1} f_1 + \cdots + \Phi_{i_p} f_p = 0.$$

In the matrix form, we get

$$\begin{pmatrix} B_n & \Phi_{i_1} & \dots & \Phi_{i_p} \end{pmatrix} \begin{pmatrix} f \\ f_1 \\ \vdots \\ f_p \end{pmatrix} = 0. \tag{14.31}$$

It follows from (14.31) that all minors of order $p+1$ in

$$\begin{pmatrix} x_1 & a_{1i_1} & \cdots & a_{1i_p} \\ \cdots & \cdots & \cdots & \cdots \\ x_n & a_{ni_1} & \cdots & a_{ni_p} \end{pmatrix},$$

where

$$\Phi_{i_j} = \begin{pmatrix} a_{1i_j} & \cdots & a_{ni_j} \end{pmatrix}^{\top}, \quad j = 1, \dots, p, \ a_{si_j} \in \mathbb{F}$$

are zero. It is obvious that then all minors of order p in

$$\begin{pmatrix} a_{1i_1} & \dots & a_{1i_p} \\ \vdots & \ddots & \vdots \\ a_{ni_1} & \dots & a_{ni_p} \end{pmatrix}$$

are zero, contradicting the linear independence of $\Phi_{i_1}, \dots, \Phi_{i_p}$. Therefore, $p = 0$ and (14.21) is trivial. The proof is complete. □

One of the consequences of the techniques suggested above is the following.

Theorem 14.53. *Let $M_{n,c}$ be the free metabelian and nilpotent of class c Lie algebra of rank n over a commutative ring $\mathbb{F}$. Then*

$$\operatorname{Aut} M_{n,c} \cong \mathrm{GL}(n, \mathbb{F}) \rtimes \mathrm{I}\operatorname{Aut} M_{n,c},$$

where $\mathrm{I}\operatorname{Aut} M_{n,c}$ is isomorphic to the group of all matrices of the form $I + QA_n$ with all entries in $\mathbb{F}[x_1, \dots, x_n]/\Delta^c$. The form of A_n is given in Section 14.5.2.2 *and Δ is the ideal of polynomials in $\mathbb{F}[x_1, \dots, x_n]$ with constant term zero.*

14.5.2.4 Algebras of derivation

Our approach via wreath product enables us to obtain analogues of the results of the preceding sections, in the case of derivation algebras. We give necessary definitions and formulations without proofs since these follow the same profile as previously. Note that derivations of free Lie algebras have been discussed in Section 2.5.5.

Given an ideal R in a free Lie algebra $L = L(x_1, \dots, x_n)$ over $\mathbb{F}$ as above such that L/R is a free $\mathbb{F}$-module, we again consider the canonical embedding μ of L/R^2 in the wreath product $W = B \oplus F$. By $\overline{\mathrm{Der}\,W}$ we denote the subalgebra in $\mathrm{Der}\,W$ whose elements leave $B = L/R$ and F invariant. If $\overline{\mathrm{Der}\,L/R^2}$ stands for the set of all derivations of $M = L/R^2$ which leave R invariant then with each derivation $\delta \in \mathrm{Der}\,L/R^2$ one can uniquely associate $\overline{\delta} \in \overline{\mathrm{Der}\,W}$ in the same way as in Proposition 14.47. If we define $\mathrm{I\,Der}\,W$ as the subset in $\mathrm{Der}\,W$ mapping B to zero and by $\mathrm{I\,Der}\,L/R^2$ the set of derivations in $\overline{\mathrm{Der}\,L/R^2}$ mapping M into R then, under the mapping $\delta \mapsto \overline{\delta}$ mentioned just above, it follows from $\delta \in I\,\mathrm{Der}\,M$ that $\overline{\delta} \in I\,\mathrm{Der}\,W$.

Theorem 14.54. *If L/R is finitely related, free as an $\mathbb{F}$-module, and can be defined by m relations with $n = \mathrm{rank}\,L$ then $\mathrm{I\,Der}\,L/R^2$ is isomorphic to the algebra of all matrices of the form QA_n where Q is an arbitrary $n \times m$ matrix with coefficients in $U(L/R)$ and A_n as in Theorem 14.48 relative to the ordinary commutator of matrices.*

Theorem 14.55. *The derivation algebra of the free metabelian algebra of rank n over a ring $\mathbb{F}$ decomposes as the semidirect product of $\mathfrak{gl}(n, \mathbb{F})$, full matrix Lie algebra of order n over $\mathbb{F}$, and an ideal isomorphic to the algebra of all $n \times n$ matrices of the form QA_n where Q is an arbitrary $n \times n(n-1)/2$ matrix, A_n a fixed $n(n-1)/2 \times n$ matrix as in Section 14.5.2.2, both with entries from $\mathbb{F}[x_1, \dots, x_n]$.*

An interesting corollary is this.

Theorem 14.56. *The derivation algebra $\mathrm{Der}\,M_n$ of the free metabelian algebra of arbitrary rank $n > 1$ is not finitely generated.*

Bibliography

[1] Acciarri, C., Khukhro, E. I., Shumyatsky, P.: Profinite groups with an automorphism whose fixed points are right Engel. Proc. Am. Math. Soc. 147(9), 3691–3703 (2019)
[2] Adian, S. I., Razborov, A. A.: Periodical groups and Lie algebras. Usp. Mat. Nauk 42, 3–68 (1987). English translation: Russian Mathematical Surveys, 42, 3–68 (1987)
[3] Aljadeff, E., Giambruno, A., Karasik, Y.: Polynomial identities with involution, superinvolutions and the Grassmann envelope. Proc. Am. Math. Soc. 145, 1843–1857 (2017)
[4] Aljadeff, E., Haile, D.: Simple G-graded algebras and their polynomial identities. Trans. Am. Math. Soc. 366, 1749–1771 (2014)
[5] Amayo, R. K., Stewart, I. N.: Infinite-dimensional Lie algebras. Noordhood, Leyden (1974)
[6] Amitsur, S. A.: A generalization of Hilbert's Nullstellensatz. Proc. Am. Math. Soc. 8, 649–656 (1957)
[7] Andreev, K. K.: Nilpotent groups and Lie algebras. Algebra Log. 7, 4–14 (1968)
[8] Andreev, K. K.: Nilpotent groups and Lie algebras, II. Algebra Log. 8, 625–635 (1969). Letter to the Editor. Algebra i Logika 10, 226–228 (1971)
[9] Andreev, K. K., Shabel'nikova, D. G.: Characteristic subalgebras of relatively free algebras. Sib. Mat. Zh. 14, 1336–1337 (1973)
[10] Andrews, G. E.: The Theory of Partitions. Addison-Wesley (1976)
[11] Artamonov, V. A.: Chain varieties of linear algebras. Tr. Mosk. Mat. Obŝ. 29, 51–78 (1973)
[12] Artamonov, V. A.: Varieties of restricted Lie algebras. Sib. Mat. Zh. 15, 1197–1212 (1974)
[13] Artamonov, V. A.: On chain varieties of Lie algebras. Trans. Am. Math. Soc. 221, 323–338 (1976)
[14] Artamonov, V. A.: The categories of free metabelian groups and Lie algebras. Comment. Math. Univ. Carol. 18, 143–159 (1977)
[15] Artamonov, V. A.: Lattices of varieties of linear algebras. Usp. Mat. Nauk 33, 135–168 (1978)
[16] Artamonov, V. A.: Projective metabelian groups and Lie algebras. Izv. Akad. Nauk SSSR, Ser. Mat. 42, 135–168 (1978)
[17] Artin, E.: Geometric Algebra. Interscience, New York (1957)
[18] Bachmuth, S., Mochizuki, H. Y., Walkup, D. W.: A nonsolvable group of exponent 5. Bull. Am. Soc. 76, 638–640 (1970)
[19] Bahturin, Yu.A.: Two remarks on the varieties of Lie algebras. Mat. Zametki 4, 387–398 (1968)
[20] Bahturin, Yu.A.: The approximation of Lie algebras. Mat. Zametki 12, 713–716 (1972)
[21] Bahturin, Yu.A.: Identities in Lie algebras, I. Vestn. Mosk. Univ. Mat. Mech. 1, 12–18 (1973)
[22] Bahturin, Yu.A.: Identities in Lie algebras, II. Vestn. Mosk. Univ. Mat. Mech. 2, 26–33 (1973)
[23] Bahturin, Yu.A.: Identities in the universal envelopes of Lie algebras. J. Aust. Math. Soc. 12–18 (1974)
[24] Bahturin, Yu.A.: Identities in metabelian Lie algebras. Tr. Semin. Im. I.G. Petrovskogo 1, 12–18 (1975)
[25] Bahturin, Yu.A.: Simple Lie algebras satisfying a nontrivial identity. Serdica, Bulg. Math. Publi. 2, 241–246 (1976)
[26] Bahturin, Yu.A.: Lectures on Lie algebras. Akademie-Verlag, Berlin (1978)
[27] Bahturin, Yu.A.: A remark on irreducible representations of Lie algebras. J. Aust. Math. Soc. 27, 332–336 (1979)
[28] Bahturin, Yu.A.: On homomorphisms of soluble Lie algebras. J. Lond. Math. Soc. 20, 415–422 (1979)
[29] Bahturin, Yu.A.: On identical relations in free polynilpotent Lie algebras. J. Lond. Math. Soc. 20, 39–52 (1979)
[30] Bahturin, Yu.A.: Two-variables identities in $\mathfrak{sl}(2, \mathbf{C})$. Tr. Semin. Im. I.G. Petrovskogo 5, 205–208 (1979)

https://doi.org/10.1515/9783110566659-015

[31] Bahturin, Yu.A.: On Lie subalgebras of associative PI-algebras. J. Algebra 67, 257–271 (1980)
[32] Bahturin, Yu.A.: On special Lie algebras. Algebra, Izdat. Mosk. Univ. 26–30 (1980)
[33] Bahturin, Yu.A.: Special varieties of Lie algebras. Algebra Log. 20, 522–530 (1981)
[34] Bahturin, Yu.A.: Artinian special Lie algebras. Algebra, Izdat. Mosk. Univ. 24–26 (1982)
[35] Bakhturin, Yu.A.: The structure of a PI-envelope of a finite-dimensional Lie algebra. Izv. Vysš. Učebn. Zaved., Mat. 11, 60–62, 85 (1985). (Russian)
[36] Bakhturin, Yu.A.: Identities in Lie Algebras [in Russian]. Nauka, Moscow (1985)
[37] Bahturin, Yu.A.: Identical Relations in Lie Algebras. VNU Press, Utrecht (1987). 309 pp.
[38] Bahturin, Yu.A.: Basic Structures of Modern Algebra. Mathematics and Its Applications, vol. 265. Kluwer AP (1993). 422 pp.
[39] Bahturin, Y., Drensky, V.: Graded polynomial identities of matrices. Linear Algebra Appl. 357, 15–34 (2002)
[40] Bahturin, Y., Fischman, D., Montgomery, S.: On the generalized Lie structure of associative algebras. Isr. J. Math. 96, part A, 27–48 (1996)
[41] Bahturin, Yu.A., Giambruno, A., Riley, D.: Group-graded algebras with polynomial identity. Isr. J. Math. 104, 146–155 (1998)
[42] Bahturin, Yu.A., Linchenko, V.: Identities of algebras with actions of Hopf algebras. J. Algebra 202, 634–654 (1998)
[43] Bahturin, Yu.A., Kegel, O.: Universal sums of abelian subalgebras. Commun. Algebra 23, 2975–2990 (1995)
[44] Bahturin, Yu.A., Kostrikin, A. I.: II International conference of the foreign alumnae of Moscow University. Usp. Mat. Nauk 40, 231–235 (1985)
[45] Bahturin, Y., Linchenko, V.: Identities of algebras with actions of Hopf algebras. J. Algebra 202, 634–654 (1998)
[46] Bahturin, Y., Mikhalev, A. A., Petrogradsky, V. M., Zaicev, M. V.: Infinite-dimensional Lie superalgebras. De Gruyter Expositions in Mathematics, vol. 7. Walter De Gruyter, Berlin (1992). 250 pp.
[47] Bahturin, Y., Mishchenko, S., Regev, A.: On the Lie and associative codimensions growth. Commun. Algebra 27(10), 4901–4908 (1999)
[48] Bahturin, Y., Montgomery, S.: PI-envelopes of Lie superalgebras. Proc. Am. Math. Soc. 127, 2829–2839 (1999)
[49] Bahturin, Y., Nabiyev, S.: Automorphisms and derivations of abelian extensions of some Lie algebras. Abh. Math. Semin. Univ. Hamb. 62, 43–57 (1992)
[50] Bahturin, Yu. A., Olshansky, A. Yu.: Identical relations in finite Lie rings. Mat. Sb. 96, 543–559 (1975)
[51] Bahturin, Yu. A., Olshansky, A. Yu.: On approximation and characteristic subalgebras of free Lie algebras. Tr. Semin. Im. I.G. Petrovskogo 2, 145–150 (1976)
[52] Bahturin, Yu. A., Olshansky, A. Yu.: Soluble just-non-Cross varieties of Lie rings. Mat. Sb. 100, 384–399 (1976)
[53] Bahturin, Yu. A., Olshanskii, A. Yu.: Large restricted Lie algebras. J. Algebra 310, 413–427 (2007)
[54] Bahturin, Y., Olshanskii, A.: Actions of maximal growth. Proc. Lond. Math. Soc. (3) 101, 27–72 (2010)
[55] Bahturin, Y., Olshanskii, A.: Filtrations and distortion in infinite-dimensional algebras. J. Algebra 327, 251–291 (2011)
[56] Bahturin, Y., Olshanskii, A.: Growth of subalgebras and subideals in free Lie algebras. J. Algebra 422, 277–305 (2015)
[57] Bahturin, Y., Shpilrain, V.: On Lie algebras with wild automorphisms. Results Math. 28, 209–213 (1995)

[58] Bahturin, Yu.A., Strade, H.: Locally finite-dimensional simple Lie algebras. Mat. Sb. 185, 2–31 (1994)

[59] Bahturin, Y., Witherspoon, S.: Delta sets and polynomial identities in pointed Hopf algebras, Algebras and Representation Theory. submitted

[60] Bahturin, Yu.A., Zaicev, M. V.: Identities of graded algebras. J. Algebra 205, 1–12 (1998)

[61] Bahturin, Yu.A., Yasumura, Felipe Yukihide: Graded polynomial identities as identities of universal algebras. Linear Algebra Appl. 562, 1–14 (2019)

[62] Bahturin, Yu.A., Yasumura, Felipe Yukihide: Distinguishing simple algebras by means of polynomial identities.. São Paulo J. Math. Sci. 13, 39–72 (2019)

[63] Barnes, D. W.: The Frattini argument for Lie algebras. Math. Z. 133, 277–283 (1973)

[64] Barnes, D. W., Newell, M. L.: Some theorems on saturated homomorphs of soluble Lie algebras. Math. Z. 115, 179–187 (1970)

[65] Bartholdi, L.: Self-similar Lie algebras. J. Eur. Math. Soc. 17(12), 3113–3151 (2015)

[66] Baumslag, G.: Lecture Notes on Nilpotent Groups. CBMS Reg. Conf. Ser. Math. AMS, vol. 2 (1971)

[67] Benediktovich, I. I., Zalesskii, A. E.: *T*-ideals of free Lie algebra with polynomial growth of the codimension sequence. Vescì Akad. Navuk BSSR 3, 5–10 (1980)

[68] Berele, A.: Cocharacter sequences for algebras with Hopf algebra actions. J. Algebra 185, 869–885 (1996)

[69] Berele, A., Regev, A.: On the codimensions of the verbally prime P.I. algebras. Isr. J. Math. 91, 239–247 (1995)

[70] Berezin, F. A.: Introduction to Algebra and Analysis with Anti-commuting Variables. Izdat. Mosk Univ., Moscow (1983)

[71] Bergen, G., Cohen, M.: Actions of commutative Hopf algebras. Bull. Lond. Math. Soc. 18, 159–164 (1986)

[72] Bianchi, A., Diniz, D.: Identities and isomorphisms of finite-dimensional graded simple algebras. J. Algebra 526, 333–344 (2019)

[73] Billig, Y.: On the homomorphic image of a special Lie algebra. Mat. Sb. 136, 320–323 (1988). Engl. Transl.: Yu. V. Billing, "On a homomorphic image of a special Lie algebra", Math. USSR-Sb., 64(2), 319–322 (1989)

[74] Birkhoff, G.: Lattice Theory, vol. XXV. AMS Colloq. Publi. (1973)

[75] Block, R. E.: Differentiably simple algebras. Bull. Am. Math. Soc. 74(6), 1086–1090 (1968)

[76] Bokut', L. A.: A basis for free polynilpotent Lie algebras. Algebra Log. 2, 13–19 (1963)

[77] Bokut', L. A.: Unsolvability of the problems of equation and subalgebra in finitely presented Lie algebras. Izv. Akad. Nauk SSSR, Ser. Mat. 36, 1173–1219 (1972)

[78] Bokut', L. A., Kukin, G. P.: Algorithmic and Combinatorial Algebra. Mathematics and Its Applications, vol. 255. Kluwer AP (1994). 472 pp.

[79] Bokut, L. A., Latyshev, V., Shestakov, I., Zelmanov, E. (eds.): Selected Works of A.I. Shirshov, vol. VIII. Birkhäuser Mathematics (2009). 250 pp. Translated by Bremner, Murray and Kotchetov, Mikhail V.

[80] Bourbaki, N.: Algèbre, Ch. 7–9. Hermann, Paris (1958)

[81] Bourbaki, N.: Groupes et Algèbres de Lie, Ch. 1. Hermann, Paris (1960)

[82] Bourbaki, N.: Groupes et Algèbres de Lie, Ch. 4–6. Hermann, Paris (1968).

[83] Bourbaki, N.: Groupes et Algèbres de Lie, Ch. 7–8. Hermann, Paris (1975)

[84] Bourbaki, N.: Groupes et algèbres de Lie, Chapitres 2 et 3, Hermann, (1972) (rèimpr. 2006), 1re èd., 314 pp.

[85] Braun, A.: Lie rings and the Engel condition. J. Algebra 31, 287–292 (1974)

[86] Braun, A.: The radical in a finitely generated P.I.-algebra. Bull. Am. Math. Soc. 7, 385–386 (1982)

[87] Bryant, R. M., Drensky, V.: Dense subgroups of the automorphism groups of free algebras. Can. J. Math. 45, 1135–1154 (1993)
[88] Bryant, R. M., Vaughan-Lee, M. R.: Soluble varieties of Lie algebras. Q. J. Math. 23, 107–112 (1972)
[89] Cohen, D. E.: On the laws of a metabelian variety. J. Algebra 5, 267–273 (1967)
[90] Cohn, P. M.: A non-nilpotent Lie ring satisfying the Engel condition and a non-nilpotent Engel group. Proc. Camb. Phil. Soc. Math. Phys. Sci. 51, 401–405 (1955)
[91] Cohn, P. M.: On the embeddings of rings in skew fields. Proc. Lond. Math. Soc. 11, 511–530 (1961)
[92] Cohn, P. M.: Subalgebras or free associative algebras. Proc. Lond. Math. Soc. 14, 618–632 (1964)
[93] Cohn, P. M.: Universal Algebra. Harper & Row, New York (1965)
[94] Curtis, C. W.: Noncommutative extensions of Hilbert rings. Proc. Am. Math. Soc. 4, 945–955 (1953)
[95] Curtis, C. W., Reiner, I.: Representation Theory of Finite Groups and Associative Algeþras. Wiley-Interscience, New York (1962)
[96] de Morais Costa, O. A., Petrogradsky, V.: Fractal just infinite nil Lie superalgebra of finite width. J. Algebra 504, 291–335 (2018)
[97] Di Vincenzo, O., Koshlukov, P., Valenti, A.: Gradings on the algebra of upper triangular matrices and their graded identities. J. Algebra 275, 550–566 (2004)
[98] Di Vincenzo, O., Koshlukov, P.: On the ∗-polynomial identities of $M_{1,1}(E)$. J. Pure Appl. Algebra 215, 262–275 (2011)
[99] Di Vincenzo, O., Koshlukov, P., La Scala, R.: Involutions for upper triangular matrix algebras. Adv. Appl. Math. 37(4), 541–568 (2006)
[100] Di Vincenzo, O., Spinelli, E.: Graded polynomial identities on upper block triangular matrix algebras. J. Algebra 415, 50–64 (2014)
[101] Dixmier, J.: Algèbres Enveloppantes. Gauthier-Villars, Paris (1974)
[102] Dniester Notebook: Unsolved Problems of the Theory of Rings and Modules, 2nd edn. Novosibirsk (1976)
[103] Drensky, V. S.: Identities in Lie algebras. Algebra Log. 13, 265–290 (1974)
[104] Drensky, V. S.: Soluble varieties of Lie algebras. Ph.D. Thesis, Moscow University (1979)
[105] Drensky, V. S.: Codimensions of T-ideals and Hilbert series of relatively free algebras. C. R. Bulg. Acad. Sci. 34, 1201–1204 (1981)
[106] Drensky, V. S.: Identities in matrix Lie algebras. Tr. Semin. Im. I.G. Petrovskogo 6, 47–55 (1981)
[107] Drensky, V. S.: Representation of the symmetric group and varieties of linear algebras. Mat. Sb. 115, 98–115 (1981)
[108] Drensky, V. S.: Automorphisms of relatively free algebras. Commun. Algebra 18(l2), 4323–4351 (1990)
[109] Drensky, V.: Endomorphisms and automorphisms of relatively free algebras. Supplemento ai Rendiconti del Circolo Matematico di Palermo 31, 97–132 (1993)
[110] Drensky, V.: Automorphisms of polynomial, free and generic matrix algebras. In: Dlaþ, V., Márki, L. (eds.) Trends in Ring Theory, Proc. Conf. Miskolc, 1996. CMS Conference Proceedings, vol. 22, pp. 13–26. AMS, Providence (1998)
[111] Drensky, V. S., Gupta, C. K.: Automorphisms of free nilpotent Lie algebras. Can. J. Math. 11, 259–279 (1990)
[112] Drensky, V., Racine, M.: Distinguishing simple Jordan algebras by means of polynomial identities. Commun. Algebra 20, 309–327 (1992)
[113] Eĭdel'kind, D. I.: Exact representations of relatively free groups. Algebra Log. 10, 449–473 (1971)

[114] Eĭdel'kind, D. I.: Magnus groups. Mat. Sb. 92, 209–223 (1973)
[115] Elduque, A., Kochetov, M.: Gradings on Simple Lie Algebras. Mathematical Surveys and Monographs, vol. 189. American Mathematical Society (2013)
[116] Fox, R. H.: Free differential calculus. I. Derivation in the free group ring. Ann. Math. 57, 547–560 (1953)
[117] Freitas, J. A., Koshlukov, P., Krasilnikov, A.: Z-graded identities of the Lie algebra W_1. J. Algebra 427, 226–251 (2015)
[118] Genov, G. K.: On verbal ideals of free Lie algebras. Godišn. Sofiĭsk. Univ. Mat. Fak. (1971–72) 66, 177–189 (1974)
[119] Giambruno, A., Milies, C., Valenti, A.: Star-polynomial identities: computing the exponential growth of the codimensions. J. Algebra 469, 302–322 (2017)
[120] Giambruno, A., Ioppolo, A., La Mattina, D.: Varieties of algebras with superinvolution of almost polynomial growth. Algebr. Represent. Theory 19, 599–611 (2016)
[121] Giambruno, A., Mishchenko, S., Zaicev, M.: Algebras with intermediate growth of the codimensions. Adv. Appl. Math. 37, 360–377 (2006)
[122] Giambruno, A., Regev, A., Zaicev, M.: Simple and semi simple Lie algebras and codimension growth. Trans. Am. Math. Soc. 352, 1935–1946 (2000)
[123] Giambruno, A., Zaicev, M.: Polynomial Identities and Asymptotic Methods. AMS Math. Surv. Monogr., vol. 123, (2005) xiii+352 pp.
[124] Giambruno, A., Zaicev, M.: On codimension growth of finite-dimensional Lie superalgebras. J. Lond. Math. Soc. (2) 85, 534–548 (2012)
[125] Giambruno, A., Zaicev, M.: Non integrality of the PI-exponent of special Lie algebras. Adv. Appl. Math. 51, 619–634 (2013)
[126] Gilmore, R.: Lie Groups, Lie Algebras and Some of Their Applications. Wiley, New York (1974)
[127] Golod, E. S.: On nil-algebras and finitely approximable p-groups. Izv. Akad. Nauk SSSR, Ser. Mat. 28, 273–276 (1964)
[128] Gordienko, A.: Amitsur's Conjecture for associative algebras with a generalized Hopf action. J. Pure Appl. Algebra 217, 1395–1411 (2013)
[129] Gordienko, A., Kochetov, M.: Derivations, gradings, actions of algebraic groups, and codimension growth of polynomial identities. Algebr. Represent. Theory 17(2), 539–563 (2014)
[130] Goto, M., Grosshans, F.: Semisimple Lie Algebras. Lecture Notes in Pure and Applied Mathematics, vol. 38. Marcel Dekker, New York (1978)
[131] Grigorchuk, R. I.: On the Burnside problem for periodic groups. Funkc. Anal. Prilozh. 14, 53–54 (1980)
[132] Grishkov, A. N.: On the growth of varieties of Lie algebras. Mat. Zametki 44, 51–54 (1988). Engl. transl.: Math. Notes, 44(1–2), 515–517 (1988)
[133] Gupta, C. K., Krasilnikov, A. N.: A simple example of a non-finitely based system of polynomial identities. Comm. Algebra 30(10), 4851–4866 (2002)
[134] Gupta, N., Sidki, S.: On the Burnside problem for periodic groups. Math. Z. 182(3), 385–388 (1983)
[135] Hall, M. Jr.: Coset representations in free groups. Trans. Am. Math. Soc. 67, 421–432 (1949)
[136] Hall, M.: A basis for free Lie algebras and higher commutators in free groups. Proc. Am. Math. Soc. 1, 575–581 (1950)
[137] Hall, M.: The Theory of Groups. Macmillan, New York (1959)
[138] Hall, M.: Combinatorial Theory. Blaisdell, London (1967)
[139] Hall, P.: A contribution to the theory of groups of prime power order. Proc. Lond. Math. Soc. 4, 29–95 (1932)
[140] Herstein, I. N.: Noncommutative Rings, Carus Monograph, vol. 15. Wiley, New York (1968)

[141] Higgins, P. J.: Lie rings satisfying the Engel condition. Proc. Camb. Phil. Soc. Math. Phys. Sci. 50, 8–15 (1954)
[142] Higman, G.: Ordering by divisibility in abstract algebras. Proc. Lond. Math. Soc. 2, 326–336 (1952)
[143] Higman, G.: Groups and rings having automorphisms without nontrivial fixed elements. J. Lond. Math. Soc. 32, 321–334 (1958)
[144] Hochschild, G. P.: Basic Theory of Algebraic Groups and Lie algebras. Graduate Texts in Mathematics, vol. 75. Springer-Verlag, New York (1981)
[145] Humphreys, J. E.: Introduction to Lie Algebras and Representation Theory. Graduate Texts in Mathematics, vol. 9. Springer-Verlag, New York (1972)
[146] Iltyakov, A. V.: On finite basis of identities of Lie algebra representations. Nova J. Algebra Geom. 1, 207–259 (1992)
[147] Ioppolo, A., La Mattina, D.: Polynomial codimension growth of algebras with involutions and superinvolutions. J. Algebra 472, 519–545 (2017)
[148] Ioppolo, A., Martino, F.: Superinvolutions on upper-triangular matrix algebras. J. Pure Appl. Algebra 222(8), 2022–2039 (2018)
[149] Ito, N.: Über das Produkt von zwei abelschen Gruppen. Math. Z. 63, 401–404 (1955)
[150] Jacobson, N.: Lie Algebras. Wiley-Interscience, New York (1962)
[151] Jacobson, N.: Structure of Rings. AMS Coll. Publ., vol. 37. AMS, Providence (1968)
[152] Jacobson, N.: PI-algebras. An Introduction. Lect. Notes Math., vol. 441 (1975)
[153] James, G.: The Representation Theory of the Symmetric Groups. Lect. Notes Math., vol. 682 (1978)
[154] Kac, V. G.: Simple graded Lie algebras of finite growth. Math. USSR, Izv. 2, 1271–1311 (1968)
[155] Kac, V.: Lie superalgebras. Adv. Math. 26, 8–96 (1977)
[156] Kac, V.: Infinite dimensional Lie algebras, 3rd edn. Cambridge University Press (1994). 424 pp.
[157] Kaloujnine, L., Krasner, M.: Produit complet des groupes de permutations et problème d'extension des groupes. Acta Math. Szeged 13, 208–230 (1951). 14, 39–66 (1951), 14, 69–82 (1951)
[158] Kaplansky, I.: Rings with a polynomial identity. Bull. Amer. Math. Soc 54, 575–580 (1948)
[159] Kaplansky, I.: Lie Algebras and Locally Compact Groups. University of Chicago Press, Chicago (1971)
[160] Kargapolov, M. I., Merzlayakov, Yu.I.: Fundamentals of the Theory of Groups. Graduate Texts in Mathematics, vol. 62. Springer-Verlag, New York (1979)
[161] Kegel, O. H.: Zur Nilpotenz gewisser assoziativer ringe. Math. Ann. 149, 258–260 (1963)
[162] Kemer, A. R.: Capelli identities and the nilpotence of the radical of finitely generated PI-algebra. Dokl. Akad. Nauk SSSR 245, 793–797 (1980)
[163] Kemer, A. R.: Ideals of Identities of Associative Algebras. Transl. of Monographs, vol. 87. Amer. Math. Soc. (1991) 81pp.
[164] Kerber, A.: Representations of Permutation Groups, I. Lect. Notes Math., vol. 240 (1971). II, vol. 495 (1975)
[165] Kharchenko, V. K.: Galois extensions and rings of fractions. Algebra Log. 13, 460–484 (1974)
[166] Kharchenko, V.: Galois extensions and rings of quotients. Algebra i Logika 4, 460–484, 488
[167] Klementyev, S. G., Petrogradsky, V. M.: Growth of solvable Lie superalgebras. Commun. Algebra 33(3), 865–895 (2005)
[168] Klementyev, S. G., Petrogradsky, V. M.: On growth of almost solvable Lie algebras. Usp. Mat. Nauk 60(5), 165–166 (2005). English translation in Russian Math. Surveys, 60(5), 970–972 (2005)
[169] Klementyev, S. G., Petrogradsky, V. M.: On growth of almost polynilpotent Lie algebras. In: Giambruno, A., Polcino Milies, C., Sehgal, S. K. (eds.) Groups, Rings and Group Rings. Contemp. Math., vol. 499, pp. 173–180. AMS, RI (2009)

[170] Klyačko, A. A.: A connection between the varieties of groups and Lie rings. Issled. po alg., Saratov, Izdat. Saratov. Univ. 4, 43–50 (1974)
[171] Klyačko, A. A.: Homogeneous Lie elements and varieties of nilpotent groups. Usp. Mat. Nauk 39, 171–172 (1974)
[172] Klyačko, A. A.: Lie elements in tensor algebra. Sib. Mat. Zh. 15, 1296–1304 (1974)
[173] Koshlukov, P.: Graded polynomial identities for the Lie algebra $sl_2(K)$. Int. J. Algebra Comput. 18, 825–836 (2008)
[174] Koshlukov, P., Zaicev: Identities and isomorphisms of graded simple algebras. Linear Algebra Appl. 432, 3141–3148 (2010)
[175] Kostrikin, A. I.: On the connection between periodic groups and Lie rings. Izv. Akad. Nauk SSSR, Ser. Mat. 21, 289–310 (1957)
[176] Kostrikin, A. I.: Lie rings satisfying the Engel condition. Izv. Akad. Nauk SSSR, Ser. Mat. 21, 515–540 (1957)
[177] Kostrikin, A. I.: The Burnside problem. Izv. Akad. Nauk SSSR, Ser. Mat. 23, 3–34 (1959)
[178] Kostrikin, A. I.: Introduction to Algebra, Universitext. Springer-Verlag, New York (1982)
[179] Kostrikin, A. I.: On Burnside's problem. Izv. Akad. Nauk SSSR, Ser. Mat. 23, 3–34 (1959)
[180] Kostrikin, A. I.: Lie algebras and finite groups. Tr. Mat. Inst. V.A. Steklova, Akad. Nauk SSSR 168, 132–154 (1984)
[181] Kostrikin, A. I.: Around Burnside, M. Nauka (1986) 232 pp. Engl. Transl.: Springer-Verlag, 1991, 216 pp.
[182] Kovács, L. G.: Varieties of nilpotent groups of small class. In: Proc. 18th SRI, Canberra, pp. 205–299 (1978)
[183] Kovács, L. G., Newman, M. F.: Cross varieties of groups. Proc. R. Soc., Ser. A 292, 530–536 (1966)
[184] Krasil'nikiov, A. N.: Finite basis property for some varieties of Lie algebras. Vestn. Mosk. Univ. Mat. Mech. 2, 34–38 (1982)
[185] Krasilnikov, A. N.: Identities of Lie algebras with nilpotent commutator ideal over a field of finite characteristic. Mat. Zametki 51, 47–52 (1992)
[186] Krause, G., Lenagan, T. H.: Growth of Algebras and Gelfand-Kirillov Dimension, Revised edn. Graduate Studies in Mathematics, vol. 22 (1999). 212 pp.
[187] Kruse, R. L.: Identities satisfied by a finite ring. J. Algebra 26, 298–318 (1973)
[188] Kukin, G. P.: Primitive elements of free Lie algebras. Algebra Log. 9, 458–472 (1970). (Russian)
[189] Kukin, G. P.: Subalgebras of free Lie p-algebras. Algebra Log. 11, 535–550 (1972)
[190] Kukin, G. P.: Subalgebras of free p-Lie algebras. Algebra Log. 11, 535–550 (1972)
[191] Kukin, G. P.: Free products of restricted Lie algebras. Mat. Sb. 95, 53–83 (1974)
[192] Kukin, G. P.: Intersection of subalgebras in free Lie algebras. Algebra Log. 16, 577–587 (1977)
[193] Kukin, G. P.: On the equation problem for Lie algebras. Sib. Mat. Zh. 18, 1194–1197 (1977)
[194] Kukin, G. P.: Algorithmic problems for soluble Lie algebras. Algebra Log. 17, 402–415 (1978)
[195] Kukin, G. P.: Bases of free Lie algebras. Mat. Zametki 24, 375–382 (1978)
[196] Kurosh, A. G.: The Theory of Groups, 2nd edn. Chelsea, New York (1960)
[197] Kushkulei, A., Razmyslov, Yu.: Varieties generated by irreducible representations of Lie algebras. Vestn. Mosk. Univ. Ser. I Mat. Mech. 5, 4–7 (1983). (Russian)
[198] Lambek, J.: Lectures on Rings and Modules. Blaisdell, London (1966)
[199] Lang, S.: Algebra. Addison-Wesley, Reading (1965)
[200] Latyshev, V. N.: Lie algebras with identical defining relations. Sib. Mat. Zh. 4, 821–829 (1963)
[201] Latyshev, V. N.: Two remarks on PI-algebras. Sib. Mat. Zh. 4, 1120–1121 (1963)
[202] Latyšev, V. N.: Regev's Theorem on identities of tensor products of PI-algebras. Usp. Mat. Nauk 27, 213–214 (1972)
[203] Lazard, M.: Sur les groupes nilpotents et les anneaux de Lie. Ann. Sci. Éc. Norm. Supér. 71, 101–190 (1954)

[204] Levitzki, J.: On a problem of A. Kurosch. Bull. Amer. Math. Soc 52, 1033–1035 (1946)
[205] Lin, K.: Quantum central extensions, C. R. Math. Acad. Sci. Canada 13(14), 135–140 (1991)
[206] Linchenko, V. V.: Identities of Lie algebras with actions of Hopf algebras. Commun. Algebra 25, 3179–3187 (1997)
[207] L'vov, I. V.: Varieties of associative rings. I. Algebra Log. 12, 269–297 (1973). II, 12, 667–688
[208] L'vov, I. V.: Finite-dimensional algebras with infinite bases for identical relations. Sib. Mat. Zh. 19, 91–99 (1978)
[209] L'vov, I. V.: Varieties generated by finite alternative rings. Algebra Log. 17, 282–286 (1978)
[210] Lyndon, R. C., Schupp, P.: Combinatorial Group Theory. Ergeb. d. Math. u. ihrer Grenzgeb., vol. 89. Springer-Verlag, New York (1977)
[211] Lysenok, I. G.: A system of defining relations for a Grigorchuk group. Math. Notes 38, 784–792 (1985). Translation from Mat. Zametki 38(4), 503–516 (1985)
[212] Macdonald, S. O.: Various varieties. J. Aust. Math. Soc. 16, 363–367 (1973)
[213] Magnus, W.: Beziehungen zwischen Gruppen und Idealen in einen speziellen Ring. Math. Ann. 111, 259–280 (1935)
[214] Magnus, W.: Über Beziehungen zwischen höheren Kommutatoren. J. Crelle 177, 105–115 (1937)
[215] Magnus, W., Karrass, A., Solitar, D.: Combinatorial Group Theory. Wiley, New York (1966)
[216] Malcev, A. I.: Generalized nilpotent algebras and their adjoint groups. Mat. Sb. 25, 347–366 (1949)
[217] Malcev, A. I.: Nilpotent torsion free groups. Izv. Akad. Nauk SSSR, Ser. Mat. 13, 201–212 (1949)
[218] Malcev, A. I.: On algebras with identical defining relations. Mat. Sb. 26, 19–33 (1950)
[219] Malcev, A. I.: Algebraic Systems. Grundl. d. Math. Wiss., vol. 192. Springer-Verlag, New York (1973)
[220] Mal'cev, Yu.N., Parfenov, V. A.: An example of a nonassociative algebra admiting no finite basis for identities. Sib. Mat. Zh. 18, 1420–1421 (1977)
[221] Martínez, C., Zelmanov, E.: Nil algebras and unipotent groups of finite width. Adv. Math. 147, 328–344 (1999)
[222] Mathieu, O.: Classification of simple graded Lie algebras of finite growth. Invent. Math. 108, 455–519 (1992)
[223] Medvedev, Yu.A.: Identities of finite Jordan algebras. Algebra Log. 18, 723–748 (1979)
[224] Medvedev, Yu.A.: Cross varieties of algebras. Mat. Sb. 115, 391–425 (1980)
[225] Mikhalev, A. A.: Free color Lie superalgebras Dokl. Akad. Nauk SSSR 286, 551–554 (1986). (Russian). Engl. transl., Soviet. Math. Dokl., 33, 136–139 (1986)
[226] Mikhalev, A. A.: Subalgebras of free Lie p-superalgebras. Mat. Zametki 43, 178–191 (1988). (Russian) Engl. transl., Math. Notes, 43, 99–106 (1988)
[227] Mikhalev, A. A., Zolotykh, A. A.: Combinatorial Aspects of Lie Superalgebras. CRC Press, Boca Raton, FL (1995). viii+260 pp.
[228] Mishchenko, S. P.: Variety of centre-by-metabelian Lie algebras over a field of characteristic zero. Mat. Zametki 30, 649–657 (1981)
[229] Mishchenko, S. P.: Varieties of Lie algebras with weakly growing codimension series. Vestn. Mosk. Univ. Mat. Mech. 5, 63–66 (1982)
[230] Mishchenko, S. P.: Engel identity and its applications. Mat. Sb. 121, 424–430 (1983)
[231] Mishchenko, S. P.: Engel problem. Mat. Sb. 124, 56–67 (1983)
[232] Mishchenko, S. P.: Varieties of hypercentrally metabelian Lie algebras over a field of characteristic zero. Vestn. Mosk. Univ. Mat. Mech. 5, 33–37 (1983)
[233] Mishchenko, S. P.: Engel's problem. Mat. Sb. 124, 56–67 (1984)
[234] Mishchenko, S. P.: On varieties of polynomial growth of Lie algebras over a field of characteristic zero. Mat. Zametki 40, 713–721, 828 (1986).

[235] Mishchenko, S. P.: Growth in varieties of Lie algebras. Usp. Mat. Nauk 45, 25–45 (1990). Engl. Transl.: Russian Math. Surveys 45, 27-52 (1990)
[236] Mishchenko, S. P.: On varieties of Lie algebras of intermediate growth. Vescî Akad. Navuk BSSR 2, 42–45 (1987). (Russian)
[237] Mishchenko, S. P.: A version of the height theorem for Lie algebras. Math. Notes 47, 83–89 (1990)
[238] Mishchenko, S. P.: On certain classes of Lie algebras. Vestn. Mosk. Univ., Ser Mat. Mech. 3, 55–57 (1992)
[239] Mishchenko, S. P.: Color Young diagrams, Vestn. Mosk. Univ., Ser Mat. Mech. 1, 90–91 (1993)
[240] Mishchenko, S. P.: Lower bounds on the dimensions of irreducible representations of symmetric groups and on the exponents of varieties of Lie algebras. Mat. Sb. 187(1), 83–94 (1996). English transl., in Sbornik: Math. 187(1), 81–92
[241] Mishchenko, S. P., Petrogradsky, V. M.: Exponents of varieties of Lie algebras with a nilpotent commutator subalgebra. Commun. Algebra 27, 2223–2330 (1999)
[242] Mishchenko, S. P., Regev, A., Zaicev, M.: Integrality of exponents of some abelian-by-nilpotent varieties of Lie algebras. Commun. Algebra 28(9), 4105–4130 (2000)
[243] Mishchenko, S. P., Zaicev, M. V.: Sequences of codimensions of identities and their asymptotic behavior. Fundam. Appl. Mat. 2(2), 115–127 (2019)
[244] Mishchenko, S. S.: New example of a variety of Lie algebras with fractional exponent. Vestn. Mosk. Univ., Ser. 1 Mat. Mech. 6, 44–47 (2011). Engl. Transl.: Moscow Univ. Math. Bull. 66(6), 264–266 (2011)
[245] Montgomery, S.: Fixed Rings of Finite Automorphism Groups of Associative Rings. Lect. Notes Math., vol. 818. Springer–Verlag, Berlin (1980)
[246] Montgomery, S.: Hopf Algebras and Their Actions on Rings. CBMS Regional Conference Series in Mathematics, vol. 82 (1993). 238 pp.
[247] Neumann, H.: Varieties of Groups. Ergeb. d. Math. u. ihrer Grenzgeb, vol. 37. Springer-Verlag, New York (1967)
[248] Neumann, P. M.: An improved bound for BFC-groups. J. Aust. Math. Soc. 11, 19–27 (1970)
[249] Oates, S., Powell, M. B.: Identical relations in finite groups. J. Algebra 1, 11–39 (1964)
[250] Olshansky, A. Yu: Soluble just-non-Cross varieties of groups. Mat. Sb. 85, 115–131 (1971)
[251] Olshansky, A. Yu: Conditional identities in finite groups. Sib. Mat. Zh. 15, 1409–1413 (1974)
[252] Olshansky, A. Yu: Some infinite systems of identities. Tr. Semin. Im. I.G. Petrovskogo 3, 139–145 (1978)
[253] Panyukov, V. V.: On the solubility of Lie algebras of positive characteristic. Russ. Math. Surv. 45, 181–182 (1990)
[254] Papistas, A. I.: Automorphisms or free polynilpotent Lie algebras. Commun. Algebra 21(12), 4391–4395 (1993)
[255] Parfenov, V. A.: Varieties of Lie algebras. Algebra Log. 6, 61–73 (1967)
[256] Parfenov, V. A.: A property of ideals in a free Lie algebra. Sib. Mat. Zh. 10, 940–944 (1969)
[257] Passman, D.: Group rings satisfying a polynomial identity. J. Algebra 20, 103–117 (1972)
[258] Passman, D. S.: Enveloping algebras satisfying a polynomial identity. J. Algebra 134, 469–490 (1990)
[259] Petravchuk, A. P.: Lie algebras which are the sum of an abelian and a nilpotent subalgebra. Ukr. Math. J. 40, 385–388 (1988)
[260] Petravchuk, A. P.: The solvability of a Lie algebra which decomposes into a direct sum of an abelian and a nilpotent subalgebra. Ukr. Math. J. 43, 986–991 (1991)
[261] Petravchuk, A. P.: On the sum of two Lie algebras with finite-dimensional commutants. Ukr. Math. J. 47, 1244–1252 (1996)
[262] Petravchuk, A. P.: On the sum of an almost abelian Lie algebra and a Lie algebra that is finite-dimensional over its center. Ukr. Math. J. 51, 707–715 (2000)

[263] Petrogradsky, V. M.: Existence of identities in the restricted enveloping algebra. Math. Notes 49, 60–66 (1991)
[264] Petrogradsky, V. M.: On types of overexponential growth of identities in Lie PI-algebras. Fundam. Prikl. Mat. 1(4), 989–1007 (1995). (Russian)
[265] Petrogradsky, V. M.: Intermediate growth in Lie algebras and their enveloping algebras. J. Algebra 179, 459–482 (1996)
[266] Petrogradsky, V. M.: Growth of polynilpotent varieties of Lie algebras and rapidly growing entire functions. Mat. Sb. 188(6), 119–138 (1997). English translation in Russian Acad. Sci. Sb. Math., 188(6), 913–931 (1997)
[267] Petrogradskii, V. M.: On numerical characteristics of subvarieties for three varieties of Lie algebras. Mat. Sb. 188, 111–126 (1999). Engl. transl.: Sbornik: Math. 190, 887–902 (1999)
[268] Petrogradsky, V. M.: Growth of finitely generated polynilpotent Lie algebras and groups, generalized partitions, and functions analytic in the unit circle. Int. J. Algebra Comput. 9, 179–212 (1999)
[269] Petrogradsky, V. M.: Exponential generating functions and complexity of Lie varieties. Isr. J. Math. 113, 323–339 (1999)
[270] Petrogradsky, V. M.: Growth of finitely generated polynilpotent Lie algebras and groups, generalized partitions, and functions analytic in the unit circle. Int. J. Algebra Comput. 9(2), 179–212 (1999)
[271] Petrogradsky, V. M.: Exponential Schreier's formula for free Lie algebras and its applications. J. Math. Sci. (N.Y.) 93(6), 939–950 (1999). Algebra, 11
[272] Petrogradsky, V. M.: Schreier's formula for free Lie algebras. Arch. Math. (Basel) 75(1), 16–28 (2000)
[273] Petrogradsky, V. M.: On growth of Lie algebras, generalized partitions, and analytic functions. (Formal power series and algebraic combinatorics, Vienna, 1997). Discrete Math. 217(1–3), 337–351 (2000)
[274] Petrogradsky, V. M.: On generating functions for subalgebras of free Lie superalgebras. Discrete Math. 246(1–3), 269–284 (2002)
[275] Petrogradsky, V.: Scale for codimension growth of Poisson PI-algebras. preprint
[276] Petrogradsky, V. M.: On some types of intermediate growth in Lie algebras. Usp. Mat. Nauk 48, 181–182 (1993). Engl. transl., Russian Math. Surveys, 48(5), 181–182 (1993)
[277] Petrogradsky, V. M.: Intermediate growth in Lie algebras and their enveloping algebras. J. Algebra 179, 459–482 (1996)
[278] Petrogradsky, V. M.: On Lie algebras with nonintegral q-dimensions. Proc. Am. Math. Soc. 125, 649–656 (1997)
[279] Petrogradsky, V. M.: Examples of self-iterating Lie algebras. J. Algebra 302(2), 881–886 (2006). MR2293788
[280] Petrogradsky, V. M.: Fractal nil graded Lie superalgebras. J. Algebra 466, 229–283 (2016). MR3541686
[281] Petrogradsky V., N.: Lie p-algebras of slow growth. Commun. Algebra 45(7), 2912–2941 (2017)
[282] Petrogradsky, V.: Clover nil restricted Lie algebras of quasi-linear growth. J. Algebra Appl. to appear
[283] Petrogradsky, V.: Nil restricted Lie algebras of oscillating intermediate growth. arXiv:2004.05157
[284] Petrogradsky, V. M., Razmyslov, Yu.P., Shishkin, E. O.: Wreath products and Kaluzhnin-Krasner embedding for Lie algebras. Proc. Am. Math. Soc. 135, 625–636 (2007). MR2262857
[285] Petrogradsky, V. M., Shestakov, I. P.: Examples of self-iterating Lie algebras, 2. J. Lie Theory 19(4), 697–724 (2009). MR2599000
[286] Petrogradsky, V. M., Shestakov, I. P.: On properties of Fibonacci restricted Lie algebra. J. Lie Theory 23(2), 407–431 (2013). MR3113516

[287] Petrogradsky, V., Shestakov, I. P.: Fractal nil graded Lie, associative, Poisson, and Jordan superalgebras. J. Algebra 574, 453–513 (2021)
[288] Petrogradsky, V. M., Shestakov, I. P., Zelmanov, E. I.: Nil graded self-similar algebras. Groups Geom. Dyn. 4, 873–900 (2010)
[289] Pihtil'kov, S. A.: On special Lie algebras. Usp. Mat. Nauk 36, 225–226 (1981)
[290] Pillen, C.: Die Summe einer abelschen und einer nilpotenten Lie-Algebra ist ausflosbar. Results Math. 11, 117–121 (1987)
[291] Pogudin, G., Razmyslov, Y. P.: Prime Lie algebras satisfying the standard Lie identity of degree 5. J. Algebra 468, 182–192 (2016)
[292] Polin, S. V.: Identities of finite algebras. Sib. Mat. Zh. 17, 1356–1366 (1976)
[293] Postnikov, M. M.: Lectures on Geometry, 5th Term. Lie groups and Lie algebras. Nauka, Moscow (1982)
[294] Premet, A., Strade, H.: Simple Lie algebras of small characteristic. VI. Completion of the classification. J. Algebra 320, 3559–3604 (2008)
[295] Procesi, C.: The invariant theory of $n \times n$ matrices. Adv. Math. 19, 306–381 (1976)
[296] Procesi, C., Small, L.: Endomorphism rings of modules over PI-algebras. Math. Z. 106, 178–180 (1968)
[297] Radford, D. E.: Divided power structures on Hopf algebras and embedding Lie algebras into special-derivation algebras. J. Algebra 98, 143–170 (1986)
[298] Razmyslov, Yu.P.: On Engel Lie algebras. Algebra Log. 10, 33–44 (1971)
[299] Razmyslov, Yu.P.: An example of an insoluble just-non-Cross variety of groups. Algebra Log. 11, 186–205 (1972)
[300] Razmyslov, Yu.P.: On Lie algebras satisfying the Engel condition. Algebra Log. 10 33–44 (1972)
[301] Razmyslov, Yu.P.: Finite basis property for identities of matrix algebra of order two over a field of characteristic zero. Algebra Log. 12, 81–113 (1973)
[302] Razmyslov, Y. P.: Trace identities of full matrix algebras over a field of characteristic zero. Math. USSR, Izv. 8(4), 727–760 (1974)
[303] Razmyslov, Yu.P.: Finite basis property for some varieties of algebras. Algebra Log. 13, 685–693 (1974)
[304] Razmyslov, Yu.P.: Jacobson's radical in PI-algebras. Algebra Log. 13, 337–360 (1974)
[305] Razmyslov, Yu.P.: Hall-Higman problem. Izv. Akad. Nauk SSSR, Ser. Mat. 42, 833–847 (1978)
[306] Razmyslov, Yu.P.: Algebras satisfying identical relations of the Capelli type. Izv. Akad. Nauk SSSR, Ser. Mat. 45, 143–166 (1981)
[307] Razmyslov, Yu.P.: Algebras satisfying Capelli type identities. Izv. Akad. Nauk SSSR, Ser. Mat. 45, 143–166 (1981)
[308] Razmyslov, Yu.P.: Varieties of representations of finite-dimensional algebras. Vestn. Mosk. Univ., Ser. 1 Mat. Mech. 6, 31–37 (1982)
[309] Razmyslov, Yu.P.: Varieties of representations of finite-dimensional Lie algebras in prime algebras. Vestn. Mosk. Univ. Mech. 6, 31–37 (1982)
[310] Razmyslov, Yu.P.: Simple Lie algebras satisfying the standard Lie identity of degree 5. Math. USSR, Izv. 26, 553–590 (1986)
[311] Razmyslov, Yu.P.: Simple Lie algebras in the varieties generated by simple Lie algebras of Cartan type. Izv. Akad. Nauk SSSR, Ser. Mat. 51, 1228–1264 (1987)
[312] Razmyslov, Yu.P.: On inclusion of varieties generated by simple Lie algebras. Vestn. Mosk. Univ., Ser. 1 Mat. Mech. 2, 34–37 (1989)
[313] Razmyslov, Yu.P.: Identities of Algebras and their Representations. AMS Transl. of Math. Mon., vol. 138 (1994). 318 pp.
[314] Regev, A.: Existence of polynomial identities in $A \otimes B$. Bull. Am. Math. Soc. 77(6), 1067–1069 (1971)

[315] Regev, A.: Existence of identities in $A \otimes B$. Isr. J. Math. 11, 131–152 (1972)
[316] Regev, A.: The polynomial identities of matrices in characteristic zero. Commun. Algebra 8, 1417–1467 (1980)
[317] Regev, A.: Asymptotic values for degrees associated with stripes of Young diagrams. Adv. Math. 41, 115–136 (1981)
[318] Regev, A.: Codimensions and trace codimensions of matrices are asymptotically equal. Isr. J. Math. 47, 246–250 (1984)
[319] Reutenauer, C.: Applications of a noncommutative Jacobian matrix. J. Pure Appl. Algebra 77, 169–181 (1992)
[320] Reutenauer, C.: Free Lie algebras, LMS Mon., New Ser., vol. 7. Clarendon Press, Oxford, New York (1993) xiii+269 pp.
[321] Ritter, E. V.: Skeletons of soluble varieties of Lie algebras. Izv. Vysš. Učebn. Zaved. Mat. 3, 50–61 (1980)
[322] Rowen, L.: Polynomial Identities in Ring Theory. Academic Press, New York (1980). xx+365 pp.
[323] Sanov, I. N.: Solution of Burnside's problem for exponent 4. Učen. Zap. Leningrad Univ., Ser. Mat. 55, 166–170 (1940)
[324] Scheunert, M.: The theory of Lie superalgebras. Lect. Notes Math., vol. 734 (1979)
[325] Schreier, O.: Die Untergruppen der freien Gruppen. Abh. Math. Semin. Univ. Hamb. 5, 161–183 (1927)
[326] Seligman, G.: Modular Lie Algebras. Engeb. d. Math. u. ihrer Grenzgeb, vol. 40. Springer-Verlag, New York (1967)
[327] Seminaire "Sophus Lie". *Théorie des Algèbres de Lie. Topologie des Groupes de Lie*. École Norm. Super., Paris (1954).
[328] Serre, J.-P.: Lie Algebras and Lie Groups. Benjamin, New York (1965)
[329] Serre, J.-P.: Algèbres de Lie Semi-simples Complexes. Benjamin, New York (1966)
[330] Sheina, G. V.: Some varieties of Lie algebras. Sib. Mat. Zh. 17, 194–199 (1976)
[331] Sheina, G. V.: Metabelian varieties of Lie algebras. Usp. Mat. Nauk 33, 209–210 (1978)
[332] Sheina, G. V.: Varieties of metabelian Lie A-algebras, I. Vestn. Mosk. Univ. Mat. Mech. 4, 37–46 (1977). II, 3, 52–59, 1978
[333] Shestakov, I., Zaicev, M.: Polynomial identities of finite dimensional simple algebras. Commun. Algebra 39(3), 929–932 (2011)
[334] Shestakov, I. P., Zelmanov, E.: Some examples of nil Lie algebras. J. Eur. Math. Soc. 10(2), 391–398 (2008)
[335] Shirshov, A. I.: On representation of Lie rings in associative rings. Usp. Mat. Nauk 8, 173–175 (1953)
[336] Shirshov, A. I.: Subalgebras of free Lie algebras. Mat. Sb. 33, 441–452 (1953)
[337] Shirshov, A. I.: Rings with identical relations. Mat. Sb. 43, 277–283 (1957)
[338] Shirshov, A. I.: Free Lie rings. Mat. Sb. 45, 113–122 (1958)
[339] Shirshov, A. I.: A conjecture in the theory of Lie algebras. Sib. Mat. Zh. 3, 297–301 (1962)
[340] Shirshov, A. I.: Bases of free Lie algebras. Algebra Log. 1, 14–19 (1962)
[341] Shmelkin, A. L.: Free polynilpotent groups. Izv. Akad. Nauk SSSR, Ser. Mat. 28, 91–122 (1964). English transl. Amer. Math. Soc. Transl. ser., Amer. Math. Soc., Providence, RI, 55, 270–304 (1966)
[342] Shmelkin, A. L.: Wreath products of Lie algebras and their applications in the theory of groups. Tr. Mosk. Mat. Obŝ. 29, 247–260 (1973)
[343] Shmelkin, A. L.: A connection between Lie algebras and groups. Usp. Mat. Nauk 33, 193–194 (1978)
[344] Skornyakov, L. A.: Elements of Lattice Theory. 2nd edn. Nauka, Moscow (1982)
[345] Skornyakov, L. A.: Basic General Algebra. Nauka, Moscow (1983)
[346] Stanley, R. P.: Enumerative Combinatorics. Brooks/Coole, Monterey, California (1986)

[347] Stovba, V. V.: Finite basis property for some varieties of Lie and associative algebras. Vestn. Mosk. Univ. Mat. Mech. 2, 54–58 (1982)

[348] Stovba, V. V.: On the property of being a finite basis of solvable varieties of Lie algebras of finite axiomatic rank. Mat. Sb. (N.S.) 129(171), 104–120 (1986) Translation in Math. USSR-Sb. 57 (1987), no. 1, 111–129

[349] Towers, D. A.: A Frattini theory for algebras. Proc. Lond. Math. Soc. 27, 440–462 (1973)

[350] Traustason, G.: A constructive approach to Zel'manov's global nilpotency theorem for n-Engel Lie algebras over a field of characteristic zero. Int. J. Algebra Comput. 8, 317–326 (1998)

[351] Ufnarovskiy, V. A.: Combinatorial and Asymptotic Methods in Algebra. Itogi Nauki i Tekhniki, Sovrem. Probl. Mat. Fund. Naprav., vol. 57. Moscow (1989). Engl. transl., Encyclopaedia Math. Sci., vol. 57, Algebra, Springer, Berlin, 1995

[352] Vais, A. Ya: Special varieties of Lie algebras. Algebra Log. 28, 17–24 (1989)

[353] Vaughan-Lee, M. R.: Varieties of Lie algebras. D. Phil. Thesis, Oxford (1970)

[354] Vaughan-Lee, M. R.: Varieties of Lie algebras. Q. J. Math. 21, 297–308 (1970)

[355] Vaughan-Lee, M. R.: Centre-by-metabelian Lie algebras. J. Aust. Math. Soc. 15, 259–264 (1973)

[356] Vaughan-Lee, M. R.: Abelian-by-nilpotent varieties of Lie algebras. J. Lond. Math. Soc. 11, 263–266 (1975)

[357] Vaughan-Lee, M. R.: The Restricted Burnside Problem, 2nd edn. London Math. Soc. Monographs (N.S.), vol. 5. Oxford University Press (1993). 256 pp.

[358] Verevkin, A. B., Zaicev, M. V., Mishchenko, S. P.: On sufficient condition for existence of the exponent of the linear algebras variety. Vestn. Mosk. Univ. Ser. I Mat. Mech. 2, 36–39 (2011). Translation in Moscow Univ. Math. Bull. 66(2), 86–89 (2011)

[359] Viennot, J.: Algèbres de Lie libres et monoïdes libres. Lect. Notes Math., vol. 691 (1978)

[360] Volichenko, I. B.: A variety of Lie algebras connected with standard identities, I. Vesci Akad. Navuk BSSR, Ser. Fiz.-Mat. Navuk 1, 23–30 (1980). II, 2, 22–30 (1980)

[361] Volichenko, I. B.: Varieties of centre-by-metabelian Lie algebras. Prepr. 1969, Inst. Mat. AN BSSR (1980)

[362] Volichenko, I. B.: A study of some extremal varieties of Lie algebras. Ph.D. Thesis, Minsk (1981)

[363] Volichenko, I. B.: Variety of Lie algebras $\mathbf{AN}_2$ over a field of characteristic zero. Dokl. Akad. Navuk BSSR 25, 1063–1066 (1981)

[364] Volichenko, I. B.: Varieties of Lie algebras with identity $[[X_1,X_2,X_3],[X_4,X_5,X_6]] = 0$ over a field of characteristic zero. Sib. Mat. Zh. 25, 40–54 (1984) (in Russian)

[365] Volkov, M. V.: Lattices of varieties of algebras. Mat. Sb. 109, 60–79 (1979)

[366] Volkov, M. V., Geĭn, A. G.: Identities of virtually nilpotent Lie rings. Mat. Sb. 118, 132–142 (1982)

[367] van der Waerden, B. L.: Algebra II. Springer-Verlag, New York (1982)

[368] Wam, Zue-Xian: Lie Algebras. Pergamon Press, Oxford (1975)

[369] Weyl, H.: The Classical Groups. Princeton University Press, Princeton (1946)

[370] Winter, D.: Abstract Lie Algebras. MIT Press, Cambridge, MA (1972)

[371] Witt, E.: Treue Darstellung Lieschen Ringe. J. Crelle 177, 152–160 (1937)

[372] Witt, E.: Die Unterringe der freien Lieschen Ringe. Math. Z. 64, 195–216 (1956)

[373] Zaicev, M. V.: Finite basis property for some varieties of Lie algebras. Mat. Sb. 106, 499–506 (1978)

[374] Zaicev, M. V.: Schreier varieties of Lie algebras. Mat. Zametki 28, 119–126 (1980)

[375] Zaicev, M. V.: Decomposability of commutators of varieties of Lie algebras and groups. Mat. Sb. 116, 312–330 (1981)

[376] Zaicev, M. V.: Identities of affine Kac-Moody algebras. Vestn. Mosk. Univ., Ser. 1 Mat. Mech. 2, 34–38 (1982)

[377] Zaicev, M. V.: Special Lie algebras. Usp. Mat. Nauk 48, 103–140 (1993)
[378] Zaicev, M. V.: Standard identity and Capelli identities in Lie algebras. Usp. Mat. Nauk 49, 153–154 (1994) Engl. Transl.: Russian Math. Surveys, 49(1994), 186–187 (2002)
[379] Zaicev, M. V.: Integrality of exponents of growth of identities of finite-dimensional Lie algebras. Izv. Ross. Akad. Nauk Ser. Mat. 66, 23–48 (2002). Engl. Transl.: Izv. Math. 66, 463–487 (2002)
[380] Zaicev, M. V., Mishchenko, S. P.: An example of a variety of Lie algebras with a fractional exponent. Algebra J. Math. Sci. (New York) 93, 977–982 (1999)
[381] Zel'manov, E. I.: Engel Lie-algebras. Dokl. Akad. Nauk SSSR 292, 265–268 (1987)
[382] Zel'manov, E. I.: Engelian Lie algebras. Sib. Mat. Zh. 29, 112–117 (1988)
[383] Zel'manov, E. I.: The solution of the restricted Burnside problem for groups of odd exponent. Math. USSR, Izv. 36, 41–60 (1991)
[384] Zel'manov, E. I.: A solution of the restricted Burnside problem for 2-groups. Mat. Sb. 18(2), 568–592 (1991). Engl. Transl.: Math. USSR-Sb., 72, 543–565 (1992)
[385] Zel'manov, E. I.: Lie algebras and torsion groups with identity. J. Comb. Algebra 1, 289–340 (2017)
[386] Zhevlakov, K. A., Slin'ko, A. M., Shestakov, I. P., Shirshov, A. I.: Rings That Are Nearly Associative. Academic Press, New York (1982)
[387] Zusmanovich, P. A.: A Lie algebra that can be written as a sum of two nilpotent subalgebras is solvable. Math. Notes 50, 909–912 (1991)

Subject index

Index of authors

Index of notation

De Gruyter Expositions in Mathematics

Volume 34
Marek Jarnicki, Peter Pflug
Extension of Holomorphic Functions, 2nd Edition, 2020
ISBN 978-3-11-062766-4, e-ISBN 978-3-11-063027-5

Volume 67
Alcides Lins Neto, Bruno Scárdua
Complex Algebraic Foliations, 2020
ISBN 978-3-11-060107-7, e-ISBN 978-3-11-060205-0

Volume 66
Anke Kalauch, Onno van Gaans
Pre-Riesz Spaces, 2018
ISBN 978-3-11-047539-5, e-ISBN 978-3-11-047629-3

Volume 65
Yakov Berkovich, Zvonimir Janko
Groups of Prime Power Order. Volume 6, 2018
ISBN 978-3-11-053097-1, e-ISBN 978-3-11-053314-9

Volume 64
Yakov Berkovich, Lev S. Kazarin, Emmanuel M. Zhmud'
Characters of Finite Groups. Volume 2, 2019
ISBN 978-3-11-022408-5, e-ISBN 978-3-11-022409-2

Volume 63
Yakov Berkovich, Lev S. Kazarin, Emmanuel M. Zhmud'
Characters of Finite Groups. Volume 1, 2017
ISBN 978-3-11-022406-1, e-ISBN 978-3-11-022407-8

Volume 62
Yakov Berkovich, Zvonimir Janko
Groups of Prime Power Order. Volume 5, 2016
ISBN 978-3-11-029534-4, e-ISBN 978-3-11-029535-1

Volume 61
Yakov Berkovich, Zvonimir Janko
Groups of Prime Power Order. Volume 4, 2016
ISBN 978-3-11-028145-3, e-ISBN 978-3-11-028147-7